FINITE MATHEMATICS AND ITS APPLICATIONS

Fourth Edition

FINITE MATHEMATICS AND ITS APPLICATIONS

Larry J. Goldstein

David I. Schneider
Department of Mathematics
University of Maryland

Martha J. Siegel
Department of Mathematics
Towson State University

PRENTICE HALL, ENGLEWOOD CLIFFS, NEW JERSEY, 07632

Library of Congress Cataloging-in-Publication Data

Goldstein, Larry Joel.
    Finite mathematics and its applications / Larry J. Goldstein,
David I. Schneider, Martha J. Siegel. — 4th ed.
        p.   cm.
    Includes index.

        1. Mathematics.   I. Schneider, David I.   II. Siegel, Martha J.
III. Title.
QA39.2.G643   1991
510—dc20   90–25331
                                                          CIP

Editorial/production: Thomas Aloisi/Nicholas Romanelli
Interior design: Donna M. Wickes
Cover design: Lee Cohen
Cover art: Salem Krieger
Prepress buyer: Paula Massenaro
Manufacturing buyer: Lori Bulwin
Photo research: Lori Morris–Nantz/Barbara Scott
Acquisition editor: Steve Comny

Photo/art credits: p. 1, Salem Krieger; p. 3, Margot Granitaas/The Image Works; p. 11, Cary Wolinsky/Stock, Boston; p. 144, Peter Southwick/Stock, Boston; p. 202, Christa Armstrong, Photo Researchers, Inc.; p. 260, Chrysler Corporation; p. 322, IBM Corp.; p. 389, Nita Winter/The Image Works; p. 421, Rapho/Photo Researchers, Inc.; p. 451, Thelma Shumsky/The Image Works; p. 481, Hazel Hankin/Stock, Boston; p. 522, Jim Pickerell/Stock, Boston; p. 581, Stuart Rosner/Stock, Boston.

Finite Mathematics and Its Applications, fourth edition
Larry J. Goldstein/David I. Schneider/Martha J. Siegel

Copyright 1991, 1988, 1984, 1980 by Prentice-Hall, Inc.
a Simon & Schuster Company
Englewood Cliffs, New Jersey 07632

Printed in the United States of America

10   9   8   7   6   5   4   3   2   1

ISBN 0-13-318221-5

Prentice-Hall International (UK) Limited, *London*
Prentice-Hall of Australia Piy. Limited, *Sydney*
Prentice-Hall Canada Inc., *Toronto*
Prentice-Hall Hispanoamericana, S.A., *Mexico*
Prentice-Hall of India Private Limited, *New Delhi*
Prentice-Hall of Japan, Inc., *Tokyo*
Simon & Schuster Asia Pte. Ltd., *Singapore*
Editora Prentice-Hall do Brasil, Ltda., *Rio de Janeiro*

Mathematics and Its Applications

This volume is one of a collection of texts for freshman and sophomore college mathematics courses. Included in this collection are the following.

Calculus and Its Applications, fifth edition by L. Goldstein, D. Lay, and D. Schneider. A text designed for a two-semester course in calculus for students of business and the social and life sciences. Emphasizes an intuitive approach and integrates applications into the development.

Brief Calculus and Its Applications, fifth edition by L. Goldstein, D. Lay, and D. Schneider. Consists of the first eight chapters of the above book. Suitable for shorter courses with a slight reordering of some topics

Finite Mathematics and Its Applications, fourth edition by L. Goldstein, D. Schnieder, and M. Siegel. A traditional finite mathematics text for students of business and the social and life sciences. Allows courses to begin with either linear mathematics (linear programming, matrices) or probability and statistics.

Mathematics for the Management, Life, and Social Sciences, second edition by L. Goldstein, D. Lay, and D. Schneider. A text for a two-semester course covering finite mathematics, precalculus, and calculus.

CONTENTS

Preface   xi

Index of Applications   xv

Introduction   1

1 **Linear Equations and Straight Lines**   3

    1.1 Coordinate Systems and Graphs   4
    1.2 Linear Inequalities   13
    1.3 The Intersection Point of a Pair of Lines   23
    1.4 The Slope of a Straight Line   29

2 **Matrices**   49

    2.1 Solving Systems of Linear Equations, I   50
    2.2 Solving Systems of Linear Equations, II   61
    2.3 Arithmetic Operations on Matrices   69
    2.4 The Inverse of a Matrix   85
    2.5 The Gauss–Jordan Method for Calculating
        Inverses   97
    2.6 Input–Output Analysis   102

3 **Linear Programming, A Geometric Approach**   111

    3.1 A Linear Programming Problem   112
    3.2 Linear Programming, I   119
    3.3 Linear Programming, II   129

4 The Simplex Method   144

4.1 Slack Variables and the Simplex Tableau   145
4.2 The Simplex Method, I—Maximum Problems   155
4.3 The Simplex Method, II—Minimum Problems   168
4.4 Marginal Analysis and Matrix Formulations of Linear Programming Problems   176
4.5 Duality   185

5 Sets and Counting   202

5.1 Sets   203
5.2 A Fundamental Principle of Counting   211
5.3 Venn Diagrams and Counting   218
5.4 The Multiplication Principle   226
5.5 Permutations and Combinations   232
5.6 Further Counting Problems   238
5.7 The Binomial Theorem   245
5.8 Multinomial Coefficients and Partitions   251

6 Probability   260

6.1 Introduction   261
6.2 Experiments, Outcomes, and Events   262
6.3 Assignment of Probabilities   272
6.4 Calculating Probabilities of Events   283
6.5 Conditional Probability and Independence   292
6.6 Tree Diagrams   305
6.7 Bayes' Theorem   312

7 Probability and Statistics   322

7.1 Frequency and Probability Distributions   323
7.2 The Mean   335
7.3 The Variance and Standard Deviation   346
7.4 Binomial Trials   356
7.5 The Normal Distribution   363
7.6 Normal Approximation to the Binomial Distribution   378

8 Markov Processes   389

8.1 The Transition Matrix   390
8.2 Regular Stochastic Matrices   401
8.3 Absorbing Stochastic Matrices   409

9 The Theory of Games   421

    9.1 Games and Strategies   422
    9.2 Mixed Strategies   430
    9.3 Determining Optimal Mixed Strategies   437

10 The Mathematics of Finance   451

    10.1 Interest   452
    10.2 Annuities   461
    10.3 Amortization of Loans   471

11 Difference Equations and Mathematical Models   481

    11.1 Introduction to Difference Equations, I   482
    11.2 Introduction to Difference Equations, II   489
    11.3 Graphing Difference Equations   497
    11.4 Mathematics of Personal Finance   509
    11.5 Modelling with Difference Equations   514

12 Logic   522

    12.1 Introduction to Logic   523
    12.2 Truth Tables   528
    12.3 Implication   537
    12.4 Logical Implication and Equivalence   547
    12.5 Valid Argument   537
    12.6 Predicate Calculus   565

13 Graphs   581

    13.1 Graphs as Models   582
    13.2 Paths and Circuits   598
    13.3 Hamiltonian Circuits and Spanning Trees   612
    13.4 Directed Graphs   626
    13.5 Matrices and Graphs   648
    13.6 Trees   663

Appendix Tables   A1
    Table 1   Areas under the Standard Normal Curve   A1
    Table 2 $(1 + i)^n$ Compound Amount of \$1 Invested for n Interest
                Periods at Interest Rate i per Period   A2

Table 3 $\dfrac{1}{(1 + i)^n}$ Present Value of \$1 Principal That Will
Accumulate to \$1 in n Interest Periods at a Compound
Rate of i per Period   A3

Table 4 $s_{\overline{n}|\,i}$ Future Value of an Ordinary Annuity of n \$1
Payments Each, Immediately after the Last Payment at
Compound Interest Rate of i per Period   A4

Table 5 $\dfrac{1}{s_{\overline{n}|\,i}}$ Rent per Period for an Ordinary Annuity of n
Payments, Compound Interest Rate i per Period, and
Future Value \$1   A5

Table 6 $a_{\overline{n}|\,i}$ Present Value of an Ordinary Annuity of n
Payments of \$1 One Period Before the First Payment
with Interest Compounded at i per Period   A6

Table 7 $\dfrac{1}{a_{\overline{n}|\,i}}$ Rent per Period for an Ordinary Annuity of n
Payments Whose Present Value is \$1 with Interest
Compounded at i per Period   A7

Answers to Exercises   **A9**

Index   **I–1**

PREFACE

This work is the fourth edition of our text for the traditional finite mathematics course taught to first- and second-year college students, especially those majoring in business and the social and biological sciences. Finite mathematics courses exhibit tremendous diversity with respect to both content and approach. Therefore, in revising this book, we have incorporated a wide range of topics from which an instructor may design a curriculum, as well as a high degree of flexibility in the order in which the topics may be presented. In the case of the mathematics of finance, we have even allowed for flexibility in the approach of the presentation.

In this edition, we have attempted to maintain our popular student-oriented approach. This approach manifests itself throughout and, in particular, in the following features:

Applications   We provide realistic applications that illustrate the uses of finite mathematics in other disciplines. The reader may survey the variety of applications by turning to the Index of Applications on page xv. Wherever possible, we have attempted to use applications to motivate the mathematics. For example, the idea of linear programming is introduced in Chapter 3 via a discussion of production options for a factory with a labor limitation.

Examples   We have included many more worked examples than is customary (549). Furthermore, we have included computational details to enhance readability by students whose basic skills are weak.

Exercises   The more than 1900 exercises comprise about one-quarter of the text—the most important part of the text in our opinion. The exercises at the ends of the sections are usually arranged in the order in which the text proceeds, so that the homework assignments may easily be made after only part of a section is discussed. Interesting applications and more challenging problems tend to be located near the ends of the exercise sets. Supplementary exercises at the end of each chapter expand the other exercise sets and provide cumulative exercises that require skills from earlier chapters.

Practice Problems The practice problems have proved to be a popular and useful feature of the book. The practice problems are carefully selected exercises located at the end of each section, just before the exercise set. Complete solutions are given following the exercise set. The practice problems often focus on points that are potentially confusing or are likely to be overlooked. We recommend that the reader seriously attempt the practice problems and study their solutions before moving on to the exercises. In effect, the practice problems constitute a built-in workbook.

Minimal Prerequisites Because of great variation in student preparation, we have kept the formal prerequisites to a minimum. We assume only a first year of high school algebra. Furthermore, we review, as needed, those topics which are typically weak spots for students.

NEW IN THIS EDITION

Among the changes in this edition, the following are the most significant.

1. *Additional Exercises.* The stock of exercises from the previous edition has been significantly expanded. Among the new exercises are some that test understanding and others that challenge the exceptional students.

2. *New Chapters on Logic and Graphs.* The recent recognition of the value of discrete mathematics in the curriculum has produced a shift in content in many finite mathematics courses. These chapters have been added to serve courses being modified in that direction. With the inclusion of this material, the book now covers nearly all the topics recommended by the CUPM Subcommittee on Discrete Mathematics. (See "Discrete Mathematics in the First Two Years," *MAA Notes* 15, 1989.)

(a) The chapter on logic emphasizes the elements of propositional calculus, implication, and valid argument. The section on the predicate calculus pays particular attention to the precision of mathematical language and relates to sets and computing.

(b) The chapter on graph theory and its applications includes basic definitions, elementary theorems, and applications to business, social sciences, and computing. The introduction to PERT in the digraph section is enhanced by optional material on stochastic scheduling methods, thus linking this material to the chapters on probability and statistics. Matrix representation of graphs and digraphs provides a further application of matrix algebra.

This edition has more material than can be covered in most one-semester courses. Therefore, the instructor can structure the course to the students' needs and interests. The book divides naturally into four parts. The first part consists of linear mathematics: linear equations, matrices, and linear programming (Chapters 1–4); the second part is devoted to probability and statistics (Chapters 5–7); the third part covers topics utilizing the ideas of the other parts (Chapters 8–10); and the fourth part explores key topics from discrete mathematics that are sometimes included in the modern finite mathematics curriculum (Chapters 11–13). We prefer to begin with linear mathematics since it makes for a smooth transition from high school mathematics and leads quickly to interesting applications, especially linear programming. Our preference notwithstanding, the instructor may begin this book with Chapter 5 (Sets and Counting) and then do either the linear mathematics or the probability and statistics.

Answers to the odd-numbered exercises are included at the back of the book.

If you have comments or suggestions, we would like to hear from you. We hope that you enjoy using this book as much as we have enjoyed writing it.

ACKNOWLEDGMENTS

While writing this book, we have received assistance from many persons. And our heartfelt thanks goes out to them all. Especially, we should like to thank the following reviewers, who took the time and energy to share their ideas, preferences, and often their enthusiasm, with us.

Reviewers of the first edition: James F. Hurley, University of Connecticut; Sam Councilman, California State University, Long Beach; Carl D. Meyer, Jr., North Carolina State University; Stephen H. Brown, Auburn University; Bart Braden, Northern Kentucky University; James C. Thorpe, University of Missouri, St. Louis; Joseph Stampfli, Indiana University; Martin C. Tangora, University of Illinois, Chicago Circle; William D. Blair, Northern Illinois University; Richard Pellerin, Northern Virginia Community College; Roger Osborn, University of Texas, Austin; Thomas J. Hill, University of Oklahoma, Norman; Donald E. Myers, University of Arizona, Tempe.

Reviewers of the second edition: D.R. Dunninger, Michigan State University; Hiram Paley, University of Illinois, Urbana-Champaign; Joan McCarter, Arizona State University; Robert Carmignani, University of Missouri; Philip Kutzko, University of Iowa; Juan Gatica, University of Iowa; Frank Warner, University of Pennsylvania; Richard Porter, Northeastern University; James Hurley, University of Connecticut; William Ramaley, Fort Lewis College; Donald E. Myers, University of Arizona, Tempe.

Reviewers of the third edition: Robin G. Symonds, Indiana University at Kokomo; Elizabeth Teles, Montgomery College; Charles J. Miller, Foothill Community College.

Reviewers of the fourth edition: Phil Steitz, Beloit College. Barry Cipra.

The authors thank the many people at Prentice Hall who contributed to the success of our books. We appreciate the efforts of the production, art, manufacturing, marketing, and sales departments. An extra special thanks to Steve Comny, acquisitions editor at Prentice Hall, for his help in planning and executing this new edition. His partnership and friendship have added a warm personal dimension to the writing process.

Larry J. Goldstein
David I. Schneider
Martha J. Siegel

INDEX OF APPLICATIONS

Business and Economics

Amortization 471ff, 476, 479

Analyzing advertising strategies 448

Analyzing business and investment options 60, 354, 437, 447, 450

Annuity 461ff, 458, 466, 467, 511, 512, 513, 521

Auditing of income tax returns 320

Break-even point 44

Business competition, strategies 422, 429

Changes in stock prices 208

Closed Leontief model 106

Communication systems 583, 598, 600, 624

Comparing car dealerships 323, 336, 338, 349

Comparing cost of car rentals 48

Compound interest 453, 454, 456, 457, 458, 459, 460, 479, 488, 490, 491, 496, 513, 521

Conglomerates 107, 108

Consumer habits 422

Consumer loans 474, 475, 476, 479, 488, 492, 493, 496, 507

Consumer price index 204

Cost and revenue curves 34, 39

Counting number of ways of assigning employees to tasks 253, 254

Counting possibilities for management positions 228, 234

Deciding whether to diversify 194, 200, 201

Deferred annuity 469

Depreciation 34

Determining production schedules 50

Double-declining balance method of depreciation 496, 497

Earnings of investment trust 76

Effectiveness of advertising campaign 36, 385

Fixed cost 34, 39

Fluctuations in agricultural prices and production 507, 517

Input–output analysis 102, 103, 105, 107, 108, 110

Insurance probabilities 301, 315, 343, 345

Inventory planning 372

Investment planning 60, 479, 480

Labor costs 83

Linear depreciation 9, 490

Marginal cost of production 34, 39

Marginal revenue of production 43

Maximizing investment returns 129, 138

Maximizing profits 166, 176, 179

Meeting schedules 620, 625

Minimizing transportation costs 133, 135, 137, 171, 183, 184

Mortgage 472, 476, 479, 480, 513, 521

Optimal scheduling and critical paths 629, 630, 631, 632, 641, 642, 645, 646, 647, 675

Optimizing production schedules 112, 117, 119, 126, 127, 139, 140, 166, 167, 176, 183, 184, 190, 583

Organizing economic and business data 204, 205, 208, 209

Perpetuity 469

Pipelines 616, 617

Quality control 270, 278, 282, 283, 285, 299, 301, 309, 310, 311, 314, 315, 320, 351, 354, 358, 376, 377, 381, 385, 387, 419

Revenue of a business 38, 73, 77

Sales commission 41, 48

Sales performance 362, 385

Salvage value 10, 35

Savings account transactions 452, 454, 457, 458, 459, 460, 482, 483, 488, 506

Shadow prices 191

Shipping 116, 126, 132

Simple interest 456, 457, 490, 493

Sinking fund 477

Straight-line method of depreciation 34

Supply and demand curves 25, 28, 39

Unemployment 205

Volume of sales 373, 377

Social Sciences

Accidental nuclear war 311

Acquaintance tables 586, 593, 624

Census statistics 317, 357

Common language usage 542, 572

Commuter trends 408

Counting options for Senate committee selection 237, 243

Counting ways to cast ballots 249

Crime statistics 316

Demographic analysis 91, 93, 357

Effects of immigration and emigration 487, 488, 518, 521

Effects of maternal influence 392, 393, 401

Evaluating performance in tests 362, 368, 384

Exam scores 348, 377

Exam strategies 388

I.Q. scores 377

Jury decisions 363

Learning process 397, 519

Marriage trends 91, 93

Opinion sampling 289, 306

Political polling 235, 266, 302, 362, 582
Population growth 487
Positions for political candidates 428
Psychology experiments 280, 311, 399, 407
Social mobility 419
Spread of information 507, 516, 521
Test for discrimination 362, 387
Voter analysis 83
Voter patterns 400, 408

Biology and Medicine
Birth weights of infants 371, 373
Counting possible outcomes of a medical experiment 226
Designing an optimum diet 68, 122, 127, 139, 143, 175, 184, 193
Epidemiology 83, 94, 411, 417
Frequency of cricket chirps 40
Genetics 272, 363, 408, 519
Glucose infusion 518
Heights of adult males 387
Heights of elephants 376
Life expectancy of wildlife 338
Medical diagnosis 261, 301
Medical screening 285, 301, 308, 311, 313
Movement of solutes through a cell membrane 518
Nutrition planning 60, 68
Organizing medical data 261
Recovery from disease 358, 382, 384
Removal of a drug from the bloodstream 518
Which group had fewer cavities? 344

General Interest
Absorption of light 518
Aircraft service 626
Bridge 253
Bridges of Konigsberg 606
Carnival games 332, 334, 345
Coin tossing 229, 238, 239, 242, 266, 270, 271, 273, 280, 289, 298, 319, 327, 333, 334, 359, 362, 383
Coloring maps and houses 618, 621, 622, 676
Comparing sports performances 344, 353
Computer languages and programming 533, 541, 542, 544, 545, 546, 579, 642, 647
Counting options for everyday tasks 216, 217, 219, 221, 222, 223, 224, 227, 228, 229, 235, 236, 237, 240, 241, 242, 243, 249, 250, 254, 255, 256, 258, 259
Counting possible contest outcomes 235, 255, 256, 257
Counting possible outcomes of sports events 223, 228, 242, 243, 253
Counting travel options 227, 229, 230, 240
Dice 263, 271, 273, 276, 280, 297, 318, 330, 341, 345, 355, 363, 383
Dictionaries 667, 668
Energy needs 519
Floor plans 610, 621, 675
Gambler's ruin 415
Games 277, 296, 298, 428, 429, 450

Logical puzzles 557, 613, 663
Nobel laureates 219
Odds 279
Optimizing exercise routine 118
Party planning 619, 625
Planning political campaign 176
Poker 282, 289
Pollution control 264, 268
Predicting sports performance 320, 344, 353, 357, 384, 385, 387, 388
Preference listing 637, 640
Press 550
Probabilities of everyday occurrences 264, 269, 270, 271, 276, 280, 281, 286, 287, 288, 289, 290, 291, 292, 296, 299, 301, 302, 303, 304, 320, 321, 362, 363
Radioactive decay 507, 515
Railroad networks 593
Road inspector's path 610, 611
Roulette 281, 341, 344, 384
School bus route 612
Sets and logic 571
Skydiving 506
Snow plow path 597, 605
Tournaments 636, 637, 638, 639
Traffic and street patterns 633, 634

FINITE MATHEMATICS
AND ITS
APPLICATIONS

INTRODUCTION

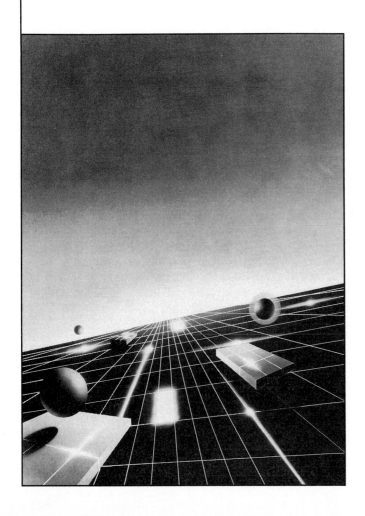

Finite mathematics has broad applications in business, the biological sciences, psychology, sociology, and political science. As we shall see, its mathematical content includes matrices and their applications to linear programming, game theory, and Markov processes as well as elementary probability and statistics. Before discussing these topics, let us look at some examples of the applications we will be considering.

PROBLEM 1 (*Business*)   A pension fund has $30 million to invest. The money is to be divided among Treasury notes, bonds, and stocks. The rules for administration of the fund require that at least $3 million be invested in each type of investment, at least half the money be invested in Treasury notes and bonds, and the amount invested in bonds not exceed twice the amount invested in Treasury notes. The annual yields for the various investments are 7% for Treasury notes, 8% for bonds, and 9% for stocks. How should the money be allocated among the various investments to produce the largest return?

PROBLEM 2 (*Medicine*)   Suppose that the reliability of a skin test for detecting active pulmonary tuberculosis is as follows: If a person has tuberculosis, the test will be positive with probability .98. If the person does not have tuberculosis, the test will be negative with probability .99. The incidence of tuberculosis in a certain population is 2 cases per 10,000. Suppose that someone is chosen at random and tested and the test turns out to be positive. What is the probability that the person actually has active pulmonary tuberculosis?

PROBLEM 3 (*Political Science*)   It is observed that, in a certain group of states, 70% of Democratic governors are succeeded by Democrats and 30% by Republicans. Also, 40% of Republican governors are succeeded by Democrats and 60% by Republicans. Assume that this trend remains unchanged for many years. Currently, 25% of the governors are Republicans. In the long run, what percentage of governors in these states will be Democratic?

PROBLEM 4 (*Sociology*)   Suppose that at 8 A.M. one Saturday a town's radio and TV stations start broadcasting news of the mayor's resignation. Each hour 30% of those who have not yet heard the news become informed. How many hours elapse before 98% of the population has heard the news?

The problems above exhibit great diversity in their respective fields of application, and we shall require an equally diverse collection of mathematical tools to solve them. All the tools required belong to finite mathematics and will be discussed in the chapters that follow.

1

LINEAR EQUATIONS AND STRAIGHT LINES

1.1 Coordinate Systems and Graphs

1.2 Linear Inequalities

1.3 The Intersection Point of a Pair of Lines

1.4 The Slope of a Straight Line

Many applications considered later involve linear equations and their geometric counterparts—straight lines. So let us begin by studying the basic facts about these two important notions.

1.1. Coordinate Systems and Graphs

Often we can display numerical data by using a *Cartesian coordinate system* on either a line or a plane. We construct a Cartesian coordinate system on a line by choosing an arbitrary point O (the *origin*) on the line and a unit of distance along the line. We then assign each point on the line a number that reflects its directed distance from the origin. Positive numbers refer to points on the right of the origin, negative numbers to points on the left. In Fig. 1 we have drawn a Cartesian coordinate system on the line and have labeled a number of points with their corresponding numbers. Each point on the line corresponds to a number (positive, negative, or zero). Conversely, every number corresponds to a point on the line.

FIGURE 1

A Cartesian coordinate system may be used to numerically describe points on a line. In a similar fashion, we can construct a Cartesian coordinate system to numerically locate points on a plane. Such a system consists of two perpendicular lines called the *coordinate axes*. These lines are usually drawn so that one is horizontal and one is vertical. The horizontal line is called the *x-axis*, the vertical line the *y-axis*. Their point of intersection is called the *origin* (Fig. 2). Each point of the plane is identified by a pair of numbers (a, b). The first number, a, tells the number of units from the point to the *y*-axis (Fig. 3). When a is positive, the point is to the right of the *y*-axis; when a is negative, the point is to the left of the *y*-axis. The second number, b, gives the number of units from the point to the *x*-axis (Fig. 4). When b is positive, the point is above the *x*-axis; when b is negative, the point is below. The numbers a and b are called, respectively, the *x- and y-coordinates of the point*. In order to plot the point (a, b), begin at the origin, move a units in the *x*-direction and then b units in the *y*-direction.

FIGURE 2

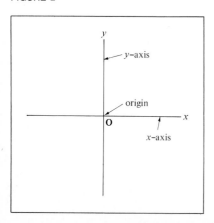

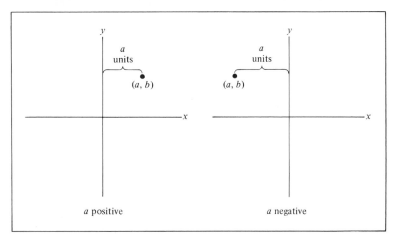

FIGURE 3

FIGURE 4

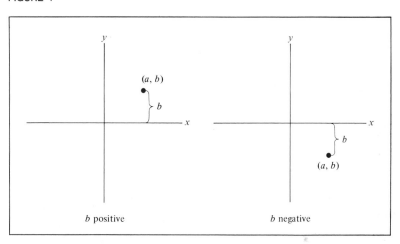

EXAMPLE 1 Plot the following points.

(a) (2, 1)   (b) (−1, 3)   (c) (−2, −1)   (d) (0, −3)

Solution (a)

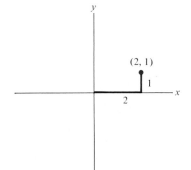

(b)

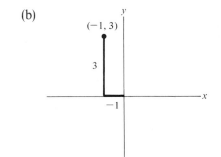

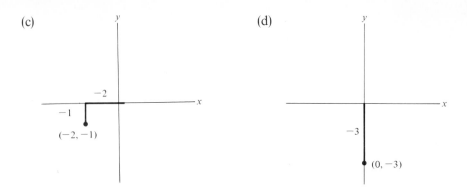

(c)

(−2, −1)

(d)

(0, −3)

In many applications one encounters equations expressing relationships between variables x and y. Some typical equations are

$$y = 2x - 1$$

$$5x^2 + 3y^3 = 7$$

$$y = 2x^2 - 5x + 8.$$

To any equation in x and y one can associate a certain collection of points in the plane. Namely, the point (a, b) belongs to the collection provided that the equation is satisfied when we substitute a for each occurrence of x and substitute b for each occurrence of y. This collection of points is usually a curve of some sort and is called the *graph of the equation.*

EXAMPLE 2   Are the following points on the graph of the equation $8x - 4y = 4$?

(a) (3, 5)   (b) (5, 17)

Solution   (a) Substitute 3 for each occurrence of x and 5 for each occurrence of y in the equation.

$$8x - 4y = 4$$

$$8 \cdot 3 - 4 \cdot 5 = 4$$

$$24 - 20 = 4$$

$$4 = 4.$$

Since the equation is satisfied, the point (3, 5) is on the graph of the equation.

(b) Replace x by 5 and y by 17 in the equation.

$$8x - 4y = 4$$

$$8 \cdot 5 - 4 \cdot 17 = 4$$

$$40 - 68 = 4$$

$$-28 = 4.$$

The equation is clearly not satisfied, so the point (5, 17) is *not* on the graph of the equation.

EXAMPLE 3   Sketch the graph of the equation $y = 2x - 1$.

Solution   First find some points on the graph by choosing various values for x and determining the corresponding values for y:

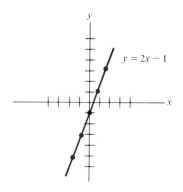

x	y
-2	$2 \cdot (-2) - 1 = -5$
-1	$2 \cdot (-1) - 1 = -3$
0	$2 \cdot 0 - 1 = -1$
1	$2 \cdot 1 - 1 = 1$
2	$2 \cdot 2 - 1 = 3$

Thus the points $(-2, -5), (-1, -3), (0, -1), (1, 1)$, and $(2, 3)$ are all on the graph. Plot these points. It appears that the points lie on a straight line. By taking more values for x and plotting the corresponding points, it is easy to become convinced that the graph of $y = 2x - 1$ is, indeed, a straight line.

EXAMPLE 4   Sketch the graph of the equation $x = 3$.

Solution   It is clear that the x-coordinate of any point on the graph must be 3. The y-coordinate can be anything. So some points on the graph are $(3, 0), (3, 5), (3, -4), (3, -2)$. Again the graph is a straight line, as in the accompanying sketch.

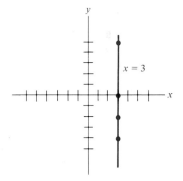

EXAMPLE 5   Sketch the graph of the equation $x = a$, where a is any given number.

Solution   The x-coordinate of any point on the graph must be a. Reasoning as in Example 4, the graph is a vertical line a units away from the y-axis. (Of course, if a is negative, then the line lies to the left of the y-axis.)

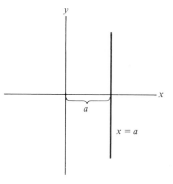

The equations in Examples 3 to 5 are all special cases of the general equation

$$cx + dy = e,$$

corresponding to particular choices of the

constants c, d, e. Any such equation is called a *linear equation* (in the two variables x and y).

The *standard form* of a linear equation is obtained by solving for y if y appears, and for x if y does not appear. In the former case the standard form looks like

$$y = mx + b \qquad (m, b \text{ constants}),$$

whereas in the latter it looks like

$$x = a \qquad (a \text{ constant}).$$

EXAMPLE 6   Find the standard form of the following equations.

(a) $8x - 4y = 4$ (b) $2x + 3y = 3$ (c) $2x = 6$

Solution   (a) Since y appears, we obtain the standard form by solving for y in terms of x:

$$8x - 4y = 4$$
$$-4y = -8x + 4$$
$$y = 2x - 1.$$

Thus the standard form of $8x - 4y = 4$ is $y = 2x - 1$—that is, $y = mx + b$ with $m = 2$, $b = -1$.

(b) Again y occurs, so we solve for y:

$$2x + 3y = 3$$
$$3y = -2x + 3$$
$$y = -\tfrac{2}{3}x + 1.$$

So the standard form is $y = -\tfrac{2}{3}x + 1$. Here $m = -\tfrac{2}{3}$ and $b = 1$.

(c) Here y does not occur, so we solve for x:

$$2x = 6$$
$$x = 3.$$

Thus the standard form of $2x = 6$ is $x = 3$—that is, $x = a$ where a is 3.

We have seen that any linear equation has one of the two standard forms $y = mx + b$, $x = a$. From Example 5, the graph of $x = a$ is a vertical line, a units from the y-axis. What can be said about the graph of $y = mx + b$? In Example 3, we saw that the graph is a straight line in the special case $y = 2x - 1$. Actually, the graph of $y = mx + b$ is always a straight line. To sketch the graph, we need only locate two points. Two convenient points to locate are the *intercepts*, the points where the line crosses the x- and y-axes. When x is 0, $y = m \cdot 0 + b = b$.

Thus $(0, b)$ is on the graph of $y = mx + b$ and is the y-intercept of the line. The x-intercept is found as follows: A point on the x-axis has y-coordinate 0. So the x-coordinate of the x-intercept can be found by setting $y = 0$—that is, $mx + b = 0$—and solving this equation for x.

EXAMPLE 7   Sketch the graph of the equation $y = 2x - 1$.

Solution   Here $m = 2, b = -1$. The y-intercept is $(0, b) = (0, -1)$. To find the x-intercept, we must solve $2x - 1 = 0$. But then $2x = 1$ and $x = \frac{1}{2}$. So the x-intercept is $(\frac{1}{2}, 0)$. Plot the two points $(0, -1)$ and $(\frac{1}{2}, 0)$ and draw the straight line through them.

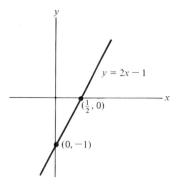

Note The line in Example 7 had different x- and y-intercepts. Two other circumstances can occur, however. First, the two intercepts may be the same, as in $y = 3x$. Second, there may be no x-intercept, as in $y = 1$. In both of these circumstances, we must plot some point other than an intercept in order to graph the line.

To summarize, then:

> **To graph the equation $y = mx + b$:**
>
> 1. Plot the y-intercept $(0, b)$.
> 2. Plot some other point. [The most convenient choice is often the x-intercept $(x, 0)$, where x is determined by solving $mx + b = 0$.]
> 3. Draw a line through the two points.

The next example gives an application of linear equations.

EXAMPLE 8   (*Linear Depreciation*)   For tax purposes, businesses must keep track of the current values of each of their assets. A common mathematical model is to assume that the current value y is related to the age x of the asset by a linear equation. A moving company buys a 40-foot van with a useful lifetime of 5 years. After x months of use, the value y of the van is estimated by the linear equation

$$y = 25{,}000 - 400x.$$

(a) Sketch the graph of this linear equation.

(b) What is the value of the van after 5 years?

(c) What economic interpretation can be given to the y-intercept of the graph?

Solution (a) The y-intercept is $(0, 25{,}000)$. The x-intercept is found from the equation

$$25{,}000 - 400x = 0$$

$$x = \frac{125}{2},$$

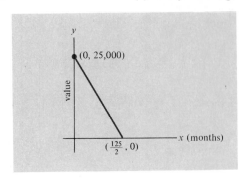

FIGURE 5

so the x-intercept is $(125/2, 0)$. The graph of the linear equation is sketched in Fig. 5. Note how the value decreases as the age of the truck increases. The value of the truck reaches 0 after $125/2$ months. Note also that we have only sketched the portion of the graph which has physical meaning, namely the portion for x between 0 and $125/2$.

(b) After 5 years (or 60 months), the value of the van is

$$y = 25{,}000 - 400(60)$$

$$= 25{,}000 - 24{,}000$$

$$= 1000.$$

Since the useful life of the van is 5 years, this value represents the *salvage value* of the van.

(c) The y-intercept corresponds to the value of the truck at $x = 0$ months, that is, the initial value of the truck, $25{,}000.

PRACTICE PROBLEMS 1

1. Plot the point $(500, 200)$.

2. Is the point $(4, -7)$ on the graph of the linear equation $2x - 3y = 1$? Is the point $(5, 3)$?

3. Graph the linear equation $5x + y = 10$.

4. Graph the straight line $y = 3x$.

EXERCISES 1

In Exercises 1–8, plot the given point.

1. $(2, 3)$	**2.** $(-1, 4)$	**3.** $(0, -2)$	**4.** $(2, 0)$
5. $(-2, 1)$	**6.** $(-1, -\frac{5}{2})$	**7.** $(-20, 40)$	**8.** $(25, 30)$

In Exercises 9–12, each linear equation is in the standard form $y = mx + b$. Identify m and b.

9. $y = 5x + 8$	**10.** $y = -2x - 6$	**11.** $y = 3$	**12.** $y = \frac{2}{3}x$

In Exercises 13–16, put the linear equations into standard form.

13. $14x + 7y = 21$ **14.** $x - y = 3$ **15.** $3x = 5$

16. $-\frac{1}{2}x + \frac{2}{3}y = 10$

In Exercises 17–20, find the x-intercept and the y-intercept of each line.

17. $y = -4x + 8$ **18.** $y = 5$ **19.** $x = 7$ **20.** $y = -8x$

In Exercises 21–26, graph the given linear equation.

21. $y = \frac{1}{3}x - 1$ **22.** $y = 2x$ **23.** $y = \frac{5}{2}$

24. $x = 0$ **25.** $3x + 4y = 24$ **26.** $x + y = 3$

27. Which of the following equations describe the same line as the equation $2x + 3y = 6$?

(a) $4x + 6y = 12$ (b) $y = -\frac{2}{3}x + 2$

(c) $x = 3 - \frac{3}{2}y$ (d) $6 - 2x - y = 0$

(e) $y = 2 - \frac{2}{3}x$ (f) $x + y = 1$

28. Which of the following equations describe the same line as the equation $\frac{1}{2}x - 5y = 1$?

(a) $2x - \frac{1}{5}y = 1$ (b) $x = 5y + 2$

(c) $2 - 5x + 10y = 0$ (d) $y = .1(x - 2)$

(e) $10y - x = -2$ (f) $1 + .5x = 2 + 5y$

29. Each of the lines L_1, L_2, and L_3 in Fig. 6 is the graph of one of the equations (a), (b), and (c). Match each of the following equations with its corresponding line.

(a) $x + y = 3$ (b) $2x - y = -2$ (c) $x = 3y + 3$

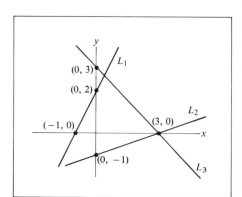

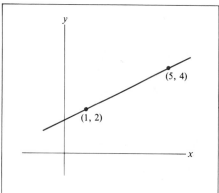

FIGURE 6   FIGURE 7

30. Which one of the following equations is graphed in Fig. 7?

(a) $x + y = 3$ (b) $y = x - 1$ (c) $2y = x + 3$

31. The temperature of water in a heating tea kettle rises according to the equation $y = 30x + 72$, where y is the temperature (in degrees Fahrenheit) x minutes after the kettle was put on the burner.

   (a) After how many minutes will the water boil?

   (b) What physical interpretation can be given to the y-intercept of the graph?

   (c) Does the x-intercept have a meaningful physical interpretation?

32. Find an equation of the line having y-intercept $(0, 5)$ and x-intercept $(4, 0)$.

33. Consider an equation of the form $y = mx + b$. When the value of m remains fixed and the value of b changes, the graph is translated vertically. As the value of b increases, does the graph move up or down?

34. Can a line have more than one x-intercept?

35. What is the equation of the x-axis?

36. What is the general form of the equation of a line that is parallel to the x-axis?

37. Does every line have a y-intercept?

38. Does every line have an x-intercept?

39. When is the x-intercept of a line the same as the y-intercept?

40. Give the x- and y-intercepts of the line $\dfrac{x}{a} + \dfrac{y}{b} = 1$.

SOLUTIONS TO PRACTICE PROBLEMS 1

1. Since the numbers are large, make each hatchmark correspond to 100. Then the point $(500, 200)$ is found by starting at the origin, moving 500 units to the right and then 200 units up.

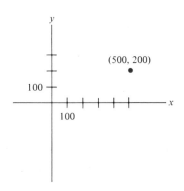

2. Substitute the point $(4, -7)$ into the equation $2x - 3y = 1$.

$$2(4) - 3(-7) = 1$$
$$8 + 21 = 1$$
$$29 = 1.$$

The equation does not hold, so $(4, -7)$ is not on the graph. Similarly, if we substitute $(5, 3)$ into the equation, we find that

$$2 \cdot 5 - 3 \cdot 3 = 1$$

$$10 - 9 = 1$$

$$1 = 1.$$

So the equation holds and $(5, 3)$ is on the graph.

3. The standard form is obtained by solving for y:

$$y = -5x + 10.$$

Thus $m = -5$ and $b = 10$. The y-intercept is $(0, 10)$. To get the x-intercept, set $y = 0$:

$$0 = -5x + 10$$

$$5x = 10$$

$$x = 2.$$

So the x-intercept is $(2, 0)$.

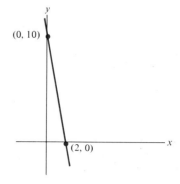

4. To find the y-intercept, set $x = 0$; then $y = 3 \cdot 0 = 0$. To find the x-intercept, set $y = 0$; then $3x = 0$ or $x = 0$. The two intercepts are the same point $(0, 0)$. We must therefore plot some other point. Setting $x = 1$, $y = 3 \cdot 1 = 3$, so another point on the line is $(1, 3)$.

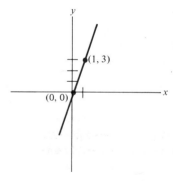

1.2. Linear Inequalities

In this section we study the properties of inequalities. Start with a Cartesian coordinate system on the line (Fig. 1). Recall that it is possible to associate with each point of the line a number; and conversely, with each number (positive, negative, or zero) it is possible to associate a point on the line.

FIGURE 1

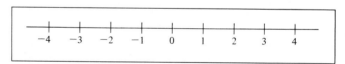

Let a and b be any numbers. We say that *a is less than b* if a lies to the left of b on the line. When a is less than b, we write $a < b$ (read: *a is less than b*). Thus, for example, $2 < 3$, $-1 < 2$, $-3 < -1$, and $0 < 4$. When a is less than b, we also say that *b is greater than a*, and we write $b > a$ (read: *b is greater than a*). Thus, for example, $3 > 2$, $2 > -1$, $-1 > -3$, and $4 > 0$. Sometimes it is convenient to have a notation that means that a is no larger than b. The notation used for this is $a \leq b$ (read: *a is less than or equal to b*). Similarly, the notation $a \geq b$ (read: *a is greater than or equal to b*) means that a is no smaller than b. The symbols $>$, $<$, $\geq$, $\leq$ are called *inequality signs*. It is easiest to remember the meaning of these various symbols by noting that the symbols $>$ and $<$ always *point toward the smaller* of a and b.

An inequality expresses a relationship between the quantities on both its sides. This relationship is very similar to the relationship expressed by an equation. And just as some problems require solving equations, others require solving inequalities. Our next task will be to state and illustrate the arithmetic operations permissible in dealing with inequalities. These permissible operations form the basis of a technique for solving the inequalities occurring in applications.

☐ **Inequality Property 1** Suppose that $a < b$ and c is any number. Then $a + c < b + c$ and $a - c < b - c$. In other words, the same number can be added to or subtracted from both sides of the inequality.

For example, start with the inequality $2 < 3$ and add 4 to both sides to get

$$2 + 4 < 3 + 4.$$

That is,

$$6 < 7,$$

a correct inequality.

EXAMPLE 1 Solve the inequality $x + 4 > 3$. That is, determine all values of x for which the inequality holds.

Solution We proceed as we would in dealing with an equation. Isolate x on the left by subtracting 4 from both sides, which is permissible by Inequality Property 1:

$$x + 4 > 3$$
$$(x + 4) - 4 > 3 - 4$$
$$x > -1.$$

That is, the values of x for which the inequality holds are exactly those x greater than -1.

In dealing with an equation, both sides may be multiplied or divided by a number. However, multiplying or dividing an inequality by a number requires some care. The result depends on whether the number is positive or negative. More precisely:

□ Inequality Property 2

2A. If $a < b$ and c is positive, then $ac < bc$.

2B. If $a < b$ and c is negative, then $ac > bc$.

In other words, an inequality may be multiplied by a positive number, just as in the case of equations. But to multiply an inequality by a negative number, it is necessary to reverse the inequality sign. For example, the inequality $-1 < 2$ can be multiplied by 4 to get $-4 < 8$, a correct statement. But if we were to multiply by -4, it would be necessary to reverse the inequality sign, since -4 is negative. In this latter case we would get $4 > -8$, a correct statement.

Note Properties 1 and 2 of inequalities are stated using only $<$. However, exactly the same properties hold if $<$ is replaced by $>$, $\leq$, or $\geq$.

EXAMPLE 2 Solve the inequality $-3x + 2 \geq -1$.

Solution Treat the inequality as if it were an equation. The goal is to isolate x on one side. To this end, first subtract 2 from both sides (Property 1). This gives

$$(-3x + 2) - 2 \geq -1 - 2$$

$$-3x \geq -3.$$

Next, we multiply by $-\frac{1}{3}$. (This gives x on the left.) But $-\frac{1}{3}$ is negative, so by Property 2B we must reverse the inequality sign. Thus

$$-\tfrac{1}{3}(-3x) \leq -\tfrac{1}{3}(-3)$$

$$x \leq 1.$$

Therefore, the values of x satisfying the inequality are precisely those ≤ 1.

The inequalities of greatest interest to us* are those in two variables x and y and having the form $cx + dy \leq e$ or $cx + dy \geq e$, where c, d, and e are given numbers, with c and d not both 0. We will call such inequalities *linear inequalities*. When $d \neq 0$ (that is, when y actually appears), the inequality can be put into one of the *standard forms* $y \leq mx + b$ or $y \geq mx + b$. When $d = 0$, the inequality can be put into one of the *standard forms* $x \leq a$ or $x \geq a$. The procedure for putting a linear inequality into standard form is analogous to that for putting a linear equation into standard form.

EXAMPLE 3 Put the linear inequality $2x - 3y \geq -9$ into standard form.

---

* Such inequalities arise in our discussion of linear programming in Chapter 3.

Solution   Since we want the y term on the left and all other terms on the right, begin by subtracting $2x$ from both sides:

$$-3y \geq -2x - 9.$$

Next, multiply by $-\frac{1}{3}$, remembering to change the inequality sign, since $-\frac{1}{3}$ is negative:

$$y \leq -\tfrac{1}{3}(-2x - 9)$$

$$y \leq \tfrac{2}{3}x + 3.$$

The last inequality is in standard form.

EXAMPLE 4   Find the standard form of the inequality $\frac{1}{2}x \geq 4$.

Solution   Note that y does not appear in the inequality. Just as was the case in finding the standard form of a linear equation when y does not appear, solve for x. To do this multiply by 2 to get

$$x \geq 8,$$

the standard form.

Graphing Linear Inequalities   Associated with every linear inequality, there is a set of points of the plane, the set of all those points which satisfy the inequality. This set of points is called the *graph* of the inequality.

EXAMPLE 5   Determine whether or not the given point satisfies the inequality $y \geq -\frac{2}{3}x + 4$.

(a) $(3, 4)$ (b) $(0, 0)$

Solution   Substitute the x-coordinate of the point for x and the y-coordinate for y and determine if the resulting inequality is correct or not.

(a) $4 \geq -\frac{2}{3}(3) + 4$ (b) $0 \geq -\frac{2}{3}(0) + 4$
    $4 \geq -2 + 4$ $0 \geq 0 + 4$
    $4 \geq 2$ (correct) $0 \geq 4$ (not correct)

Therefore, the point $(3, 4)$ satisfies the inequality and the point $(0, 0)$ does not.

It is easiest to determine the graph of a given inequality after it has been written in standard form. Therefore, let us describe the graphs of each of the standard forms. The easiest to handle are the forms $x \geq a$ and $x \leq a$.

A point satisfies the inequality $x \geq a$ if and only if its x-coordinate is greater than or equal to a. The y-coordinate can be anything. Therefore, the graph of $x \geq a$ consists of all points to the right of and on the vertical line $x = a$. We will display the graph by crossing out the portion of the plane to the left of the line. (See Fig. 2.) Similarly, the graph of $x \leq a$ consists of the points to the left of and on the line $x = a$. This graph is shown in Fig. 3.

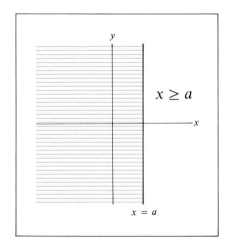

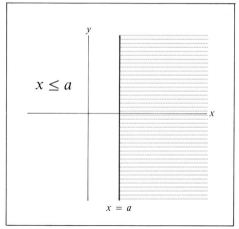

FIGURE 2   FIGURE 3

Here is a simple procedure for graphing the other two standard forms.

To graph $y \geq mx + b$ or $y \leq mx + b$:

1. Draw the graph of $y = mx + b$.
2. Throw away, that is, "cross out" the portion of the plane not satisfying the inequality. The graph of $y \geq mx + b$ consists of all points above or on the line. The graph of $y \leq mx + b$ consists of all points below or on the line.

The graphs of the inequalities $y \geq mx + b$ and $y \leq mx + b$ are shown in Fig. 4. Some simple reasoning suffices to show why these graphs are correct. Draw

FIGURE 4

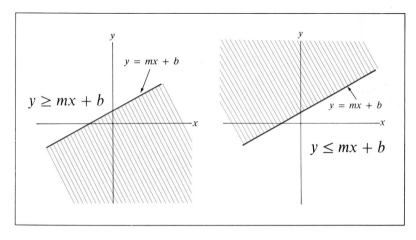

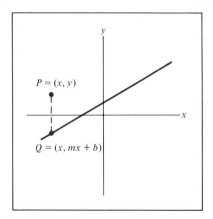

FIGURE 5

the line $y = mx + b$, as in Fig. 5, and pick any point P above the line. Suppose that P has coordinates (x, y). Let Q be the point on the line that lies directly below P. Then Q has the same first coordinate as P, that is, x. Since Q lies on the line, the second coordinate of Q is $mx + b$. Since the second coordinate of P is clearly larger than the second coordinate of Q, we must have $y \geq mx + b$. Similarly, any point below the line satisfies $y \leq mx + b$. Thus the two graphs are as given in Fig. 4.

EXAMPLE 6 Graph the inequality $2x + 3y \geq 15$.

Solution In order to apply the procedure above, the inequality must first be put into standard form:

$$2x + 3y \geq 15$$

$$3y \geq -2x + 15$$

$$y \geq -\tfrac{2}{3}x + 5.$$

The last inequality is in standard form. Next, we graph the line $y = -\tfrac{2}{3}x + 5$. Its intercepts are $(0, 5)$ and $(\tfrac{15}{2}, 0)$. Since the inequality is "$y \geq$" we cross out the region below the line and label the region above with the inequality. The graph consists of all points above or on the line (Fig. 6).

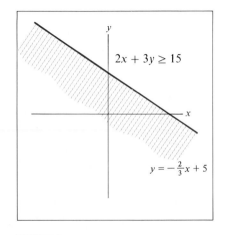

FIGURE 6

FIGURE 7

EXAMPLE 7 Graph the inequality $4x - 2y \geq 12$.

Solution First put the inequality in standard form:

$$4x - 2y \geq 12$$

$$-2y \geq -4x + 12$$

$$y \leq 2x - 6$$

(note the change in the inequality sign!). Next, graph $y = 2x - 6$. The intercepts are $(0, -6)$ and $(3, 0)$. Since the inequality is "$y \leq$" the graph consists of all points below or on the line (Fig. 7).

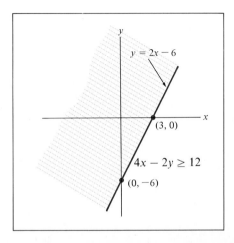

So far, we have been concerned only with graphing single inequalities. The next example concerns graphing a system of inequalities. That is, it asks us to determine all points of the plane which *simultaneously* satisfy all inequalities of a system.

EXAMPLE 8   Graph the system of inequalities

$$\begin{cases} 2x + 3y \geq 15 \\ 4x - 2y \geq 12 \\ \qquad\quad y \geq 0. \end{cases}$$

Solution   The first two inequalities have already been graphed in Examples 6 and 7. The graph of $y \geq 0$ consists of all points above the x-axis. In Fig. 8, any point that is crossed out is *not* on the graph of at least one inequality. So the points which simultaneously satisfy all three inequalities are those in the remaining clear region.

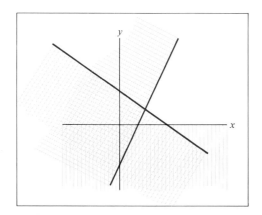

Remark At first, our convention of crossing out those points *not* on the graph of an inequality (instead of shading the points *on* the graph) may have seemed odd. However, the real advantage of this convention becomes apparent when graphing a system of inequalities. Imagine trying to find the graph of the system of Example 8 if the points *on* the graph of each inequality had been shaded. It would have been necessary to locate the points that had been shaded three times. This is hard to do.

FIGURE 8

The graph of a system of inequalities is called a *feasible set*. The feasible set associated to the system of Example 8 is a three-sided, unbounded region.

Given a specific point, we should be able to decide whether or not the point lies in the feasible set. The next example shows how this is done.

EXAMPLE 9   Determine whether the points (5, 3) and (4, 2) are in the feasible set of the system of inequalities of Example 8.

Solution   If we had very accurate measuring devices, we could plot the points in the graph of Figure 8 and determine whether or not they lie in the feasible set. However, there is a simpler and more reliable algebraic method. Just substitute the coordinates of the points into each of the inequalities of the system and see whether or not *all* of the inequalities are satisfied. So doing we find that (5, 3) is in the feasible set and (4, 2) is not.

$$(5, 3) \quad \begin{cases} 2(5) + 3(3) \geq 15 \\ 4(5) - 2(3) \geq 12 \\ \qquad\quad (3) \geq 0 \end{cases} \qquad \begin{cases} 19 \geq 15 \\ 14 \geq 12 \\ \;\; 3 \geq 0 \end{cases}$$

$$(4, 2) \quad \begin{cases} 2(4) + 3(2) \geq 15 \\ 4(4) - 2(2) \geq 12 \\ \qquad\quad (2) \geq 0 \end{cases} \qquad \begin{cases} 14 \geq 15 \\ 12 \geq 12 \\ \;\; 2 \geq 0 \end{cases}$$

PRACTICE PROBLEMS 2

1. Graph the inequality $3x - y \geq 3$.

2. Graph the feasible set for the system of inequalities

$$\begin{cases} x \geq 0, \quad y \geq 0 \\ x + 2y \leq 4 \\ 4x - 4y \geq -4. \end{cases}$$

EXERCISES 2

In Exercises 1–4, tell whether the inequality is true or false.

1. $2 \leq -3$ 2. $-2 \leq 0$ 3. $7 \leq 7$ 4. $0 \geq \frac{1}{2}$

In Exercises 5–7, solve for x.

5. $2x - 5 \geq 3$ 6. $3x - 7 \leq 2$ 7. $-5x + 13 \leq -2$

8. Which of the following results from solving $-x + 1 \leq 3$ for x?

    (a) $x \leq 4$ (b) $x \leq 2$ (c) $x \geq -4$ (d) $x \geq -2$

In Exercises 9 –14, put the linear inequality into standard form.

9. $2x + y \leq 5$ 10. $-3x + y \geq 1$ 11. $5x - \frac{1}{3}y \leq 6$

12. $\frac{1}{2}x - y \leq -1$ 13. $4x \geq -3$ 14. $-2x \leq 4$

In Exercises 15–22, determine whether or not the given point satisfies the given inequality.

15. $3x + 5y \leq 12, (2, 1)$ 16. $-2x + y \geq 9, (3, 15)$

17. $y \geq -2x + 7, (3, 0)$ 18. $y \leq \frac{1}{2}x + 3, (4, 6)$

19. $y \leq 3x - 4, (3, 5)$ 20. $y \geq x, (-3, -2)$

21. $x \geq 5, (7, -2)$ 22. $x \leq 7, (0, 0)$

In Exercises 23–26, graph the given inequality by crossing out (i.e., discarding) the points not satisfying the inequality.

23. $y \leq \frac{1}{3}x + 1$

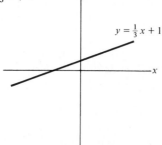

24. $y \geq -x + 1$

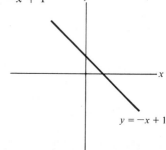

25. $x \geq 4$ **26.** $y \leq 2$

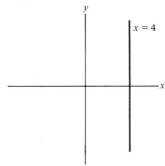 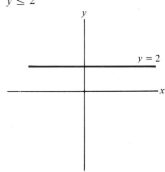

In Exercises 27–36, graph the given inequality.

27. $y \leq 2x + 1$ **28.** $y \geq -3x + 6$ **29.** $x \geq 2$

30. $x \geq 0$ **31.** $x + 4y \geq 12$ **32.** $4x - 4y \geq 8$

33. $4x - 5y + 25 \geq 0$ **34.** $.2y - x \geq .4$ **35.** $\frac{1}{2}x - \frac{1}{3}y \leq 1$

36. $3y - 3x \leq 2y + x + 1$

In Exercises 37–42, graph the feasible set for the system of inequalities.

37. $\begin{cases} y \leq 2x - 4 \\ y \geq 0 \end{cases}$ **38.** $\begin{cases} y \geq -\frac{1}{3}x + 1 \\ x \geq 0 \end{cases}$ **39.** $\begin{cases} x + 2y \geq 2 \\ 3x - y \geq 3 \end{cases}$

40. $\begin{cases} 3x + 6y \geq 24 \\ 3x + y \geq 6 \end{cases}$ **41.** $\begin{cases} x + 5y \leq 10 \\ x + y \leq 3 \\ x \geq 0, \ y \geq 0 \end{cases}$ **42.** $\begin{cases} x + 2y \geq 6 \\ x + y \geq 5 \\ x \geq 1 \end{cases}$

In Exercises 43–46, determine whether the given point is in the feasible set of this system of inequalities:

$$\begin{cases} 6x + 3y \leq 96 \\ x + y \leq 18 \\ 2x + 6y \leq 72 \\ x \geq 0, \quad y \geq 0. \end{cases}$$

43. $(8, 7)$ **44.** $(14, 3)$ **45.** $(9, 10)$ **46.** $(16, 0)$

In Exercises 47–50, determine whether the given point is above or below the given line.

47. $y = 2x + 5, (3, 9)$ **48.** $3x - y = 4, (2, 3)$

49. $7 - 4x + 5y = 0, (0, 0)$ **50.** $x = 2y + 5, (6, 1)$

51. Give a system of inequalities for which the graph is the region between the pair of lines $8x - 4y - 4 = 0$ and $8x - 4y = 0$.

52. Find a system of linear inequalities having the feasible set of Fig. 9.

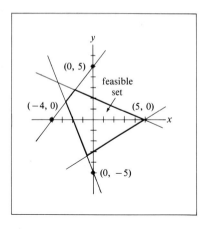

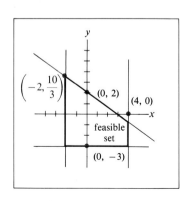

FIGURE 9

FIGURE 10

53. Find a system of linear inequalities having the feasible set of Fig. 10.

SOLUTIONS TO PRACTICE PROBLEMS 2

1. Linear inequalities are easiest to graph if they are first put into standard form. Subtract $3x$ from both sides and multiply by -1:

$$3x - y \geq 3$$

$$-y \geq -3x + 3$$

$$y \leq 3x - 3.$$

Now graph the line $y = 3x - 3$. The graph of the inequality is the portion of the plane below and on the line ("$\leq$" corresponds to below), so throw away (that is, cross out) the portion above the line.

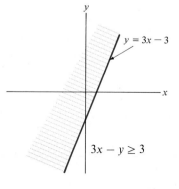

2. Begin by putting the linear inequalities into standard form and then graphing them all on the same coordinate system.

$$\begin{cases} x \geq 0, \quad y \geq 0 \\ x + 2y \leq 4 \\ 4x - 4y \geq -4 \end{cases} \quad \text{has standard form} \quad \begin{cases} x \geq 0, \quad y \geq 0 \\ y \leq -\tfrac{1}{2}x + 2 \\ y \leq x + 1. \end{cases}$$

A good procedure to follow is to graph all of the linear equations and then cross out the regions to be thrown away one at a time.

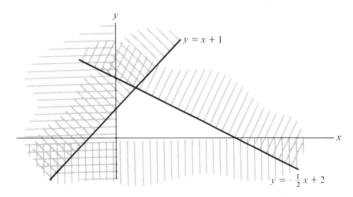

The inequalities $x \geq 0$ and $y \geq 0$ arise frequently in applications. The first has the form $x \geq a$, where $a = 0$, and the second has the form $y \geq mx + b$, where $m = 0$ and $b = 0$. To graph them, just cross out all points to the left of the y-axis and all points below the x-axis, respectively.

1.3. The Intersection Point of a Pair of Lines

Suppose that we are given a pair of intersecting straight lines L and M. Let us consider the problem of determining the coordinates of the point of intersection $S = (x, y)$ (see Fig. 1). We may as well assume that the equations of L and M are given in standard form. First, let us assume that both lines are in the first standard form— that is, that the equations are

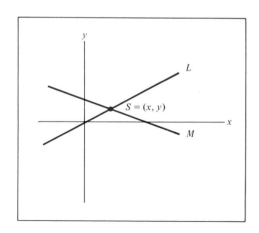

FIGURE 1

$$L: \ y = mx + b, \qquad M: \ y = nx + c.$$

Since the point S is on both lines, its coordinates satisfy both equations. In particular, we have two expressions for its y-coordinate:

$$y = mx + b = nx + c.$$

The last equality gives an equation from which x can easily be determined. Then the value of y can be determined as $mx + b$. Let us see how this works in a particular example.

EXAMPLE 1 Find the point of intersection of the two lines $y = 2x - 3$, $y = x + 1$.

Solution To find the x-coordinate of the point of intersection, equate the two expressions for y and solve for x:

$$2x - 3 = x + 1$$
$$2x - x = 1 + 3$$
$$x = 4.$$

To find the value of y, set $x = 4$ in either equation, say the first. Then

$$y = 2 \cdot 4 - 3 = 5.$$

So the point of intersection is $(4, 5)$.

EXAMPLE 2 Find the point of intersection of the two lines $x + 2y = 6$, $5x + 2y = 18$.

Solution To use the method described, the equations must be in standard form. Solving both equations for y, we get the standard forms

$$y = -\tfrac{1}{2}x + 3$$
$$y = -\tfrac{5}{2}x + 9.$$

Equating the expressions for y gives

$$-\tfrac{1}{2}x + 3 = -\tfrac{5}{2}x + 9$$
$$\tfrac{5}{2}x - \tfrac{1}{2}x = 9 - 3$$
$$2x = 6$$
$$x = 3.$$

Setting $x = 3$ in the first equation gives

$$y = -\tfrac{1}{2}(3) + 3 = \tfrac{3}{2}.$$

So the intersection point is $(3, \tfrac{3}{2})$.

The method above works when both equations have the first standard form. In case one equation has the standard form $x = a$, things are much simpler. The value of x is then given directly without any work, namely $x = a$. The value of y can be found by substituting a for x in the other equation.

EXAMPLE 3 Find the point of intersection of the lines $y = 2x - 1$, $x = 2$.

Solution The x-coordinate of the intersection point is 2, and the y-coordinate is $y = 2 \cdot 2 - 1 = 3$. Therefore, the intersection point is $(2, 3)$.

The method introduced above may be used to solve systems of two equations in two variables.

EXAMPLE 4 Solve the following system of linear equations:

$$\begin{cases} 2x + 3y = 7 \\ 4x - 2y = 9. \end{cases}$$

Solution First convert the equations to standard form:

$$2x + 3y = 7$$
$$3y = -2x + 7$$
$$y = -\tfrac{2}{3}x + \tfrac{7}{3};$$
$$4x - 2y = 9$$
$$-2y = -4x + 9$$
$$y = 2x - \tfrac{9}{2}.$$

Now equate the two expressions for y:

$$2x - \tfrac{9}{2} = -\tfrac{2}{3}x + \tfrac{7}{3}$$
$$\tfrac{8}{3}x = \tfrac{7}{3} + \tfrac{9}{2} = \tfrac{14}{6} + \tfrac{27}{6} = \tfrac{41}{6}$$
$$x = \tfrac{3}{8} \cdot \tfrac{41}{6} = \tfrac{41}{16}$$
$$y = 2x - \tfrac{9}{2} = 2(\tfrac{41}{16}) - \tfrac{9}{2} = \tfrac{5}{8}.$$

So the solution of the given system is $x = \tfrac{41}{16}$, $y = \tfrac{5}{8}$.

Supply and Demand Curves Let p denote the price of a commodity and q the quantity. Economists study two sorts of graphs which express relationships between p and q. To describe these graphs, let us plot price along the horizontal axis and quantity along the vertical axis. The first graph relating p and q is called a *supply curve* and expresses the relationship between p and q from a manufacturer's point of view. For every price p, the supply curve specifies the quantity q which the manufacturer is willing to produce at the market price p. The higher the price, the more the manufacturer is willing to supply. So supply curves rise when viewed from left to right.

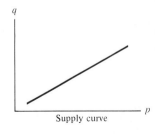
Supply curve

   The second curve relating p and q is called a *demand curve* and expresses the relationship between p and q from the consumers' viewpoint. For each price p, the demand curve expresses the quantity q which consumers will buy if the market price is p. The higher the price, the less consumers will buy. So demand curves fall when viewed from left to right.

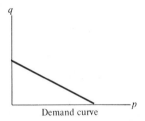
Demand curve

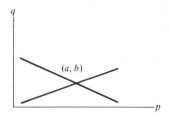
(a, b)

Suppose that the supply and demand curves for a commodity are drawn on a single coordinate system. The intersection point (a, b) of the two curves has an economic significance: prices will stabilize at a dollars per unit and the quantity produced will be b units.

EXAMPLE 5 Suppose that the supply curve for a certain commodity is the straight line whose equation is $q = 5000p - 10,000$ (p in dollars). Suppose that the demand curve for the same commodity is the straight line whose equation is $q = -2000p + 11,000$. Determine the price at which the commodity will sell and determine the quantity produced.

Solution We must solve the system of linear equations

$$\begin{cases} q = 5000p - 10,000 \\ q = -2000p + 11,000. \end{cases}$$

$$5000p - 10,000 = -2000p + 11,000$$

$$7000p = 21,000$$

$$p = 3$$

$$q = 5000(3) - 10,000$$

$$= 5000.$$

Thus the commodity will sell for $3 and 5000 units will be produced.

PRACTICE PROBLEMS 3

Figure 2 shows the feasible set of a system of linear inequalities; its four vertices are labeled $A, B, C,$ and D.

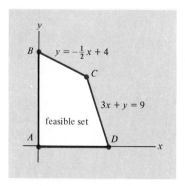

FIGURE 2

1. Use the method of this section to find the coordinates of the point C.

2. Determine the coordinates of the points A and B by inspection.

3. Find the coordinates of the point D.

EXERCISES 3

In Exercises 1–6, find the point of intersection of the given pair of straight lines.

1. $\begin{cases} y = 4x - 5 \\ y = -2x + 7 \end{cases}$

2. $\begin{cases} y = 3x - 15 \\ y = -2x + 10 \end{cases}$

3. $\begin{cases} x - 4y = -2 \\ x + 2y = 4 \end{cases}$

4. $\begin{cases} 2x - 3y = 3 \\ y = 3 \end{cases}$

5. $\begin{cases} y = \frac{1}{3}x - 1 \\ x = 12 \end{cases}$

6. $\begin{cases} 2x - 3y = 3 \\ x = 6 \end{cases}$

7. Does $(6, 4)$ satisfy the following system of linear equations?

$$\begin{cases} x - 3y = -6 \\ 3x - 2y = 10 \end{cases}$$

8. Does $(12, 4)$ satisfy the following system of linear equations?

$$\begin{cases} y = \frac{1}{3}x - 1 \\ x = 12 \end{cases}$$

Solve the following systems of linear equations.

9. $\begin{cases} 2x + y = 7 \\ x - y = 3 \end{cases}$

10. $\begin{cases} x + 2y = 4 \\ \frac{1}{2}x + \frac{1}{2}y = 3 \end{cases}$

11. $\begin{cases} 5x - 2y = 1 \\ 2x + y = -4 \end{cases}$

12. $\begin{cases} x + 2y = 6 \\ x - \frac{1}{3}y = 4 \end{cases}$

Find the coordinates of the vertices of the following feasible sets.

13.

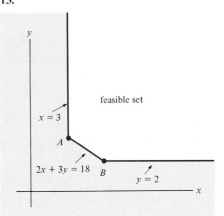

14.

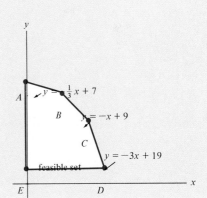

15.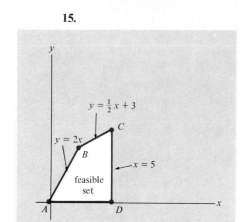

16.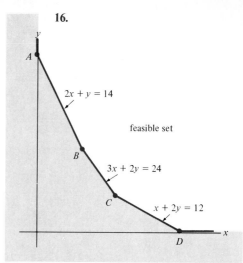

In Exercises 17–22, graph the feasible set for the system of inequalities and find the co-ordinates of the vertices.

17. $\begin{cases} 2y - x \le 6 \\ x + 2y \ge 10 \\ x \le 6 \end{cases}$

18. $\begin{cases} 2x + y \ge 10 \\ x \ge 2 \\ y \ge 2 \end{cases}$

19. $\begin{cases} x + 3y \le 18 \\ 2x + y \le 16 \\ x \ge 0, \quad y \quad 0 \end{cases}$

20. $\begin{cases} 5x + 2y \ge 14 \\ x + 3y \ge 8 \\ x \ge 0, \quad y \ge 0 \end{cases}$

21. $\begin{cases} 4x + y \ge 8 \\ x + y \ge 5 \\ x + 3y \ge 9 \\ x \ge 0, \quad y \ge 0 \end{cases}$

22. $\begin{cases} x + 4y \le 28 \\ x + y \le 10 \\ 3x + y \le 24 \\ x \ge 0, \quad y \ge 0 \end{cases}$

23. The supply curve for a certain commodity is $q = 10{,}000p - 500$.

    (a) Suppose that the market price is \$2 per unit. How many units of the commodity will be sold?

    (b) What is the highest price at which a quantity of 0 will be supplied?

24. The demand curve for a certain commodity is $q = -1000p + 32{,}500$.

    (a) Suppose that the market price is \$1 per unit. How many units of the commodity will be demanded?

    (b) What is the lowest price at which a quantity of 0 will be demanded?

25. Suppose that supply and demand are described by the supply and demand curves of Exercises 23 and 24.

    (a) Determine the equilibrium selling price of the commodity.

    (b) Determine the number of units that will be produced at that price.

26. The x- and y-intercepts of line L_1 are 1 unit to the left and 2 units below the x- and y-intercepts of line L_2, respectively. The equation for L_1 is $y = 3x + 6$. Find the point of intersection of L_1 and L_2.

27. A plant supervisor must apportion her 40-hour work week between hours working on the assembly line and hours supervising the work of others. She is paid $12 per hour for working and $15 per hour for supervising. If her earnings for a certain week are $504, how much time does she spend on each task?

SOLUTIONS TO PRACTICE PROBLEMS 3

1. The point C is the point of intersection of the lines with equations $y = -\frac{1}{2}x + 4$ and $3x + y = 9$. In order to use the method of this section, the second equation must first be put into its standard form $y = -3x + 9$. Now equate the two expressions for y and solve.

$$-\tfrac{1}{2}x + 4 = -3x + 9$$
$$\tfrac{5}{2}x = 5$$
$$x = \tfrac{2}{5} \cdot 5 = 2$$
$$y = -\tfrac{1}{2}(2) + 4 = 3.$$

Therefore, $C = (2, 3)$.

2. $A = (0, 0)$, since the point A is the origin. $B = (0, 4)$, since it is the y-intercept of the line with equation $y = -\frac{1}{2}x + 4$.

3. D is the x-intercept of the line $3x + y = 9$. Its first coordinate is found by setting $y = 0$ and solving for x.

$$3x + (0) = 9$$
$$x = 3.$$

Therefore, $D = (3, 0)$.

1.4. The Slope of a Straight Line

As we have seen, any linear equation can be put into one of the two standard forms $y = mx + b$ or $x = a$. In this section, let us exclude linear equations whose standard form is of the latter type. *Geometrically, this means that we will consider only nonvertical lines.*

Suppose that we are given a nonvertical line L whose equation is $y = mx + b$. The number m is called the *slope of L*. That is, the slope is the coefficient of x in the standard form of the equation of the line.

EXAMPLE 1 Find the slopes of the lines having the following equations:

(a) $y = 2x + 1$ (b) $y = -\frac{3}{4}x + 2$ (c) $y = 3$ (d) $-8x + 2y = 4$

Solution (a) $m = 2$.

(b) $m = -\frac{3}{4}$.

(c) When we write the equation in the form $y = 0 \cdot x + 3$, we see that $m = 0$.

(d) First, we put the equation in standard form:

$$-8x + 2y = 4$$
$$2y = 8x + 4$$
$$y = 4x + 2.$$

Thus $m = 4$.

The definition of the slope given is in terms of the standard form of the equation of the line. Let us give an alternative definition.

Geometric Definition of Slope   Let L be a line passing through the points (x_1, y_1) and (x_2, y_2). Then the slope of L is given by the formula

$$m = \frac{y_2 - y_1}{x_2 - x_1}. \tag{1}$$

That is, the slope is the difference in the y-coordinates divided by the difference in the x-coordinates, with both differences formed in the same order.

Before proving this definition equivalent to the first one given, let us show how it can be used.

EXAMPLE 2   Find the slope of the line passing through the points $(1, 3)$ and $(4, 6)$.

Solution   We have

$$m = \frac{[\text{difference in } y\text{-coordinates}]}{[\text{difference in } x\text{-coordinates}]} = \frac{6 - 3}{4 - 1} = \frac{3}{3} = 1.$$

Thus $m = 1$. [Note that if we reverse the order of the points and use formula (1) to compute the slope, then we get

$$m = \frac{3 - 6}{1 - 4} = \frac{-3}{-3} = 1,$$

which is the same answer. The order of the points is immaterial. The important concern is to make sure that the differences in the x- and y-coordinates are formed in the same order.]

The slope of a line does not depend on which pair of points are chosen as (x_1, y_1) and (x_2, y_2). Consider the line $y = 4x - 3$ and two points $(1, 1)$ and $(3, 9)$, which are on the line. Using these two points, we calculate the slope to be

$$m = \frac{9 - 1}{3 - 1} = \frac{8}{2} = 4.$$

Now let's choose two other points on the line, say $(2, 5)$ and $(-1, -7)$, and use these points to determine m. We obtain

$$m = \frac{-7 - 5}{-1 - 2} = \frac{-12}{-3} = 4.$$

The two pairs of points give the same slope, 4.

Justification of Formula 1   Since (x_1, y_1) and (x_2, y_2) are both on the line, both points satisfy the equation of the line, which has the form $y = mx + b$. Thus

$$y_2 = mx_2 + b$$

$$y_1 = mx_1 + b.$$

Subtracting these two equations gives

$$y_2 - y_1 = mx_2 - mx_1 = m(x_2 - x_1).$$

Dividing by $x_2 - x_1$, we have

$$m = \frac{y_2 - y_1}{x_2 - x_1},$$

which is formula (1). So the two definitions of slope lead to the same number.

Let us now study four of the most important properties of the slope of a straight line. We begin with the so-called steepness property, since it provides us with a geometric interpretation for the number m.

> **Steepness Property**   Let the line L have slope m. If we start at any point on the line and move 1 unit to the right, then we must move m units vertically in order to return to the line (Fig. 1). (Of course, if m is positive, then we move up; and if m is negative, we move down.)

FIGURE 1

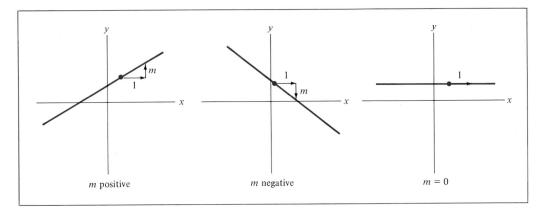

m positive m negative $m = 0$

EXAMPLE 3 Illustrate the steepness property for each of the lines.

(a) $y = 2x + 1$ (b) $y = -\frac{3}{4}x + 2$ (c) $y = 3$

Solution (a) Here $m = 2$. So starting from any point on the line, proceeding 1 unit to the right, we must go 2 units up to return to the line (Fig. 2).

(b) Here $m = -\frac{3}{4}$. So starting from any point on the line, proceeding 1 unit to the right, we must go $\frac{3}{4}$ unit down to return to the line (Fig. 3).

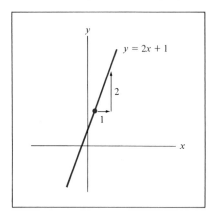

FIGURE 2

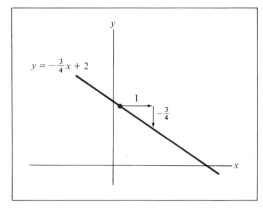

FIGURE 3

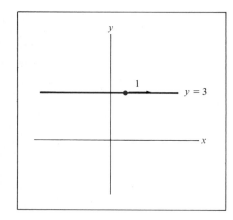

FIGURE 4

(c) Here $m = 0$. So going 1 unit to the right requires going 0 units vertically to return to the line (Fig. 4).

In the next example we introduce a new method for graphing a linear equation. This method relies on the steepness property and is much more efficient than finding two points on the line (e.g., the two intercepts).

EXAMPLE 4 Use the steepness property to draw the graph of $y = \frac{1}{2}x + \frac{3}{2}$.

Solution The y-intercept is $(0, \frac{3}{2})$, as we read off from the equation. We can find another point on the line using the steepness property. Start at $(0, \frac{3}{2})$. Go 1 unit to the right. Since the slope is $\frac{1}{2}$, we must move vertically $\frac{1}{2}$ unit to return to the line. But this locates a second point on the line. So we draw the line through the two points. The entire procedure is illustrated in Fig. 5.

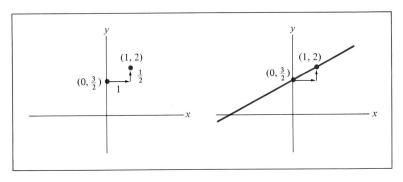

FIGURE 5

Actually, in order to use the steepness property to graph an equation, all that is needed is the slope plus *any* point (not necessarily the y-intercept).

EXAMPLE 5   Graph the line of slope -1 and passing through the point (2, 2).

Solution   Start at (2, 2), move 1 unit to the right and then -1 unit vertically, that is, 1 unit down. The line through (2, 2) and the resulting point is the desired line (see Fig. 6).

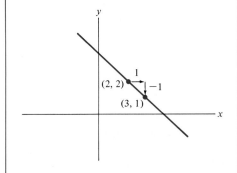

Slope measures the *steepness* of a line. Namely, the slope of a line tells whether it is rising or falling, and how fast. Specifically, lines of positive slope rise as we move from left to right. Lines of negative slope fall, and lines of zero slope stay level. The larger the magnitude of the slope, the steeper the ascent or descent. These facts are directly implied by the steepness property (see Fig. 7).

FIGURE 6

FIGURE 7

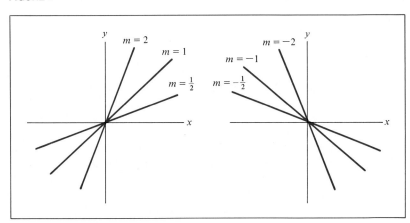

Consider a line with equation $y = mx + b$, and let (x_1, y_1) be any point on the line. If we start from this point and move 1 unit to the right, the first coordinate of the new point is $x_1 + 1$, since the x-coordinate is increased by 1. Now go far enough vertically to return to the line. Denote the y-coordinate of this new point by y_2 (see Fig. 8). We must show that to get y_2, we add m to y_1. That is, $y_2 = y_1 + m$. By equation (1), we can compute m as

$$m = \frac{[\text{difference in } y\text{-coordinates}]}{[\text{difference in } x\text{-coordinates}]}$$

$$= \frac{y_2 - y_1}{1} = y_2 - y_1.$$

In other words, $y_2 = y_1 + m$, which is what we desired to show.

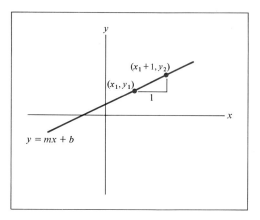

FIGURE 8

Often the slopes of the straight lines which occur in applications have interesting and significant interpretations. An application in the field of economics is illustrated in the next example.

EXAMPLE 6 Suppose a manufacturer finds that the cost y of producing x units of a certain commodity is given by the formula $y = 2x + 5000$. What interpretation can be given the slope of the graph of this equation?

Solution Suppose that the firm is producing at a certain level and increases production by 1 unit. That is, x is increased by 1 unit. By the steepness property, the value of y then increases by 2, which is the slope of the line whose equation is $y = 2x + 5000$. Thus, *each additional unit of production costs $2*. (The graph of $y = 2x + 5000$ is called a *cost curve*. It relates the size of production to total cost. The graph is a straight line, and economists call its slope the *marginal cost of production*. The y-coordinate of the y-intercept is called the *fixed cost*. In this case the fixed cost is $5000, and it includes costs such as rent and insurance which are incurred even if no units are produced.)

In applied problems having time as a variable, the letter t is often used in place of the letter x. If so, straight lines have equations of the form $y = mt + b$ and are graphed on a ty-coordinate system.

EXAMPLE 7 (*Depreciation*) The federal government allows an income tax deduction for the decrease in value (or *depreciation*) of capital assets (such as buildings and equipment). One method of calculating the depreciation is to take equal amounts over the expected lifetime of the asset. This method is called *straight-line depreciation*. For tax purposes the value V of an asset t years after purchase is figured according to the equation $V = -100{,}000t + 700{,}000$. The expected life of the asset is 5 years.

(a) How much did the asset originally cost?

(b) What is the annual deduction for depreciation?

(c) What is the salvage value of the asset? (That is, what is the value of the asset after 5 years?)

Solution   (a) The original cost is the value of V at $t = 0$, namely

$$V = -100{,}000(0) + 700{,}000 = 700{,}000.$$

That is, the asset originally cost $700,000.

(b) By the steepness property, each increase of 1 in t causes a decrease in V of 100,000. That is, the value is decreasing by $100,000 per year. So the depreciation deduction is $100,000 each year.

(c) After 5 years, the value V is given by

$$V = -100{,}000(5) + 700{,}000 = 200{,}000.$$

The salvage value is $200,000.

We have seen in Example 5 how to sketch a straight line when given its slope and one point on it. Let us now see how to find the equation of the line from this same data.

Point-Slope Formula   The equation of the straight line passing through (x_1, y_1) and having slope m is given by $y - y_1 = m(x - x_1)$.

EXAMPLE 8   Find the equation of the line that passes through $(2, 3)$ and has slope $\frac{1}{2}$.

Solution   Here $x_1 = 2$, $y_1 = 3$, $m = \frac{1}{2}$. So the equation is

$$y - 3 = \tfrac{1}{2}(x - 2)$$
$$y - 3 = \tfrac{1}{2}x - 1$$
$$y = \tfrac{1}{2}x + 2.$$

EXAMPLE 9   Find the equation of the line through the points $(3, 1)$ and $(6, 0)$.

Solution   We can compute the slope from equation (1):

$$[\text{slope}] = \frac{[\text{difference in } y\text{-coordinates}]}{[\text{difference in } x\text{-coordinates}]} = \frac{1 - 0}{3 - 6} = -\tfrac{1}{3}.$$

Now we can determine the equation from the point-slope formula with $(x_1, y_1) = (3, 1)$ and $m = -\frac{1}{3}$:

$$y - 1 = -\tfrac{1}{3}(x - 3)$$

$$y = -\tfrac{1}{3}x + 2.$$

[Question: What would the equation be if we had chosen $(x_1, y_1) = (6, 0)$?]

EXAMPLE 10 For each unit in monthly advertising expenditure, a store experiences a 6-unit increase in sales revenue. Even without advertising, the store has $30,000 in sales revenue per month. Let x be the number of dollars of advertising expenditures per month and let y be the number of dollars in sales revenue per month.

(a) Find the equation of the line that expresses the relationship between x and y.

(b) If the store spends $10,000 in advertising, what will be the sales revenue for the month?

(c) How much would the store have to spend on advertising to attain $150,000 in sales revenue for the month?

Solution (a) The steepness property tells us that the line has slope $m = 6$. Since $x = 0$ (no advertising expenditures) yields $y = \$30,000$, the y-intercept of the line is $(0, 30,000)$. Therefore, the standard form of the equation of a line is

$$y = 6x + 30,000.$$

(b) If $x = 10,000$, then $y = 6(10,000) + 30,000 = 90,000$. Therefore, the sales revenue for the month will be $90,000.

(c) We are given that $y = 150,000$, and we must find the value of x for which

$$150,000 = 6x + 30,000.$$

Solving for x, we obtain $6x = 120,000$, and hence $x = \$20,000$. In order to attain $150,000 in sales revenue, the store should invest $20,000 in advertising.

Verification of the Point-Slope Formula Let (x, y) be any point on the line passing through the point (x_1, y_1) and having slope m. Then, by equation (1), we have

$$m = \frac{y - y_1}{x - x_1}.$$

Multiplying through by $x - x_1$ gives

$$y - y_1 = m(x - x_1). \tag{2}$$

Thus, every point (x, y) on the line satisfies equation (2). So (2) gives the equation of the line through (x_1, y_1) and having slope m.

The next property of slope relates the slopes of two perpendicular lines.

Perpendicular Property   When two lines are perpendicular, their slopes are negative reciprocals of one another. That is, if two lines with slopes m and n are perpendicular to one another, then*

$$m = -\frac{1}{n}.$$

Conversely, if two lines have slopes that are negative reciprocals of one another, then they are perpendicular.

A proof of the perpendicular property is outlined in Exercise 83. Let us show how it can be used to help find equations of lines.

EXAMPLE 11   Find the equation of the line perpendicular to the graph of $y = 2x - 5$ and passing through $(1, 2)$.

Solution   The slope of the graph of $y = 2x - 5$ is 2. By the perpendicular property, the slope of a line perpendicular to it is $-\frac{1}{2}$. If a line has slope $-\frac{1}{2}$ and passes through $(1, 2)$, then it has equation

$$y - 2 = -\tfrac{1}{2}(x - 1)$$

or

$$y = -\tfrac{1}{2}x + \tfrac{5}{2}$$

(by the point-slope formula).

The final property of slopes gives the relationship between slopes of parallel lines. A proof is outlined in Exercise 82.

Parallel Property   Parallel lines have the same slope. Conversely, if two lines have the same slope, then they are parallel.

EXAMPLE 12   Find the equation of the line through $(2, 0)$ and parallel to the line whose equation is $y = \frac{1}{3}x - 11$.

Solution   The slope of the line having equation $y = \frac{1}{3}x - 11$ is $\frac{1}{3}$. Therefore, any line parallel to it also has slope $\frac{1}{3}$. Therefore, the desired line passes through $(2, 0)$

---

* If $n = 0$, this formula does not say anything, since $1/0$ is undefined. However, in this case, one line is horizontal and one vertical, the vertical one having an undefined slope.

and has slope $\frac{1}{3}$, so its equation is

$$y - 0 = \tfrac{1}{3}(x - 2)$$

or

$$y = \tfrac{1}{3}x - \tfrac{2}{3}.$$

PRACTICE PROBLEMS 4

Suppose that the revenue y from selling x units of a certain commodity is given by the formula $y = 4x$. (Revenue is the amount of money received from the sale of the commodity.)

1. What interpretation can be given to the slope of the graph of this equation?

2. (See Example 6.) Find the coordinates of the point of intersection of $y = 4x$ and $y = 2x + 5000$.

3. What interpretation can be given to the value of the x-coordinate of the point found in Problem 2?

EXERCISES 4

In Exercises 1–4, find the slope of the line having the given equation.

1. $y = \frac{2}{3}x + 7$ 2. $y = -4$

3. $y - 3 = 5(x + 4)$ 4. $7x + 5y = 10$

In Exercises 5–8, plot each pair of points, draw the straight line between them, and find the slope.

5. $(3, 4), (7, 9)$ 6. $(-2, 1), (3, -3)$

7. $(0, 0), (5, 4)$ 8. $(4, 17), (-2, 17)$

9. What is the slope of any line parallel to the y-axis?

10. Why doesn't it make sense to talk about the slope of the line between the two points $(2, 3)$ and $(2, -1)$?

In Exercises 11–14, graph the given linear equation by beginning at the y-intercept, moving 1 unit to the right and m units in the y-direction.

11. $y = -2x + 1$ 12. $y = 4x - 2$ 13. $y = 3x$ 14. $y = -2$

In Exercises 15–22, find the equation of line L.

15.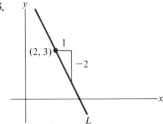

16.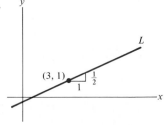

17.

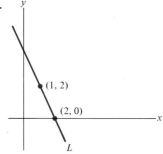

18.

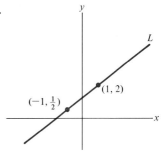

19.

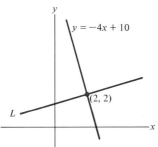

L perpendicular to $y = -4x + 10$

20.

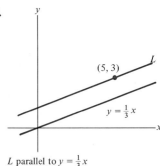

L parallel to $y = \frac{1}{3}x$

21.

L parallel to $y = -x + 2$

22.

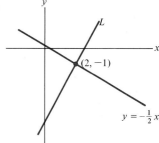

L perpendicular to $y = -\frac{1}{2}x$

23. Find the equation of the line passing through the point $(2, 3)$ and parallel to the x-axis.

24. Find the equation of the line passing through $(0, 0)$ and having slope 1.5.

25. Find the y-intercept of the line passing through the point $(5, 6)$ and having slope $\frac{3}{5}$.

26. Find the slope of the line passing through the point $(1, 4)$ and having y-intercept $(0, 4)$.

27. A salesperson's weekly pay depends on the volume of sales. If she sells x units of goods, then her pay is $y = 5x + 60$ dollars. Give an interpretation to the slope and the y-intercept of this straight line.

28. A manufacturer has fixed costs (such as rent and insurance) of $2000 per month. The cost of producing each unit of goods is $4. Give the linear equation for the cost of producing x units per month.

29. Suppose the price that must be set in order to sell q items is given by the equation $p = -3q + 1200$.

(a) Find and interpret the p-intercept of the graph of the equation.

(b) Find and interpret the q-intercept of the graph of the equation.

(c) Find and interpret the slope of the graph of the equation.

(d) What price must be set in order to sell 350 items?

(e) What quantity will be sold if the price is $300?

(f) Sketch the graph of the equation.

30. Biologists have found that the number of chirps that crickets of a certain species make per minute is related to the temperature. The relationship is very close to linear. At $68°F$ those crickets chirp about 124 times a minute. At $80°F$ they chirp about 172 times a minute.

(a) Find the linear equation relating Fahrenheit temperature F and the number of chirps c.

(b) If you only count chirps for 15 seconds, how can you quickly estimate the temperature?

31. Suppose the cost of making 20 radios is $6800 and the cost of making 50 radios is $9500.

(a) Find the cost equation.

(b) What is the fixed cost?

(c) What is the marginal cost of production?

(d) Sketch the graph of the equation.

Exercises 32–34 are related.

32. Suppose the total cost y of making x coats is given by the formula $y = 40x + 2400$.

(a) What is the cost of making 100 coats?

(b) How many coats can be made for $3600?

(c) Find and interpret the y-intercept of the graph of the equation.

(d) Find and interpret the slope of the graph of the equation.

33. Suppose the total revenue y from the sale of x coats is given by the formula $y = 100x$.

(a) What is the revenue if 300 coats are sold?

(b) How many coats must be sold to have a revenue of $6000?

(c) Find and interpret the y-intercept of the graph of the equation.

(d) Find and interpret the slope of the graph of the equation.

34. Consider a coat factory with the cost and revenue equations given in Exercises 32 and 33.

(a) Find the equation giving the profit y resulting from making and selling x coats.

(b) Find and interpret the y-intercept of the graph of the equation.

(c) Find and interpret the x-intercept of the graph of the equation.

(d) Find and interpret the slope of the graph of the equation.

(e) How much profit will be made if 80 coats are sold?

(f) How many coats must be sold to have a profit of $6000?

(g) Sketch the graph of the equation found in part (a).

An apartment complex has a storage tank to hold its heating oil. The tank was filled on January 1, but no more deliveries of oil will be made until sometime in March. Let t denote the number of days after January 1 and let y denote the number of gallons of fuel oil in the tank. Current records show that y and t will be related by the equation $y = 30{,}000 - 400t$.

35. Graph the equation $y = 30{,}000 - 400t$.

36. How much oil will be in the tank on February 1?

37. How much oil will be in the tank on February 15?

38. Determine the y-intercept of the graph. Explain its significance.

39. Determine the t-intercept of the graph. Explain its significance.

A corporation receives payment for a large contract on July 1, bringing its cash reserves to $2.3 million. Let y denote its cash reserves (in millions) t days after July 1. The corporation's accountants estimate that y and t will be related by the equation $y = 2.3 - .15t$.

40. Graph the equation $y = 2.3 - .15t$.

41. How much cash does the corporation have on the morning of July 16?

42. Determine the y-intercept of the graph. Explain its significance.

43. Determine the t-intercept of the graph. Explain its significance.

44. Determine the cash reserves on July 4.

45. When will the cash reserves be $.8 million?

In a certain factory, each day the expected number of accidents is related to the number of overtime hours by a linear equation. Suppose that on one day there were 1000 overtime hours logged and 8 accidents reported, and on another day there were 400 overtime hours logged and 5 accidents.

46. Find the equation relating the number of accidents, y, to the number of overtime hours, x.

47. What are the expected number of accidents when no overtime hours are logged?

48. What are the expected number of accidents when 2000 hours of overtime are logged?

49. To what level should the number of overtime hours be restricted if the company feels it cannot tolerate more than 10 accidents a day.

A furniture salesperson earns $160 a week plus 10% commission on her sales. Let x denote her sales and y her income for a week.

50. Express y in terms of x.

51. Determine her week's income if she sells $1000 in merchandise that week.

52. How much must she sell in a week in order to earn $500?

Find the equations of the following lines.

53. Slope is 3; y-intercept is $(0, -1)$

54. Slope is $-\frac{1}{2}$; y-intercept is $(0, 0)$

55. Slope is 1; $(1, 2)$ on line

56. Slope is $-\frac{1}{3}$; $(6, -2)$ on line

57. Slope is -7; $(5, 0)$ on line

58. Slope is $\frac{1}{2}$; $(2, -3)$ on line

59. Slope is 0; $(7, 4)$ on line

60. Slope is $-\frac{2}{5}$; $(0, 5)$ on line

61. $(2, 1)$ and $(4, 2)$ on line

62. $(5, -3)$ and $(-1, 3)$ on line

63. $(0, 0)$ and $(1, -2)$ on line

64. $(2, -1), (3, -1)$ on line

In each of Exercises 61–64 we specify a line by giving the slope and one point on the line. We give the first coordinate of some points on the line. Without deriving the equation of the line, find the second coordinate of each of the points.

65. Slope is 2, $(1, 3)$ on line; $(2, \quad)$; $(0, \quad)$; $(-1, \quad)$

66. Slope is -3, $(2, 2)$ on line; $(3, \quad)$; $(4, \quad)$; $(1, \quad)$

67. Slope is $-\frac{1}{4}$, $(-1, -1)$ on line; $(0, \quad)$; $(1, \quad)$; $(-2, \quad)$

68. Slope is $\frac{1}{3}$; $(-5, 2)$ on line; $(-4, \quad)$; $(-3, \quad)$; $(-2, \quad)$

For each pair of lines in the figures below, determine the one with the greater slope.

69.

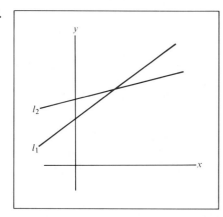

70.

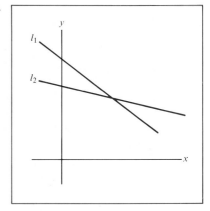

71. Centigrade and Fahrenheit temperatures are related by a linear equation. Use the fact that $0°C = 32°F$ and $100°C = 212°F$ to find an equation.

72. An archeologist dates a bone fragment discovered at a depth of 4 feet as approximately 1500 B.C. and dates a pottery shard at a depth of 8 feet as approximately 2100 B.C. Assuming there is a linear relationship between depths and dates at this archeological site, find the equation that relates depth to date. How deep should the archeologist dig to look for relics from 3000 B.C.?

73. Consider a linear equation of the form $y = mx + b$. When the value of b remains fixed and the value of m changes, the graph rotates about the point $(0, b)$. As the value of m increases, will the graph rotate in a clockwise or counterclockwise direction?

74. Write an inequality whose graph consists of the points on or below the straight line passing through the two points $(2, 5)$ and $(4, 9)$.

75. Write an inequality whose graph consists of the points on or above the line with slope .4 and y-intercept $(0, 3)$.

76. Find a system of inequalities having the feasible set in Fig. 9.

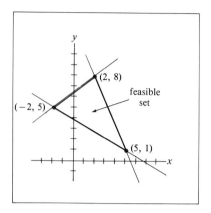

FIGURE 9   FIGURE 10

77. Find a system of inequalities having the feasible set in Fig. 10.

78. Show that the points $(1, 3)$, $(2, 4)$, and $(3, -1)$ are *not* on the same line.

79. For what value of k will the three points $(1, 5)$, $(2, 7)$, and $(3, k)$ be on the same line?

80. Find the value of a for which the line through the points $(a, 1)$ and $(2, -3.1)$ is parallel to the line through the points $(-1, 0)$ and $(3.8, 2.4)$?

81. Rework Exercise 80 where the word "parallel" is replaced by the word "perpendicular."

82. Prove the parallel property. [*Hint:* If $y = mx + b$ and $y = m'x + b'$ are the equations of two lines, then the two lines have a point in common if and only if the equation $mx + b = m'x + b'$ has a solution x.]

83. Prove the perpendicular property. [*Hint:* Without loss of generality, assume that both lines pass through the origin. Use the point-slope formula, the Pythagorean theorem, and the accompanying figure.]

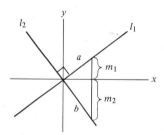

SOLUTIONS TO PRACTICE PROBLEMS 4

1. By the steepness property, whenever x is increased by 1 unit, the value of y is increased by 4 units. Therefore, *each additional unit of production brings in $4 of revenue.* (The graph of $y = 4x$ is called a *revenue curve* and its slope is called the *marginal revenue of production.*)

2. $\begin{cases} y = 4x \\ y = 2x + 5000. \end{cases}$ Equate expressions for y: $4x = 2x + 5000$

$$2x = 5000$$
$$x = 2500$$
$$y = 4(2500) = 10{,}000.$$

3. When producing 2500 units, the revenue equals the cost. This value of x is called the *break-even point*. Since (profit) = (revenue) − (cost), the company will make a profit only if its level of production is greater than the break-even point.

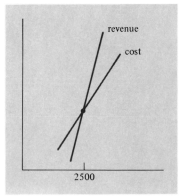

Chapter 1: CHECKLIST

☐ Cartesian coordinate system for the
☐ Cartesian coordinate system for the plane
☐ Origin, x-axis, y-axis
☐ Graph of an equation
☐ Linear equation in two variables
☐ Standard form of a linear equation in two variables
☐ y-intercept, x-intercept of a line
☐ Inequality properties for addition, subtraction, and multiplication
☐ Linear inequalities in two variables
☐ Graph of a linear inequality
☐ Feasible set of a system of linear inequalities
☐ Method to find point of intersection of pair of straight lines
☐ Slope of a line
☐ $m = \dfrac{y_2 - y_1}{x_2 - x_1}$
☐ Steepness property of slope
☐ Point-slope formula: $y - y_1 = m(x - x_1)$
☐ Perpendicular property
☐ Parallel property

Chapter 1: SUPPLEMENTARY EXERCISES

1. What is the equation of the y-axis?

2. Graph the linear equation $y = -\frac{1}{2}x$.

3. Find the point of intersection of the pair of straight lines $x - 5y = 6$ and $3x = 6$.

4. Find the slope of the line having the equation $3x - 4y = 8$.

5. Find the equation of the line having y-intercept $(0, 5)$ and x-intercept $(10, 0)$.

6. Graph the linear inequality $x - 3y \geq 12$.

7. Does the point $(1, 2)$ satisfy the linear inequality $3x + 4y \geq 11$?

8. Find the point of intersection of the pair of straight lines $2x - y = 1$ and $x + 2y = 13$.

9. Find the equation of the straight line passing through the point $(15, 16)$ and parallel to the line $2x - 10y = 7$.

10. Find the y-coordinate of the point having x-coordinate 1 and lying on the line $y = 3x + 7$.

11. Find the x-intercept of the straight line with equation $x = 5$.

12. Graph the linear inequality $y \leq 6$.

13. Solve the following system of linear equations:
$$\begin{cases} 3x - 2y = 1 \\ 2x + y = 24. \end{cases}$$

14. Graph the feasible set for the following system of inequalities:
$$\begin{cases} 2y + 7x \geq 30 \\ 4y - x \geq 0 \\ \qquad y \leq 8. \end{cases}$$

15. Find the y-intercept of the line passing through the point $(4, 9)$ and having slope $\frac{1}{2}$.

16. The fee charged by a local moving company depends on the amount of time required for the move. If t hours is required, then the fee is $y = 35t + 20$ dollars. Give an interpretation of the slope and y-intercept of this line.

17. Are the points $(1, 2)$, $(2, 0)$, and $(3, 1)$ on the same line?

18. Write an equation of the line with x-intercept $(3, 0)$ and y-intercept $(0, -2)$.

19. For what values of a and b does the following system of linear equations have a solution?
$$\begin{cases} 4x + 6y = a \\ 2x + 3y = b \end{cases}$$

20. Write the inequality whose graph is the half-plane below the line with slope $\frac{2}{3}$ and y-intercept $\frac{3}{2}$.

21. Write the inequality whose graph is the half-plane above the line through $(2, -1)$ and $(6, 8.6)$.

22. Solve the system of linear equations
$$\begin{cases} 1.2x + 2.4y = .6 \\ 4.8y - 1.6x = 2.4. \end{cases}$$

23. Find the equation of the line through $(1, 1)$ and the intersection of the lines $y = -x + 1$ and $y = 2x + 3$.

24. Find all numbers x such that $2x + 3(x - 2) \geq 0$.

25. Graph the equation $x + \frac{1}{2}y = 4$ and give the slope and both intercepts.

26. Do the three graphs of the linear equations $2x - 3y = 1$, $5x + 2y = 0$, and $x + y = 1$ contain a common point?

27. Show that the lines with equations $2x - 3y = 1$ and $3x + 2y = 4$ are perpendicular.

28. Each of the half-planes (A), (B), (C), and (D) is the graph of one of the linear inequalities (a), (b), (c), and (d). Match each half-plane with its inequality.

(A)
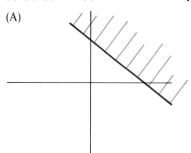

(B)
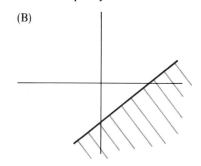

(C)
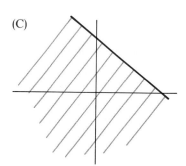

(D)
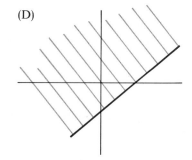

(a) $x + y \geq 1$ (b) $x + y \leq 1$

(c) $x - y \leq 1$ (d) $y - x \leq -1$

29. Each of the lines, L_1, L_2, and L_3 in Fig. 1 is the graph of one of the equations (a), (b), and (c). Match each of the following equations with its corresponding line.

(a) $4x + y = 17$ (b) $y = x + 2$

(c) $2x + 3y = 11$

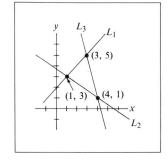

FIGURE 1

30. Find a system of inequalities having the feasible set in Fig. 2 and find the coordinates of the unspecified vertex. [*Note:* There is a right angle at the vertex $(4, \frac{3}{2})$.]

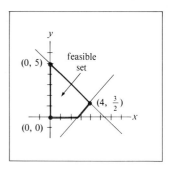

FIGURE 2

31. Consider the following four equations:

$$q = 100p + 500$$
$$q = 150p - 100$$
$$q = -75p - 750$$
$$q = -75p + 500.$$

One is the equation of a supply curve and another is the equation of a demand curve. Find the intersection point of those two curves.

32. Find the vertices of the following feasible set.

$$\begin{cases} x \geq 0 \\ y \geq 0 \\ 5x + y \leq 50 \\ 2x + 3y \leq 33 \\ x - 2y \geq -8 \end{cases}$$

33. It is not possible to draw a straight line through all three of the points $(2, 4)$, $(5, 8)$, and $(7, 9)$. However, it *is* possible for a straight line to *miss* all three points by the same amount; that is, there is a line that makes the vertical distances d_1, d_2, and d_3 in Fig. 3 equal. Find the equation of this line. [*Hint:* Let the equation of the line be $y = mx + b$. Then the point P has coordinates $(2, 2m + b)$.]

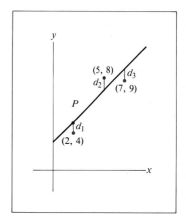

FIGURE 3

34. For a certain manufacturer, the production and sale of each additional unit yields an additional profit of $10. The sale of 1000 units yields a profit of $4000.

   (a) Find the profit equation.

   (b) Find the y- and x-intercepts of the graph of the profit equation.

   (c) Sketch the graph of the profit equation.

35. One-day car rentals cost $50 plus 10¢ per mile from company A and $40 plus 20¢ per mile from company B.

   (a) For each company, give the linear equations for the cost, y, when x miles are driven.

   (b) Which company offers the best value when the car is driven for 80 miles?

   (c) Which company offers the best value when the car is driven for 160 miles?

   (d) For what mileage do the two companies offer the same value?

36. (*Linear Depreciation*) For accounting purposes, the value of certain items is depreciated linearly over time from the purchase price to the salvage value. Suppose that a computer was purchased in 1980 for $5000 and sold in 1988 for $1000.

   (a) Find the linear equation giving its value, y, after x years.

   (b) How much was the computer worth in 1984?

   (c) When was the computer valued at $2000?

37. Graph the linear inequality $x \leq 3y + 2$.

38. A furniture store offers its new employees a weekly salary of $200 plus a 3% commission on sales. After one year, employees receive $100 per week plus a 5% commission. For what weekly sales level will the two scales produce the same salary?

2

MATRICES

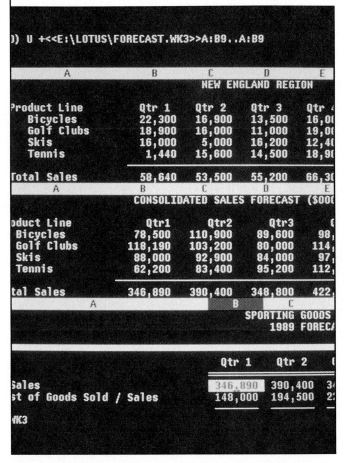

2.1 Solving Systems of Linear Equations, I

2.2 Solving Systems of Linear Equations, II

2.3 Arithmetic Operations on Matrices

2.4 The Inverse of a Matrix

2.5 The Gauss-Jordan Method for Calculating Inverses

2.6 Input-Output Analysis

We begin this chapter by developing a method for solving systems of linear equations in any number of variables. Our discussion of this method will lead naturally into the study of mathematical objects called *matrices*. The arithmetic and applications of matrices are the main topics of the chapter. We will discuss in detail the application of matrix arithmetic to input-output analysis, which can be (and is) used to make production decisions for large businesses and entire economies.

2.1. Solving Systems of Linear Equations, I

In the preceding chapter we presented a method for solving systems of linear equations in two variables. The method of Chapter 1 is very efficient for determining the solutions. Unfortunately, it works only for systems of linear equations having *two* variables. In many applications we meet systems having more than two variables, as the following example illustrates.

EXAMPLE 1 The Upside Down company specializes in making down jackets, ski vests, and comforters. The requirements for down and labor and the profits earned are given in the accompanying chart.

	Down (pounds)	Labor (person-hours)	Profit ($)
Jacket	3	2	6
Vest	2	1	6
Comforter	4	1	2

Each week the company has available 600 pounds of down and 275 person-hours of labor, and wants to earn a profit of $1150. How many of each item should the company make each week?

Solution The requirements and earnings can be expressed by a system of equations. Let x be the number of jackets, y the number of vests, and z the number of comforters. If 600 pounds of down are used, then

$$[\text{down in jackets}] + [\text{down in vests}] + [\text{down in comforters}] = 600.$$
$$3\,[\#\text{ jackets}] \;+\; 2\,[\#\text{ vests}] \;+\; 4\,[\#\text{ comforters}] \;= 600.$$

That is,

$$3x + 2y + 4z = 600.$$

Similarly, the equation for labor is

$$2x + y + z = 275.$$

And the equation for the profit is

$$6x + 6y + 2z = 1150.$$

The numbers x, y, and z must simultaneously satisfy a system of three linear equations in three variables.

$$\begin{cases} 3x + 2y + 4z = 600 \\ 2x + y + z = 275 \\ 6x + 6y + 2z = 1150 \end{cases} \tag{1}$$

Later we present a method for determining the solution to this system. This method yields the solution $x = 50$, $y = 125$, and $z = 50$. It is easy to confirm that these values of x, y, z satisfy all three equations.

$$3(50) + 2(125) + 4(50) = 600$$

$$2(50) + (125) + (50) = 275$$

$$6(50) + 6(125) + 2(50) = 1150.$$

Thus, the Upside Down Company can make a profit of $1150 by producing 50 jackets, 125 vests, and 50 comforters.

In this section we develop a step-by-step procedure for solving systems of linear equations such as (1). The procedure, called the *Gaussian elimination method*, consists of repeatedly simplifying the system, using so-called elementary row operations, until the solution stares us in the face!

In the system of linear equations (1) the equations have been written in such a way that the x-terms, the y-terms, and the z-terms lie in different columns. We shall always be careful to display systems of equations with separate columns for each variable. One of the key ideas of the Gaussian elimination method is to think of the solution as a system of linear equations in its own right. For example, we can write the solution of the system (1) as:

$$\begin{cases} x = 50 \\ y = 125 \\ z = 50 \end{cases} \tag{2}$$

This is just a system of linear equations in which the coefficients of most terms are zero! Since the only terms with nonzero coefficients are arranged on a diagonal, such a system is said to be in *diagonal form.*

Our method for solving a system of linear equations consists of repeatedly using three operations which alter the system but do not change the solutions. The operations are used to transform the system into a system in diagonal form. Since the operations involve only elementary arithmetic and are applied to entire equations (i.e., rows of the system), they are called *elementary row operations*. Let us begin our study of the Gaussian elimination method by introducing these operations.

> **Elementary Row Operation 1**   Rearrange the equations in any order.

This operation is harmless enough. It certainly does not change the solutions of the system.

> **Elementary Row Operation 2**   Multiply an equation by a nonzero number.

For example, if we are given the system of linear equations

$$\begin{cases} 2x - 3y + 4z = 11 \\ 4x - 19y + z = 31 \\ 5x + 7y - z = 12, \end{cases}$$

then we may replàce it by a new system obtained by leaving the last two equations unchanged and multiplying the first equation by 3. To accomplish this, multiply each term of the first equation by 3. The transformed system is

$$\begin{cases} 6x - 9y + 12z = 33 \\ 4x - 19y + z = 31 \\ 5x + 7y - z = 12. \end{cases}$$

The operation of multiplying an equation by a nonzero number does not change the solutions of the system. For if a particular set of values of the variables satisfy the original equation, they satisfy the resulting equation, and vice versa.

Elementary row operation 2 may be used to make the coefficient of a particular variable 1.

EXAMPLE 2   Replace the system

$$\begin{cases} -5x + 10y + 20z = 4 \\ x - 12z = 1 \\ x + y + z = 0 \end{cases}$$

by an equivalent system in which the coefficient of x in the first equation is 1.

Solution   The coefficient of x in the first equation is -5, so we use elementary row operation 2 to multiply the first equation by $-\frac{1}{5}$. Multiplying each term of the first equation by $-\frac{1}{5}$ gives

$$\begin{cases} x - 2y - 4z = -\frac{4}{5} \\ x - 12z = 1 \\ x + y + z = 0. \end{cases}$$

Another operation that can be performed on a system without changing its solutions is to replace one equation by its sum with some other equation. For example, consider this system of equations:

$$A: \begin{cases} x + y - 2z = 3 \\ x + 2y - 5z = 4 \\ 5x + 8y - 18z = 14. \end{cases}$$

We can replace the second equation by the sum of the first and the second. Since

$$\begin{array}{r} x + y - 2z = 3 \\ + \quad x + 2y - 5z = 4 \\ \hline 2x + 3y - 7z = 7, \end{array}$$

the resulting system is

$$B: \begin{cases} x + y - 2z = 3 \\ 2x + 3y - 7z = 7 \\ 5x + 8y - 18z = 14. \end{cases}$$

If a particular choice of x, y, z satisfies system A, it also satisfies system B. This is because system B results from adding equations. Similarly, system A can be derived from system B by subtracting equations. So any particular solution of system A is a solution of system B, and vice versa.

The operation of adding equations is usually used in conjunction with elementary row operation 2. That is, an equation is changed by adding to it a nonzero multiple of another equation. For example, consider the system

$$\begin{cases} x + y - 2z = 3 \\ x + 2y - 5z = 4 \\ 5x + 8y - 18z = 14. \end{cases}$$

Let us change the second equation by adding to it twice the first. Since

$$\begin{array}{rr} 2(\text{first}) & 2x + 2y - 4z = 6 \\ + (\text{second}) & x + 2y - 5z = 4 \\ \hline & 3x + 4y - 9z = 10, \end{array}$$

the new second equation is

$$3x + 4y - 9z = 10$$

and the transformed system is

$$\begin{cases} x + y - 2z = 3 \\ 3x + 4y - 9z = 10 \\ 5x + 8y - 18z = 14. \end{cases}$$

Since addition of equations and elementary row operation 2 are often used together, let us define a third elementary row operation to be the combination:

Elementary Row Operation 3 Change an equation by adding to it a multiple of another equation.

For reference, let us summarize the elementary row operations we have just defined.

Elementary Row Operations

1. Rearrange the equations in any order.
2. Multiply an equation by a nonzero number.
3. Change an equation by adding to it a multiple of another equation.

The idea of the Gaussian elimination method is to transform an arbitrary system of linear equations into diagonal form by repeated application of the three elementary row operations. To see how the method works, consider the following example.

EXAMPLE 3 Solve the following system by the Gaussian elimination method.

$$\begin{cases} x - 3y = 7 \\ -3x + 4y = -1. \end{cases}$$

Solution Let us transform this system into diagonal form by examining one column at a time, starting from the left. Examine the first column:

$$\begin{array}{c} x \\ -3x \end{array}$$

The coefficient of the top x is 1, which is exactly what it should be for the system to be in diagonal form. So we do nothing to this term. Now examine the next term in the column, $-3x$. In diagonal form this term must be absent. In order to accomplish this we add a multiple of the first equation to the second. Since the coefficient of x in the second is -3, we add three times the first equation to the second equation in order to cancel the x-term. (Abbreviation: $[2] + 3[1]$. The $[2]$ means that we are changing equation 2. The expression $[2] + 3[1]$

means that we are replacing equation 2 by the original equation plus three times equation 1.)

$$\begin{cases} x - 3y = 7 \\ -3x + 4y = -1 \end{cases} \xrightarrow{\;[2] + 3[1]\;} \begin{cases} x - 3y = 7 \\ - 5y = 20. \end{cases}$$

The first column now has the proper form, so we proceed to the second column. In diagonal form that column will have one nonzero term, namely the second, and the coefficient of y in that term must be 1. To bring this about, multiply the second equation by $-\frac{1}{5}$ (abbreviation: $-\frac{1}{5}[2]$):

$$\begin{cases} x - 3y = 7 \\ - 5y = 20 \end{cases} \xrightarrow{\;-\frac{1}{5}[2]\;} \begin{cases} x - 3y = 7 \\ y = -4. \end{cases}$$

The second column still does not have the correct form. We must get rid of the $-3y$ term in the first equation. We do this by adding a multiple of the second equation to the first. Since the coefficient of the term to be canceled is -3, we add three times the second equation to the first:

$$\begin{cases} x - 3y = 7 \\ y = -4 \end{cases} \xrightarrow{\;[1] + 3[2]\;} \begin{cases} x = -5 \\ y = -4. \end{cases}$$

The system is now in diagonal form and the solution can be read off: $x = -5$, $y = -4$.

EXAMPLE 4 Use the Gaussian elimination method to solve the system

$$\begin{cases} 2x - 6y = -8 \\ -5x + 13y = 1. \end{cases}$$

Solution We can perform the calculations in a mechanical way, proceeding column by column from the left:

$$\begin{cases} 2x - 6y = -8 \\ -5x + 13y = 1 \end{cases} \xrightarrow{\;\frac{1}{2}[1]\;} \begin{cases} x - 3y = -4 \\ -5x + 13y = 1 \end{cases}$$

$$\xrightarrow{\;[2] + 5[1]\;} \begin{cases} x - 3y = -4 \\ -2y = -19 \end{cases}$$

$$\xrightarrow{\;-\frac{1}{2}[2]\;} \begin{cases} x - 3y = -4 \\ y = \frac{19}{2} \end{cases}$$

$$\xrightarrow{\;[1] + 3[2]\;} \begin{cases} x = \frac{49}{2} \\ y = \frac{19}{2}. \end{cases}$$

So the solution of the system is $x = \frac{49}{2}$, $y = \frac{19}{2}$.

The calculation becomes easier to follow if we omit writing down the variables at each stage and work only with the coefficients. At each stage of the computation the system is represented by a rectangular array of numbers. For instance, the original system is written*

$$\begin{bmatrix} 2 & -6 & | & -8 \\ -5 & 13 & | & 1 \end{bmatrix}.$$

The elementary row operations are performed on the rows of this rectangular array just as if the variables were there. So, for example, the first step above is to multiply the first equation by $\frac{1}{2}$. This corresponds to multiplying the first row of the array by $\frac{1}{2}$ to get

$$\begin{bmatrix} 1 & -3 & | & -4 \\ -5 & 13 & | & 1 \end{bmatrix}.$$

The diagonal form just corresponds to the array

$$\begin{bmatrix} 1 & 0 & | & \frac{49}{2} \\ 0 & 1 & | & \frac{19}{2} \end{bmatrix}.$$

Note that this array has ones down the diagonal and zeros everywhere else on the left. The solution of the system appears on the right.

A rectangular array of numbers is called a *matrix* (plural: *matrices*). In the next example, we use matrices to carry out the Gaussian elimination method.

EXAMPLE 5   Use the Gaussian elimination method to solve the system

$$\begin{cases} 3x - 6y + 9z = 0 \\ 4x - 6y + 8z = -4 \\ -2x - y + z = 7. \end{cases}$$

Solution   The initial array corresponding to the system is

$$\begin{bmatrix} 3 & -6 & 9 & | & 0 \\ 4 & -6 & 8 & | & -4 \\ -2 & -1 & 1 & | & 7 \end{bmatrix}.$$

We must use elementary row operations to transform this array into diagonal form—that is, with ones down the diagonal and zeros everywhere else on the left:

$$\begin{bmatrix} 1 & 0 & 0 & | & * \\ 0 & 1 & 0 & | & * \\ 0 & 0 & 1 & | & * \end{bmatrix}.$$

---

* The vertical line between the second and third columns is a placemarker which separates the data obtained from the right- and left-hand sides of the equations. It is inserted for visual convenience.

We proceed one column at a time.

$$\begin{bmatrix} 3 & -6 & 9 & | & 0 \\ 4 & -6 & 8 & | & -4 \\ -2 & -1 & 1 & | & 7 \end{bmatrix} \xrightarrow{\frac{1}{3}[1]} \begin{bmatrix} 1 & -2 & 3 & | & 0 \\ 4 & -6 & 8 & | & -4 \\ -2 & -1 & 1 & | & 7 \end{bmatrix} \xrightarrow{[2]+(-4)[1]}$$

$$\begin{bmatrix} 1 & -2 & 3 & | & 0 \\ 0 & 2 & -4 & | & -4 \\ -2 & -1 & 1 & | & 7 \end{bmatrix} \xrightarrow{[3]+2[1]} \begin{bmatrix} 1 & -2 & 3 & | & 0 \\ 0 & 2 & -4 & | & -4 \\ 0 & -5 & 7 & | & 7 \end{bmatrix} \xrightarrow{\frac{1}{2}[2]}$$

$$\begin{bmatrix} 1 & -2 & 3 & | & 0 \\ 0 & 1 & -2 & | & -2 \\ 0 & -5 & 7 & | & 7 \end{bmatrix} \xrightarrow{[1]+2[2]} \begin{bmatrix} 1 & 0 & -1 & | & -4 \\ 0 & 1 & -2 & | & -2 \\ 0 & -5 & 7 & | & 7 \end{bmatrix} \xrightarrow{[3]+5[2]}$$

$$\begin{bmatrix} 1 & 0 & -1 & | & -4 \\ 0 & 1 & -2 & | & -2 \\ 0 & 0 & -3 & | & -3 \end{bmatrix} \xrightarrow{(-\frac{1}{3})[3]} \begin{bmatrix} 1 & 0 & -1 & | & -4 \\ 0 & 1 & -2 & | & -2 \\ 0 & 0 & 1 & | & 1 \end{bmatrix} \xrightarrow{[1]+1[3]}$$

$$\begin{bmatrix} 1 & 0 & 0 & | & -3 \\ 0 & 1 & -2 & | & -2 \\ 0 & 0 & 1 & | & 1 \end{bmatrix} \xrightarrow{[2]+2[3]} \begin{bmatrix} 1 & 0 & 0 & | & -3 \\ 0 & 1 & 0 & | & 0 \\ 0 & 0 & 1 & | & 1 \end{bmatrix}.$$

The last array is in diagonal form, so we just put back the variables and read off the solution:

$$x = -3, \qquad y = 0, \qquad z = 1.$$

Because so much arithmetic has been performed, it is a good idea to check the solution by substituting the values for x, y, z into each of the equations of the original system. This will uncover any arithmetic errors that may have occurred.

$$\begin{cases} 3x - 6y + 9z = 0 \\ 4x - 6y + 8z = -4 \\ -2x - y + z = 7 \end{cases} \qquad \begin{cases} 3(-3) - 6(0) + 9(1) = 0 \\ 4(-3) - 6(0) + 8(1) = -4 \\ -2(-3) - (0) + (1) = 7 \end{cases}$$

$$\begin{cases} -9 - 0 + 9 = 0 \\ -12 - 0 + 8 = -4 \\ 6 - 0 + 1 = 7 \end{cases}$$

$$\begin{cases} 0 = 0 \\ -4 = -4 \\ 7 = 7. \end{cases}$$

So we have indeed found a solution of the system.

Remark Note that so far we have not had to use elementary row operation 1, which allows interchange of equations. But in some examples it is definitely needed.

Consider this system:

$$\begin{cases} y + z = 0 \\ 3x - y + z = 6 \\ 6x \quad\;\; - z = 3. \end{cases}$$

The first step of the Gaussian elimination method consists of making the x-coefficient 1 in the first equation. But we cannot do this, since the first equation does not involve x. To remedy this difficulty, just interchange the first two equations to guarantee that the first equation involves x. Now proceed as before. Of course, in terms of the matrix of coefficients, interchanging equations corresponds to interchanging rows of the matrix.

PRACTICE PROBLEMS 1

1. Determine whether the following systems of linear equations are in diagonal form.

(a) $\begin{cases} x \quad\;\; + z = 3 \\ \quad y \quad\quad = 2 \\ \quad\quad\;\; z = 7 \end{cases}$
(b) $\begin{cases} x \quad\quad = 3 \\ \quad y = 5 \\ \quad z \quad = 7 \end{cases}$
(c) $\begin{cases} x \quad\quad = -1 \\ \quad y \quad = 0 \\ \quad 3z = 4 \end{cases}$

2. Perform the indicated elementary row operation.

(a) $\begin{cases} x - 3y = 2 \\ 2x + 3y = 5 \end{cases} \xrightarrow{\;[2] + (-2)[1]\;}$

(b) $\begin{cases} x + y = 3 \\ -x + 2y = 5 \end{cases} \xrightarrow{\;[2] + (1)[1]\;}$

3. State the next elementary row operation which should be performed when applying the Gaussian elimination method.

(a) $\left[\begin{array}{ccc|c} 0 & 2 & 4 & 1 \\ 0 & 3 & -7 & 0 \\ 3 & 6 & -3 & 3 \end{array}\right]$

(b) $\left[\begin{array}{ccc|c} 1 & -3 & 4 & 5 \\ 0 & 2 & 3 & 4 \\ -6 & 5 & -7 & 0 \end{array}\right]$

EXERCISES 1

In Exercises 1–8, perform the indicated elementary row operations and give their abbreviations.

1. Operation 2: multiply the first equation by 2.

$$\begin{cases} \frac{1}{2}x - 3y = 2 \\ 5x + 4y = 1. \end{cases}$$

2. Operation 2: multiply the second equation by -1.

$$\begin{cases} x + 4y = 6 \\ \quad\;\; -y = 2. \end{cases}$$

3. Operation 3: change the second equation by adding to it 5 times the first equation.

$$\begin{cases} x + 2y = 3 \\ -5x + 4y = 1. \end{cases}$$

4. Operation 3: change the second equation by adding to it $(-\frac{1}{2})$ times the first equation.

$$\begin{cases} x - 6y = 4 \\ \frac{1}{2}x + 2y = 1. \end{cases}$$

5. Operation 3: change the third equation by adding to it (-4) times the first equation.

$$\begin{cases} x - 2y + z = 0 \\ y - 2z = 4 \\ 4x + y + 3z = 5. \end{cases}$$

6. Operation 3: change the third equation by adding to it 3 times the second equation.

$$\begin{cases} x + 6y - 4z = 1 \\ y + 3z = 1 \\ -3y + 7z = 2. \end{cases}$$

7. Operation 3: change the first row by adding to it $\frac{1}{2}$ times the second row.

$$\left[\begin{array}{cc|c} 1 & -\frac{1}{2} & 3 \\ 0 & 1 & 4 \end{array}\right].$$

8. Operation 3: change the third row by adding to it (-4) times the second row.

$$\left[\begin{array}{ccc|c} 1 & 0 & 7 & 9 \\ 0 & 1 & -2 & 3 \\ 0 & 4 & 8 & 5 \end{array}\right].$$

In Exercises 9–16, state the next elementary row operation which must be performed in order to put the matrix into diagonal form. Do not perform the operation.

9. $\left[\begin{array}{cc|c} 1 & -5 & 1 \\ -2 & 4 & 6 \end{array}\right]$ 10. $\left[\begin{array}{cc|c} 1 & 3 & 4 \\ 0 & 2 & 6 \end{array}\right]$ 11. $\left[\begin{array}{cc|c} 1 & 2 & 3 \\ 0 & 1 & 4 \end{array}\right]$

12. $\left[\begin{array}{ccc|c} 1 & -2 & 5 & 7 \\ 0 & -3 & 6 & 9 \\ 4 & 5 & -6 & 7 \end{array}\right]$ 13. $\left[\begin{array}{ccc|c} 0 & 5 & -3 & 6 \\ 2 & -3 & 4 & 5 \\ 4 & 1 & -7 & 8 \end{array}\right]$

14. $\left[\begin{array}{ccc|c} 1 & 4 & -2 & 5 \\ 0 & -3 & 6 & 9 \\ 0 & 4 & 3 & 1 \end{array}\right]$ 15. $\left[\begin{array}{ccc|c} 1 & 0 & 3 & 4 \\ 0 & 1 & 2 & 5 \\ 0 & 0 & 1 & 6 \end{array}\right]$

16. $\left[\begin{array}{ccc|c} 1 & 2 & 4 & 5 \\ 0 & 0 & 3 & 6 \\ 0 & 1 & 1 & 7 \end{array}\right]$

Solve the following linear systems by using the Gaussian elimination method.

17. $\begin{cases} 3x + 9y = 6 \\ 2x + 8y = 6 \end{cases}$ 18. $\begin{cases} \frac{1}{3}x + 2y = 1 \\ -2x - 4y = 6 \end{cases}$

19. $\begin{cases} x - 3y + 4z = 1 \\ 4x - 10y + 10z = 4 \\ -3x + 9y - 5z = -6 \end{cases}$

20. $\begin{cases} \frac{1}{2}x + y = 4 \\ -4x - 7y + 3z = -31 \\ 6x + 14y + 7z = 50 \end{cases}$

21. $\begin{cases} 2x - 2y + 4 = 0 \\ 3x + 4y - 1 = 0 \end{cases}$

22. $\begin{cases} 2x + 3y = 4 \\ -x + 2y = -2 \end{cases}$

23. $\begin{cases} 4x - 4y + 4z = -8 \\ x - 2y - 2z = -1 \\ 2x + y + 3z = 1 \end{cases}$

24. $\begin{cases} x + 2y + 2z - 11 = 0 \\ x - y - z + 4 = 0 \\ 2x + 5y + 9z - 39 = 0 \end{cases}$

25. $\begin{cases} .2x + .3y = 4 \\ .6x + 1.1y = 15 \end{cases}$

26. $\begin{cases} \frac{3}{2}x + 6y = 9 \\ \frac{1}{2}x - \frac{2}{3}y = 11 \end{cases}$

27. $\begin{cases} x + y + 4z = 3 \\ 4x + y - 2z = -6 \\ -3x + 2z = 1 \end{cases}$

28. $\begin{cases} -2x - 3y + 2z = -2 \\ x + y = 3 \\ -x - 3y + 5z = 8 \end{cases}$

29. $\begin{cases} -x + y = -1 \\ x + z = 4 \\ 6x - 3y + 2z = 10 \end{cases}$

30. $\begin{cases} x + 2z = 9 \\ y + z = 1 \\ 3x - 2y = 9 \end{cases}$

31. A 600-seat movie theater charges $5.50 admission for adults and $2.50 for children. If the theater is full and $1911 is collected, how many adults and how many children are in the audience?

32. A baseball player's batting average is determined by dividing the number of hits by the number of times at bat and multiplying by 1000. (Batting averages are usually, but not necessarily, rounded to the nearest whole number.) For instance, if a player gets 2 hits in 5 times at bat, his batting average is 400: ($\frac{2}{5} \times 1000 = 400$). Partway through the season, a player thinks to himself: If I get a hit in my next time at bat, my average will go up to 250; if I don't get a hit, it will drop to 187.5. How many times has this player batted, how many hits has he had, and what is his current batting average?

33. A bank wishes to invest a $100,000 trust fund in three sources—bonds paying 8%, certificates of deposit paying 7%, and first mortgages paying 10%. The bank wishes to realize an $8000 annual income from the investment. A condition of the trust is that the total amount invested in bonds and certificates of deposit must be triple the amount invested in mortgages. How much should the bank invest in each possible category? Let x, y, and z, respectively, be the amounts invested in bonds, certificates of deposit, and first mortgages. Solve the system of equations by the Gaussian elimination method.

34. A dietician wishes to plan a meal around three foods. Each ounce of food I contains 10% of the daily requirements for carbohydrates, 10% for protein, and 15% for vitamin C. Each ounce of food II contains 10% of the daily requirements for carbohydrates, 5% for protein, and 0% for vitamin C. Each ounce of food III contains 10% of the daily requirements for carbohydrates, 25% for protein, and 10% for vitamin C. How many ounces of each food should be served in order to supply exactly the daily requirements for each of carbohydrates, protein and vitamin C? Let x, y, and z, respectively, be the number of ounces of foods I, II, and III.

SOLUTIONS TO PRACTICE PROBLEMS 1

1. (a) Not in diagonal form, since the first equation contains both x and z.

   (b) Not in diagonal form, since the variables are not arranged in diagonal fashion.

   (c) Not in diagonal form, since the coefficient of z is not 1.

2. (a) Change the system into another system in which the second equation is altered by having (-2) (first equation) added to it. The new system is

$$\begin{cases} x - 3y = 2 \\ 9y = 1. \end{cases}$$

   The equation $9y = 1$ was obtained as follows:

$$\begin{array}{rl} (-2)(\text{first equation}) & -2x + 6y = -4 \\ + (\text{second equation}) & \underline{2x + 3y = 5} \\ & 9y = 1. \end{array}$$

   (b) Replace the second equation by the first equation multiplied by 1 and added to the second. This is the same as adding the first equation to the second. The result is

$$\begin{cases} x + y = 3 \\ 3y = 8. \end{cases}$$

3. (a) The first row should contain a nonzero number as its first entry. This can be accomplished by interchanging the first and third rows.

   (b) The first column can be put into proper form by eliminating the -6. To accomplish this, multiply the first row by 6 and add this product to the third row. The notation for this operation is

$$\xrightarrow{[3] + 6[1]}$$

2.2. Solving Systems of Linear Equations, II

In this section we introduce the operation of pivoting and consider systems of linear equations which do not have unique solutions.

Roughly speaking, the Gaussian elimination method applied to a matrix proceeds as follows: Consider the columns one at a time, from left to right. For each column use the elementary row operations to transform the appropriate entry to a one and the remaining entries in the column to zeros. (The "appropriate" entry is the first entry in the first column, the second entry in the second column, and so forth.) This sequence of elementary row operations performed for each column is called *pivoting*. More precisely:

Method *To pivot a matrix about a given nonzero entry:*

1. Transform the given entry into a one.
2. Transform all other entries in the same column into zeros.

Pivoting is used in solving problems other than systems of linear equations. As we shall see in Chapter 4, it is the basis for the simplex method of solving linear programming problems.

EXAMPLE 1 Pivot the matrix about the circled element.

$$\left[\begin{array}{cc|c} 18 & \boxed{-6} & 15 \\ 5 & -2 & 4 \end{array}\right]$$

Solution The first step is to transform the -6 to a 1. We do this by multiplying the first row by $-\frac{1}{6}$:

$$\left[\begin{array}{cc|c} 18 & -6 & 15 \\ 5 & -2 & 4 \end{array}\right] \xrightarrow{-\frac{1}{6}[1]} \left[\begin{array}{cc|c} -3 & 1 & -\frac{5}{2} \\ 5 & -2 & 4 \end{array}\right].$$

Next, we transform the -2 (the only remaining entry in column 2) into a 0:

$$\left[\begin{array}{cc|c} -3 & 1 & -\frac{5}{2} \\ 5 & -2 & 4 \end{array}\right] \xrightarrow{[2] + 2[1]} \left[\begin{array}{cc|c} -3 & 1 & -\frac{5}{2} \\ -1 & 0 & -1 \end{array}\right].$$

The last matrix is the result of pivoting the original matrix about the circled entry.

In terms of pivoting, we can give the following summary of the Gaussian elimination method.

> **Gaussian Elimination Method to Transform a System of Linear Equations into Diagonal Form**
>
> 1. Write down the matrix corresponding to the linear system.
> 2. Make sure that the first entry in the first column is nonzero. Do this by interchanging the first row with one of the rows below it, if necessary.
> 3. Pivot the matrix about the first entry in the first column.
> 4. Make sure that the second entry in the second column is nonzero. Do this by interchanging the second row with one of the rows below it, if necessary.
> 5. Pivot the matrix about the second entry in the second column.
> 6. Continue in this manner.

All the systems considered in the preceding section had only a single solution. In this case we say that the solution is *unique*. Let us now use the Gaussian elimination method to study the various possibilities other than a unique solution. We first experiment with an example.

EXAMPLE 2 Determine all solutions of the system

$$\begin{cases} 2x + 2y + 4z = 8 \\ x - y + 2z = 2 \\ -x + 5y - 2z = 2. \end{cases}$$

Solution We set up the matrix corresponding to the system and perform the appropriate pivoting operations. (The elements pivoted about are circled.)

$$
\begin{bmatrix} ② & 2 & 4 & 8 \\ 1 & -1 & 2 & 2 \\ -1 & 5 & -2 & 2 \end{bmatrix}
\begin{array}{l} \frac{1}{2}[1] \\ [2]+(-1)[1] \\ \hline [3]+(1)[1] \end{array} \rightarrow
\begin{bmatrix} 1 & 1 & 2 & 4 \\ 0 & ⊖② & 0 & -2 \\ 0 & 6 & 0 & 6 \end{bmatrix}
$$

$$
\begin{array}{l} (-\frac{1}{2})[2] \\ [1]+(-1)[2] \\ \hline [3]+(-6)[2] \end{array} \rightarrow
\begin{bmatrix} 1 & 0 & 2 & 3 \\ 0 & 1 & 0 & 1 \\ 0 & 0 & 0 & 0 \end{bmatrix}.
$$

Note that our method must terminate here, since there is no way to transform the third entry in the third column into a 1 without disturbing the columns already in appropriate form. The equations corresponding to the last matrix read

$$
\begin{cases} x & + 2z = 3 \\ & y & = 1 \\ & 0 & = 0. \end{cases}
$$

The last equation does not involve any of the variables and so may be omitted. This leaves the two equations

$$
\begin{cases} x & + 2z = 3 \\ & y & = 1. \end{cases}
$$

Now, taking the $2z$ term in the first equation to the right side, we can write the equations

$$
\begin{cases} x = 3 - 2z \\ y = 1. \end{cases}
$$

The value of y is given: $y = 1$. The value of x is given in terms of z. To find a solution to this system, assign any value to z. Then the first equation gives a value for x and thereby a specific solution to the system. For example, if we take $z = 1$, then the corresponding specific solution is

$$
z = 1
$$
$$
x = 3 - 2(1) = 1
$$
$$
y = 1.
$$

If we take $z = -3$, the corresponding specific solution is

$$
z = -3
$$
$$
x = 3 - 2(-3) = 9
$$
$$
y = 1.
$$

Thus, we see that the original system has infinitely many specific solutions, corresponding to the infinitely many possible different choices for z.

We say that the *general solution* of the system is

$$z = \text{any value}$$
$$x = 3 - 2z$$
$$y = 1.$$

When a linear system cannot be *completely* diagonalized:

1. Apply the Gaussian elimination method to as many columns as possible. Proceed from left to right, but do not disturb columns that have already been put into proper form.
2. Variables corresponding to columns not in proper form can assume any value.
3. The other variables can be expressed in terms of the variables of step 2.

EXAMPLE 3 Find all solutions of the linear system

$$\begin{cases} x + 2y - z + 3w = 5 \\ y + 2z + w = 7. \end{cases}$$

Solution The Gaussian elimination method proceeds as follows:

$$\begin{bmatrix} 1 & 2 & -1 & 3 & | & 5 \\ 0 & \textcircled{1} & 2 & 1 & | & 7 \end{bmatrix} \quad \text{(The first column is already in proper form.)}$$

$$\xrightarrow{[1] + (-2)[2]} \begin{bmatrix} 1 & 0 & -5 & 1 & | & -9 \\ 0 & 1 & 2 & 1 & | & 7 \end{bmatrix}.$$

We cannot do anything further with the last two columns (without disturbing the first two columns), so the corresponding variables, z and w, can assume any values. Writing down the equations corresponding to the last matrix yields

$$\begin{cases} x - 5z + w = -9 \\ y + 2z + w = 7 \end{cases}$$

or

$$z = \text{any value}$$
$$w = \text{any value}$$
$$x = -9 + 5z - w$$
$$y = 7 - 2z - w.$$

To determine an example of a specific solution, let $z = 1$, $w = 2$. Then a specific solution of the original system is

$$z = 1$$
$$w = 2$$
$$x = -9 + 5(1) - (2) = -6$$
$$y = 7 - 2(1) - (2) = 3.$$

EXAMPLE 4   Find all solutions of the system of equations

$$\begin{cases} x - 7y + z = 3 \\ 2x - 14y + 3z = 4. \end{cases}$$

Solution   The first pivot operation is routine:

$$\left[\begin{array}{ccc|c} \circled{1} & -7 & 1 & 3 \\ 2 & -14 & 3 & 4 \end{array}\right] \xrightarrow{[2] + (-2)[1]} \left[\begin{array}{ccc|c} 1 & -7 & 1 & 3 \\ 0 & 0 & 1 & -2 \end{array}\right].$$

However, it is impossible to pivot about the zero in the second column. So skip the second column and pivot about the second entry in the third column to get

$$\left[\begin{array}{ccc|c} 1 & -7 & 0 & 5 \\ 0 & 0 & 1 & -2 \end{array}\right].$$

This is as far as we can go. The variable corresponding to the second column, namely y, can assume any value, and the general solution of the system is obtained from the equations

$$\begin{cases} x - 7y = 5 \\ z = -2. \end{cases}$$

Therefore, the general solution of the system is

$$y = \text{any value}$$
$$x = 5 + 7y$$
$$z = -2.$$

We have seen that a linear system may have a unique solution or it may have infinitely many solutions. But another phenomenon can occur: A system may have no solutions at all, as the next example shows.

EXAMPLE 5   Find all solutions of the system

$$\begin{cases} x - y + z = 3 \\ x + y - z = 5 \\ -2x + 4y - 4z = 1. \end{cases}$$

Solution We apply the Gaussian elimination method to the matrix of the system.

$$\begin{bmatrix} ① & -1 & 1 & | & 3 \\ 1 & 1 & -1 & | & 5 \\ -2 & 4 & -4 & | & 1 \end{bmatrix} \xrightarrow[\substack{[2] + (-1)[1] \\ [3] + 2[1]}]{} \begin{bmatrix} 1 & -1 & 1 & | & 3 \\ 0 & ② & -2 & | & 2 \\ 0 & 2 & -2 & | & 7 \end{bmatrix}$$

$$\xrightarrow[\substack{\frac{1}{2}[2] \\ [1] + (1)[2] \\ [3] + (-2)[2]}]{} \begin{bmatrix} 1 & 0 & 0 & | & 4 \\ 0 & 1 & -1 & | & 1 \\ 0 & 0 & 0 & | & 5 \end{bmatrix}.$$

We cannot pivot about the last zero in the third column, so we have carried the method as far as we can. Let us write out the equations corresponding to the last matrix:

$$\begin{cases} x & = 4 \\ y - z & = 1 \\ 0 & = 5. \end{cases}$$

matter what the values of x, y, z. Thus, the original system has no solutions. Systems with no solutions can always be detected by the presence of inconsistent equations in the last matrix resulting from the Gaussian elimination method.

At first it might seem strange that some systems have no solutions, some have one, and yet others have infinitely many. The reason for the difference can be explained geometrically. For simplicity, consider the case of systems of two equations in two variables. Each equation in this case has a graph in the xy-plane, and the graph is a straight line. As we have seen, solving the system corresponds to finding the points lying on both lines. There are three possibilities. First, the two lines may intersect. In this case the solution is unique. Second, the two lines may be parallel. Then the two lines do not intersect and the system has no solutions. Finally, the two equations may represent the same line, as, for example, do the equations $2x + 3y = 1$, $4x + 6y = 2$. In this case every point on the line is a solution of the system; that is, there are infinitely many solutions (Fig. 1).

FIGURE 1

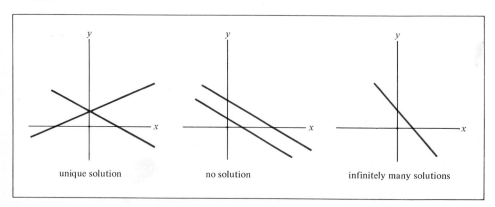

unique solution   no solution   infinitely many solutions

PRACTICE PROBLEMS 2

1. Find a specific solution to a system of linear equations whose general solution is

$$w = \text{any value}$$

$$y = \text{any value}$$

$$z = 7 + 6w$$

$$x = 26 - 2y + 14w.$$

2. Find all solutions of this system of linear equations.

$$\begin{cases} 2x + 4y - 4z - 4w = 24 \\ -3x - 6y + 10z - 18w = -8 \\ -x - 2y + 4z - 10w = 2. \end{cases}$$

EXERCISES 2

Pivot each of the following matrices about the circled element.

1. $\begin{bmatrix} ② & -4 & 6 \\ 3 & 7 & 1 \end{bmatrix}$

2. $\begin{bmatrix} 1 & 2 & 3 \\ 4 & ⑧ & -12 \end{bmatrix}$

3. $\begin{bmatrix} 7 & 1 & 4 & 5 \\ -1 & 1 & ② & 6 \\ 4 & 0 & 2 & 3 \end{bmatrix}$

4. $\begin{bmatrix} 5 & 10 & -10 & 12 \\ 4 & 3 & 6 & 12 \\ 4 & ④ & 4 & -16 \end{bmatrix}$

5. $\begin{bmatrix} ② & 3 \\ 6 & 0 \\ 1 & 5 \end{bmatrix}$

6. $\begin{bmatrix} 2 & 1 \\ ⊖① & 0 \end{bmatrix}$

7. $\begin{bmatrix} 4 & 3 & 0 \\ \frac{2}{3} & 0 & -2 \\ 1 & 3 & ⑥ \end{bmatrix}$

8. $\begin{bmatrix} 1 & 0 & 2 \\ -1 & 1 & ⊖② \\ 1 & 2 & 6 \end{bmatrix}$

Use the Gaussian elimination method to find all solutions of the following systems of linear equations.

9. $\begin{cases} 2x - 4y = 6 \\ -x + 2y = -3 \end{cases}$

10. $\begin{cases} -\frac{1}{2}x + y = \frac{3}{2} \\ -3x + 6y = 10 \end{cases}$

11. $\begin{cases} x + 2y = 5 \\ 3x - y = 1 \\ -x + 3y = 5 \end{cases}$

12. $\begin{cases} x - 6y = 12 \\ -\frac{1}{2}x + 3y = -6 \\ \frac{1}{3}x - 2y = 4 \end{cases}$

13. $\begin{cases} x - y + 3z = 3 \\ -2x + 3y - 11z = -4 \\ x - 2y + 8z = 6 \end{cases}$

14. $\begin{cases} x - 3y + z = 5 \\ -2x + 7y - 6z = -9 \\ x - 2y - 3z = 6 \end{cases}$

15. $\begin{cases} x + y + z = -1 \\ 2x + 3y + 2z = 3 \\ 2x + y + 2z = -7 \end{cases}$

16. $\begin{cases} x - 3y + 2z = 10 \\ -x + 3y - z = -6 \\ -x + 3y + 2z = 6 \end{cases}$

17. $\begin{cases} x + 2y + 3z = 4 \\ 5x + 6y + 7z = 8 \\ x + 2y + 3z = 5 \end{cases}$

18. $\begin{cases} x + 3y = 7 \\ x + 2y = 5 \\ -x + y = 2 \end{cases}$

19. $\begin{cases} x + y - 2z + 2w = 5 \\ 2x + y - 4z + w = 5 \\ 3x + 4y - 6z + 9w = 20 \\ 4x + 4y - 8z + 8w = 20 \end{cases}$

20. $\begin{cases} 2y + z - w = 1 \\ x - y + z + w = 14 \\ -x - 9y - z + 4w = 11 \\ x + y + z = 9 \end{cases}$

21. $\begin{cases} 6x - 4y = 2 \\ -3x + 3y = 6 \\ 5x + 2y = 39 \end{cases}$

22. $\begin{cases} 3x + 2y = 5 \\ -x + 3y = 2 \\ 5x + 2y = 6 \\ 6x + y = 39 \end{cases}$

In Exercises 23–25, find three solutions to the system of equations.

23. $\begin{cases} x + 2y + z = 5 \\ y + 3z = 9 \end{cases}$

24. $\begin{cases} x + 5y + 3z = 9 \\ 2x + 9y + 7z = 5 \end{cases}$

25. $\begin{cases} x + 7y - 3z = 8 \\ z = 5 \end{cases}$

26. An out-of-shape athlete runs 6 miles per hour, swims 1 mile per hour, and bikes 10 miles per hour. He entered a triathalon, which requires all three events, and finished 5 hours and 40 minutes later. A friend who runs 8 miles per hour, swims 2 miles per hour, and bikes 15 miles per hour finished the same course in 3 hours and 35 minutes. The total course was 32 miles long. How many miles was each segment (running, swimming, and biking)?

27. In a laboratory experiment, a researcher wants to provide a rabbit with exactly 1000 units of vitamin A, exactly 1600 units of vitamin C, and exactly 2400 units of vitamin E. The rabbit is fed a mixture of three foods. Each gram of food 1 contains 2 units of vitamin A, 3 units of vitamin C, and 5 units of vitamin E. Each gram of food 2 contains 4 units of vitamin A, 7 units of vitamin C, and 9 units of vitamin E. Each gram of food 3 contains 6 units of vitamin A, 10 units of vitamin C, and 14 units of vitamin E. How many grams of each food should the rabbit be fed?

28. Rework Exercise 27 with the requirement for vitamin E changed to 2000 units.

29. Find all solutions to the following system of linear equations.

$$\begin{cases} x^2 + y^2 + z^2 = 14 \\ x^2 - y^2 + 2z^2 = 15 \\ x^2 + 2y^2 + 3z^2 = 36 \end{cases}$$

30. For what value of k will the following system of linear equations have a solution?

$$\begin{cases} 2x + 6y = 4 \\ x + 7y = 10 \\ kx + 8y = 4 \end{cases}$$

31. For what value(s) of k will the following system of linear equations have no solution? Infinitely many solutions?

$$\begin{cases} 2x - 3y = 4 \\ -6x + 9y = k \end{cases}$$

SOLUTIONS TO PRACTICE PROBLEMS 2

1. Since w and y can each assume any value, select any numbers, say $w = 1$ and $y = 2$. Then $z = 7 + 6(1) = 13$ and $x = 26 - 2(2) + 14(1) = 36$. So $x = 36$, $y = 2$, $z = 13$, $w = 1$ is a specific solution. There are infinitely many different specific solutions, since there are infinitely many different choices for w and y.

2. We apply the Gaussian elimination method to the matrix of the system.

$$\begin{bmatrix} ② & 4 & -4 & -4 & | & 24 \\ -3 & -6 & 10 & -18 & | & -8 \\ -1 & -2 & 4 & -10 & | & 2 \end{bmatrix} \xrightarrow[\substack{\frac{1}{2}[1] \\ [2] + 3[1] \\ [3] + 1[1]}]{} \begin{bmatrix} 1 & 2 & -2 & -2 & | & 12 \\ 0 & 0 & ④ & -24 & | & 28 \\ 0 & 0 & 2 & -12 & | & 14 \end{bmatrix}$$

$$\xrightarrow[\substack{\frac{1}{4}[2] \\ [1] + 2[2] \\ [3] + (-2)[2]}]{} \begin{bmatrix} 1 & 2 & 0 & -14 & | & 26 \\ 0 & 0 & 1 & -6 & | & 7 \\ 0 & 0 & 0 & 0 & | & 0 \end{bmatrix}.$$

The corresponding system of equation is

$$\begin{cases} x + 2y & - 14w = 26 \\ & z - 6w = 7. \end{cases}$$

The general solution is

$$w = \text{any value}$$
$$y = \text{any value}$$
$$z = 7 + 6w$$
$$x = 26 - 2y + 14w.$$

2.3. Arithmetic Operations on Matrices

We introduced matrices in the preceding sections in order to display the coefficients of a system of linear equations. For example, the linear system

$$\begin{cases} 5x - 3y = \frac{1}{2} \\ 4x + 2y = -1 \end{cases}$$

is represented by the matrix

$$\begin{bmatrix} 5 & -3 & | & \frac{1}{2} \\ 4 & 2 & | & -1 \end{bmatrix}.$$

After we have become accustomed to using such matrices in solving linear systems, we may omit the vertical line which separates the left and the right sides of the equations. We need only remember that the right side of the equations is recorded in the right column. So, for example, we would write the matrix above in the form

$$\begin{bmatrix} 5 & -3 & \frac{1}{2} \\ 4 & 2 & -1 \end{bmatrix}$$

A matrix is *any* rectangular array of numbers and may be of any size. Here are some examples of matrices of various sizes:

$$\begin{bmatrix} 3 & 7 \\ 0 & -1 \end{bmatrix}, \quad \begin{bmatrix} 1 \\ 2 \end{bmatrix}, \quad \begin{bmatrix} 2 & 1 \end{bmatrix}, \quad \begin{bmatrix} 6 \end{bmatrix}, \quad \begin{bmatrix} 5 & 7 & -1 \\ 0 & 3 & 5 \\ 6 & 0 & 5 \end{bmatrix}$$

Examples of matrices abound in everyday life. For example, the newspaper stock-market report is a large matrix with several thousand rows, one for each listed stock. The columns of the matrix give various data about each stock, such as opening and closing price, number of shares traded, and so forth. Another example of a matrix is a mileage charge on a road map. The rows and columns are labeled with the names of cities. The number at a given row and column gives the distance between the corresponding cities.

In these everyday examples, matrices are used only to display data. However, the most important applications involve arithmetic operations on matrices—namely, addition, subtraction, and multiplication of matrices. The major goal of this section is to discuss these operations. Before we can do so, however, we need some vocabulary with which to describe matrices.

A matrix is described by the number of rows and columns it contains. For example, the matrix

$$\begin{bmatrix} 7 & 5 \\ \frac{1}{2} & -2 \\ 2 & -11 \end{bmatrix}$$

has three rows and two columns and is referred to as a 3 × 2 (read: "three-by-two") *matrix*. The matrix $\begin{bmatrix} 4 & 5 & 0 \end{bmatrix}$ has one row and three columns and is a 1 × 3 *matrix*. A matrix with only one row is often called a *row matrix* (sometimes also called a *row vector*). A matrix, such as $\begin{bmatrix} 2 \\ 7 \end{bmatrix}$, which has only one column is called a *column matrix* or *column vector*. If a matrix has the same number of rows and columns, it is called a *square matrix*. Here are some square matrices of various sizes:

$$\begin{bmatrix} 5 \end{bmatrix}, \quad \begin{bmatrix} 1 & 2 \\ 3 & 4 \end{bmatrix}, \quad \begin{bmatrix} 2 & -1 & 0 \\ 3 & 5 & 4 \\ 0 & 3 & -7 \end{bmatrix}$$

The rows of a matrix are numbered from the top down, and the columns are numbered from left to right. For example, the first row of the matrix

$$\begin{bmatrix} 1 & -1 & 0 \\ 2 & 1 & 7 \\ -3 & 2 & 4 \end{bmatrix}$$

is $\begin{bmatrix} 1 & -1 & 0 \end{bmatrix}$, and its third column is

$$\begin{bmatrix} 0 \\ 7 \\ 4 \end{bmatrix}.$$

The numbers in a matrix, called *entries*, may be identified in terms of the row and column containing the entry in question. For example, the entry in the first row, third column of the following matrix is 0:

$$\begin{bmatrix} 1 & -1 & \boxed{0} \\ 2 & 1 & 7 \\ -3 & 2 & 4 \end{bmatrix};$$

the entry in the second row, first column is 2:

$$\begin{bmatrix} 1 & -1 & 0 \\ \boxed{2} & 1 & 7 \\ -3 & 2 & 4 \end{bmatrix};$$

and the entry in the third row, third column is 4:

$$\begin{bmatrix} 1 & -1 & 0 \\ 2 & 1 & 7 \\ -3 & 2 & \boxed{4} \end{bmatrix}.$$

We use double-subscripted letters to indicate the locations of the entries of a matrix. We denote the entry in the ith row, jth column by a_{ij}. For instance, in the above matrix we have $a_{13} = 0$, $a_{21} = 2$, and $a_{33} = 4$.

We say that two matrices A and B are *equal*, denoted $A = B$, provided that they have the same size and that all their corresponding entries are equal.

Addition and Subtraction of Matrices   We define the sum $A + B$ of two matrices A and B only if A and B are two matrices of the size—that is, if A and B have the same number of rows and the same number of columns. In this case $A + B$ is the matrix formed by adding the corresponding entries of A and B. Thus, for example,

$$\begin{bmatrix} 2 & 0 \\ 1 & 1 \\ 5 & 3 \end{bmatrix} + \begin{bmatrix} 5 & 4 \\ 0 & 2 \\ 2 & 6 \end{bmatrix} = \begin{bmatrix} 2+5 & 0+4 \\ 1+0 & 1+2 \\ 5+2 & 3+6 \end{bmatrix} = \begin{bmatrix} 7 & 4 \\ 1 & 3 \\ 7 & 9 \end{bmatrix}.$$

We subtract matrices of the same size by subtracting corresponding entries. Thus, we have

$$\begin{bmatrix} 7 \\ 1 \end{bmatrix} - \begin{bmatrix} 3 \\ 2 \end{bmatrix} = \begin{bmatrix} 7 - 3 \\ 1 - 2 \end{bmatrix} = \begin{bmatrix} 4 \\ -1 \end{bmatrix}.$$

Multiplication of Matrices   It might seem that to define the product of two matrices, one would start with two matrices of like size and multiply the corresponding entries. But this definition is not useful, since the calculations that arise in applications require a somewhat more complex multiplication. In the interests of simplicity, we start by defining the product of a row matrix times a column matrix.

If A is a row matrix and B is a column matrix, then we can form the product $A \cdot B$ provided that the two matrices have the same length. The product $A \cdot B$ is the 1×1 matrix obtained by multiplying corresponding entries of A and B and then forming the sum.

We may put this definition into algebraic terms as follows. Suppose that A is the row matrix

$$A = [a_1 \quad a_2 \quad \cdots \quad a_n],$$

and B is the column matrix

$$B = \begin{bmatrix} b_1 \\ b_2 \\ \vdots \\ b_n \end{bmatrix}.$$

Note that A and B are both of the same length, namely n. Then

$$A \cdot B = [a_1 \quad a_2 \quad \cdots \quad a_n] \cdot \begin{bmatrix} b_1 \\ b_2 \\ \vdots \\ b_n \end{bmatrix}$$

is calculated by multiplying corresponding entries of A and B and forming the sum; that is,

$$A \cdot B = [a_1 b_1 + a_2 b_2 + \cdots + a_n b_n].$$

Notice that the product is a 1×1 matrix, namely a single number in brackets.

Here are some examples of the product of a row matrix times a column matrix:

$$[3 \quad \tfrac{1}{2}] \cdot \begin{bmatrix} 1 \\ 4 \end{bmatrix} = [3 \cdot 1 + \tfrac{1}{2} \cdot 4] = [5],$$

$$[2 \quad 0 \quad -1] \cdot \begin{bmatrix} 6 \\ 5 \\ 3 \end{bmatrix} = [2 \cdot 6 + 0 \cdot 5 + (-1) \cdot 3] = [9].$$

In multiplying a row matrix times a column matrix, it helps to use both hands. Use your left index finger to point to an element of the row matrix and your right to point to the corresponding element of the column. Multiply the elements you are pointing to and keep a running total of the products in your head. After each multiplication move your fingers to the next elements of each matrix. With a little practice you should be able to multiply a row times a column quickly and accurately.

The definition of multiplication above may seem strange. But products of this sort occur in many down-to-earth problems. Consider, for instance, the next example.

EXAMPLE 1   A dairy farm produces three items—milk, eggs, and cheese. The prices of these items are \$1.50 per gallon, \$.80 per dozen, and \$2 per pound, respectively. In a certain week the dairy farm sells 30,000 gallons of milk, 2000 dozen eggs, and 5000 pounds of cheese. Represent its total revenue as a matrix product.

Solution   The total revenue equals

$$(1.50)(30,000) + (.80)(2000) + (2)(5000).$$

This suggests that we define two matrices: the first displays the prices of the various produce:

$$[1.50 \quad .80 \quad 2].$$

The second represents the production:

$$\begin{bmatrix} 30,000 \\ 2000 \\ 5000 \end{bmatrix}.$$

Then the revenue for the week, when placed in a 1×1 matrix, equals

$$[1.50 \quad .80 \quad 2] \begin{bmatrix} 30,000 \\ 2000 \\ 5000 \end{bmatrix} = [56,600].$$

The principle behind Example 1 is this: any sum of products of the form $a_1b_1 + a_2b_2 + \cdots + a_nb_n$, when placed in a 1×1 matrix, can be written as the matrix product

$$[a_1b_1 + a_2b_2 + \cdots + a_nb_n] = [a_1 \quad a_2 \quad \cdots \quad a_n] \begin{bmatrix} b_1 \\ b_2 \\ \vdots \\ b_n \end{bmatrix}.$$

Let us illustrate the procedure for multiplying more general matrices by working out a typical product:

$$\begin{bmatrix} 2 & 1 \\ 0 & 1 \\ 1 & 0 \end{bmatrix} \cdot \begin{bmatrix} 1 & 1 \\ 4 & 2 \end{bmatrix}.$$

To obtain the entries of the product, we multiply the rows of the left matrix by the columns of the right matrix, taking care to arrange the products in a specific way to yield a matrix, as follows. Start with the first row on the left, $[2 \quad 1]$, and the first column on the right, $\begin{bmatrix} 1 \\ 4 \end{bmatrix}$. Their product is $[6]$, so we enter 6 as the element in the first row, first column of the product:

$$\begin{bmatrix} 2 & 1 \\ 0 & 1 \\ 1 & 0 \end{bmatrix} \cdot \begin{bmatrix} 1 & 1 \\ 4 & 2 \end{bmatrix} = \begin{bmatrix} 6 & \\ & \end{bmatrix}.$$

The product of the first row of the left matrix and the second column of the right matrix is $[4]$, so we put a 4 in the first row, second column of the product:

$$\begin{bmatrix} 2 & 1 \\ 0 & 1 \\ 1 & 0 \end{bmatrix} \cdot \begin{bmatrix} 1 & 1 \\ 4 & 2 \end{bmatrix} = \begin{bmatrix} 6 & 4 \\ & \end{bmatrix}.$$

There are no more columns which can be multiplied by the first row, so let us move to the second row and shift back to the first column. Correspondingly, we move one row down in the product:

$$\begin{bmatrix} 2 & 1 \\ 0 & 1 \\ 1 & 0 \end{bmatrix} \cdot \begin{bmatrix} 1 & 1 \\ 4 & 2 \end{bmatrix} = \begin{bmatrix} 6 & 4 \\ 4 & \end{bmatrix}$$

$$\begin{bmatrix} 2 & 1 \\ 0 & 1 \\ 1 & 0 \end{bmatrix} \cdot \begin{bmatrix} 1 & 1 \\ 4 & 2 \end{bmatrix} = \begin{bmatrix} 6 & 4 \\ 4 & 2 \end{bmatrix}.$$

We have now exhausted the second row of the left matrix, so we shift to the third row and correspondingly move down one row in the product.

$$\begin{bmatrix} 2 & 1 \\ 0 & 1 \\ 1 & 0 \end{bmatrix} \cdot \begin{bmatrix} 1 & 1 \\ 4 & 2 \end{bmatrix} = \begin{bmatrix} 6 & 4 \\ 4 & 2 \\ 1 & \end{bmatrix}$$

$$\begin{bmatrix} 2 & 1 \\ 0 & 1 \\ 1 & 0 \end{bmatrix} \cdot \begin{bmatrix} 1 & 1 \\ 4 & 2 \end{bmatrix} = \begin{bmatrix} 6 & 4 \\ 4 & 2 \\ 1 & 1 \end{bmatrix}.$$

Note that we have now multiplied every row of the left matrix by every column of the right matrix. This completes computation of the product:

$$\begin{bmatrix} 2 & 1 \\ 0 & 1 \\ 1 & 0 \end{bmatrix} \cdot \begin{bmatrix} 1 & 1 \\ 4 & 2 \end{bmatrix} = \begin{bmatrix} 6 & 4 \\ 4 & 2 \\ 1 & 1 \end{bmatrix}.$$

EXAMPLE 2   Calculate the following product:

$$\begin{bmatrix} 1 & 5 \\ 3 & 2 \end{bmatrix} \cdot \begin{bmatrix} 1 & 2 \\ 1 & 0 \end{bmatrix}.$$

Solution

$$\begin{bmatrix} 1 & 5 \\ 3 & 2 \end{bmatrix} \cdot \begin{bmatrix} 1 & 2 \\ 1 & 0 \end{bmatrix} = \begin{bmatrix} 6 & \\ & \end{bmatrix}$$

$$\begin{bmatrix} 1 & 5 \\ 3 & 2 \end{bmatrix} \cdot \begin{bmatrix} 1 & 2 \\ 1 & 0 \end{bmatrix} = \begin{bmatrix} 6 & 2 \\ & \end{bmatrix}$$

$$\begin{bmatrix} 1 & 5 \\ 3 & 2 \end{bmatrix} \cdot \begin{bmatrix} 1 & 2 \\ 1 & 0 \end{bmatrix} = \begin{bmatrix} 6 & 2 \\ 5 & \end{bmatrix}$$

$$\begin{bmatrix} 1 & 5 \\ 3 & 2 \end{bmatrix} \cdot \begin{bmatrix} 1 & 2 \\ 1 & 0 \end{bmatrix} = \begin{bmatrix} 6 & 2 \\ 5 & 6 \end{bmatrix}.$$

Thus

$$\begin{bmatrix} 1 & 5 \\ 3 & 2 \end{bmatrix} \cdot \begin{bmatrix} 1 & 2 \\ 1 & 0 \end{bmatrix} = \begin{bmatrix} 6 & 2 \\ 5 & 6 \end{bmatrix}.$$

Notice that we cannot use the method above to compute the product $A \cdot B$ of *any* matrices A and B. For the procedure to work, it is crucial that the number of entries of each row of A be the same as the number of entries of each column of B. (Or, to put it another way, the number of columns of the left matrix must equal the number of rows of the right matrix.) Therefore, in order for us to form the product $A \cdot B$, the sizes of A and B must match up in a special way. If A is $m \times n$ and B is $p \times q$, then the product $A \cdot B$ is defined only in case the "inner" dimensions n and p are equal. In that case, the product is determined by the "outer" dimensions m and q. It is an $m \times q$ matrix.

$$A \quad \cdot \quad B \quad = \quad C.$$
$$m \times n \quad p \times q \quad m \times q$$
$$\underbrace{\qquad}_{\text{equal}}$$

So, for example,

$$\begin{bmatrix} \end{bmatrix}\begin{bmatrix} \end{bmatrix} = \begin{bmatrix} \end{bmatrix}$$
$$3 \times 4 \qquad 4 \times 2 \qquad 3 \times 2$$

$$\begin{bmatrix} \end{bmatrix}\begin{bmatrix} \end{bmatrix} = \begin{bmatrix} \end{bmatrix}.$$
$$2 \times 2 \quad 2 \times 1 \qquad 2 \times 1$$

If the sizes of A and B do not match up in the way just described, the product $A \cdot B$ is not defined.

EXAMPLE 3 Calculate the following products, if defined.

(a) $\begin{bmatrix} 3 & -1 \\ 2 & 0 \\ 1 & 5 \end{bmatrix} \begin{bmatrix} 1 & 0 \\ 5 & -4 \\ 2 & -1 \end{bmatrix}$
(b) $\begin{bmatrix} 3 & -1 \\ 2 & 0 \\ 1 & 5 \end{bmatrix} \begin{bmatrix} 5 & 4 \\ -2 & 3 \end{bmatrix}$

Solution (a) The matrices to be multiplied are 3×2 and 3×2. The inner dimensions do not match, so the product is undefined.

(b) We are asked to multiply a 3×2 matrix times a 2×2 matrix. The inner dimensions match, so the product is defined and has size determined by the outer dimensions, that is, 3×2.

$$\begin{bmatrix} 3 & -1 \\ 2 & 0 \\ 1 & 5 \end{bmatrix} \cdot \begin{bmatrix} 5 & 4 \\ -2 & 3 \end{bmatrix} = \begin{bmatrix} 3 \cdot 5 + (-1) \cdot (-2) & 3 \cdot 4 + (-1) \cdot 3 \\ 2 \cdot 5 + 0 \cdot (-2) & 2 \cdot 4 + 0 \cdot 3 \\ 1 \cdot 5 + 5 \cdot (-2) & 1 \cdot 4 + 5 \cdot 3 \end{bmatrix}$$

$$= \begin{bmatrix} 17 & 9 \\ 10 & 9 \\ -5 & 19 \end{bmatrix}.$$

Multiplication of matrices has many properties in common with multiplication of ordinary numbers. However, there is at least one important difference. With matrix multiplication, the order of the factors is usually important. For example, the product of a 2×3 matrix times a 3×2 matrix is defined and the product is a 2×2 matrix. If the order is reversed to a 3×2 matrix times a 2×3 matrix, the product is a 3×3 matrix. So reversing the order may change the size of the product. Even when it does not, reversing the order may still change the entries in the product, as the following two products demonstrate.

$$\begin{bmatrix} 1 & 5 \\ 3 & 2 \end{bmatrix} \begin{bmatrix} 1 & 2 \\ 1 & 0 \end{bmatrix} = \begin{bmatrix} 6 & 2 \\ 5 & 6 \end{bmatrix}$$

$$\begin{bmatrix} 1 & 2 \\ 1 & 0 \end{bmatrix} \begin{bmatrix} 1 & 5 \\ 3 & 2 \end{bmatrix} = \begin{bmatrix} 7 & 9 \\ 1 & 5 \end{bmatrix}.$$

EXAMPLE 4 An investment trust has investments in three states. Its deposits in each state are divided among bonds, mortgages, and consumer loans. On January 1 the amount (in millions of dollars) of money invested in each category by state is given by the matrix

	Bonds	Mortgages	Consumer loans
State A	10	5	20
State B	30	12	10
State C	15	6	25

The current average yields are 7% for bonds, 9% for mortgages, and 15% for consumer loans. Determine the earnings of the trust from its investments in each state.

Solution Define the matrix of investment yields by

$$\begin{bmatrix} .07 \\ .09 \\ .15 \end{bmatrix} \begin{matrix} \text{Bonds} \\ \text{Mortgages} \\ \text{Consumer loans.} \end{matrix}$$

The amount earned in state A, for instance, is

[amount of bonds][yield of bonds]

+ [amount of mortgages][yield of mortgages]

+ [amount of consumer loans][yield of consumer loans]

$$= (10)(.07) + (5)(.09) + (20)(.15).$$

And this is just the first entry of the product

$$\begin{bmatrix} 10 & 5 & 20 \\ 30 & 12 & 10 \\ 15 & 6 & 25 \end{bmatrix} \begin{bmatrix} .07 \\ .09 \\ .15 \end{bmatrix}.$$

Similarly, the earnings for the other states are the second and third entries of the product. Carrying out the arithmetic, we find that

$$\begin{bmatrix} 10 & 5 & 20 \\ 30 & 12 & 10 \\ 15 & 6 & 25 \end{bmatrix} \begin{bmatrix} .07 \\ .09 \\ .15 \end{bmatrix} = \begin{bmatrix} 4.15 \\ 4.68 \\ 5.34 \end{bmatrix}.$$

Therefore, the trust earns $4.15 million in state A, $4.68 million in state B, and $5.34 million in state C.

EXAMPLE 5 A clothing manufacturer has factories in Los Angeles, San Antonio, and Newark. Sales (in thousands) during the first quarter of last year are summarized in the production matrix

	Los Angeles	San Antonio	Newark
Coats	12	13	38
Shirts	25	5	26
Sweaters	11	8	8
Ties	5	0	12

During this period the selling price of a coat was $100, of a shirt $10, of a sweater $25, and of a tie $5.

(a) Use a matrix calculation to determine the total revenue produced by each of the factories.

(b) Suppose that the prices had been $110, $8, $20, and $10, respectively. How would this have affected the revenue of each factory?

Solution   (a) For each factory, we wish to multiply the price of each item by the number produced to arrive at revenue. Since the production figures for the various items of clothing are arranged down the columns, we arrange the prices in a row matrix, ready for multiplication. The price matrix is

$$[100 \quad 10 \quad 25 \quad 5].$$

The revenues of the various factories are then the entries of the product

$$[100 \quad 10 \quad 25 \quad 5] \begin{bmatrix} 12 & 13 & 38 \\ 25 & 5 & 26 \\ 11 & 8 & 8 \\ 5 & 0 & 12 \end{bmatrix} = \begin{matrix} \text{Los Angeles} & \text{San Antonio} & \text{Newark} \\ [1750 & 1550 & 4320]. \end{matrix}$$

Since the production figures are in thousands, the revenue figures are in thousands of dollars. That is, the Los Angeles factory has revenues of $1,750,000, the San Antonio factory $1,550,000, and the Newark factory $4,320,000.

(b) In a similar way, we determine the revenue of each factory if the price matrix had been $[110 \quad 8 \quad 20 \quad 10]$.

$$[110 \quad 8 \quad 20 \quad 10] \begin{bmatrix} 12 & 13 & 38 \\ 25 & 5 & 26 \\ 11 & 8 & 8 \\ 5 & 0 & 12 \end{bmatrix} = \begin{matrix} \text{Los Angeles} & \text{San Antonio} & \text{Newark} \\ [1790 & 1630 & 4668]. \end{matrix}$$

The change in revenue at each factory can be read from the difference of the revenue matrices:

$$[1790 \quad 1630 \quad 4668] - [1750 \quad 1550 \quad 4320] = [40 \quad 80 \quad 348].$$

That is, if prices had been as given in part (b), then revenues of the Los Angeles factory would have increased by 40, revenues at San Antonio would have increased by 80, and revenues at Newark would have increased by 348.

There are special matrices analogous to the number 1. Such matrices are called *identity matrices*. The identity matrix I_n of size n is the $n \times n$ square matrix with all zeros, except for ones down the upper-left-to-lower-right diagonal. Here are the identity matrices of sizes 2, 3, and 4:

$$I_2 = \begin{bmatrix} 1 & 0 \\ 0 & 1 \end{bmatrix}, \qquad I_3 = \begin{bmatrix} 1 & 0 & 0 \\ 0 & 1 & 0 \\ 0 & 0 & 1 \end{bmatrix}, \qquad I_4 = \begin{bmatrix} 1 & 0 & 0 & 0 \\ 0 & 1 & 0 & 0 \\ 0 & 0 & 1 & 0 \\ 0 & 0 & 0 & 1 \end{bmatrix}.$$

The characteristic property of an identity matrix is that it plays the role of the number 1; that is,

$$I_n \cdot A = A \cdot I_n = A$$

for all $n \times n$ matrices A.

One of the principal uses of matrices is in dealing with systems of linear equations. Matrices provide a compact way of writing systems, as the next example shows.

EXAMPLE 6   Write the system of linear equations

$$\begin{cases} -2x + 4y = 2 \\ -3x + 7y = 7 \end{cases}$$

as a matrix equation.

Solution   The system of equations can be written in the form

$$\begin{bmatrix} -2x + 4y \\ -3x + 7y \end{bmatrix} = \begin{bmatrix} 2 \\ 7 \end{bmatrix}$$

So consider the matrices

$$A = \begin{bmatrix} -2 & 4 \\ -3 & 7 \end{bmatrix}, \qquad X = \begin{bmatrix} x \\ y \end{bmatrix}, \qquad B = \begin{bmatrix} 2 \\ 7 \end{bmatrix}$$

Notice that

$$AX = \begin{bmatrix} -2 & 4 \\ -3 & 7 \end{bmatrix} \begin{bmatrix} x \\ y \end{bmatrix} = \begin{bmatrix} -2x + 4y \\ -3x + 7y \end{bmatrix}$$

Thus AX is a 2×1 column matrix whose entries correspond to the left side of the given system of linear equations. Since the entries of B correspond to the right side of the system of equations, we can rewrite the given system in the form

$$AX = B$$

—that is,

$$\begin{bmatrix} -2 & 4 \\ -3 & 7 \end{bmatrix} \begin{bmatrix} x \\ y \end{bmatrix} = \begin{bmatrix} 2 \\ 7 \end{bmatrix}$$

The matrix A of the example above displays the coefficients of the variables x and y, so it is called the *coefficient matrix* of the system.

PRACTICE PROBLEMS 3

1. Compute

$$\begin{bmatrix} 3 & 1 & 2 \\ -1 & 0 & \frac{1}{2} \\ 0 & 4 & 1 \end{bmatrix} \begin{bmatrix} 7 & -1 & 0 \\ 5 & 4 & 2 \\ -6 & 0 & 4 \end{bmatrix}$$

2. Give the system of linear equations which is equivalent to the matrix equation

$$\begin{bmatrix} 3 & -6 \\ 2 & 1 \end{bmatrix} \begin{bmatrix} x \\ y \end{bmatrix} = \begin{bmatrix} 5 \\ 0 \end{bmatrix}$$

3. Give a matrix equation equivalent to this system of equations:

$$\begin{cases} 8x + 3y = 7 \\ 9x - 2y = -5. \end{cases}$$

EXERCISES 3

In Exercises 1–6, give the size and special characteristics of the given matrix (such as square, column, row, identity).

1. $\begin{bmatrix} 3 & 2 & .4 \\ \frac{1}{2} & 0 & 6 \end{bmatrix}$

2. $\begin{bmatrix} 3 \\ -1 \end{bmatrix}$

3. $[2 \quad \frac{1}{3} \quad 0]$

4. $\begin{bmatrix} 1 & 0 \\ 0 & 1 \end{bmatrix}$

5. $[-2]$

6. $\begin{bmatrix} 0 & 0 & 0 & 0 \\ 0 & 0 & 0 & 0 \end{bmatrix}$

In Exercises 7–14, perform the indicated matrix calculations.

7. $\begin{bmatrix} 4 & -2 \\ 3 & 0 \end{bmatrix} + \begin{bmatrix} 5 & 5 \\ 4 & -1 \end{bmatrix}$

8. $\begin{bmatrix} 8 \\ -3 \end{bmatrix} + \begin{bmatrix} 5 \\ 6 \end{bmatrix}$

9. $\begin{bmatrix} 2 & 8 \\ \frac{4}{3} & 4 \\ 1 & -2 \end{bmatrix} - \begin{bmatrix} 1 & 5 \\ \frac{1}{3} & 2 \\ -3 & 0 \end{bmatrix}$

10. $\begin{bmatrix} 1 & 0 \\ 0 & 1 \end{bmatrix} - \begin{bmatrix} .8 & .5 \\ .2 & .5 \end{bmatrix}$

11. $[5 \quad 3]\begin{bmatrix} 1 \\ 2 \end{bmatrix}$

12. $[1 \quad 0 \quad 0]\begin{bmatrix} \frac{1}{2} \\ 6 \\ 2 \end{bmatrix}$

13. $[6 \quad 1 \quad 5]\begin{bmatrix} \frac{1}{2} \\ -3 \\ 2 \end{bmatrix}$

14. $[0 \quad 0]\begin{bmatrix} 5 \\ -3 \end{bmatrix}$

In Exercises 15–20, the sizes of two matrices are given. Tell whether or not the product AB is defined. If so, give its size.

15. $A, 3 \times 4; B, 4 \times 5$

16. $A, 3 \times 3; B, 3 \times 4$

17. $A, 3 \times 2; B, 3 \times 2$

18. $A, 1 \times 1; B, 1 \times 1$

19. $A, 3 \times 3; B, 3 \times 1$

20. $A, 4 \times 2; B, 3 \times 4$

In Exercises 21–30, perform the multiplication.

21. $\begin{bmatrix} 3 & 1 \\ 0 & 2 \end{bmatrix}\begin{bmatrix} 1 & 4 \\ 3 & 5 \end{bmatrix}$

22. $\begin{bmatrix} 4 & -1 \\ 2 & \frac{1}{2} \end{bmatrix}\begin{bmatrix} 3 \\ 2 \end{bmatrix}$

23. $\begin{bmatrix} 4 & 1 & 0 \\ -2 & 0 & 3 \\ 1 & 5 & -1 \end{bmatrix}\begin{bmatrix} 5 \\ 1 \\ 2 \end{bmatrix}$

24. $\begin{bmatrix} 0 & 0 \\ 0 & 0 \\ 0 & 0 \end{bmatrix}\begin{bmatrix} 1 & 2 \\ 3 & 4 \end{bmatrix}$

25. $\begin{bmatrix} 1 & 0 \\ 0 & 1 \end{bmatrix}\begin{bmatrix} 5 & 6 \\ 7 & 8 \end{bmatrix}$

26. $\begin{bmatrix} 1 & 2 \\ 1 & 3 \end{bmatrix}\begin{bmatrix} 3 & -2 \\ -1 & 1 \end{bmatrix}$

27. $\begin{bmatrix} .6 & .3 \\ .4 & .7 \end{bmatrix}\begin{bmatrix} .6 & .3 \\ .4 & .7 \end{bmatrix}$

28. $\begin{bmatrix} 0 & 1 & 2 \\ -1 & 4 & \frac{1}{2} \\ 1 & 3 & 0 \end{bmatrix}\begin{bmatrix} 3 & -1 & 5 \\ 0 & 2 & 2 \\ 4 & -6 & 0 \end{bmatrix}$

29. $\begin{bmatrix} 2 & -1 & 4 \\ 0 & 1 & 0 \\ \frac{1}{2} & 3 & -2 \end{bmatrix} \begin{bmatrix} 4 & 8 & 0 \\ 3 & -1 & 2 \\ 5 & 0 & 1 \end{bmatrix}$

30. $\begin{bmatrix} 1 & 0 & 0 \\ 0 & 1 & 0 \\ 0 & 0 & 1 \end{bmatrix} \begin{bmatrix} 1 \\ 2 \\ 3 \end{bmatrix}$

In Exercises 31–34, give the system of linear equations that is equivalent to the matrix equation. Do not solve.

31. $\begin{bmatrix} 2 & 3 \\ 4 & 5 \end{bmatrix} \begin{bmatrix} x \\ y \end{bmatrix} = \begin{bmatrix} 6 \\ 7 \end{bmatrix}$

32. $\begin{bmatrix} -3 & 4 \\ 0 & 1 \end{bmatrix} \begin{bmatrix} x \\ y \end{bmatrix} = \begin{bmatrix} 1 \\ 1 \end{bmatrix}$

33. $\begin{bmatrix} 1 & 2 & 3 \\ 4 & 5 & 6 \\ 7 & 8 & 9 \end{bmatrix} \begin{bmatrix} x \\ y \\ z \end{bmatrix} = \begin{bmatrix} 10 \\ 11 \\ 12 \end{bmatrix}$

34. $\begin{bmatrix} 1 & 0 & 0 \\ 0 & 1 & 0 \\ 0 & 0 & 1 \end{bmatrix} \begin{bmatrix} x \\ y \\ z \end{bmatrix} = \begin{bmatrix} 1 \\ 2 \\ 3 \end{bmatrix}$

In Exercises 35–38, write the given system of linear equations in matrix form.

35. $\begin{cases} 3x + 2y = -1 \\ 7x - y = 2 \end{cases}$

36. $\begin{cases} 5x - 2y = 6 \\ -3x + 4y = 0 \end{cases}$

37. $\begin{cases} x - 2y + 3z = 5 \\ y + z = 6 \\ z = 2 \end{cases}$

38. $\begin{cases} -2x + 4y - z = 5 \\ x + 6y + 3z = -1 \\ 7x + 4z = 8 \end{cases}$

The distributive law says that $(A + B)C = AC + BC$. That is, adding A and B and then multiplying on the right by C gives the same result as first multiplying each of A and B on the right by C and then adding. In Exercises 39 and 40, verify the distributive law for the given matrices.

39. $A = \begin{bmatrix} 1 & 2 \\ 0 & 3 \end{bmatrix}, B = \begin{bmatrix} 3 & -2 \\ 4 & 5 \end{bmatrix}, C = \begin{bmatrix} 1 & 6 \\ 2 & 0 \end{bmatrix}$

40. $A = \begin{bmatrix} 1 & 0 & 0 \\ 0 & 1 & 0 \\ 0 & 0 & 1 \end{bmatrix}, B = \begin{bmatrix} 2 & 1 & 3 \\ 0 & 5 & -1 \\ 3 & 6 & 0 \end{bmatrix}, C = \begin{bmatrix} 0 \\ 3 \\ -4 \end{bmatrix}$

Two $n \times n$ matrices A and B are called *inverses* (of one another) if both products AB and BA equal I_n. Check that the pairs of matrices in Exercises 41 and 42 are inverses.

41. $\begin{bmatrix} 3 & -1 \\ -1 & \frac{1}{2} \end{bmatrix}, \begin{bmatrix} 1 & 2 \\ 2 & 6 \end{bmatrix}$

42. $\begin{bmatrix} 2 & 8 & -11 \\ -1 & -5 & 7 \\ 1 & 2 & -3 \end{bmatrix}, \begin{bmatrix} 1 & 2 & 1 \\ 4 & 5 & -3 \\ 3 & 4 & -2 \end{bmatrix}$

43. The quantities of pants, shirts, and jackets owned by Mike and Don are given by the matrix A, and the costs of these items are given by matrix B.

$$\begin{array}{c} \\ \text{Mike} \\ \text{Don} \end{array} \begin{array}{ccc} \text{Pants} & \text{Shirts} & \text{Jackets} \\ \begin{bmatrix} 6 & 8 & 2 \\ 2 & 5 & 3 \end{bmatrix} \end{array} = A, \qquad \begin{array}{c} \text{Pants} \\ \text{Shirts} \\ \text{Jackets} \end{array} \begin{bmatrix} 20 \\ 15 \\ 50 \end{bmatrix} = B.$$

(a) Calculate the matrix AB.

(b) Interpret the entries of the matrix AB.

44. A company has three appliance stores that sell washers, dryers, and stoves. Matrices A and B give the wholesale and retail prices of these items, respectively. Matrices C

and *D* give the quantities of these items sold by the three stores in September and October, respectively.

$$A = \begin{bmatrix} \overset{\text{Washers}}{300} & \overset{\text{Dryers}}{250} & \overset{\text{Stoves}}{450} \end{bmatrix} \qquad B = \begin{bmatrix} \overset{\text{Washers}}{500} & \overset{\text{Dryers}}{450} & \overset{\text{Stoves}}{750} \end{bmatrix}$$

$$C = \begin{bmatrix} \overset{\text{Store 1}}{30} & \overset{\text{Store 2}}{40} & \overset{\text{Store 3}}{20} \\ 20 & 30 & 10 \\ 10 & 5 & 35 \end{bmatrix} \begin{matrix} \text{Washers} \\ \text{Dryers} \\ \text{Stoves} \end{matrix}$$

$$D = \begin{bmatrix} \overset{\text{Store 1}}{20} & \overset{\text{Store 2}}{50} & \overset{\text{Store 3}}{30} \\ 30 & 10 & 20 \\ 10 & 20 & 30 \end{bmatrix} \begin{matrix} \text{Washers} \\ \text{Dryers} \\ \text{Stoves} \end{matrix}$$

Determine and interpret the following matrices.

(a) *AC*   (b) *AD*   (c) *BC*

(d) *BD*   (e) *B* − *A*   (f) (*B* − *A*)*C*

(g) (*B* − *A*)*D*   (h) *C* + *D*   (i) (*B* − *A*)(*C* + *D*)

45. Three professors teaching the same course have entirely different grading policies. The percentage of each grade given by the professors is summarized in the matrix

	Grade				
	A	B	C	D	F
Prof. I	25	35	30	10	0
Prof. II	10	20	40	20	10
Prof. III	5	10	20	40	25

(a) The point values of the grades are A = 4, B = 3, C = 2, D = 1, and F = 0. Use matrix multiplication to determine the average grade given by each professor.

(b) Professor I has 240 students, Professor II has 120 students, and Professor III has 40 students. Use matrix multiplication to determine the numbers of A's, B's, C's, D's, and F's given.

46. A professor bases semester grades on four 100-point items: homework, quizzes, a midterm exam, and a final exam. Students may choose one of three schemes summarized in the matrix below for weighting the points from the four items. Use matrix multiplication to determine the most advantageous weighting scheme for a student who earned 97 points on homework, 72 points on the quizzes, 83 points on the midterm exam, and 75 points on the final exam.

	Items			
	Hw	Qu	ME	FE
Scheme I	.10	.10	.30	.50
Scheme II	.10	.20	.30	.40
Scheme III	.15	.15	.35	.35

47. In a certain town the percentages of voters voting Democratic and Republican by various age groups is summarized by this matrix:

$$
\begin{array}{c}
\\
\text{Under 30} \\
\text{30–50} \\
\text{Over 50}
\end{array}
\begin{array}{cc}
\text{Dem.} & \text{Rep.} \\
\left[\begin{array}{cc}
.65 & .35 \\
.55 & .45 \\
.45 & .55
\end{array}\right] & = A.
\end{array}
$$

The population of voters in the town by age group is given by the matrix

$$
B = \begin{bmatrix} 6000 & 8000 & 4000 \end{bmatrix}.
$$
$$
\underbrace{}_{\substack{\text{Under} \\ 30}} \ \underbrace{}_{\text{30–50}} \ \underbrace{}_{\substack{\text{Over} \\ 50}}
$$

Interpret the entries of the matrix product BA.

48. Refer to Exercise 47.

(a) Using the given data, which party would win and what would be the percentage of the winning vote?

(b) Suppose that the population of the town shifted toward older residents as reflected in the population matrix $B = \begin{bmatrix} 2000 & 4000 & 12{,}000 \end{bmatrix}$. What would be the result of the election now?

49. Suppose that a contractor employs carpenters, bricklayers, and plumbers, working three shifts per day. The number of man-hours employed in each of the shifts is summarized in the matrix

$$
\begin{array}{c}
\\
\\
\text{Carpenters} \\
\text{Bricklayers} \\
\text{Plumbers}
\end{array}
\begin{array}{c}
\text{Shift} \\
\begin{array}{ccc}
1 & 2 & 3 \\
\end{array} \\
\left[\begin{array}{ccc}
50 & 20 & 10 \\
30 & 30 & 15 \\
20 & 20 & 5
\end{array}\right].
\end{array}
$$

Labor in shift 1 costs $10 per hour, in shift 2 $15 per hour, and in shift 3 $20 per hour. Use matrix multiplication to compute the amount spent on each type of labor.

50. A flu epidemic hits a large city. Each resident of the city is either sick, well, or a carrier. The proportion of people in each of the categories is expressed by the matrix

$$
\begin{array}{c}
\\
\text{Well} \\
\text{Sick} \\
\text{Carrier}
\end{array}
\begin{array}{c}
\text{Age} \\
\begin{array}{ccc}
0\text{–}10 & 10\text{–}30 & \text{Over 30}
\end{array} \\
\left[\begin{array}{ccc}
.70 & .70 & .60 \\
.10 & .20 & .30 \\
.20 & .10 & .10
\end{array}\right] = A.
\end{array}
$$

The population of the city is distributed by age and sex as follows:

$$
\text{Age} \quad
\begin{array}{c}
0\text{–}10 \\
10\text{–}30 \\
\text{Over }30
\end{array}
\left[
\begin{array}{cc}
\overset{\text{Male}}{60{,}000} & \overset{\text{Female}}{65{,}000} \\
100{,}000 & 110{,}000 \\
200{,}000 & 230{,}000
\end{array}
\right] = B.
$$

(a) Compute AB.

(b) How many sick males are there?

(c) How many female carriers are there?

51. The matrix equation $AX = B$, where A is a 3×3 matrix, has the solution $x = 5$, $y = 4$, $z = 3$. What will be the new solution to this equation if the first and third columns of A are interchanged?

52. Find the values of a and b for which $A \cdot B = I_3$, where

$$
A = \begin{bmatrix} 3 & 2 & 0 \\ 1 & 1 & 0 \\ 0 & 0 & 1 \end{bmatrix} \text{ and } B = \begin{bmatrix} a & b & 0 \\ -1 & 3 & 0 \\ 0 & 0 & 1 \end{bmatrix}.
$$

SOLUTIONS TO PRACTICE PROBLEMS 3

1. Answer:

$$
\begin{bmatrix} 3 & 1 & 2 \\ -1 & 0 & \frac{1}{2} \\ 0 & 4 & 1 \end{bmatrix}
\begin{bmatrix} 7 & -1 & 0 \\ 5 & 4 & 2 \\ -6 & 0 & 4 \end{bmatrix}
=
\begin{bmatrix} 14 & 1 & 10 \\ -10 & 1 & 2 \\ 14 & 16 & 12 \end{bmatrix}.
$$

The systematic steps to be taken are:

(a) Determine the size of the product matrix.
Since we have a $\underset{\text{outer dimensions}}{\textcircled{3} \times 3 \text{ times a } 3 \times \textcircled{3}}$, the size of the product is given by the outer dimensions or 3×3. Begin by drawing a 3×3 rectangular array.

(b) Find the entries one at a time.
To find the entry in the first row, first column of the product, look at the first row of the left given matrix and the first column of the right given matrix and form their product.

$$
\begin{bmatrix} 3 & 1 & 2 \\ -1 & 0 & \frac{1}{2} \\ 0 & 4 & 1 \end{bmatrix}
\begin{bmatrix} 7 & -1 & 0 \\ 5 & 4 & 2 \\ -6 & 0 & 4 \end{bmatrix}
=
\begin{bmatrix} 14 & & \\ & & \\ & & \end{bmatrix},
$$

since $3 \cdot 7 + 1 \cdot 5 + 2(-6) = 14$. In general, to find the entry in the ith row, jth column of the product, put one finger on the ith row of the left given matrix and another finger on the jth column of the right given matrix. Then multiply the row matrix times the column matrix to get the desired entry.

2. Denote the three matrices by A, X, and B, respectively. Since b_{11} (the entry of the first row, first column of B) is 5, this means that

$$[\text{first row of } A]\begin{bmatrix} \text{first} \\ \text{column} \\ \text{of } X \end{bmatrix} = [b_{11}].$$

That is,

$$[3 \quad -6]\begin{bmatrix} x \\ y \end{bmatrix} = [5]$$

or

$$3x - 6y = 5.$$

Similarly, $b_{21} = 0$ says that $2x + y = 0$. Therefore, the corresponding system of linear equations is

$$\begin{cases} 3x - 6y = 5 \\ 2x + \ y = 0. \end{cases}$$

3. The coefficient matrix is

$$\begin{bmatrix} 8 & 3 \\ 9 & -2 \end{bmatrix}.$$

So the system of equations is equivalent to the matrix equation

$$\begin{bmatrix} 8 & 3 \\ 9 & -2 \end{bmatrix}\begin{bmatrix} x \\ y \end{bmatrix} = \begin{bmatrix} 7 \\ -5 \end{bmatrix}.$$

2.4. The Inverse of a Matrix

In the preceding section we introduced the operations of addition, subtraction, and multiplication of matrices. In this section let us pursue the algebra of matrices a bit further and consider equations involving matrices. Specifically, we shall consider equations of the form

$$AX = B, \tag{1}$$

where A and B are given matrices and X is an unknown matrix whose entries are to be determined. Such equations among matrices are intimately bound up with the theory of systems of linear equations. Indeed, we described the connection in a special case in Example 6 of the preceding section. In that example we wrote the system of linear equations

$$\begin{cases} -2x + 4y = 2 \\ -3x + 7y = 7 \end{cases}$$

as a matrix equation of the form (1), where

$$A = \begin{bmatrix} -2 & 4 \\ -3 & 7 \end{bmatrix}, \qquad B = \begin{bmatrix} 2 \\ 7 \end{bmatrix}, \qquad X = \begin{bmatrix} x \\ y \end{bmatrix}.$$

Note that by determining the entries (x and y) of the unknown matrix X, we solve the system of linear equations. We will return to this example after we have made a complete study of the matrix equation (1).

As motivation for our solution of equation (1), let us consider the analogous equation among numbers:

$$ax = b,$$

where a and b are given numbers* and x is to be determined. Let us examine its solution in great detail. Multiply both sides by $1/a$. (Note that $1/a$ makes sense, since $a \neq 0$.)

$$\left(\frac{1}{a}\right) \cdot (ax) = \frac{1}{a} \cdot b$$

$$\left(\frac{1}{a} \cdot a\right) \cdot x = \frac{1}{a} \cdot b$$

$$1 \cdot x = \frac{1}{a} \cdot b$$

$$x = \frac{1}{a} \cdot b.$$

Let us model our solution of equation (1) on the calculation above. To do so, we wish to multiply both sides of the equation by a matrix that plays the same role in matrix arithmetic as $1/a$ plays in ordinary arithmetic. Our first task then will be to introduce this matrix and study its properties.

The number $1/a$ has the following relationship to the number a:

$$\frac{1}{a} \cdot a = a \cdot \frac{1}{a} = 1. \tag{2}$$

The matrix analog of the number 1 is an identity matrix I. This prompts us to generalize equation (2) to matrices as follows. Suppose that we are given a square matrix A. Then the *inverse* of A, denoted A^{-1}, is a square matrix with the property

$$A^{-1}A = I \qquad \text{and} \qquad AA^{-1} = I, \tag{3}$$

where I is an identity matrix of the same size as A. The matrix A^{-1} is the matrix analogue of the number $1/a$. It can be shown that a matrix A has at most one inverse. (However, A may not have an inverse at all; see below.)

If we are given a matrix A, then it is easy to determine whether or not a given matrix is its inverse. Merely check equation (3) with the given matrix substituted for A^{-1}. For example, if

$$A = \begin{bmatrix} -2 & 4 \\ -3 & 7 \end{bmatrix},$$

---

* We may as well assume that $a \neq 0$. Otherwise, x does not occur.

then

$$A^{-1} = \begin{bmatrix} -\frac{7}{2} & 2 \\ -\frac{3}{2} & 1 \end{bmatrix}.$$

Indeed, we have

$$\underset{A^{-1}}{\begin{bmatrix} -\frac{7}{2} & 2 \\ -\frac{3}{2} & 1 \end{bmatrix}} \underset{A}{\begin{bmatrix} -2 & 4 \\ -3 & 7 \end{bmatrix}} = \begin{bmatrix} 7 - 6 & -14 + 14 \\ 3 - 3 & -6 + 7 \end{bmatrix} = \underset{I_2}{\begin{bmatrix} 1 & 0 \\ 0 & 1 \end{bmatrix}}$$

and

$$\underset{A}{\begin{bmatrix} -2 & 4 \\ -3 & 7 \end{bmatrix}} \underset{A^{-1}}{\begin{bmatrix} -\frac{7}{2} & 2 \\ -\frac{3}{2} & 1 \end{bmatrix}} = \begin{bmatrix} 7 - 6 & -4 + 4 \\ \frac{21}{2} - \frac{21}{2} & -6 + 7 \end{bmatrix} = \underset{I_2}{\begin{bmatrix} 1 & 0 \\ 0 & 1 \end{bmatrix}}.$$

The inverse of a matrix can be calculated using Gaussian elimination, as the next example illustrates.

EXAMPLE 1 Let $A = \begin{bmatrix} 3 & 1 \\ 5 & 2 \end{bmatrix}$. Determine A^{-1}.

Solution Since A is a 2×2 matrix, A^{-1} is also a 2×2 matrix and satisfies

$$AA^{-1} = I_2 \qquad \text{and} \qquad A^{-1}A = I_2, \tag{4}$$

where $I_2 = \begin{bmatrix} 1 & 0 \\ 0 & 1 \end{bmatrix}$ is a 2×2 identity matrix. Suppose that

$$A^{-1} = \begin{bmatrix} x & y \\ z & w \end{bmatrix}.$$

Then the first equation of (4) reads

$$\begin{bmatrix} 3 & 1 \\ 5 & 2 \end{bmatrix} \begin{bmatrix} x & y \\ z & w \end{bmatrix} = \begin{bmatrix} 1 & 0 \\ 0 & 1 \end{bmatrix}.$$

Multiplying out the matrices on the left gives

$$\begin{bmatrix} 3x + z & 3y + w \\ 5x + 2z & 5y + 2w \end{bmatrix} = \begin{bmatrix} 1 & 0 \\ 0 & 1 \end{bmatrix}.$$

Now equate corresponding elements in the two matrices to obtain the equations

$$\begin{cases} 3x + z = 1 \\ 5x + 2z = 0, \end{cases} \qquad \begin{cases} 3y + w = 0 \\ 5y + 2w = 1. \end{cases}$$

Notice that the equations break up into two pairs of linear equations, each pair involving only two variables. Solving these two systems of linear equations yields $x = 2$, $z = -5$, $y = -1$, $w = 3$. Therefore,

$$A^{-1} = \begin{bmatrix} 2 & -1 \\ -5 & 3 \end{bmatrix}.$$

Indeed, we may readily verify that

$$\begin{bmatrix} 3 & 1 \\ 5 & 2 \end{bmatrix}\begin{bmatrix} 2 & -1 \\ -5 & 3 \end{bmatrix} = \begin{bmatrix} 1 & 0 \\ 0 & 1 \end{bmatrix}$$

$$\begin{bmatrix} 2 & -1 \\ -5 & 3 \end{bmatrix}\begin{bmatrix} 3 & 1 \\ 5 & 2 \end{bmatrix} = \begin{bmatrix} 1 & 0 \\ 0 & 1 \end{bmatrix}.$$

The method above can be used to calculate the inverse of matrices of any size, although it involves considerable calculation. We shall provide a rather efficient computational method for calculating A^{-1} in the next section. For now, however, let us be content with the above method. Using it, we can derive a general formula for A^{-1} in case A is a 2×2 matrix.

> *To determine the inverse of a 2×2 matrix* **Let**
>
> $$A = \begin{bmatrix} a & b \\ c & d \end{bmatrix}.$$
>
> Let $\Delta = ad - bc$ and assume that $\Delta \neq 0$. Then A^{-1} is given by the formula
>
> $$A^{-1} = \begin{bmatrix} \dfrac{d}{\Delta} & -\dfrac{b}{\Delta} \\ -\dfrac{c}{\Delta} & \dfrac{a}{\Delta} \end{bmatrix}. \qquad (5)$$

We will omit the derivation of this formula. It proceeds along lines similar to those of Example 1. Notice that formula (5) involves division by Δ. Since division by 0 is not permissible, it is necessary that $\Delta \neq 0$ for formula (5) to be applied. We will discuss the case $\Delta = 0$ below.

Equation (5) can be reduced to a simple step-by-step procedure.

> *To determine the inverse of* $\begin{bmatrix} a & b \\ c & d \end{bmatrix}$ *if* $\Delta = ad - bc \neq 0$:
>
> **1.** Interchange a and d to get $\begin{bmatrix} d & b \\ c & a \end{bmatrix}$.
>
> **2.** Change the signs of b and c to get $\begin{bmatrix} d & -b \\ -c & a \end{bmatrix}$.
>
> **3.** Divide all entries by Δ to get $\begin{bmatrix} \dfrac{d}{\Delta} & -\dfrac{b}{\Delta} \\ -\dfrac{c}{\Delta} & \dfrac{a}{\Delta} \end{bmatrix}$.

EXAMPLE 2 Calculate the inverse of $\begin{bmatrix} -2 & 4 \\ -3 & 7 \end{bmatrix}$.

Solution $\Delta = (-2) \cdot 7 - 4 \cdot (-3) = -2$, so $\Delta \neq 0$, and we may use the computation above.

1. Interchange a and d:

$$\begin{bmatrix} 7 & 4 \\ -3 & -2 \end{bmatrix}$$

2. Change signs of b and c:

$$\begin{bmatrix} 7 & -4 \\ 3 & -2 \end{bmatrix}$$

3. Divide all entries by $\Delta = -2$:

$$\begin{bmatrix} -\frac{7}{2} & 2 \\ -\frac{3}{2} & 1 \end{bmatrix}$$

Thus

$$\begin{bmatrix} -2 & 4 \\ -3 & 7 \end{bmatrix}^{-1} = \begin{bmatrix} -\frac{7}{2} & 2 \\ -\frac{3}{2} & 1 \end{bmatrix}.$$

Not every square matrix has an inverse. Indeed, it may be impossible to satisfy equations (3) for any choice of A^{-1}. This phenomenon can even occur in the case of 2×2 matrices. Here, one can show that *if* $\Delta = 0$, *then the matrix does not have an inverse*. The next example illustrates this phenomenon in a special case.

EXAMPLE 3 Show that $\begin{bmatrix} 1 & 1 \\ 1 & 1 \end{bmatrix}$ does not have an inverse.

Solution Note first that $\Delta = 1 \cdot 1 - 1 \cdot 1 = 0$, so the inverse cannot be computed via equation (5). Suppose that the given matrix did have an inverse, say

$$\begin{bmatrix} s & t \\ u & v \end{bmatrix}.$$

Then the following equation would hold:

$$\begin{bmatrix} s & t \\ u & v \end{bmatrix}\begin{bmatrix} 1 & 1 \\ 1 & 1 \end{bmatrix} = \begin{bmatrix} 1 & 0 \\ 0 & 1 \end{bmatrix}.$$

On multiplying out the two matrices on the left, we get the equation

$$\begin{bmatrix} s + t & s + t \\ u + v & u + v \end{bmatrix} = \begin{bmatrix} 1 & 0 \\ 0 & 1 \end{bmatrix},$$

or, on equating entries in the first row:

$$s + t = 1, \qquad s + t = 0.$$

But $s + t$ cannot equal both 1 and 0. So we reach a contradiction, and therefore the original matrix cannot have an inverse.

We were led to introduce the inverse of a matrix from a discussion of the matrix equation $AX = B$. Let us now return to that discussion. Suppose that A and B are given matrices and that we wish to solve the matrix equation

$$AX = B$$

for the unknown matrix X. Suppose further that A has an inverse A^{-1}. Multiply both sides of the equation on the left by A^{-1} to obtain

$$A^{-1} \cdot AX = A^{-1}B.$$

Because $A^{-1} \cdot A = I$, we have

$$IX = A^{-1}B$$
$$X = A^{-1}B.$$

Thus the matrix X is found by simply multiplying B on the left by A^{-1}, and we can summarize our findings as follows:

> *Solving a Matrix Equation*   If the matrix A has an inverse, then the solution of the matrix equation
>
> $$AX = B$$
>
> is given by
>
> $$X = A^{-1}B.$$

Matrix equations can be used to solve systems of linear equations, as illustrated in the next example.

EXAMPLE 4   Use a matrix equation to solve the system of linear equations

$$\begin{cases} -2x + 4y = 2 \\ -3x + 7y = 7. \end{cases}$$

Solution   In Example 6 of the preceding section we saw that the system could be written as a matrix equation:

$$\underset{A}{\begin{bmatrix} -2 & 4 \\ -3 & 7 \end{bmatrix}} \underset{X}{\begin{bmatrix} x \\ y \end{bmatrix}} = \underset{B}{\begin{bmatrix} 2 \\ 7 \end{bmatrix}}.$$

We happen to know A^{-1} from Example 2, namely

$$A^{-1} = \begin{bmatrix} -\frac{7}{2} & 2 \\ -\frac{3}{2} & 1 \end{bmatrix}.$$

So we may compute the matrix $X = A^{-1}B$:

$$X = \begin{bmatrix} x \\ y \end{bmatrix} = \begin{bmatrix} -\frac{7}{2} & 2 \\ -\frac{3}{2} & 1 \end{bmatrix}\begin{bmatrix} 2 \\ 7 \end{bmatrix} = \begin{bmatrix} 7 \\ 4 \end{bmatrix}.$$

Thus the solution of the system is $x = 7$, $y = 4$.

EXAMPLE 5 Let x and y denote the number of married and single adults in a certain town as of January 1. Let m and s denote the corresponding numbers for the following year. A statistical survey shows that x, y, m, and s are related by the equations

$$.9x + .2y = m$$

$$.1x + .8y = s.$$

In a given year there were found to be 490,000 married adults and 147,000 single adults.

(a) How many married adults were there in the preceding year?

(b) How many married adults were there 2 years ago?

Solution (a) The given equations can be written in the matrix form

$$AX = B,$$

where

$$A = \begin{bmatrix} .9 & .2 \\ .1 & .8 \end{bmatrix}, \qquad X = \begin{bmatrix} x \\ y \end{bmatrix}, \qquad B = \begin{bmatrix} m \\ s \end{bmatrix}.$$

We are given that $B = \begin{bmatrix} 490,000 \\ 147,000 \end{bmatrix}$. So, since

$$X = A^{-1}B$$

and

$$A^{-1} = \begin{bmatrix} \frac{8}{7} & -\frac{2}{7} \\ -\frac{1}{7} & \frac{9}{7} \end{bmatrix},$$

we have

$$X = \begin{bmatrix} \frac{8}{7} & -\frac{2}{7} \\ -\frac{1}{7} & \frac{9}{7} \end{bmatrix}\begin{bmatrix} 490,000 \\ 147,000 \end{bmatrix} = \begin{bmatrix} 518,000 \\ 119,000 \end{bmatrix}.$$

Thus last year there were 518,000 married adults and 119,000 single adults.

(b) We deduce x and y for two years ago from the values of m and s for last year, namely $m = 518,000$, $s = 119,000$.

$$X = A^{-1}B = \begin{bmatrix} \frac{8}{7} & -\frac{2}{7} \\ -\frac{1}{7} & \frac{9}{7} \end{bmatrix} \begin{bmatrix} 518{,}000 \\ 119{,}000 \end{bmatrix} = \begin{bmatrix} 558{,}000 \\ 79{,}000 \end{bmatrix}.$$

That is, 2 years ago there were 558,000 married adults and 79,000 single adults.

EXAMPLE 6 In the next section we will show that if

$$A = \begin{bmatrix} 4 & -2 & 3 \\ 8 & -3 & 5 \\ 7 & -2 & 4 \end{bmatrix}, \quad \text{then} \quad A^{-1} = \begin{bmatrix} -2 & 2 & -1 \\ 3 & -5 & 4 \\ 5 & -6 & 4 \end{bmatrix}.$$

(a) Use this fact to solve the system of linear equations

$$\begin{cases} 4x - 2y + 3z = 1 \\ 8x - 3y + 5z = 4 \\ 7x - 2y + 4z = 5. \end{cases}$$

(b) Solve the system of equations

$$\begin{cases} 4x - 2y + 3z = 4 \\ 8x - 3y + 5z = 7 \\ 7x - 2y + 4z = 6. \end{cases}$$

Solution (a) The system can be written in the matrix form

$$\underset{A}{\begin{bmatrix} 4 & -2 & 3 \\ 8 & -3 & 5 \\ 7 & -2 & 4 \end{bmatrix}} \underset{X}{\begin{bmatrix} x \\ y \\ z \end{bmatrix}} = \underset{B}{\begin{bmatrix} 1 \\ 4 \\ 5 \end{bmatrix}}.$$

The solution of this matrix equation is $X = A^{-1}B$ or

$$\begin{bmatrix} x \\ y \\ z \end{bmatrix} = \begin{bmatrix} -2 & 2 & -1 \\ 3 & -5 & 4 \\ 5 & -6 & 4 \end{bmatrix} \begin{bmatrix} 1 \\ 4 \\ 5 \end{bmatrix} = \begin{bmatrix} 1 \\ 3 \\ 1 \end{bmatrix}.$$

Thus the solution of the system is $x = 1, y = 3, z = 1$.

(b) This system has the same left-hand side as the preceding system, so its solution is

$$\begin{bmatrix} x \\ y \\ z \end{bmatrix} = \begin{bmatrix} -2 & 2 & -1 \\ 3 & -5 & 4 \\ 5 & -6 & 4 \end{bmatrix} \begin{bmatrix} 4 \\ 7 \\ 6 \end{bmatrix} = \begin{bmatrix} 0 \\ 1 \\ 2 \end{bmatrix}.$$

That is, the solution of the system is $x = 0, y = 1, z = 2$.

Using the method of matrix equations to solve a system of linear equations is especially efficient if one wishes to solve a number of systems all having the same left-hand sides but different right-hand sides. For then A^{-1} must be computed only once for all the systems under consideration. (This point is useful in Exercises 17–20.)

PRACTICE PROBLEMS 4

1. Show that the inverse of

$$\begin{bmatrix} -4 & 1 & 2 \\ 7 & -1 & -4 \\ -\frac{1}{2} & 0 & \frac{1}{2} \end{bmatrix} \text{ is } \begin{bmatrix} 1 & 1 & 4 \\ 3 & 2 & 4 \\ 1 & 1 & 6 \end{bmatrix}.$$

2. Use the method of this section to solve the system of linear equations

$$\begin{cases} .8x + .6y = 5 \\ .2x + .4y = 2. \end{cases}$$

EXERCISES 4

In Exercises 1 and 2, use the fact that

$$\begin{bmatrix} 2 & 2 \\ \frac{1}{2} & 1 \end{bmatrix}^{-1} = \begin{bmatrix} 1 & -2 \\ -\frac{1}{2} & 2 \end{bmatrix}.$$

1. Solve $\begin{cases} 2x + 2y = 4 \\ \frac{1}{2}x + y = 1. \end{cases}$

2. Solve $\begin{cases} 2x + 2y = 14 \\ \frac{1}{2}x + y = 4. \end{cases}$

In Exercises 3–10, find the inverse of the given matrix.

3. $\begin{bmatrix} 7 & 2 \\ 3 & 1 \end{bmatrix}$

4. $\begin{bmatrix} 2 & 3 \\ 5 & 7 \end{bmatrix}$

5. $\begin{bmatrix} 6 & 2 \\ 5 & 2 \end{bmatrix}$

6. $\begin{bmatrix} 1 & .5 \\ 0 & .5 \end{bmatrix}$

7. $\begin{bmatrix} .7 & .2 \\ .3 & .8 \end{bmatrix}$

8. $\begin{bmatrix} 0 & 1 \\ 1 & 0 \end{bmatrix}$

9. $[3]$

10. $[.2]$

In Exercises 11–14, use the method of this section to solve the system of linear equations.

11. $\begin{cases} x + 2y = 3 \\ 2x + 6y = 5 \end{cases}$

12. $\begin{cases} 5x + 3y = 1 \\ 7x + 4y = 2 \end{cases}$

13. $\begin{cases} \frac{1}{2}x + 2y = 4 \\ 3x + 16y = 0 \end{cases}$

14. $\begin{cases} .8x + .6y = 2 \\ .2x + .4y = 1 \end{cases}$

15. It is found that the number of married and single adults in a certain town are subject to the following statistics. Suppose that x and y denote the number of married and single adults, respectively, in a given year (say as of January 1) and let m, s denote the corresponding numbers for the following year. Then

$$.8x + .3y = m$$

$$.2x + .7y = s.$$

(a) Write this system of equations in matrix form.

(b) Solve the resulting matrix equation for $X = \begin{bmatrix} x \\ y \end{bmatrix}$.

(c) Suppose that in a given year there were found to be 100,000 married adults and 50,000 single adults. How many married (resp. single) adults were there the preceding year?

(d) How many married (resp. single) adults were there 2 years ago?

16. A flu epidemic is spreading through a town of 48,000 people. It is found that if x and y denote the numbers of people sick and well in a given week, respectively, and if s and w denote the corresponding numbers for the following week, then

$$\tfrac{1}{3}x + \tfrac{1}{4}y = s$$

$$\tfrac{2}{3}x + \tfrac{3}{4}y = w.$$

(a) Write this system of equations in matrix form.

(b) Solve the resulting matrix equation for $X = \begin{bmatrix} x \\ y \end{bmatrix}$.

(c) Suppose that 13,000 people are sick in a given week. How many were sick the preceding week?

(d) Same question as part (c), except assume that 14,000 are sick.

In Exercises 17 and 18, use the fact that

$$\begin{bmatrix} 1 & 2 & 2 \\ 1 & 3 & 2 \\ 1 & 2 & 3 \end{bmatrix}^{-1} = \begin{bmatrix} 5 & -2 & -2 \\ -1 & 1 & 0 \\ -1 & 0 & 1 \end{bmatrix}.$$

17. Solve $\begin{cases} x + 2y + 2z = 1 \\ x + 3y + 2z = -1 \\ x + 2y + 3z = -1. \end{cases}$

18. Solve $\begin{cases} x + 2y + 2z = 1 \\ x + 3y + 2z = 0 \\ x + 2y + 3z = 0. \end{cases}$

In Exercises 19 and 20, use the fact that

$$\begin{bmatrix} 9 & 0 & 2 & 0 \\ -20 & -9 & -5 & 5 \\ 4 & 0 & 1 & 0 \\ -4 & -2 & -1 & 1 \end{bmatrix}^{-1} = \begin{bmatrix} 1 & 0 & -2 & 0 \\ 0 & 1 & 0 & -5 \\ -4 & 0 & 9 & 0 \\ 0 & 2 & 1 & -9 \end{bmatrix}.$$

19. Solve $\begin{cases} 9x \quad\quad + 2z \quad\quad = 1 \\ -20x - 9y - 5z + 5w = 0 \\ 4x \quad\quad + z \quad\quad = 0 \\ -4x - 2y - z + w = -1. \end{cases}$

20. Solve $\begin{cases} 9x \quad\quad + 2z \quad\quad = 2 \\ -20x - 9y - 5z + 5w = 1 \\ 4x \quad\quad + z \quad\quad = 3 \\ -4x - 2y - z + w = 0. \end{cases}$

21. Without computing Δ, show that the matrix $\begin{bmatrix} 6 & 3 \\ 2 & 1 \end{bmatrix}$ does not have an inverse.

22. If $A^{-1} = \begin{bmatrix} 2 & 7 \\ 1 & -3 \end{bmatrix}$, what is the matrix A?

23. There are two age groups for a particular species of organism. Group I consists of all organisms aged under 1 year, while group II consists of all organisms aged from 1 to 2 years. No organism survives more than 2 years. The average number of offspring per year born to each member of group I is 1, while the average number of organisms per year born to each member of group II is 2. Nine-tenths of group I survive to enter group II each year.

   (a) Let x and y represent the initial number of organisms in groups I and II, respectively. Let a and b represent the number of organisms in groups I and II, respectively, after one year. Write a matrix equation relating $\begin{bmatrix} x \\ y \end{bmatrix}$ to $\begin{bmatrix} a \\ b \end{bmatrix}$.

   (b) If there are initially 450,000 organisms in group I and 360,000 organisms in group II, calculate the number of organisms in each of the groups after 1 year and after 2 years.

   (c) Suppose that at a certain time there were 810,000 organisms in group I and 630,000 organisms in group II. Determine the population of each group 1 year earlier.

24. If $A^2 = \begin{bmatrix} -2 & -1 \\ 2 & -1 \end{bmatrix}$ and $A^3 = \begin{bmatrix} -2 & 1 \\ -2 & -3 \end{bmatrix}$, what is A?

SOLUTIONS TO PRACTICE PROBLEMS 4

1. To see if this matrix is indeed the inverse, multiply it by the original matrix and find out if the products are identity matrices.

$$\begin{bmatrix} 1 & 1 & 4 \\ 3 & 2 & 4 \\ 1 & 1 & 6 \end{bmatrix}\begin{bmatrix} -4 & 1 & 2 \\ 7 & -1 & -4 \\ -\frac{1}{2} & 0 & \frac{1}{2} \end{bmatrix} = \begin{bmatrix} 1 & 0 & 0 \\ 0 & 1 & 0 \\ 0 & 0 & 1 \end{bmatrix}, \quad \text{an identity matrix.}$$

$$\begin{bmatrix} -4 & 1 & 2 \\ 7 & -1 & -4 \\ -\frac{1}{2} & 0 & \frac{1}{2} \end{bmatrix}\begin{bmatrix} 1 & 1 & 4 \\ 3 & 2 & 4 \\ 1 & 1 & 6 \end{bmatrix} = \begin{bmatrix} 1 & 0 & 0 \\ 0 & 1 & 0 \\ 0 & 0 & 1 \end{bmatrix}.$$

2. The matrix form of this system is

$$\begin{bmatrix} .8 & .6 \\ .2 & .4 \end{bmatrix}\begin{bmatrix} x \\ y \end{bmatrix} = \begin{bmatrix} 5 \\ 2 \end{bmatrix}.$$

Therefore, the solution is

$$\begin{bmatrix} x \\ y \end{bmatrix} = \begin{bmatrix} .8 & .6 \\ .2 & .4 \end{bmatrix}^{-1}\begin{bmatrix} 5 \\ 2 \end{bmatrix}.$$

To compute the inverse of the 2×2 matrix, first compute Δ.

$$\Delta = ad - bc = (.8)(.4) - (.6)(.2) = .32 - .12 = .2.$$

Thus

$$\begin{bmatrix} .8 & .6 \\ .2 & .4 \end{bmatrix}^{-1} = \begin{bmatrix} .4/.2 & -.6/.2 \\ -.2/.2 & .8/.2 \end{bmatrix} = \begin{bmatrix} 2 & -3 \\ -1 & 4 \end{bmatrix}.$$

Therefore,

$$\begin{bmatrix} x \\ y \end{bmatrix} = \begin{bmatrix} 2 & -3 \\ -1 & 4 \end{bmatrix} \begin{bmatrix} 5 \\ 2 \end{bmatrix} = \begin{bmatrix} 4 \\ 3 \end{bmatrix}.$$

So the solution is $x = 4$, $y = 3$.

2.5. The Gauss-Jordan Method for Calculating Inverses

Of the several popular methods for finding the inverse of a matrix, the Gauss-Jordan method is probably the easiest to describe. It can be used on square matrices of any size. Also, the mechanical nature of the computations allows this method to be programmed for a computer with relative ease. We shall illustrate the procedure with a 2×2 matrix, whose inverse can also be calculated using the method of the previous section. Let

$$A = \begin{bmatrix} \frac{1}{2} & 1 \\ 1 & 3 \end{bmatrix}.$$

It is simple to check that

$$A^{-1} = \begin{bmatrix} 6 & -2 \\ -2 & 1 \end{bmatrix}.$$

Let us now derive this result using the Gauss-Jordan method.

Step 1   Write down the matrix A, and on its right an identity matrix of the same size.

This is most conveniently done by placing I_2 beside A in a single matrix.

$$\begin{bmatrix} \frac{1}{2} & 1 & \vline & 1 & 0 \\ 1 & 3 & \vline & 0 & 1 \end{bmatrix}.$$
$$\underbrace{}_{A} \quad \underbrace{}_{I_2}$$

Step 2   Perform elementary row operations on the left-hand matrix so as to transform it into an identity matrix. Each operation performed on the left-hand matrix is also performed on the right-hand matrix.

This step proceeds exactly like the Gaussian elimination method and may be most conveniently expressed in terms of pivoting.

$$\left[\begin{array}{cc|cc} \textcircled{$\frac{1}{2}$} & 1 & 1 & 0 \\ 1 & 3 & 0 & 1 \end{array}\right], \quad \left[\begin{array}{cc|cc} 1 & 2 & 2 & 0 \\ 0 & \textcircled{1} & -2 & 1 \end{array}\right], \quad \left[\begin{array}{cc|cc} 1 & 0 & 6 & -2 \\ 0 & 1 & -2 & 1 \end{array}\right].$$

> *Step 3* When the matrix on the left becomes an identity matrix, the matrix on the right is the desired inverse.

So, from the last matrix of our calculation above, we have

$$A^{-1} = \left[\begin{array}{cc} 6 & -2 \\ -2 & 1 \end{array}\right].$$

This is the same result obtained earlier.

We will demonstrate why the method above works after some further examples.

EXAMPLE 1 Find the inverse of the matrix

$$A = \left[\begin{array}{ccc} 4 & -2 & 3 \\ 8 & -3 & 5 \\ 7 & -2 & 4 \end{array}\right].$$

Solution

$$\left[\begin{array}{ccc|ccc} \textcircled{4} & -2 & 3 & 1 & 0 & 0 \\ 8 & -3 & 5 & 0 & 1 & 0 \\ 7 & -2 & 4 & 0 & 0 & 1 \end{array}\right]$$

$$\left[\begin{array}{ccc|ccc} 1 & -\frac{1}{2} & \frac{3}{4} & \frac{1}{4} & 0 & 0 \\ 0 & \textcircled{1} & -1 & -2 & 1 & 0 \\ 0 & \frac{3}{2} & -\frac{5}{4} & -\frac{7}{4} & 0 & 1 \end{array}\right]$$

$$\left[\begin{array}{ccc|ccc} 1 & 0 & \frac{1}{4} & -\frac{3}{4} & \frac{1}{2} & 0 \\ 0 & 1 & -1 & -2 & 1 & 0 \\ 0 & 0 & \textcircled{$\frac{1}{4}$} & \frac{5}{4} & -\frac{3}{2} & 1 \end{array}\right]$$

$$\left[\begin{array}{ccc|ccc} 1 & 0 & 0 & -2 & 2 & -1 \\ 0 & 1 & 0 & 3 & -5 & 4 \\ 0 & 0 & 1 & 5 & -6 & 4 \end{array}\right].$$

Therefore,

$$A^{-1} = \left[\begin{array}{ccc} -2 & 2 & -1 \\ 3 & -5 & 4 \\ 5 & -6 & 4 \end{array}\right].$$

Not all square matrices have inverses. If a matrix does not have an inverse, this will become apparent when applying the Gauss-Jordan method. At some point there will be no way to continue transforming the left-hand matrix into an identity matrix. This is illustrated in the next example.

EXAMPLE 2 Find the inverse of the matrix

$$A = \begin{bmatrix} 1 & 3 & 2 \\ 0 & 1 & 4 \\ 1 & 5 & 10 \end{bmatrix}.$$

Solution

$$\left[\begin{array}{ccc|ccc} 1 & 3 & 2 & 1 & 0 & 0 \\ ⓪ & 1 & 4 & 0 & 1 & 0 \\ 1 & 5 & 10 & 0 & 0 & 1 \end{array}\right]$$

$$\left[\begin{array}{ccc|ccc} 1 & 3 & 2 & 1 & 0 & 0 \\ 0 & ① & 4 & 0 & 1 & 0 \\ 0 & 2 & 8 & -1 & 0 & 1 \end{array}\right]$$

$$\left[\begin{array}{ccc|ccc} 1 & 0 & -10 & 1 & -3 & 0 \\ 0 & 1 & 4 & 0 & 1 & 0 \\ 0 & 0 & 0 & -1 & -2 & 1 \end{array}\right].$$

Since the third row of the left-hand matrix has only zero entries, it is impossible to complete the Gauss-Jordan method. Therefore, the matrix A has no inverse matrix.

Verification of the Gauss-Jordan Method for Calculating Inverses

In the preceding section we showed how to calculate the inverse by solving several systems of linear equations. Actually, the Gauss-Jordan method is just an organized way of going about the calculation. To see why, let us consider a concrete example:

$$A = \begin{bmatrix} 4 & -2 & 3 \\ 8 & -3 & 5 \\ 7 & -2 & 4 \end{bmatrix}.$$

We wish to determine A^{-1}, so regard it as a matrix of unknowns:

$$A^{-1} = \begin{bmatrix} x_1 & x_2 & x_3 \\ y_1 & y_2 & y_3 \\ z_1 & z_2 & z_3 \end{bmatrix}.$$

The statement $AA^{-1} = I_3$ is

$$\begin{bmatrix} 4 & -2 & 3 \\ 8 & -3 & 5 \\ 7 & -2 & 4 \end{bmatrix} \begin{bmatrix} x_1 & x_2 & x_3 \\ y_1 & y_2 & y_3 \\ z_1 & z_2 & z_3 \end{bmatrix} = \begin{bmatrix} 1 & 0 & 0 \\ 0 & 1 & 0 \\ 0 & 0 & 1 \end{bmatrix}.$$

Multiplying out the matrices on the left and comparing the result with the matrix on the right gives us nine equations, namely

$$\begin{cases} 4x_1 - 2y_1 + 3z_1 = 1 \\ 8x_1 - 3y_1 + 5z_1 = 0 \\ 7x_1 - 2y_1 + 4z_1 = 0 \end{cases}$$

$$\begin{cases} 4x_2 - 2y_2 + 3z_2 = 0 \\ 8x_2 - 3y_2 + 5z_2 = 1 \\ 7x_2 - 2y_2 + 4z_2 = 0 \end{cases}$$

$$\begin{cases} 4x_3 - 2y_3 + 3z_3 = 0 \\ 8x_3 - 3y_3 + 5z_3 = 0 \\ 7x_3 - 2y_3 + 4z_3 = 1. \end{cases}$$

Notice that each system of equations corresponds to one column of unknowns in A^{-1}. More precisely, if we set

$$X_1 = \begin{bmatrix} x_1 \\ y_1 \\ z_1 \end{bmatrix}, \qquad X_2 = \begin{bmatrix} x_2 \\ y_2 \\ z_2 \end{bmatrix}, \qquad X_3 = \begin{bmatrix} x_3 \\ y_3 \\ z_3 \end{bmatrix},$$

then the three systems above have the respective matrix forms

Now imagine the process of applying Gaussian elimination to solve these three systems. We apply elementary row operations to the matrices

$$\left[\begin{array}{c|c} A & \begin{matrix} 1 \\ 0 \\ 0 \end{matrix} \end{array}\right], \qquad \left[\begin{array}{c|c} A & \begin{matrix} 0 \\ 1 \\ 0 \end{matrix} \end{array}\right], \qquad \left[\begin{array}{c|c} A & \begin{matrix} 0 \\ 0 \\ 1 \end{matrix} \end{array}\right].$$

The process ends when we convert A into the identity matrix, at which point the solutions may be read off the right column. So the procedure ends with the matrices

$$[I_3 \mid X_1], \qquad [I_3 \mid X_2], \qquad [I_3 \mid X_3].$$

Realize, however, that at each step of the three Gaussian eliminations we are performing the same operations, since all three start with the matrix A on the left. So, in order to save calculation, perform the three Gaussian eliminations simultaneously by performing the row operations on the composite matrix

$$\left[\begin{array}{c|c} A & \begin{matrix} 1 & 0 & 0 \\ 0 & 1 & 0 \\ 0 & 0 & 1 \end{matrix} \end{array}\right] = [A \mid I_3].$$

The procedure ends when this matrix is converted into

$$[I_3 \mid X_1 \ X_2 \ X_3].$$

That is, since $A^{-1} = [X_1 \quad X_2 \quad X_3]$, the procedure ends with A^{-1} on the right. This is the reasoning behind the Gauss-Jordan method of calculating inverses.

PRACTICE PROBLEMS 5

1. Use the Gauss-Jordan method to calculate the inverse of the matrix
$$\begin{bmatrix} 1 & 0 & 2 \\ 0 & 1 & -4 \\ 0 & 0 & 2 \end{bmatrix}.$$

2. Solve the system of linear equations
$$\begin{cases} x & + 2z = 4 \\ & y - 4z = 6 \\ & 2z = 9. \end{cases}$$

EXERCISES 5

Use the Gauss-Jordan method to compute the inverses of the following matrices.

1. $\begin{bmatrix} 7 & 3 \\ 5 & 2 \end{bmatrix}$
2. $\begin{bmatrix} 5 & -2 \\ 6 & 2 \end{bmatrix}$

3. $\begin{bmatrix} 10 & 12 \\ 3 & -4 \end{bmatrix}$
4. $\begin{bmatrix} 1 & -3 \\ 0 & 1 \end{bmatrix}$

5. $\begin{bmatrix} 2 & -4 \\ -1 & 2 \end{bmatrix}$
6. $\begin{bmatrix} 1 & 3 & 1 \\ -1 & 2 & 0 \\ 2 & 11 & 3 \end{bmatrix}$

7. $\begin{bmatrix} 1 & 2 & -2 \\ 1 & 1 & 1 \\ 0 & 0 & 1 \end{bmatrix}$
8. $\begin{bmatrix} 2 & 2 & 0 \\ 0 & -2 & 0 \\ 3 & 0 & 1 \end{bmatrix}$

9. $\begin{bmatrix} -2 & 5 & 2 \\ 1 & -3 & -1 \\ -1 & 2 & 1 \end{bmatrix}$
10. $\begin{bmatrix} 1 & 0 & 0 \\ 2 & 1 & -2 \\ -1 & 2 & 1 \end{bmatrix}$

11. $\begin{bmatrix} 1 & 6 & 0 & 0 \\ 1 & 5 & 0 & 0 \\ 0 & 0 & 4 & 2 \\ 0 & 0 & 50 & 2 \end{bmatrix}$
12. $\begin{bmatrix} 6 & 0 & 2 & 0 \\ -6 & 1 & 0 & 1 \\ 1 & 0 & 1 & 0 \\ -9 & 0 & -1 & 1 \end{bmatrix}$

In Exercises 13-16, use matrix inversion to solve the system of linear equations.

13. $\begin{cases} x + y + 2z = 3 \\ 3x + 2y + 2z = 4 \\ x + y + 3z = 5 \end{cases}$
14. $\begin{cases} x + 2y + 3z = 4 \\ 3x + 5y + 5z = 3 \\ 2x + 4y + 2z = 4 \end{cases}$

15. $\begin{cases} x & -2z - 2w = 0 \\ y & -5w = 1 \\ -4x & +9z + 9w = 2 \\ 2y + z - 8w = 3 \end{cases}$

16. $\begin{cases} y + 2z = 1 \\ 2x + y + 3z = 2 \\ x + y + 2z = 3 \end{cases}$

17. Suppose that we try to solve the matrix equation $AX = B$ by using matrix inversion, but find that even though the matrix A is a square matrix, it has no inverse. What can be said about the outcome from solving the associated system of linear equations by the Gaussian elimination method?

18. Find the 2×2 matrix C for which $A \cdot C = B$, where

$$A = \begin{bmatrix} 3 & 2 \\ 4 & 3 \end{bmatrix} \text{ and } B = \begin{bmatrix} 5 & 6 \\ 1 & 7 \end{bmatrix}.$$

$4 + 7 - 6$

19. Find a 2×2 matrix A for which

$$A \cdot \begin{bmatrix} 2 \\ 1 \end{bmatrix} = \begin{bmatrix} -1 \\ 4 \end{bmatrix} \text{ and } A \cdot \begin{bmatrix} 5 \\ 3 \end{bmatrix} = \begin{bmatrix} 0 \\ 2 \end{bmatrix}.$$

20. Let $A = \begin{bmatrix} 7 & 4 \\ 3 & 2 \end{bmatrix}$ and $B = \begin{bmatrix} 1 & 5 \\ -3 & 4 \end{bmatrix}$. Each of the equations $AX = B$ and $XA = B$ has a 2×2 matrix X as a solution. Find X in each case and explain why the two answers are different.

$12 - 4 - 1$

$-6 + 2 + 1$

$7 - 6 = 1$

SOLUTIONS TO PRACTICE PROBLEMS 5

1. First write the given matrix beside an identity matrix of the same size

$$\left[\begin{array}{ccc|ccc} 1 & 0 & 2 & 1 & 0 & 0 \\ 0 & 1 & -4 & 0 & 1 & 0 \\ 0 & 0 & 2 & 0 & 0 & 1 \end{array} \right].$$

The object is to use elementary row operations to transform the 3×3 matrix on the left into the identity matrix. The first two columns are already in the correct form.

$$\left[\begin{array}{ccc|ccc} 1 & 0 & 2 & 1 & 0 & 0 \\ 0 & 1 & -4 & 0 & 1 & 0 \\ 0 & 0 & 2 & 0 & 0 & 1 \end{array} \right]$$

$$\xrightarrow{\frac{1}{2}[3]} \left[\begin{array}{ccc|ccc} 1 & 0 & 2 & 1 & 0 & 0 \\ 0 & 1 & -4 & 0 & 1 & 0 \\ 0 & 0 & 1 & 0 & 0 & \frac{1}{2} \end{array} \right]$$

$$\xrightarrow{[1] + (-2)[3]} \left[\begin{array}{ccc|ccc} 1 & 0 & 0 & 1 & 0 & -1 \\ 0 & 1 & -4 & 0 & 1 & 0 \\ 0 & 0 & 1 & 0 & 0 & \frac{1}{2} \end{array} \right]$$

$$\xrightarrow{[2] + (4)[3]} \left[\begin{array}{ccc|ccc} 1 & 0 & 0 & 1 & 0 & -1 \\ 0 & 1 & 0 & 0 & 1 & 2 \\ 0 & 0 & 1 & 0 & 0 & \frac{1}{2} \end{array} \right].$$

SEC. 2.5: *The Gauss-Jordan Method for Calculating Inverses* **101**

Thus the inverse of the given matrix is

$$\begin{bmatrix} 1 & 0 & -1 \\ 0 & 1 & 2 \\ 0 & 0 & \frac{1}{2} \end{bmatrix}.$$

2. The matrix form of this system of equations is $AX = B$, where A is the matrix whose inverse was found in Problem 1, and

$$B = \begin{bmatrix} 4 \\ 6 \\ 9 \end{bmatrix}.$$

Therefore, $X = A^{-1}B$, so that

$$\begin{bmatrix} x \\ y \\ z \end{bmatrix} = \begin{bmatrix} 1 & 0 & -1 \\ 0 & 1 & 2 \\ 0 & 0 & \frac{1}{2} \end{bmatrix} \begin{bmatrix} 4 \\ 6 \\ 9 \end{bmatrix} = \begin{bmatrix} -5 \\ 24 \\ \frac{9}{2} \end{bmatrix}.$$

So the solution of the system is $x = -5$, $y = 24$, $z = \frac{9}{2}$.

2.6. Input-Output Analysis

In recent years matrix arithmetic has played an ever-increasing role in economics, especially in that branch of economics called *input-output analysis*. Pioneered by the Harvard economist Vassily Leontieff, input-output analysis is used to analyze an economy in order to determine how much output must be produced by each segment of the economy in order to meet given consumption and export demands. As we shall see, such analysis leads into matrix calculations and in particular to inverses of matrices. Input-output analysis has been of such great significance that Leontieff was awarded the 1973 Nobel prize in economics for his fundamental work in the subject.

Suppose that we divide an economy up into a number of industries—transportation, agriculture, steel, and so on. Each industry produces a certain output using certain raw materials (or input). The input of each industry is made up in part by the outputs of other industries. For example, in order to produce food, agriculture uses as input the output of many industries, such as transportation (tractors and trucks) and oil (gasoline and fertilizers). This interdependence among the industries of the economy is summarized in a matrix—a so-called *input-output matrix*. There is one column for each industry's input requirements. The entries in the column reflect the amount of input required from each of the industries. A typical input-output matrix looks like this:

Input requirements of:

	Industry 1	Industry 2	Industry 3	...
Industry 1				
Industry 2				
Industry 3				

From

$\vdots$

It is most convenient to express the entries of this matrix in monetary terms. That is, each column gives the dollar values of the various inputs needed by an industry in order to produce $1 worth of output.

There are consumers (other than the industries themselves) who want to purchase some of the output of these industries. The quantity of goods that these consumers want (or demand) is called the *final demand* on the economy. The final demand can be represented by a column matrix, with one entry for each industry, indicating the amount of consumable output demanded from the industry:

$$[\text{final demand}] = \begin{bmatrix} \text{amount from industry 1} \\ \text{amount from industry 2} \\ \vdots \end{bmatrix}.$$

We shall consider the situation in which the final-demand matrix is given and it is necessary to determine how much output should be produced by each industry in order to provide the needed inputs of the various industries and also to satisfy the final demand. The proper level of output can be computed using matrix calculations as illustrated in the next example.

EXAMPLE 1   Suppose that an economy is composed of only three industries—coal, steel, and electricity. Each of these industries depends on the others for some of its raw materials. Suppose that to make $1 of coal, it takes no coal, but $.02 of steel and $.01 of electricity; to make $1 of steel, it takes $.15 of coal, $.03 of steel, and $.08 of electricity; and to make $1 of electricity, it takes $.43 of coal, $.20 of steel, and $.05 of electricity. How much should each industry produce to allow for consumption (not used for production) at these levels: $2 billion coal, $1 billion steel, $3 billion electricity?

Solution   Put all the data indicating the interdependence of the industries in a matrix. In each industry's column put the amount of input from each of the industries needed to produce $1 of output in that particular industry:

$$\begin{array}{c} \\ \text{Coal} \\ \text{Steel} \\ \text{Electricity} \end{array} \begin{array}{ccc} \text{Coal} & \text{Steel} & \text{Electricity} \\ \begin{bmatrix} 0 & .15 & .43 \\ .02 & .03 & .20 \\ .01 & .08 & .05 \end{bmatrix} \end{array} = A.$$

This matrix is the *input-output matrix* corresponding to the economy. Let D denote the final-demand matrix. Then, letting the numbers in D stand for billions of dollars, we have

$$D = \begin{bmatrix} 2 \\ 1 \\ 3 \end{bmatrix}.$$

Suppose that the coal industry produces x billion dollars of output, the steel industry y billion dollars, and the electrical industry z billion dollars. Our problem is to determine the x, y, and z that yield the desired amounts left over from the production process. As an example, consider coal. The amount of coal that can be consumed or exported is just

$$x - [\text{amount of coal used in production}].$$

To determine the amount of coal used in production, refer to the input-output matrix. Production of x billion dollars of coal takes $0 \cdot x$ billion dollars of coal, production of y billion dollars of steel takes $.15y$ billion dollars of coal, and production of z billion dollars of electricity takes $.43z$ billion dollars of coal. Thus,

$$[\text{amount of coal used in production}] = 0 \cdot x + .15y + .43z.$$

This quantity should be recognized as the first entry of a matrix product. Namely, if we let

$$X = \begin{bmatrix} x \\ y \\ z \end{bmatrix},$$

then

$$\begin{bmatrix} \text{coal} \\ \text{steel} \\ \text{electricity} \end{bmatrix}_{\text{used in production}} = \begin{bmatrix} 0 & .15 & .43 \\ .02 & .03 & .20 \\ .01 & .08 & .05 \end{bmatrix} \begin{bmatrix} x \\ y \\ z \end{bmatrix}.$$

$$= AX.$$

But then the amount of each output available for purposes other than production is $X - AX$. That is, we have the matrix equation

$$X - AX = D.$$

To solve this equation for X, proceed as follows. Since $IX = X$, write the equation in the form

$$IX - AX = D$$

$$(I - A)X = D$$

$$\boxed{X = (I - A)^{-1}D.} \tag{1}$$

So, in other words, X may be found by multiplying D on the left by $(I - A)^{-1}$. Let us now do the arithmetic.

$$I - A = \begin{bmatrix} 1 & 0 & 0 \\ 0 & 1 & 0 \\ 0 & 0 & 1 \end{bmatrix} - \begin{bmatrix} 0 & .15 & .43 \\ .02 & .03 & .20 \\ .01 & .08 & .05 \end{bmatrix} = \begin{bmatrix} 1 & -.15 & -.43 \\ -.02 & .97 & -.20 \\ -.01 & -.08 & .95 \end{bmatrix}.$$

Applying the Gauss-Jordan method, we find

$$(I - A)^{-1} = \begin{bmatrix} 1.01 & .20 & .50 \\ .02 & 1.05 & .23 \\ .01 & .09 & 1.08 \end{bmatrix},$$

where all figures are carried to two decimal places (Exercise 5). Therefore,

$$X = (I - A)^{-1}D = \begin{bmatrix} 1.01 & .20 & .50 \\ .02 & 1.05 & .23 \\ .01 & .09 & 1.08 \end{bmatrix} \begin{bmatrix} 2 \\ 1 \\ 3 \end{bmatrix} = \begin{bmatrix} 3.72 \\ 1.78 \\ 3.35 \end{bmatrix}.$$

In other words, coal should produce $3.72 billion worth of output, steel $1.78 billion, and electricity $3.35 billion. This output will meet the required final demands from each industry.

The analysis above is useful in studying not only entire economies but also segments of economies and even individual companies.

EXAMPLE 2 A conglomerate has three divisions, which produce computers, semiconductors, and business forms. For each $1 of output, the computer division needs $.02 worth of computers, $.20 worth of semiconductors, and $.10 worth of business forms. For each $1 of output, the semiconductor division needs $.02 worth of computers, $.01 worth of semiconductors, and $.02 worth of business forms. For each $1 of output, the business forms division requires $.10 worth of computers and $.01 worth of business forms. The conglomerate estimates the sales demand to be $300,000,000 for the computer division, $100,000,000 for the semiconductor division, and $200,000,000 for the business forms division. At what level should each division produce in order to satisfy this demand?

Solution The conglomerate can be viewed as a miniature economy and its sales as the final demand. The input-output matrix for this "economy" is

	Computers	Semiconductors	Business forms
Computers	.02	.02	.10
Semiconductors	.20	.01	0
Business forms	.10	.02	.01

$$= A.$$

The final-demand matrix is

$$D = \begin{bmatrix} 3 \\ 1 \\ 2 \end{bmatrix},$$

where the demand is expressed in hundreds of millions of dollars. By equation (1) the matrix X, giving the desired levels of production for the various divisions, is given by

$$X = (I - A)^{-1}D.$$

But

$$I - A = \begin{bmatrix} .98 & -.02 & -.10 \\ -.20 & .99 & 0 \\ -.10 & -.02 & .99 \end{bmatrix},$$

so that (Exercise 6)

$$(I - A)^{-1} = \begin{bmatrix} 1.04 & .02 & .10 \\ .21 & 1.01 & .02 \\ .11 & .02 & 1.02 \end{bmatrix},$$

and

$$(I - A)^{-1}D = \begin{bmatrix} 3.34 \\ 1.68 \\ 2.39 \end{bmatrix}.$$

Therefore,

$$X = \begin{bmatrix} 3.34 \\ 1.68 \\ 2.39 \end{bmatrix}.$$

That is, the computer division should produce $334,000,000, the semiconductor division $168,000,000, and the business forms division $239,000,000.

Input-output analysis is usually applied to the entire economy of a country having hundreds of industries. The resulting matrix equation $(I - A)X = D$ could be solved by the Gaussian elimination method. However, it is best to find the inverse of $I - A$ and solve for X as we have done in the examples of this section. Over a short period, D might change but A is unlikely to change. Therefore, the proper outputs to satisfy the new demand can easily be determined by using the already computed inverse of $I - A$.

The Closed Leontieff Model   The foregoing description of an economy is usually called the *Leontieff open model*, since it views exports as an activity that takes place external to the economy. However, it is possible to consider export as but another industry in the economy. Instead of describing exports by a demand column D, we describe it by a column in the input-output matrix. That is, the export column describes how each dollar of exports is divided among the various industries. Since exports are now regarded as another industry, each of the original columns has an additional entry, namely the amount of output from the export industry (that is, imports) used to produce $1 of goods (of the industry corresponding to the column). If A denotes the expanded input-output matrix and X the production matrix (as before), then AX is the matrix describing the total demand experienced by each of the industries. In order for the economy to function efficiently, the total amount demanded by the various industries should equal the amount produced. That is, the production matrix must satisfy the equation

$$AX = X.$$

By studying the solutions to this equation, it is possible to determine the equilibrium states of the economy—that is, the production matrices X for which the amounts produced exactly equal the amounts needed by the various industries. The model just described is called *Leontieff's closed model*.

We may expand the Leontieff closed model to include the effects of labor and monetary phenomena by considering labor and banking as yet further industries to be incorporated in the input-output matrix.

PRACTICE PROBLEMS 6

1. Let

$$I = \begin{bmatrix} 1 & 0 & 0 \\ 0 & 1 & 0 \\ 0 & 0 & 1 \end{bmatrix}, \quad A = \begin{bmatrix} .1 & 0 & .1 \\ .2 & .1 & .1 \\ .1 & .2 & 0 \end{bmatrix}, \quad X = \begin{bmatrix} x \\ y \\ z \end{bmatrix}, \quad D = \begin{bmatrix} 100 \\ 200 \\ 50 \end{bmatrix}.$$

Solve the matrix equation

$$(I - A)X = D.$$

2. Let I, A, X be as in Problem 1, but let

$$D = \begin{bmatrix} 300 \\ 100 \\ 100 \end{bmatrix}.$$

Solve the matrix equation $(I - A)X = D$.

EXERCISES 6

1. Suppose that in the economy of Example 1 the demand for electricity triples and the demand for coal doubles, whereas the demand for steel increases only by 50%. At what levels should the various industries produce in order to satisfy the new demand?

2. Suppose that the conglomerate of Example 2 is faced with an increase of 50% in demand for computers, a doubling in demand for semiconductors, and a decrease of 50% in demand for business forms. At what levels should the various divisions produce in order to satisfy the new demand?

3. Suppose that the conglomerate of Example 2 experiences a doubling in the demand for business forms. At what levels should the computer and semiconductor divisions produce?

4. A multinational corporation does business in the United States, Canada, and England. Its branches in one country purchase goods from the branches in other countries according to the matrix

		Branch in:		
		United States	Canada	England
Purchase from	United States	.02	0	.02
	Canada	.01	.03	.01
	England	.03	0	.01

where the entries in the matrix represent percentages of total sales by the respective branch. The external sales by each of the offices are $800,000,000 for the U.S. branch,

$300,000,000 for the Canadian branch, and $1,400,000,000 for the English branch. At what level should each of the branches produce in order to satisfy the total demand?

5. Show that to two decimal places

$$\begin{bmatrix} 1 & -.15 & -.43 \\ -.02 & .97 & -.20 \\ -.01 & -.08 & .95 \end{bmatrix}^{-1} = \begin{bmatrix} 1.01 & .20 & .50 \\ .02 & 1.05 & .23 \\ .01 & .09 & 1.08 \end{bmatrix}.$$

(A hand calculator would help a great deal here.)

6. Show that to two decimal places

$$\begin{bmatrix} .98 & -.02 & -.10 \\ -.2 & .99 & 0 \\ -.10 & -.02 & .99 \end{bmatrix}^{-1} = \begin{bmatrix} 1.04 & .02 & .10 \\ .21 & 1.01 & .02 \\ .11 & .02 & 1.02 \end{bmatrix}.$$

7. A corporation has a plastics division and an industrial equipment division. For each $1 worth of output, the plastics division needs $.02 worth of plastics and $.10 worth of equipment. For each $1 worth of output, the industrial equipment division needs $.01 worth of plastics and $.05 worth of equipment. At what level should the divisions produce to meet a demand for $930,000 worth of plastics and $465,000 worth of industrial equipment?

8. Rework Exercise 7 under the condition that the demand for plastics is $1,860,000 and the demand for industrial equipment is $2,790,000.

SOLUTIONS TO PRACTICE PROBLEMS 6

1. The equation $(I - A)X = D$ has the form $CX = D$, where C is the matrix $I - A$. From Section 4 we know that $X = C^{-1}D$. That is,

$$X = (I - A)^{-1}D.$$

Now

$$I - A = \begin{bmatrix} 1 & 0 & 0 \\ 0 & 1 & 0 \\ 0 & 0 & 1 \end{bmatrix} - \begin{bmatrix} .1 & 0 & .1 \\ .2 & .1 & .1 \\ .1 & .2 & 0 \end{bmatrix} = \begin{bmatrix} .9 & 0 & -.1 \\ -.2 & .9 & -.1 \\ -.1 & -.2 & 1 \end{bmatrix}.$$

Using the Gauss-Jordan method to find the inverse of this matrix, we have (to two decimal places)

$$(I - A)^{-1} = \begin{bmatrix} 1.13 & .03 & .12 \\ .27 & 1.14 & .14 \\ .17 & .23 & 1.04 \end{bmatrix}.$$

Therefore, rounding to the nearest integer, we have

$$X = (I - A)^{-1}D = \begin{bmatrix} 1.13 & .03 & .12 \\ .27 & 1.14 & .14 \\ .17 & .23 & 1.04 \end{bmatrix} \begin{bmatrix} 100 \\ 200 \\ 50 \end{bmatrix} = \begin{bmatrix} 125 \\ 262 \\ 115 \end{bmatrix}.$$

2. We have $X = (I - A)^{-1}D$, where $(I - A)^{-1}$ is as computed in Problem 1. So

$$X = \begin{bmatrix} 1.13 & .03 & .12 \\ .27 & 1.14 & .14 \\ .17 & .23 & 1.04 \end{bmatrix} \begin{bmatrix} 300 \\ 100 \\ 100 \end{bmatrix} = \begin{bmatrix} 354 \\ 209 \\ 178 \end{bmatrix}.$$

Chapter 2: CHECKLIST

☐ System of linear equations
☐ Elementary row operations
☐ Diagonal form
☐ Gaussian elimination method
☐ Matrix
☐ Pivoting
☐ Row matrix
☐ Column matrix
☐ Square matrix
☐ The ijth entry, a_{ij}
☐ Addition and subtraction of matrices
☐ Multiplication of matrices
☐ Identity matrix, I_n
☐ Inverse of a matrix, A^{-1}
☐ Formula for inverse of a 2×2 matrix
☐ Solution of matrix equation $AX = B$
☐ Use of inverse matrix to solve system of linear equations
☐ Gauss-Jordan method for calculating inverse of a matrix
☐ Input-output analysis

Chapter 2: SUPPLEMENTARY EXERCISES

Pivot each of the following matrices around the circled element.

1. $\begin{bmatrix} ③ & -6 & 1 \\ 2 & 4 & 6 \end{bmatrix}$

2. $\begin{bmatrix} -5 & -3 & 1 \\ 4 & ② & 0 \\ 0 & 6 & 7 \end{bmatrix}$

Use the Gaussian elimination method to find all solutions of the following systems of linear equations.

3. $\begin{cases} \frac{1}{2}x - y = -3 \\ 4x - 5y = -9 \end{cases}$

4. $\begin{cases} 3x \qquad + 9z = 42 \\ 2x + y + 6z = 30 \\ -x + 3y - 2z = -20 \end{cases}$

5. $\begin{cases} 3x - 6y + 6z = -5 \\ -2x + 3y - 5z = \frac{7}{3} \\ x + y + 10z = 3 \end{cases}$

6. $\begin{cases} 3x + 6y - 9z = 1 \\ 2x + 4y - 6z = 1 \\ 3x + 4y + 5z = 0 \end{cases}$

$$7. \begin{cases} x + 2y - 5z + 3w = 16 \\ -5x - 7y + 13z - 9w = -50 \\ -x + y - 7z + 2w = 9 \\ 3x + 4y - 7z + 6w = 33 \end{cases} \qquad 8. \begin{cases} 5x - 10y = 5 \\ 3x - 8y = -3 \\ -3x + 7y = 0 \end{cases}$$

Perform the indicated matrix operations.

$$9. \begin{bmatrix} 2 \\ -1 \\ 0 \end{bmatrix} + \begin{bmatrix} 3 \\ 4 \\ 7 \end{bmatrix} \qquad 10. \begin{bmatrix} 1 & 3 & -2 \\ 4 & 0 & -1 \end{bmatrix} \begin{bmatrix} 3 & 5 \\ 1 & 0 \\ 0 & -6 \end{bmatrix}$$

11. Find the inverse of the appropriate matrix and use it to solve the system of equations

$$\begin{cases} 3x + 2y = 0 \\ 5x + 4y = 2. \end{cases}$$

12. The matrices

$$\begin{bmatrix} 4 & -2 & 3 \\ 8 & -3 & 5 \\ 7 & -2 & 4 \end{bmatrix} \quad \text{and} \quad \begin{bmatrix} -2 & 2 & -1 \\ 3 & -5 & 4 \\ 5 & -6 & 4 \end{bmatrix}$$

are inverses of each other. Use these matrices to solve the following systems of linear equations.

$$\text{(a)} \begin{cases} -2x + 2y - z = 1 \\ 3x - 5y + 4z = 0 \\ 5x - 6y + 4z = 3 \end{cases} \qquad \text{(b)} \begin{cases} 4x - 2y + 3z = 0 \\ 8x - 3y + 5z = -1 \\ 7x - 2y + 4z = 2 \end{cases}$$

Use the Gauss-Jordan method to calculate the inverses of the following matrices.

$$13. \begin{bmatrix} 2 & 6 \\ 1 & 2 \end{bmatrix} \qquad 14. \begin{bmatrix} 1 & 1 & 1 \\ 3 & 4 & 3 \\ 1 & 1 & 2 \end{bmatrix}$$

15. The economy of a small country can be regarded as consisting of two industries, I and II, whose input-output matrix is

$$A = \begin{bmatrix} .4 & .2 \\ .1 & .3 \end{bmatrix}.$$

How many units should be produced by each industry in order to meet a demand for 8 units from industry I and 12 units from industry II?

16. In an arms race between two superpowers, each nation takes stock of its own and its enemy's nuclear arsenal each year. Each nation has the policy of dismantling a certain percentage of its stockpile each year and adding that same percentage of its competitor's stockpile. Nation A uses 20% and nation B uses 10%. Suppose that the current stockpiles of nations A and B are 10,000 and 7000 weapons, respectively.
    (a) What will the stockpiles be in each of the next two years?
    (b) What were the stockpiles in each of the previous two years?
    (c) Show that the "missile gap" between the superpowers decreases by 30% each year under these policies. Show that the total number of weapons decreases each year if nation A begins with the most weapons, and the total number of weapons increases if nation B begins with the most weapons.

3

LINEAR PROGRAMMING,
A GEOMETRIC APPROACH

3.1 A Linear Programming Problem

3.2 Linear Programming, I

3.3 Linear Programming, II

Linear programming is a method for solving problems in which a linear function (representing cost, profit, distance, weight, or the like) is to be maximized or minimized. Such problems are called *optimization problems.* As we shall see, these problems, when translated into mathematical language, involve systems of linear inequalities, systems of linear equations, and eventually (in Chapter 4) matrices.

3.1. A Linear Programming Problem

Let us begin with a detailed discussion of a typical problem that can be solved by linear programming.

Furniture Manufacturing Problem

A furniture manufacturer makes two types of furniture—chairs and sofas. For simplicity, divide the production process into three distinct operations—carpentry, finishing, and upholstery. The amount of labor required for each operation varies. Manufacture of a chair requires 6 hours of carpentry, 1 hour of finishing, and 2 hours of upholstery. Manufacture of a sofa requires 3 hours of carpentry, 1 hour of finishing, and 6 hours of upholstery. Owing to limited availability of skilled labor as well as of tools and equipment, the factory has available each day 96 man-hours for carpentry, 18 man-hours for finishing, and 72 man-hours for upholstery. The profit per chair is $80 and the profit per sofa $70. How many chairs and how many sofas should be produced each day in order to maximize the profit?

It is often helpful to tabulate data given in verbal problems. Our first step, then, is to construct a chart.

	Chair	Sofa	Available labor
Carpentry	6 hours	3 hours	96 man-hours
Finishing	1 hour	1 hour	18 man-hours
Upholstery	2 hours	6 hours	72 man-hours
Profit	$80	$70	

The next step is to translate the problem into mathematical language. As you know, this is done by identifying what is unknown and denoting the unknown quantities by letters. Since the problem asks for the optimum number of chairs and sofas to be produced each day, there are two unknowns—the number of chairs produced each day and the number of sofas produced each day. Let x denote the former and y the latter.

In order to achieve a large profit, one need only manufacture a large number of chairs and sofas. But, owing to restricted availability of tools and labor, the factory cannot manufacture an unlimited quantity of furniture. Let us translate the restrictions into mathematical language. Each row of the chart gives one restriction. The first row says that the amount of carpentry required is 6 hours for each chair and 3 hours for each sofa. Also, there are available only 96 man-hours of carpentry per day. We can compute the total number of man-hours of carpentry required per day to produce x chairs and y sofas as follows:

[number of man-hours per day of carpentry]

$$= \text{(number of hours carpentry per chair)} \cdot \text{(number of chairs per day)}$$
$$+ \text{(number of hours carpentry per sofa)} \cdot \text{(number of sofas per day)}$$
$$= 6 \cdot x + 3 \cdot y.$$

The requirement that at most 96 man-hours of carpentry be used per day means that x and y must satisfy the inequality

$$6x + 3y \leq 96. \tag{1}$$

The second row of the chart gives a restriction imposed by finishing. Since 1 hour of finishing is required for each chair and sofa, and since at most 18 man-hours of finishing are available per day, the same reasoning as used to derive inequality (1) yields

$$x + y \leq 18. \tag{2}$$

Similarly, the third row of the chart gives the restriction due to upholstery:

$$2x + 6y \leq 72. \tag{3}$$

A further restriction is given by the fact that the numbers of chairs and sofas must be nonnegative:

$$x \geq 0, \quad y \geq 0. \tag{4}$$

Now that we have written down the restrictions which constrain x and y, let us express the profit (which is to be maximized) in terms of x and y. The profit comes from two sources—chairs and sofas. Therefore,

$$[\text{profit}] = [\text{profit from chairs}] + [\text{profit from sofas}]$$
$$= [\text{profit per chair}] \cdot [\text{number of chairs}]$$
$$+ [\text{profit per sofa}] \cdot [\text{number of sofas}]$$
$$= 80x + 70y \tag{5}$$

Combining (1) to (5), we arrive at the following:

☐ **Furniture Manufacturing Problem—Mathematical Formulation** Find numbers x and y for which $80x + 70y$ is as large as possible, and for which all the following inequalities hold simultaneously:

$$\begin{cases} 6x + 3y \leq 96 \\ x + y \leq 18 \\ 2x + 6y \leq 72 \\ x \geq 0, \quad y \geq 0. \end{cases} \tag{6}$$

We may describe this mathematical problem in the following general way. We are required to maximize an expression in a certain number of variables, where the variables are subject to restrictions in the form of one or more in-equalities. Problems of this sort are called *mathematical programming problems.* Actually, general mathematical programming problems can be quite involved, and their solutions may require very sophisticated mathematical ideas. However, this is not the case with the furniture manufacturing problem. What makes it a rather simple mathematical programming problem is that both the expression to be maximized and the inequalities are linear. For this reason the furniture manufacturing problem is called a *linear programming problem.* The theory of linear programming is a fairly recent advance in mathematics. It was developed over the last 40 years to deal with the increasingly more complicated problems of our technological society. The 1975 Nobel prize in economics was awarded to Kantorovich and Koopmans for their pioneering work in the field of linear programming.

We will solve the furniture manufacturing problem in Section 2, where we will develop a general technique for handling similar linear programming prob-lems. At this point it is worthwhile to attempt to gain some insights into the problem and possible methods for attacking it.

It seems clear that a factory will operate most efficiently when its labor is fully utilized. Let us therefore take the operations one at a time and determine the conditions on x and y that fully utilize the three kinds of labor. The restriction on carpentry asserts that

$$6x + 3y \leq 96.$$

If x and y were chosen so that $6x + 3y$ is actually *less* than 96, we would leave the carpenters idle some of the time, a waste of labor. Thus, it would seem reasonable to choose x and y to satisfy

$$6x + 3y = 96.$$

Similarly, to utilize all the finishers' time, x and y must satisfy

$$x + y = 18,$$

and to utilize all the upholsterers' time, we must have

$$2x + 6y = 72.$$

Thus, if no labor is to be wasted, then x and y must satisfy the system of equations

$$\begin{cases} 6x + 3y = 96 \\ x + y = 18 \\ 2x + 6y = 72. \end{cases} \tag{7}$$

Let us now graph the three equations of (7), which represent the conditions for full utilization of all forms of labor. (See chart below and Fig. 1.)

Equation	Standard form	x-intercept	y-intercept
$6x + 3y = 96$	$y = -2x + 32$	$(16, 0)$	$(0, 32)$
$x + y = 18$	$y = -x + 18$	$(18, 0)$	$(0, 18)$
$2x + 6y = 72$	$y = -\frac{1}{3}x + 12$	$(36, 0)$	$(0, 12)$

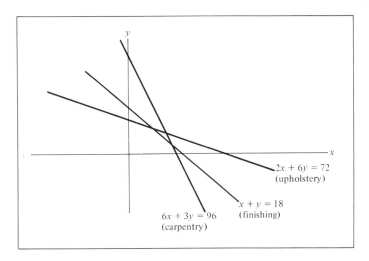

$2x + 6y = 72$
(upholstery)

$x + y = 18$
(finishing)

$6x + 3y = 96$
(carpentry)

FIGURE 1

What does Fig. 1 say about the furniture manufacturing problem? Each particular pair of numbers (x, y) is called a *production schedule*. Each of the lines in Fig. 1 gives the production schedules which fully utilize one of the types of labor. Notice that the three lines do not have a common intersection point. This means that there is *no* production schedule which *simultaneously* makes full use of all three types of labor. In any production schedule at least some of the man-hours of labor must be wasted. This is not a solution to the furniture manufacturing problem, but is a valuable insight. It says that in the inequalities of (6) not all of the corresponding equations can hold. This suggests that we take a closer look at the system of inequalities.

The standard forms of the inequalities (6) are

$$\begin{cases} y \le -2x + 32 \\ y \le -x + 18 \\ y \le -\frac{1}{3}x + 12 \\ x \ge 0, \quad y \ge 0. \end{cases}$$

By using the techniques of Section 1.2, we arrive at a feasible set for the system of inequalities above, as shown in Fig. 2.

The feasible set for the furniture manufacturing problem is a bounded, five-sided region. The points on and inside the boundary of this feasible set give the production schedules which satisfy all the restrictions. In the next section we will show how to pick out the particular point of the feasible set that corresponds to a maximum profit.

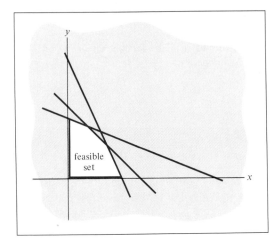

feasible set

FIGURE 2

PRACTICE PROBLEMS 1

1. Determine whether the following points are in the feasible set of the furniture manufacturing problem: (a) $(10, 9)$; (b) $(14, 4)$.

2. A physical fitness enthusiast decides to devote her exercise time to a combination of jogging and cycling. She wants to earn aerobic points (a measure of the benefit of the exercise to strengthening the heart and lungs) and also to achieve relaxation and enjoyment. She jogs at 6 miles per hour and cycles at 18 miles per hour. An hour of jogging earns 12 aerobic points and an hour of cycling earns 9 aerobic points. Each week she would like to earn at least 36 aerobic points, cover at least 54 miles, and cycle at least as much as she jogs.

   (a) Fill in the chart below.

	One hour of jogging	*One hour of cycling*	*Requirement*
Miles covered			
Aerobic points			
Time required			

   (b) Let x be the number of hours of jogging and y the number of hours of cycling each week. Referring to the chart, give the inequalities that x and y must satisfy due to miles covered and aerobic points.

   (c) Give the inequalities that x and y must satisfy due to her preference for cycling and also due to the fact that x and y cannot be negative.

   (d) Express the time required as a linear function of x and y.

   (e) Graph the feasible set for the system of linear inequalities.

EXERCISES 1

In Exercises 1–4, determine whether the given point is in the feasible set of the furniture manufacturing problem. (The inequalities are given below.)

$$\begin{cases} 6x + 3y \le 96 \\ x + y \le 18 \\ 2x + 6y \le 72 \\ x \ge 0 \quad y \ge 0. \end{cases}$$

1. $(8, 7)$ 2. $(14, 3)$ 3. $(9, 10)$ 4. $(16, 0)$

5. (*Shipping Problem*) A truck traveling from New York to Baltimore is to be loaded with two types of cargo. Each crate of cargo A is 4 cubic feet in volume, weighs 100 pounds, and earns \$13 for the driver. Each crate of cargo B is 3 cubic feet in volume, weighs 200 pounds, and earns \$9 for the driver. The truck can carry no more than 300 cubic feet of crates and no more than 10,000 pounds. Also, the number of crates of cargo B must be less than or equal to twice the number of crates of cargo A.

(a) Fill in the chart below.

	A	B	Truck capacity
Volume	4	3	300
Weight	100	200	10 000
Earnings	$13	$9	

$$y \leq 2x$$

(b) Let x be the number of crates of cargo A and y the number of crates of cargo B. Referring to the chart, give the two inequalities that x and y must satisfy because of the truck's capacity for volume and weight.

(c) Give the inequalities that x and y must satisfy because of the last sentence of the problem and also because x and y cannot be negative.

(d) Express the earnings from carrying x crates of cargo A and y crates of cargo B.

(e) Graph the feasible set for the shipping problem.

6. (*Mining Problem*) A coal company owns mines in two different locations. Each day mine 1 produces 4 tons of anthracite (hard coal), 4 tons of ordinary coal, and 7 tons of bituminous (soft) coal. Each day mine 2 produces 10 tons of anthracite, 5 tons of ordinary coal, and 5 tons of bituminous coal. It costs the company $150 per day to operate mine 1 and $200 per day to operate mine 2. An order is received for 80 tons of anthracite, 60 tons of ordinary coal, and 75 tons of bituminous coal.

(a) Fill in the chart below.

	Mine 1	Mine 2	Ordered
Anthracite			
Ordinary			
Bituminous			
Daily cost			

(b) Let x be the number of days mine 1 should be operated and y the number of days mine 2 should be operated. Refer to the chart and give three inequalities that x and y must satisfy to fill the order.

(c) Give other requirements that x and y must satisfy.

(d) Find the cost of operating mine 1 for x days and mine 2 for y days.

(e) Graph the feasible set for the mining problem.

7. A dairy farmer concludes that his small herd of cows will need at least 4550 pounds of protein in their winter feed, at least 26,880 pounds of total digestible nutrients (TDN), and at least 43,200 international units (IUs) of vitamin A. Each pound of alfalfa hay provides .13 pound of protein, .48 pound of TDN, and 2.16 IUs of vitamin A. Each pound of ground ear corn supplies .065 pound of protein, .96 pound of TDN, and no

vitamin A. Alfalfa hay costs $1 per 100-pound sack. Ground ear corn costs $1.60 per 100-pound sack.

(a) Fill in the chart below.

	Alfalfa	Corn	Requirements
Protein			
TDN			
Vitamin A			
Cost			

(b) Let x be the number of pounds of alfalfa hay, and y be the number of pounds of ground ears of corn to be bought. Give the inequalities that x and y must satisfy.

(c) Graph the feasible set of the system of linear inequalities.

(d) Express the cost of buying x pounds of alfalfa hay and y pounds of ground ears of corn.

SOLUTIONS TO PRACTICE PROBLEMS 1

1. A point is in the feasible set of a system of inequalities if it satisfies every inequality. Either the original form or the standard form of the inequalities may be used. The original form of the inequalities of the furniture manufacturing problem is

$$\begin{cases} 6x + 3y \le 96 \\ x + y \le 18 \\ 2x + 6y \le 72 \\ x \ge 0, \quad y \ge 0. \end{cases}$$

(a) $(10, 9)$

$$\begin{cases} 6(10) + 3(9) \le 96 & 87 \le 96 & \text{true} \\ 10 + 9 \le 18 & 19 \le 18 & \text{false} \\ 2(10) + 6(9) \le 72 & 74 \le 72 & \text{false} \\ 10 \ge 0, \quad 9 \ge 0; & 10 \ge 0, \quad 9 \ge 0. & \text{true} \end{cases}$$

(b) $(14, 4)$

$$\begin{cases} 6(14) + 3(4) \le 96 & 96 \le 96 & \text{true} \\ 14 + 4 \le 18 & 18 \le 18 & \text{true} \\ 2(14) + 6(4) \le 72 & 52 \le 72 & \text{true} \\ 14 \ge 0, \quad 4 \ge 0; & 14 \ge 0, \quad 4 \ge 0. & \text{true} \end{cases}$$

Therefore, $(14, 4)$ is in the feasible set and $(10, 9)$ is not.

2. (a)

	One hour of jogging	One hour of cycling	Requirement
Miles covered	6	18	54
Aerobic points	12	9	36
Time required	1	1	

(b) Miles covered: $6x + 18y \geq 54$.
Aerobic points: $12x + 9y \geq 36$.

(c) $y \geq x, x \geq 0$. It is not necessary to list $y \geq 0$ since this is automatically assured if the other two inequalities hold.

(d) $x + y$. (An objective of the exercise program might be to minimize $x + y$.)

(e)

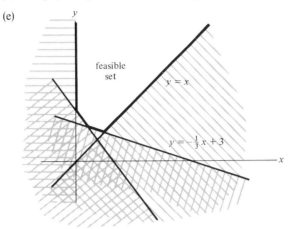

3.2. Linear Programming, I

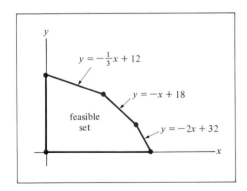

FIGURE 1

We have shown that the feasible set for the furniture manufacturing problem—that is, the set of points corresponding to production schedules satisfying all five restriction inequalities—consists of the points in and on the interior and boundary of the five-sided region drawn in Fig. 1. For reference, we have labeled each line segment with the equation of the line to which it belongs. The line segments intersect in five points, each of which is a corner of the feasible set. Such a corner is called a *vertex*. Somehow, we must pick out of the feasible set an *optimum point*—that is, a point corresponding to a production schedule which yields a maximum profit. To assist us in this task, we have the following result:*

□ **Fundamental Theorem of Linear Programming** The maximum (or minimum) value of the objective function is achieved at one of the vertices of the feasible set.

This result does not completely solve the furniture manufacturing problem for us, but it comes close. It tells us that an optimum production schedule (a, b)

---

* For a proof, see Section 3.3.

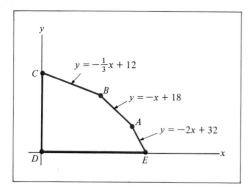

FIGURE 2

corresponds to one of the five points labeled A–E in Fig. 2. So to complete the solution of the furniture manufacturing problem, it suffices to find the coordinates of the five points, evaluate the profit at each, and then choose the point corresponding to the maximum profit.

Solution of the Furniture Manufacturing Problem

Let us begin by determining the coordinates of the points A–E in Fig. 2. Remembering that the x-axis has the equation $y = 0$ and the y-axis the equation $x = 0$, we see from Fig. 2 that the coordinates of A–E can be found as intersections of the following lines:

$$A: \begin{cases} y = -x + 18 \\ y = -2x + 32 \end{cases} \qquad B: \begin{cases} y = -x + 18 \\ y = -\tfrac{1}{3}x + 12 \end{cases}$$

$$C: \begin{cases} y = -\tfrac{1}{3}x + 12 \\ x = 0 \end{cases} \qquad D: \begin{cases} y = 0 \\ x = 0 \end{cases}$$

$$E: \begin{cases} y = 0 \\ y = -2x + 32. \end{cases}$$

The point D is clearly $(0, 0)$, and C is clearly the point $(0, 12)$. We obtain A from

$$-x + 18 = -2x + 32$$

$$x = 14$$

$$y = -14 + 18 = 4.$$

Hence $A = (14, 4)$. Similarly, we obtain B from

$$-x + 18 = -\tfrac{1}{3}x + 12$$

$$-\tfrac{2}{3}x = -6$$

$$x = 9$$

$$y = -9 + 18 = 9,$$

so $B = (9, 9)$. Finally, E is obtained from

$$0 = -2x + 32$$

$$x = 16$$

$$y = 0,$$

and thus $E = (16, 0)$. We have displayed the vertices in Fig. 3 and listed them in Table 1. In the second column we have evaluated the profit, which is given by

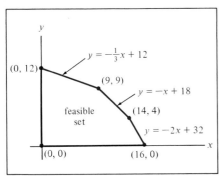

FIGURE 3

TABLE 1

Vertex	Profit $= 80x + 70y$
(14, 4)	$80(14) + 70(4) = 1400$
(9, 9)	$80(9) + 70(9) = 1350$
(0, 12)	$80(0) + 70(12) = 840$
(0, 0)	$80(0) + 70(0) = 0$
(16, 0)	$80(16) + 70(0) = 1280$

$80x + 70y$, at each of the vertices. Note that the largest profit occurs at the vertex (14, 4), so the solution of the linear programming problem is $x = 14$, $y = 4$. In other words, the factory should produce 14 chairs and 4 sofas each day in order to achieve maximum profit, and the maximum profit is $1400 per day.

The furniture manufacturing problem is one particular example of a linear programming problem. Generally, such problems involve finding the values of x and y which maximize (or minimize) a particular linear expression in x and y and where x and y are chosen so as to satisfy one or more restrictions in the form of linear inequalities. The expression that is to be maximized (or minimized) is called the *objective function*. On the basis of our experience with the furniture manufacturing problem, we can summarize the steps to be followed in approaching *any* linear programming problem.

Step 1   Translate the problem into mathematical language.

A. Organize the data.
B. Identify the unknown quantities and define corresponding variables.
C. Translate the restrictions into linear inequalities.
D. Form the objective function.

Step 2   Graph the feasible set.

A. Put the inequalities in standard form.
B. Graph the straight line corresponding to each inequality.
C. Determine the side of the line belonging to the graph of each inequality. Cross out the other side. The remaining region is the feasible set.

Step 3   Determine the vertices of the feasible set.
Step 4   Evaluate the objective function at each vertex. Determine the optimum point.

Linear programming can be applied to many problems. The Army Corps of Engineers has used linear programming to plan the location of a series of dams so as to maximize the resulting hydroelectric power production. The restrictions were to provide adequate flood control and irrigation. Public transit companies have used linear programming to plan routes and schedule buses in order to maximize services. The restrictions in this case arose from the limitations on manpower, equipment, and funding. The petroleum industry uses linear programming in the refining and blending of gasoline. Profit is maximized subject to restrictions on availability of raw materials, refining capacity, and product specifications. Some large advertising firms have used linear programming in media selection. The problem consists of determining how much to spend in each medium in order to maximize the number of consumers reached. The restrictions come from limitations on the budget and the relative costs of different media. Linear programming has been also used by psychologists to design an optimum battery of tests. The problem is to maximize the correlation between test scores and the characteristic that is to be predicted. The restrictions are imposed by the length and cost of the testing.

Linear programming is also used by dieticians in planning meals for large numbers of people. The object is to minimize the cost of the diet, and the restrictions reflect the minimum daily requirements of the various nutrients considered in the diet. The next example is representative of this type of problem. Whereas in actual practice many nutritional factors are considered, we shall simplify the problem by considering only three: protein, calories, and riboflavin.

EXAMPLE 1   *(Nutrition Problem)*   Suppose that in a developing nation the government wants to encourage everyone to make rice and soybeans part of his staple diet. The object is to design a lowest-cost diet which provides certain minimum levels of protein, calories, and vitamin B_2 (riboflavin). Suppose that one cup of uncooked rice costs 21 cents and contains 15 grams of protein, 810 calories, and $\frac{1}{9}$ milligram of riboflavin. On the other hand, one cup of uncooked soybeans costs 14 cents and contains 22.5 grams of protein, 270 calories, and $\frac{1}{3}$ milligram of riboflavin. Suppose that the minimum daily requirements are 90 grams of protein, 1620 calories, and 1 milligram of riboflavin. Design the lowest-cost diet meeting these specifications.

Solution   We solve the problem by following steps 1–4. The first step is to translate the problem into mathematical language, and the first part of this step is to organize the data, preferably into a chart (Table 2).

Now that we have organized the data, we ask for the unknowns. We wish to know how many cups each of rice and soybeans should comprise the diet, so we identify appropriate variables:

$$x = \text{number of cups of rice per day}$$

$$y = \text{number of cups of soybeans per day.}$$

Next, we obtain the restrictions on the variables. There is one restriction corresponding to each nutrient. That is, there is one restriction for each row of the

TABLE 2

	Rice	Soybeans	Required level per day
Protein (grams/cup)	15	22.5	90
Calories (per cup)	810	270	1620
Riboflavin (milligrams/cup)	$\frac{1}{9}$	$\frac{1}{3}$	1
Cost (cents/cup)	21	14	

chart. If x cups of rice and y cups of soybeans are consumed, then the amount of protein is $15x + 22.5y$ grams. Thus, from the first row of the chart, $15x + 22.5y \geq 90$, a restriction expressing the fact that there must be at least 90 grams of protein per day. Similarly, the restrictions for calories and riboflavin lead to the inequalities $810x + 270y \geq 1620$ and $\frac{1}{9}x + \frac{1}{3}y \geq 1$, respectively. As in the furniture manufacturing problem, x and y cannot be negative, so there are two further restrictions: $x \geq 0$, $y \geq 0$. In all there are five restrictions:

$$\begin{cases} 15x + 22.5y \geq 90 \\ 810x + 270y \geq 1620 \\ \frac{1}{9}x + \frac{1}{3}y \geq 1 \\ x \geq 0, \quad y \geq 0. \end{cases} \tag{1}$$

Now that we have the restrictions, we form the objective function, which tells what we are out to maximize or minimize. Since we wish to minimize cost, we express cost in terms of x and y. Now x cups of rice costs $21x$ cents and y cups of soybeans costs $14y$ cents, so the objective function is given by

$$[\text{cost}] = 21x + 14y. \tag{2}$$

And the problem can finally be stated in mathematical form: Minimize the objective function (2) subject to the restrictions (1). This completes the first step of the solution process.

The second step requires that we graph each of the inequalities (1). In Table 3 we have summarized all the steps necessary to obtain the information from which to draw the graphs. We have sketched the graphs in Fig. 4. From Fig. 4(b) we see

TABLE 3

			Intercepts		
Inequality	Standard form	Line	x	y	Graph
$15x + 22.5y \geq 90$	$y \geq -\frac{2}{3}x + 4$	$y = -\frac{2}{3}x + 4$	$(6, 0)$	$(0, 4)$	above
$810x + 270y \geq 1620$	$y \geq -3x + 6$	$y = -3x + 6$	$(2, 0)$	$(0, 6)$	above
$\frac{1}{9}x + \frac{1}{3}y \geq 1$	$y \geq -\frac{1}{3}x + 3$	$y = -\frac{1}{3}x + 3$	$(9, 0)$	$(0, 3)$	above
$x \geq 0$	$x \geq 0$	$x = 0$	$(0, 0)$	—	right
$y \geq 0$	$y \geq 0$	$y = 0$	—	$(0, 0)$	above

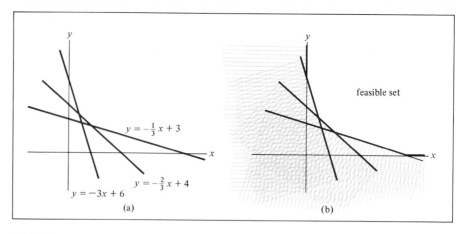

FIGURE 4

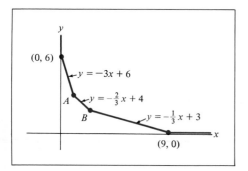

FIGURE 5

that the feasible set is an unbounded, five-sided region. There are four vertices, two of which are known from Table 3, since they are intercepts of boundary lines. Label the remaining two vertices A and B (Fig. 5).

The third step of the solution process consists of determining the coordinates of A and B. From Fig. 5, these coordinates can be found by solving the following systems of equations:

$$A: \begin{cases} y = -3x + 6 \\ y = -\tfrac{2}{3}x + 4 \end{cases} \qquad B: \begin{cases} y = -\tfrac{2}{3}x + 4 \\ y = -\tfrac{1}{3}x + 3. \end{cases}$$

To solve the first system, equate the two expressions for y:

$$-\tfrac{2}{3}x + 4 = -3x + 6$$
$$3x - \tfrac{2}{3}x = 6 - 4$$
$$\tfrac{7}{3}x = 2$$
$$x = \tfrac{6}{7}$$
$$y = -3x + 6 = -3(\tfrac{6}{7}) + 6 = \tfrac{24}{7}$$
$$A = (\tfrac{6}{7}, \tfrac{24}{7}).$$

Similarly, we find B:

$$-\tfrac{2}{3}x + 4 = -\tfrac{1}{3}x + 3$$
$$-\tfrac{1}{3}x = -1$$
$$x = 3$$
$$y = -\tfrac{2}{3}(3) + 4 = 2$$
$$B = (3, 2).$$

TABLE 4

Vertex	$Cost = 21x + 14y$
$(0, 6)$	$21 \cdot 0 + 14 \cdot 6 = 84$
$(\frac{6}{7}, \frac{24}{7})$	$21 \cdot \frac{6}{7} + 14 \cdot \frac{24}{7} = 66$
$(3, 2)$	$21 \cdot 3 + 14 \cdot 2 = 91$
$(9, 0)$	$21 \cdot 9 + 14 \cdot 0 = 189$

The fourth step consists of evaluating the objective function, in this case $21x + 14y$, at each vertex. From Table 4, we see that the minimum cost is achieved at the vertex $(\frac{6}{7}, \frac{24}{7})$. So the optimum diet—that is, the one which gives nutrients at the desired levels but at minimum cost—is the one which has $\frac{6}{7}$ cup of rice per day and $\frac{24}{7}$ cups of soybeans per day.

Note: We have assumed that all linear programming problems have solutions. Although every linear programming problem presented in this text has a solution, there are problems that have no optimal feasible solution. This can happen in two ways. First, there might be no points in the feasible set. Second, feasible solutions to the system of inequalities might exist, but the objective function might not have a maximum (or minimum) value within the feasible set. See Exercises 17 and 18.

PRACTICE PROBLEMS 2

1. The feasible set for the nutrition problem is shown in the accompanying sketch. The cost is $21x + 14y$. *Without* using the fundamental theorem of linear programming, explain why the cost could not possibly be minimized at the point (4, 4).

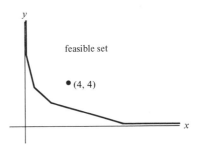

2. Rework the nutrition problem assuming that the cost of rice is changed to 7 cents per cup.

EXERCISES 2

For each of the feasible sets in Exercises 1–4, determine x and y so that the objective function $4x + 3y$ is maximized.

1.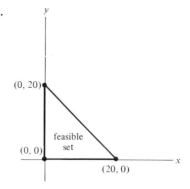

2.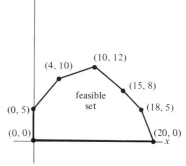

3.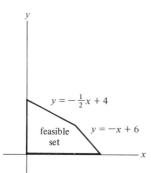

4.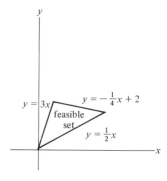

5. *(Shipping Problem)*   Refer to Exercises 1, Problem 5. How many crates of each cargo should be shipped in order to satisfy the shipping requirements and yield the greatest earnings? (See the graph of the feasible set below.)

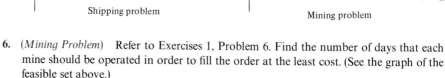

6. *(Mining Problem)*   Refer to Exercises 1, Problem 6. Find the number of days that each mine should be operated in order to fill the order at the least cost. (See the graph of the feasible set above.)

In Exercises 7 and 8, rework the furniture manufacturing problem, where everything is the same except that the profit per chair is changed to the given value. (See Table 1 for vertices.)

7. $150

8. $60

9. Minimize the objective function $7x + 4y$ subject to the restrictions

$$\begin{cases} y \geq -2x + 11 \\ y \leq -x + 10 \\ y \leq -\frac{1}{3}x + 6 \\ y \geq -\frac{1}{4}x + 4. \end{cases}$$

10. Maximize the objective function $x + 2y$ subject to the restrictions

$$\begin{cases} y \leq -x + 100 \\ y \geq \frac{1}{3}x + 20 \\ y \leq x. \end{cases}$$

11. Maximize the objective function $100x + 150y$ subject to the constraints

$$\begin{cases} x + 3y \leq 120 \\ 35x + 10y \leq 780 \\ x \leq 20 \\ x \geq 0, \quad y \geq 0. \end{cases}$$

12. Minimize the objective function $\frac{1}{2}x + \frac{3}{4}y$ subject to the constraints

$$\begin{cases} 2x + 2y \geq 8 \\ 3x + 5y \geq 16 \\ x \geq 0, \quad y \geq 0. \end{cases}$$

13. A contractor builds two types of homes. The first type requires one lot, $12,000 capital, and 150 man-days of labor to build and is sold for a profit of $2400. The second type of home requires one lot, $32,000 capital, and 200 man-days of labor to build and is sold for a profit of $3400. The contractor owns 150 lots and has available for the job $2,880,000 capital and 24,000 man-days of labor. How many homes of each type should she build in order to realize the greatest profit?

14. A nutritionist, working for NASA, must meet certain nutritional requirements and yet keep the weight of the food at a minimum. He is considering a combination of two foods which are packaged in tubes. Each tube of food A contains 4 units of protein, 2 units of carbohydrate, 2 units of fat, and weighs 3 pounds. Each tube of food B contains 3 units of protein, 6 units of carbohydrate, 1 unit of fat, and weighs 2 pounds. The requirement calls for 42 units of protein, 30 units of carbohydrate, and 18 units of fat. How many tubes of each food should be supplied to the astronauts?

15. The Beautiful Day Fruit Juice Company makes two varieties of fruit drink. Each can of Fruit Delight contains 10 ounces of pineapple juice, 3 ounces of orange juice, and 1 ounce of apricot juice, and makes a profit of 20 cents. Each can of Heavenly Punch contains 10 ounces of pineapple juice, 2 ounces of orange juice, and 2 ounces of apricot juice, and makes a profit of 30 cents. Each week, the company has available 9000 ounces of pineapple juice, 2400 ounces of orange juice, and 1400 ounces of apricot juice. How many cans of Fruit Delight and of Heavenly Punch should be produced each week in order to maximize profits?

16. The Bluejay Lacrosse Stick Company makes two kinds of lacrosse sticks. Type A sticks require 2 man-hours for cutting, 1 man-hour for stringing, and 2 man-hours for finishing, and are sold for a profit of $8. Type B sticks require 1 man-hour for cutting, 3 man-hours

for stringing, and 2 man-hours for finishing, and are sold for a profit of $10. Each day the company has available 120 man-hours for cutting, 150 man-hours for stringing, and 140 man-hours for finishing. How many lacrosse sticks of each kind should be manufactured each day in order to maximize profits?

17. A farmer has 100 acres on which to plant oats or corn. Each acre of oats requires $18 capital and 2 hours of labor. Each acre of corn requires $36 capital and 6 hours of labor. Labor costs are $8 per hour. The farmer has $2100 available for capital and $2400 available for labor. If the revenue is $55 from each acre of oats and $125 from each acre of corn, what planting combination will produce the greatest total profit? (Profit here is revenue plus leftover capital and labor cash reserve.) What is the maximum profit?

18. Suppose the farmer of Exercise 17 can allocate the $4500 available for capital and labor however he or she wants.

    (a) Without solving the linear programming problem, explain why the optimal profit cannot be less than what was found in Exercise 17.

    (b) Find the optimal solution in the new situation. Does it provide more profit than in Exercise 17?

19. A company makes two items I_1 and I_2 from three raw materials M_1, M_2, and M_3. Item I_1 uses 3 ounces of M_1, 2 ounces of M_2, and 2 ounces of M_3. Item I_2 uses 4 ounces of M_1, 1 ounce of M_2, and 3 ounces of M_3. The profit on item I_1 is $8 and on item I_2 is $6. The company has a daily supply of 40 ounces of M_1, 20 ounces of M_2, and 60 ounces of M_3.

    (a) How many of items I_1 and I_2 should be made each day to maximize profit?

    (b) What is the maximum profit?

    (c) How many ounces of each raw material is used?

    (d) If the profit on item I_1 increases to $13, how many of items I_1 and I_2 should be made each day to maximize profit?

20. Consider the following linear programming problem:
    Maximize $M = 10x + 6y$ subject to the constraints

    $$\begin{cases} x + y \geq 6 \\ x \geq 0, \quad y \geq 0. \end{cases}$$

    (a) Sketch the feasible set.

    (b) Determine three points in the feasible set and calculate M at each of them.

    (c) Show that the objective function attains no maximum value for points in the feasible set.

21. Consider the following linear programming problem:
    Minimize $M = 10x + 6y$ subject to the constraints

    $$\begin{cases} x + y \geq 6 \\ 4x + 3y \leq 4 \\ x \geq 0, \quad y \geq 0. \end{cases}$$

    $3y = -\frac{4}{3}y + 14$

    $y = -\frac{4}{4}x + \frac{14}{3}$

(a) Sketch the feasible set for the linear programming problem.

(b) Determine a point of the feasible set.

SOLUTIONS TO PRACTICE PROBLEMS 2

1. The point P has a smaller value of x and a smaller value of y than $(4, 4)$ and is still in the feasible set. It therefore corresponds to a lower cost than $(4, 4)$ and still meets the requirements. We conclude that no interior point of the feasible set could possibly be an optimum point. This geometric argument indicates that an optimum point might be one that juts out far—that is, a vertex.

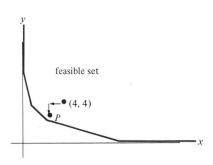

2. The system of linear inequalities, feasible set, and vertices will all be the same as before. Only the objective function changes. The new objective function is $7x + 14y$. The minimum cost occurs when using 3 cups of rice and 2 cups of soybeans.

Vertex	Cost $= 7x + 14y$
$(0, 6)$	84
$(\frac{6}{7}, \frac{24}{7})$	54
$(3, 2)$	49
$(9, 0)$	63

3.3. Linear Programming, II

In this section we apply the technique of linear programming to the design of a portfolio for a retirement fund and to the transportation of goods from warehouses to retail outlets. The significant new feature of each of these problems is that, on the surface, they appear to involve more than two variables. However, they can be translated into mathematical language so that only two variables are required.

EXAMPLE 1 (*Investment Analysis*) A pension fund has $30 million to invest. The money is to be divided among Treasury notes, bonds, and stocks. The rules for administration of the fund require that at least $3 million be invested in each type of investment, at least half the money be invested in Treasury notes and bonds, and the amount invested in bonds not exceed twice the amount invested in Treasury notes. The annual yields for the various investments are 7% for Treasury notes, 8% for bonds, and 9% for stocks. How should the money be allocated among the various investments to produce the largest return?

Solution First, let us agree that all numbers stand for millions. That is, we write 30 to stand for 30 million. This will save us from writing too many zeros. In examining the problem, we find that very little organization needs to be done. The rules for administration of the fund are written in a form from which inequalities can be read right off. Let us just summarize the remaining data in the first row of a chart (Table 1).

There appear to be three variables—the amounts to be invested in each of the three categories. However, since the three investments must total 30, we need only two variables. Let x = the amount to be invested in Treasury notes and y = the amount to be invested in bonds. Then the amount invested in stocks is $30 - (x + y)$. We have displayed the variables in Table 1.

TABLE 1

	~~Treasury notes~~ Real Estate	Bonds	Stocks
Yield	.07	.08	.09
Variables	x	y	$30 - (x + y)$

Now for the restrictions. Since at least 3 (million dollars) must be invested in each category, we have the three inequalities

$$x \geq 3$$

$$y \geq 3$$

$$30 - (x + y) \geq 3.$$

Moreover, since at least half the money, or 15, must be invested in ~~Treasury notes~~ Real Estate and bonds, we must have

$$x + y \geq 15.$$

Finally, since the amount invested in bonds must not exceed twice the amount invested in ~~Treasury notes~~ Real Estate, we must have

$$y \leq 2x.$$

(In this example we do not need to state that $x \geq 0$, $y \geq 0$, since we have already required that they be greater than or equal to 3.) Thus there are five restriction inequalities:

$$\begin{cases} x \geq 3, \quad y \geq 3 \\ 30 - (x + y) \geq 3 \\ \qquad x + y \geq 15 \\ \qquad\qquad y \leq 2x. \end{cases} \tag{1}$$

Next, we form the objective function, which in this case equals the total return on the investment. Since x dollars is invested at 7%, y dollars at 8%, and

TABLE 2

Inequality	Standard form	Equation	Intercepts x	Intercepts y	Graph
$x \geq 3$	$x \geq 3$	$x = 3$	$(3, 0)$	—	Right of line
$y \geq 3$	$y \geq 3$	$y = 3$	—	$(0, 3)$	Above line
$30 - (x + y) \geq 3$	$y \leq -x + 27$	$y = -x + 27$	$(27, 0)$	$(0, 27)$	Below line
$x + y \geq 15$	$y \geq -x + 15$	$y = -x + 15$	$(15, 0)$	$(0, 15)$	Above line
$y \leq 2x$	$y \leq 2x$	$y = 2x$	$(0, 0)$	$(0, 0)$	Below line

$30 - (x + y)$ dollars at 9%, the total return is

$$[\text{return}] = .07x + .08y + .09[30 - (x + y)]$$

$$= .07x + .08y + 2.7 - .09x - .09y$$

$$= 2.7 - .02x - .01y. \tag{2}$$

So the mathematical statement of the problem is: Maximize the objective function (2) subject to the restrictions (1).

The next step of the solution is to graph the inequalities (1). The necessary information is tabulated in Table 2.

One point about the chart is worth noting: It contains enough data to graph each of the lines, with the exception of $y = 2x$. The reason is that the x- and y-intercepts of this line are the same, $(0, 0)$. So to graph $y = 2x$, we must find an additional point on the line. For example, if we set $x = 2$, then $y = 4$, so $(2, 4)$ is on the line. In Fig. 1(a) we have drawn the various lines, and in Fig. 1(b) we have crossed out the appropriate regions to produce the graph of the system. The

FIGURE 1

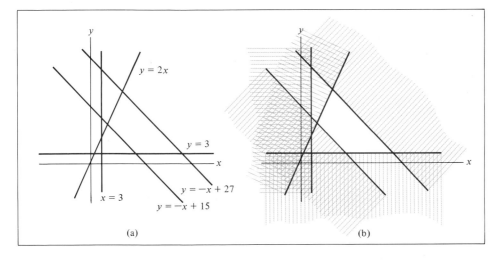

(a)   (b)

feasible set, as well as the equations of the various lines that make up its boundary, are shown in Fig. 2. From Fig. 2 we find the pairs of equations that determine each of the vertices A–D. This is the third step of the solution procedure.

$$A: \begin{cases} y = 3 \\ y = -x + 15 \end{cases} \qquad B: \begin{cases} y = 3 \\ y = -x + 27 \end{cases}$$

$$C: \begin{cases} y = -x + 27 \\ y = 2x \end{cases} \qquad D: \begin{cases} y = 2x \\ y = -x + 15. \end{cases}$$

A and B are the easiest to determine. To find A, we must solve

$$3 = -x + 15$$
$$x = 12$$
$$y = 3$$
$$A = (12, 3).$$

Similarly, $B = (24, 3)$. To find C, we must solve

$$2x = -x + 27$$
$$3x = 27$$
$$x = 9$$
$$y = 2(9) = 18$$
$$C = (9, 18).$$

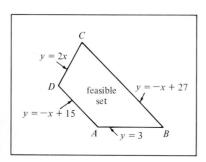

FIGURE 2

Similarly, $D = (5, 10)$.

Finally, we list the four vertices A, B, C, D and evaluate the objective function (2) at each one. The results are summarized in Table 3.

It is clear that the largest return occurs when $x = 5$, $y = 10$. In other words, $5 million should be invested in Treasury notes, $10 million in bonds, and $30 - (x + y) = 30 - (5 + 10) = \$15$ million in stocks.

Linear programming is of use not only in analyzing investments but in the fields of transportation and shipping. It is often used to plan routes, determine locations of warehouses, and develop efficient procedures for getting goods to people. Many linear programming problems of this variety can be formulated as *transportation problems*. A typical transportation problem involves determining

TABLE 3

Vertex	Return = $2.7 - .02x - .01y$
(5, 10)	$2.7 - .02(5) - .01(10) = \$2.5$ million
(9, 18)	$2.7 - .02(9) - .01(18) = \$2.34$ million
(24, 3)	$2.7 - .02(24) - .01(3) = \$2.19$ million
(12, 3)	$2.7 - .02(12) - .01(3) = \$2.43$ million

the least-cost scheme for delivering a commodity stocked in a number of different warehouses to a number of different locations, say retail stores. Of course, in practical applications, it is necessary to consider problems involving perhaps dozens or even hundreds of warehouses, and possibly just as many delivery locations. For problems on such a grand scale, the methods developed so far are inadequate. For one thing, the number of variables required is usually more than two. We must wait until Chapter 4 for methods that apply to such problems. However, the next example gives an instance of a transportation problem which does not involve too many warehouses or too many delivery points. It gives the flavor of general transportation problems.

EXAMPLE 2 Suppose that a Maryland TV dealer has stores in Annapolis and Rockville and warehouses in College Park and Baltimore. The cost of shipping a set from College Park to Annapolis is $6; from College Park to Rockville, $3; from Baltimore to Annapolis, $9; and from Baltimore to Rockville, $5. Suppose that the Annapolis store orders 25 TV sets and the Rockville store 30. Suppose further that the College Park warehouse has a stock of 45 sets and the Baltimore warehouse 40. What is the most economical way to supply the requested TV sets to the two stores?

Solution The first step in solving a linear programming problem is to translate it into mathematical language. And the first part of this step is to organize the information given, preferably in the form of a chart. In this case, since the problem is geographic, we draw a schematic diagram, as in Fig. 3, which shows the flow of goods between warehouses and retail stores. By each route, we have written the cost. Below each warehouse we have written down its stock and below each retail store the number of TV sets it ordered.

Next, let us determine the variables. It appears initially that four variables are required, namely the number of TV sets to be shipped over each route. However, a closer look shows that only two variables are required. For if x denotes the number of TV sets to be shipped from College Park to Rockville, then since Rockville ordered 30 sets, the number shipped from Baltimore to Rockville is $30 - x$. Similarly, if y denotes the number of sets shipped from College Park to Annapolis, then the number shipped from Baltimore to Annapolis is $25 - y$. We have written the appropriate shipment sizes beside the various routes in Fig. 3.

As the third part of the translation process let us write down the restrictions on the variables. Basically, there are two kinds of restrictions: none of x, y, $30 - x$, $25 - y$ can be negative, and a warehouse cannot ship more TV sets than it has in stock. Referring to Fig. 3, we see that College Park ships $x + y$ sets, so that

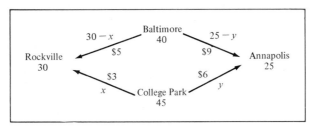

FIGURE 3

$x + y \leq 45$. Similarly, Baltimore ships $(30 - x) + (25 - y)$ sets, so that $(30 - x) + (25 - y) \leq 40$. Simplifying this inequality, we get

$$55 - x - y \leq 40$$

$$-x - y \leq -15$$

$$x + y \geq 15.$$

The inequality $30 - x \geq 0$ can be simplified to $x \leq 30$, and the inequality $25 - y \geq 0$ can be written $y \leq 25$. So our restriction inequalities are these:

$$\begin{cases} x \geq 0, & y \geq 0 \\ x \leq 30, & y \leq 25 \\ x + y \geq 15 \\ x + y \leq 45. \end{cases} \tag{3}$$

The final step in the translation process is to form the objective function. In this problem we are attempting to minimize cost, so the objective function must express the cost in terms of x and y. Refer again to Fig. 3. There are x sets going from College Park to Rockville, and each costs \$3 to transport, so the cost of delivering these x sets is $3x$. Similarly, the costs of making the other deliveries are $6y$, $5(30 - x)$, and $9(25 - y)$, respectively. Thus the objective function is

$$[\text{cost}] = 3x + 6y + 5(30 - x) + 9(25 - y)$$

$$= 3x + 6y + 150 - 5x + 225 - 9y$$

$$= 375 - 2x - 3y. \tag{4}$$

So the mathematical problem we must solve is: Find x and y that minimize the objective function (4) and satisfy the restrictions (3).

FIGURE 4

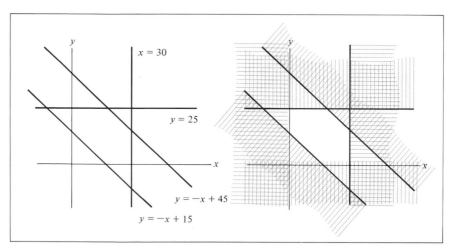

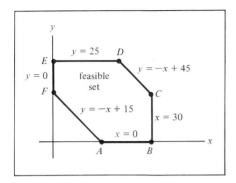

FIGURE 5

To solve the mathematical problem, we must graph the system of inequalities in (3). Four of the inequalities have graphs determined by horizontal and vertical lines. The only inequalities involving any work are $x + y \geq 15$ and $x + y \leq 45$. And even these are very easy to graph. The result is the graph in Fig. 4.

In Fig. 5 we have drawn the feasible set and have labeled each boundary line with its equation. The vertices A–F are now simple to determine. First, A and F are the intercepts of the line $y = -x + 15$. Therefore, $A = (15, 0)$ and $F = (0, 15)$. Since B is the x-intercept of the line $x = 30$, we have $B = (30, 0)$. Similarly, $E = (0, 25)$. Since C is on the line $x = 30$, its x-coordinate is 30. Its y-coordinate is $y = -30 + 45 = 15$, so $C = (30, 15)$. Similarly, since D has y-coordinate 25, its x-coordinate is given by $25 = -x + 45$ or $x = 20$. Thus $D = (20, 25)$.

We have listed in Table 4 the vertices A–F, as well as the cost corresponding to each one. The minimum cost occurs at the vertex $(20, 25)$. So $x = 20$, $y = 25$

TABLE 4

Vertex	Cost $= 375 - 2x - 3y$
(0, 25)	300
(0, 15)	330
(15, 0)	345
(30, 0)	315
(30, 15)	270
(20, 25)	260

yields the minimum of the objective function. In other words, 20 TV sets should be shipped from College Park to Rockville and 25 from College Park to Annapolis, $30 - x = 10$ from Baltimore to Rockville, and $25 - y = 0$ from Baltimore to Annapolis. This solves our problem.

Remarks Concerning the Transportation Problem Note that the highest-cost route is the one from Baltimore to Annapolis. The solution we have obtained eliminates any shipments over this route. One might infer from this that one should always avoid the most expensive route. But this is not correct reasoning. To see why, reconsider Example 2, except change the cost of transporting a TV set from Baltimore to Annapolis from $9 to $7. The Baltimore-Annapolis route is still the most expensive. However, in this case the minimum cost is not obtained by eliminating the Baltimore-Annapolis route. For the revised problem, the linear inequalities stay the same. So the feasible set and the vertices remain the same. The only change is in the objective function, which now is given by

$$[\text{cost}] = 3x + 6y + 5(30 - x) + 7(25 - y)$$
$$= 325 - 2x - y.$$

Therefore, the costs at the various vertices are as given in Table 5. So the minimum cost of \$250 is achieved when $x = 30$, $y = 15$, $30 - x = 0$, and $25 - y = 10$. Note that 10 sets are being shipped from Baltimore to Annapolis, even though this is the most expensive route.

It is even possible for the cost function to be optimized simultaneously at two different vertices. For example, if the cost from Baltimore to Annapolis is \$8 and all other data are the same as in Example 2, then the optimum cost is \$260 and is achieved at both vertices (30, 15) and (20, 25).

TABLE 5

Vertex	Cost $= 325 - 2x - y$
(0, 25)	300
(0, 15)	310
(15, 0)	295
(30, 0)	265
(30, 15)	250
(20, 25)	260

Verification of the Fundamental Theorem

The fundamental theorem of linear programming asserts that the objective function assumes its optimum value at a vertex of the feasible set. Let us verify this fact. For simplicity, we give the argument only in a special case, namely for the furniture manufacturing problem. However, this is for convenience of exposition only. The same argument as given below may be used to prove the fundamental theorem in general. Our argument relies on the parallel property for straight lines, which asserts that parallel lines have the same slope.

EXAMPLE 3 Prove the fundamental theorem of linear programming in the special case of the furniture manufacturing problem.

Solution The profit derived from producing x chairs and y sofas is $80x + 70y$ dollars. Let us examine all those production schedules having a given profit. As an example, consider a profit of \$2800. Then x and y must satisfy $80x + 70y = 2800$. That is, (x, y) must lie on the line whose equation is $80x + 70y = 2800$, or in standard form, $y = -\frac{8}{7}x + 40$. The slope of this line is $-\frac{8}{7}$ and its y-intercept is (0, 40). We have drawn this line in Fig. 6(a), in which we have also drawn the feasible set for the furniture manufacturing problem. Note two fundamental facts: (1) Every production schedule on the line corresponds to a profit of \$2800. (2) The line lies above the feasible set. In particular, no production schedule on the line satisfies all the restrictions of the problem. The difficulty is that \$2800 is too high a profit for which to ask.

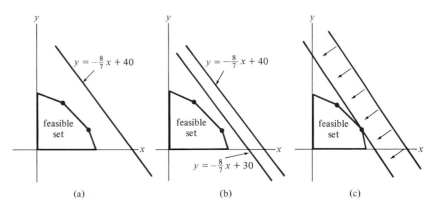

FIGURE 6

So now lower the profit, say, to $2100. In this case the production schedule (x, y) lies on the line $80x + 70y = 2100$, or in standard form, $y = -\frac{8}{7}x + 30$. This line is drawn in Fig. 6(b). Note that since both lines have slope $-\frac{8}{7}$, they are parallel by the parallel property. Actually, if we look at the production schedules yielding any fixed profit p, then they will lie along a line of slope $-\frac{8}{7}$, which is then parallel to the two lines already drawn. For if the production schedule (x, y) yields a profit p, then $80x + 70y = p$ or $y = -\frac{8}{7}x + p/70$. In other words, (x, y) lies on a line of slope $-\frac{8}{7}$ and y-intercept $(0, p/70)$. In particular, all the "lines of constant profit" are parallel to one another. So let us go back to the line of $2800 profit. It does not touch the feasible set. So now lower the profit and therefore translate the line downward parallel to itself. Next lower the profit until we first touch the feasible set. This line now touches the feasible set at a vertex [Fig. 6(c)]. And this vertex corresponds to the optimum production schedule, since any other point of the feasible set lies on a "line of constant profit" corresponding to an even lower profit. This shows why the fundamental theorem of linear programming is true.

If the objective function is parallel to one of the boundary lines of the feasible set, there might be infinitely many solutions—all points on that boundary might provide optimum values for the objective function. Two such points are vertices of the feasible set.

EXAMPLE 4   Reconsider the TV shipping problem discussed in Example 2 with the cost of shipping from Baltimore to Annapolis now $8 and all other data are the same as in Example 2. Minimize the shipping costs.

Solution   The cost function to be minimized is now

$$[\text{cost}] = 350 - 2x - 2y$$

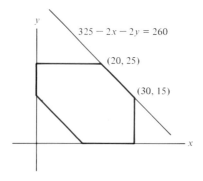

$325 - 2x - 2y = 260$

$(20, 25)$

$(30, 15)$

FIGURE 7

TABLE 6

Vertex	$Cost = 350 - 2x - 2y$
$(0, 25)$	300
$(0, 15)$	320
$(15, 0)$	320
$(30, 0)$	290
$(30, 15)$	260
$(20, 25)$	260

Note that for any fixed value of the cost, the slope of the objective function is -1. The feasible set is identical to that of Example 2 (Fig. 5). The slope of the boundary line $y = -x + 45$ is also -1. A check of the vertices of the feasible set (Table 6) shows that the minimal cost of $260 is achieved at the two vertices $C = (30, 15)$ and $D = (20, 25)$. In fact, the cost of $260 is achieved at every point on the boundary line joining these two vertices. Let's verify this in two cases. The points $(25, 20)$ and $(28, 17)$ both lie on the line $y = -x + 45$ and produce a cost of $260. Figure 7 illustrates the result.

PRACTICE PROBLEMS 3

Problems 1 and 2 refer to Example 1. Translate the statement into an inequality.

1. The amount to be invested in bonds is at most $5 million more than the amount to be invested in Treasury notes.

2. No more than $25 million should be invested in stocks and bonds.

3. Rework Example 1, assuming that the yield for Treasury notes goes up to 8%.

4. A linear programming problem has objective function: $[cost] = 5x + 10y$, which is to be minimized. Figure shows the feasible set and the straight line of all combinations of x and y for which $[cost] = \$20$.

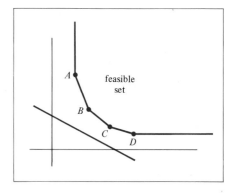

FIGURE 8

(a) Give the linear equation (in standard form) of the line of constant cost c.

(b) As c increases, does the line of constant cost c move up or down?

(c) By inspection, find the vertex of the feasible set that gives the optimum solution.

EXERCISES 3

1. Mr. Smith decides to feed his pet Doberman pinscher a combination of two dog foods. Each can of brand A contains 3 units of protein, 1 unit of carbohydrates, and 2 units of fat and costs 80 cents. Each can of brand B contains 1 unit of protein, 1 unit of carbohydrates, and 6 units of fat and costs 50 cents. Mr. Smith feels that each day his dog should have at least 6 units of protein, 4 units of carbohydrates, and 12 units of fat. How many cans of each dog food should he give to his dog each day in order to provide the minimum requirements at the least cost?

2. An oil company owns two refineries. Refinery I produces each day 100 barrels of high-grade oil, 200 barrels of medium-grade oil, and 300 barrels of low-grade oil and costs $10,000 to operate. Refinery II produces each day 200 barrels of high-grade, 100 barrels of medium-grade, and 200 barrels of low-grade oil and costs $9000 to operate. An order is received for 1000 barrels of high-grade oil, 1000 barrels of medium-grade oil, and 1800 barrels of low-grade oil. How many days should each refinery be operated in order to fill the order at the least cost?

3. A produce dealer in Florida ships oranges, grapefruits, and avocados to New York by truck. Each truckload consists of 100 crates, of which at least 20 crates must be oranges, at least 10 crates must be grapefruits, at least 30 crates must be avocados, and there must be at least as many crates of oranges as grapefruits. The profit per crate is $5 for oranges, $6 for grapefruits, and $4 for avocados. How many crates of each type should be shipped in order to maximize the profit? [$Hint$: Let x = number of crates of oranges, y = number of crates of grapefruit. Then $100 - x - y$ = number of crates of avocados.]

4. Mr. Jones has $9000 to invest in three types of stocks: low-risk, medium-risk, and high-risk. He invests according to three principles. The amount invested in low-risk stocks will be at most $1000 more than the amount invested in medium-risk stocks. At least $5000 will be invested in low- and medium-risk stocks. No more than $7000 will be invested in medium- and high-risk stocks. The expected yields are 6% for low-risk stocks, 7% for medium-risk stocks, and 8% for high-risk stocks. How much money should Mr. Jones invest in each type of stock in order to maximize his total expected yield?

5. An automobile manufacturer has assembly plants in Detroit and Cleveland, each of which can assemble cars and trucks. The Detroit plant can assemble at most 800 vehicles in one day at a cost of $1200 per car and $2100 per truck. The Cleveland plant can assemble at most 500 vehicles in one day at a cost of $1000 per car and $2000 per truck. A rush order is received for 600 cars and 300 trucks. How many vehicles of each type should each plant produce in order to fill the order at the least cost? [$Hint$: Let x = number of cars to be produced in Detroit, y = number of trucks to be produced in Detroit, $600 - x$ = number of cars to be produced in Cleveland, $300 - y$ = number of trucks to be produced in Cleveland.]

6. A foreign car dealer with warehouses in New York and Baltimore receives orders from dealers in Philadelphia and Trenton. The dealer in Philadelphia needs 4 cars and the

dealer in Trenton needs 7. The New York warehouse has 6 cars and the Baltimore warehouse has 8. The cost of shipping cars from Baltimore to Philadelphia is $120 per car, from Baltimore to Trenton $90 per car, from New York to Philadelphia $100 per car, from New York to Trenton $70 per car. Find the number of cars to be shipped from each warehouse to each dealer in order to minimize the shipping cost.

7. Figure 9(a) shows the feasible set of the nutrition problem of Section 2 and the straight line of all combinations of rice and soybeans for which the cost is 42 cents.

   (a) The objective function is $21x + 14y$. Give the linear equation (in standard form) of the line of constant cost c.

   (b) As c increases, does the line of constant cost move up or down?

   (c) By inspection, find the vertex of the feasible set that gives the optimum solution.

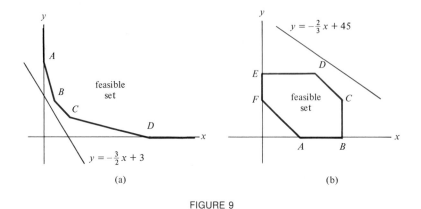

(a)   (b)

FIGURE 9

8. Figure 9(b) shows the feasible set of the transportation problem of Example 2 and the straight line of all combinations of shipments for which the transportation cost is $240.

   (a) The objective function is $[\text{cost}] = 375 - 2x - 3y$. Give the linear equation (in standard form) of the line of constant cost c.

   (b) As c increases, does the line of constant cost move up or down?

   (c) By inspection, find the vertex of the feasible set that gives the optimum solution.

9. An oil refinery produces gasoline, jet fuel, and diesel fuel. The profits per gallon from the sale of these fuels are $.15, $.12, and $.10, respectively. The refinery has a contract with an airline to deliver a minimum of 20,000 gallons per day of jet fuel and/or gasoline (or some of each). It has a contract with a trucking firm to deliver a minimum of 50,000 gallons per day of diesel fuel and/or gasoline (or some of each). The refinery can produce 100,000 gallons of fuel per day, distributed among the fuels in any fashion. It wishes to produce at least 5000 gallons per day of each fuel. How many gallons of each should be produced in order to maximize the profit?

10. Suppose that a price war reduces the profits of gasoline in Problem 9 to $.05 per gallon and that the profits on jet fuel and diesel fuel are unchanged. How many gallons of each fuel should now be produced to maximize the profit?

11. Mr Jones hires three workers, Tom, Dick, and Harry, for one 8-hour job entailing 9 hours of gardening and 15 hours of painting. Tom charges $7 per hour for gardening and $10 per hour for painting. Dick charges $8 per hour for gardening and $9 per hour for painting. Harry charges $8.50 for either job. (Note: Read all parts of exercise before solving.)

   (a) How should Mr. Jones apportion the work in order to minimize the amount he must pay the workers?

   (b) How should the workers apportion the work in order to maximize their total wages?

   (c) Answer part (b) under the additional assumption that the workers want to make within $10 of each other.

   (d) Answer part (a) under the additional assumption that the workers make within $10 of each other.

   (e) Suppose that the maximum or minimum total wages in parts (a) through (d) are A, B, C, and D respectively. Before working parts (a) through (d), order the numbers A, B, C, and D. [For instance, one (incorrect) ordering would be $A \leq B \leq C \leq D$.]

12. Explain why the region on the right cannot be the feasible set of a linear programming problem.

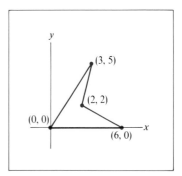

13. Consider the feasible set in Figure 10. For what values of k will the objective function $x + ky$ be maximized at the vertex $(3, 4)$?

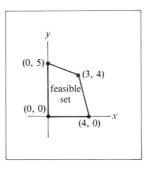

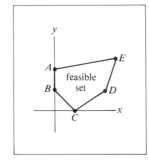

FIGURE 10   FIGURE 11

14. Consider the feasible set in Figure 11. Explain why the objective function $ax + by$, with a and b positive, must have its maximum value at point E.

SOLUTIONS TO PRACTICE PROBLEMS 3

1. Amount invested in bonds $= y$. Five million dollars more than the amount invested in Treasury notes is $x + 5$. Therefore, $y \leq x + 5$.

2. Amount invested in stocks $= 30 - (x + y)$. Amount invested in bonds $= y$. Therefore,

$$30 - (x + y) + y \leq 25$$
$$30 - x \leq 25$$
$$x \geq 5.$$

3. The feasible set stays the same but the return becomes

$$[\text{return}] = .08x + .08y + .09[30 - (x + y)]$$
$$= .08x + .08y + 2.7 - .09x - .09y$$
$$= 2.7 - .01x - .01y.$$

When the return is evaluated at each of the vertices of the feasible set, the greatest return is achieved at two vertices. Either of these vertices yields an optimum solution.

	$2.7 - .01x - .01y$
(5, 10)	2.55
(12, 3)	2.55
(24, 3)	2.43
(9, 18)	2.43

4. (a) The values of x and y for which the cost is c dollars satisfy $5x + 10y = c$. The standard form of this linear equation is $y = -\frac{1}{2}x + c/10$.

   (b) The line $y = -\frac{1}{2}x + c/10$ has slope $-\frac{1}{2}$ and y-intercept $(0, c/10)$. As c increases, the slope stays the same, but the y-intercept moves up. Therefore, the line moves up.

   (c) The line of constant cost $20 does not contain any points of the feasible set, so such a low cost cannot be achieved. Increase the cost until the line of constant cost just touches the feasible set. As c increases, the line moves up (keeping the same slope) and first touches the feasible set at vertex C. Therefore, taking x and y to be the coordinates of C yields the minimum cost.

Chapter 3: CHECKLIST

☐ Objective function
☐ Fundamental theorem of linear programming
☐ Four-step procedure for solving linear programming problems

Chapter 3: SUPPLEMENTARY EXERCISES

1. Terrapin Airlines wants to fly 1400 members of a ski club to Colorado. The airline owns two types of planes. Type A can carry only 50 passengers, requires three stewards, and costs $14,000 for the trip. Type B can carry 300 passengers, requires four stewards, and

costs $90,000 for the trip. If the airline must use at least as many type A planes as type B and has available only 42 stewards, how many planes of each type should be used to minimize the cost for the trip?

2. A nutritionist is designing a new breakfast cereal using wheat germ and enriched oat flour as the basic ingredients. Each ounce of wheat germ contains 2 milligrams of niacin, 3 milligrams of iron, .5 milligram of thiamin, and costs 3 cents. Each ounce of enriched oat flour contains 3 milligrams of niacin, 3 milligrams of iron, .25 milligram of thiamin, and costs 4 cents. The nutritionist wants the cereal to have at least 7 milligrams of niacin, 8 milligrams of iron, and 1 milligram of thiamin. How many ounces of wheat germ and how many ounces of enriched oat flour should be used in each serving in order to meet the nutritional requirements at the least cost?

3. An automobile manufacturer makes hardtops and sports cars. Each hardtop requires 8 man-hours to assemble, 2 man-hours to paint, 2 man-hours to upholster, and is sold for a profit of $90. Each sports car requires 18 man-hours to assemble, 2 man-hours to paint, 1 man-hour to upholster, and is sold for a profit of $100. During each day 360 man-hours are available to assemble, 50 man-hours to paint, and 40 man-hours to upholster automobiles. How many hardtops and sports cars should be produced each day in order to maximize the profit?

4. A confectioner makes two raisin-nut mixtures. A box of mixture A contains 6 ounces of peanuts, 1 ounce of raisins, 4 ounces of cashews, and sells for 50 cents. A box of mixture B contains 12 ounces of peanuts, 3 ounces of raisins, 2 ounces of cashews, and sells for 90 cents. He has available 5400 ounces of peanuts, 1200 ounces of raisins, and 2400 ounces of cashews. How many boxes of each mixture should he make in order to maximize revenue?

5. A textbook publisher puts out 72 new books each year, which are classified as elementary, intermediate, and advanced. The company's policy for new books is to publish at least four advanced books, at least three times as many elementary books as intermediate books, and at least twice as many intermediate books as advanced books. On the average, the annual profits are $8000 for each elementary book, $7000 for each intermediate book, and $1000 for each advanced book. How many new books of each type should be published in order to maximize the annual profit while conforming to company policy?

4

THE SIMPLEX METHOD

4.1 Slack Variables and the Simplex Tableau

4.2 The Simplex Method, I—Maximum Problems

4.3 The Simplex Method, II—Minimum Problems

4.4 Marginal Analysis and Matrix Formulations of Linear Programming Problems

4.5 Duality

In Chapter 3 we introduced a graphical method for solving linear programming problems. This method, although very simple, is of limited usefulness, since it applies only to problems which involve (or can be reduced to) two variables. On the other hand, linear programming applications in business and economics can involve dozens or even hundreds of variables. In this chapter we describe a method for handling such applications. This method, called the *simplex method* (or *simplex algorithm*), was developed by the mathematician George B. Dantzig in the late 1940s and today is the principal method used in solving complex linear programming problems. The simplex method can be used for problems in any number of variables and is easily adapted to computer calculations.

4.1. Slack Variables and the Simplex Tableau

In this section and the next, we explain how the simplex method can be used to solve linear programming problems. Let us reconsider the furniture manufacturing problem of Chapter 3. You may recall that the problem is to determine the number of chairs and the number of sofas that should be produced each day in order to maximize the profit. The requirements and availability of resources for carpentry, finishing, and upholstery determine the constraints on the production schedule. Thus, we try to find numbers x and y for which

$$80x + 70y$$

is as large as possible subject to the constraints:

$$\begin{cases} 6x + 3y \leq 96 \\ x + y \leq 18 \\ 2x + 6y \leq 72 \\ x \geq 0, \quad y \geq 0. \end{cases}$$

Here x is the number of chairs to be produced each day and y is the number of sofas to be produced each day.

This problem exhibits certain features that make it particularly convenient to work with.

1. The objective function is to be maximized.
2. Each variable is constrained to be ≥ 0.
3. All other constraints are of the form

$$\text{[linear polynomial]}^* \leq \text{[nonnegative constant]}.$$

A linear programming problem satisfying these conditions is said to be in *standard form*. Our initial discussion of the simplex method will involve only such problems. Then, in Section 3, we will consider problems in nonstandard form.

---

* A linear polynomial is an expression of the form $ax + by + cz + \cdots + dw$, where $a, b, c, \ldots, d$ are specific numbers and $x, y, z, \ldots, w$ are variables. Some examples are $2x - 3y + z, x + 2y + 3z - 4w$, and $-x + 3z - 2w$.

The essential feature of the simplex method is that it provides a systematic method of testing selected vertices of the feasible set until an optimal vertex is reached. The method usually begins at the origin, if it is in the feasible set, and then considers the adjacent vertex that most improves the value of the objective function. This process continues until the optimal vertex is found.

The first step of the simplex method is to convert the given linear programming problem into a system of linear *equations*. To see how this is done, consider the furniture manufacturing problem. It specifies that the variables x and y are subject to the constraint

$$6x + 3y \leq 96.$$

Let us introduce another variable, u, which turns the inequality into an equation:

$$6x + 3y + u = 96.$$

The variable u "takes up the slack" between $6x + 3y$ and 96 and is therefore called a *slack variable*. Moreover, since $6x + 3y$ is at most 96, the variable u must be ≥ 0. In a similar way the constraint

$$x + y \leq 18$$

can be turned into the equation

$$x + y + v = 18,$$

where v is a slack variable and $v \geq 0$. The third constraint

$$2x + 6y \leq 72$$

becomes the equation

$$2x + 6y + w = 72$$

where w is also a slack variable and $w \geq 0$. Let us even turn our objective function $80x + 70y$ into an equation by introducing the new variable M defined by $M = 80x + 70y$. Then M is the variable we want to maximize. Moreover, it satisfies the equation

$$-80x - 70y + M = 0.$$

Thus, the furniture manufacturing problem can be restated in terms of a system of linear equations as discussed below.

Furniture Manufacturing Problem

Among all the solutions of the system of linear equations

$$\left\{ \begin{array}{rcl} 6x + 3y + u & = & 96 \\ x + y + v & = & 18 \\ 2x + 6y + w & = & 72 \\ -80x - 70y + M & = & 0 \end{array} \right.$$

find one for which $x \geq 0$, $y \geq 0$, $u \geq 0$, $v \geq 0$, $w \geq 0$, and for which M is as large as possible.

In a similar way, any linear programming problem in standard form can be reduced to that of determining a certain type of solution of a system of linear equations.

EXAMPLE 1   Formulate the following linear programming problem in terms of a system of linear equations.

Maximize the objective function $3x + 4y$ subject to the constraints

$$\begin{cases} x + y \le 20 \\ x + 2y \le 25 \\ x \ge 0 \\ y \ge 0. \end{cases}$$

Solution   The two constraints $x + y \le 20$ and $x + 2y \le 25$ yield the equations

$$x + y + u \quad\quad = 20$$
$$x + 2y \quad\quad + v = 25$$

where u and v are slack variables and $u \ge 0$ and $v \ge 0$. The objective function gives the equation $M = 3x + 4y$ or

$$-3x - 4y + M = 0.$$

So the problem can be reformulated: Among all the solutions of the system of linear equations

$$\begin{cases} x + y + u \quad\quad\quad\quad = 20 \\ x + 2y \quad + v \quad\quad = 25 \\ -3x - 4y \quad\quad\quad + M = 0, \end{cases}$$

find one for which $x \ge 0$, $y \ge 0$, $u \ge 0$, $v \ge 0$, and M is as large as possible.

EXAMPLE 2   Formulate the following linear programming problem in terms of a system of linear equations.

Maximize the objective function $x + 2y + z$ subject to the constraints

$$\begin{cases} x - y + 2z \le 10 \\ 2x + y + 3z \le 12 \\ x \ge 0, \quad y \ge 0, \quad z \ge 0. \end{cases}$$

Solution   The two constraints $x - y + 2z \le 10$ and $2x + y + 3z \le 12$ yield the equations

$$x - y + 2z + u \quad\quad = 10$$
$$2x + y + 3z \quad\quad + v = 12.$$

The objective function yields the equation $M = x + 2y + z$—that is,

$$-x - 2y - z + M = 0.$$

So the problem can be reformulated: Among all solutions of the system of linear equations

$$\begin{cases} x - y + 2z + u & = 10 \\ 2x + y + 3z \quad\; + v & = 12 \\ -x - 2y - z \qquad\qquad + M = 0, \end{cases}$$

find one for which $x \geq 0, y \geq 0, z \geq 0, u \geq 0, v \geq 0$, and M is as large as possible.

We shall now discuss a scheme for solving systems of equations like those just encountered. For the moment, we will not worry about maximizing M or keeping the variables ≥ 0. Rather, let us concentrate on a particular method for determining solutions. In order to be concrete, consider the system of linear equations from the furniture manufacturing problem:

$$\begin{cases} 6x + 3y + u & = 96 \\ x + y \quad\; + v & = 18 \\ 2x + 6y \qquad\; + w & = 72 \\ -80x - 70y \qquad\qquad + M = 0. \end{cases} \tag{1}$$

This system of equations has an infinite number of solutions. We can rewrite the equations as

$$u = 96 - 6x - 3y$$
$$v = 18 - x - y$$
$$w = 72 - 2x - 6y$$
$$M = \quad\quad 80x + 70y.$$

Given any values of x and y, we can determine corresponding values for u, v, w, and M. For example, if $x = 0$ and $y = 0$, then $u = 96, v = 18, w = 72$, and $M = 0$. These values for u, v, w, and M are precisely the numbers that appear to the right of the equality signs in our system of linear equations. Therefore, this particular solution could have been read off directly from system (1) without any computation. This method of generating solutions is used in the simplex method, so let us explore further the special properties of the system which allowed us to read off a specific solution so easily.

Note that the system of linear equations has six variables: x, y, u, v, w, and M. These variables can be divided into two groups. Group I consists of those which were set equal to 0, namely, x and y. Group II consists of those whose particular values were read off from the right-hand sides of the equations, namely, u, v, w, and M. Note also that the system has a special form which allows the particular values of the group II variables to be read off: Each of the equations involves exactly one of the group II variables, and these variables always appear with coefficient 1. Thus, for example, the first equation involves the group II variable u:

$$6x + 3y + u = 96.$$

Therefore, when all group I variables (x and y) are set equal to 0, only the u term remains on the left, and the particular value of u can be read off from the right-hand side.

The special form of the system can best be described in matrix form. Write the system in the usual way as a matrix, but add column headings corresponding to the variables:

$$
\begin{array}{cccccc}
x & y & u & v & w & M \\
\end{array}
$$

$$
\left[
\begin{array}{cccccc|c}
6 & 3 & 1 & 0 & 0 & 0 & 96 \\
1 & 1 & 0 & 1 & 0 & 0 & 18 \\
2 & 6 & 0 & 0 & 1 & 0 & 72 \\
-80 & -70 & 0 & 0 & 0 & 1 & 0
\end{array}
\right].
$$

Note closely the columns corresponding to the group II variables u, v, w, and M:

$$
\begin{array}{cccccc}
x & y & u & v & w & M \\
\end{array}
$$

$$
\left[
\begin{array}{cccccc|c}
6 & 3 & 1 & 0 & 0 & 0 & 96 \\
1 & 1 & 0 & 1 & 0 & 0 & 18 \\
2 & 6 & 0 & 0 & 1 & 0 & 72 \\
-80 & -70 & 0 & 0 & 0 & 1 & 0
\end{array}
\right].
$$

The presence of these columns gives the system the special form discussed above. Indeed, the u column asserts that u appears only in the first equation and its coefficient there is 1, and similarly for the v, w, and M columns.

The property of allowing us to read off a particular solution from the right-hand column is shared by all linear systems whose matrices contain the columns

$$
\begin{array}{cccc}
1 & 0 & 0 & \cdots & 0 \\
0 & 1 & 0 & \cdots & 0 \\
0 & 0 & 1 & \cdots & 0 \\
\vdots & \vdots & \vdots & & \vdots \\
0 & 0 & 0 & & 1.
\end{array}
$$

(These columns need not appear in exactly the order shown.) The variables corresponding to these columns are called the group II variables. The group I variables consist of all the others. To get one particular solution to the system, set all the group I variables equal to zero and read off the values of the group II variables from the right-hand side of the system. This procedure is illustrated in the following example.

EXAMPLE 3   Determine by inspection one set of solutions to each of these systems of linear equations:

(a) $\begin{cases} x - 5y + u & = 3 \\ -2x + 8y + v & = 11 \\ -\frac{1}{2}x + M = 0 \end{cases}$ (b) $\begin{cases} -y + 2u + v & = 12 \\ x + \frac{1}{2}y - 6u & = -1 \\ 3y + 8u + M = 4 \end{cases}$

Solution (a) The matrix of the system is

$$
\begin{array}{ccccc}
x & y & u & v & M \\
\end{array}
$$
$$
\left[\begin{array}{ccccc|c}
1 & -5 & 1 & 0 & 0 & 3 \\
-2 & 8 & 0 & 1 & 0 & 11 \\
-\frac{1}{2} & 0 & 0 & 0 & 1 & 0
\end{array}\right].
$$

We look for each variable whose column contains one entry of 1 and all the other entries 0.

$$
\begin{array}{ccccc}
x & y & u & v & M \\
\end{array}
$$
$$
\left[\begin{array}{ccccc|c}
1 & -5 & 1 & 0 & 0 & 3 \\
-2 & 8 & 0 & 1 & 0 & 11 \\
-\frac{1}{2} & 0 & 0 & 0 & 1 & 0
\end{array}\right].
$$

The group II variables should be u, v, M, with x, y as the group I variables. Set all group I variables equal to 0. The corresponding values of the group II variables may then be read off the last column: $u = 3$, $v = 11$, $M = 0$. So one solution of the system is

$$x = 0, \qquad y = 0, \qquad u = 3, \qquad v = 11, \qquad M = 0.$$

(b) The matrix of the system is

$$
\begin{array}{ccccc}
x & y & u & v & M \\
\end{array}
$$
$$
\left[\begin{array}{ccccc|c}
0 & -1 & 2 & 1 & 0 & 12 \\
1 & \frac{1}{2} & -6 & 0 & 0 & -1 \\
0 & 3 & 8 & 0 & 1 & 4
\end{array}\right].
$$

The shaded columns show that the group II variables should be v, x, M, with y, u, as the group I variables. So the corresponding solution is

$$x = -1, \qquad y = 0, \qquad u = 0, \qquad v = 12, \qquad M = 4.$$

A *simplex tableau* is a matrix (corresponding to a linear system) in which each of the columns

$$
\begin{array}{cccc}
1 & 0 & \cdots & 0 \\
0 & 1 & \cdots & 0 \\
\vdots & \vdots & & \vdots \\
0 & 0 & & 1
\end{array}
$$

is present (in some order) to the left of the vertical line. We have seen how to construct a simplex tableau corresponding to a linear programming problem in standard form. From this initial simplex tableau we can read off one particular solution of the linear system by using the method described above. This particular solution may or may not correspond to the solution of the original optimization problem. If it does not, we replace the initial tableau with another one whose corresponding solution is "closer" to the optimum. How do we replace the initial simplex tableau with another? Just pivot it about a nonzero entry! Indeed, one of the key reasons the simplex method works is that pivoting transforms one simplex tableau into another. Note also that since pivoting consists of elementary row operations, the solution corresponding to a transformed tableau is a solution of the original linear system. The next example illustrates how pivoting transforms a tableau into another one.

EXAMPLE 4 Consider the simplex tableau obtained from the furniture manufacturing problem:

$$
\begin{array}{cccccc}
x & y & u & v & w & M \\
\end{array}
$$
$$
\left[\begin{array}{cccccc|c}
⑥ & 3 & 1 & 0 & 0 & 0 & 96 \\
1 & 1 & 0 & 1 & 0 & 0 & 18 \\
2 & 6 & 0 & 0 & 1 & 0 & 72 \\
-80 & -70 & 0 & 0 & 0 & 1 & 0
\end{array}\right]
$$

(a) Pivot this tableau around the circled entry, 6.

(b) Calculate the particular solution corresponding to the transformed tableau which results from setting the new group I variables equal to 0.

Solution (a) The first step in pivoting is to replace the pivot element 6 by a 1. To do this, multiply the first row of the tableau by $\frac{1}{6}$ to get

$$
\begin{array}{cccccc}
x & y & u & v & w & M \\
\end{array}
$$
$$
\left[\begin{array}{cccccc|c}
1 & \frac{1}{2} & \frac{1}{6} & 0 & 0 & 0 & 16 \\
1 & 1 & 0 & 1 & 0 & 0 & 18 \\
2 & 6 & 0 & 0 & 1 & 0 & 72 \\
-80 & -70 & 0 & 0 & 0 & 1 & 0
\end{array}\right].
$$

Next we must replace all nonpivot elements in the first column by zeros. Do this by adding to the second row (-1) times the first row:

$$
\begin{array}{cccccc}
x & y & u & v & w & M \\
\end{array}
$$
$$
\left[\begin{array}{cccccc|c}
1 & \frac{1}{2} & \frac{1}{6} & 0 & 0 & 0 & 16 \\
0 & \frac{1}{2} & -\frac{1}{6} & 1 & 0 & 0 & 2 \\
2 & 6 & 0 & 0 & 1 & 0 & 72 \\
-80 & -70 & 0 & 0 & 0 & 1 & 0
\end{array}\right].
$$

and by adding to the third row (-2) times the first row:

$$
\begin{array}{cccccc}
x & y & u & v & w & M \\
\end{array}
$$
$$
\left[\begin{array}{cccccc|c}
1 & \frac{1}{2} & \frac{1}{6} & 0 & 0 & 0 & 16 \\
0 & \frac{1}{2} & -\frac{1}{6} & 1 & 0 & 0 & 2 \\
0 & 5 & -\frac{1}{3} & 0 & 1 & 0 & 40 \\
-80 & -70 & 0 & 0 & 0 & 1 & 0
\end{array}\right].
$$

and, finally, by adding to the fourth row 80 times the first row:

$$
\begin{array}{cccccc}
x & y & u & v & w & M \\
\end{array}
$$

$$
\begin{bmatrix}
1 & \frac{1}{2} & \frac{1}{6} & 0 & 0 & 0 & 16 \\
0 & \frac{1}{2} & -\frac{1}{6} & 1 & 0 & 0 & 2 \\
0 & 5 & -\frac{1}{3} & 0 & 1 & 0 & 40 \\
0 & -30 & \frac{40}{3} & 0 & 0 & 1 & 1280
\end{bmatrix}.
$$

Note that we indeed get a new simplex tableau. The new group II variables are x, v, w, and M. The group I variables are y and u.

$$
\begin{array}{cccccc}
x & y & u & v & w & M \\
\end{array}
$$

$$
\begin{bmatrix}
1 & \frac{1}{2} & \frac{1}{6} & 0 & 0 & 0 & 16 \\
0 & \frac{1}{2} & -\frac{1}{6} & 1 & 0 & 0 & 2 \\
0 & 5 & -\frac{1}{3} & 0 & 1 & 0 & 40 \\
0 & -30 & \frac{40}{3} & 0 & 0 & 1 & 1280
\end{bmatrix}.
$$

(b) Set the group I variables equal to 0:

$$y = 0, \qquad u = 0.$$

Read off the particular values of the group II variables from the right-hand column:

$$x = 16, \qquad v = 2, \qquad w = 40, \qquad M = 1280.$$

So the particular solution corresponding to the transformed tableau is

$$x = 16, \qquad y = 0, \qquad u = 0, \qquad v = 2, \qquad w = 40, \qquad M = 1280.$$

We see that the simplex tableau leads to the vertex $(16, 0)$ that was listed and tested in the graphical solution of the problem presented in Chapter 3.

PRACTICE PROBLEMS 1

1. Determine by inspection a particular solution of the following system of linear equations:

$$
\begin{cases}
x + 2y + 3u & = 6 \\
y & + v & = 4 \\
5y + 2u & + M = 0.
\end{cases}
$$

2. Pivot the simplex tableau about the circled element:

$$
\begin{bmatrix}
2 & 4 & 1 & 0 & 0 & 6 \\
3 & \textcircled{1} & 0 & 1 & 0 & 0 \\
1 & 1 & 0 & 0 & 1 & 1
\end{bmatrix}.
$$

EXERCISES 1

For each of the following linear programming problems, determine the corresponding linear system and restate the linear programming problem in terms of the linear system.

1. Maximize $8x + 13y$ subject to the constraints

$$\begin{cases} 20x + 30y \leq 3500 \\ 50x + 10y \leq 5000 \\ x \geq 0 \\ y \geq 0. \end{cases}$$

2. Maximize $x + 15y$ subject to the constraints

$$\begin{cases} 3x + 2y \leq 10 \\ x \quad\quad \leq 15 \\ \quad\quad y \leq 3 \\ x + \ y \leq 5 \\ x \geq 0 \\ y \geq 0. \end{cases}$$

3. Maximize $x + 2y - 3z$ subject to the constraints

$$\begin{cases} x + \quad y + z \leq 100 \\ 3x \quad\quad + z \leq 200 \\ 5x + 10y \quad\quad \leq 100 \\ x \geq 0 \\ y \geq 0 \\ z \geq 0. \end{cases}$$

4. Maximize $2x + y + 50$ subject to the constraints

$$\begin{cases} x + 3y \leq 24 \\ y \leq 5 \\ x + 7y \leq 10 \\ x \geq 0 \\ y \geq 0. \end{cases}$$

5-8. For each of the linear programming problems in Exercises 1–4:

   (a) Set up the simplex tableau.
   (b) Determine the particular solution corresponding to the tableau.

Find the particular solutions corresponding to these tableaux.

9.

x	y	u	v	M	
0	2	1	0	0	10
1	3	0	12	0	15
0	−1	0	17	1	20

10.

x	y	u	v	M	
1	0	3	11	0	6
0	1	10	17	0	16
0	0	5	-1	1	3

11.

x	y	z	u	v	w	M	
0	3	1	0	1	15	0	15
1	-1	0	0	2	-5	0	10
0	2	0	1	-5	4	0	23
0	11	0	0	11	6	1	-11

12.

x	y	z	u	v	w	M	
6	0	1	0	5	-1	0	$\frac{1}{4}$
5	1	0	0	3	$\frac{1}{3}$	0	100
4	0	0	1	8	$\frac{1}{2}$	0	11
2	0	0	0	6	$\frac{1}{7}$	1	$-\frac{1}{2}$

13. Pivot the simplex tableau

x	y	u	v	M	
2	3	1	0	0	12
①	1	0	1	0	10
-10	-20	0	0	1	0

about the indicated element and compute the particular solution corresponding to the new tableau.

(a) 2   (b) 3

(c) 1 (second row, first column)   (d) 1 (second row, second column).

14. Pivot the simplex tableau

x	y	u	v	M	
5	4	1	0	0	100
10	6	0	1	0	1200
-1	2	0	0	1	0

about the indicated element and compute the solution corresponding to the new tableau.

(a) 5   (b) 4   (c) 10   (d) 6.

15. Determine which of the pivot operations in Exercise 13 increases M the most.

16. Determine which of the pivot operations in Exercise 14 increases M the most.

SOLUTIONS TO PRACTICE PROBLEMS 1

1. The matrix of the system is

$$\begin{array}{ccccc} x & y & u & v & M \\ \begin{bmatrix} 1 & 2 & 3 & 0 & 0 \\ 0 & 1 & 0 & 1 & 0 \\ 0 & 5 & 2 & 0 & 1 \end{bmatrix} & \begin{array}{c} 6 \\ 4 \\ 0 \end{array} \end{array},$$

from which we see that the group II variables are x, v, M, and the group I variables y, u. To obtain a solution, we set the group I variables equal to 0. We obtain from the first equation that $x = 6$, from the second that $v = 4$, and from the third that $M = 0$. Thus a solution of the system is $x = 6$, $y = 0$, $u = 0$, $v = 4$, $M = 0$.

2. We must use elementary row operations to transform the second column into

$$\begin{bmatrix} 0 \\ 1 \\ 0 \end{bmatrix}.$$

$$\begin{bmatrix} 2 & 4 & 1 & 0 & 0 \\ 3 & 1 & 0 & 1 & 0 \\ 1 & 1 & 0 & 0 & 1 \end{bmatrix} \begin{array}{c} 6 \\ 0 \\ 1 \end{array} \xrightarrow{[1] + (-4)[2]} \begin{bmatrix} -10 & 0 & 1 & -4 & 0 \\ 3 & 1 & 0 & 1 & 0 \\ 1 & 1 & 0 & 0 & 1 \end{bmatrix} \begin{array}{c} 6 \\ 0 \\ 1 \end{array}$$

$$\xrightarrow{[3] + (-1)[2]} \begin{bmatrix} -10 & 0 & 1 & -4 & 0 \\ 3 & 1 & 0 & 1 & 0 \\ -2 & 0 & 0 & -1 & 1 \end{bmatrix} \begin{array}{c} 6 \\ 0 \\ 1 \end{array}.$$

4.2. The Simplex Method, I—Maximum Problems

We can now describe the simplex method for solving linear programming problems. The procedure will be illustrated as we solve the furniture manufacturing problem of Section 4.1. Recall that we must maximize the objective function $80x + 70y$ subject to the constraints

$$\begin{cases} 6x + 3y \le 96 \\ x + y \le 18 \\ 2x + 6y \le 72 \\ x \ge 0, \quad y \ge 0. \end{cases}$$

> *Step 1* Introduce slack variables and state the problem in terms of a system of linear equations.

We carried out this step in the preceding section. The result was the following restatement of the problem.

Among all the solutions of the system of linear equations

$$\begin{cases} 6x + 3y + u & = 96 \\ x + y + v & = 18 \\ 2x + 6y + w & = 72 \\ -80x - 70y + M = 0 \end{cases}$$

find one for which $x \geq 0$, $y \geq 0$, $u \geq 0$, $v \geq 0$, $w \geq 0$, and for which M is as large as possible.

Step 2 Construct the simplex tableau corresponding to the linear system.

This step was also carried out in Section 4.1. The tableau is

	x	y	u	v	w	M	
u	6	3	1	0	0	0	96
v	1	1	0	1	0	0	18
w	2	6	0	0	1	0	72
M	-80	-70	0	0	0	1	0

Note that we have made two additions to the previously found tableau. First, we have separated the last row from the others by means of a horizontal line. This is because the last row, which corresponds to the objective function in the original problem, will play a special role in what follows. The second addition is that we have labeled each row with one of the group II variables—namely, the variable whose value is determined by the row. Thus, for example, the first row gives the particular value of u, which is 96, so the row is labeled with a u. We will find these labels convenient.

Corresponding to this tableau, there is a particular solution to the linear system, namely, the one obtained by setting all group I variables equal to 0. Reading the values of the group II variables from the last column, we obtain

$$x = 0, \quad y = 0, \quad u = 96, \quad v = 18, \quad w = 72, \quad M = 0.$$

Our object is to make M as large as possible. How can the value of M be increased? Look at the equation corresponding to the last row of the tableau. It reads

$$-80x - 70y + M = 0.$$

Note that two of the coefficients, -80 and -70, are negative. Or, what amounts to the same thing, if we solve for M and get

$$M = 80x + 70y,$$

then the coefficients on the right-hand side are *positive*. This fact is significant. It says that M can be increased by increasing either the value of x or the value of

y. A unit change in x will increase M by 80 units, whereas a unit change in y will increase M by 70 units. And since we wish to increase M by as much as possible, it is reasonable to attempt to increase the value of x. Let us indicate this by drawing an arrow pointing to the x column of the tableau:

$$
\begin{array}{c}
\begin{array}{ccccccc}
 & x & y & u & v & w & M
\end{array} \\
\begin{array}{c} u \\ v \\ w \\ M \end{array}
\left[
\begin{array}{cccccc|c}
6 & 3 & 1 & 0 & 0 & 0 & 96 \\
1 & 1 & 0 & 1 & 0 & 0 & 18 \\
2 & 6 & 0 & 0 & 1 & 0 & 72 \\
\hline
-80 & -70 & 0 & 0 & 0 & 1 & 0
\end{array}
\right]
\end{array}
\qquad (1)
$$

$\uparrow$

In order to increase x (from its present value, zero), we will pivot about one of the entries (above the horizontal line) in the x column. In this way, x will become a group II variable and hence will not necessarily be zero in our next particular solution. But which entry should we pivot around? To find out, let us experiment. The results from pivoting about the 6, the 1, and the 2 in the x column are, respectively:

$$
\begin{array}{c}
\begin{array}{ccccccc}
 & x & y & u & v & w & M
\end{array} \\
\begin{array}{c} x \\ v \\ w \\ M \end{array}
\left[
\begin{array}{cccccc|c}
1 & \frac{1}{2} & \frac{1}{6} & 0 & 0 & 0 & 16 \\
0 & \frac{1}{2} & -\frac{1}{6} & 1 & 0 & 0 & 2 \\
0 & 5 & -\frac{1}{3} & 0 & 1 & 0 & 40 \\
\hline
0 & -30 & \frac{40}{3} & 0 & 0 & 1 & 1280
\end{array}
\right]
\end{array}
$$
$$\text{Pivot about 6}$$

$$
\begin{array}{c}
\begin{array}{ccccccc}
 & x & y & u & v & w & M
\end{array} \\
\begin{array}{c} u \\ x \\ w \\ M \end{array}
\left[
\begin{array}{cccccc|c}
0 & -3 & 1 & -6 & 0 & 0 & -12 \\
1 & 1 & 0 & 1 & 0 & 0 & 18 \\
0 & 4 & 0 & -2 & 1 & 0 & 36 \\
\hline
0 & 10 & 0 & 80 & 0 & 1 & 1440
\end{array}
\right]
\end{array}
$$
$$\text{Pivot about 1}$$

$$
\begin{array}{c}
\begin{array}{ccccccc}
 & x & y & u & v & w & M
\end{array} \\
\begin{array}{c} u \\ v \\ x \\ M \end{array}
\left[
\begin{array}{cccccc|c}
0 & -15 & 1 & 0 & -3 & 0 & -120 \\
0 & -2 & 0 & 1 & -\frac{1}{2} & 0 & -18 \\
1 & 3 & 0 & 0 & \frac{1}{2} & 0 & 36 \\
\hline
0 & 170 & 0 & 0 & 40 & 1 & 2880
\end{array}
\right]
\end{array}
$$
$$\text{Pivot about 2}$$

Note that the labels on the rows have changed because the group II variables are now different. The solutions corresponding to these tableaux are, respectively:

$x = 16$,	$y = 0$,	$u = 0$,	$v = 2$,	$w = 40$,	$M = 1280$,
$x = 18$,	$y = 0$,	$u = -12$,	$v = 0$,	$w = 36$,	$M = 1440$,
$x = 36$,	$y = 0$,	$u = -120$,	$v = -18$,	$w = 0$,	$M = 2880$.

The second and third solutions violate the requirement that all variables be ≥ 0. Thus, we use the first solution in which we pivoted about 6. Using this solution, we have increased the value of M to 1280 and have replaced our original tableau by

	x	y	u	v	w	M	
x	1	$\frac{1}{2}$	$\frac{1}{6}$	0	0	0	16
v	0	$\frac{1}{2}$	$-\frac{1}{6}$	1	0	0	2
w	0	5	$-\frac{1}{3}$	0	1	0	40
M	0	-30	$\frac{40}{3}$	0	0	1	1280

Can M be increased further? To answer this question, look at the last row of the tableau, corresponding to the equation

$$-30y + \tfrac{40}{3}u + M = 1280.$$

There is a negative coefficient for the variable y in this equation. Correspondingly, when the equation is solved for M, there is a positive coefficient for y:

$$M = 1280 + 30y - \tfrac{40}{3}u.$$

Now it is clear we should try to increase y. So we pivot about one of the entries in the y column. A calculation for each of the possible pivots shows that pivoting about the first or the third entries leads to solutions having some negative values. Therefore, we pivot about the second entry in the y column. The result is

	x	y	u	v	w	M	
x	1	0	$\frac{1}{3}$	-1	0	0	14
y	0	1	$-\frac{1}{3}$	2	0	0	4
w	0	0	$\frac{4}{3}$	-10	1	0	20
M	0	0	$\frac{10}{3}$	60	0	1	1400

The corresponding solution is

$$x = 14, \qquad y = 4, \qquad u = 0, \qquad v = 0, \qquad w = 20, \qquad M = 1400.$$

Note that with this pivot operation we have increased M from 1280 to 1400.

Can we increase M any further? Let us reason as before. Use the last row of the current tableau to write M in terms of the other variables:

$$\tfrac{10}{3}u + 60v + M = 1400, \qquad M = -\tfrac{10}{3}u - 60v.$$

Note, however, that in contrast to the previous expressions for M, this one has *no positive coefficients*. And since u and v are ≥ 0, this means that M can be *at most* 1400. But M is already 1400. So M cannot be increased further. Thus we have shown that the maximum value of M is 1400, and this occurs when $x = 14$ and $y = 4$. Thus, to maximize profits, the furniture manufacturer should be making 14 chairs and 4 sofas each day. The maximum profit is \$1400. From the tableau we can read off the values of the slack variables: $u = 0$, $v = 0$, and $w = 20$. This shows that we have no slack resulting from the first inequality, so we have used all the man-hours available for carpentry. Similarly, since $v = 0$, we have used all of the man-hours available for finishing. But since $w = 20$, we have 20 man-hours of upholstery remaining when we manufacture the optimal number of chairs and sofas.

Let us compare the simplex method solution of the furniture manufacturing problem with the geometric solution carried out in Chapter 3. Both solutions yielded the same optimal production schedule. In the geometric solution, we found *all* of the vertices of the feasible set and then evaluated the objective function at every one of these vertices. The following table was obtained:

Vertex	Profit $= 80x + 70y$
(14, 4)	$80(14) + 70(4) = 1400$
(9, 9)	$80(9) + 70(9) = 1350$
(0, 12)	$80(0) + 70(12) = 840$
(0, 0)	$80(0) + 70(0) = 0$
(16, 0)	$80(16) + 70(0) = 1280$

We selected the optimal solution ($x = 14$, $y = 4$) because it produced the greatest profit.

With the simplex method, we only had to consider *some* of the vertices. In the initial tableau we first considered the vertex $(0, 0)$—that is, both x and y were 0. M was also 0. In the second tableau we looked at the vertex $(16, 0)$—that is $x = 16$ and $y = 0$, and the tableau showed that $M = 1280$. Finally, as a result of the last pivot operation, we came to the vertex $(14, 4)$. This meant that $x = 14$ and $y = 4$. The value of the objective function was read from the tableau: $M = 1400$. Since we could not increase M any more, we did not have to consider any other vertices. In larger linear programming problems, the time saved from looking at just *some* of the vertices, rather than *all* of the vertices, can be substantial.

Based on the discussion above, we can state several general principles. First of all, the following criterion determines when a simplex tableau yields a maximum.

> *Condition for a Maximum* The particular solution derived from a simplex tableau is a maximum if and only if the bottom row contains no negative entries except perhaps the entry in the last column.*

We saw this condition illustrated in the example above. Each of the first two tableaux had negative entries in the last row, and, as we showed, their corresponding solutions were not maxima. However, the third tableau, with no negative entries in the last row, did yield a maximum.

---

* In Section 3 we shall encounter maximum problems whose final tableaux have a negative number in the lower right-hand corner.

The crucial point of the simplex method is the correct choice of a pivot element. In the example above we decided to choose a pivot element from the column corresponding to the most negative entry in the last row. It can be proved that this is the proper choice in general; that is, we have the following rule:

> *Choosing the Pivot Column*   The pivot element should be chosen from that column to the left of the vertical line which has the most negative entry in the last row.*

Choosing the correct pivot element from the designated column is somewhat more complicated. Our approach above was to calculate the tableau associated with each element and observe that only one corresponded to a solution with nonnegative elements. However, there is a simpler way to make the choice. As an illustration, let us reconsider tableau (1). We have already decided to pivot around some entry in the first column. For each *positive* entry in the pivot column we compute a ratio: the corresponding entry in the right-hand column divided by the entry in the pivot column. So for example, for the first entry the ratio is $\frac{96}{6}$, for the second entry the ratio is $\frac{18}{1}$, and for the third entry the ratio is $\frac{72}{2}$. We write these ratios to the right of the matrix as follows:

$$
\begin{array}{c}
 \\
u \\
v \\
w \\
M
\end{array}
\begin{array}{c}
\begin{array}{cccccc}
x & y & u & v & w & M
\end{array} \\
\left[
\begin{array}{cccccc|c}
6 & 3 & 1 & 0 & 0 & 0 & 96 \\
1 & 1 & 0 & 1 & 0 & 0 & 18 \\
2 & 6 & 0 & 0 & 1 & 0 & 72 \\
-80 & -70 & 0 & 0 & 0 & 1 & 0
\end{array}
\right]
\end{array}
\begin{array}{c}
96/6 \\
18/1 \\
72/2 \\
\end{array}
$$

It is possible to prove the following rule, which allows us to determine the pivot element from the above display:

> *Choosing the Pivot Element*   For each positive entry of the pivot column, compute the appropriate ratio. Choose as pivot element the one corresponding to the least nonnegative ratio.

For instance, consider the choice of pivot element in the example above. The least of the ratios is 16. So we choose 6 as the pivot element.

At first, this method for choosing the pivot element might seem very odd. However, it is just a way of guaranteeing that the last column of the new tableau will have entries ≥ 0. And that is just the basis on which we chose the pivot element earlier. To obtain further insight, let us analyze the example above yet further.

---

* In case two or more columns are tied for the honor of being the pivot column, an arbitrary choice among them may be made.

Suppose that we pivot our tableau about the 6 in column 1. The first step in pivoting is to divide the pivot row by the pivot element (in this case 6). This gives the array

$$\begin{array}{cccccc}x & y & u & v & w & M\end{array}$$

$$\begin{bmatrix} 1 & \frac{1}{2} & \frac{1}{6} & 0 & 0 & 0 & \frac{96}{6} \\ 1 & 1 & 0 & 1 & 0 & 0 & 18 \\ 2 & 6 & 0 & 0 & 1 & 0 & 72 \\ \hline -80 & -70 & 0 & 0 & 0 & 1 & 0 \end{bmatrix},$$

where we have written $\frac{96}{6}$ rather than 16 to emphasize that we have divided by the pivot element. The next step in the pivot procedure is to replace the second row by the second row plus (-1 times the first row). The result is

$$\begin{array}{cccccc}x & y & u & v & w & M\end{array}$$

$$\begin{bmatrix} 1 & \frac{1}{2} & \frac{1}{6} & 0 & 0 & 0 & \frac{96}{6} \\ 0 & \frac{1}{2} & -\frac{1}{6} & 1 & 0 & 0 & 18 - \frac{96}{6} \\ 2 & 6 & 0 & 0 & 1 & 0 & 72 \\ \hline -80 & -70 & 0 & 0 & 0 & 1 & 0 \end{bmatrix}.$$

The next step in the pivot process is to replace the third row by $[3] + (-2)[1]$ to obtain

$$\begin{array}{cccccc}x & y & u & v & w & M\end{array}$$

$$\begin{bmatrix} 1 & \frac{1}{2} & \frac{1}{6} & 0 & 0 & 0 & \frac{96}{6} \\ 0 & \frac{1}{2} & -\frac{1}{6} & 1 & 0 & 0 & 18 - \frac{96}{6} \\ 0 & 5 & -\frac{1}{3} & 0 & 1 & 0 & 72 - 2(\frac{96}{6}) \\ \hline -80 & -70 & 0 & 0 & 0 & 1 & 0 \end{bmatrix}.$$

The final step of the pivot process is to replace the fourth row by $[4] + 80[1]$ to obtain

$$\begin{array}{cccccc}x & y & u & v & w & M\end{array}$$

$$\begin{bmatrix} 1 & \frac{1}{2} & \frac{1}{6} & 0 & 0 & 0 & \frac{96}{6} \\ 0 & \frac{1}{2} & -\frac{1}{6} & 1 & 0 & 0 & 18 - \frac{96}{6} \\ 0 & 5 & -\frac{1}{3} & 0 & 1 & 0 & 72 - 2(\frac{96}{6}) \\ \hline 0 & -30 & \frac{40}{3} & 0 & 0 & 1 & 1280 \end{bmatrix}.$$

The entries in the upper part of the right-hand column may be written

$$\frac{96}{6}, \qquad \frac{18}{1} - \frac{96}{6}, \qquad 2(\frac{72}{2} - \frac{96}{6}).$$

If we had pivoted about the 1 or the 2 in the first column of the original tableau, the upper entries in the last column of the tableau would have been

$$6(\frac{96}{6} - \frac{18}{1}), \qquad \frac{18}{1}, \qquad 2(\frac{72}{2} - \frac{18}{1})$$

and

$$6(\frac{96}{6} - \frac{72}{2}), \qquad \frac{18}{1} - \frac{72}{2}, \qquad \frac{72}{2}$$

respectively. Notice that all of the combinations of the differences of the pairs of ratios appear in these triples. In the first case, the ratio $\frac{96}{6}$ is subtracted from each of the other two ratios, whereas in the next two cases, the ratios $\frac{18}{1}$ and $\frac{72}{2}$ are subtracted. In order that the upper entries in the last column be nonnegative, we must subtract off the smallest of the ratios. That is, we should pivot about the entry corresponding to the smallest ratio. This is the rationale governing our choice of pivot element!

Now that we have assembled all the components of the simplex method, we can summarize it as follows:

The Simplex Method for Problems in Standard Form

1. Introduce slack variables and state the problem in terms of a system of linear equations.
2. Construct the simplex tableau corresponding to the system.
3. Determine if the left part of the bottom row contains negative entries. If none are present, the solution corresponding to the tableau yields a maximum and the problem is solved.
4. If the left part of the bottom row contains negative entries, construct a new simplex tableau.
   (a) Choose the pivot column by inspecting the entries of the last row of the current tableau, excluding the right-hand entry. The pivot column is the one containing the most negative of these entries.
   (b) Choose the pivot element by computing ratios associated with the positive entries of the pivot column. The pivot element is the one corresponding to the smallest nonnegative ratio.
   (c) Construct the new simplex tableau by pivoting around the selected element.
5. Return to step 3. Steps 3 and 4 are repeated as many times as necessary to find a maximum.

Let us now work some problems to see how this method is applied.

EXAMPLE 1   Maximize the objective function $10x + y$ subject to the constraints

$$\begin{cases} x + 2y \le 10 \\ 3x + 4y \le 6 \\ x \ge 0, \quad y \ge 0. \end{cases}$$

Solution   The corresponding system of linear equations with slack variables is

$$\begin{cases} x + 2y + u & = 10 \\ 3x + 4y \quad + v & = 6 \\ -10x - y \quad\quad\quad + M = 0, \end{cases}$$

and we must find that solution of the system for which $x \geq 0$, $y \geq 0$, $u \geq 0$, $v \geq 0$, and M is as large as possible. Here is the initial simplex tableau:

$$
\begin{array}{c}
 \\
u \\
v \\
M
\end{array}
\begin{array}{c}
\begin{array}{ccccc}
x & y & u & v & M
\end{array} \\
\left[
\begin{array}{ccccc|c}
1 & 2 & 1 & 0 & 0 & 10 \\
3 & 4 & 0 & 1 & 0 & 6 \\
-10 & -1 & 0 & 0 & 1 & 0
\end{array}
\right].
\end{array}
$$

Note that this tableau does not correspond to a maximum, since the left part of the bottom row has negative entries. So we pivot to create a new tableau. Since -10 is the most negative entry in the last row, we choose the first column as the pivot column. To determine the pivot element, we compute ratios:

$$
\begin{array}{c}
 \\
u \\
v \\
M
\end{array}
\begin{array}{c}
\begin{array}{ccccc}
x & y & u & v & M
\end{array} \hspace{1cm} \text{Ratios} \\
\left[
\begin{array}{ccccc|c}
1 & 2 & 1 & 0 & 0 & 10 \\
③ & 4 & 0 & 1 & 0 & 6 \\
-10 & -1 & 0 & 0 & 1 & 0
\end{array}
\right]
\begin{array}{l}
10/1 = 10 \\
6/3 = 2 \\

\end{array}
\end{array}
$$

The smallest ratio is 2, so we pivot about 3, which we have circled. The new tableau is therefore

$$
\begin{array}{c}
 \\
u \\
x \\
M
\end{array}
\begin{array}{c}
\begin{array}{ccccc}
x & y & u & v & M
\end{array} \\
\left[
\begin{array}{ccccc|c}
0 & \frac{1}{3} & 1 & -\frac{1}{3} & 0 & 8 \\
1 & \frac{4}{3} & 0 & \frac{1}{3} & 0 & 2 \\
0 & \frac{37}{3} & 0 & \frac{10}{3} & 1 & 20
\end{array}
\right].
\end{array}
$$

Note that this tableau corresponds to a maximum, since there are no negative entries in the left part of the last row. The solution corresponding to the tableau is

$$x = 2, \qquad y = 0, \qquad u = 8, \qquad v = 0, \qquad M = 20.$$

Therefore, the objective function assumes its maximum value of 20 when $x = 2$ and $y = 0$.

The simplex method can be used to solve problems in any number of variables. Let us illustrate the method for three variables.

EXAMPLE 2   Maximize the objective function $x + 2y + z$ subject to the constraints

$$
\begin{cases}
x - y + 2z \leq 10 \\
2x + y + 3z \leq 12 \\
x \geq 0, \quad y \geq 0, \quad z \geq 0.
\end{cases}
$$

Solution We determined the corresponding linear system in Example 1 of Section 1:

$$\begin{cases} x - y + 2z + u & = 10 \\ 2x + y + 3z & + v & = 12 \\ -x - 2y - z & + M = 0. \end{cases}$$

So the simplex method works as follows:

$$\begin{array}{c} \begin{array}{ccccccc} & x & y & z & u & v & M \end{array} \\ \begin{array}{c} u \\ v \\ M \end{array} \left[\begin{array}{cccccc|c} 1 & -1 & 2 & 1 & 0 & 0 & 10 \\ 2 & ① & 3 & 0 & 1 & 0 & 12 \\ -1 & -2 & -1 & 0 & 0 & 1 & 0 \end{array} \right] \end{array} \quad 12/1 = 12 \quad \text{(smallest ratio)}$$

$$\begin{array}{c} \begin{array}{ccccccc} & x & y & z & u & v & M \end{array} \\ \begin{array}{c} u \\ y \\ M \end{array} \left[\begin{array}{cccccc|c} 3 & 0 & 5 & 1 & 1 & 0 & 22 \\ 2 & 1 & 3 & 0 & 1 & 0 & 12 \\ 3 & 0 & 5 & 0 & 2 & 1 & 24 \end{array} \right]. \end{array}$$

Thus the solution of the original problem $(x = 0, y = 12, z = 0)$ yields the maximum value of the objective function $x + 2y + z$. The maximum value is 24.

PRACTICE PROBLEMS 2

1. Which of these simplex tableaux has a solution which corresponds to a maximum for the associated linear programming problem?

(a)
$$\begin{array}{c} \begin{array}{cccccc} & x & y & u & v & M \end{array} \\ \left[\begin{array}{ccccc|c} 3 & 1 & 0 & 1 & 0 & 5 \\ 2 & 0 & 0 & 0 & 1 & 0 \\ -1 & -2 & 1 & 0 & 0 & 3 \end{array} \right] \end{array}$$

(b)
$$\begin{array}{c} \begin{array}{cccccc} x & y & u & v & M \end{array} \\ \left[\begin{array}{ccccc|c} 2 & 1 & 0 & 11 & 0 & 10 \\ 1 & 0 & 1 & 7 & 0 & 1 \\ 1 & 0 & 0 & 4 & 1 & -2 \end{array} \right] \end{array}$$

2. Suppose that in the solution of a linear programming problem by the simplex method we encounter the following simplex tableau. What is the next step in the solution?

$$\begin{array}{c} \begin{array}{cccccc} x & y & u & v & M \end{array} \\ \left[\begin{array}{ccccc|c} 0 & 4 & 1 & 2 & 0 & 4 \\ 1 & 5 & 0 & 1 & 0 & 9 \\ 0 & 2 & 0 & -3 & 1 & 6 \end{array} \right] \end{array}$$

EXERCISES 2

For each of the following simplex tableaux:

(a) Compute the next pivot element.

(b) Determine the next tableau.

(c) Determine the particular solution corresponding to the tableau of part (b).

1.
x	y	u	v	M	
6	2	1	0	0	10
1	3	0	1	0	6
-4	-12	0	0	1	0

2.
x	y	u	v	M	
1	0	3	1	0	5
0	1	2	0	0	12
-6	0	5	0	1	10

3.
x	y	u	v	M	
5	12	1	0	0	12
15	10	0	1	0	5
4	-2	0	0	1	0

4.
x	y	u	v	M	
0	6	3	1	0	5
1	-5	2	0	0	8
0	20	-10	0	1	22

Solve the following linear programming problems using the simplex method.

5. Maximize $x + 3y$ subject to the constraints

$$\begin{cases} x + y \le 7 \\ x + 2y \le 10 \\ x \ge 0, \quad y \ge 0. \end{cases}$$

6. Maximize $x + 2y$ subject to the constraints

$$\begin{cases} -x + y \le 100 \\ 6x + 6y \le 1200 \\ x \ge 0, \quad y \ge 0. \end{cases}$$

7. Maximize $4x + 2y$ subject to the constraints

$$\begin{cases} 5x + y \le 80 \\ 3x + 2y \le 76 \\ x \ge 0, \quad y \ge 0. \end{cases}$$

8. Maximize $2x + 6y$ subject to the constraints

$$\begin{cases} -x + 8y \le 160 \\ 3x - y \le 3 \\ x \ge 0, \quad y \ge 0. \end{cases}$$

9. Maximize $x + 3y + 5z$ subject to the constraints

$$\begin{cases} x \quad\quad + 2z \le 10 \\ \quad 3y + z \le 24 \\ x \ge 0, \quad y \ge 0, \quad z \ge 0. \end{cases}$$

10. Maximize $-x + 8y + z$ subject to the constraints

$$\begin{cases} x - 2y + 9z \le 10 \\ \quad y + 4z \le 12 \\ x \ge 0, \quad y \ge 0, \quad z \ge 0. \end{cases}$$

11. Maximize $2x + 3y$ subject to the constraints

$$\begin{cases} 5x + y \le 30 \\ 3x + 2y \le 60 \\ x + y \le 50 \\ x \ge 0, \quad y \ge 0. \end{cases}$$

12. Maximize $10x + 12y + 10z$ subject to the constraints

$$\begin{cases} x - 2y \quad\quad \le 6 \\ 3x \quad\quad + z \le 9 \\ \quad y + 3z \le 12 \\ x \ge 0, \quad y \ge 0, \quad z \ge 0. \end{cases}$$

13. Maximize $6x + 7y + 300$ subject to the constraints

$$\begin{cases} 2x + 3y \le 400 \\ x + y \le 150 \\ x \ge 0, \quad y \ge 0. \end{cases}$$

14. Maximize $10x + 20y + 50$ subject to the constraints

$$\begin{cases} x + y \le 10 \\ 5x + 2y \le 20 \\ x \ge 0, \quad y \ge 0. \end{cases}$$

15. Suppose that a furniture manufacturer makes chairs, sofas, and tables. The amounts of labor of various types as well as the relative availability of each type are summarized by the following chart:

	Chair	Sofa	Table	Daily labor available (man-hours)
Carpentry	6	3	8	768
Finishing	1	1	2	144
Upholstery	2	5	0	216
profit	80	70	120	

The profit per chair is $80, per sofa $70, and per table $120. How many pieces of each type of furniture should be manufactured each day in order to maximize the profit?

16. A stereo store sells three brands of stereo system, brands A, B, and C. It can sell a total of 100 stereo systems per month. Brands A, B, and C take up, respectively, 5, 4, and 4 cubic feet of warehouse space and a maximum of 480 cubic feet of warehouse space is available. Brands A, B, and C generate sales commissions of $40, $20, and $30, respectively, and $3200 is available to pay the sales commissions. The profit generated from the sale of each brand is $70, $210, and $140, respectively. How many of each brand of stereo system should be sold to maximize the profit?

17. A furniture manufacturer produces small sofas, large sofas, and chairs. The profits per item are, respectively, $60, $60, and $50. The pieces of furniture require the following numbers of man-hours for their manufacture:

	Carpentry	Upholstery	Finishing
Small sofas	10	30	20
Large sofas	10	30	0
Chairs	10	10	10

The following amounts of labor are available per month: carpentry at most 1200 hours, upholstery at most 3000 hours, and finishing at most 1800 hours. How many each of small sofas, large sofas, and chairs should be manufactured to maximize the profit?

18. Maximize $60x + 90y + 300z$ subject to the constraints

$$\begin{cases} x + y + z \le 600 \\ x + 3y \quad\;\; \le 600 \\ 2x \quad\;\; + z \le 900 \\ x \ge 0, \quad y \ge 0, \quad z \ge 0. \end{cases}$$

19. Maximize $200x + 500y$ subject to the constraints

$$\begin{cases} x + 4y \le 300 \\ x + 2y \le 200. \end{cases}$$

SOLUTIONS TO PRACTICE PROBLEMS 2

1. (a) Does not correspond to a maximum, since among the entries $-1, -2, 1, 0, 0$ on the last row, at least one is negative.

   (b) Corresponds to a maximum since none of the entries $1, 0, 0, 4, 1$ of the last row is negative. Note that it does not matter that the entry -2 in the right-hand corner of the matrix is negative. This number gives the value of M. In this example -2 is as large as M can become.

2. First choose the column corresponding to the most negative entry of the final row, that is, the fourth column. For each entry in the fourth column which is above the horizontal

line, compute the ratio with the sixth column. The smallest ratio is 2 and appears in the first row, so the next operation is to pivot around the 2 in the first row of the fourth column.

$$\begin{bmatrix} 0 & 4 & 1 & ② & 0 & 4 \\ 1 & 5 & 0 & 1 & 0 & 9 \\ 0 & 2 & 0 & -3 & 1 & 6 \end{bmatrix} \quad \begin{matrix} 4/2 = 2 \\ 9/1 = 9 \\ \end{matrix}$$

$$\uparrow$$

4.3. The Simplex Method, II—Minimum Problems

In the preceding section we developed the simplex method and applied it to a number of problems. However, throughout we restricted ourselves to linear programming problems in standard form. Recall that such problems satisfied three properties: (1) the objective function is to be maximized; (2) each variable must be ≥ 0; and (3) all constraints other than those implied by (2) must be of the form

[linear polynomial] $\leq$ [nonnegative constant].

In this section we shall do what we can to relax these restrictions.

Let us begin with restriction (3). This could be violated in two ways. First, the constant on the right-hand side of one or more constraints could be negative. Thus, for example, one constraint might be

$$x - y \leq -2.$$

A second way in which restriction (3) can be violated is for some constraints to involve $\geq$ rather than $\leq$. An example of such a constraint is

$$2x + 3y \geq 5.$$

However, we can convert such a constraint into one involving $\leq$ by multiplying both sides of the inequality by -1:

$$-2x - 3y \leq -5.$$

Of course, the right-hand constant is no longer nonnegative. Thus, if we allow negative constants on the right, we can write all constraints in the form

[linear polynomial] $\leq$ [constant].

Henceforth, the first step in solving a linear programming problem will be to write the constraints in this form. Let us now see how to deal with the phenomenon of negative constants.

EXAMPLE 1 Maximize the objective function $5x + 10y$ subject to the constraints

$$\begin{cases} x + y \leq 20 \\ 2x - y \geq 10 \\ x \geq 0, \quad y \geq 0. \end{cases}$$

Solution The first step is to put the second constraint into $\leq$ form. Multiply the second inequality by -1 to obtain

$$\begin{cases} x + y \leq 20 \\ -2x + y \leq -10 \\ x \geq 0, \quad y \geq 0. \end{cases}$$

Just as before, write this as a linear system:

$$\begin{cases} x + y + u = 20 \\ -2x + y + v = -10 \\ -5x - 10y + M = 0. \end{cases}$$

From the linear system construct the simplex tableau:

$$\begin{array}{c} \\ u \\ v \\ M \end{array} \begin{array}{ccccc} x & y & u & v & M \\ \end{array}$$

	x	y	u	v	M	
u	1	1	1	0	0	20
v	-2	1	0	1	0	-10
M	-5	-10	0	0	1	0

Everything would proceed exactly as before, except that the right-hand column has a -10 in it. This means that the initial value for v is -10, which violates the condition that all variables be ≥ 0. Before we can apply the simplex method of Section 2, we must first put the tableau into standard form. This can be done by pivoting so as to remove the negative entry in the right column.

We choose the pivot element as follows. Look along the left side of the -10 row of the tableau and locate any negative entry. There is only one: -2. Use the column containing the -2—column 1—as the pivot column. Now compute ratios as before:*

	x	y	u	v	M		
u	1	1	1	0	0	20	$20/1 = 20$
v	$\boxed{-2}$	1	0	1	0	-10	$-10/-2 = 5$
M	-5	-10	0	0	1	0	

The smallest positive ratio is 5, so we choose -2 as the pivot element. The new tableau is

	x	y	u	v	M	
u	0	$\frac{3}{2}$	1	$\frac{1}{2}$	0	15
x	1	$-\frac{1}{2}$	0	$-\frac{1}{2}$	0	5
M	0	$-\frac{25}{2}$	0	$-\frac{5}{2}$	1	25

---

* Note, however, that in this circumstance we compute ratios corresponding to both positive *and* negative entries (except the last) in the pivot column, considering further only those ratios which are positive.

Note that all entries in the right-hand column are now nonnegative;* that is, the corresponding solution has all variables ≥ 0. From here on we follow the simplex method for tableaux in standard form:

$$
\begin{array}{c}
\quad\quad\ \ x \quad\ \ y \quad\ u \quad\ v \quad\ M \\
\begin{array}{c} u \\ x \\ M \end{array}
\left[
\begin{array}{ccccc|c}
0 & \tfrac{3}{2} & 1 & \tfrac{1}{2} & 0 & 15 \\
1 & -\tfrac{1}{2} & 0 & -\tfrac{1}{2} & 0 & 5 \\
0 & -\tfrac{25}{2} & 0 & -\tfrac{5}{2} & 1 & 25
\end{array}
\right]
\quad 15/\tfrac{3}{2} = 10
\end{array}
$$

$$\uparrow$$

$$
\begin{array}{c}
\quad\quad\ \ x \quad\ \ y \quad\ u \quad\ v \quad\ M \\
\begin{array}{c} y \\ x \\ M \end{array}
\left[
\begin{array}{ccccc|c}
0 & 1 & \tfrac{2}{3} & \tfrac{1}{3} & 0 & 10 \\
1 & 0 & \tfrac{1}{3} & -\tfrac{1}{3} & 0 & 10 \\
0 & 0 & \tfrac{25}{3} & \tfrac{5}{3} & 1 & 150
\end{array}
\right].
\end{array}
$$

So the maximum value of M is 150, which is attained for $x = 10$, $y = 10$.

In summary:

The Simplex Method for Problems in Nonstandard Form

1. If necessary, convert all inequalities (except $x \geq 0$, $y \geq 0$), into the form

$$[\text{linear polynomial}] \leq [\text{constant}].$$

2. If a negative number appears in the upper part of the last column of the simplex tableau, remove it by pivoting.
   (a) Select one of the negative entries in its row. The column containing the entry will be the pivot column.
   (b) Select the pivot element by determining the least of the positive ratios associated to entries in the pivot column (except the bottom entry).
   (c) Pivot.
3. Repeat step 2 until there are no negative entries in the upper part of the right-hand column of the simplex tableau.
4. Proceed to apply the simplex method for tableaux in standard form.

The method we have just developed can be used to solve *minimum* problems as well as maximum problems. Minimizing the objective function f is the same as maximizing $(-1) \cdot f$. This is so since multiplying an inequality by -1 reverses the direction of the inequality sign. Thus, in order to apply our method to a minimum problem, we merely multiply the objective function by -1 and turn the problem into a maximum problem.

---

* In general, it may be necessary to pivot several times before all elements in the last column are ≥ 0.

EXAMPLE 2   Minimize the objective function $3x + 2y$ subject to the constraints

$$\begin{cases} x + y \geq 10 \\ x - y \leq 15 \\ x \geq 0, \quad y \geq 0. \end{cases}$$

Solution   First transform the problem so that the first two constraints are in $\leq$ form:

$$\begin{cases} -x - y \leq -10 \\ x - y \leq 15 \\ x \geq 0, \quad y \geq 0. \end{cases}$$

Instead of minimizing $3x + 2y$, let us maximize $-3x - 2y$. Let $M = -3x - 2y$. Then our initial simplex tableau reads

	x	y	u	v	M		
u	-1	$\boxed{-1}$	1	0	0	-10	$-10/-1 = 10$
v	1	-1	0	1	0	15	
M	3	2	0	0	1	0	

We first eliminate the -10 in the right-hand column. We have a choice of two negative entries in the -10 row. Let us choose the one in the y column. The ratios are then tabulated as above, and we pivot around the circled element. The new tableau is*

	x	y	u	v	M	
y	1	1	-1	0	0	10
v	2	0	-1	1	0	25
M	1	0	2	0	1	-20

Since all entries in the bottom row, except the last, are positive, this tableau corresponds to a maximum. Thus the maximum value of $-3x - 2y$ (subject to the constraints) is -20 and this value occurs for $x = 0$, $y = 10$. Thus the *minimum* value of $3x + 2y$ subject to the constraints is 20.

Let us now rework an applied problem previously treated (see Example 2 in Section 3 of Chapter 3), this time using the simplex method. For easy reference we restate the problem.

EXAMPLE 3   (*Transportation Problem*)   Suppose that a TV dealer has stores in Annapolis and Rockville and warehouses in College Park and Baltimore. The cost of shipping sets from College Park to Annapolis is $6 per set; from College Park to

---

* Note that we do not need the last entry in the last column positive. We require *only* that x, y, u, and v be ≥ 0.

Rockville, $3; from Baltimore to Annapolis, $9; and from Baltimore to Rockville, $5. Suppose that the Annapolis store orders 25 TV sets and the Rockville store 30. Further suppose that the College Park warehouse has a stock of 45 sets, and the Baltimore warehouse 40. What is the most economical way to supply the requested TV sets to the two stores?

Solution via the Simplex Method

As in the previous solution, let x be the number of sets shipped from College Park to Rockville, and y the number shipped from College Park to Annapolis. The flow of sets is depicted in Fig. 1.

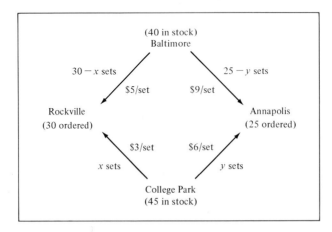

FIGURE 1

Exactly as in our previous solution, we reduce the problem to the following algebraic form: Minimize $375 - 2x - 3y$ subject to the constraints

$$\begin{cases} x \le 30, \quad y \le 25 \\ x + y \ge 15 \\ x + y \le 45 \\ x \ge 0, \quad y \ge 0. \end{cases}$$

Two changes are needed. First, instead of minimizing $375 - 2x - 3y$ we maximize $-(375 - 2x - 3y) = 2x + 3y - 375$. Second, we write the constraint $x + y \ge 15$ in the form

$$-x - y \le -15.$$

With these changes made, we can write down the linear system:

$$\begin{cases} x & +\ t & & & & = 30 \\ & y & +\ u & & & = 25 \\ -x\ - & y & & +\ v & & = -15 \\ x\ + & y & & & +\ w & = 45 \\ -2x\ - & 3y & & & +\ M & = -375. \end{cases}$$

From here on we follow our routine procedure in a mechanical way:

	x	y	t	u	v	w	M		
t	1	0	1	0	0	0	0	30	30/1 = 30
u	0	1	0	1	0	0	0	25	
v	(-1)	-1	0	0	1	0	0	-15	-15/-1 = 15
w	1	1	0	0	0	1	0	45	45/1 = 45
M	-2	-3	0	0	0	0	1	-375	

↑ (under x)

	x	y	t	u	v	w	M		
t	0	-1	1	0	(1)	0	0	15	15/1 = 15
u	0	1	0	1	0	0	0	25	
x	1	1	0	0	-1	0	0	15	
w	0	0	0	0	1	1	0	30	30/1 = 30
M	0	-1	0	0	-2	0	1	-345	

↑ (under v)

	x	y	t	u	v	w	M		
v	0	-1	1	0	1	0	0	15	
u	0	1	0	1	0	0	0	25	25/1 = 25
x	1	0	1	0	0	0	0	30	
w	0	(1)	-1	0	0	1	0	15	15/1 = 15
M	0	-3	2	0	0	0	1	-315	

↑ (under y)

	x	y	t	u	v	w	M		
v	0	0	0	0	1	1	0	30	
u	0	0	(1)	1	0	-1	0	10	10/1 = 10
x	1	0	1	0	0	0	0	30	30/1 = 30
y	0	1	-1	0	0	1	0	15	
M	0	0	-1	0	0	3	1	-270	

↑ (under t)

	x	y	t	u	v	(w)	M		
v	0	0	0	0	1	1	0	30	
t	0	0	1	1	0	-1	0	10	20 + 0(c)
x	1	0	0	-1	0	1	0	20	.
y	0	1	0	1	0	0	0	25	
M	0	0	0	1	0	2	1	-260	

This last tableau corresponds to a maximum. So $2x + 3y - 375$ has a maximum value -260, and therefore $375 - 2x - 3y$ has a minimum value 260. This value occurs when $x = 20$ and $y = 25$. This is in agreement with our previous graphical solution of the problem.

The calculations used in the preceding example are not that much simpler than those in the original solution. Why, then, should we concern ourselves with the simplex method? For one thing, the simplex method is so mechanical in its execution that it is much easier to program for a computer. For another, our previous method was restricted to problems in two variables. However, suppose that the two warehouses were to deliver their TV sets to three or four or perhaps even 100 stores? Our previous method could not be applied. However, the simplex method, although yielding very large matrices and very tedious calculations, is applicable. Indeed, this is the method used by many industrial concerns to optimize distribution of their products.

Some Further Comments on the Simplex Method   Our discussion has omitted some of the technical complications arising in the simplex method. A complete discussion of these is beyond the scope of this book. However, let us mention three. First of all, it is possible that a given linear programming problem has more than one solution. This can occur, for example, if there are ties for the choice of pivot column. For instance, if the bottom row of the simplex tableau is

$$[-3 \quad -7 \quad 4 \quad -7 \quad 1 \quad 3],$$
$$\qquad \uparrow \qquad\qquad \uparrow$$

then -7 is the most negative entry and we may choose as pivot column either the second or fourth. In such a circumstance the pivot column may be chosen arbitrarily. Different choices, however, may lead to different solutions of the problem.

A second difficulty is that a given linear programming problem may have no solution at all. In this case the method will break down at some point. For example, among the ratios at a given stage there may be no nonnegative ones to consider. Then we cannot choose a pivot element. Such a breakdown of the method indicates that the associated linear programming problem has no solution.

Finally, whenever there is a tie for the choice of a pivot element, we choose one of the candidates arbitrarily. Occasionally, this may lead to a loop in which the simplex algorithm leads back to a previously encountered tableau. To avoid the loop from reoccuring, we should then make a different selection from the tied pivot possibilities.

PRACTICE PROBLEMS 3

1. Convert the following minimum problem into a maximum problem in standard form:
   Minimize $3x + 4y$ subject to the constraints

$$\begin{cases} x - y \geq 0 \\ 3x - 4y \geq 0 \\ x \geq 0, \quad y \geq 0. \end{cases}$$

2. Suppose that the solution of a minimum problem yields the final simplex tableau

x	y	u	v	M	
1	6	−1	0	0	11
0	5	3	1	0	16
0	2	4	0	1	−40

What is the minimum value sought in the original problem?

EXERCISES 3

Solve the following linear programming problems by the simplex method.

1. Maximize $40x + 30y$ subject to the constraints
$$\begin{cases} x + y \le 5 \\ -2x + 3y \ge 12 \\ x \ge 0, \quad y \ge 0. \end{cases}$$

2. Maximize $3x - y$ subject to the constraints
$$\begin{cases} 2x + 5y \le 100 \\ x \qquad \ge 10 \\ \qquad y \ge 0. \end{cases}$$

3. Minimize $3x + y$ subject to the constraints
$$\begin{cases} x + y \ge 3 \\ 2x \qquad \ge 5 \\ x \ge 0, \quad y \ge 0. \end{cases}$$

4. Minimize $3x + 5y + z$ subject to the constraints
$$\begin{cases} x + y + z \ge 20 \\ y + 2z \ge 10 \\ x \ge 0, \quad y \ge 0, \quad z \ge 0. \end{cases}$$

5. Minimize $13x + 4y$ subject to the constraints
$$\begin{cases} y \ge -2x + 11 \quad \to \; y + 2x \ge 11 \;\to\; -y - 2x \le -11 \\ y \le -x + 10 \quad \to \; y + x \le 10 \\ y \le -\frac{1}{3}x + 6 \quad \to \; y + \frac{1}{3}x \le 6 \\ y \ge -\frac{1}{4}x + 4 \quad \to \; y + \frac{1}{4}x \ge 4 \;\to\; -y - \frac{1}{4}x \le -4 \\ x \ge 0, \quad y \ge 0. \end{cases}$$

6. Minimize $500 - 10x - 3y$ subject to the constraints
$$\begin{cases} x + y \le 20 \\ 3x + 2y \ge 50 \\ x \ge 0, \quad y \ge 0. \end{cases}$$

7. A dietician is designing a daily diet that is to contain at least 60 units of protein, 40 units of carbohydrate, and 120 units of fat. The diet is to consist of two types of foods. One serving of food A contains 30 units of protein, 10 units of carbohydrate, and 20 units of fat and costs $3. One serving of food B contains 10 units of protein, 10 units of carbohydrate, and 60 units of fat and costs $1.50. Design the diet that provides the daily requirements at the least cost.

8. A manufacturing company has two plants, each capable of producing radios, television sets, and stereo systems. The daily production capacities of each plant are as follows:

	Plant I	Plant II
Radios	10	20
Television sets	30	20
Stereo systems	20	10

Plant I costs $1500 per day to operate, whereas plant II costs $1200. How many days should each plant be operated to fill an order for 1000 radios, 1800 television sets, and 1000 stereo systems at the minimum cost?

9. An appliance store sells three brands of color television sets, brands A, B, and C. The profit per set is $30 for brand A, $50 for brand B, and $60 for brand C. The total warehouse space allotted to all brands is sufficient for 600 sets and the inventory is delivered only once per month. At least 100 customers per month will demand brand A, at least 50 will demand brand B, and at least 200 will demand either brand B or brand C. How can the appliance store satisfy all these constraints and earn maximum profit?

10. A citizen decides to campaign for the election of a candidate for city council. Her goal is to generate at least 210 votes by a combination of door-to-door canvassing, letter writing, and phone calls. She figures that each hour of door-to-door canvassing will generate four votes, each hour of letter writing will generate two votes, and each hour on the phone will generate three votes. She would like to devote at least seven hours to phone calls and spend at most half of her time at door-to-door canvassing. How much time should she allocate to each task in order to achieve her goal in the least amount of time?

SOLUTIONS TO PRACTICE PROBLEMS 3

1. To minimize $3x + 4y$, we maximize $-(3x + 4y) = -3x - 4y$. So the associated maximum problem is: Maximize $-3x - 4y$ subject to the constraints

$$\begin{cases} -x + y \leq 0 \\ -3x + 4y \leq 0 \\ x \geq 0, \quad y \geq 0. \end{cases}$$

2. The value -40 in the lower right corner gives the solution of the associated *maximum* problem. The minimum value originally sought is the negative of the maximum value— that is, $-(-40) = 40$.

4.4. Marginal Analysis and Matrix Formulations of Linear Programming Problems

Marginal Analysis of Linear Programming Problems   The simplex method not only provides the optimal solutions to linear programming problems but gives other useful information. As a matter of fact, each number appearing in the final simplex tableau has an interpretation that not only sheds light on the current situation but can be used to analyze the benefits of small changes in the available resources.

Consider the final simplex tableau of the furniture manufacturing problem. Since the variable u was introduced to take up the slack in the carpentry inequality, the u-column has been labeled *carpentry*. Similarly, the v-column and the w-column have been labeled *finishing* and *upholstery*, respectively:

	x	y	(Carpentry) u	(Finishing) v	(Upholstery) w	M	
x	1	0	$\frac{1}{3}$	-1	0	0	14
y	0	1	$-\frac{1}{3}$	2	0	0	4
w	0	0	$\frac{4}{3}$	-10	1	0	20
M	0	0	$\frac{10}{3}$	60	0	1	1400

The optimum profit is $1400, which occurs when $x = 14$ chairs and $y = 4$ sofas.

We now consider the following question. If additional labor becomes available, how will this change the production level and the profit? To be specific, suppose we had 3 more man-hours of labor available for carpentry. The initial tableau of the furniture manufacturing problem would become

	x	y	(Carpentry) u	(Finishing) v	(Upholstery) w	M	
	6	3	1	0	0	0	$96 + 3$
	1	1	0	1	0	0	$18 + 0$
	2	6	0	0	1	0	$72 + 0$
	-80	-70	0	0	0	1	$0 + 0$

Note that only the first entry of the right-hand column has been changed. The increment to the right-hand column can be written as the column

$$\begin{matrix} 3 \\ 0 \\ 0 \\ \overline{0}, \end{matrix}$$

which is three times the u-column. Now, when the simplex method is performed on the new initial tableau, all of the row operations will affect the increment column exactly as they affected the u-column in the original initial tableau. Therefore, the final increment column will be three times the final u-column. Hence, the new final tableau will be

	x	y	(Carpentry) u	(Finishing) v	(Upholstery) w	M	
x	1	0	$\frac{1}{3}$	-1	0	0	$14 + 3(\frac{1}{3})$
y	0	1	$-\frac{1}{3}$	2	0	0	$4 + 3(-\frac{1}{3})$
w	0	0	$\frac{4}{3}$	-10	1	0	$20 + 3(\frac{4}{3})$
M	0	0	$\frac{10}{3}$	60	0	1	$1400 + 3(\frac{10}{3})$

Thus, when 3 additional man-hours of carpentry are available, $x = 14 + 3(\frac{1}{3}) = 15$ chairs and $y = 4 + 3(-\frac{1}{3}) = 3$ sofas should be produced. The maximum profit will increase to the new value of $M = 1400 + 3(\frac{10}{3}) = 1410$ dollars.

The number 3 was arbitrary. If h is a suitable number, then adding h man-hours of labor for carpentry to the original problem results in the final tableau:

	x	y	(Carpentry) u	(Finishing) v	(Upholstery) w	M	
x	1	0	$\frac{1}{3}$	-1	0	0	$14 + h(\frac{1}{3})$
y	0	1	$-\frac{1}{3}$	2	0	0	$4 + h(-\frac{1}{3})$
w	0	0	$\frac{4}{3}$	-10	1	0	$20 + h(\frac{4}{3})$
M	0	0	$\frac{10}{3}$	60	0	1	$1400 + h(\frac{10}{3})$

The number h can be positive or negative. For instance, if one less man-hour of labor is available for carpentry, then setting $h = -1$ yields the optimal production schedule $x = 13\frac{2}{3}$, $y = 4\frac{1}{3}$ and a profit of $\$1396\frac{2}{3}$. The only restriction on h is that the numbers in the upper part of the right-hand column of the final tableau must all be nonnegative.

Similarly, if in the original problem, the amount of labor available for finishing is increased by h man-hours, the initial tableau becomes

	x	y	(Carpentry) u	(Finishing) v	(Upholstery) w	M	
u	6	3	1	0	0	0	$96 + 0$
v	1	1	0	1	0	0	$18 + h$
w	2	6	0	0	1	0	$72 + 0$
M	-80	-70	0	0	0	1	$0 + 0$

The right-hand column of the original tableau was changed by adding an increment column that is h times the v-column, and therefore the right-hand column of the new final tableau will be the original final right-hand column plus h times the final v-column:

	x	y	(Carpentry) u	(Finishing) v	(Upholstery) w	M	
x	1	0	$\frac{1}{3}$	-1	0	0	$14 + h(-1)$
y	0	1	$-\frac{1}{3}$	2	0	0	$4 + h(2)$
w	0	0	$\frac{4}{3}$	-10	1	0	$20 + h(-10)$
M	0	0	$\frac{10}{3}$	60	0	1	$1400 + h(60)$

Hence, if one additional man-hour of labor were available for finishing, the optimal production schedule would be $x = 13$ chairs, $y = 6$ sofas, and the profit would be $\$1460$.

Finally, if in the original problem, the amount of labor available for upholstery is increased by h man-hours, the initial and final tableaux become:

| | | | (Carpentry) | (Finishing) | (Upholstery) | | |
	x	y	u	v	w	M	
u	6	3	1	0	0	0	$96 + 0$
v	1	1	0	1	0	0	$18 + 0$
w	2	6	0	0	1	0	$72 + h$
M	-80	-70	0	0	0	1	$0 + 0$

| | | | (Carpentry) | (Finishing) | (Upholstery) | | |
	x	y	u	v	w	M	
x	1	0	$\frac{1}{3}$	-1	0	0	$14 + h(0)$
y	0	1	$-\frac{1}{3}$	2	0	0	$4 + h(0)$
w	0	0	$\frac{4}{3}$	-10	1	0	$20 + h(1)$
M	0	0	$\frac{10}{3}$	60	0	1	$1400 + h(0)$

Therefore, a change in the amount of labor available for upholstery has no effect on the production schedule or the profit. This makes sense, since we had excess labor available for upholstery in the solution to the original problem. The slack in carpentry and finishing was used up (u and v were 0), but there was slack in the labor available for upholstery (w was 20).

In summary, each of the slack variable columns in the final tableau of the original furniture manufacturing problem gives the sensitivity to change in the production schedule and in the profit due to a suitable change in one of the factors of production. The final values in each of these columns ($u = \frac{10}{3}$, $v = 60$, and $w = 0$) are called the *marginal* values of the three factors of production—carpentry, finishing, and upholstery.

The following example does a complete analysis of a new linear programming problem.

EXAMPLE 1 The Cutting Edge Knife Company manufactures paring knives and pocket knives. Each paring knife requires 3 man-hours of labor, 7 units of steel, and 4 units of wood. Each pocket knife requires 6 man-hours of labor, 5 units of steel, and 3 units of wood. The profit on each paring knife is $3 and the profit on each pocket knife is $5. Each day the company has available 90 man-hours of labor, 138 units of steel, and 120 units of wood.

(a) How many of each type of knife should the Cutting Edge Knife Company manufacture daily in order to maximize its profits?

(b) Suppose that an additional 18 units of steel were available each day. What effect would this have on the optimal solution?

(c) Generalize the result in (b) to the case where the increase in the number of units of steel available each day is h. (The value of h can be positive or negative.) For what range of values will be result be valid?

Solution We need to find the number of paring knives, x, and pocket knives, y, that will maximize the profit, $M = 3x + 5y$, subject to the constraints

$$\begin{cases} 3x + 6y \leq 90 \\ 7x + 5y \leq 138 \\ 4x + 3y \leq 120 \\ x \geq 0, \quad y \geq 0. \end{cases}$$

The initial tableau with slack variables u, v, and w added for labor, steel, and wood is

	x	y	(Labor) u	(Steel) v	(Wood) w	M		
u	3	6	1	0	0	0	90	$90/6 = 15$
v	7	5	0	1	0	0	138	$138/5 = 27.6$
w	4	3	0	0	1	0	120	$120/3 = 40$
M	-3	-5	0	0	0	1	0	

The proper pivot element is the entry 6 in the y-column. The next tableau is

	x	y	(Labor) u	(Steel) v	(Wood) w	M		
y	$\frac{1}{2}$	1	$\frac{1}{6}$	0	0	0	15	$15/\frac{1}{2} = 30$
v	$\frac{9}{2}$	0	$-\frac{5}{6}$	1	0	0	63	$63/\frac{9}{2} = 14$
w	$\frac{5}{2}$	0	$-\frac{1}{2}$	0	1	0	75	$75/\frac{5}{2} = 30$
M	$-\frac{1}{2}$	0	$\frac{5}{6}$	0	0	1	75	

The proper pivot element is the entry $\frac{9}{2}$ in the x-column. The next tableau is

	x	y	(Labor) u	(Steel) v	(Wood) w	M	
y	0	1	$\frac{7}{27}$	$-\frac{1}{9}$	0	0	8
x	1	0	$-\frac{5}{27}$	$\frac{2}{9}$	0	0	14
w	0	0	$-\frac{1}{27}$	$-\frac{5}{9}$	1	0	40
M	0	0	$\frac{20}{27}$	$\frac{1}{9}$	0	1	82

Since there are no negative entries in the last row of this tableau, the simplex method is complete.

(a) The Cutting Edge Knife Company should produce 14 paring knives and 8 pocket knives each day for a profit of \$82. (Since the slack variable w has the value 40, there will be 40 excess units of wood each day.)

(b) Since 18 additional units of steel are available, the final tableau of the revised problem can be obtained from the final tableau of the original problem by adding 18 times the v-column to the right-hand column:

		(Labor)	(Steel)	(Wood)			
	x	y	u	v	w	M	
y	0	1	$\frac{7}{27}$	$-\frac{1}{9}$	0	0	$8 + 18(-\frac{1}{9})$
x	1	0	$-\frac{5}{27}$	$\frac{2}{9}$	0	0	$14 + 18(\frac{2}{9})$
w	0	0	$-\frac{1}{27}$	$-\frac{5}{9}$	1	0	$40 + 18(-\frac{5}{9})$
M	0	0	$\frac{20}{27}$	$\frac{1}{9}$	0	1	$82 + 18(\frac{1}{9})$

The company should make 4 more paring knives $[18(\frac{2}{9}) = 4]$ and 2 fewer pocket knives $[18(-\frac{1}{9}) = -2]$. So doing will increase the profits by \$2 $[18(\frac{1}{9}) = 2]$.

(c) With h additional units of steel available, the right-hand column of the final tableau will be similar to the tableau above but with 18 replaced by h.

		(Labor)	(Steel)	(Wood)			
	x	y	u	v	w	M	
y	0	1	$\frac{7}{27}$	$-\frac{1}{9}$	0	0	$8 + h(-\frac{1}{9})$
x	1	0	$-\frac{5}{27}$	$\frac{2}{9}$	0	0	$14 + h(\frac{2}{9})$
w	0	0	$-\frac{1}{27}$	$-\frac{5}{9}$	1	0	$40 + h(-\frac{5}{9})$
M	0	0	$\frac{20}{27}$	$\frac{1}{9}$	0	1	$82 + h(\frac{1}{9})$

Therefore, the number of paring knives made should be $14 + h(\frac{2}{9})$ and the number of pocket knives made should be $8 + h(-\frac{1}{9})$. The new profit will be $82 + h(\frac{1}{9})$ dollars. This analysis is valid provided that each of the entries in the upper part of the right-hand column of the tableau is nonnegative. The restrictions on h given by each of these entries are

Entry	Restriction
$8 + h(-\frac{1}{9})$	$h \leq 72$
$14 + h(\frac{2}{9})$	$h \geq -63$
$40 + h(-\frac{5}{9})$	$h \leq 72$

All of these restrictions will be satisfied if h is between -63 and 72.

Matrix Formulations of Linear Programming Problems   Linear programming problems can be neatly stated in terms of matrices. Such formulations provide a convenient way to define the *dual* of a linear programming problem, an important concept that is studied in the next section. To introduce the matrix formulation of a linear programming problem, we first need the concept of inequality for matrices.

Let A and B be two matrices of the same size. We say that A is less than or equal to B (denoted $A \le B$) if each entry of A is less than or equal to the corresponding entry of B. For instance, we have the following matrix inequalities:

$$\begin{bmatrix} 2 & -3 \\ \frac{1}{2} & 0 \end{bmatrix} \le \begin{bmatrix} 5 & -1 \\ 1 & 0 \end{bmatrix} \quad \text{and} \quad \begin{bmatrix} 5 \\ 6 \end{bmatrix} \le \begin{bmatrix} 8 \\ 9 \end{bmatrix}.$$

The symbol $\ge$ has an analogous meaning for matrices.

EXAMPLE 2 Let

$$A = \begin{bmatrix} 6 & 3 \\ 1 & 1 \\ 2 & 6 \end{bmatrix}, \quad B = \begin{bmatrix} 96 \\ 18 \\ 72 \end{bmatrix}, \quad C = [80 \quad 70], \quad X = \begin{bmatrix} x \\ y \end{bmatrix}.$$

Carry out the indicated matrix multiplications in the following statement: Maximize CX subject to the constraints $AX \le B$, $X \ge 0$. (Here 0 is the zero matrix.)

Solution $CX = [80 \quad 70] \begin{bmatrix} x \\ y \end{bmatrix} = [80x + 70y].$

$$AX = \begin{bmatrix} 6 & 3 \\ 1 & 1 \\ 2 & 6 \end{bmatrix} \begin{bmatrix} x \\ y \end{bmatrix} = \begin{bmatrix} 6x + 3y \\ x + y \\ 2x + 6y \end{bmatrix}.$$

$$AX \le B \text{ means } \begin{bmatrix} 6x + 3y \\ x + y \\ 2x + 6y \end{bmatrix} \le \begin{bmatrix} 96 \\ 18 \\ 72 \end{bmatrix} \quad \text{or} \quad \begin{cases} 6x + 3y \le 96 \\ x + y \le 18 \\ 2x + 6y \le 72. \end{cases}$$

$$X \ge 0 \text{ means } \begin{bmatrix} x \\ y \end{bmatrix} \ge \begin{bmatrix} 0 \\ 0 \end{bmatrix} \quad \text{or} \quad \begin{cases} x \ge 0 \\ y \ge 0. \end{cases}$$

Hence the statement "Maximize CX subject to the constraints $AX \le B$, $X \ge 0$" is a matrix formulation of the furniture manufacturing problem.

Another concept that is needed for the definition of the dual of a linear programming problem is the *transpose of a matrix*. If A is an $m \times n$ matrix, then the matrix A^T (pronounced A transpose) is the $n \times m$ matrix whose ij^{th} entry is the ji^{th} entry of A. The rows of A^T are the columns of A, and vice versa.

EXAMPLE 3 Find the transpose of

(a) $\begin{bmatrix} 3 & -2 & 4 \\ 6 & 5 & 0 \end{bmatrix}$

(b) $\begin{bmatrix} 5 \\ 2 \\ 1 \end{bmatrix}.$

Solution   (a) Since the given matrix has two rows and three columns, its transpose will have three rows and two columns. The entries of the first row of the transpose will be the entries in the first column of the original matrix:

$$\begin{bmatrix} 3 & 6 \\ & \\ & \end{bmatrix}.$$

The entries of the second and third rows are obtained in a similar manner from the second and third columns of the original matrix. Therefore

$$\begin{bmatrix} 3 & -2 & 4 \\ 6 & 5 & 0 \end{bmatrix}^T = \begin{bmatrix} 3 & 6 \\ -2 & 5 \\ 4 & 0 \end{bmatrix}.$$

(b) Since the given matrix has three rows and one column, its transpose has one row and three columns. That row consists of the single column of the original matrix:

$$\begin{bmatrix} 5 \\ 2 \\ 1 \end{bmatrix}^T = \begin{bmatrix} 5 & 2 & 1 \end{bmatrix}.$$

PRACTICE PROBLEMS 4

Consider the furniture manufacturing problem, whose final simplex tableau appears on page 152.

1. Suppose that the number of man-hours of labor for finishing that are available each day are decreased by 2. What will be the effect on the optimal number of chairs and sofas produced and on the profit?

2. For what range of values of h will a marginal analysis on the effect of a change of h man-hours in the amount of available labor for finishing be valid?

EXERCISES 4

Exercises 1 and 2 refer to the Cutting Edge Knife Company problem of Example 1.

1. Suppose that the number of man-hours of labor that are available each day are increased by 54. What will be the effect on the optimal numbers of knives produced and on the profit?

2. For what range of values of h will a marginal analysis on the effect of a change of h man-hours in the amount of available labor be valid?

Exercises 3 and 4 refer to the transportation problem of Example 3 in Section 4.3.

3. Suppose that the number of TV sets stocked in the College Park warehouse are increased to 50. What will be the effect on the optimal numbers of TV sets shipped from each warehouse to each store and what will be the change in the cost?

4. For what range of values of h will a marginal analysis on the effect of a change of h in the number of TV sets stocked in the College Park store be valid?

5. In the furniture manufacturing problem, for what range of values of h will a marginal analysis on the effect of a change of h man-hours in the amount of available labor for carpentry be valid?

6. Consider the nutrition problem in Example 1 of Section 3.2. Solve the problem by the simplex method and then determine the optimal quantities of soybeans and rice in the diet, and the new cost, if the daily requirement for calories is increased to 1700. For what range of values of h will a marginal analysis on the effects of a change of h calories be valid?

In Exercises 7–10, find the transpose of the given matrix.

7. $\begin{bmatrix} 9 & 4 \\ 1 & 8 \\ 1 & -3 \end{bmatrix}$

   8. $\begin{bmatrix} 4 \\ 0 \\ 6 \end{bmatrix}$

9. $\begin{bmatrix} 7 & 6 & 5 & 1 \end{bmatrix}$

   10. $\begin{bmatrix} 5 & 2 \\ 3 & -1 \end{bmatrix}$

11. Is it true that the transpose of the transpose of a matrix is the original matrix?

12. Find an example of a matrix that is its own transpose.

In Exercises 13 and 14, give the matrix formulation of the linear programming problem.

13. Minimize $7x + 5y + 4z$ subject to
$$\begin{cases} 3x + 8y + 9z \geq 75 \\ x + 2y + 5z \geq 80 \\ 4x + y + 7z \geq 67 \\ x \geq 0, \quad y \geq 0, \quad z \geq 0. \end{cases}$$

14. Maximize $20x + 30y$ subject to
$$\begin{cases} 7x + 8y \leq 55 \\ x + 2y \leq 78 \\ x \quad\quad \leq 25 \\ x \geq 0, \quad y \geq 0. \end{cases}$$

15. Give a matrix formulation of the Cutting Edge Knife Company problem of Example 1.

16. Does every linear programming problem have a matrix formulation? If not, under what conditions will a linear programming problem have a matrix formulation?

In Exercises 17 and 18, let

$$C = \begin{bmatrix} 2 & 3 \end{bmatrix}, \quad X = \begin{bmatrix} x \\ y \end{bmatrix}, \quad A = \begin{bmatrix} 7 & 4 \\ 5 & 8 \\ 1 & 3 \end{bmatrix}, \quad B = \begin{bmatrix} 33 \\ 44 \\ 55 \end{bmatrix}, \quad \text{and} \quad U = \begin{bmatrix} u \\ v \\ w \end{bmatrix}.$$

17. Give the linear programming problem whose matrix formulation is "Minimize CX subject to the constraints $AX \geq B$, $X \geq \mathbf{0}$."

18. Give the linear programming problem whose matrix formulation is: "Maximize $B^T U$ subject to the constraints $A^T U \leq C^T$, $U \geq \mathbf{0}$."

SOLUTIONS TO PRACTICE PROBLEMS 4

1. Since the finishing column in the original final tableau is

$$\frac{\begin{array}{c} -1 \\ 2 \\ -10 \end{array}}{60,}$$

the right-hand column in the new tableau is

$$\frac{\begin{array}{c} 14 + (-2)(-1) \\ 4 + (-2)(2) \\ 20 + (-2)(-10) \end{array}}{1400 + (-2)(60).}$$

Therefore, the new values of x, y, and M are 16, 0, and 1280.

2. Using h instead of -2 in the marginal analysis, we find that the right-hand column of the new final tableau is

$$\frac{\begin{array}{c} 14 + h(-1) \\ 4 + h(2) \\ 20 + h(-10) \end{array}}{1400 + h(60).}$$

Of course, this analysis is only valid if the three numbers above the line are not negative. That is, $14 + h(-1) \geq 0$, $4 + h(2) \geq 0$, and $20 + h(-10) \geq 0$. These three inequalities can be simplified to $h \leq 14$, $h \geq -2$, and $h \leq 2$. Therefore, in order to satisfy all three inequalities, h must be in the range $-2 \leq h \leq 2$.

4.5. Duality

Each linear programming problem may be converted into a related linear programming problem called its *dual*. The dual problem is sometimes easier to solve than the original problem and moreover has the same optimum value. Furthermore, the solution of the dual problem often can provide valuable insights into the original problem. In order to understand the relationship between a linear programming problem and its dual, it is best to begin with a concrete example.

PROBLEM A Maximize the objective function $6x + 5y$ subject to the constraints

$$\begin{cases} 4x + 8y \leq 32 \\ 3x + 2y \leq 12 \\ x \geq 0, \quad y \geq 0. \end{cases}$$

The dual of Problem A is the following problem.

PROBLEM B   Minimize the objective function $32u + 12v$ subject to the constraints

$$\begin{cases} 4u + 3v \geq 6 \\ 8u + 2v \geq 5 \\ u \geq 0, \quad v \geq 0. \end{cases}$$

The relationship between the two problems is easiest to see if we write them in their matrix formulations. Problem A is:

$$\text{Maximize } \begin{bmatrix} 6 & 5 \end{bmatrix} \begin{bmatrix} x \\ y \end{bmatrix} \text{ subject to the constraints}$$

$$\begin{bmatrix} 4 & 8 \\ 3 & 2 \end{bmatrix} \begin{bmatrix} x \\ y \end{bmatrix} \leq \begin{bmatrix} 32 \\ 12 \end{bmatrix} \quad \text{and} \quad \begin{bmatrix} x \\ y \end{bmatrix} \geq \begin{bmatrix} 0 \\ 0 \end{bmatrix}.$$

Problem B is:

$$\text{Minimize } \begin{bmatrix} 32 & 12 \end{bmatrix} \begin{bmatrix} u \\ v \end{bmatrix} \text{ subject to the constraints}$$

$$\begin{bmatrix} 4 & 3 \\ 8 & 2 \end{bmatrix} \begin{bmatrix} u \\ v \end{bmatrix} \geq \begin{bmatrix} 6 \\ 5 \end{bmatrix} \quad \text{and} \quad \begin{bmatrix} u \\ v \end{bmatrix} \geq \begin{bmatrix} 0 \\ 0 \end{bmatrix}.$$

Each of the numeric matrices in Problem B is the transpose of one of the matrices in Problem A. Let

$$C = \begin{bmatrix} 6 & 5 \end{bmatrix}, \quad X = \begin{bmatrix} x \\ y \end{bmatrix}, \quad A = \begin{bmatrix} 4 & 8 \\ 3 & 2 \end{bmatrix}, \quad B = \begin{bmatrix} 32 \\ 12 \end{bmatrix}, \quad U = \begin{bmatrix} u \\ v \end{bmatrix},$$

$$\text{and} \quad \mathbf{0} = \begin{bmatrix} 0 \\ 0 \end{bmatrix}.$$

Problem A is: "Maximize CX subject to the constraints $AX \leq B$, $X \geq \mathbf{0}$."
Problem B is: "Minimize $B^T U$ subject to the constraints $A^T U \geq C^T$, $U \geq \mathbf{0}$."
Problem A is referred to as the *primal* problem and Problem B as its *dual*.

The Dual of a Linear Programming Problem

1. If the original (primal) problem has the form:

    Maximize CX subject to the constraints $AX \leq B$, $X \geq \mathbf{0}$,

    then the dual problem is:

    Minimize $B^T U$ subject to the constraints $A^T U \geq C^T$, $U \geq \mathbf{0}$.

2. If the original (primal) problem has the form:

    Minimize CX subject to the constraints $AX \geq B$, $X \geq \mathbf{0}$,

    then the dual problem is:

    Maximize $B^T U$ subject to the constraints $A^T U \leq C^T$, $U \geq \mathbf{0}$.

Note that Problem A is a standard maximization problem. That is, all of the inequalities involve ≤, except for $x \geq 0$ and $y \geq 0$. Problem B is a standard minimization problem in that all of its inequalities are ≥. The coefficients of the objective function for Problem A are the numbers on the right-hand side of the inequalities of Problem B, and vice versa. The coefficient matrices for the left-hand sides of the inequalities are transposes of one another. In an analogous way, we can start with a standard minimization problem and define its dual to be a standard maximization problem. Any linear programming problem can be put into one of these two standard forms. (If an inequality points in the wrong direction, we need only multiply it by -1.) Therefore, every linear programming problem has a dual.

EXAMPLE 1 Determine the dual of the following linear programming problem. Minimize $18x + 20y + 2z$ subject to the constraints

$$\begin{cases} 3x - 5y - 2z \leq 4 \\ 6x \quad\quad - 8z \geq 9 \\ x \geq 0, \quad y \geq 0, \quad z \geq 0. \end{cases}$$

Solution We first put the problem into standard form. Since the primal problem is a minimization problem, we must write all constraints with the inequality sign ≥. To put the first inequality in this form, we multiply by -1 to obtain

$$-3x + 5y + 2z \geq -4.$$

We now write the problem in matrix form:

$$\text{Minimize } \begin{bmatrix} 18 & 20 & 2 \end{bmatrix} \begin{bmatrix} x \\ y \\ z \end{bmatrix} \text{ subject to}$$

$$\begin{bmatrix} -3 & 5 & 2 \\ 6 & 0 & -8 \end{bmatrix} \begin{bmatrix} x \\ y \\ z \end{bmatrix} \geq \begin{bmatrix} -4 \\ 9 \end{bmatrix} \quad \text{and} \quad \begin{bmatrix} x \\ y \\ z \end{bmatrix} \geq \begin{bmatrix} 0 \\ 0 \\ 0 \end{bmatrix}.$$

The dual is:

$$\text{Maximize } \begin{bmatrix} -4 & 9 \end{bmatrix} \begin{bmatrix} u \\ v \end{bmatrix} \text{ subject to the constraints}$$

$$\begin{bmatrix} -3 & 6 \\ 5 & 0 \\ 2 & -8 \end{bmatrix} \begin{bmatrix} u \\ v \end{bmatrix} \leq \begin{bmatrix} 18 \\ 20 \\ 2 \end{bmatrix} \quad \text{and} \quad \begin{bmatrix} u \\ v \end{bmatrix} \geq \begin{bmatrix} 0 \\ 0 \end{bmatrix}.$$

Multiplying the matrices, we obtain the following:

$$\text{Maximize } -4u + 9v \text{ subject to}$$

$$\begin{cases} -3u + 6v \leq 18 \\ 5u \quad\quad \leq 20 \\ 2u - 8v \leq 2 \\ u \geq 0, \quad v \geq 0. \end{cases}$$

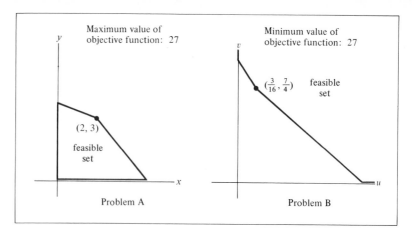

FIGURE 1

Let us now return to Problems A and B in order to examine the connection between the solutions of a linear programming problem and its dual problem. Problems A and B both involve two variables and hence can be solved by the geometric method of Chapter 3. Figure 1 shows their respective feasible sets and the vertices that yield the optimum values of the objective functions. The feasible sets do not look alike and the optimal vertices are different. However, both problems have the same optimum value, 27. The relationship between the two problems is brought into even sharper focus by looking at the final tableaux that arise when the two problems are solved by the simplex method. (*Note:* In Problem B, the original variables are u and v and the slack variables have been named x and y.)

FINAL TABLEAU

	x	y	u	v	M	
y	0	1	$\frac{3}{16}$	$-\frac{1}{4}$	0	3
x	1	0	$-\frac{1}{8}$	$\frac{1}{2}$	0	2
	0	0	$\frac{3}{16}$	$\frac{7}{4}$	1	27

Problem A

	u	v	x	y	M	
v	0	1	$-\frac{1}{2}$	$\frac{1}{4}$	0	$\frac{7}{4}$
u	1	0	$\frac{1}{8}$	$-\frac{3}{16}$	0	$\frac{3}{16}$
	0	0	2	3	1	-27

Problem B

The final tableau for Problem A contains the solution to Problem B ($u = \frac{3}{16}$, $v = \frac{7}{4}$) in the final entries of the u and v columns. Similarly, the final tableau for Problem B gives the solution to Problem A ($x = 2$, $y = 3$) in the final entries of its x and y columns. This situation always occurs. The solutions to a linear programming problem and its dual problem may be obtained simultaneously by solving just one of the problems using the simplex method and applying the following theorem .

Fundamental Theorem of Duality

1. If either the primal problem or the dual problem has an optimal solution, then they both have an optimal solution and their objective functions have the same values at these optimal points.

2. If both the primal and the dual problems have a feasible solution, then they both have optimal solutions and their objective functions have the same value at these optimal points.

3. The solution of one of these problems by the simplex method yields the solution of the other problem as the final entries in the columns associated with the slack variables.

EXAMPLE 2 Solve the linear programming problem of Example 1 by applying the simplex method to its dual problem.

Solution In the solution to Example 1, the dual problem is: Maximize $-4u + 9v$ subject to the constraints

$$\begin{cases} -3u + 6v \leq 18 \\ 5u \quad\quad \leq 20 \\ 2u - 8v \leq 2 \\ u \geq 0, \quad v \geq 0. \end{cases}$$

Since there are three nontrivial inequalities, the simplex method calls for three slack variables. Denote the slack variables by x, y, and z. Let $M = -4u + 9v$ and apply the simplex method.

	u	v	x	y	z	M	
x	-3	⑥	1	0	0	0	18
y	5	0	0	1	0	0	20
z	2	-8	0	0	1	0	2
	4	-9	0	0	0	1	0

	u	v	x	y	z	M	
v	$-\frac{1}{2}$	1	$\frac{1}{6}$	0	0	0	3
y	⑤	0	0	1	0	0	20
z	-2	0	$\frac{4}{3}$	0	1	0	26
	$-\frac{1}{2}$	0	$\frac{3}{2}$	0	0	1	27

	u	v	x	y	z	M	
v	0	1	$\frac{1}{6}$	$\frac{1}{10}$	0	0	5
u	1	0	0	$\frac{1}{5}$	0	0	4
z	0	0	$\frac{4}{3}$	$\frac{2}{5}$	1	0	34
	0	0	$\frac{3}{2}$	$\frac{1}{10}$	0	1	29

Since the maximum value of the dual problem is 29, we know that the minimum value of the original problem is also 29. Looking at the last row of the final tableau, we conclude that this minimum value is assumed when $x = \frac{3}{2}$, $y = \frac{1}{10}$, and $z = 0$.

In Example 2, the dual problem was easier to solve than the original problem given in Example 1. Thus, we see how consideration of the dual problem may simplify the solution of linear programming problems in some cases.

An Economic Interpretation of the Dual Problem   To illustrate the interpretation of the dual problem, let's reconsider the furniture manufacturing problem. Recall that this problem asked us to maximize the profit from the sale of x chairs and y sofas subject to limitations on the amount of labor available for carpentry, finishing, and upholstery. In mathematical terms, the problem required us to maximize $80x + 70y$ subject to the constraints

$$\begin{cases} 6x + 3y \leq 96 \\ x + y \leq 18 \\ 2x + 6y \leq 72 \\ x \geq 0, \quad y \geq 0. \end{cases}$$

Its dual problem is to minimize $96u + 18v + 72w$ subject to the constraints

$$\begin{cases} 6u + v + 2w \geq 80 \\ 3u + v + 6w \geq 70 \\ u \geq 0, \quad v \geq 0, \quad w \geq 0. \end{cases}$$

The variables u, v, and w can be assigned a meaning so that the dual problem has a significant interpretation in terms of the original problem.

First, recall the following table of data (man-hours except as noted):

	Chair	Sofa	Available labor
Carpentry	6	3	96
Finishing	1	1	18
Upholstery	2	6	72
Profit	$80	$70	

Suppose that we have an opportunity to hire out all our workers. Suppose that hiring out the carpenters will yield a profit of u dollars per hour, the finishers v dollars per hour, and the upholsterers w dollars per hour. Of course, u, v, and w all must be ≥ 0. However, there are other constraints that we should reasonably impose. Any scheme for hiring out the workers should generate at least as much profit as is currently being generated in the construction of chairs and sofas. In terms of the potential profits from hiring the workers out, the labor involved in constructing a chair will generate

$$6u + v + 2w$$

dollars of profit. And this amount should be at least equal to the $80 profit that could be earned by using the labor to construct a chair. That is, we have the constraint

$$6u + v + 2w \geq 80.$$

Similarly, considering the labor involved in building a sofa, we derive the constraint

$$3u + v + 6w \geq 70.$$

Since there are available 96 hours of carpentry, 18 hours of finishing, and 72 hours of upholstery, the total profit from hiring out the workers would be

$$96u + 18v + 72w.$$

Thus the problem of determining the least acceptable profit from hiring out the workers is equivalent to the following: Minimize $96u + 18v + 72w$ subject to the constraints

$$\begin{cases} 6u + v + 2w \geq 80 \\ 3u + v + 6w \geq 70 \\ u \geq 0, \quad v \geq 0, \quad w \geq 0. \end{cases}$$

This is just the dual of the furniture manufacturing problem. The values u, v, and w are measures of the value of an hour's labor by each type of worker. Economists often refer to them as *shadow prices*. The fundamental theorem of duality asserts that the minimum acceptable profit that can be achieved by hiring the workers out is equal to the maximum profit that can be generated if they make furniture.

As we saw in the furniture manufacturing problem, the maximum profit, $M = 1400$, is achieved when

$$x = 14, \qquad y = 4, \qquad u = 0, \qquad v = 0, \qquad w = 20.$$

For the dual problem, we can read from the final tableau of the primal problem that the minimum acceptable profit, $M = 1400$, is achieved when

$$u = \tfrac{10}{3}, \qquad v = 60, \qquad w = 0, \qquad x = 0, \qquad y = 0.$$

The solution of the primal problem gives the activity level which meets the constraints imposed by the resources and provides the maximum profit. Such a model is called an *allocation problem*. On the other hand, the solution of the dual problem assigns values to each of the resources of production. The solution of the dual problem might be used by an insurance salesperson or by an accountant to impute a value to each resource. It is referred to as a *valuation problem*.

We summarize the situation briefly below:
If the original problem is a maximization problem:

$$\text{Maximize } CX \text{ subject to } AX \leq B \text{ and } X \geq \mathbf{0},$$

then we interpret the solution matrix X as the *activity matrix* in which each entry gives the optimal level of each activity. The matrix B is the *capacity* or *resources matrix* where each entry represents the available amount of a (scarce) resource. The matrix C is the *profit matrix* whose entries are the unit profits for each activity represented in X. The dual solution matrix U is the *imputed value matrix* which gives the imputed value of each of the resources in the production process.

If the original problem is a minimization problem:

$$\text{Minimize } CX \text{ subject to } AX \geq B \text{ and } X \geq \mathbf{0},$$

then we interpret X as the *activity matrix* in which each entry gives the optimal level of an activity. B is the *requirements matrix* in which each entry is a minimum required level of production for some commodity. C is the *cost matrix* where each entry is the unit cost of the corresponding activity in X. The dual solution matrix U is the *imputed cost matrix* whose entries are the costs imputed to the required commodities.

We see that $x = 14 > 0$ in the solution to the primal furniture manufacturing problem, and $x = 0$ in the dual. Also, $u = \frac{10}{3} > 0$ in the dual and $u = 0$ in the primal problem. In general, this is the complementary nature of the solutions to the primal and dual problems. Each variable that has a positive value in the solution of the primal problem has the value 0 in the solution of the dual. Likewise, if a variable has a positive value in the solution to the dual problem, then it has the value 0 in the solution of the primal problem. This result is called the *principle of complementary slackness*.

The economic interpretation of complementary slackness is this: If a slack variable from the primal problem is positive, then having available more of the corresponding resource cannot improve the value of the objective function. For instance, since $w > 0$ in the solution of the primal problem and $2x + 6y + w = 72$, not all of the labor available for upholstery will be used in the optimal production schedule. Therefore, the marginal value of the resource is zero. As we saw in Section 4.4, the marginal value of a resource is the final value in the column of its slack variable in the primal problem. But this number is the value of the main variable in the solution of the dual problem.

When we gave an economic interpretation to the dual of the furniture manufacturing problem, we first had to assign units to each of the variables of the dual problem. For instance, u was in units of profit per man-hours of carpentry, or

$$\frac{\text{profit}}{\text{man-hours of carpentry}}.$$

The following systematic procedure can often be used to obtain the units for the variables of the dual problem from units appearing in the primal problem. Assume that the primal problem is stated in one of the two forms given on page 178.

1. Replace each entry of the matrix A by its units, written in fraction form. Label each column and row with the corresponding variable.
2. Replace each entry of the matrix C by its units, written in fraction form.

3. To find the units for a variable of the dual problem, select any entry in its row in A, divide the corresponding entry in C by the entry chosen in A, and simplify the fraction.

For the furniture manufacturing problem, the matrices are:

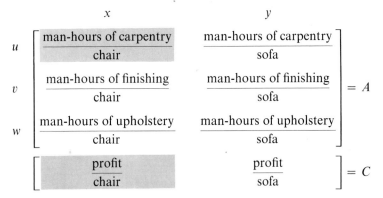

Using the first entry in the row labeled u, the units for u in the dual problem are:

$$\frac{\text{profit}}{\text{chair}} \div \frac{\text{man-hours of carpentry}}{\text{chair}} = \frac{\text{profit}}{\text{chair}} \cdot \frac{\text{chair}}{\text{man-hours of carpentry}}$$

$$= \frac{\text{profit}}{\text{man-hours of carpentry}}.$$

Hence, u is measured in dollars per man-hour of carpentry.

EXAMPLE 3 A rancher needs to provide a daily dietary supplement to the minks on his ranch. He needs 6 units of protein and 5 units of carbohydrates to add to their regular feed each day. Bran X costs 32 cents per ounce and supplies 4 units of protein and 8 units of carbohydrates per ounce. Wheatchips cost 12 cents per ounce and each ounce supplies 3 units of protein and 2 units of carbohydrates.

(a) Determine the mixture of Bran X and Wheatchips that will meet the daily requirements at minimum cost.

(b) Solve and interpret the dual problem.

Solution (a) Let x be the number of ounces of Bran X and y the number of ounces of Wheatchips to be added daily to the feed. Then we must find the values of x and y that minimize $32x + 12y$ subject to the constraints

$$\begin{cases} 4x + 3y \geq 6 \\ 8x + 2y \geq 5 \\ x \geq 0, \quad y \geq 0. \end{cases}$$

For simplicity, we solve the dual problem. Let u and v be the dual variables. We must find the values of u and v that maximize $6u + 5v$ subject to the constraints

$$\begin{cases} 4u + 8v \le 32 \\ 3u + 2v \le 12 \\ u \ge 0, \quad v \ge 0. \end{cases}$$

The final tableau for the dual problem (with slack variables x and y) is

	u	v	x	y	M	
v	0	1	$\frac{3}{16}$	$-\frac{1}{4}$	0	3
u	1	0	$-\frac{1}{8}$	$\frac{1}{2}$	0	2
M	0	0	$\frac{3}{16}$	$\frac{7}{4}$	1	27

The solution to the rancher's problem is to add $x = \frac{3}{16}$ ounces of Bran X and $y = \frac{7}{4} = 1\frac{3}{4}$ ounces of Wheatchips to the daily feed at a minimum cost of 27 cents per day.

(b) The solution of the dual problem can be read from the tableau above: $u = 2$ and $v = 3$. To find the units of u and v, we construct a chart from the original problem:

$$\begin{array}{c c}
 & \quad x \qquad\qquad\qquad\qquad y \\
\begin{matrix} u \\[2.6em] v \\[2.6em] {} \end{matrix} &
\left[\begin{matrix}
\dfrac{\text{units of protein}}{\text{ounce Bran } X} & \dfrac{\text{units of protein}}{\text{ounce Wheatchips}} \\[1.5em]
\dfrac{\text{units of carbohydrates}}{\text{ounce Bran } X} & \dfrac{\text{units of carbohydrates}}{\text{ounce Wheatchips}} \\[1.5em]
\dfrac{\text{cents}}{\text{ounce Bran } X} & \dfrac{\text{cents}}{\text{ounce Wheatchips}}
\end{matrix}\right]
\end{array}$$

The units for u in the dual problem are

$$\frac{\text{cents}}{\text{ounce Bran } X} \div \frac{\text{units of protein}}{\text{ounce Bran } X} = \frac{\text{cents}}{\text{ounce Bran } X} \cdot \frac{\text{ounce Bran } X}{\text{units of protein}}$$

$$= \frac{\text{cents}}{\text{units of protein}}.$$

We can interpret the dual this way: If someone were to provide the perfect one daily supplement for the minks with 6 units of protein and 5 units of carbohydrates, the rancher would expect to pay 27 cents. He certainly would pay no more, since he could mix his own supplement for that price per day. The value of the protein is 2 cents per unit and the value of the carbohydrates is 3 cents per unit.

A Useful Application of the Dual One type of decision that businesses must make is whether or not to introduce new products. The following example uses the matrix forms of a linear programming problem and its dual to determine the proper course of action.

EXAMPLE 4 Consider the furniture manufacturing problem once again. Suppose that the manufacturer has the same resources but is considering adding a new product to his line, love seats. The manufacture of a love seat requires 3 hours of carpentry, 2 hours of finishing, and 4 hours of upholstery. What profit must he gain per love seat in order to justify adding love seats to his product line?

Solution Let z denote the number of love seats to be produced each day and let p be the profit per love seat. The new furniture manufacturing problem is:

Maximize $80x + 70y + pz$ subject to

$$\begin{cases} 6x + 3y + 3z \leq 96 \\ x + y + 2z \leq 18 \\ 2x + 6y + 4z \leq 72 \\ x \geq 0, \quad y \geq 0. \end{cases}$$

What is the dual of this problem? All inequalities are $\leq$ in the original maximization problem so we may proceed directly to write the dual.

Minimize $96u + 18v + 72w$ subject to

$$\begin{cases} 6u + v + 2w \geq 80 \\ 3u + v + 6w \geq 70 \\ 3u + 2v + 4w \geq p \\ u \geq 0, \quad v \geq 0, \quad w \geq 0 \end{cases}$$

Whereas the new furniture manufacturing problem has one more variable than the original problem, its dual has one more constraint than the dual of the original problem,

$$3u + 2v + 4w \geq p.$$

If the optimal solution to the original problem (just making chairs and sofas) were to remain optimal in the new problem, the variable z would not enter the set of group II variables. If that were the case, the solution to the dual would also remain optimal. But that solution was

$$u = \tfrac{10}{3}, \qquad v = 60, \qquad w = 0.$$

And the new dual would require that $3u + 2v + 4w \geq p$. Since

$$3(\tfrac{10}{3}) + 2(60) + 4(0) = 130,$$

if the profit per love seat is at most $130, the previous solution will remain optimal. That is, the manufacturer should make 14 chairs and 4 sofas and 0 love seats. However, if the profit per love seat exceeds $130, the variable z will enter the set of group II variables and we will find an optimal production schedule in which $z > 0$.

PRACTICE PROBLEMS 5

A linear programming problem involving three variables and four nontrivial inequalities has the number 52 as the maximum value of its objective function.

1. How many variables and nontrivial inequalities will the dual problem have?

2. What is the optimum value for the objective function of the dual problem?

EXERCISES 5

In Exercises 1–6, determine the dual problem of the given linear programming problem.

1. Maximize $4x + 2y$ subject to the constraints

$$\begin{cases} 5x + y \le 80 \\ 3x + 2y \le 76 \\ x \ge 0, \quad y \ge 0. \end{cases}$$

2. Minimize $30x + 60y + 50z$ subject to the constraints

$$\begin{cases} 5x + 3y + z \ge 2 \\ x + 2y + z \ge 3 \\ x \ge 0, \quad y \ge 0, \quad z \ge 0. \end{cases}$$

3. Minimize $10x + 12y$ subject to the constraints

$$\begin{cases} x + 2y \ge 1 \\ -x + y \ge 2 \\ 2x + 3y \ge 1 \\ x \ge 0, \quad y \ge 0. \end{cases}$$

4. Maximize $80x + 70y + 120z$ subject to the constraints

$$\begin{cases} 6x + 3y + 8z \le 768 \\ x + y + 2z \le 144 \\ 2x + 5y \quad\quad \le 216 \\ x \ge 0, \quad y \ge 0, \quad z \ge 0. \end{cases}$$

5. Minimize $3x + 5y + z$ subject to the constraints

$$\begin{cases} 2x - 4y - 6z \le 7 \\ y \ge 10 - 8x - 9z \\ x \ge 0, \quad y \ge 0, \quad z \ge 0. \end{cases}$$

6. Maximize $2x - 3y + 4z - 5w$ subject to the constraints

$$\begin{cases} x + y + z + w - 6 \le 10 \\ 7x + 9y - 4z - 3w \quad\quad \ge 5 \\ x \ge 0, \quad y \ge 0, \quad z \ge 0, \quad w \ge 0. \end{cases}$$

7. The final simplex tableau for the linear programming problem of Exercise 1 above appears below. Give the solution to the problem and to its dual.

$$\begin{array}{c@{\quad}c@{\quad}c@{\quad}c@{\quad}c@{\quad}c}
 & x & y & u & v & M \\
x & \begin{bmatrix} 1 & 0 & \frac{2}{7} & -\frac{1}{7} & 0 \\ 0 & 1 & -\frac{3}{7} & \frac{5}{7} & 0 \\ 0 & 0 & \frac{2}{7} & \frac{6}{7} & 1 \end{bmatrix} & & & & \begin{matrix} 12 \\ 20 \\ 88 \end{matrix}
\end{array}$$

	x	y	u	v	M	
x	1	0	$\frac{2}{7}$	$-\frac{1}{7}$	0	12
y	0	1	$-\frac{3}{7}$	$\frac{5}{7}$	0	20
M	0	0	$\frac{2}{7}$	$\frac{6}{7}$	1	88

8. The final simplex tableau for the *dual* of the linear programming problem of Exercise 2 above appears below. Give the solution to the problem and to its dual.

	u	v	x	y	z	M	
v	5	1	1	0	0	0	30
y	-7	0	-2	1	0	0	0
z	-4	0	-1	0	1	0	20
M	13	0	3	0	0	1	90

9. The final simplex tableau for the *dual* of the linear programming problem of Exercise 3 above appears below. Give the solution to the problem and to its dual.

	u	v	w	x	y	M	
x	3	0	5	1	1	0	22
v	2	1	3	0	1	0	12
M	3	0	5	0	2	1	24

10. The final simplex tableau for the linear programming problem of Exercise 4 above appears below. Give the solution to the problem and to its dual.

	x	y	z	u	v	w	M	
x	1	0	0	$\frac{5}{12}$	$-\frac{5}{3}$	$\frac{1}{12}$	0	98
z	0	0	1	$-\frac{1}{8}$	1	$-\frac{1}{8}$	0	21
y	0	1	0	$-\frac{1}{6}$	$\frac{2}{3}$	$\frac{1}{6}$	0	4
M	0	0	0	$\frac{20}{3}$	$\frac{100}{3}$	$\frac{10}{3}$	1	10,640

In Exercises 11–14, determine the dual problem. Solve either the original problem or its dual by the simplex method and then give the solutions to both.

11. Minimize $3x + y$ subject to the constraints

$$\begin{cases} x + y \geq 3 \\ 2x \quad\;\; \geq 5 \\ x \geq 0, \quad y \geq 0. \end{cases}$$

12. Minimize $3x + 5y + z$ subject to the constraints

$$\begin{cases} x + y + z \geq 20 \\ \quad\;\; y + 2z \geq 0 \\ x \geq 0, \quad y \geq 0, \quad z \geq 0. \end{cases}$$

13. Maximize $10x + 12y + 10z$ subject to the constraints

$$\begin{cases} x - 2y \quad\quad\; \leq 6 \\ 3x \quad\;\; + z \leq 9 \\ \quad\;\; y + 3z \leq 12 \\ x \geq 0, \quad y \geq 0, \quad z \geq 0. \end{cases}$$

14. Maximize $x + 3y$ subject to the constraints

$$\begin{cases} x + y \le 7 \\ x + 2y \le 10 \\ x \ge 0, \quad y \ge 0. \end{cases}$$

15. Give an economic interpretation to the dual of the Cutting Edge Knife Company problem of Example 1 of Section 4.4.

16. Give an economic interpretation to the dual of Exercise 8 of Section 4.3.

17. Give an economic interpretation to the dual of the mining problem of Exercise 6 of Sections 3.1 and 3.2.

18. Give an economic interpretation to the dual of the nutrition problem of Example 1 of Section 3.2.

19. Consider the Cutting Edge Knife Company problem of Example 1 of Section 4.4. Suppose that the company is thinking of also making table knives. If each table knife requires 4 hours of labor, 6 units of steel, and 2 units of wood, what profit must be realized per knife in order to justify adding this product?

SOLUTIONS TO PRACTICE PROBLEMS 5

1. Four variables and three nontrivial inequalities. The number of variables in the dual problem is always the same as the number of nontrivial inequalities in the original problem. The number of nontrivial inequalities in the dual problem is the same as the number of variables in the original problem.

2. Minimum value of 52. The original problem and the dual problem always have the same optimum values. However, if this value is a maximum for one of the problems, it will be a minimum for the other.

Chapter 4: CHECKLIST

- ☐ Standard form of linear programming problem
- ☐ Slack variable
- ☐ Group I, group II variables
- ☐ Simplex tableau
- ☐ Simplex method for problems in standard form
- ☐ Converting minimization problems to maximization problems
- ☐ Reduction of linear programming problems to standard form
- ☐ Dual problem
- ☐ Fundamental theorem of duality
- ☐ Matrix formulation of linear programming problem
- ☐ Marginal analysis
- ☐ Economic interpretation of the dual

Chapter 4: SUPPLEMENTARY EXERCISES

Use the simplex method to solve the following linear programming problems.

1. Maximize $3x + 4y$ subject to the constraints

$$\begin{cases} 2x + y \le 7 \\ -x + y \le 1 \\ x \ge 0, \quad y \ge 0. \end{cases}$$

2. Maximize $2x + 5y$ subject to the constraints

$$\begin{cases} x + y \le 7 \\ 4x + 3y \le 24 \\ x \ge 0, \quad y \ge 0. \end{cases}$$

3. Maximize $2x + 3y$ subject to the constraints

$$\begin{cases} x + 2y \le 14 \\ x + y \le 9 \\ 3x + 2y \le 24 \\ x \ge 0, \quad y \ge 0. \end{cases}$$

4. Maximize $3x + 7y$ subject to the constraints

$$\begin{cases} x + 2y \le 10 \\ 4x + 3y \le 30 \\ -2x + y \le 0 \\ x \ge 0, \quad y \ge 0. \end{cases}$$

5. Minimize $x + y$ subject to the constraints

$$\begin{cases} 7x + 5y \ge 40 \\ x + 4y \ge 9 \\ x \ge 0, \quad y \ge 0. \end{cases}$$

6. Minimize $3x + 2y$ subject to the constraints

$$\begin{cases} x + y \ge 6 \\ x + 2y \ge 0 \\ x \ge 0, \quad y \ge 0. \end{cases}$$

7. Minimize $20x + 30y$ subject to the constraints

$$\begin{cases} x + 4y \ge 8 \\ x + y \ge 5 \\ 2x + y \ge 7 \\ x \ge 0, \quad y \ge 0. \end{cases}$$

8. Minimize $5x + 7y$ subject to the constraints

$$\begin{cases} 2x + y \geq 10 \\ 3x + 2y \geq 18 \\ x + 2y \geq 10 \\ x \geq 0, \quad y \geq 0. \end{cases}$$

9. Maximize $36x + 48y + 70z$ subject to the constraints

$$\begin{cases} x \leq 4 \\ y \leq 6 \\ z \leq 8 \\ 4x + 3y + 2z \leq 38 \\ x \geq 0, \quad y \geq 0, \quad z \geq 0. \end{cases}$$

10. Maximize $3x + 4y + 5z + 4w$ subject to the constraints

$$\begin{cases} 6x + 9y + 12z + 15w \leq 672 \\ x - y + 2z + 2w \leq 92 \\ 5x + 10y - 5z + 4w \leq 280 \\ x \geq 0, \quad y \geq 0, \quad z \geq 0, \quad w \geq 0. \end{cases}$$

11. Determine the dual problem of the linear programming problem in Exercise 3 above.

12. Determine the dual problem of the linear programming problem in Exercise 7 above.

13. The final simplex tableau for the linear programming problem of Exercise 3 appears below. Give the solution to the problem and to its dual.

	x	y	u	v	w	M	
y	0	1	1	-1	0	0	5
x	1	0	-1	2	0	0	4
w	0	0	1	-4	1	0	2
M	0	0	1	1	0	1	23

14. The final simplex tableau for the *dual* of the linear programming problem of Exercise 7 appears below. Give the solution to the problem and to its dual.

	u	v	w	x	y	M	
v	0	1	$\frac{7}{3}$	$\frac{4}{3}$	$-\frac{1}{3}$	0	$\frac{50}{3}$
u	1	0	$-\frac{1}{3}$	$-\frac{1}{3}$	$\frac{1}{3}$	0	$\frac{10}{3}$
	0	0	2	4	1	1	110

15, 16. For each of the linear programming problems in Exercises 3 and 7, identify the matrices A, B, C, X, U and state the problem and its dual in terms of matrices.

17. Consider Exercise 16 of Section 3.2.

(a) Solve the problem by the simplex method.

(b) The Bluejay Lacrosse Stick Company is considering diversifying by also making tennis rackets. A tennis racket requires 1 man-hour for cutting, 4 man-hours for stringing, and 2 man-hours for finishing. How much profit must the company be able to make on each tennis racket in order to justify the diversification?

18. Consider the stereo store of Exercise 16 in Section 4.2. A fourth brand of stereo system has appeared on the market. Brand D takes up 3 cubic feet of storage space and generates a sales commission of $30. What profit would the store have to realize on the sale of each brand D stereo set in order to justify carrying it?

5

SETS AND COUNTING

5.1   **Sets**

5.2   **A Fundamental Principle of Counting**

5.3   **Venn Diagrams and Counting**

5.4   **The Multiplication Principle**

5.5   **Permutations and Combinations**

5.6   **Further Counting Problems**

5.7   **The Binomial Theorem**

5.8   **Multinomial Coefficients and Partitions**

In this chapter we introduce some ideas useful in the study of probability (Chapter 6). Our first topic, the theory of sets, will provide a convenient language and notation in which to discuss probability. Using set theory, we will develop a number of counting principles which can also be applied to computing probabilities.

5.1. Sets

In many applied problems one must consider collections of various sorts of objects. For example, a survey of unemployment might consider the collection of all U.S. cities with current unemployment greater than 7%. A study of birthrates might consider the collection of countries with a current birthrate less than 20 per 1000 population. Such collections are examples of sets. A *set* is any collection of objects. The objects, which may be countries, cities, years, numbers, letters, or anything else, are called the *elements* of the set. A set is often specified by a listing of its elements inside a pair of braces. For example, the set whose elements are the first six letters of the alphabet is written

$$\{a, b, c, d, e, f\}.$$

Similarly, the set whose elements are the even numbers between 1 and 11 is written

$$\{2, 4, 6, 8, 10\}.$$

We can also specify a set by giving a description of its elements (without actually listing the elements). For example, the set $\{a, b, c, d, e, f\}$ can also be written

$$\{\text{the first six letters of the alphabet}\},$$

and the set $\{2, 4, 6, 8, 10\}$ can be written

$$\{\text{all even numbers between 1 and 11}\}.$$

For convenience, we usually denote sets by capital letters A, B, C, and so on.
The great diversity of sets is illustrated by the following examples:

1. In a linear programming problem, the feasible set is the set of all points satisfying a system of linear inequalities. The feasible set of the furniture manufacturing problem is the set of all points on or inside the five-sided region in Fig. 1.

FIGURE 1

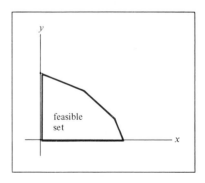

2. Let $B = \{$license plate numbers consisting of three letters followed by three digits$\}$. Some typical elements of B are

$$\text{SBG } 602, \quad \text{GXZ } 179, \quad \text{YHJ } 006.$$

The number of elements in B is sufficiently large so that listing all of them is impractical. However, in this chapter we will develop a technique that allows us to calculate the number of elements of B.

3. Let $C = \{$possible sequences of outcomes of tossing a coin three times$\}$. If we let H denote "heads" and T denote "tails," the various sequences can be easily described:

$$C = \{\text{HHH, THH, HTH, HHT, TTH, THT, HTT, TTT}\},$$

where, for instance, HTH means "first toss heads, second toss tails, third toss heads."

Sets arise in many practical contexts, as the next example shows.

EXAMPLE 1 The following table gives the rate of inflation, as measured by the percentage change in the consumer price index, for the years 1971–1986.

Year	Inflation (%)	Year	Inflation (%)
1971	4.3	1979	11.3
1972	3.3	1980	13.5
1973	6.2	1981	10.4
1974	11.0	1982	6.1
1975	9.1	1983	3.2
1976	5.8	1984	4.3
1977	6.5	1985	3.6
1978	7.7	1986	1.9

Let

$$A = \{\text{years from 1971 to 1986 in which inflation was above 10\%}\},$$

$$B = \{\text{years from 1971 to 1986 in which inflation was below 5\%}\}.$$

Determine the elements of A and B.

Solution By reading the table, we see that

$$A = \{1974, 1979, 1980, 1981\},$$

$$B = \{1971, 1972, 1983, 1984, 1985, 1986\}.$$

Suppose we are given two sets, A and B. Then it is possible to form new sets from A and B as follows: The *union* of A and B, denoted $A \cup B$, is the set consisting of all those elements which belong to either A *or* B (or both). The *intersection* of A and B, denoted $A \cap B$, is the set consisting of those elements which belong to both A *and* B. For example, let

$$A = \{1, 2, 3, 4\}, \quad B = \{1, 3, 5, 7, 11\}.$$

Then, since $A \cup B$ consists of those elements belonging to either A or B (or both), we have

$$A \cup B = \{1, 2, 3, 4, 5, 7, 11\}.$$

Moreover, since $A \cap B$ consists of those elements that belong to both A and B, we have

$$A \cap B = \{1, 3\}.$$

EXAMPLE 2 Here is a table giving the rates of unemployment and inflation for the years 1976–1986.

Year	Unemployment (%)	Inflation (%)
1976	7.6	5.8
1977	6.9	6.5
1978	6.0	7.7
1979	5.8	11.3
1980	7.0	13.5
1981	7.5	10.4
1982	9.5	6.1
1983	9.5	3.2
1984	7.4	4.3
1985	7.1	3.6
1986	6.9	1.9

Let

$A = \{$years from 1976 to 1986 in which unemployment is at least 7%$\}$,

$B = \{$years from 1976 to 1986 in which the inflation rate is at least 7%$\}$.

(a) Describe the sets $A \cap B$ and $A \cup B$.

(b) Determine the elements of A, B, $A \cap B$, and $A \cup B$.

Solution (a) From the descriptions of A and B, we have

$A \cap B = \{$years from 1976 to 1986 in which unemployment is at least 7% and inflation is at least 7%$\}$

$A \cup B = \{$years from 1976 to 1986 in which either unemployment or inflation (or both) is at least 7%$\}$.

(b) From the table, we see that

$$A = \{1976, 1980, 1981, 1982, 1983, 1984, 1985\}$$

$$B = \{1978, 1979, 1980, 1981\}$$

$$A \cap B = \{1980, 1981\}$$

$$A \cup B = \{1976, 1978, 1979, 1980, 1981, 1982, 1983, 1984, 1985\}$$

We have defined the union and the intersection of two sets. In a similar manner, we can define the union and intersection of any number of sets. For example, if A, B, C are three sets, then their union, denoted $A \cup B \cup C$, is the set whose elements are precisely those which belong to at least one of the sets A, B, C. Similarly, the intersection of A, B, C, denoted $A \cap B \cap C$, is the set consisting of those elements which belong to all the sets A, B, C. In a similar way, we may define the union and intersection of more than three sets.

Suppose that we are given a set A. We may form new sets by selecting elements from A. Sets formed in this way are called *subsets* of A. More precisely, a set B is called a *subset* of A provided that every element of B is also an element of A. For example, let $A = \{1, 2, 3\}$, $B = \{1, 3\}$. Since the elements of B (1 and 3) are also elements of A, B is a subset of A.

One set which is very often considered is the set which contains no elements at all. This set is called the *empty set* (or *null set*) and is written $\varnothing$. The empty set is a subset of every set.*

EXAMPLE 3 Let $A = \{a, b, c\}$. Find all subsets of A.

Solution Since A contains three elements, every subset of A has at most three elements. We look for subsets according to the number of elements:

Number of elements in subset	Possible subsets
0	$\varnothing$
1	$\{a\}$, $\{b\}$, $\{c\}$
2	$\{a, b\}$, $\{a, c\}$, $\{b, c\}$
3	$\{a, b, c\}$

Thus, we see that A has eight subsets, namely those listed on the right. (Note that we count A as a subset of itself.)

It is usually convenient to regard all sets involved in a particular discussion as subsets of a single larger set. Thus, for example, if a problem involves the sets $\{a, b, c\}$, $\{e, f\}$, $\{g\}$, $\{b, x, y\}$, then we can regard all of these as subsets of the set

$$U = \{\text{all letters of the alphabet}\}.$$

Since U contains all sets being discussed, it is called a *universal set* (for the particular problem). In this book we shall always specify the particular universal set we have in mind.

Suppose that U is a universal set and A is a subset of U. The set of elements of U which are not in A is called the *complement* of A, denoted A'. For example, if

$$U = \{1, 2, 3, 4, 5, 6, 7, 8, 9\}$$

and

$$A = \{2, 4, 6, 8\},$$

---

* Here is why: Let A be any set. Every element of $\varnothing$ also belongs to A. If you do not agree, then you must produce an element of $\varnothing$ which does not belong to A. But you cannot, since $\varnothing$ has no elements. So $\varnothing$ is a subset of A.

then
$$A' = \{1, 3, 5, 7, 9\}.$$

EXAMPLE 4 Let $U = \{a, b, c, d, e, f, g\}$, $S = \{a, b, c\}$, $T = \{a, c, d\}$. List the elements of the sets

(a) S' (b) T' (c) $(S \cap T)'$ (d) $S' \cap T'$

Solution (a) S' consists of those elements of U which are not in S, so $S' = \{d, e, f, g\}$.

(b) Similarly, $T' = \{b, e, f, g\}$.

(c) To determine $(S \cap T)'$ we must first determine $S \cap T$:
$$S \cap T = \{a, c\}.$$

Then we determine the complement of this set:
$$(S \cap T)' = \{b, d, e, f, g\}.$$

(d) We determined S' and T' in parts (a) and (b). The set $S' \cap T'$ consists of the elements that belong to both S' and T'. Therefore, referring to (a) and (b), we have
$$S' \cap T' = \{e, f, g\}.$$

PRACTICE PROBLEMS 1

1. Let $U = \{a, b, c, d, e, f, g\}$, $R = \{a, b, c, d\}$, $S = \{c, d, e\}$, $T = \{c, e, g\}$. List the elements of the sets

   (a) R' (b) $R \cap S$

   (c) $(R \cap S) \cap T$ (d) $R \cap (S \cap T)$

2. Let $U = \{$all Nobel prize winners$\}$, $W = \{$women who have won Nobel prizes$\}$, $A = \{$Americans who have won Nobel prizes$\}$, $L = \{$winners of the Nobel prize in literature$\}$. Describe the following sets.

   (a) W' (b) $A \cap L'$ (c) $W \cap A \cap L'$

3. Refer to Problem 2. Use set-theoretic notation to describe $\{$Nobel prize winners who are American men or recipients of the prize in literature$\}$.

EXERCISES 1

1. Let $U = \{1, 2, 3, 4, 5, 6, 7\}$, $S = \{1, 2, 3, 4\}$, $T = \{1, 3, 5, 7\}$. List the elements of the sets
   (a) S' (b) $S \cup T$ (c) $S \cap T$ (d) $S' \cap T$

2. Let $U = \{1, 2, 3, 4, 5\}$, $S = \{1, 2, 3\}$, $T = \{5\}$. List the elements of the sets
   (a) S' (b) $S \cup T$ (c) $S \cap T$ (d) $S' \cap T$

3. Let $U = \{$all letters of the alphabet$\}$, $R = \{a, b, c\}$, $S = \{c, d, e, f\}$, $T = \{x, y, z\}$. List the elements of the sets

(a) $R \cup S$ (b) $R \cap S$ (c) $S \cap T$

4. Let $U = \{a, b, c, d, e, f, g\}$, $R = \{a\}$, $S = \{a, b\}$, $T = \{b, d, e, f, g\}$. List the elements of the sets

(a) $R \cup S$ (b) $R \cap S$ (c) T' (d) $T' \cup S$

5. List all subsets of the set $\{1, 2\}$.

6. List all subsets of the set $\{1\}$.

7. Let $U = \{$all college students$\}$, $M = \{$all male college students$\}$, $F = \{$all college students who like football$\}$. Describe the elements of the sets

(a) $M \cap F$ (b) M' (c) $M' \cap F'$ (d) $M \cup F$

8. Let $U = \{$all corporations$\}$, $S = \{$all corporations with headquarters in New York City$\}$, $T = \{$all privately owned corporations$\}$. Describe the elements of the sets

(a) S' (b) T' (c) $S \cap T$ (d) $S \cap T'$

9. The Standard and Poor's Index measures the price of a certain collection of 500 stocks. Table 1 compares the percentage change in the index during the first 5 days of certain years with the percentage change for the entire year. Let $U = \{$all years from 1950 to 1977$\}$, $S = \{$all years during which the Index increased by 2% or more during the first 5 days$\}$, $T = \{$all years for which the index increased by 16% or more during the entire year$\}$. List the elements of the sets

(a) S (b) T (c) $S \cap T$ (d) $S' \cap T$ (e) $S \cap T'$

10. Refer to Table 1. Let $U = \{$all years from 1950 to 1977$\}$, $A = \{$all years during which the Index declined during the first 5 days$\}$, $B = \{$all years during which the Index declined

TABLE 1   Percentage Change in the Standard and Poor's Index

	Percent change for first 5 days	Percent change for year		Percent change for first 5 days	Percent change for year
1977	−2.3	−12.36	1963	2.6	18.9
1976	4.9	19.1	1962	−3.4	−11.8
1975	2.2	31.5	1961	1.2	23.1
1974	−1.5	−29.7	1960	−0.7	−3.0
1973	1.5	−17.4	1959	0.3	8.5
1972	1.4	15.6	1958	2.5	38.1
1971	0.0	10.8	1957	−0.9	−14.3
1970	0.7	−0.1	1956	−2.1	2.6
1969	−2.9	−11.4	1955	−1.8	26.4
1968	0.2	−7.7	1954	0.5	45.0
1967	3.1	20.1	1953	−0.9	−6.6
1966	0.8	−13.1	1952	0.6	11.8
1965	0.7	9.1	1951	2.3	16.5
1964	1.3	13.0	1950	2.0	21.8

for the entire year}. List the elements of the sets

(a) A (b) B (c) $A \cap B$ (d) $A' \cap B$ (e) $A \cap B'$

11. Refer to Exercise 9. Describe verbally the fact that $S \cap T' = \emptyset$.

12. Refer to Exercise 10. Describe verbally the fact that $A \cap B'$ has two elements.

13. Let $U = \{a, b, c, d, e, f\}$, $R = \{a, b, c\}$, $S = \{a, c, e\}$, $T = \{e, f\}$. List the elements of the following sets.

(a) $(R \cup S)'$ (b) $R \cup S \cup T$ (c) $R \cap S \cap T$

(d) $R \cap S \cap T'$ (e) $R' \cap S \cap T$ (f) $S \cup T$

(g) $(R \cup S) \cap (R \cup T)$ (h) $(R \cap S) \cup (R \cap T)$ (i) $R' \cap T'$

14. Let $U = \{1, 2, 3, 4, 5\}$, $R = \{1, 3, 5\}$, $S = \{3, 4, 5\}$, $T = \{2, 4\}$. List the elements of the following sets.

(a) $R \cap S \cap T$ (b) $R \cap S \cap T'$ (c) $R \cap S' \cap T$

(d) $R' \cap T$ (e) $R \cup S$ (f) $R' \cup R$

(g) $(S \cap T)'$ (h) $S' \cup T'$

In Exercises 15–20, simplify the given expression.

15. $(S')'$ 16. $S \cap S'$ 17. $S \cup S'$

18. $S \cap \emptyset$ 19. $T \cap S \cap T'$ 20. $S \cup \emptyset$

A large corporation classifies its many divisions by their performance in the preceding year. Let $P = \{$divisions that made a profit$\}$, $L = \{$divisions that had an increase in labor costs$\}$, and $T = \{$divisions whose total revenue increased$\}$. Describe the following sets using set-theoretic notation.

21. {divisions that had increases in labor costs or total revenue}

22. {divisions that did not make a profit}

23. {divisions that made a profit despite an increase in labor costs}

24. {divisions that had an increase in labor costs and were either unprofitable or did not increase their total revenue}

25. {profitable divisions with increases in labor costs and total revenue}

26. {divisions that were unprofitable or did not have increases in either labor costs or total revenue}

An automobile insurance company classifies applicants by their driving records for the previous three years. Let $S = \{$applicants who have received speeding tickets$\}$, $A = \{$applicants who have caused accidents$\}$, and $D = \{$applicants who have been arrested for driving while intoxicated$\}$. Describe the sets in Exercises 27–32 using set-theoretic notation.

27. {applicants who have not received speeding tickets}

28. {applicants who have caused accidents and been arrested for drunk driving}

29. {applicants who received speeding tickets, caused accidents, or were arrested for drunk driving}

30. {applicants who have not been arrested for drunk driving but have received speeding tickets or have caused accidents}

31. {applicants who have not both caused accidents and received speeding tickets but who have been arrested for drunk driving}

32. {applicants who have not caused accidents or have not been arrested for drunk driving}

Let U = {people at Mount College}, A = {students at Mount College}, B = {teachers at Mount College}, C = {females at Mount College}, and D = {males at Mount College}. Describe verbally the sets in Exercises 33 to 40.

33. $A \cap D$ 34. $B \cap C$ 35. $A \cap B$ 36. $B \cup C$

37. $A \cup C'$ 38. $(A \cap D)'$ 39. D' 40. $D \cap U$

Let U = {all people}, S = {people who like strawberry ice cream}, V = {people who like vanilla ice cream}, and C = {people who like chocolate ice cream). Describe the sets in Exercises 41 to 46 using set-theoretical notation.

41. {people who don't like strawberry ice cream}

42. {people who like vanilla but not chocolate ice cream}

43. {people who like vanilla or chocolate but not strawberry ice cream}

44. {people who don't like any of the three flavors of ice cream}

45. {people who don't like either chocolate or vanilla ice cream}

46. {people who like only strawberry ice cream}

47. Let U be the set of vertices in the feasible set on the right. Let R = {vertices (x, y) with $x > 0$}, S = {vertices (x, y) with $y > 0$}, and T = {vertices (x, y) with $x \le y$}. List the elements of the following sets:

     (a) R (b) S (c) T

     (d) $R' \cup S$ (e) $R' \cap T$ (f) $R \cap S \cap T$

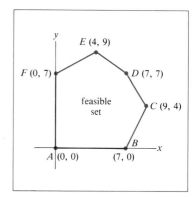

48. Sam ordered a baked potato at a restaurant. The waitress offered him butter, cheese, and chives as toppings. How many different ways can he have his potato? List them.

49. Let S = {1, 3, 5, 7} and T = {2, 5, 7}. Give an example of a subset of T that is not a subset of S.

50. Suppose that S and T are subsets of the set U. Under what circumstance will $S \cap T = T$?

51. Suppose that S and T are subsets of the set U. Under what circumstance will $S \cup T = T$?

52. Find three subsets of the set of integers from 1 through 10, R, S, and T, such that $R \cup (S \cap T)$ is different from $(R \cup S) \cap T$.

SOLUTIONS TO PRACTICE PROBLEMS 1

1. (a) $\{e, f, g\}$.

   (b) $\{c, d\}$.

   (c) $\{c\}$. This problem asks for the intersection of two sets. The first set is $R \cap S = \{c, d\}$ and the second set is $T = \{c, e, g\}$. The intersection of these sets is $\{c\}$.

   (d) $\{c\}$. Here again the problem asks for the intersection of two sets. However, now the first set is $R = \{a, b, c, d\}$ and the second set is $S \cap T = \{c, e\}$. The intersection of these sets is $\{c\}$.

   [*Note:* It should be expected that the set $(R \cap S) \cap T$ is the same as the set $R \cap (S \cap T)$, for each set consists of those elements that are in all three sets. Therefore, each of these sets equals the set $R \cap S \cap T$.]

2. (a) $W' = \{$men who have won Nobel prizes$\}$. This is so since W' consists of those elements of U that are not in W—that is, those Nobel prize winners who are not women.

   (b) $A \cap L' = \{$Americans who have received Nobel prizes in fields other than literature$\}$.

   (c) $W \cap A \cap L' = \{$American women who have received Nobel prizes in fields other than literature$\}$. This is so since to qualify for $W \cap A \cap L'$, a Nobel prize winner must simultaneously be in W, in A, and in L'—that is, a woman, an American, and not a winner of the Nobel prize in literature.

3. $(A \cap W') \cup L$.

5.2. A Fundamental Principle of Counting

A counting problem is one that requires us to determine the number of elements in a set S. Counting problems arise in many applications of mathematics and comprise the mathematical field of *combinatorics*. We shall study a number of different sorts of counting problems in the remainder of this chapter.

If S is any set, we will denote the number of elements in S by $n(S)$. Thus, for example, if $S = \{1, 7, 11\}$, then $n(S) = 3$, and if $S = \{a, b, c, d, e, f, g, h, i\}$, then $n(S) = 9$. Of course, if $S = \varnothing$, the empty set, then $n(S) = 0$. (The empty set contains no elements.)

Let us begin by stating one of the fundamental principles of counting, the so-called *inclusion-exclusion principle*.

> *Inclusion-Exclusion Principle*   Let S and T be sets. Then
> $$n(S \cup T) = n(S) + n(T) - n(S \cap T). \tag{1}$$

Note that formula (1) connects the four quantities $n(S \cup T)$, $n(S)$, $n(T)$, $n(S \cap T)$. Given any three, the remaining quantity can be determined by using this formula.

To test the plausibility of the inclusion-exclusion principle, consider this example. Let $S = \{a, b, c, d, e\}$, $T = \{a, c, g, h\}$. Then

$$S \cup T = \{a, b, c, d, e, g, h\} \qquad n(S \cup T) = 7$$

$$S \cap T = \{a, c\} \qquad n(S \cap T) = 2.$$

In this case the inclusion-exclusion principle reads

$$n(S \cup T) = n(S) + n(T) - n(S \cap T),$$

$$7 \quad = 5 + 4 - 2$$

which is correct.

Here is the reason for the validity of the inclusion-exclusion principle: The left side of formula (1) is $n(S \cup T)$, the number of elements in either S or T (or both). As a first approximation to this number, add the number of elements in S to the number of elements in T, obtaining $n(S) + n(T)$. However, if an element lies in both S and T, it is counted twice—once in $n(S)$ and again in $n(T)$. To make up for this double counting we must subtract off the number of elements counted twice, namely $n(S \cap T)$. This gives us $n(S) + n(T) - n(S \cap T)$ as the number of elements in $S \cup T$.

The next example illustrates a typical use of the inclusion-exclusion principle in an applied problem.

EXAMPLE 1 In 1986, *Executive* magazine surveyed the presidents of the 500 largest corporations in the United States. Of these 500 people, 310 had degrees (of any sort) in business, 238 had undergraduate degrees in business, and 184 had graduate degrees in business. How many presidents had both undergraduate and graduate degrees in business?

Solution Let

$$S = \{\text{presidents with an undergraduate degree in business}\},$$

$$T = \{\text{presidents with a graduate degree in business}\}.$$

Then

$$S \cup T = \{\text{presidents with at least one degree in business}\},$$

$$S \cap T = \{\text{presidents with both undergraduate and graduate degrees in business}\}.$$

From the data given we have

$$n(S) = 238, \qquad n(T) = 184, \qquad n(S \cup T) = 310.$$

The problem asks for $n(S \cap T)$. By the inclusion-exclusion principle we have

$$n(S \cup T) = n(S) + n(T) - n(S \cap T)$$

$$310 = 238 + 184 - n(S \cap T)$$

$$n(S \cap T) = 112.$$

That is, exactly 112 of the presidents had both undergraduate and graduate degrees in business.

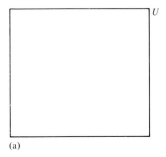

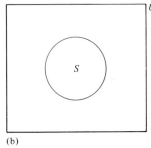

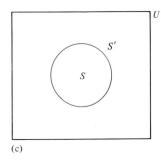

(a) (b) (c)

FIGURE 1

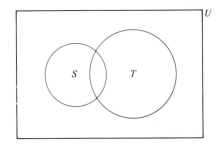

FIGURE 2

It is possible to visualize sets geometrically by means of drawings known as *Venn diagrams*. Such graphical representations of sets are very useful tools in solving counting problems. In order to describe Venn diagrams, let us begin with a single set S contained in a universal set U. Draw a rectangle and view its points as the elements of U [Fig. 1(a)]. To show that S is a subset of U we draw a circle inside the rectangle and view S as the set of points in the circle [Fig. 1(b)]. The resulting diagram is called a *Venn diagram* of S. It illustrates the proper relationship between S and U. Since S' consists of those elements of U which are not in S, we may view the portion of the rectangle which is outside of the circle as representing S' [Fig. 1(c)].

Venn diagrams are particularly useful for visualizing the relationship between two or more sets. Suppose that we are given two sets S and T in a universal set U. As before, we represent each of the sets by means of a circle inside the rectangle (Fig. 2).

We can now illustrate a number of sets by shading in appropriate regions of the rectangle. For instance, in Fig. 3 we have shaded the regions corresponding to T, $S \cup T$, and $S \cap T$.

FIGURE 3

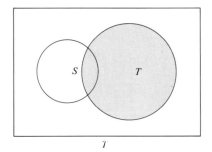

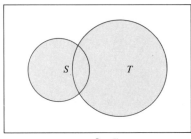

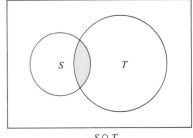

T $S \cup T$ $S \cap T$

EXAMPLE 2 Shade the portions of the rectangle corresponding to the sets

(a) $S \cap T'$ (b) $(S \cap T')'$

Solution (a) $S \cap T'$ consists of the points in S and in T', that is, the points in S and not in T. So we shade the points that are in the circle S but are not in the circle T [Fig. 4(a)].

(b) $(S \cap T')'$ is the complement of the set $(S \cap T')$. Therefore, it consists of exactly those points not shaded in Fig. 4(a). [See Fig. 4(b).]

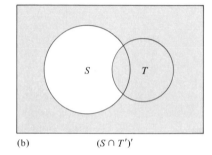

(a) $S \cap T'$ (b) $(S \cap T')'$

FIGURE 4

In a similar manner, Venn diagrams can illustrate intersections and unions of three sets. Some representative regions are shaded in Fig. 5.

FIGURE 5

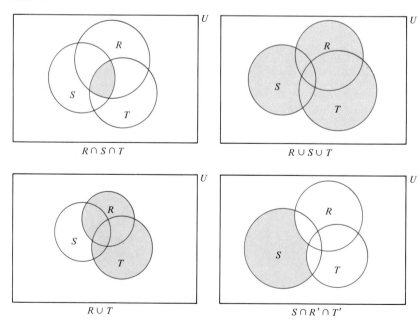

$R \cap S \cap T$ $R \cup S \cup T$

$R \cup T$ $S \cap R' \cap T'$

There are many formulas expressing relationships between intersections and unions of sets. Possibly the most fundamental are the two formulas known as De Morgan's laws.

> *De Morgan's Laws* Let S and T be sets. Then
> $$(S \cup T)' = S' \cap T'$$
> and
> $$(S \cap T)' = S' \cup T'.$$

In other words, De Morgan's laws state that to form the complement of a union (or intersection) form the complements of the individual sets and change unions to intersections (or intersections to unions).

Verification of De Morgan's Laws Let us utilize Venn diagrams to describe $(S \cup T)'$. In Fig. 6(a) we have shaded the area corresponding to $S \cup T$. In Fig. 6(b) we have shaded the area corresponding to $(S \cup T)'$. On the other hand, in Fig. 6(c) we have shaded the area corresponding to S' and in Fig. 6(d) the area corresponding to T'. By considering the common shaded areas of Fig. 6(c) and (d) we arrive at the shaded area corresponding to $S' \cap T'$ [Fig. 6(e)]. Note that this is the same region as shaded in Fig. 6(b).

FIGURE 6

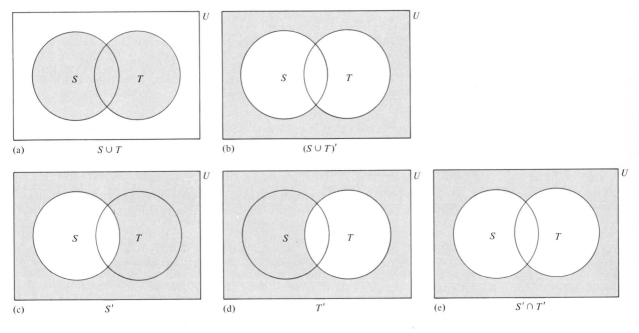

(a) $S \cup T$ (b) $(S \cup T)'$

(c) S' (d) T' (e) $S' \cap T'$

Therefore,

$$(S \cup T)' = S' \cap T'.$$

This verifies the first of De Morgan's laws. The proof of the second law is similar.

PRACTICE PROBLEMS 2

1. Draw a two-circle Venn diagram and shade the portion corresponding to the set $(S \cap T') \cup (S \cap T)$.

2. Suppose that $n(S) + n(T) = n(S \cup T)$. What can you conclude about S and T?

EXERCISES 2

1. Find $n(S \cup T)$, given that $n(S) = 5$, $n(T) = 4$, and $n(S \cap T) = 2$.

2. Find $n(S \cup T)$, given that $n(S) = 17$, $n(T) = 13$, and $n(S \cap T) = 9$.

3. Find $n(S \cap T)$, given that $n(S) = 7$, $n(T) = 8$, and $n(S \cup T) = 15$.

4. Find $n(S \cap T)$, given that $n(S) = 4$, $n(T) = 12$, and $n(S \cup T) = 15$.

5. Find $n(S)$, given that $n(T) = 7$, $n(S \cap T) = 5$, and $n(S \cup T) = 13$.

6. Find $n(T)$, given that $n(S) = 14$, $n(S \cap T) = 6$, and $n(S \cup T) = 14$.

7. If $n(S) = n(S \cap T)$, what can you conclude about S and T?

8. If $n(S) = n(S \cup T)$, what can you conclude about S and T?

9. Suppose that each of the 180 million adults in South America is fluent in Portuguese or Spanish. If 99 million are fluent in Portuguese and 95 million are fluent in Spanish, how many are fluent in both languages?

10. Suppose that all of the 1000 freshmen at a certain college are enrolled in a math or an English course. Suppose that 400 are taking both math and English and 600 are taking English. How many are taking a math course?

11. The combined membership of the MAA (Mathematical Association of America) and the AMS (American Mathematical Society) is approximately 43,000. Twenty-three thousand people belong to the AMS and 7000 of them also belong to the MAA. How many people belong to the MAA?

12. A survey of employees in a certain company revealed that 300 people subscribe to *Newsweek*, 200 subscribe to *Time*, and 50 subscribe to both. How many people subscribe to at least one of these magazines?

13. Motors Inc. manufactured 325 cars with automatic transmissions, 216 with power steering, and 89 with both of these options. How many cars were manufactured with at least one of the two options?

14. A survey of 100 investors in stocks and bonds revealed that 80 investors owned stocks and 70 owned bonds. How many investors owned both stocks and bonds?

In Exercises 15–26, draw a two-circle Venn diagram and shade the portion corresponding to the set.

15. $S' \cap T'$

16. $S' \cap T$

17. $S \cup T'$

18. $S' \cup T'$

19. $(S \cap T)'$

20. $(S' \cap T)'$

21. $(S \cap T') \cup (S' \cap T)$

22. $(S \cap T) \cup (S' \cap T')$

23. $S \cup (S \cap T)$

24. $S \cup (T' \cup S)$

25. $S \cup S'$

26. $S \cap S'$

In Exercises 27–38, draw a three-circle Venn diagram and shade the portion corresponding to the set.

27. $R \cap S \cap T'$

28. $R' \cap S' \cap T$

29. $R \cup (S \cap T)$

30. $R \cap (S \cup T)$

31. $R \cap (S' \cup T)$

32. $R' \cup (S \cap T')$

33. $R \cap T$

34. $S \cap T'$

35. $R' \cap S' \cap T'$

36. $(R \cup S \cup T)'$

37. $(R \cap T) \cup (S \cap T')$

38. $(R \cup S') \cap (R \cup T')$

In Exercises 39–44, use De Morgan's laws to simplify the given expression.

39. $S' \cup (S \cap T)'$

40. $T \cap (S \cup T)'$

41. $(S' \cup T)'$

42. $(S' \cap T')'$

43. $T \cup (S \cap T)'$

44. $(S' \cap T)' \cup S$

In Exercises 45–50, give a set-theoretic expression that describes the shaded portion of the Venn diagram.

45.

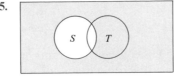

46.

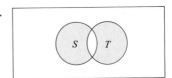

47.

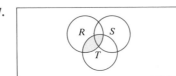

48.

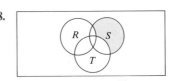

49.

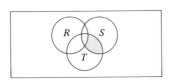

50.

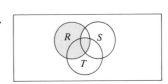

By drawing a Venn diagram, replace each of the expressions in Exercises 51–53 with one involving at most one union and the complement symbol applied only to R, S, and T.

51. $(T \cap S) \cup (T \cap R) \cup (R \cap S') \cup (T \cap R' \cap S')$

52. $(R \cap S) \cup (S \cap T) \cup (R \cap S' \cap T')$

53. $((R \cap S') \cup (S \cap T') \cup (T \cap R'))'$

1. $(S \cap T') \cup (S \cap T)$ is given as a union of two sets, $S \cap T'$ and $S \cap T$. The Venn diagrams for these two sets are given in Fig. 7(a) and (b). The desired set consists of the elements which are in one or the other (or both) of the two sets. Therefore, its Venn diagram is obtained by shading everything that is shaded in either Fig. 7(a) or (b). [See Fig. 7(c).] [*Note:* Looking at Fig. 7(c) reveals that $(S \cap T') \cup (S \cap T)$ and S are the same set. Often Venn diagrams can be used to simplify complicated set-theoretic expressions.]

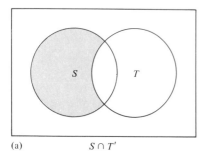

(a) $S \cap T'$

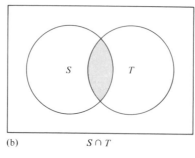

(b) $S \cap T$

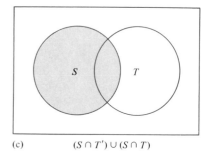

(c) $(S \cap T') \cup (S \cap T)$

FIGURE 7

2. From the inclusion-exclusion principle, we obtain $n(S \cap T) = 0$—that is, $S \cap T = \varnothing$. We conclude that S and T have no elements in common.

5.3. Venn Diagrams and Counting

In this section we discuss the use of Venn diagrams in solving counting problems. The techniques developed are especially useful in analyzing survey data.

Each Venn diagram divides the universal set U into a certain number of regions. For example, the Venn diagram for a single set divides U into two regions—the inside and outside of the circle [Fig. 1(a)]. The Venn diagram for two sets divides U into four regions [Fig. 1(b)]. And the Venn diagram for three sets divides U into eight regions [Fig. 1(c)]. Each of the regions is called a *basic region* for the Venn diagram. Knowing the number of elements in each basic

FIGURE 1

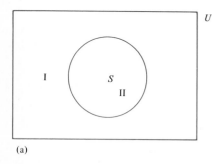

(a)

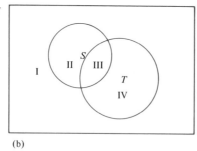

(b)

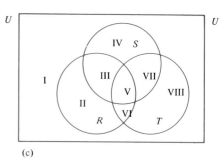

(c)

region is of great use in many applied problems. As an illustration consider the following example.

EXAMPLE 1 Let

$U = \{$winners of the Nobel prize during the period 1901–1986$\}$,

$A = \{$American winners of the Nobel prize during the period 1901–1986$\}$,

$C = \{$winners of the Nobel prize in chemistry during the period 1901–1986$\}$,

$P = \{$winners of the Nobel Peace Prize during the period 1901–1986$\}$.

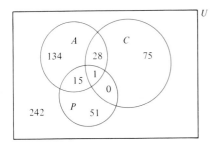

FIGURE 2

These sets are illustrated in the Venn diagram of Fig. 2 in which each basic region has been labeled with the number of elements in it.

(a) How many Americans received a Nobel prize during the period 1901–1986?

(b) How many Americans received Nobel prizes in fields other than chemistry and peace during this period?

(c) How many Americans received the Nobel Peace Prize during this period?

(d) How many people received Nobel prizes during this period?

Solution

(a) The number of Americans who received a Nobel prize is the total number contained in the circle A, which is $134 + 15 + 1 + 28 = 178$ [Fig. 3(a)].

(b) The question asks for the number of Nobel laureates in A but not in C and not in P. So start with the A circle and eliminate those basic regions belonging to C or P [Fig. 3(b)]. There remains a single basic region with 134 Nobel laureates. Note that this region corresponds to $A \cap C' \cap P'$.

(c) The question asks for the number of elements in both A and P—that is, $n(A \cap P)$. But $A \cap P$ comprises two basic regions [Fig. 3(c)]. Thus, to compute $n(A \cap P)$ we add the numbers in these basic regions to obtain $15 + 1 = 16$ Americans who have won the Nobel Peace Prize.

FIGURE 3

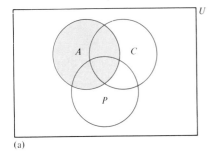

(a)

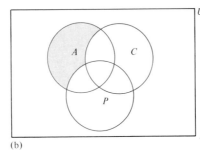

(b)

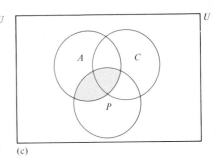

(c)

(d) The number of recipients is just $n(U)$, and we obtain it by adding together the numbers corresponding to the basic regions. We obtain $242 + 134 + 28 + 1 + 15 + 75 + 0 + 51 = 546$.

One need not always be given the number of elements in each of the basic regions of a Venn diagram. Very often these data can be deduced from given information.

EXAMPLE 2 Consider the set of 500 corporate presidents of Example 1, Section 2.

(a) Draw a Venn diagram displaying the given data and determine the number of elements in each basic region.

(b) Determine the number of presidents having exactly one degree (graduate or undergraduate) in business.

Solution (a) Recall that we defined the following sets:

$$S = \{\text{presidents with an undergraduate degree in business}\}$$

$$T = \{\text{presidents with a graduate degree in business}\}.$$

We were given the following data:

$$n(S) = 238, \qquad n(T) = 184, \qquad n(S \cup T) = 310.$$

FIGURE 4

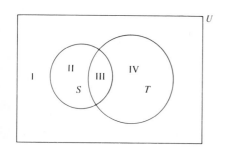

We draw a Venn diagram corresponding to S and T (Fig. 4). Notice that none of the given information corresponds to a basic region of the Venn diagram. So we must use our wits to determine the number of presidents in each of the regions I–IV. Region I is the complement of $S \cup T$, so it contains

$$n(U) - n(S \cup T) = 500 - 310 = 190$$

presidents. Region III is just $S \cap T$. By using the inclusion exclusion principle, in Example 1, Section 2, we determined that $n(S \cap T) = 112$. Now the total number of presidents in II and III combined equals $n(S)$, or 238. Therefore, the number of presidents in II is

$$238 - 112 = 126.$$

Similarly, the number of presidents in IV is

$$184 - 112 = 72.$$

Thus, we may fill in the data determined to obtain a completed Venn diagram (Fig. 5).

(b) The number of people with exactly one business degree corresponds to the shaded region in Fig. 6. Adding together the number of presidents in each of these regions gives $126 + 72 = 198$ presidents with exactly one business degree.

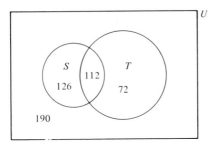

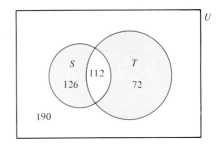

FIGURE 5

FIGURE 6

Here is another example illustrating the procedure for determining the number of elements in each of the basic regions of a Venn diagram.

EXAMPLE 3 An advertising agency finds that of its 170 clients, 115 use television (T), 100 use radio (R), 130 use magazines (M), 75 use television and radio, 95 use radio and magazines, 85 use television and magazines, and 70 use all three. Use these data to complete a Venn diagram displaying the use of mass media (Fig. 7).

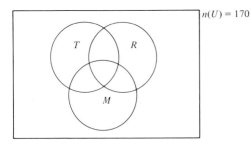

$n(U) = 170$

FIGURE 7

Solution Of the various data given, only the last item corresponds to one of the eight basic regions of the Venn diagram, namely the "70" corresponding to the use of all three media. So we begin by entering this number in the diagram [Fig. 8(a)]. We can fill in the rest of the Venn diagram by working with the remaining information one piece at a time in the reverse order that it is given. Since 85 clients advertise in television and magazines, $85 - 70 = 15$ advertise in television and magazines but not on radio. The appropriate region is labeled in Fig. 8(b). In Fig. 8(c) the

FIGURE 8

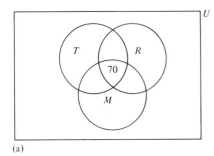

(a)

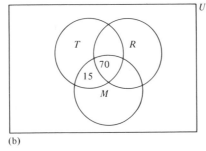

(b)

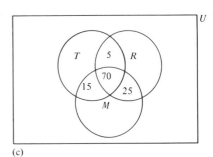

(c)

next two pieces of information have been used in the same way to fill in two more basic regions. In Fig. 8(c) we observe that three of the four basic regions comprising M have been filled in. Since $n(M) = 130$, we deduce that the number of clients advertising only in magazines is $130 - (15 + 70 + 25) = 130 - 110 = 20$ [Fig. 9(a)]. By similar reasoning the number of clients using only radio advertising and the number using only television advertising can be determined [Fig. 9(b)]. Adding together the numbers in the three circles gives the number of clients utilizing television, radio, or magazines as $25 + 5 + 0 + 15 + 70 + 25 + 20 = 160$. Since there were 170 clients in total, the remainder—or $170 - 160 = 10$ clients— use none of these media. Figure 9(c) gives a complete display of the data.

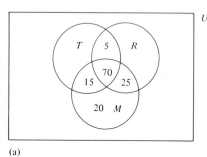

(a)

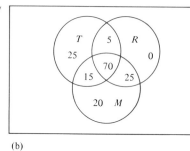

(b)

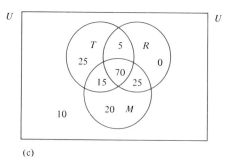

(c)

FIGURE 9

PRACTICE PROBLEMS 3

1. Of the 1000 freshmen at a certain college, 700 take mathematics courses, 300 take mathematics and economics courses, and 200 do not take any mathematics or economics courses. Represent these data in a Venn diagram.

2. Refer to the Venn diagram from Problem 1.

   (a) How many of the freshmen take an economics course?

   (b) How many take an economics course but not a mathematics course?

EXERCISES 3

In Exercises 1–8, draw an appropriate Venn diagram and use the given data to determine the number of elements in each basic region.

1. $n(U) = 14$, $n(S) = 5$, $n(T) = 6$, $n(S \cap T) = 2$.

2. $n(U) = 20$, $n(S) = 11$, $n(T) = 7$, $n(S \cap T) = 7$.

3. $n(U) = 44$, $n(R) = 17$, $n(S) = 17$, $n(T) = 17$, $n(R \cap S) = 7$, $n(R \cap T) = 6$, $n(S \cap T) = 5$, $n(R \cap S \cap T) = 2$.

4. $n(U) = 29$, $n(R) = 10$, $n(S) = 12$, $n(T) = 10$, $n(R \cap S) = 1$, $n(R \cap T) = 5$, $n(S \cap T) = 4$, $n(R \cap S \cap T) = 1$.

5. $n(U) = 75, n(S) = 15, n(T) = 25, n(S' \cap T') = 40.$

6. $n(S) = 9, n(T) = 11, n(S \cap T) = 5, n(S') = 13.$

7. $n(R') = 22, \quad n(R \cup S) = 21, \quad n(S) = 14, \quad n(T) = 22, \quad n(R \cap S) = 7, \quad n(S \cap T) = 9,$
   $n(R \cap T) = 11, n(R \cap S \cap T) = 5.$

8. $n(U) = 64, n(R \cup S \cup T) = 45, n(R) = 22, n(T) = 26, n(R \cap S) = 4, n(S \cap T) = 6,$
   $n(R \cap T) = 8, n(R \cap S \cap T) = 1.$

9. A survey of 70 high school students revealed that 35 like folk music, 15 like classical music, and 5 like both. How many of the students surveyed do not like either folk or classical music?

10. A total of 546 Nobel prizes had been awarded by 1986. Fourteen of the 84 prizes in literature were awarded to Scandinavians. Scandinavians received a total of 43 awards. How many Nobel prizes outside of literature have been awarded to non-Scandinavians?

11. Out of 35 students in a finite math class, 22 are male, 19 are business majors, 27 are freshmen, 14 are male business students, 17 are male freshmen, 15 are freshmen business majors, and 11 are male freshmen business majors. How many upperclass women nonbusiness majors are in the class? How many women business majors are in the class?

12. A survey of 100 college faculty who exercise regularly found that 45 jog, 30 swim, 20 cycle, 6 jog and swim, 1 jogs and cycles, 5 swim and cycle, and 1 does all three. How many of the faculty members do not do any of these three activities? How many just jog?

13. One hundred college students were surveyed after voting in an election involving a Democrat and a Republican. Fifty of the students were freshmen, 55 voted Democratic, and 25 were nonfreshmen who voted Republican. How many freshmen voted Democratic?

14. A group of 100 workers were asked if they were college graduates and if they belonged to a union. Sixty were not college graduates, 20 were nonunion college graduates, and 30 were union members. How many of the workers were neither college graduates nor union members?

15. A class of 30 students was given a diagnostic test on the first day of a mathematics course. At the end of the semester, only 2 of the 21 students who had passed the diagnostic test failed the course. A total of 23 students passed the course. How many students managed to pass the course even though they had failed the diagnostic test?

16. A group of applicants for training as air-traffic controllers consisted of 35 pilots, 20 veterans, 30 pilots who were not veterans, and 50 people who were neither veterans nor pilots. How large was the group?

17. One of Shakespeare's sonnets has a verb in 11 of its 14 lines, an adjective in 9 lines, and both in 7 lines. How many lines have a verb but no adjective? An adjective and no verb? Neither an adjective nor a verb?

Of the 130 students who took a math exam, 90 correctly answered the first question, 62 correctly answered the second question, and 50 correctly answered both questions. Exercises 18–22 refer to these students.

18. How many students correctly answered either the first or second question?

19. How many students did not answer either of the two questions correctly?

20. How many students answered either the first or the second question correctly, but not both?

21. How many students answered the second question correctly, but not the first?

22. How many students missed the second question?

A collector of football cards has 2200 cards. He has 1500 players from the National Football League (NFL), 900 who played defense, and 400 who played defense for the NFL. Exercises 23–28 refer to these football players.

23. How many players were either in the NFL or played defense?

24. How many players played defense, but were not in the NFL?

25. How many players played offense, but were not in the NFL?

26. How many players played offense in the NFL?

27. How many players either played defense or were in the NFL, but not both?

28. How many players did not play defense for the NFL?

A campus radio station surveyed 190 students to determine the types of music they liked. The survey revealed that 114 liked rock, 50 liked country, and 41 liked classical music. Moreover, 14 liked rock and country, 15 liked rock and classical, 11 liked classical and country, and 5 liked all three types of music. Exercises 29–36 refer to the students in this survey.

29. How many students like rock only?

30. How many students like country but not rock?

31. How many students like classical and country, but not rock?

32. How many students like classical or country, but not rock?

33. How many students like exactly one of the three types of music?

34. How many students do not like any of the three types of music?

35. How many students like at least two of the three types of music?

36. How many students do not like either rock or country?

A merchant surveyed 400 people to determine the way they learned about an upcoming sale. The survey showed that 180 learned about the sale from the radio, 190 from television, 190 from the newspaper, 80 from radio and television, 90 from radio and newspaper, 50 from television and newspaper, and 30 from all three sources. Exercises 37–42 refer to the people in this survey.

37. How many people learned of the sale from newspapers or radio, but not both?

38. How many people learned of the sale only from newspapers?

39. How many people learned of the sale from radio or television, but not the newspaper?

40. How many people learned of the sale from at least two of the three media?

41. How many people learned of the sale from exactly one of the three media?

42. How many people learned of the sale from radio and television, but not the newspaper?

43. One hundred and eighty business executives were surveyed to determine if they regularly read *Fortune*, *Time*, or *Money* magazines. Seventy-five read *Fortune*, 70 read *Time*, 55 read *Money*, 45 read exactly two of the three magazines, 25 read *Fortune* and *Time*, 25 read *Time* and *Money*, and 5 read all three magazines. How many read none of the three magazines?

44. A survey of the characteristics of 100 small businesses which had failed revealed that 95 of them either were undercapitalized, had inexperienced management, or had a poor location. Four of the businesses had all three of these characteristics. Forty businesses were undercapitalized but had experienced management and good location. Fifteen businesses had inexperienced management but sufficient capitalization and good location. Seven were undercapitalized and had inexperienced management. Nine were undercapitalized and had poor location. Ten had inexperienced management and poor location. How many of the businesses had poor location? Which of the three characteristics was most prevalent in the failed businesses?

45. A shopping mall has three large "anchor" department stores—Grady's, Price's, and Zack's. A customer survey shows that 97% of the shoppers at this mall go into at least one of the three anchors, 50% go into exactly one, and 30% go into all three. The survey shows that everyone who goes into Grady's also goes into Price's, which draws 80% of all shoppers. Determine the possible percentage of shoppers who go into Zack's.

SOLUTIONS TO PRACTICE PROBLEMS 3

1. Draw a Venn diagram with two circles, one for mathematics (*M*) and one for economics (*E*) [Fig. 10(a)]. This Venn diagram has four basic regions, and our goal is to label each basic region with the proper number of students. The numbers for two of the basic regions are given directly. Since "300 take mathematics and economics," $n(M \cap E) = 300$. Since

FIGURE 10

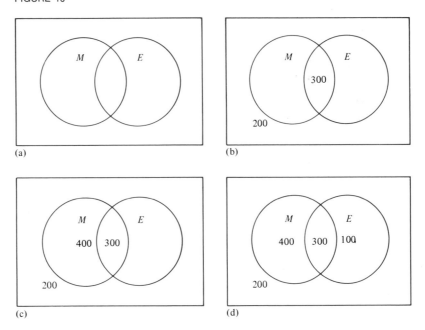

(a)

(b)

(c)

(d)

"200 do not take any mathematics or economics courses," $n((M \cup E)') = 200$ [Fig. 10(b)]. Now "700 take mathematics courses." Since M is made up of two basic regions and one region has 300 elements, the other basic region of M must contain 400 elements [Fig. 10(c)]. At this point all but one of the basic regions have been labeled and $400 + 300 + 200 = 900$ students have been accounted for. Since there is a total of 1000 students, the remaining basic region has 100 students [Fig. 10(d)].

2. (a) 400. "Economics" refers to the entire circle E which is made up of two basic regions, one having 300 elements and the other 100. (A common error is to interpret the question as asking for the number of freshmen who take economics exclusively and therefore give the answer 100. To say that a person takes an economics course does not imply anything about the person's enrollment in mathematics courses.)

(b) 100.

5.4. The Multiplication Principle

In this section we introduce a second fundamental principle of counting, the so-called *multiplication principle*. By way of motivation, consider the following example.

EXAMPLE 1 A medical researcher wishes to test the effect of a drug on a rat's perception by studying its ability to run a maze while under the influence of the drug. The maze is constructed so that, in order to arrive at the exit point C, the rat must pass through a central point B. There are five paths from the entry point A to B, and three paths from B to C. In how many different ways can the rat run the maze from A to C? (See Fig. 1.)

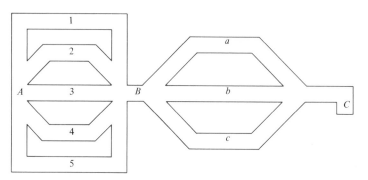

FIGURE 1   Paths through a maze

Solution The paths from A to B have been labeled 1 through 5, and the paths from B to C have been labeled a through c. The various paths through the maze can be schematically represented as in Fig. 2. The diagram shows that there are five ways to go from A to B. For each of these there are three ways to go from B to C. So there are five groups of three paths each, and therefore $5 \cdot 3 = 15$ possible

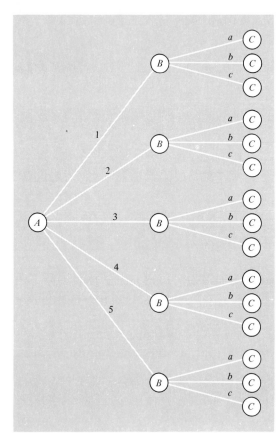

FIGURE 2

paths from *A* to *C*. (A diagram such as Fig. 2, called a *tree diagram*, is useful in enumerating the various possibilities in counting problems.)

In the problem above, choosing a path is a task that can be broken up into two consecutive operations.

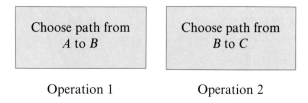

| Choose path from *A* to *B* | Choose path from *B* to *C* |

Operation 1   Operation 2

The first operation can be performed in five ways and, after the first operation has been carried out, the second can be performed in three ways. And we determined that the entire task can be performed in $5 \cdot 3 = 15$ ways. The same reasoning as used above yields the following useful counting principle:

Multiplication Principle   Suppose that a task is composed of two consecutive operations. If operation 1 can be performed in *m* ways and, for each of these, operation 2 can be performed in *n* ways, then the complete task can be performed in $m \cdot n$ ways.

EXAMPLE 2   An airline passenger must fly from New York to Frankfurt via London. There are 8 flights leaving New York for London. All of these provide connections on any one of 19 flights from London to Frankfurt. In how many different ways can the passenger book reservations?

Solution   The task "fly from New York to Frankfurt" is composed of two consecutive operations:

| Fly from New York to London | Fly from London to Frankfurt |

Operation 1   Operation 2

From the data given the multiplication principle implies that the task can be accomplished in $8 \cdot 19 = 152$ ways.

It is possible to generalize the multiplication principle to tasks consisting of more than two operations.

Generalized Multiplication Principle Suppose that a task consists of t operations performed consecutively. Suppose that operation 1 can be performed in m_1 ways; for each of these, operation 2 in m_2 ways; for each of these, operation 3 in m_3 ways; and so forth. Then the task can be performed in

$$m_1 \cdot m_2 \cdot m_3 \cdots m_t$$

ways.

EXAMPLE 3   A corporation has a board of directors consisting of 10 members. The board must select from among its members a chairman, vice-chairman, and secretary. In how many ways can this be done?

Solution   The task "select the three officers" can be divided into three consecutive operations:

| Select chairman | Select vice-chairman | Select secretary |

Since there are 10 directors, operation 1 can be performed in 10 ways. After the chairman has been selected, there are 9 directors left as possible candidates for vice-chairman, so that for each way of performing operation 1, operation 2 can be performed in 9 ways. After this has been done, there are 8 directors who are possible candidates for secretary, so operation 3 can be performed in 8 ways. By the generalized multiplication principle the number of possible ways to perform the sequence of three operations equals $10 \cdot 9 \cdot 8$ or 720. So the officers of the board can be selected in 720 ways.

In Example 3 we made important use of the phrase "for each of these" in the generalized multiplication principle. The operation "select a vice-chairman" can be performed in 10 ways, since any member of the board is eligible. However, when we view the selection process as a sequence of operations of which "select a vice-chairman" is the second operation, the situation has changed. *For each way* that the first operation is performed one person will have been used up; hence there will be only 9 possibilities for choosing the vice-chairman.

EXAMPLE 4   In how many ways can a baseball team of nine members arrange themselves in a line for a group picture?

Solution   Choose the players by their place in the picture, say from left to right. The first can be chosen in nine ways; for each of these choices the second can be chosen in eight ways; for each of these choices the third can be chosen in seven ways; and so forth. So the number of possible arrangements is

$$9 \cdot 8 \cdot 7 \cdot 6 \cdot 5 \cdot 4 \cdot 3 \cdot 2 \cdot 1 = 362,880.$$

EXAMPLE 5   A certain state uses automobile license plates which consist of three letters followed by three digits. How many such license plates are there?

Solution   The task in this case, "form a license plate," consists of a sequence of six operations: three for choosing letters and three for choosing digits. Each letter can be chosen in 26 ways and each digit in 10 ways. So the number of license plates is

$$26 \cdot 26 \cdot 26 \cdot 10 \cdot 10 \cdot 10 = 17,576,000.$$

PRACTICE PROBLEMS 4

1. There are six seats available in a sedan. In how many ways can six people be seated if only three can drive?

2. A multiple-choice exam contains 10 questions, each having 3 possible answers. How many different ways are there of completing the exam?

EXERCISES 4

1. If there are three routes from College Park to Baltimore and five routes from Baltimore to New York, how many routes are there from College Park to New York via Baltimore?

2. How many different outfits consisting of a coat and a hat can be chosen from two coats and three hats?

3. How many different two-letter words can be formed allowing repetition of letters?

4. How many different two-letter words can be formed such that the two letters are distinct?

5. A railway has 20 stations. If the names of the point of departure and the destination are printed on each ticket, how many different kinds of single tickets must be printed?

6. Refer to Exercise 5. How many different kinds of tickets are needed if each ticket may be used in either direction between two stations?

7. A man has five different pairs of gloves. In how many ways can he select a right-hand glove and a left-hand glove that do not match?

8. How many license plates consisting of two letters followed by four digits are possible?

9. How many ways can five people be arranged in a line for a group picture?

10. In how many different ways can four books be arranged on a bookshelf?

11. Toss a coin six times and observe the sequence of heads or tails that results. How many different sequences are possible?

12. Refer to Exercise 11. In how many of the sequences are the first and last tosses identical?

13. Twenty athletes enter an Olympic event. How many different possibilities are there for winning the Gold Medal, Silver Medal, and Bronze Medal?

14. A sportswriter is asked to rank eight teams. How many different orderings are possible?

15. In how many different ways can a 30-member football team select a captain and a cocaptain?

16. How many different outfits can be selected from two coats, three hats, and two scarves?

17. How many different words can be formed using the four letters of the word "MATH"?

18. If you can travel from Frederick, Maryland, to Baltimore, Maryland, by car, bus, or train and from Baltimore to London by airplane or ship, how many different ways are there to go from Frederick to London?

19. An exam contains five "true or false" questions. In how many different ways can the exam be completed?

20. A company has 700 employees. Explain why there must be two people with the same pair of initials.

21. A computer manufacturer assigns serial numbers to its computers. The first symbol of a serial number is either A, B, or C, indicating the manufacturing plant. The second and third symbols taken together are one of the numbers $01, 02, \ldots, 12$ indicating the month of manufacture. The final four symbols are digits. How many possible serial numbers are there?

22. How many four-letter words (including nonsense words) can be made from the letters of "statistics" assuming that each word may not have repeated letters?

23. How many four-letter words (including nonsense words) can be made from the letters h, o, t, s, m, x, and e for each of the following conditions?

    (a) Letters can be repeated.

    (b) Letters cannot be repeated.

    (c) Words must begin with an h and repetitions are allowed.

    (d) Words must end with a vowel and repetitions are not allowed.

24. A group of five boys and three girls is to be photographed.

    (a) How many ways can they be arranged in one row?

    (b) How many ways can they be arranged with the girls in the front row and the boys in the back row?

25. The manager of a little league baseball team has picked the nine starting players for a game. How many different batting orders are possible under each of the following conditions?

    (a) There are no restrictions.

    (b) The pitcher must bat last.

    (c) The pitcher must bat last, the catcher eighth, and the shortstop first.

26. How many different ways can a Venn diagram with two circles be shaded?

27. How many different ways can a Venn diagram with three circles be shaded?

28. A club can elect a member as president and a different member as treasurer in 506 different ways. How many members does the club have?

29. An exam contains six true and false statements. In how many ways can the exam be completed if leaving the answer blank is also an option?

30. A physiologist wants to test the effects of exercise and meditation on blood pressure. She devises four different exercise programs and three different meditation programs. If she wants 10 subjects for each combination of exercise and meditation program, how many volunteers must she recruit?

31. A college student eats all his meals at a restaurant offering six breakfast specials, seven lunch specials, and four dinner specials. How many weeks can he go without repeating an entire day's menu selection?

32. An area code is a three-digit number where the first digit cannot be 0 or 1. How many different area codes are possible?

33. A circular combination lock requires three numbers to be opened, where each dial can be set at any one of the numbers 1 through 40. The lock can be opened even if the values tried are off by 1. For example, 5-40-17 will open the lock if the correct combination is 6-1-16. How many tries must a thief make before he can be certain of opening the lock?

SOLUTIONS TO PRACTICE PROBLEMS 4

1. 360. Pretend that you are given the task of seating the six people. This task consists of six operations performed consecutively, as shown in the accompanying chart.

Operation	Number of ways operation can be performed
1: select person to drive	3
2: select person for middle front seat	5
3: select person for right front seat	4
4: select person for left rear seat	3
5: select person for middle rear seat	2
6: select person for right rear seat	1

After you have performed operation 1, five people will remain, and any one of these five can be seated in the middle front seat. After operation 2, four people remain, and so on. By the generalized multiplication principle, the task can be performed in $3 \cdot 5 \cdot 4 \cdot 3 \cdot 2 \cdot 1 = 360$ ways.

2. 3^{10}. The task of answering the questions consists of 10 consecutive operations, each of which can be performed in three ways. Therefore, by the generalized multiplication principle, the task can be performed in $\underbrace{3 \cdot 3 \cdot 3 \cdot \ldots \cdot 3}_{10 \text{ terms}}$ ways.

[*Note:* The answer can be left as 3^{10} or can be multiplied out to 59,049.]

5.5. Permutations and Combinations

In preceding sections we have solved a variety of counting problems using Venn diagrams and the generalized multiplication principle. Let us now turn our attention to two types of counting problems which occur very frequently and which can be solved using formulas derived from the generalized multiplication principle. These problems involve what are called permutations and combinations, each of which are particular types of arrangements of elements of a set. The sort of arrangements we have in mind are illustrated in two problems:

PROBLEM A   How many words (by which we mean strings of letters) of two distinct letters can be formed from the letters $\{a, b, c\}$?

PROBLEM B   A construction crew has three members. A team of two must be chosen for a particular job. In how many ways can the team be chosen?

Each of the two problems can be solved by enumerating all possibilities.

Solution of Problem A

There are six possible words, namely

$$ab, ac, ba, bc, ca, cb.$$

Solution of Problem B

Designate the three crew members by a, b, and c. Then there are three possible two-man teams, namely

$$ab, ac, bc.$$

(Note that ba, the team consisting of b and a, is the same as the team ab.)

We deliberately set up both problems using the same letters in order to facilitate comparison. Both problems are concerned with counting the number of arrangements of the elements of the set $\{a, b, c\}$, taken two at a time, without allowing repetition (for example, aa was not allowed). However, in Problem A the order of the arrangement mattered, whereas in Problem B it did not. Arrangements of the sort considered in Problem A are called *permutations*, whereas those in Problem B are called *combinations*.

More precisely, suppose that we are given a set of n objects.* Then a *permutation of n objects taken r at a time* is an arrangement of r of the n objects in a specific order. So, for example, Problem A was concerned with permutations of the three objects a, b, c ($n = 3$), taken two at a time ($r = 2$). A *combination of n objects taken r at a time* is a selection of r objects from among the n, with order disregarded. Thus, for example, in Problem B we considered combinations of the three objects a, b, c ($n = 3$), taken two at a time ($r = 2$).

It is convenient to introduce the following notation for counting permutations and combinations. Let

$P(n, r)$ = the number of permutations of n objects taken r at a time,

$C(n, r)$ = the number of combinations of n objects taken r at a time.

* All assumed to be different.

Thus, for example, from our solutions to Problems A and B we have

$$P(3, 2) = 6, \qquad C(3, 2) = 3.$$

There are very simple formulas for $P(n, r)$ and $C(n, r)$ which allow us to calculate these quantities for any n and r. Let us begin by stating the formula for $P(n, r)$. For $r = 1, 2, 3$ we have, respectively,

$$P(n, 1) = n$$

$$P(n, 2) = n(n - 1) \qquad \text{(two factors)}$$

$$P(n, 3) = n(n - 1)(n - 2) \qquad \text{(three factors)},$$

and, in general,

$$P(n, r) = n(n - 1)(n - 2) \cdots (n - r + 1) \qquad (r \text{ factors}). \qquad (1)$$

This formula is verified at the end of this section.

EXAMPLE 1 Compute the following numbers.

(a) $P(100, 2)$ (b) $P(6, 4)$ (c) $P(5, 5)$

Solution (a) Here $n = 100$, $r = 2$. So we take the product of two factors, beginning with 100:

$$P(100, 2) = 100 \cdot 99 = 9900.$$

(b) $P(6, 4) = 6 \cdot 5 \cdot 4 \cdot 3 = 360$

(c) $P(5, 5) = 5 \cdot 4 \cdot 3 \cdot 2 \cdot 1 = 120$

In order to state the formula for $C(n, r)$, we must introduce some further notation. Suppose that r is any positive integer. We denote by $r!$ (read "r factorial") the product of all positive integers from r down to 1:

$$r! = r \cdot (r - 1) \cdot \ldots \cdot 2 \cdot 1.$$

Thus, for instance,

$$1! = 1$$

$$2! = 2 \cdot 1 = 2$$

$$3! = 3 \cdot 2 \cdot 1 = 6$$

$$4! = 4 \cdot 3 \cdot 2 \cdot 1 = 24$$

$$5! = 5 \cdot 4 \cdot 3 \cdot 2 \cdot 1 = 120.$$

In terms of this notation we can state a very simple formula for $C(n, r)$, the number of combinations of n things taken r at a time.

$$C(n, r) = \frac{P(n, r)}{r!} = \frac{n(n-1) \cdot \ldots \cdot (n-r+1)}{r(r-1) \cdot \ldots \cdot 1}. \qquad (2)$$

This formula is verified at the end of the section.

EXAMPLE 2   Compute the following numbers.

(a) $C(100, 2)$ (b) $C(6, 4)$ (c) $C(5, 5)$

Solution   (a) $C(100, 2) = \dfrac{P(100, 2)}{2!} = \dfrac{100 \cdot 99}{2 \cdot 1} = 4950$

(b) $C(6, 4) = \dfrac{P(6, 4)}{4!} = \dfrac{6 \cdot 5 \cdot 4 \cdot 3}{4 \cdot 3 \cdot 2 \cdot 1} = 15$

(c) $C(5, 5) = \dfrac{P(5, 5)}{5!} = \dfrac{5 \cdot 4 \cdot 3 \cdot 2 \cdot 1}{5 \cdot 4 \cdot 3 \cdot 2 \cdot 1} = 1$

EXAMPLE 3   Solve Problems A and B using formulas (1) and (2).

Solution   The number of two-letter words which can be formed from the three letters a, b, c is equal to $P(3, 2) = 3 \cdot 2 = 6$, in agreement with our previous solution.

     The number of two-worker teams which can be formed from three individuals is equal to $C(3, 2)$, and

$$C(3, 2) = \frac{P(3, 2)}{2!} = \frac{3 \cdot 2}{2 \cdot 1} = 3,$$

in agreement with our previous result.

EXAMPLE 4   The board of directors of a corporation has 10 members. In how many ways can they choose a committee of 3 board members to negotiate a merger?

Solution   Since the committee of three involves no ordering of its members, we are concerned here with combinations. The number of combinations of 10 people taken 3 at a time is $C(10, 3)$, which is

$$C(10, 3) = \frac{10 \cdot 9 \cdot 8}{3 \cdot 2 \cdot 1} = 120.$$

Thus there are 120 choices for the committee.

EXAMPLE 5   Eight horses are entered in a race in which a first, second, and third prize will be awarded. Assuming no ties, how many different outcomes are possible?

Solution   In this example we are considering ordered arrangements of three horses, so we are dealing with permutations. The number of permutations of eight horses taken three at a time is

$$P(8, 3) = 8 \cdot 7 \cdot 6 = 336,$$

so the number of possible outcomes of the race is 336.

EXAMPLE 6   A political pollster wishes to survey 1500 individuals chosen from a sample of 5,000,000 adults. In how many ways can the 1500 individuals be chosen?

Solution   No ordering of the 1500 individuals is involved, so we are dealing with combinations. So the number in question is $C(5,000,000, 1500)$, a number too large to be written down in digit form. (It has several thousand digits!) But it could be calculated with the aid of a computer.

EXAMPLE 7   A club has 10 members. In how many ways can they choose a slate of four officers, consisting of a president, vice-president, secretary, and treasurer?

Solution   In this problem we are dealing with an ordering of four members. (The first is the president, the second the vice-president, and so on.) So we are dealing with permutations, and the number of ways of choosing the officers is

$$P(10, 4) = 10 \cdot 9 \cdot 8 \cdot 7 = 5040.$$

Verification of the Formulas for $P(n, r)$ and $C(n, r)$ Let us first derive the formula for $P(n, r)$, the number of permutations of n objects taken r at a time. The task of choosing r objects (in a given order) consists of r consecutive operations (Fig. 1). The first operation can be performed in n ways. For each way that the first operation is performed one object will have been used up and so we can perform the second operation in $n - 1$ ways, and so on. For each way of performing the sequence of operations $1, 2, 3, \ldots, r - 1$ the rth operation can be performed in $n - (r - 1) = n - r + 1$ ways. By the generalized multiplication principle, the task of choosing the r objects from among the n can be performed in

$$n(n - 1) \cdot \ldots \cdot (n - r + 1)$$

FIGURE 1

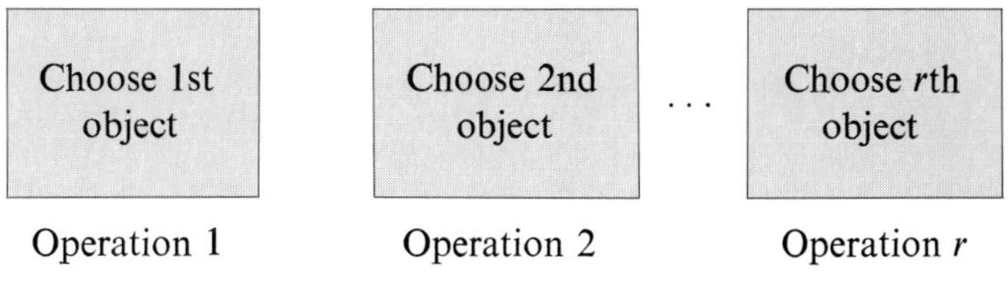

Choose 1st object	Choose 2nd object	. . .	Choose rth object
Operation 1	Operation 2		Operation r

ways. That is,

$$P(n, r) = n(n - 1) \cdot \ldots \cdot (n - r + 1),$$

which is formula (1).

Let us now verify the formula for $C(n, r)$, the number of combinations of n objects taken r at a time. Each such combination is a set of r objects and therefore can be ordered in

$$P(r, r) = r(r - 1) \cdot \ldots \cdot 2 \cdot 1 = r!$$

ways by formula (1). In other words, each different combination of r objects gives rise to $r!$ permutations of the same r objects. On the other hand, each permutation of n objects taken r at a time gives rise to a combination of n objects taken r at a time, by simply ignoring the order of the permutation. Thus, if we start with the $P(n, r)$ permutations, we will have all the combinations of n objects taken r at a time, with each combination repeated $r!$ times. Thus

$$P(n, r) = r! \, C(n, r).$$

On dividing both sides of the equation by $r!$, we obtain formula (2).

PRACTICE PROBLEMS 5

1. Calculate:

   (a) 3!   (b) 7!   (c) $C(7, 3)$ (d) $P(7, 3)$

2. A newborn child is to be given a first name and a middle name from a selection of 10 names. How many different possibilities are there?

EXERCISES 5

Evaluate:

1. $P(4, 2)$	2. $P(5, 1)$	3. $P(6, 3)$	4. $P(5, 4)$
5. $C(10, 3)$	6. $C(12, 2)$	7. $C(5, 4)$	8. $C(6, 3)$
9. 6!	10. 10!/4!		

11. How many different selections of two books can be made from a set of nine books?

12. A pizza parlor offers five toppings for the plain cheese base of the pizzas. How many different pizzas are possible which use three of the toppings?

13. In how many ways can four people line up in a row for a group picture?

14. How many different outcomes of "winner" and "runner-up" are possible if there are six contestants in a pie-eating contest?

15. If you are going on a trip and decide to take three of your seven sweaters, how many different possibilities are there?

16. A student must choose five courses out of seven that he would like to take. How many possibilities are there?

17. How many different three-letter words are there having no repetition of letters?

18. A sportswriter makes a preseason guess of the top 10 football teams (in order) from among 40 major teams. How many different possibilities are there?

19. In how many different ways can a committee of 15 senators be selected from the 100 members of the U.S. Senate?

20. Theoretically, how many possibilities are there for first, second, and third place in a marathon race with 1000 entries?

21. How many different poker hands are there? (A poker hand consists of five cards.)

22. How many different poker hands consist entirely of aces and kings?

23. Two children, Moe and Joe, are allowed to select candy from a plate of nine pieces of candy. Moe, being younger, is allowed to choose first, but can only take two candies. Joe is then allowed to take three of the remaining candies. Joe complains that he has fewer choices than Moe. Is Joe correct? How many choices will each child have?

24. How many different ways can six jazz albums, four blues albums, and three swing albums be arranged on a shelf if the records for each of the three categories must be grouped together?

25. A truck driver has to deliver bread to five grocery stores. In how many different ways can he schedule the order of his stops?

26. Suppose the stores in Exercise 25 are located in two towns far enough apart that the driver wants to make all stops in one town before going on to the next. In how many different ways can he schedule the order of his stops if the larger town has three stores?

SOLUTIONS TO PRACTICE PROBLEMS 5

1. (a) $3! = 3 \cdot 2 \cdot 1 = 6$ (b) $7! = 7 \cdot 6 \cdot 5 \cdot 4 \cdot 3 \cdot 2 \cdot 1 = 5040$

    (c) $C(7, 3) = \dfrac{7 \cdot 6 \cdot 5}{3 \cdot 2 \cdot 1} = \dfrac{7 \cdot \cancel{6} \cdot 5}{\cancel{3} \cdot \cancel{2} \cdot 1} = 35$

    [A convenient procedure to follow when calculating $C(n, r)$ is to first write the product expansion of $r!$ in the denominator and then to write in the numerator an integer from the descending expansion of $n!$ above each integer in the denominator.]

    (d) $P(7, 3) = \underbrace{7 \cdot 6 \cdot 5}_{3 \text{ factors}} = 210$

    [In general, $P(n, r)$ is the product of the first r factors in the descending expansion of $n!$]

2. 90. The first question to be asked here is whether permutations or combinations are involved. Two names are to be selected, and the order of the names is important. (The name Amanda Beth is different from the name Beth Amanda.) Since the problem asks for arrangements of 10 names taken 2 at a time in a *specific order*, the number of arrangements if $P(10, 2) = 10 \cdot 9 = 90$. In general, order is important if a different outcome results when two items in the selection are interchanged.

5.6. Further Counting Problems

In the last section we introduced permutations and combinations and developed formulas for counting all permutations (or combinations) of a given type. Many counting problems can be formulated in terms of permutations or combinations. But in order to successfully use the formulas of Section 5, we must be able to recognize these problems when they occur and to translate them into a form in which the formulas may be applied. In this section we practice doing that. We consider three typical applications giving rise to permutations or combinations. The first two applications may seem, at first glance, to have little practical significance. However, they suggest a common way to "model" outcomes of real-life situations having two equally likely results.

As our first application, consider a coin-tossing experiment in which we toss a coin a fixed number of times. We can describe the outcome of the experiment as a sequence of "heads" and "tails." For instance, if a coin is tossed three times, then one possible outcome is "head on first toss, tail on second toss, and tail on third toss." This outcome can be abbreviated as HTT. We can use the methods of the preceding section to count the number of possible outcomes having various prescribed properties.

EXAMPLE 1   Suppose that an experiment consists of tossing a coin 10 times and observing the sequence of heads and tails.

(a) How many different outcomes are possible?

(b) How many different outcomes have exactly four heads?

Solution   (a) Visualize each outcome of the experiment as a sequence of 10 boxes, where each box contains one letter, H or T, with the first box recording the result of the first toss, the second box recording the result of the second toss, and so

H	T	H	T	T	T	H	T	H	T
1	2	3	4	5	6	7	8	9	10

forth. Each box can be filled in two ways. So by the generalized multiplication principle the sequence of 10 boxes can be filled in

$$\underbrace{2 \cdot 2 \cdot \ldots \cdot 2}_{10 \text{ factors}} = 2^{10}$$

ways. So there are $2^{10} = 1024$ different possible outcomes.

(b) An outcome with four heads corresponds to filling the boxes with 4 H's and 6 T's. A particular outcome is determined as soon as we decide where to place the H's. The 4 boxes to receive H's can be selected from the 10 boxes in $C(10, 4)$ ways. So the number of outcomes with 4 heads is

$$C(10, 4) = \frac{10 \cdot 9 \cdot 8 \cdot 7}{4 \cdot 3 \cdot 2 \cdot 1} = 210.$$

Ideas similar to those applied in Example 1 are useful in counting even more complicated sets of outcomes of coin-tossing experiments. The second part of our next example highlights a trick that can often save time and effort.

EXAMPLE 2 Consider the coin-tossing experiment of Example 1.

(a) How many different outcomes have at most two heads?

(b) How many different outcomes have at least three heads?

Solution (a) The outcomes with at most two heads are those having 0, 1, or 2 heads. Let us count the number of these outcomes separately:

0 heads: There is 1 outcome, namely TTTTTTTTTT.

1 head: To determine such an outcome we just select the box in which to put the single H. And this can be done in $C(10, 1) = 10$ ways.

2 heads: To determine such an outcome we just select the boxes in which to put the two H's. And this can be done in $C(10, 2) = (10 \cdot 9)/(2 \cdot 1) = 45$ ways.

Adding up all the possible outcomes, we see that the number of outcomes with at most two heads is equal to $1 + 10 + 45 = 56$.

(b) "At least three heads" refers to an outcome with either 3, 4, 5, 6, 7, 8, 9, or 10 heads. And the total number of such outcomes is

$$C(10, 3) + C(10, 4) + \cdots + C(10, 10).$$

This sum can, of course, be calculated, but there is a less tedious way to solve the problem. Just start with all outcomes [1024 of them by Example 1(a)] and subtract off those with at most two heads [56 of them by part (a)]. So the number of outcomes with at least three heads is $1024 - 56 = 968$.

Let us now turn to a different sort of counting problem, namely one that involves counting the number of paths between two points.

EXAMPLE 3 In Fig. 1 we have drawn a partial map of the streets in a certain city. A tourist wishes to walk from point A to point B. We have drawn two possible routes from A to B. What is the total number of routes (with no backtracking) from A to B?

Solution Any particular route can be described by giving the directions of each block walked in the appropriate order. For instance, the route on the left of Fig. 1 is described as "a block south, a block south, a block east, a block east, a block east, a block

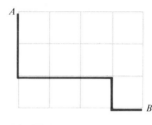 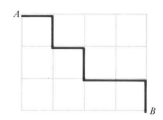

FIGURE 1

south, a block east." Using S for south and E for east, this route can be designated by the string of letters SSEEESE. Similarly, the route on the right is ESESEES. Note that each route is then described by a string of seven letters, of which three are S's (we must go three blocks south) and four E's (we must go four blocks east). Selecting a route is thus the same as placing three S's in a string of seven boxes:

	S		S			S

The three boxes to receive S's can be selected in $C(7, 3) = 35$ ways. So the number of paths from A to B is 35.

Let us now move on to a third type of counting problem. Suppose that we have an urn in which there are a certain number of red balls and a certain number of white balls. We perform an experiment which consists of selecting a number of balls from the urn and observing the color distribution of the sample selected. (This model may be used, for example, to describe the process of selecting people to be polled in a survey. The different colors correspond to different opinions.) By using familiar counting techniques we can calculate the number of possible samples having a given color distribution. The next example illustrates a typical computation.

EXAMPLE 4 An urn contains 25 numbered balls, of which 15 are red and 10 are white. A sample of 5 balls is to be selected.

(a) How many different samples are possible?

(b) How many samples contain all red balls?

(c) How many samples contain 3 red balls and 2 white balls?

(d) How many samples contain at least 4 red balls?

Solution (a) A sample is just an unordered selection of 5 balls out of 25. There are $C(25, 5)$ such samples. Numerically, we have

$$C(25, 5) = \frac{25 \cdot 24 \cdot 23 \cdot 22 \cdot 21}{5 \cdot 4 \cdot 3 \cdot 2 \cdot 1} = 53{,}130$$

samples.

(b) To form a sample of all red balls we must select 5 balls from the 15 red ones. This can be done in $C(15, 5)$ ways—that is, in

$$C(15, 5) = \frac{15 \cdot 14 \cdot 13 \cdot 12 \cdot 11}{5 \cdot 4 \cdot 3 \cdot 2 \cdot 1} = 3003$$

ways.

(c) To answer this question we use both the multiplication principle and the formula for $C(n, r)$. We form a sample of 3 red balls and 2 white balls using a sequence of two operations:

Select 3 red balls	Select 2 white balls
Operation 1	Operation 2

The first operation can be performed in $C(15, 3)$ ways and the second in $C(10, 2)$ ways. Thus, the total number of samples having 3 red and 2 white balls is $C(15, 3) \cdot C(10, 2)$. However,

$$C(15, 3) = \frac{15 \cdot 14 \cdot 13}{3 \cdot 2 \cdot 1} = 455$$

$$C(10, 2) = \frac{10 \cdot 9}{2 \cdot 1} = 45$$

$$C(15, 3) \cdot C(10, 2) = 455 \cdot 45 = 20{,}475.$$

So the number of possible samples is 20,475.

(d) A sample with at least 4 red balls has either 4 or 5 red balls. By part (b) the number of samples with 5 red balls is 3003. Using the same reasoning as in part (c), the number of samples with 4 red balls is $C(15, 4) \cdot C(10, 1) = 1365 \cdot 10 = 13{,}650$. Thus the total number of samples having at least 4 red balls is $13{,}650 + 3003 = 16{,}653$.

PRACTICE PROBLEMS 6

1. A newspaper reporter wants an indication of how the 15 members of the school board feel about a certain proposal. She decides to question a sample of 6 of the board members.

   (a) How many different samples are possible?

   (b) Suppose that 10 of the board members support the proposal and 5 oppose it. How many of the samples reflect the distribution of the board? That is, in how many of the samples do 4 people support the proposal and 2 oppose it?

2. A basketball player shoots eight free throws and lists the sequence of results of each trial in order. Let S represent "success" and F represent "failure." Then, for instance, FFSSSSSS represents the outcome of missing the first two shots and hitting the rest.

(a) How many different outcomes are possible?

(b) How many of the outcomes have six successes?

EXERCISES 6

1. An experiment consists of tossing a coin six times and observing the sequence of heads and tails.

(a) How many different outcomes are possible?

(b) How many different outcomes have exactly three heads?

(c) How many different outcomes have more heads than tails?

(d) How many different outcomes have at least two heads?

2. Refer to the map in Fig. 2. How many routes are there from A to B?

FIGURE 2   FIGURE 3

3. Refer to the map in Fig. 3. How many routes are there from A to B?

4. An urn contains 12 numbered balls, of which 8 are red and 4 are white. A sample of 4 balls is to be selected.

(a) How many different samples are possible?

(b) How many samples contain all red balls?

(c) How many samples contain 2 red balls and 2 white balls?

(d) How many samples contain at least 3 red balls?

(e) How many samples contain a different number of red balls than white balls?

5. A bag of 10 apples contains 2 rotten apples and 8 good apples. A shopper selects a sample of 3 apples from the bag.

(a) How many different samples are possible?

(b) How many samples contain all good apples?

(c) How many samples contain at least 1 rotten apple?

6. An experiment consists of tossing a coin 8 times and observing the sequence of heads and tails.

   (a) How many different outcomes are possible?

   (b) How many different outcomes have exactly 3 heads?

   (c) How many different outcomes have at least 2 heads?

   (d) How many different outcomes have 4 heads or 5 heads?

7. How many ways can a group of 100 students be assigned to dorms A, B, and C, with 25 assigned to dorm A, 40 to dorm B, and 35 to dorm C?

8. In the World Series the American League team ("A") and the National League team ("N") play until one team wins four games. If the sequence of winners is designated by letters (NAAAA means National League won the first game and lost the next four), how many different sequences are possible?

9. Refer to the map in Fig. 3. How many of the routes from A to B pass through the point C?

10. A package contains 100 fuses, of which 10 are defective. A sample of 5 fuses is selected at random.

    (a) How many different samples are there?

    (b) How many of the samples contain 2 defective fuses?

    (c) How many of the samples contain at least 1 defective fuse?

11. In how many ways can a committee of 15 senators be selected from the 100 members of the U.S. Senate so that no two committee members are from the same state?

12. An exam contains five "true or false" questions. How many of the 32 different ways of answering these questions contain 3 or more correct answers?

13. How many poker hands consist of 3 aces and 2 kings?

14. How many poker hands consist of 2 aces, 2 cards of another denomination, and 1 card of a third denomination?

15. How many poker hands consist of 3 cards of one denomination and 2 cards of another denomination? (Such a poker hand is called a "full house.")

16. How many poker hands consist of 2 cards of one denomination, 2 cards of another (different) denomination, and 1 card of a third denomination? (Such a poker hand is called "two pairs.")

17. A five-digit ZIP code is said to be detour-prone if it looks like a valid and different ZIP code when read upside down. For instance, 68901 and 88111 are detour-prone, whereas 32145 and 10801 are not. How many of the 10^5 possible zip code numbers are detour-prone?

18. Nine college students are interested in playing a pickup basketball game. How many different ways are there of choosing two teams of four people each from this group?

19. A singer wants to produce a "survey" record consisting of 2 songs from each of her previous 5 albums. If each album contained 10 songs, in how many different ways can songs be chosen for the survey record?

20. In how many different ways can 16 tennis players be paired off for the first round of a tournament?

SOLUTIONS TO PRACTICE PROBLEMS 6

1. (a) $C(15, 6)$. Each sample is an unordered selection of 15 objects taken 6 at a time.

   (b) $C(10, 4) \cdot C(5, 2)$. Asking for the number of samples of a certain type is the same as asking for the number of ways that the task of forming such a sample can be performed. This task is composed of two consecutive operations. Operation 1, selecting 4 people from among the 10 that support the proposal, can be performed in $C(10, 4)$ ways. Operation 2, selecting 2 people from among the 5 people that oppose the proposal, can be performed in $C(5, 2)$ ways. Therefore, by the multiplication principle, the complete task can be performed in $C(10, 4) \cdot C(5, 2)$ ways.

   [*Note:* $C(15, 6) = 5005$ and $C(10, 4) \cdot C(5, 2) = 2100$. Therefore, less than half of the possible samples reflect the true distribution of the school board.]

2. (a) 2^8. Apply the generalized multiplication principle.

   (b) $C(8, 6)$ or 28. Each outcome having 6 successes corresponds to a sequence of 8 letters of which 6 are S's and 2 are F's. Such an outcome is specified by selecting the 6 locations for the S's from among the 8 locations, and this has $C(8, 6)$ possibilities.

5.7. The Binomial Theorem

In the preceding two sections we have dealt with permutations and combinations and, in particular, have derived a formula for $C(n, r)$, the number of combinations of n objects taken r at a time. Namely, we have

$$C(n, r) = \frac{P(n, r)}{r!} = \frac{n(n-1) \cdot \ldots \cdot (n-r+1)}{r!}. \tag{1}$$

Actually, formula (1) was verified in case both n and r are positive integers. But it is useful to consider $C(n, r)$ also in case $r = 0$. In this case we are considering the number of combinations of n things taken 0 at a time. There is clearly only one such combination: the one containing no elements. Therefore,

$$C(n, 0) = 1. \tag{2}$$

Here is another convenient formula for $C(n, r)$:

$$C(n, r) = \frac{n!}{r! (n-r)!}. \tag{3}$$

For instance, according to formula (3),

$$C(8, 3) = \frac{8!}{3!(8 - 3)!} = \frac{8!}{3!5!} = \frac{8 \cdot 7 \cdot 6 \cdot \cancel{5} \cdot \cancel{4} \cdot \cancel{3} \cdot \cancel{2} \cdot \cancel{1}}{3 \cdot 2 \cdot 1 \cdot \cancel{5} \cdot \cancel{4} \cdot \cancel{3} \cdot \cancel{2} \cdot \cancel{1}} = \frac{8 \cdot 7 \cdot 6}{3 \cdot 2 \cdot 1}$$

which agrees with the result given by formula (1).

Note that

$$n(n - 1) \cdot \ldots \cdot (n - r + 1)$$

$$\frac{n(n - 1) \cdot \ldots \cdot (n - r + 1)\cancel{(n - r)}\cancel{(n - r - 1)} \cdot \ldots \cdot \cancel{2} \cdot \cancel{1}}{\cancel{(n - r)}\cancel{(n - r - 1)} \cdot \ldots \cdot \cancel{2} \cdot \cancel{1}} = \frac{n!}{(n - r)!}.$$

Then, by formula (1), we have

$$C(n, r) = \frac{n(n - 1) \cdot \ldots \cdot (n - r + 1)}{r!} = \frac{\dfrac{n!}{(n - r)!}}{r!} = \frac{n!}{r!(n - r)!},$$

which is formula (3).

Note that for $r = 0$, formula (3) reads

$$C(n, 0) = \frac{n!}{0!(n - 0)!} = \frac{n!}{0!n!} = \frac{1}{0!}.$$

Let us agree that the value of $0!$ is 1. Then the right-hand side of the equation above is 1, so that formula (3) holds also for $r = 0$.

Formula (3) can be used to prove many facts about $C(n, r)$. For example, here is a formula which is useful in calculating $C(n, r)$ for large values of r:

$$C(n, r) = C(n, n - r). \tag{4}$$

For example, suppose that we wish to calculate $C(100, 98)$. If we apply formula (4), we have

$$C(100, 98) = C(100, 100 - 98) = C(100, 2) = \frac{100 \cdot 99}{2 \cdot 1} = 4950.$$

Apply formula (3) to evaluate $C(n, n - r)$:

$$C(n, n - r) = \frac{n!}{(n - r)!(n - (n - r))!} = \frac{n!}{(n - r)!r!}$$

$$= C(n, r) \quad \text{[by formula (3) again]}.$$

The formula is intuitively reasonable since each time we select a subset of r elements we are excluding a subset of $n - r$ elements. Thus there are as many subsets of $n - r$ elements as there are subsets of r elements.

An alternative notation for $C(n, r)$ is $\binom{n}{r}$. Thus, for example,

$$\binom{5}{2} = C(5, 2) = \frac{5 \cdot 4}{2 \cdot 1} = 10.$$

The symbol $\binom{n}{r}$ is called a *binomial coefficient*. To discover why, let us tabulate the values of $\binom{n}{r}$ for some small values of n and r.

$$n = 2: \quad \binom{2}{0} = 1 \quad \binom{2}{1} = 2 \quad \binom{2}{2} = 1$$

$$n = 3: \quad \binom{3}{0} = 1 \quad \binom{3}{1} = 3 \quad \binom{3}{2} = 3 \quad \binom{3}{3} = 1$$

$$n = 4: \quad \binom{4}{0} = 1 \quad \binom{4}{1} = 4 \quad \binom{4}{2} = 6 \quad \binom{4}{3} = 4 \quad \binom{4}{4} = 1$$

$$n = 5: \quad \binom{5}{0} = 1 \quad \binom{5}{1} = 5 \quad \binom{5}{2} = 10 \quad \binom{5}{3} = 10 \quad \binom{5}{4} = 5 \quad \binom{5}{5} = 1.$$

Each row of the table above consists of the coefficients that arise in expanding $(x + y)^n$. To see this, inspect the results of expanding $(x + y)^n$ for $n = 2, 3, 4, 5$:

$$(x + y)^2 = x^2 + 2xy + y^2$$
$$(x + y)^3 = x^3 + 3x^2y + 3xy^2 + y^3$$
$$(x + y)^4 = x^4 + 4x^3y + 6x^2y^2 + 4xy^3 + y^4$$
$$(x + y)^5 = x^5 + 5x^4y + 10x^3y^2 + 10x^2y^3 + 5xy^4 + y^5.$$

Compare the coefficients in any row with the values in the corresponding row of binomial coefficients. Note that they are the same. Thus, we see that the binomial coefficients arise as coefficients in multiplying out powers of the binomial $x + y$; whence the name *binomial coefficient*.

What we observed above for the exponents $n = 2, 3, 4, 5$ holds true for any positive integer n. We have the following result, a proof of which is given at the end of this section.

Binomial Theorem

$$(x+y)^n = \binom{n}{0}x^n + \binom{n}{1}x^{n-1}y + \binom{n}{2}x^{n-2}y^2 + \cdots + \binom{n}{n-1}xy^{n-1} + \binom{n}{n}y^n.$$

EXAMPLE 1 Expand $(x + y)^6$.

Solution By the binomial theorem

$$(x + y)^6 = \binom{6}{0}x^6 + \binom{6}{1}x^5y + \binom{6}{2}x^4y^2 + \binom{6}{3}x^3y^3$$

$$+ \binom{6}{4}x^2y^4 + \binom{6}{5}xy^5 + \binom{6}{6}y^6.$$

Furthermore,

$$\binom{6}{0} = 1 \qquad \binom{6}{1} = \frac{6}{1} = 6 \qquad \binom{6}{2} = \frac{6 \cdot 5}{2 \cdot 1} = 15$$

$$\binom{6}{3} = \frac{6 \cdot 5 \cdot 4}{3 \cdot 2 \cdot 1} = 20 \qquad \binom{6}{4} = \frac{6 \cdot 5 \cdot \cancel{4} \cdot \cancel{3}}{\cancel{4} \cdot \cancel{3} \cdot 2 \cdot 1} = 15$$

$$\binom{6}{5} = \frac{6 \cdot \cancel{5} \cdot \cancel{4} \cdot \cancel{3} \cdot \cancel{2}}{\cancel{5} \cdot \cancel{4} \cdot \cancel{3} \cdot \cancel{2} \cdot 1} = 6 \qquad \binom{6}{6} = \frac{\cancel{6} \cdot \cancel{5} \cdot \cancel{4} \cdot \cancel{3} \cdot \cancel{2} \cdot \cancel{1}}{\cancel{6} \cdot \cancel{5} \cdot \cancel{4} \cdot \cancel{3} \cdot \cancel{2} \cdot \cancel{1}} = 1.$$

Thus,

$$(x + y)^6 = x^6 + 6x^5y + 15x^4y^2 + 20x^3y^3 + 15x^2y^4 + 6xy^5 + y^6.$$

The binomial theorem can be used to count the number of subsets of a set, as shown in the next example.

EXAMPLE 2 Determine the number of subsets of a set with five elements.

Solution Let us count the number of subsets of each possible size. A subset of r elements can be chosen in $\binom{5}{r}$ ways, since $C(5, r) = \binom{5}{r}$. So the set has $\binom{5}{0}$ subsets with 0 elements, $\binom{5}{1}$ subsets with 1 element, $\binom{5}{2}$ subsets with 2 elements, and so on. Therefore, the total number of subsets is

$$\binom{5}{0} + \binom{5}{1} + \binom{5}{2} + \binom{5}{3} + \binom{5}{4} + \binom{5}{5}.$$

On the other hand, the binomial theorem for $n = 5$ gives

$$(x + y)^5 = \binom{5}{0}x^5 + \binom{5}{1}x^4y + \binom{5}{2}x^3y^2 + \binom{5}{3}x^2y^3 + \binom{5}{4}xy^4 + \binom{5}{5}y^5.$$

Set $x = 1$ and $y = 1$ in this formula, deriving

$$(1 + 1)^5 = \binom{5}{0}1^5 + \binom{5}{1}1^4 \cdot 1 + \binom{5}{2}1^3 \cdot 1^2 + \binom{5}{3}1^2 \cdot 1^3 + \binom{5}{4}1 \cdot 1^4 + \binom{5}{5}1^5$$

$$2^5 = \binom{5}{0} + \binom{5}{1} + \binom{5}{2} + \binom{5}{3} + \binom{5}{4} + \binom{5}{5}.$$

Thus the total number of subsets of a set with five elements (the right side) equals $2^5 = 32$.

There is nothing special about the number five in the preceding example. An analogous argument gives the following result:

> A set of n elements has 2^n subsets.

EXAMPLE 3 A pizza parlor offers a plain cheese pizza to which any number of six possible toppings can be added. How many different pizzas can be ordered?

Solution Ordering a pizza requires selecting a subset of the six possible toppings. Since the set of six toppings has 2^6 different subsets, there are 2^6 or 64 different pizzas. (Note that the plain cheese pizza corresponds to selecting the empty subset of toppings.)

Proof of the Binomial Theorem Note that

$$(x + y)^n = \underbrace{(x + y)(x + y) \cdot \ldots \cdot (x + y)}_{n \text{ factors}}.$$

Multiplying out these factors involves forming all products, where one term is selected from each factor, and then combining like products. For instance,

$$(x + y)(x + y)(x + y) = x \cdot x \cdot x + x \cdot x \cdot y + x \cdot y \cdot x + y \cdot x \cdot x + x \cdot y \cdot y$$
$$+ y \cdot x \cdot y + y \cdot y \cdot x + y \cdot y \cdot y.$$

The first product on the right, $x \cdot x \cdot x$, is obtained by selecting the x-term from each of the three factors. The next term, $x \cdot x \cdot y$, is obtained by selecting the x-terms from the first two factors and the y-term from the third. The next product, $x \cdot y \cdot x$, is obtained by selecting the x-terms from the first and third factors and the y-term from the second. And so on. There are as many products containing two x's and one y as there are ways of selecting the factor from which to pick the y-term—namely $\binom{3}{1}$.

In general, when multiplying the n factors $(x + y)(x + y) \cdots (x + y)$, the number of products having k y's (and therefore $n - k$ x's) is equal to the number of different ways of selecting the k factors from which to take the y-term—that is, $\binom{n}{k}$. Therefore, the coefficient of $x^{n-k}y^k$ is $\binom{n}{k}$. This proves the binomial theorem.

PRACTICE PROBLEMS 7

1. Calculate $\binom{12}{8}$.

2. An ice cream parlor offers 10 flavors of ice cream and 5 toppings. How many different servings are possible if each choice consists of one flavor of ice cream and as many toppings as desired?

EXERCISES 7

Calculate:

1. $\binom{6}{2}$ 2. $\binom{20}{18}$ 3. $0!$ 4. $\binom{9}{0}$ 5. $\binom{18}{15}$

6. $1!$ 7. $\binom{500}{0}$ 8. $\binom{25}{24}$ 9. $(x + y)^7$ 10. $(x + y)^8$

11. Determine the first three terms in the binomial expansion of $(x + y)^{10}$.

12. Determine the first three terms in the binomial expansion of $(x + y)^{20}$.

13. How many different subsets can be chosen from a set of six elements?

14. How many different subsets can be chosen from a set of 100 elements?

15. How many different tips could you leave in a restaurant if you had a nickel, a dime, a quarter, and a half-dollar?

16. A pizza parlor offers mushroom, green peppers, onions, and sausage as topping for the plain cheese base. How many different types of pizzas can be made?

17. An ice cream parlor offers four flavors of ice cream, three toppings, and two sizes of glasses. How many different servings consisting of a single flavor of ice cream are possible?

18. A salad bar offers a base of lettuce to which tomatoes, chickpeas, beets, pinto beans, olives, and green peppers can be added. There are five salad dressings available. How many different salads are possible? (Assume that each salad contains lettuce and at most one salad dressing.)

19. How many batting orders are possible in a nine-member baseball team if the catcher must bat fourth and the pitcher last?

20. How many ways can a selection of at least one book be made from a set of eight books?

21. In how many ways can a committee of 5 people be chosen from 12 married couples if:

    (a) The committee must consist of two men and three women?

    (b) A man and his wife cannot both serve on the committee?

22. Suppose that you are voting in an election for state delegate. Two state delegates are to be elected from among seven candidates. In how many different ways can you cast your ballot? (*Note:* You may vote for two candidates. However, some people "single-shoot" and others don't pull any levers.)

23. Show that, for n a positive integer, we have

$$\binom{n}{0} + \binom{n}{1} + \cdots + \binom{n}{n} = 2^n$$

[*Hint:* Write down the binomial theorem for $(1 + 1)^n$.]

24. Show that, for n a positive integer, we have

$$\binom{n}{0} - \binom{n}{1} + \cdots \pm \binom{n}{n} = 0.$$

[*Hint:* Write down the binomial theorem for $(1 + (-1))^n$.]

25. Show that, for n, k positive integers, we have

$$\binom{n}{k - 1} + \binom{n}{k} = \binom{n + 1}{k}.$$

$$\left[\textit{Hint:} \binom{n}{k - 1} + \binom{n}{k} = \frac{n!}{(k - 1)!(n - k)!} \left(\frac{1}{n - k + 1} + \frac{1}{k} \right). \right]$$

26. In the following table (known as Pascal's triangle) the entries in the nth row are $\binom{n}{0}$, $\binom{n}{1}$, $\binom{n}{2}$, $\ldots$, $\binom{n}{n}$. Use Exercise 25 to explain how we can use each row to calculate the row below it by using only addition. Complete the first eight rows of Pascal's triangle.

$$
\begin{array}{ccccccc}
1 & 1 & & & & & \\
1 & 2 & 1 & & & & \\
1 & 3 & 3 & 1 & & & \\
1 & 4 & 6 & 4 & 1 & & \\
1 & 5 & 10 & 10 & 5 & 1 & \\
\vdots & & & & & &
\end{array}
$$

A feasible set is determined by choosing a subset from the set of five inequalities $\{x \geq 0,\ y \geq 0,\ x + y \leq 10,\ x + 3y \leq 21,\ x \geq y - 3\}$. See the graph on the right.

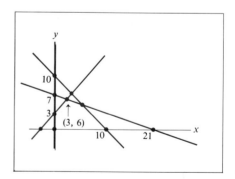

27. How many feasible sets are possible?

28. How many of the possible feasible sets have $(3, 6)$ as a vertex?

SOLUTIONS TO PRACTICE PROBLEMS 7

1. 495. $\binom{12}{8}$ is the same as $C(12, 8)$, which equals $C(12, 12 - 8)$ or $C(12, 4)$.

$$C(12, 4) = \frac{12 \cdot 11 \cdot 10 \cdot 9}{4 \cdot 3 \cdot 2 \cdot 1} = \frac{\overset{5}{\cancel{12}} \cdot 11 \cdot \cancel{10} \cdot 9}{\cancel{4} \cdot \cancel{3} \cdot \cancel{2} \cdot 1} = 495.$$

2. 320. The task of deciding what sort of serving to have consists of two operations. The first operation, selecting the flavor of ice cream, can be performed in 10 ways. The second operation, selecting the toppings, can be performed in 2^5 or 32 ways, since selecting the toppings amounts to selecting a subset from the set of 5 toppings and a set of 5 elements has 2^5 subsets. (Notice that selecting the empty subset corresponds to ordering a plain dish of ice cream.) By the multiplication principle, the task can be performed in $10 \cdot 32 = 320$ ways.

5.8. Multinomial Coefficients and Partitions

Permutation and combination problems are only two of the many types of counting problems. By appropriately generalizing the binomial coefficients to the so-called *multinomial coefficients* we can consider certain generalizations of combinations, namely *partitions*. To introduce the notion of a partition, let us return to combinations and look at them from another viewpoint. Suppose that we consider combinations of n objects taken r at a time. View the n objects as the elements of a set S. Then each combination determines an ordered division of S into two subsets, S_1 and S_2, the first containing the r elements selected and the second containing the $n - r$ elements remaining (Fig. 1). We see that $S = S_1 \cup S_2$ and $n(S_1) + n(S_2) = n$. This ordered division is called an *ordered partition* of type $(r, n - r)$. We know that the number of such partitions is just the number of ways of selecting the first subset, that is, $n!/[r!(n - r)!]$. If we let $n_1 = n(S_1) = r$ and $n_2 = n(S_2) = n - r$, then we find that the number of ordered partitions of type (n_1, n_2) is $n!/n_1!n_2!$.

We may generalize the above situation as follows: Let S be a set of n elements. An *ordered partition of S of type* $(n_1, n_2, \ldots, n_m)$ is a decomposition of S into m subsets (given in a specific order) $S_1, S_2, \ldots, S_m$, where no two of these intersect and where $n(S_1) = n_1, n(S_2) = n_2, \ldots, n(S_m) = n_m$ (Fig. 2). Since S has n elements, we clearly must have $n = n_1 + n_2 + \cdots + n_m$.

FIGURE 1

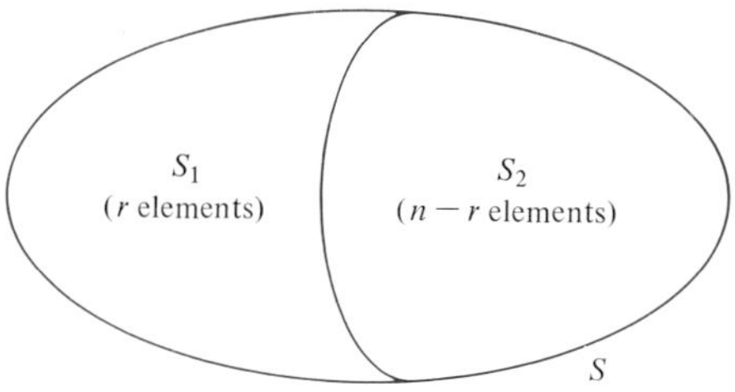

FIGURE 2

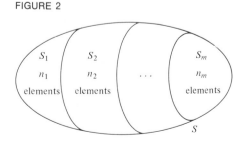

S_1 (r elements) S_2 ($n - r$ elements) S

S_1 n_1 elements S_2 n_2 elements $\cdots$ S_m n_m elements S

EXAMPLE 1 List all ordered partitions of $S = \{a, b, c, d\}$ of type (1, 1, 2).

Solution

$(\{a\}, \{b\}, \{c, d\})$ $(\{c\}, \{a\}, \{b, d\})$

$(\{a\}, \{c\}, \{b, d\})$ $(\{c\}, \{b\}, \{a, d\})$

$(\{a\}, \{d\}, \{b, c\})$ $(\{c\}, \{d\}, \{a, b\})$

$(\{b\}, \{a\}, \{c, d\})$ $(\{d\}, \{a\}, \{b, c\})$

$(\{b\}, \{c\}, \{a, d\})$ $(\{d\}, \{b\}, \{a, c\})$

$(\{b\}, \{d\}, \{a, c\})$ $(\{d\}, \{c\}, \{a, b\}).$

Note that the ordered partition $(\{a\}, \{b\}, \{c, d\})$ is different from the ordered partition $(\{b\}, \{a\}, \{c, d\})$, since in the first S_1 is $\{a\}$ whereas in the second S_1 is $\{b\}$. The order in which the subsets are given is significant.

We saw earlier that the number of ordered partitions of type (n_1, n_2) for a set of n elements is $n!/n_1! n_2!$. This result generalizes.

Number of Ordered Partitions of Type $(n_1, n_2, \ldots, n_m)$ Let S be a set of n elements. Then the number of ordered partitions of S of type $(n_1, n_2, \ldots, n_m)$ is

$$\frac{n!}{n_1! n_2! \cdots n_m!}.$$

(1)

The number of ordered partitions of type $(n_1, n_2, \ldots, n_m)$ for a set of n elements is often denoted

$$\binom{n}{n_1, n_2, \ldots, n_m}.$$

Using the notation, result (1) says that

$$\binom{n}{n_1, n_2, \ldots, n_m} = \frac{n!}{n_1! n_2! \cdots n_m!}.$$

The binomial coefficient $\binom{n}{n_1}$ can also be written $\binom{n}{n_1, n_2}$. The number

$$\binom{n}{n_1, n_2, \ldots, n_m}$$

is called a *multinomial coefficient* since it appears as the coefficient of $x_1^{n_1} x_2^{n_2} \cdots x_m^{n_m}$ in the expansion of $(x_1 + x_2 + \cdots + x_m)^n$.

EXAMPLE 2   Let S be a set of four elements. Use the formula in (1) to determine the number of ordered partitions of S of type $(1, 1, 2)$.

Solution   Here $n = 4$, $n_1 = 1$, $n_2 = 1$, and $n_3 = 2$. Therefore, the number of ordered partitions of type $(1, 1, 2)$ is

$$\binom{4}{1, 1, 2} = \frac{4!}{1!1!2!} = \frac{4 \cdot 3 \cdot 2 \cdot 1}{1 \cdot 1 \cdot 2 \cdot 1} = 12.$$

This result is the same as obtained in Example 1 by enumeration.

EXAMPLE 3   A work crew consists of 12 construction workers, all having the same skills. Construction requires four welders, three concrete workers, three heavy equipment operators, and two bricklayers. In how many ways can the 12 workers be assigned to the required tasks?

Solution   Each assignment of jobs corresponds to an ordered partition of the type $(4, 3, 3, 2)$. The number of such ordered partitions is

$$\binom{12}{4, 3, 3, 2} = \frac{12!}{4!3!3!2!} = 277{,}200.$$

Sometimes each of the m subsets of an ordered partition are required to have the same number of elements. If the set has n elements and each of the m subsets has r elements, then the number of ordered partitions of type $\underbrace{(r, r, \ldots, r)}_{m}$ is

$$\binom{n}{r, r, \ldots, r} = \frac{n!}{r!r! \cdots r!} = \frac{n!}{(r!)^m}. \tag{2}$$

EXAMPLE 4   In the game of bridge, four players seated in a specified order are each dealt 13 cards. How many different possibilities are there for the hands dealt to the players?

Solution   Each deal results in an ordered partition of the 52 cards of type $(13, 13, 13, 13)$. The number of such partitions is

$$\binom{52}{13, 13, 13, 13} = \frac{52!}{(13!)^4}.$$

This number is approximately 5.36×10^{28}.

Unordered Partitions   Determining the number of unordered partitions of a certain type is a complex matter. We still restrict our attention to the special case in which each subset is of the same size.

EXAMPLE 5   List all unordered partitions of $S = \{a, b, c, d\}$ of the type (2, 2).

Solution
$$(\{a, b\}, \{c, d\})$$
$$(\{a, c\}, \{b, d\})$$
$$(\{a, d\}, \{b, c\}).$$

Note that the partition $(\{c, d\}, \{a, b\})$ is the same as the partition $(\{a, b\}, \{c, d\})$ when order is not taken into account.

> *Number of Unordered Partitions of Type* $(r, r, \ldots, r)$ Let S be a set of n elements where $n = m \cdot r$. Then the number of unordered partitions of S of type $(r, r, \ldots, r)$ is
>
> $$\frac{1}{m!} \cdot \frac{n!}{(r!)^m}.$$

(3)

Formula (3) follows from the fact that each unordered partition of the m subsets gives rise to $m!$ ordered partitions. Therefore,

$$(m!) \, [\text{number of unordered partitions}] = [\text{number of ordered partitions}]$$

or

$$[\text{number of unordered partitions}] = \frac{1}{m!} \cdot [\text{number of ordered partitions}]$$

$$= \frac{1}{m!} \frac{n!}{(r!)^m} \qquad [\text{by formula (2)}].$$

EXAMPLE 6   Let S be a set of four elements. Use formula (3) to determine the number of unordered partitions of S of type (2, 2).

Solution   Here $n = 4$, $r = 2$, $m = 2$. Therefore, the number of unordered partitions of type (2, 2) is

$$\frac{1}{2!} \cdot \frac{4!}{(2!)^2} = \frac{1}{2} \cdot \frac{4 \cdot 3 \cdot 2 \cdot 1}{(2 \cdot 1)^2} = 3.$$

This result is the same as obtained in Example 5 by enumeration.

EXAMPLE 7   A construction crew contains 12 workers, all having similar skills. In how many ways can the workers be divided into four groups of three?

Solution   The order of the four groups is not relevant. (It does not matter which is labeled S_1 and which S_2, and so on. Only the composition of the groups is important.)

Applying formula (3) with $n = 12$, $r = 3$, and $m = 4$, we see that the number of ways is

$$\frac{1}{m!} \cdot \frac{n!}{(r!)^m} = \frac{1}{4!} \cdot \frac{12!}{(3!)^4} = \frac{1}{4 \cdot 3 \cdot 2 \cdot 1} \cdot \frac{12 \cdot 11 \cdot 10 \cdot 9 \cdot 8 \cdot 7 \cdot 6 \cdot 5 \cdot 4 \cdot 3 \cdot 2 \cdot 1}{6^4}$$

$$= \frac{2 \cdot 11 \cdot 10 \cdot 9 \cdot 8 \cdot 7 \cdot 1 \cdot 5}{6 \cdot 6} = 15,400.$$

PRACTICE PROBLEMS 8

1. A foundation wishes to award one grant of $100,000, two grants of $10,000, five grants of $5000, and five grants of $2000. Its list of potential grant recipients has been narrowed to 13 possibilities. In how many ways can the awards be made?

2. In how many different ways can six medical interns be put into three groups of two and assigned to

   (a) the radiology, neurology, and surgery departments?

   (b) share living quarters?

EXERCISES 8

Let S be a set of n elements. Determine the number of ordered partitions of the following types.

1. $n = 5; (3, 1, 1)$

2. $n = 5; (2, 1, 2)$

3. $n = 6; (2, 1, 2, 1)$

4. $n = 6; (3, 3)$

5. $n = 7; (3, 2, 2)$

6. $n = 7; (4, 1, 2)$

7. $n = 12; (4, 4, 4)$.

8. $n = 8; (3, 3, 2)$

9. $n = 12; (5, 3, 2, 2)$

10. $n = 8; (2, 2, 2, 2)$

Let S be a set of n elements. Determine the number of unordered partitions of the following types.

11. $n = 15; (3, 3, 3, 3, 3)$

12. $n = 10; (5, 5)$

13. $n = 18; (6, 6, 6)$

14. $n = 12; (4, 4, 4)$

15. A brokerage house regularly reports the behavior of a group of 20 stocks, each stock being reported as "up," "down," or "unchanged." How many different reports can show seven stocks up, five stocks down, and eight stocks unchanged?

16. An investment advisory service rates investments as A, AA, and AAA. On a certain week, it rates 15 investments. In how many ways can it rate five investments in each of the categories?

17. In a certain month (of 30 days) it rains 10 days, snows two days, and is clear 18 days. In how many ways can such weather be distributed over the month?

18. A psychology experiment observes groups of four individuals. In how many ways can an experimenter choose five groups of four from among 20 subjects?

19. During orientation, new students are divided into groups of five people. In how many ways can four groups be chosen from among 20 people?

20. Of the nine contestants in a contest, three will receive cars, three will receive TV sets, and three will receive radios. In how many different ways can prizes be awarded?

21. A corporation has four employees that it wants to place in high executive positions. One will become president, one will become vice-president, and two will be appointed to the board of directors. In how many different ways can this be accomplished?

22. The ten members of a city council decide to form two committees of six to study zoning ordinances and street-repair schedules, with an overlap of two committee members. In how many ways can the committees be formed? (*Hint:* Specify three groups, not two.)

23. In how many ways can the 14 children in a third-grade class be paired up for a trip to a museum?

24. A sales representative must travel to three cities twice each in the next 10 days. Her nontravel days are spent in the office. In how many different ways can she schedule her travel, assuming that she does not want to spend four consecutive days in the office?

25. Derive formula (1) using the generalized multiplication principle and the formula for $\binom{n}{r}$. [*Hint:* First select the elements of S_1, then the elements of S_2, and so on.]

SOLUTIONS TO PRACTICE PROBLEMS 8

1. Each choice of recipients is an ordered partition of the 13 finalists into a first subset of one ($100,000 award), a second subset of two ($10,000 award), a third subset of five ($5000 award) and a fourth subset of five ($2000 award). The number of ways to choose the recipients is thus

$$\binom{13}{1, 2, 5, 5} = \frac{13!}{1!2!5!5!}$$

$$= \frac{13 \cdot 12 \cdot 11 \cdot \cancel{10} \cdot \cancel{9} \cdot \cancel{8} \cdot 7 \cdot 6 \cdot \cancel{5 \cdot 4 \cdot 3 \cdot 2 \cdot 1}}{1 \cdot \cancel{2} \cdot 1 \cdot \cancel{3} \cdot \cancel{4} \cdot \cancel{3} \cdot \cancel{2} \cdot 1 \cdot \cancel{5 \cdot 4 \cdot 3 \cdot 2 \cdot 1}}$$

$$= 13 \cdot 12 \cdot 11 \cdot 3 \cdot 7 \cdot 6$$

$$= 216, 216.$$

2. Each partition is of the type (2, 2, 2). In part (a) the order of the subsets is important, whereas in part (b) the order is irrelevant. Consider the partitions

({Dr. A, Dr. B}, {Dr. C, Dr. D}, {Dr. E, Dr. F})

({Dr. C, Dr. D}, {Dr. A, Dr. B}, {Dr. E, Dr. F}).

With respect to part (a) these two partitions are different since in one Drs. A and B are assigned to the radiology department and in the other they are assigned to the neurology department. With respect to part (b) these two partitions are the same since, for instance, in both Drs. A and B are roommates. Therefore, the answers are

(a) $\left(\dfrac{6}{2, 2, 2}\right) = \dfrac{6!}{(2!)^3} = \dfrac{6 \cdot 5 \cdot \not{4} \cdot 3 \cdot \not{2} \cdot 1}{\not{2} \cdot \not{2} \cdot \not{2}} = 90$

(b) $\dfrac{1}{3!} \cdot \dfrac{6!}{(2!)^3} = \dfrac{1}{\not{6}} \cdot \dfrac{\not{6} \cdot 5 \cdot \not{4} \cdot 3 \cdot \not{2} \cdot 1}{\not{2} \cdot \not{2} \cdot \not{2}} = 15$

Chapter 5: CHECKLIST

- ☐ Set
- ☐ Element
- ☐ Union and intersection of sets
- ☐ Subset
- ☐ Universal set
- ☐ Complement of a set
- ☐ Inclusion-exclusion principle
- ☐ Venn diagram
- ☐ Generalized multiplication principle
- ☐ Permutations
- ☐ Combinations
- ☐ $P(n, r) = n(n - 1) \cdots (n - r + 1)$
- ☐ $n!$
- ☐ $C(n, r) = \dfrac{P(n, r)}{r!} = \dfrac{n!}{r! \, (n - r)!}$
- ☐ $C(n, r) = C(n, n - r)$
- ☐ $\dbinom{n}{r}$
- ☐ Binomial theorem
- ☐ Number of subsets of a set
- ☐ Ordered partition of a set
- ☐ Multinomial coefficient

Chapter 5: SUPPLEMENTARY EXERCISES

1. List all subsets of the set $\{a, b\}$.

2. Draw a two-circle Venn diagram and shade the portion corresponding to the set $(S \cup T')'$.

3. There are 16 contestants in a tennis tournament. How many different possibilities are there for the two people who will play in the final round?

4. In how many ways can a coach and five basketball players line up in a row for a picture if the coach insists on standing at one of the ends of the row?

5. Draw a three-circle Venn diagram and shade the portion corresponding to the set $R' \cap (S \cup T)$.

6. Calculate the first three terms in the binomial expansion of $(x + y)^{12}$.

7. An urn contains 14 numbered balls of which 8 are red and 6 are green. How many different possibilities are there for selecting a sample of 5 balls in which 3 are red and 2 are green?

8. Sixty people with a certain medical condition were given pills. Fifteen of these people received placebos. Forty people showed improvement, and 30 of these people received an actual drug. How many of the people who received the drug showed no improvement?

9. An appliance store carries seven different types of washing machines and five different types of dryers. How many different combinations are possible for a customer who wants to purchase a washing machine and a dryer?

10. There are 12 contestants in a contest. Two will receive trips around the world, 4 will receive cars, and 6 will receive color TV sets. In how many different ways can the prizes be awarded?

11. Out of a group of 115 applicants for jobs at the World Bank, 70 speak French, 65 speak Spanish, 65 speak German, 45 speak French and Spanish, 35 speak Spanish and German, 40 speak French and German, and 35 speak all three languages. How many of the people speak none of the three languages?

12. Calculate $\binom{17}{15}$.

The 100 members of the Earth Club were asked what they felt the club's priorities should be in the coming year: clean water, clean air, or recycling. The responses were 45 for clean water, 30 for clean air, 42 for recycling, 13 for both clean air and clean water, 20 for clean air and recycling, 16 for clean water and recycling, and 9 for all three. Exercises 13–20 refer to this poll.

13. How many members thought the priority should be clean air only?

14. How many members thought the priority should be clean water or clean air, but not both?

15. How many members thought the priority should be clean water or recycling, but not clean air?

16. How many members thought the priority should be clean air and recycling, but not clean water?

17. How many members thought the priority should be exactly one of the three issues?

18. How many members thought the priority should be anything but recycling?

19. How many members thought the priority should be recycling, but not clean air?

20. How many members thought the priority should be anything but one of these three issues?

21. How many different nine-letter words (i.e., sequences of letters) can be made using four S's and five T's?

22. Forty people take an exam. How many different possibilities are there for the set of people who pass the exam?

23. A survey at a small New England college showed that 400 students skied, 300 played ice hockey, and 150 did both. How many students participated in at least one of these sports?

24. A poker hand consists of five cards. How many different poker hands contain all cards of the same suit? (Such a hand is called a "flush.")

25. How many three-digit numbers are there in which no two digits are alike?

26. How many three-digit numbers are there in which exactly two digits are alike?

27. How many hands of five cards contain exactly three aces?

28. Fraternity and sorority names consist of two or three letters from the Greek alphabet. How many different names are there in which no letter appears more than once? (The Greek alphabet contains 24 letters.)

In Exercises 29–31, suppose there are three boys and three girls at a party.

29. How many different pairings of the six into three couples can be formed?

30. In how many ways can they be seated in a row such that no person is seated next to someone of the same sex?

31. In how many ways can they be seated at a round table such that no person is seated next to someone of the same sex?

32. If 10 lines are drawn in the plane so that none of them are parallel and no 3 lines intersect at the same point, how many points of intersection are there? If each of these 10 lines forms the boundary of the bounded feasible set of a system of linear inequalities, how many of the intersections occur outside the feasible set?

33. Two elementary school teachers have 24 students each. The first teacher splits his students into 4 groups of 6. The second teacher splits her students into 6 groups of 4. Which teacher has more options?

34. A consulting engineer agrees to spend three days at Widgets International, four days at Gadgets Unlimited, and three days at Doodads Incorporated in the next two work weeks. In how many different ways can she schedule her consultations?

35. A set of books can be arranged on a bookshelf in 120 different ways. How many books are in the set?

6

PROBABILITY

6.1   Introduction

6.2   Experiments, Outcomes, and Events

6.3   Assignment of Probabilities

6.4   Calculating Probabilities of Events

6.5   Conditional Probability and Independence

6.6   Tree Diagrams

6.7   Bayes' Theorem

6.1. Introduction

Many events in the world around us exhibit a random character. Yet by repeated observations of such events we can often discern long-run patterns which persist despite random, short-term fluctuations. Probability is the branch of mathematics devoted to a study of such events. To obtain a clearer idea of the sort of events considered, let us discuss a concrete example from the field of medicine.

Suppose that we wish to analyze the reliability of a skin test for active pulmonary tuberculosis. Unfortunately, such a test is not completely reliable. On the one hand, the test may be negative even for a person with tuberculosis. On the other hand, the test may be positive for a person who does not have tuberculosis. For the moment, let us concentrate on errors of the first sort and consider only individuals actually having tuberculosis. Suppose that by observing the results of the test on increasingly large populations of tuberculosis patients we accumulate the following data.

Number of tuberculosis patients (N)	Number of positive test results (m)	Relative frequency of positive test results (m/N)
100	97	.97
500	494	.988
1,000	981	.981
10,000	9,806	.9806
50,000	49,008	.98016
100,000	98,005	.98005

Note that out of each group of tuberculosis patients the test fails to identify a certain number. However, the data does exhibit a pattern. It appears that out of a very large population of tuberculosis patients the skin test will successfully identify about 98% of them. In fact, it appears that as the size of the population is increased, the relative frequency m/N more and more closely approximates the number .98. In a situation like this, we say that the skin test detects tuberculosis with a 98% likelihood or that the *probability* that the test detects tuberculosis (when present) is .98.

More generally, the *probability of an event* is a number which expresses the long-run likelihood that the event will occur. Such numbers are always chosen to lie between 0 and 1. The smaller the probability, the less likely the event is to occur. So, for example, an event having probability .1 is rather unlikely to occur; an event with probability .9 is very likely to occur; and an event with probability .5 is just as likely to occur as not.

We assign probabilities to events on the following intuitive basis: The probability of an event should represent the long-run proportion of the time that the event can be expected to occur. Thus, for example, an event with probability .9 can be expected to occur 90% of the time and an event with probability .1 can be expected to occur 10% of the time.

As we shall see, many real-life problems require us to calculate probabilities from known data. Here is one example which arises in connection with the skin test for tuberculosis.

> **MEDICAL DIAGNOSIS**   A clinic tests for active pulmonary tuberculosis. If a person has tuberculosis, the probability of a positive test result is .98. If a person does not have tuberculosis, the probability of a negative test result is .99. The incidence of tuberculosis in a certain city is 2 cases per 10,000 population. Suppose that an individual is tested and a positive result is noted. What is the probability that that individual actually has active pulmonary tuberculosis?

Before we can solve this problem, it will be necessary to do considerable preliminary work. We begin this work in Section 2, where we introduce a convenient language for discussing events and the process of observing them. In Section 3 we introduce probabilities of events, and in Sections 4, 5, 6 and 7 we develop methods for calculating probabilities of various sorts of events. The solution of the medical diagnosis problem is presented in Section 6.

6.2. Experiments, Outcomes, and Events

The events whose probabilities we wish to compute all arise as outcomes of various experiments. So as our first step in developing probability theory let us describe, in mathematical terms, the notions of experiment, outcome, and event.

For our purposes an *experiment* is an activity with an observable outcome. Here are some typical examples of experiments.

> **EXPERIMENT 1**   Flip a coin and observe the side that faces upward.
>
> **EXPERIMENT 2**   Allow a conditioned rat to run a maze and observe which one of the three possible paths it takes.
>
> **EXPERIMENT 3**   Choose a year and tabulate the amount of rainfall in New York City during that year.

We think of an experiment as being performed repeatedly. Each repetition of the experiment is called a *trial*. In each trial we observe the *outcome* of the experiment. For example, a possible outcome of Experiment 1 is "heads"; a possible outcome of Experiment 2 is "path 3"; and a possible outcome of Experiment 3 is "37.23 inches."

In order to describe an experiment in mathematical language, we construct a model of the experiment. It is most convenient to form the set consisting of all possible outcomes of the experiment. This set is called the *sample space* of the experiment. For example, if S_1, S_2, S_3 are the sample spaces for Experiments 1, 2, 3, respectively, then we immediately see that

$$S_1 = \{\text{heads, tails}\}$$

$$S_2 = \{\text{path 1, path 2, path 3}\}.$$

Moreover, since any nonnegative number is a candidate for the amount of rainfall, we have

$$S_3 = \{\text{all numbers} \geq 0\}.$$

We can describe an experiment in terms of the sample space as follows:

> Suppose that an experiment has a sample space S. Then each trial has as its outcome one of the elements of S.

Thus, for example, each trial for Experiment 1 has as its outcome one of the elements of the set

$$S_1 = \{\text{heads, tails}\}.$$

Henceforth, we shall always describe experiments in terms of their respective sample spaces. So it is important to be able to recognize the appropriate sample space in each instance. The next few examples should help you obtain the necessary facility in doing this.

EXAMPLE 1   An experiment consists of tossing a die and observing the number on the uppermost face. Describe the sample space S for this experiment.

Solution   There are six outcomes of the experiment, corresponding to the six possible numbers on the uppermost face. Therefore,

$$S = \{1, 2, 3, 4, 5, 6\}.$$

EXAMPLE 2   Once an hour a supermarket manager observes the number of people standing in checkout lines. The store has space for at most 30 customers to wait in line. What is the sample space S for this experiment?

Solution   The outcome of the experiment is the number of people standing in checkout lines. And this number may be 0, 1, 2, ..., or 30. Therefore,

$$S = \{0, 1, 2, \ldots, 30\}.$$

EXAMPLE 3   An experiment consists of throwing two dice, one red and one green, and observing the uppermost face on each. What is the associated sample space S?

Solution   Each outcome of the experiment can be regarded as an ordered pair of numbers, the first representing the number on the red die and the second the number on the green die. Thus, for example, the pair of numbers $(3, 5)$ represents the outcome "3 on the red die, 5 on the green die." The sample space consists of all possible pairs of numbers (r, g), where r and g are each one of the numbers 1, 2, 3, 4, 5, 6.

This sample space has 36 elements:

$$S = \{(1, 1), (1, 2), (1, 3), (1, 4), (1, 5), (1, 6),$$
$$(2, 1), (2, 2), (2, 3), (2, 4), (2, 5), (2, 6),$$
$$(3, 1), (3, 2), (3, 3), (3, 4), (3, 5), (3, 6),$$
$$(4, 1), (4, 2), (4, 3), (4, 4), (4, 5), (4, 6),$$
$$(5, 1), (5, 2), (5, 3), (5, 4), (5, 5), (5, 6),$$
$$(6, 1), (6, 2), (6, 3), (6, 4), (6, 5), (6, 6)\}.$$

The sample space for an experiment should be chosen so that every outcome is included and there is no overlap. Since we will be assigning probabilities to the elements of the sample space, we choose a sample space for its utility and not for its simplicity. Personal choices might differ; there is no one right answer.

Consider the experiment of counting the number of heads on three tosses of a fair coin. One possible sample space is $\{0, 1, 2, 3\}$. But we will see later that for practical purposes we might prefer

$$S = \{TTT, TTH, THT, HTT, THH, HTH, HHT, HHH\},$$

which gives the results (heads or tails) on each of the three tosses.

EXAMPLE 4   The Environmental Protection Agency orders Middle States Edison Corporation to install "scrubbers" to remove pollutants from its smokestacks. To monitor the effectiveness of the scrubbers, the corporation installs monitoring devices to record the levels of sulfur dioxide, oxides of nitrogen, and particulate matter (in parts per million) in the smokestack emissions. Consider the monitoring operation as an experiment. Describe the associated sample space.

Solution   Each reading of the instruments consists of an ordered triple of numbers (x, y, z), where $x =$ the level of sulfur dioxide, $y =$ the level of oxides of nitrogen, and $z =$ the level of particulate matter. The sample space thus consists of all possible triples (x, y, z), where $x \geq 0, y \geq 0, z \geq 0$.

The sample spaces in Examples 1 to 3 are *finite*. That is, the associated experiments have only a finite number of possible outcomes. However, the sample space of Example 4 is *infinite*, since there are infinitely many triples (x, y, z), $x \geq 0, y \geq 0, z \geq 0$.

Now that we have discussed experiments and their outcomes, let us turn our attention to the notion of "event." In connection with our preceding discussion, we can define many events whose probabilities we might wish to know. For example, in connection with experiment 2, we can consider the event

"A conditioned rat chooses either path 2 or path 3."

Here are two events associated with experiment 3:

"The annual rainfall in New York City exceeds 50 inches."
"The annual rainfall in New York City is less than 35 inches."

It is easy to describe events in terms of the sample space. For example, let us consider the die-tossing experiment of Example 1 and the following events.

I.   An even number occurs.
II. A number greater than 2 occurs.

We saw above that the sample space S for this experiment is

$$S = \{1, 2, 3, 4, 5, 6\}.$$

Assume that the experiment is performed. Then event I occurs precisely when the outcome of the experiment is 2, 4, or 6. That is, the event I occurs precisely when the outcome belongs to the set

$$E_{\mathrm{I}} = \{2, 4, 6\}.$$

Note that this set is a subset of the sample space S. Similarly, we can describe event II by the set

$$E_{\mathrm{II}} = \{3, 4, 5, 6\}.$$

Event II occurs precisely when the outcome of the experiment is an element of E_{II}.

The sets E_{I} and E_{II} contain all the information we need in order to completely describe the events I and II. This observation suggests the following definition of an event in terms of the sample space.

An *event E* is a subset of the sample space. We say that the event *occurs* when the outcome of the experiment is an element of E.

The next few examples provide some practice in describing events as subsets of the sample space.

EXAMPLE 5   Consider the supermarket of Example 2. Describe the following events as subsets of the sample space.

(a) Fewer than 5 people are waiting in line.

(b) More than 23 people are waiting in line.

(c) No people are waiting in line.

Solution   We saw that the sample space for this example is given by

$$S = \{0, 1, 2, \ldots, 30\}.$$

(a) If fewer than five people are waiting in line, then the number waiting is 0, 1, 2, 3, or 4. So the subset of S corresponding to event (a) is

$$\{0, 1, 2, 3, 4\}.$$

(b) In this case the number waiting must be 24, 25, 26, 27, 28, 29, or 30. So the event is just the subset

$$\{24, 25, 26, 27, 28, 29, 30\}.$$

(c) $\{0\}$.

EXAMPLE 6   Suppose that an experiment consists of tossing a coin three times and observing the sequence of heads and tails. (Order counts.)

(a) Determine the sample space S.

(b) Determine the event $E =$ "exactly two heads."

Solution   (a) Denote "heads" by H and "tails" by T. Then a typical outcome of the experiment is a sequence of H's and T's. So, for instance, the sequence HTT would stand for a head followed by two tails. We exhibit all such sequences and arrive at the sample space S:

$$S = \{HHH, HHT, HTH, THH, HTT, THT, TTH, TTT\}.$$

(b) Here are the outcomes in which exactly two heads occur: HHT, HTH, THH. Therefore, the event E is

$$E = \{HHT, HTH, THH\}.$$

EXAMPLE 7   A political poll surveys a group of people to determine their income levels and political affiliations. People are classified as either low-, middle-, or upper-level income, and as either Democrat, Republican, or Independent.

(a) Find the sample space corresponding to the poll.

(b) Determine the event $E_1 =$ "Independent."

(c) Determine the event $E_2 =$ "low income and not Independent."

(d) Determine the event $E_3 =$ "neither upper income nor Independent."

Solution   (a) Let us abbreviate low, middle, and upper income, respectively, by the letters L, M, and U, respectively. And let us abbreviate Democrat, Republican, and Independent by the letters D, R, and I, respectively. Then a response to the poll can be represented as a pair of letters. For example, the pair (L, D) refers to a lower-income-level Democrat. The sample space S is then given by

$$S = \{(L, D), (L, R), (L, I), (M, D), (M, R), (M, I), (U, D), (U, R), (U, I)\}.$$

(b) For the event E_1 the income level may be anything, but the political affiliation is Independent. Thus,

$$E_1 = \{(L, I), (M, I), (U, I)\}.$$

(c) For the event E_2 the income level is low and the political affiliation may be either Democrat or Republican, so that

$$E_2 = \{(L, D), (L, R)\}.$$

(d) For the event E_3 the income level may be either low or middle and the political affiliation may be Democrat or Republican. Thus,

$$E_3 = \{(L, D), (M, D), (L, R), (M, R)\}.$$

As we have seen, an event is a subset of the sample space. Two events are worthy of special mention. The first is the event corresponding to the empty set, $\emptyset$. This is called the *impossible event*, since it can never occur. The second special event is the set S itself. Every outcome is an element of S, so S always occurs. For this reason S is called the *certain event*.

One particular advantage of defining experiments and events in terms of sets is that it allows us to define new events from given ones by applying the operations of set theory. When so doing we always let the sample space S play the role of universal set. (All outcomes belong to the universal set.)

If E and F are events, then so are $E \cup F$, $E \cap F$, and E'. For example, consider the die-tossing experiment of Example 1. Then

$$S = \{1, 2, 3, 4, 5, 6\}.$$

Let E and F be the events given by

$$E = \{3, 4, 5, 6\} \qquad F = \{1, 4, 6\}.$$

Then we have

$$E \cup F = \{1, 3, 4, 5, 6\}$$
$$E \cap F = \{4, 6\}$$
$$E' = \{1, 2\}.$$

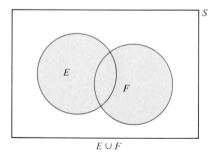

$E \cup F$

FIGURE 1

Let us interpret the events $E \cup F$, $E \cap F$, and E' using Venn diagrams. In Fig. 1 we have drawn a Venn diagram for $E \cup F$. Note that $E \cup F$ occurs precisely when the experimental outcome belongs to the shaded region—that is, to either E or F. Thus we have the following result.

The event $E \cup F$ occurs precisely when either E or F (or both) occurs.

Similarly, we can interpret the event $E \cap F$. This event occurs when the experimental outcome belongs to the shaded region of Fig. 2 — that is, to both E and F. Thus, we have an interpretation for $E \cap F$:

> The event $E \cap F$ occurs precisely when both E and F occur.

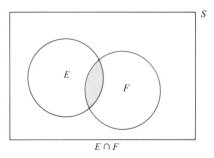

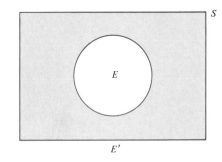

FIGURE 2

FIGURE 3

Finally, the event E' consists of all those outcomes not in E (Fig. 3). Therefore, we have:

> The event E' occurs precisely when E does not occur.

EXAMPLE 8 Consider the pollution monitoring described in Example 4. Let E, F, and G be the events

$$E = \text{"level of sulfur dioxide} \geq 100,\text{"}$$

$$F = \text{"level of particulate matter} \leq 50,\text{"}$$

$$G = \text{"level of oxides of nitrogen} \leq 30.\text{"}$$

Describe the following events.

(a) $E \cap F$ (b) E' (c) $E \cup G$ (d) $E' \cap F \cap G$

Solution   (a) $E \cap F = $ "level of sulfur dioxide ≥ 100 *and* particulate matter ≤ 50."

(b) $E' = $ "level of sulfur dioxide < 100."

(c) $E \cup G = $ "level of sulfur dioxide ≥ 100 *or* oxides of nitrogen ≤ 30."

(d) $E' \cap F \cap G = $ "level of sulfur dioxide < 100 *and* particulate matter ≤ 50 *and* oxides of nitrogen ≤ 30."

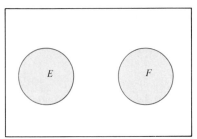

E and *F* are mutually exclusive

FIGURE 4

Suppose that E and F are events in a sample space S. We say that E and F are *mutually exclusive* (*or disjoint*) provided that $E \cap F = \varnothing$. In terms of Venn diagrams, we may represent a pair of mutually exclusive events as a pair of circles with no points in common (Fig. 4). If the events E and F are mutually exclusive, then E and F cannot simultaneously occur; if E occurs, then F does not; and if F occurs, then E does not.

EXAMPLE 9 Let $S = \{a, b, c, d, e, f, g\}$ be a sample space and let $E = \{a, b, c\}$, $F = \{e, f, g\}$, $G = \{c, d, f\}$.

(a) Are E and F mutually exclusive?

(b) Are F and G mutually exclusive?

Solution (a) $E \cap F = \varnothing$, so E and F are mutually exclusive.

(b) $F \cap G = \{f\}$, so F and G are *not* mutually exclusive.

PRACTICE PROBLEMS 2

1. A machine produces light bulbs. As part of a quality control procedure, a sample of five light bulbs is collected each hour and the number of defective light bulbs among these is observed.

    (a) What is the sample space for this experiment?

    (b) Describe the event "there are at most two defective light bulbs" as a subset of the sample space.

2. Suppose that there are two crates of citrus fruit and each crate contains oranges, grapefruit, and tangelos. An experiment consists of selecting a crate and then selecting a piece of fruit from that crate. Both the crate and the type of fruit are noted. Refer to the crates as crate I and crate II.

    (a) What is the sample space for this experiment?

    (b) Describe the event "a tangelo is selected" as a subset of the sample space.

EXERCISES 2

1. A committee of two people is to be selected from five people, R, S, T, U, and V.

    (a) What is the sample space for this experiment?

    (b) Describe the event "R is on the committee" as a subset of the sample space.

(c) Describe the event "neither R nor S is on the committee" as a subset of the sample space.

2. A letter is selected at random from the word "MISSISSIPPI."

(a) What is the sample space for this experiment?

(b) Describe the event "the letter chosen is a vowel" as a subset of the sample space.

3. An experiment consists of tossing a coin two times and observing the sequence of heads and tails.

(a) What is the sample space of this experiment?

(b) Describe the event "the first toss is a head" as a subset of the sample space.

4. A campus survey is taken to correlate the number of years that students have been on campus with their political leaning. Students are classified as freshman, sophomore, junior, or senior and as conservative or liberal.

(a) Find the sample space corresponding to the poll.

(b) Determine the event E_1 = "conservative."

(c) Determine the event E_2 = "junior and liberal."

(d) Determine the event E_3 = "neither freshman nor conservative."

5. Suppose that we have two urns—call them urn I and urn II—each containing red balls and white balls. An experiment consists of selecting an urn and then selecting a ball from that urn and noting its color.

(a) What is a suitable sample space for this experiment?

(b) Describe the event "Urn I is selected" as a subset of the sample space.

6. An experiment consists of tossing a coin four times and observing the sequence of heads and tails.

(a) What is the sample space of this experiment?

(b) Determine the event E_1 = "more heads than tails occur."

(c) Determine the event E_2 = "the first toss is a head."

(d) Determine the event $E_1 \cap E_2$.

7. A corporation efficiency expert records the time it takes an assembly line worker to perform a particular task. Let E be the event "more than 5 minutes," F the event "less than 8 minutes," and G the event "less than 4 minutes."

(a) Describe the sample space for this experiment.

(b) Describe the events $E \cap F$, $E \cap G$, E', F', $E' \cap F$, $E' \cap F \cap G$, $E \cup F$.

8. A manufacturer of kitchen appliances tests the reliability of its refrigerators by recording in a laboratory test the elapsed time between consecutive failures. Let E be the event "more than nine months," F the event "less than two years." Describe the events $E \cap F$, $E \cup F$, E', F', $(E \cup F)'$.

9. A pair of four-sided dice, each with the numbers from 1 to 4 on their sides, are rolled and the numbers facing down are observed.

   (a) List the sample space.

   (b) Describe each of the following events as a subset of the sample space.
   (i) Both numbers are even.
   (ii) At least one number is even.
   (iii) Neither number is less than or equal to 2.
   (iv) The sum of the numbers is 6.
   (v) The sum of the numbers is greater than or equal to 5.
   (vi) The numbers are the same.
   (vii) A 2 or 3 occurs, but not both.
   (viii) No 4 appears.

10. An experiment consists of selecting a car at random from a college parking lot and observing the color and make. Let E be the event "the car in red," F the event "the car is a Chevrolet," G the event "the car is a green Ford," and H the event "the car is black or a Chrysler."

    (a) Which of the following pairs of events are mutually exclusive?
    (i) E and F (ii) E and G (iii) F and G (iv) E and H
    (v) F and H (vi) G and H (vii) E' and G (viii) F' and H'

    (b) Describe each of the following events:
    (i) $E \cap F$ (ii) $E \cup F$ (iii) E' (iv) F'
    (v) G' (vi) H' (vii) $E \cup G$ (viii) $E \cap G$
    (ix) $E \cap H$ (x) $E \cup H$ (xi) $G \cap H$ (xii) $E' \cap F'$
    (xiii) $E' \cup G'$

11. Let $S = \{1, 2, 3, 4, 5, 6\}$ be a sample space, $E = \{1, 2\}$, $F = \{2, 3\}$, $G = \{1, 5, 6\}$.

    (a) Are E and F mutually exclusive?

    (b) Are F and G mutually exclusive?

12. Show that if E is any event, E' its complement, then E and E' are mutually exclusive.

13. Let $S = \{a, b, c\}$ be a sample space. Determine all possible events associated with S.

14. Let S be a sample space with n outcomes. How many events are associated with S?

15. Let $S = \{1, 2, 3, 4\}$ be a sample space, $E = \{1\}$, $F = \{2, 3\}$. Are the events $E \cup F$ and $E' \cap F'$ mutually exclusive?

16. Let S be any sample space, E, F any events associated with S. Are the events $E \cup F$ and $E' \cap F'$ mutually exclusive? [*Hint:* Apply De Morgan's laws.]

17. Suppose that 10 coins are tossed and the number of heads observed.

    (a) Describe the sample space for this experiment.

    (b) Describe the event "more heads than tails" in terms of the sample space.

18. Suppose that five nickels and five dimes are tossed and the numbers of heads from group recorded.

    (a) Describe the sample space for this experiment.

(b) Describe the event "more heads on the nickels than on the dimes" in terms of the sample space.

19. An experiment consists of observing the eye color and sex of the students at a certain school. Let E be the event "blue eyes," F the event "male," G the event "brown eyes and female."

   (a) Are E and F mutually exclusive?

   (b) Are E and G mutually exclusive?

   (c) Are F and G mutually exclusive?

20. Consider the experiment and events of Exercise 19. Describe the following events.

   (a) $E \cup F$ (b) $E \cap G$ (c) E' (d) F' (e) $(G \cup F) \cap E$ (f) $G' \cap E$

21. Suppose that you observe the length of the line at a fast-food restaurant. Describe the sample space.

22. Suppose that you observe the time to be served at a fast-food restaurant. Describe the sample space.

23. Suppose that you observe the time between customer arrivals at a fast-food restaurant. Describe the sample space.

24. Suppose that you observe the length of the line when a customer arrives and the length of time it takes for him to be served in a fast-food restaurant. Describe the sample space.

SOLUTIONS TO PRACTICE PROBLEMS 2

1. (a) $\{0, 1, 2, 3, 4, 5\}$. The sample space is the set of all outcomes of the experiment. At first glance it might seem that each outcome is a set of five light bulbs. What is observed, however, is not the specific sample but rather the number of defective bulbs in the sample. Therefore, the outcome must be a number.

   (b) $\{0, 1, 2\}$. "At most 2" means "2 or less."

2. (a) $\{$(crate I, orange), (crate I, grapefruit), (crate I, tangelo), (crate II, orange), (crate II, grapefruit), (crate II, tangelo)$\}$. Two selections are being made and should both be recorded.

   (b) $\{$(crate I, tangelo), (crate II, tangelo)$\}$. This set consists of those outcomes in which a tangelo is selected.

6.3. Assignment of Probabilities

In the last section we introduced the sample space of an experiment and used it to describe events. We complete our description of experiments by introducing probabilities associated to events. For the remainder of this chapter* let us limit our discussion to experiments with only a finite number of outcomes.

---

   * This restriction will remain in effect until our discussion of the normal distribution in Section 7.5.

Suppose that an experiment has a sample space S consisting of a finite number of outcomes $s_1, s_2, \ldots, s_N$. To each outcome we associate a number, called the *probability of the outcome*, which represents the relative likelihood that the outcome will occur. Suppose that to the outcome s_1 we associate the probability p_1, to the outcome s_2 the probability p_2, and so forth. We can summarize these data in a chart of the following sort:

Outcome	Probability
s_1	p_1
s_2	p_2
$\vdots$	$\vdots$
s_N	p_N

Such a chart is called the *probability distribution* for the experiment. The numbers $p_1, \ldots, p_N$ are chosen so that each probability represents the long-run proportion of trials in which the associated outcome can be expected to occur.

The next three examples illustrate some methods for determining probability distributions.

EXAMPLE 1 Toss an unbiased coin and observe the uppermost face. Determine the probability distribution for this experiment.

Solution Since the coin is unbiased, we expect each of the outcomes "heads" and "tails" to be equally likely. So we assign the two outcomes equal probabilities, namely $\frac{1}{2}$. The probability distribution is

Outcome	Probability
Heads	$\frac{1}{2}$
Tails	$\frac{1}{2}$

EXAMPLE 2 Toss a die and observe the uppermost face. Determine the probability distribution for this experiment.

Solution There are six possible outcomes, namely 1, 2, 3, 4, 5, 6. Assuming that the die is unbiased, these outcomes are equally likely. So we assign to each outcome the probability $\frac{1}{6}$. Here is the probability distribution for the experiment:

Outcome	Probability
1	$\frac{1}{6}$
2	$\frac{1}{6}$
3	$\frac{1}{6}$
4	$\frac{1}{6}$
5	$\frac{1}{6}$
6	$\frac{1}{6}$

Probabilities may be assigned to the elements of the sample space using common sense about the physical nature of the experiment. The fair coin has two sides, both equally likely to be face up. The balanced die has six equally probable faces. However, it may not be possible to use intuition alone to decide on a realistic probability to assign to individual sample points. Sometimes it is necessary to conduct an experiment and use the data to shed light on the long-run relative frequency with which events occur. The following example demonstrates this technique.

EXAMPLE 3   Traffic engineers measure the traffic on a major highway during rush hour. By observing the number of cars for 300 consecutive rush hours, they arrive at the following data:

Number of cars observed	Frequency observed
≤1000	30
1001–3000	45
3001–5000	135
5001–7000	75
>7000	15

(a) Describe the sample space associated to this experiment.

(b) Assign a probability distribution to this experiment.

Solution   (a) The experiment consists of counting the number of cars during rush hour and observing which one of five categories the number belongs to. The five outcomes are

$$s_1 = \text{``}\leq 1000 \text{ cars''}$$

$$s_2 = \text{``}1001\text{--}3000 \text{ cars''}$$

$$s_3 = \text{``}3001\text{--}5000 \text{ cars''}$$

$$s_4 = \text{``}5001\text{--}7000 \text{ cars''}$$

$$s_5 = \text{``}>7000 \text{ cars.''}$$

And the sample space is

$$S = \{s_1, s_2, s_3, s_4, s_5\}.$$

(b) For each outcome we use the available statistics to compute its relative frequency. For example, the relative frequency of outcome s_1 is

$$\frac{\left[\text{number of times } s_1 \text{ occurs}\right]}{\left[\text{number of trials}\right]} = \frac{30}{300} = .1.$$

That is, s_1 occurred in 10% of the observations. If we assume that the 300 observations are representative of rush hours in general, then it would seem reasonable to assign the outcome s_1 the probability .1. Similarly, we can assign probabilities to the other outcomes and arrive at this probability distribution:

Outcome	Probability
s_1	$\frac{30}{300} = .1$
s_2	$\frac{45}{300} = .15$
s_3	$\frac{135}{300} = .45$
s_4	$\frac{75}{300} = .25$
s_5	$\frac{15}{300} = .05$

This method of assigning probabilities to outcomes is valid only insofar as the observed trials are "representative." If such probability models are to be used for planning roadways or traffic patterns, they would have to be tested extensively to be sure that the probabilities are realistic representations of the long-run frequency of events.

Let an experiment have outcomes $s_1, s_2, \ldots, s_N$ with respective probabilities $p_1, p_2, p_3, \ldots, p_N$. Then the numbers $p_1, p_2, \ldots, p_N$ must satisfy two basic properties.

> **Fundamental Property 1** Each of the numbers $p_1, p_2, \ldots, p_N$ is between 0 and 1.
>
> **Fundamental Property 2** $p_1 + p_2 + \cdots + p_N = 1.$

Roughly speaking, Fundamental Property 1 says that the likelihood of each outcome lies between 0% and 100%, whereas Fundamental Property 2 says that there is a 100% likelihood that one of the outcomes $s_1, s_2, \ldots, s_N$ will occur. The two fundamental properties may be easily verified for the probability distributions of Examples 1 to 3.

Suppose that we are given an experiment with a finite number of outcomes. Let us now assign to each event E a probability, which we denote by $\Pr(E)$. If E consists of a single outcome, say $E = \{s\}$, then E is called an *elementary event*. In this case we associate to E the probability of the outcome s. If E consists of more than one outcome, we may compute $\Pr(E)$ via the so-called addition principle.

> **Addition Principle** Suppose that an event E consists of the finite number of outcomes $s, t, u, \ldots, z$. That is,
>
> $$E = \{s, t, u, \ldots, z\}.$$
>
> Then
>
> $$\Pr(E) = \Pr(s) + \Pr(t) + \Pr(u) + \cdots + \Pr(z).$$

We supplement the addition principle with the convention that the probability of the impossible event $\emptyset$ is 0. This is certainly reasonable, since the impossible event never occurs.

EXAMPLE 4   Observe two-child families. Describe the sample space for counting the number of boys and assign probabilities to each outcome.

Solution   The sample space $S = \{GG, GB, BG, BB\}$ describes the sex and birth order in two-child families. If we assume that the chance of a boy (B) [or girl (G)] on any birth is $\frac{1}{2}$ and that the sex of any child does not depend on the sex of other children born to the same parents, we would conclude that each of the four outcomes in S is equally likely. Thus we would assign probability $\frac{1}{4}$ to each outcome. Then,

$$\text{Pr(no boys)} = \text{Pr(GG)} = \tfrac{1}{4}$$

$$\text{Pr(one boy)} = \text{Pr(GB)} + \text{Pr(BG)} = \tfrac{1}{4} + \tfrac{1}{4} = \tfrac{1}{2}$$

$$\text{Pr(two boys)} = \text{Pr(GG)} = \tfrac{1}{4}$$

We are making a reasonable assignment of probabilities here, although U.S. statistics show that 51% of all live births are boys and 49% are girls. If we wish to take years of census data into account for a more accurate model, we would not assign equal probabilities to each outcome in S.

EXAMPLE 5   Suppose that we toss a die and observe the uppermost face. What is the probability that an odd number will occur?

Solution   The event "odd number occurs" corresponds to the subset of the sample space given by

$$E = \{1, 3, 5\}.$$

That is, the event occurs if a 1, 3, or 5 appears on the uppermost face. By the addition principle

$$\text{Pr}(E) = \text{Pr}(1) + \text{Pr}(3) + \text{Pr}(5).$$

As we observed in Example 2, each of the elementary outcomes in the die-tossing experiment has probability $\frac{1}{6}$. Therefore,

$$\text{Pr}(E) = \tfrac{1}{6} + \tfrac{1}{6} + \tfrac{1}{6} = \tfrac{1}{2}.$$

So we expect an odd number to occur approximately half of the time.

EXAMPLE 6   Consider the traffic study of Example 3. What is the probability that at most 5000 cars will use the highway during rush hour?

Solution   The event

"at most 5000 cars"

is the same as

$$\{s_1, s_2, s_3\},$$

where we use the same notation for the outcomes as we used in Example 3. Thus the probability of the event is

$$\Pr(s_1) + \Pr(s_2) + \Pr(s_3) = .1 + .15 + .45 = .7.$$

Therefore, we expect that traffic will involve ≤ 5000 cars in approximately 70% of the rush hours.

EXAMPLE 7 Suppose that we toss a red die and a green die and observe the numbers on the uppermost faces.

(a) Calculate the probabilities of the elementary events.

(b) Calculate the probability that the two dice show the same number.

Solution (a) As shown in Example 3 of Section 2, the sample space consists of 36 pairs of numbers:

$$S = \{(1, 1), (1, 2), \ldots, (6, 5), (6, 6)\}.$$

Each of these pairs is equally likely to occur. (How could the dice show favoritism to a particular pair?) Therefore, each outcome is expected to occur about $\frac{1}{36}$ of the time and the probability of each elementary event is $\frac{1}{36}$.

(b) The event

$$E = \text{``both dice show the same number''}$$

consists of six outcomes:

$$E = \{(1, 1), (2, 2), (3, 3), (4, 4), (5, 5), (6, 6)\}.$$

Thus, by the addition principle,

$$\Pr(E) = \tfrac{1}{36} + \tfrac{1}{36} + \tfrac{1}{36} + \tfrac{1}{36} + \tfrac{1}{36} + \tfrac{1}{36} = \tfrac{6}{36} = \tfrac{1}{6}.$$

EXAMPLE 8 A person playing a certain lottery can win $100, $10, or $1, can break even, or can lose $10. These five outcomes with their corresponding probabilities are given by the probability distribution in Table 1.

(a) What outcome has the greatest probability?

(b) What outcome has the least probability?

(c) What is the probability that the person will win some money?

Solution (a) Table 1 reveals that the outcome -10 has the greatest probability, .50. (A person playing the lottery repeatedly can expect to lose $10 about 50% of the time.) This outcome is just as likely to occur as not.

TABLE 1

Winnings	Probability
100	.02
10	.05
1	.40
0	.03
−10	.50

(b) The outcome 100 has the least probability, .02. A person playing the lottery can expect to win $100 about 2% of the time. (This outcome is quite unlikely to occur.)

(c) We are asked for the probability that the event E occurs where $E = \{100, 10, 1\}$. By the addition principle

$$\Pr(E) = \Pr(100) + \Pr(10) + \Pr(1)$$
$$= .02 + .05 + .40$$
$$= .47.$$

Here is a useful formula that relates $\Pr(E \cup F)$ to $\Pr(E \cap F)$:

Inclusion-Exclusion Principle   Let E and F be any events. Then

$$\Pr(E \cup F) = \Pr(E) + \Pr(F) - \Pr(E \cap F).$$

In particular, if E and F are mutually exclusive, then

$$\Pr(E \cup F) = \Pr(E) + \Pr(F).$$

Note the similarity of this principle to the principle of the same name that was used in Section 5.2 to count the elements in a set. The inclusion-exclusion principle will be verified after first giving an illustration of its use.

EXAMPLE 9   A factory needs two raw materials. The probability of not having an adequate supply of material A is .05, whereas the probability of not having an adequate supply of material B is .03. A study determines that the probability of a shortage of both A and B is .01. What proportion of the time can the factory operate?

Solution   Let E be the event "shortage of A," F the event "shortage of B." We are given that

$$\Pr(E) = .05, \qquad \Pr(F) = .03, \qquad \Pr(E \cap F) = .01.$$

The factory can operate only if it has both raw materials. Therefore we must calculate the proportion of the time in which there is no shortage of material A or material B. A shortage of A or B is the event $E \cup F$. By the inclusion-exclusion principle

$$Pr(E \cup F) = Pr(E) + Pr(F) - Pr(E \cap F)$$
$$= .05 + .03 - .01$$
$$= .07.$$

Thus the factory is likely to be short of one raw material or the other 7% of the time. Therefore, the factory can expect to operate 93% of the time.

Verification of the Inclusion-Exclusion Principle

From Fig. 1, we see that the set $E \cup F$ can be written as the union of disjoint sets:

$$E \cup F = (E \cap F') \cup (E \cap F) \cup (E' \cap F).$$

Therefore, $Pr(E \cup F) = Pr(E \cap F') + Pr(E \cap F) + Pr(E' \cap F)$. Let us add and subtract $Pr(E \cap F)$ to the right-hand side of the equation. Then

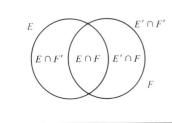

FIGURE 1

$$Pr(E \cup F) = Pr(E \cap F') + Pr(E \cap F) + Pr(E' \cap F)$$
$$+ Pr(E \cap F) - Pr(E \cap F).$$

But $(E \cap F') \cup (E \cap F) = E$ and $(E' \cap F) \cup (E \cap F) = F$ (see Fig. 1). Therefore

$$Pr(E \cup F) = Pr(E) + Pr(F) - Pr(E \cap F).$$

If E and F are mutually exclusive, then $E \cap F = \varnothing$ and $Pr(E \cap F) = 0$. Thus,

$$Pr(E \cup F) = Pr(E) + Pr(F).$$

Odds   Frequently in applications we meet statements like these:

The odds of a Republican victory are 3 to 2.
The odds of a recession next year are 1 to 3.

Such statements may be readily translated into the language of probability. For example, consider the first statement above. It means that if the election were repeated often (a theoretical possibility), then for every 2 Democratic wins there would be 3 Republican wins. That is, the Republicans would win $\frac{3}{5}$ (or 60%) of the elections. In terms of probability, this means that the probability of a Republican win is .6. In a similar way, we can translate the second statement above into probabilistic terms. The statement says that if we consider a large number of years experiencing conditions identical to this year's, then for every one that is followed by a recession three years are not. That is, the probability of having a recession next year is $\frac{1}{4}$.

We may generalize our reasoning to obtain the following result:

If the odds in favor of an event are a to b, then the probability of the event is $a/(a + b)$.

EXAMPLE 10 Suppose that the odds of rain tomorrow are 5 to 3. What is the probability that rain will occur?

Solution The probability that rain will occur is

$$\frac{5}{5+3} = \frac{5}{8}.$$

PRACTICE PROBLEMS 3

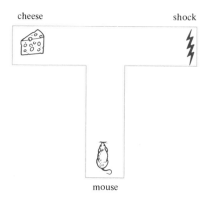

cheese   shock

mouse

FIGURE 2

1. A mouse is put into a T-maze (a maze shaped like a "T") (Fig. 2). If he turns to the left he receives cheese, and if he turns to the right he receives a mild shock. This trial is done twice with the same mouse and the directions of the turns recorded.

(a) What is the sample space for this experiment?

(b) Why would it not be reasonable to assign each outcome the same probability?

2. (a) What are the odds in favor of an event that is just as likely to occur as not?

(b) Is there any difference between the odds 6 to 4 and the odds 3 to 2?

3. Suppose that E and F are any events. Show that $\Pr(E) = \Pr(E \cap F) + \Pr(E \cap \overline{F})$.

EXERCISES 3

1. There are 774,746 words in the Bible. The word "and" occurs 46,277 times and the word "Lord" occurs 1855 times.* Suppose that a word is selected at random from the Bible.

(a) What is the probability that the word is "and"?

(b) What is the probability that the word is "and" or "Lord"?

(c) What is the probability that the word is neither "and" nor "Lord"?

2. An experiment consists of tossing a coin two times and observing the sequence of heads and tails. Each of the four outcomes has the same probability of occurring.

(a) What is the probability that "HH" is the outcome?

(b) What is the probability of the event "at least one head"?

3. Suppose that a red die and a green die are tossed and the numbers on the uppermost faces are observed. (See Example 7.)

(a) What is the probability that the numbers add up to 8?

(b) What is the probability that the sum of the numbers is less than 5?

---

* According to *The People's Almanac* by Wallechinsky and Wallace (New York: Doubleday, 1975).

4. A state is selected at random from the 50 states of the United States. What is the probability that it is one of the 6 New England states?

5. The modern American roulette wheel has 38 slots, which are labeled with 36 numbers evenly divided between red and black plus two green numbers 0 and 00. What is the probability that the ball lands on a green number?

6. An experiment consists of selecting a number at random from the set of numbers $\{1, 2, 3, 4, 5, 6, 7, 8, 9\}$. Find the probability that the number selected is:

(a) Less than 4.　　　　　　(b) Odd.　　　　　　(c) Less than 4 or odd.

7. Three horses, call them A, B, and C, are going to race against each other. The probability that A will win is $\frac{1}{3}$ and the probability that B will win is $\frac{1}{2}$. What is the probability that C will win?

8. Which of the following probabilities are feasible for an experiment having sample space $\{s_1, s_2, s_3\}$?

(a) $\Pr(s_1) = .4, \Pr(s_2) = .4, \Pr(s_3) = .4$

(b) $\Pr(s_1) = .5, \Pr(s_2) = .7, \Pr(s_3) = -.2$

(c) $\Pr(s_1) = 2, \Pr(s_2) = 1, \Pr(s_3) = \frac{1}{2}$

(d) $\Pr(s_1) = \frac{1}{4}, \Pr(s_2) = \frac{1}{2}, \Pr(s_3) = \frac{1}{4}$

9. An experiment with outcomes s_1, s_2, s_3, s_4 is described by the probability table:

Outcome	Probability
s_1	.1
s_2	.6
s_3	.2
s_4	.1

(a) What is $\Pr(\{s_1, s_2\})$?　　　　　　(b) What is $\Pr(\{s_2, s_4\})$?

10. An experiment with outcomes $s_1, s_2, s_3, s_4, s_5, s_6$ is described by the probability table:

Outcome	Probability
s_1	.05
s_2	.25
s_3	.05
s_4	.01
s_5	.63
s_6	.01

Let $E = \{s_1, s_2\}$, $F = \{s_3, s_5, s_6\}$.

(a) Determine $\Pr(E), \Pr(F)$.　　　　　　(b) Determine $\Pr(E')$.

(c) Determine $\Pr(E \cap F)$.　　　　　　(d) Determine $\Pr(E \cup F)$.

11. Convert the following odds to probabilities:

    (a) 10 to 1   (b) 1 to 2   (c) 4 to 5

12. In poker, the probability of being dealt a hand containing a pair of jacks or better is about $\frac{1}{6}$. What are the corresponding odds?

13. Nine percent of all candy bars sold in the United States are *Snickers* bars. What are the odds that a randomly selected candy bar purchased is a *Snickers* bar?

14. The odds of Americans living in the state where they were born is 16 to 9. What is the probability that an American selected at random lives in his or her birth state?

15. If the odds for Secretariat to win a horse race are $11:7$, what is the probability that Secretariat wins? Loses?

16. Four people are running for class president, Liz, Sam, Sue, and Tom. The probabilities of Sam, Sue, and Tom winning are .18, .23, and .31, respectively.

    (a) What is the probability of Liz winning?

    (b) What is the probability that a boy wins?

    (c) What is the probability that Tom loses?

    (d) What are the odds that Sue loses?

    (e) What are the odds that a girl wins?

    (f) What are the odds that Sam wins?

17. Let E and F be events for which $\Pr(E) = .6$, $\Pr(F) = .5$ and $\Pr(E \cap F) = .4$. Find

    (a) $\Pr(E \cup F)$

    (b) $\Pr(E \cap F')$ [*Hint*: See Practice Problem 3.]

18. Let E and F be events for which $\Pr(E) = .4$, $\Pr(F) = .5$ and $\Pr(E \cap F') = .3$. Find

    (a) $\Pr(E \cap F)$ (b) $\Pr(E \cup F)$

19. A statistical analysis of the wait (in minutes) at the checkout line of a certain supermarket yields the following probability distribution:

Wait	Probability
At most 3	.1
More than 3 and at most 5	.2
More than 5 and at most 10	.25
More than 10 and at most 15	.25
More than 15	.2

    (a) What is the probability of waiting more than 3 minutes but at most 15?

    (b) If you observed the waiting times of a representative sample of 10,000 supermarket customers, approximately how many would you expect to wait for more than 3 minutes but at most 15?

20. What is the probability of the certain event?

21. A computer firm analyzes the failures of circuit boards for their newest personal computer. It observes the following data.

Failures in	*Probability*
First month	.05
First 2 months	.10
First 3 months	.20
First 4 months	.25
First 5 months	.30
First 6 months	.32

Convert these data into a probability distribution.

22. Refer to the data of Exercise 21. What is the probability that a computer will not fail during its first 3 months of use? First 6 months of use?

SOLUTIONS TO PRACTICE PROBLEMS 3

1. (a) {LL, LR, RR, RL}. Here LL means that the mouse turned left both times, LR means that the mouse turned left the first time and right the second, and so on.

(b) The mouse will learn something from the first trial. If he turned left the first time and got rewarded, then he is more likely to turn left again on the second trial. Hence LL should have a greater probability than LR. Similarly, RL should be more likely than RR.

2. (a) 1 to 1. An event that is just as likely to occur as not has probability $\frac{1}{2}$. So if the odds are a to b, then we may set $a = 1$, $a + b = 2$. Thus $a = 1$, $b = 1$. (The odds could also be given as 2 to 2, 3 to 3, etc.)

(b) No. Odds of 6 to 4 correspond to a probability of $6/(6 + 4) = \frac{6}{10} = \frac{3}{5}$. Odds of 3 to 2 correspond to a probability of $3/(3 + 2) = \frac{3}{5}$. (There are always many different ways to express the same odds.)

3. The sets $(E \cap F)$ and $(E \cap F')$ have no elements in common and so are mutually exclusive. Since $E = (E \cap F) \cup (E \cap F')$, the result follows from the inclusion-exclusion principle.

6.4. Calculating Probabilities of Events

In this section we use the counting techniques of Chapter 5 to compute the probabilities of various events. In addition, we illustrate the exceptionally wide range of applications which make use of probability theory.

Experiments with Equally Likely Outcomes In the experiments associated to many common applications, all outcomes are equally likely—that is, they all have the same probability. This is the case, for example, if we toss an unbiased coin or select a person at random from a population. If a sample space has N equally likely outcomes, then the probability of each outcome is $1/N$ (since the probabilities must add up to 1). Using this fact, the probability of any event is then easy to compute. Namely, suppose that E is an event consisting of M outcomes. Then, by the addition principle,

$$\Pr(E) = \underbrace{\frac{1}{N} + \frac{1}{N} + \cdots + \frac{1}{N}}_{M \text{ times}} = \frac{M}{N}.$$

We can restate this fundamental result as follows:

Let S be a sample space consisting of N equally likely outcomes. Let E be any event. Then

$$\Pr(E) = \frac{[\text{number of outcomes in } E]}{N}. \tag{1}$$

In order to apply (1) in particular examples, it is necessary to compute N, the number of outcomes, and [number of outcomes in E]. Often these quantities can be determined using the counting techniques of Chapter 5. Some illustrative computations are provided in Examples 1 to 5 below.

We should mention that, although the urn and dice problems being considered in this section and the next might seem artificial and removed from applications, many applied problems can be described in mathematical terms as urn or dice-tossing experiments. We begin our discussion with two examples involving abstract urn problems. Then in two more examples we show the utility of urn models by applying them to quality control and medical screening problems.

EXAMPLE 1 An urn contains eight white balls and two green balls. A sample of three balls is selected at random. What is the probability of selecting only white balls?

Solution The experiment consists of selecting three balls from among the 10. Since the order in which the three balls are selected is immaterial, the samples are combinations of 10 balls taken three at a time. The total number of samples is therefore $\binom{10}{3}$, and this is N, the number of elements in the sample space. Since the selection of the sample is random, all samples are equally likely, and thus we can use formula (1) to compute the probability of any event. The problem asks us to compute the probability of the event $E =$ "all three balls selected are white." Since there are 8 white balls, the number of different samples in which all are white is $\binom{8}{3}$. Thus,

$$\Pr(E) = \frac{[\text{number of outcomes in } E]}{N} = \frac{\binom{8}{3}}{\binom{10}{3}} = \frac{56}{120} = \frac{7}{15}.$$

EXAMPLE 2   An urn contains eight white balls and two green balls. A sample of three balls is selected at random. What is the probability that the sample contains at least one green ball?

Solution   As in Example 1, $N = \binom{10}{3}$ equally likely outcomes. Let F be the event "at least one green ball is selected." Let us determine the number of different outcomes in F. These outcomes contain either one or two green balls. There are $\binom{2}{1}$ ways to select one green ball from among two; and for each of these, there are $\binom{8}{2}$ ways to select two white balls from among eight. So the number of samples containing one green ball equals $\binom{2}{1}\binom{8}{2}$. Similarly, the number of samples containing two green balls equals $\binom{2}{2}\binom{8}{1}$. Therefore, the number of outcomes in F—namely, the number of samples having at least one green ball—equals

$$\binom{2}{1}\binom{8}{2} + \binom{2}{2}\binom{8}{1} = 2 \cdot 28 + 1 \cdot 8 = 64.$$

And so

$$\Pr(F) = \frac{[\text{number of outcomes in } F]}{N} = \frac{64}{\binom{10}{3}} = \frac{64}{120} = \frac{8}{15}.$$

EXAMPLE 3   (*Quality Control*)   A toy manufacturer inspects boxes of toys before shipment. Each box contains 10 toys. The inspection procedure consists of randomly selecting three toys from the box. If any are defective, the box is not shipped. Suppose that a given box has two defective toys. What is the probability that it will be shipped?

Solution   This problem is not really new! We solved it in disguise as Example 1. The urn can be regarded as a box of toys, and the balls as individual toys. The white balls are nondefective toys and the green balls defective toys. The random selection of three balls from the urn is just the inspection procedure. And the event "all three balls selected are white" corresponds to the box being shipped. As we calculated above, the probability of this event is $\frac{7}{15}$. (Since $\frac{7}{15} \approx .47$, there is approximately a 47% chance of shipping a box with two defective toys. This inspection procedure is not particularly effective!)

EXAMPLE 4   (*Medical Screening*)   Suppose that a cruise ship returns to the United States from the Far East. Unknown to anyone, four of its 600 passengers have contracted a rare disease. Suppose that the Public Health Service screens 20 passengers, selected at random, to see whether the disease is present aboard ship. What is the probability that the presence of the disease will escape detection?

The sample space consists of samples of 20 drawn from among the 600 passengers. There are $\binom{600}{20}$ such samples. The number of samples containing none of the sick passengers is $\binom{596}{20}$. Therefore, the probability of not detecting the disease is

$$\frac{\binom{596}{20}}{\binom{600}{20}} = \frac{\dfrac{596!}{20!\,576!}}{\dfrac{600!}{20!\,580!}} = \frac{596!}{600!} \cdot \frac{580!}{576!} = \frac{580 \cdot 579 \cdot 578 \cdot 577}{600 \cdot 599 \cdot 598 \cdot 597} \approx .87.$$

So there is approximately an 87% chance that the disease will escape detection.

The Complement Rule The *complement rule* relates the probability of an event E to the probability of its complement E'. When applied together with counting techniques, it often simplifies computation of probabilities.

Complement Rule Let E be any event, E' its complement. Then

$$\Pr(E) = 1 - \Pr(E').$$

For example, recall Example 1. We determined the probability of the event

$$E = \text{"all three balls selected are white"}$$

associated to the experiment of selecting three balls from an urn containing eight white balls and two green balls. We found that $\Pr(E) = \frac{7}{15}$. On the other hand, in Example 2 we determined the probability of the event

$$F = \text{"at least one green ball is selected."}$$

The event E is the complement of F:
$$E = F'.$$
So, by the complement rule,

$$\Pr(F) = 1 - \Pr(F') = 1 - \Pr(E)$$
$$= 1 - \tfrac{7}{15} = \tfrac{8}{15},$$

in agreement with the calculations of Example 2.

The complement rule is especially useful in situations where $\Pr(E')$ is easier to compute than $\Pr(E)$. One of these situations arises in the celebrated *birthday problem*.

EXAMPLE 5 A group of five people is to be selected at random. What is the probability that two or more of them have the same birthday?

Solution For simplicity we ignore leap years. Furthermore, we assume that each of the 365 days in a year is an equally likely birthday. (Not an unreasonable assumption.) The experiment we have in mind is this. Pick out five people and observe their birthdays. The outcomes of this experiment are strings of five dates, corresponding to the birthdays. For example, one outcome of the experiment is

(June 2, April 6, Dec. 20, Feb. 12, Aug. 5).

Each date has 365 different possibilities. So, by the generalized n. principle, the total number N of possible outcomes of the experiment is

$$N = 365 \cdot 365 \cdot 365 \cdot 365 \cdot 365 = 365^5.$$

Let E be the event "at least two people have the same birthday." It is very difficult to calculate directly the number of outcomes in E. However, it is comparatively simple to compute the number of outcomes in E' and hence to compute $\Pr(E')$. This is because E' is the event "all five birthdays are different." An outcome in E' can be selected in a sequence of five steps:

Select a day	Select a different day	Select yet a different day	Select yet a different day	Select yet a different day

These five steps will result in a sequence of five different birthdays. The first step can be performed in 365 ways; for each of these, the next step in 364; for each of these, the next step in 363; for each of these, the next step in 362; and for each of these, the last step in 361. Therefore, E' contains $365 \cdot 364 \cdot 363 \cdot 362 \cdot 361$ [or $P(365, 5)$] outcomes, and

$$\Pr(E') = \frac{365 \cdot 364 \cdot 363 \cdot 362 \cdot 361}{365^5} \approx .973.$$

By the complement rule,

$$\Pr(E) = 1 - \Pr(E') = 1 - .973 = .027.$$

So the likelihood is about 2.7% that two or more of the five people will have the same birthday.

The experiment of Example 5 can be repeated using samples of 8, 10, 20, or any number of people. As before, let E be the event "at least two people have the same birthday," so that $E' =$ "all the birthdays are different." If a sample of r people is used, then the same reasoning as used above yields

$$\Pr(E') = \frac{365 \cdot 364 \cdot \ldots \cdot (365 - r + 1)}{365^r}.$$

Table 1 gives the values of $\Pr(E') = 1 - \Pr(E')$ for various values of r. You may be surprised by the numbers in the table. Even with as few as 23 people it is more likely than not that at least two people have the same birthday. With a sample of 50 people we are almost certain to have two with the same birthday. (Try this experiment in your dormitory or class.)

TABLE 1 Probability That, in a Randomly Selected Group of r People, at Least Two Will Have the Same Birthday

r	5	10	15	20	22	23	25	30	40	50
$Pr(E)$	.027	.117	.253	.411	.476	.507	.569	.706	.891	.970

Verification of the Complement Rule

If S is the sample space, then $Pr(S) = 1$, $E \cup E' = S$, and $E \cap E' = \varnothing$. Therefore, by the inclusion-exclusion principle.

$$Pr(S) = Pr(E \cup E') = Pr(E) + Pr(E').$$

So we have

$$1 = Pr(E) + Pr(E') \text{ and } Pr(E') = 1 - Pr(E).$$

PRACTICE PROBLEMS 4

1. A couple decides to have four children. What is the probability that among the children will be at least one boy and at least one girl?

2. (a) Find the probability that all the numbers are different in three spins of a roulette wheel. [*Note:* A roulette wheel has 38 numbers.]

   (b) Guess how many spins are required in order that the probability that all the numbers are different will be less than .5?

EXERCISES 4

1. An urn contains six white balls and five red balls. A sample of four balls is selected at random from the urn. What is the probability that the sample contains two white balls and two red balls?

2. A factory produces fuses, which are packaged in boxes of 10. Three fuses are selected at random from each box for inspection. The box is rejected if at least one of these three fuses is defective. What is the probability that a box containing five defective fuses will be rejected?

3. Of the nine members of the board of trustees of a college, five agree with the president on a certain issue. The president selects three trustees at random and asks for their opinions. What is the probability that at least two of them will agree with him?

4. An urn contains five red balls and four white balls. A sample of two balls is selected at random from the urn. What is the probability that at least one of the balls is red?

5. Without consultation, each of four organizations announces a one-day convention to be held during June. Find the probability that at least two organizations specify the same day for their convention.

6. Five letters are selected from the alphabet, one at a time with replacement, to form a five-letter word. What is the probability that the word has five different letters?

7. An exam contains five "true or false" questions. What is the probability that a student guessing at the answers will get three or more answers correct?

8. An airport limousine has four passengers and stops at six different hotels. What is the probability that two or more people will be staying at the same hotel? (Assume that each person is just as likely to stay in one hotel as another.)

9. In a certain agricultural region the probability of a drought during the growing season is .2, the probability of a severe cold spell is .15, and the probability of both is .1. Find the probability of

    (a) Either a drought or a severe cold spell.

    (b) Neither a drought nor a severe cold spell.

    (c) Not having a drought.

10. A coin is to be tossed seven times. What is the probability of obtaining four heads and three tails?

11. Let E and F be events such that $\Pr(E) = .3$, $\Pr(F') = .6$, and $\Pr(E \cup F) = .7$. What is $\Pr(E \cap F)$?

12. In a certain manufacturing process the probability of a type I defect is .12, the probability of a type II defect is .22, and the probability of having both types of defects is .02. Find the probability of having neither type of defect.

13. A man has six pairs of socks, from which he selects two socks at random. What is the probability that the selected socks will match?

14. Of the 15 members on a Senate committee, 10 plan to vote "yes" and 5 plan to vote "no" on an important issue. A reporter attempts to predict the outcome of the vote by questioning 6 of the senators. Find the probability that this sample is precisely representative of the final vote. That is, find the probability that four of the six senators questioned planned to vote "yes."

15. A man, a woman, and their three children randomly stand in a row for a family picture. What is the probability that the parents will be standing next to each other?

16. In poker the probabilities of being dealt a flush, a straight, and a straight flush are .0019654, .0039246, and .0000154, respectively. What is the probability of being dealt a straight or a flush?

17. Figure 1 shows a partial map of the streets in a certain city. A tourist starts at point A and selects at random a path to point B. (We shall assume that he walks only south and east.) Find the probability that

    FIGURE 1

    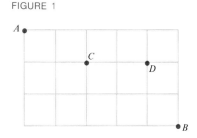

    (a) He passes through point C.

    (b) He passes through point D.

    (c) He passes through point C and point D.

    (d) He passes through point C or point D.

18. A couple decide to have four children. What is the probability that they will have more girls than boys?

19. A law firm has six senior and four junior partners. A committee of three partners is selected at random to represent the firm at a conference. What is the probability that at least one of the junior partners is on the committee?

20. A coin is to be tossed six times. What is the probability of obtaining exactly three heads?

21. A bag contains nine tomatoes, of which one is rotten. A sample of three tomatoes is selected at random. What is the probability that the sample contains the rotten tomato?

22. A vacationer has brought along four novels and four nonfiction books. One day the person selects two at random to take to the beach. What is the probability that both are novels?

Exercises 23 and 24* refer to the Illinois Lottery Lotto game. In this game, the player chooses six different integers from 1 to 40. If the six match (in any order) the six different integers drawn by the lottery, the player wins the grand prize jackpot, which starts at $1 million and grows weekly until won. Multiple winners split the pot equally. For each $1 bet, the player must pick two, presumably different, sets of six integers.

23. What is the probability of winning the Illinois Lottery Lotto with a $1 bet?

24. In the game week ending June 18, 1983, 2 million people bought $1 tickets and 78 people matched all six winning integers and split the jackpot. If all numbers were selected randomly, the likelihood of having so many joint winners would be about 10^{-115}. Can you think of any reason that such an unlikely event occurred? (*Note:* The winning numbers were 7, 13, 14, 21, 28, and 35.) What would be the best strategy in selecting the numbers to ensure that in the event you won, you would probably not have to share the jackpot with too many people?

25. Each of the regions labeled *A* through *K* in Figure 2 is a feasible set. Suppose that two distinct regions are chosen at random.

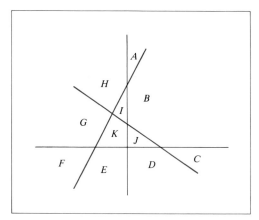

FIGURE 2

---

* The data for these exercises were taken from Allan J. Gottlieb's "Puzzle Corner" in *Technology Review*, February/March 1985.

(a) What is the probability that both regions are bounded?

(b) What is the probability that one has two edges and one has three edges?

(c) What is the probability that the regions share a edge?

(d) What is the probability that the regions lie on opposite sides of the vertical line?

26. A politician knows that a committee vote is stacked against her, 6 to 3. However, she has the option of letting a randomly selected subcommittee decide the issue. Show that the smaller the subcommittee is the better her chances of winning the vote. (Consider only subcommittees with an odd number of members, so that tie votes are precluded.)

27. A parapsychologist labels a person "clairvoyant" if they can correctly predict at least 8 out of 10 tosses of a coin.

(a) What is the probability that a person guessing heads and tails at random will be labeled clairvoyant?

(b) If 10 people are tested for clairvoyance with 10 coin tosses, what is the probability that at least one of them will be labeled clairoyant?

(c) Parapsychologists often report the "puzzling" phenomenon that people who tested "positive" for clairvoyance tend to "lose their powers" in subsequent tests. What is the probability that a person who was labeled clairvoyant after 10 tosses will "lose" that label after another 10 tosses?

28. One interpretation of a baseball player's batting average is as the probability of getting a hit each time the player goes to bat. For instance, a player with a .300 average has probability .3 of getting a hit.

(a) If a player with .3 probability of getting a hit bats four times in a game and each at-bat is an independent event, what is the probability of the player getting at least one hit in the game?

(b) What is the probability of the player in part (a) starting off the season with at least one hit in each of the first 10 games?

(c) If there are 20 players with a .300 average, what is the probability that at least one of them will start the season with a 10-game hitting streak?

29. A team got 15 hits in a baseball game, including 3 home runs, in 40 at-bats. The catcher had 3 hits in 5 at-bats. What is the probability that he had at least one of the home runs?

30. There are 7 baseball teams in each of the two divisions of the American League and 6 teams in each of the two divisions of the National League. A sportswriter "predicts" the winner of each of the four divisions by choosing a team completely at random in each division. What is the probability that the sportswriter will predict at least one winner correctly?

31. Suppose that the sportswriter in Exercise 30 eliminates one team from each division that clearly has no chance of winning, and predicts a winner at random from the remaining teams. Assuming the eliminated teams don't end up surprising anyone, what is the writer's chance of predicting at least one winner?

32. Suppose the sportswriter in Exercise 30 simply puts the 26 team names in a hat and draws 4 completely at random. Does this increase or decrease the writer's chance of picking at least one winner?

SOLUTIONS TO PRACTICE PROBLEMS 4

1. Each possible outcome is a string of four letters composed of B's and G's. By the generalized multiplication principle, there are 2^4 or 16 possible outcomes. Let E be the event "children of both sexes." Then $E' = \{BBBB, GGGG\}$, and

$$\Pr(E') = \frac{[\text{number of outcomes in } E']}{[\text{total number of outcomes}]} = \frac{2}{16} = \frac{1}{8}.$$

Therefore,

$$\Pr(E) = 1 - \Pr(E') = 1 - \frac{1}{8} = \frac{7}{8}.$$

So the probability is 87.5% that they will have children of both sexes.

2. (a) Each sequence of three numbers is just as likely to occur as any other. Therefore,

$$\Pr(\text{numbers different}) = \frac{[\text{number of outcomes with numbers different}]}{[\text{number of possible outcomes}]}$$

$$= \frac{38 \cdot 37 \cdot 36}{38^3} \approx .92.$$

(b) 8

6.5. Conditional Probability and Independence

The probability of an event depends, often in a critical way, on the sample space in question. In this section we explore this dependence in some detail by introducing what are called conditional probabilities.

To illustrate the dependence of probabilities on the sample space, consider the following example.

EXAMPLE 1 Suppose that a certain mathematics class contains 26 students. Of these, 15 are freshmen, 14 are business majors, and 7 are neither. Suppose that a person is selected at random from the class.

(a) What is the probability that the person is both a freshman and a business major?

(b) Suppose we are given the additional information that the person selected is a freshman. What is the probability that he is also a business major?

Solution Let B denote the set of business majors, F the set of freshmen. A complete Venn diagram of the class is given in Fig. 1.

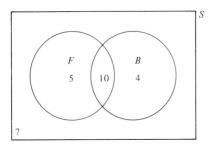

FIGURE 1

(a) In selecting a student from the class, the sample space consists of all 26 students. Since the choice is random, all students are equally likely to be selected. The event "freshman and business major" corresponds to the set $F \cap B$ of the Venn diagram. Therefore,

$$\Pr(F \cap B) = \frac{[\text{number of outcomes in } F \cap B]}{[\text{number of possible outcomes}]}$$
$$= \frac{10}{26} = \frac{5}{13}.$$

So the probability of selecting a freshman business major is $\frac{5}{13}$.

(b) If we know that the student selected is a freshman, then the possible outcomes of the experiment are restricted. They must belong to F. In other words, given the additional information we must alter the sample space from "all students" to "freshmen." Since each of the 15 freshmen is equally likely to be selected and since 10 of the 15 freshmen are business majors, the probability of choosing a business major under these circumstances is equal to $\frac{10}{15} = \frac{2}{3}$.

Example 1 may be generalized as follows. Suppose that we consider an experiment, call it experiment I, with associated events E and F. Suppose that we are given that the event F occurred. The probability that E also occurred is called the *conditional probability of E given F* and is denoted $\Pr(E|F)$.

All probabilities are defined with reference to a particular experiment. Let us be clear about the experiment associated with the conditional probability $\Pr(E|F)$. Call this one experiment II. A trial of experiment II consists of performing experiment I. If the event F occurs, then the outcome is observed. If the event F does not occur, then experiment I is repeated. And so on until F does occur. (Some trials of experiment II may last a long time!) In effect, this experiment is just experiment I, ignoring outcomes in which F does not occur. Now that we have specified the experiment, we must specify its associated probability distribution. That is, we must assign a value to the conditional probability $\Pr(E|F)$. This is done via the following formula:

Conditional Probability

$$\Pr(E|F) = \frac{\Pr(E \cap F)}{\Pr(F)},$$ (1)

provided $\Pr(F) \neq 0.$

We will provide an intuitive justification of this formula below. However, we will first give two applications.

EXAMPLE 2 Consider all families with two children (not twins). Assume that all elements of the sample space {BB, BG, GB, GG} are equally likely. (Here, for instance, BG denotes the birth sequence "boy girl.") Let E be the event {BB} and F the event "at least one boy." Calculate $\Pr(E|F)$.

Solution Each of the four birth sequences is equally likely and has probability $\frac{1}{4}$. In particular, since $F = \{BB, BG, GB\}$, $E \cap F = \{BB\}$, and we have $\Pr(E \cap F) = \frac{1}{4}$. Also, $\Pr(F) = \frac{1}{4} + \frac{1}{4} + \frac{1}{4} = \frac{3}{4}$. The conditional probability $\Pr(E|F)$ equals the probability that a two-child family has two boys given that it has at least one. This conditional probability has the value

$$\Pr(E|F) = \frac{\Pr(E \cap F)}{\Pr(F)} = \frac{\frac{1}{4}}{\frac{3}{4}} = \frac{1}{3}.$$

[Note that this result is contrary to many people's intuitions. They reason that among families with at least one boy, the other child is just as likely to be a boy as a girl. Therefore, they conclude incorrectly that the probability $\Pr(E|F)$ is $\frac{1}{2}$.]

EXAMPLE 3 Ajax Steel Company employs 20% college graduates. Of all its employees, 25% earn more than $30,000 per year, and 15% are college graduates earning more than $30,000 per year. What is the probability that an employee selected at random earns more than $30,000 per year, given that he is a college graduate?

Solution Let H and C be the events

$$H = \text{"earns more than \$30,000 per year"}$$

$$C = \text{"college graduate."}$$

We are asked to calculate $\Pr(H|C)$. The given data are

$$\Pr(H) = .25, \qquad \Pr(C) = .20, \qquad \Pr(H \cap C) = .15.$$

By formula (1), we have

$$\Pr(H|C) = \frac{\Pr(H \cap C)}{\Pr(C)} = \frac{.15}{.20} = \frac{3}{4}.$$

Thus $\frac{3}{4}$ of all college graduates at Ajax Steel earn more than $30,000 per year.

Suppose that an experiment has N equally likely outcomes. Then we may apply the following formula to calculate $\Pr(E|F)$.

> *Conditional Probability in Case of Equally Likely Outcomes*
>
> $$\Pr(E|F) = \frac{[\text{number of outcomes in } E \cap F]}{[\text{number of outcomes in } F]}, \qquad (2)$$
>
> provided [number of outcomes in F] $\neq 0$.

For instance, in Example 2, the different birth sequences were all equally likely. Moreover, the number of outcomes in $E \cap F$ (two boys) is 1, whereas the number of outcomes in F (at least one boy) is 3. Then formula (2) gives that $\Pr(E|F) = \frac{1}{3}$, in agreement with the calculation of Example 2.

Let us now justify formulas (1) and (2).

Justification of Formula (1)

Formula (1) is a definition of conditional probability and as such does not really need any justification. (We can make whatever definitions we choose!) However, let us proceed intuitively and show that the definition is a reasonable one, in the sense that formula (1) gives the expected long-run proportion of occurrences of E given that F occurs. Assume that our experiment is performed repeatedly, say for 10,000 trials. We would expect F to occur in approximately 10,000 $\Pr(F)$ trials. Among these, the trials for which E also occurs are exactly those for which *both* E and F occur. In other words, the trials for which E also occurs are exactly those for which the event $E \cap F$ occurs; and this event has probability $\Pr(E \cap F)$. Out of the original 10,000 trials, there should be approximately 10,000 $\Pr(E \cap F)$ in which E and F both occur. Thus, considering only those trials in which F occurs, the proportion in which E also occurs is

$$\frac{10{,}000 \ \Pr(E \cap F)}{10{,}000 \ \Pr(F)} = \frac{\Pr(E \cap F)}{\Pr(F)}.$$

Thus, at least intuitively, it seems reasonable to define $\Pr(E|F)$ by formula (1).

Justification of Formula (2)

Suppose that the number of outcomes of the experiment is N. Then

$$\Pr(F) = \frac{[\text{number of outcomes in } F]}{N},$$

$$\Pr(E \cap F) = \frac{[\text{number of outcomes in } E \cap F]}{N}.$$

Therefore, using formula (1), we have

$$\Pr(E|F) = \frac{\Pr(E \cap F)}{\Pr(F)}$$

$$= \frac{\dfrac{[\text{number of outcomes in } E \cap F]}{N}}{\dfrac{[\text{number of outcomes in } F]}{N}}$$

$$= \frac{[\text{number of outcomes in } E \cap F]}{[\text{number of outcomes in } F]}.$$

From formula (1), by multiplying both sides of the equation by $\Pr(F)$, we can deduce the following useful fact.

> **Product Rule**   If $\Pr(F) \neq 0$,
> $$\Pr(E \cap F) = \Pr(F) \cdot \Pr(E|F).$$
> (3)

The next example illustrates the use of this rule.

EXAMPLE 4   Assume that a certain school contains an equal number of female and male students and that 5% of the male population is color-blind. Find the probability that a randomly selected student is a color-blind male.

Solution   Let M = "male" and B = "color-blind." We wish to calculate $\Pr(B \cap M)$. From the given data

$$\Pr(M) = .5 \qquad \text{and} \qquad \Pr(B|M) = .05.$$

Therefore, by the product rule

$$\Pr(B \cap M) = \Pr(M) \cdot \Pr(B|M) = (.5)(.05) = .025.$$

Often an event G can be described as a sequence of two other events E and F. That is, G occurs if F occurs and then E occurs. The product rule allows us to compute the probability of G as the probability of F times the conditional probability $\Pr(E|F)$. The next example illustrates this point.

EXAMPLE 5   A sequence of two playing cards is drawn at random (without replacement) from a standard deck of 52 cards. What is the probability that the first card is red and the second is black?

Solution   The event in question is a sequence of two events, namely

$$F = \text{"the first card is red"}$$

$$E = \text{"the second card is black."}$$

Since half the deck consists of red cards, $\Pr(F) = \frac{1}{2}$. If we are given that F occurs, then there are only 51 cards left in the deck, of which 26 are black, so

$$\Pr(E|F) = \tfrac{26}{51}.$$

By the product rule

$$\Pr(E \cap F) = \Pr(F) \cdot \Pr(E|F) = \tfrac{1}{2} \cdot \tfrac{26}{51} = \tfrac{13}{51}.$$

The product rule may be generalized to sequences of three events E_1, E_2, E_3:

$$\Pr(E_1 \cap E_2 \cap E_3) = \Pr(E_1) \cdot \Pr(E_2|E_1) \cdot \Pr(E_3|E_1 \cap E_2).$$

Similar formulas hold for sequences of four or more events.

One of the most important applications of conditional probability is in the discussion of independent events. Intuitively, two events are independent of each other if the occurrence of one has no effect on the likelihood that the other will occur. For example, suppose that we toss a die two times. Let the events E and F be

$$F = \text{``first throw is a 6''}$$

$$E = \text{``second throw is a 3.''}$$

Then intuitively these events are independent of one another. Throwing a 6 on the first throw has no effect whatsoever on the outcome of the second throw. On the other hand, suppose that we draw a sequence of two cards at random from a deck. Then the events

$$F = \text{``first card is red''}$$

$$E = \text{``second card is black''}$$

are not independent of one another, at least intuitively. Indeed, whether or not we draw a red on the first card affects the likelihood of drawing a black on the second.

The notion of independence of events is easily formulated. If E and F are events in a sample space and $\Pr(F) \neq 0$, then the product rules states that $\Pr(E \cap F) = \Pr(E|F) \cdot \Pr(F)$. However, if the occurrence of event F does not affect the likelihood of the occurrence of event E, we would expect that $\Pr(E|F) = \Pr(E)$. Substitution then shows that $\Pr(E \cap F) = \Pr(E) \cdot \Pr(F)$.

> Let E and F be events. We say that E and F are *independent* provided that
>
> $$\Pr(E \cap F) = \Pr(E) \cdot \Pr(F).$$

If $\Pr(E) \neq 0$ and $\Pr(F) \neq 0$, then our definition is equivalent to the intuitive statement of independence stated in terms of conditional probability. The two may be used interchangeably.

> Let E and F be events with nonzero probability. E and F are *independent* provided that
>
> $$\Pr(E|F) = \Pr(E) \text{ and } \Pr(F|E) = \Pr(F).$$

EXAMPLE 6 Suppose that an experiment consists of observing the outcome of two consecutive throws of a die. Let E and F be the events

$$E = \text{``the first throw is a 3,''}$$

$$F = \text{``the second throw is a 6.''}$$

Show that these events are independent.

Solution Clearly, $\Pr(E) = \Pr(F) = \frac{1}{6}$. To compute $\Pr(E|F)$, assume that F occurs. Then there are six possible outcomes:

$$F = \{(1, 6), (2, 6), (3, 6), (4, 6), (5, 6), (6, 6)\},$$

and all outcomes are equally likely. Moreover,

$$E \cap F = \{(3, 6)\},$$

so that

$$\Pr(E|F) = \frac{[\text{number of outcomes in } E \cap F]}{[\text{number of outcomes in } F]} = \frac{1}{6} = \Pr(E).$$

Similarly, $\Pr(F|E) = \Pr(F)$. So E and F are independent events, in agreement with our intuition.

EXAMPLE 7 Suppose that an experiment consists of observing the results of drawing two consecutive cards from a 52-card deck. Let E and F be the events

$$E = \text{"second card is black,"}$$

$$F = \text{"first card is red."}$$

Are these events independent?

Solution There are the same number of outcomes with the second card red as with the second card black, so $\Pr(E) = \frac{1}{2}$. To compute $\Pr(E|F)$, note that if F occurs, then there are 51 equally likely choices for the second card, of which 26 are black, so that $\Pr(E|F) = \frac{26}{51}$. Note that $\Pr(E|F) \neq \Pr(E)$, so E and F are not independent, in agreement with our intuition.

EXAMPLE 8 Suppose that we toss a coin three times and record the sequence of heads and tails. Let E be the event "at most one head occurs" and F the event "both heads and tails occur." Are E and F independent?

Solution Using the abbreviations H for "heads" and T for "tails," we have

$$E = \{\text{TTT, HTT, THT, TTH}\}$$

$$F = \{\text{HTT, HTH, HHT, THH, THT, TTH}\}$$

$$E \cap F = \{\text{HTT, THT, TTH}\}.$$

The sample space contains eight equally likely outcomes, so that

$$\Pr(E) = \frac{1}{2}, \qquad \Pr(F) = \frac{3}{4}, \qquad \Pr(E \cap F) = \frac{3}{8}.$$

Moreover,

$$\Pr(E) \cdot \Pr(F) = \frac{1}{2} \cdot \frac{3}{4} = \frac{3}{8},$$

which equals $\Pr(E \cap F)$. So E and F are independent.

EXAMPLE 9 Suppose that a family has four children. Let E be the event "at most one boy" and F the event "at least one child of each sex." Are E and F independent?

Solution   Let B stand for "boy" and G for "girl." Then

$$E = \{GGGG, GGGB, GGBG, GBGG, BGGG\}$$

$$F = \{GGGB, GGBG, GBGG, BGGG, BBBG, BBGB, BGBB,$$

$$GBBB, BBGG, BGBG, BGGB, GBBG, GBGB, GGBB\},$$

and the sample space consists of 16 equally likely outcomes. Furthermore,

$$E \cap F = \{GGGB, GGBG, GBGG, BGGG\}.$$

Therefore,

$$\Pr(E) = \tfrac{5}{16}, \qquad \Pr(F) = \tfrac{7}{8}, \qquad \Pr(E \cap F) = \tfrac{1}{4}.$$

In this example

$$\Pr(E) \cdot \Pr(F) = \tfrac{5}{16} \cdot \tfrac{7}{8} = \tfrac{35}{128} \neq \Pr(E \cap F).$$

So E and F are *not* independent events.

EXAMPLE 10   A new hand calculator is designed to be ultrareliable by reason of its two independent calculating units. The probability that a given calculating unit fails within the first 1000 hours of operation is .001. What is the probability that at least one calculating unit will operate without failure for the first 1000 hours of operation?

Solution   Let

$$E = \text{``calculating unit 1 fails in first 1000 hours,''}$$

$$F = \text{``calculating unit 2 fails in first 1000 hours.''}$$

Then E and F are independent events, since the calculating units are independent of one another. Therefore,

$$\Pr(E \cap F) = \Pr(E) \cdot \Pr(F) = (.001)^2 = .000001$$

$$\Pr((E \cap F)') = 1 - .000001 = .999999.$$

Since $(E \cap F)' = $ "not both calculating units fail in first 1000 hours," the desired probability is .999999.

The concept of independent events can be extended to more than two events:

> A set of events is said to be *independent* if, for each collection of events chosen from among them, say $E_1, E_2, \ldots, E_n$, we have
>
> $$\Pr(E_1 \cap E_2 \cap \cdots \cap E_n) = \Pr(E_1) \cdot \Pr(E_2) \cdot \ldots \cdot \Pr(E_n).$$

EXAMPLE 11   Three events A, B, C are independent and $\Pr(A) = .5$, $\Pr(B) = .3$, $\Pr(C) = .2$.

(a) Calculate $\Pr(A \cap B \cap C)$.   (b) Calculate $\Pr(A \cap C)$.

Solution   (a) $\Pr(A \cap B \cap C) = \Pr(A) \cdot \Pr(B) \cdot \Pr(C) = (.5)(.3)(.2) = .03$.

(b) $\Pr(A \cap C) = \Pr(A) \cdot \Pr(C) = (.5)(.2) = .1$.

We shall leave as an exercise the intuitively reasonable result that if E and F are independent events, then so are E and F', E' and F, and E' and F'. This result also generalizes to any collection of independent events.

EXAMPLE 12   A company manufactures stereo components. Experience shows that defects in manufacture are independent of one another. Quality control studies reveal that

2% of turntables are defective,
3% of amplifiers are defective,
7% of speakers are defective.

A system consists of a turntable, an amplifier, and two speakers. What is the probability that the system is not defective?

Solution   Let T, A, S_1, and S_2 be the events corresponding to defective turntable, amplifier, speaker 1, and speaker 2, respectively. Then

$$\Pr(T) = .02, \qquad \Pr(A) = .03, \qquad \Pr(S_1) = \Pr(S_2) = .07.$$

We wish to calculate $\Pr(T' \cap A' \cap S_1' \cap S_2')$. By the complement rule we have

$$\Pr(T') = .98, \qquad \Pr(A') = .97, \qquad \Pr(S_1') = \Pr(S_2') = .93.$$

Since we have assumed that T, A, S_1, S_2 are independent, so are T', A', S_1', S_2'. Therefore,

$$\Pr(T' \cap A' \cap S_1' \cap S_2') = \Pr(T') \cdot \Pr(A') \cdot \Pr(S_1') \cdot \Pr(S_2')$$
$$= (.98)(.97)(.93)^2 \approx .822.$$

Thus there is an 82.2% chance that the system is not defective.

PRACTICE PROBLEMS 5

1. Suppose there are three cards: one red on both sides, one white on both sides, and one having a side of each color. A card is selected at random and placed on a table. If the up side is red, what is the probability that the down side is red? (Try guessing at the answer before working it using the formula for conditional probability.)

2. Show that if events E and F are independent of each other, then so are E and F'. [*Hint:* Since $E \cap F$ and $E \cap F'$ are mutually exclusive, we have

$$\Pr(E) = \Pr(E \cap F) + \Pr(E \cap F').]$$

EXERCISES 5

In Exercises 1–4, let S be a sample space and E and F events associated with S. Suppose that $\Pr(E) = .5$, $\Pr(F) = .3$, $\Pr(E \cap F) = .1$.

1. Calculate $\Pr(E|F)$, $\Pr(F|E)$.

2. Are E and F independent events? Explain.

3. Calculate $\Pr(E|F')$.

4. Calculate $\Pr(E'|F')$.

5. A doctor studies the known cancer patients in a certain town. The probability that a randomly chosen resident has cancer is found to be .001. It is found that 30% of the town works for Ajax Chemical Company. The probability that an employee of Ajax has cancer is equal to .003. Are the events "has cancer" and "works for Ajax" independent of one another?

6. The proportion of individuals in a certain city earning more than $25,000 per year is .25. The proportion of individuals earning more than $25,000 and having a college degree is .10. Suppose that a person is randomly chosen and he turns out to be earning more than $25,000. What is the probability that he is a college graduate?

7. A medical screening program administers three independent fitness tests. Of the persons taking the tests, 80% pass test I, 75% pass test II, and 60% pass test III. A participant is chosen at random.

   (a) What is the probability that he will pass all three tests?

   (b) What is the probability that he will pass at least two of the three tests?

8. A stereo system contains 50 transistors. The probability that a given transistor will fail in 100,000 hours of use is .0005. Assume that the failures of the various transistors are independent of one another. What is the probability that no transistor will fail during the first 100,000 hours of use?

9. A television set contains five circuit boards of type A, five of type B, and three of type C. The probability of failing in its first 5000 hours of use is .01 for a type A circuit board, .02 for a type B circuit board, and .025 for a type C circuit board. Assuming that the failures of the various circuit boards are independent of one another, compute the probability that no circuit board fails in the first 5000 hours of use.

10. A certain brand of long-life bulb has probability .01 of burning out in less than 1000 hours. Suppose that we wish to light a corridor with a number of independent bulbs in such a way that at least one of the bulbs remains lit for 1000 consecutive hours. What is the minimum number of bulbs needed to assure that the probability of success is at least .99999?

11. Let E and F be events with $\Pr(E) = \frac{1}{2}$, $\Pr(F) = \frac{1}{3}$, $\Pr(E \cap F) = \frac{1}{4}$. Compute $\Pr(E|F)$ and $\Pr(F|E)$.

12. Let E and F be events with $\Pr(E) = .3$, $\Pr(F) = .6$, and $\Pr(E \cup F) = .7$. Find:

    (a) $\Pr(E \cap F)$ (b) $\Pr(E|F)$ (c) $\Pr(F|E)$

    (d) $\Pr(E' \cap F)$ (e) $\Pr(E'|F)$

13. Of the registered voters in a certain town, 50% are Democrats, 40% favor a school loan, and 30% are Democrats who favor a school loan. Suppose that a registered voter is selected at random from the town.

   (a) What is the probability that the person is not a Democrat and opposes the school loan?

   (b) What is the conditional probability that the person favors the school loan given that he is a Democrat?

   (c) What is the conditional probability that the person is a Democrat given that he favors the school loan?

14. Of the students at a certain college, 50% regularly attend the football games, 30% are freshmen, and 40% are upperclassmen who do not regularly attend football games. Suppose that a student is selected at random.

   (a) What is the probability that the person both is a freshman and regularly attends football games?

   (b) What is the conditional probability that the person regularly attends football games given that he is a freshman?

   (c) What is the conditional probability that the person is a freshman given that he regularly attends football games?

15. A coin is tossed three times. What is the conditional probability that the outcome is HHH given that at least two heads occur?

16. Two balls are selected at random from an urn containing two white balls and three red balls. What is the conditional probability that both balls are white given that at least one of them is white?

17. The probabilities that a person A and a person B will live an additional 15 years are .8 and .7, respectively. Assuming that their lifespans are independent, what is the probability that A or B will live an additional 15 years?

18. A sample of two balls is drawn from an urn containing two white balls and three red balls. Are the events "the sample contains at least one white ball" and "the sample contains balls of both colors" independent?

19. Show that if events E and F are independent of each other, then so are E' and F'.

20. Show that if E and F are independent events then

$$\Pr(E \cup F) = 1 - \Pr(E') \cdot \Pr(F').$$

21. A basketball player with a free-throw shooting average of .5 is on the line for a one-and-one free throw. (That is, a second throw is allowed only if the first is successful.) What is the probability that the player will score 0 points? 1 point? 2 points? Assume that the two throws are independent.

22. Let $\Pr(F) > 0$.

   (a) Show that $\Pr(E'|F) = 1 - \Pr(E|F)$.

   (b) Find an example for which $\Pr(E|F') \neq 1 - \Pr(E|F)$.

23. Use the inclusion-exclusion principle for (nonconditional) probabilities to show that if E, F, and G are events in S, then

$$\Pr(E \cup F|G) = \Pr(E|G) + \Pr(F|G) - \Pr(E \cap F|G).$$

24. A hospital uses two tests to classify blood. Every blood sample is subjected to both tests. The first test correctly identifies blood type with probability .3, and the second test correctly identifies blood type with probability .5. The probability that at least one of the tests correctly identifies the blood type is .6.

(a) Find the probability that both tests correctly identify the blood type.

(b) Determine the probability that the second test is correct given that the first test is correct.

(c) Determine the probability that the first test is correct given that the second test is correct.

(d) Are the events "test I correctly identifies the blood" and "test II correctly identifies the blood" independent?

25. Sixty-five percent of the patients in the emergency room of a hospital are seen by a physician immediately. The remainder are kept waiting in the waiting room. Eighty percent of those kept waiting are seen within 2 hours. Seventy-five percent of those seen immediately are admitted to the hospital. Forty percent of those seen within 2 hours (but not immediately) are admitted to the hospital, and 10 percent of those who wait more than 2 hours are admitted to the hospital. Let A be the event "the patient is seen immediately." Let B be the event "the patient is not seen immediately but is seen within 2 hours." Let C be the event "the patient waits more than 2 hours." Let H be the event "the patient is admitted to the hospital."

(a) Find $\Pr(B)$.

(b) Find $\Pr(C)$.

(c) Find $\Pr(H)$.

26. Out of 250 students interviewed at a community college, 90 were taking mathematics but not computer science, 160 were taking mathematics, and 50 were taking neither mathematics nor computer science. Find the probability that a student chosen at random was:

(a) Taking just computer science

(b) Taking mathematics or computer science, but not both

(c) Taking computer science

(d) Not taking mathematics

(e) Taking mathematics, given the student was taking computer science

(f) Taking computer science, given the student was taking mathematics

(g) Taking mathematics, given the student was taking computer science or mathematics

(h) Taking computer science, given the student was not taking mathematics

(i) Not taking mathematics, given the student was not taking computer science.

27. Out of 250-third-grade boys, 230 played baseball, 140 played soccer, and 50 played both. Find the probability that a boy chosen at random:

   (a) Did not play either sport

   (b) Played exactly one sport

   (c) Played soccer but not baseball

   (d) Played soccer, given that he played baseball

   (e) Played baseball, given he did not play soccer

   (f) Did not play soccer, given he did not play baseball

28. Table 1 provides some information about students at a certain college.

TABLE 1

	Works Full Time	Works Part Time	Not Working
Freshman	130	460	210
Sophomore	100	500	150
Junior	80	420	100
Senior	200	300	50

Find the probability that a student selected at random is:

   (a) A senior

   (b) Working full time

   (c) Working part time, given the student is a freshman

   (d) A freshman, given the student does not work

   (e) A junior or senior, given the student does not work

   (f) Working part time or full time, given the student is a sophomore or a junior

29. Table 2 describes the voters in a certain district.

TABLE 2

	Democrat	Republican	Independent
Male	400	700	300
Female	600	300	200

Find the probability that a voter chosen at random is a:

   (a) Democrat

(b) Male

(c) Independent, given the voter is female

(d) Male, given the voter is a Republican

(e) Female, given the voter is not an Independent

(f) Democrat, given the voter is not a Republican

30. In a single-elimination tennis tournament with 16 players, each player is arbitrarily assigned a number from 1 through 16. In the first round, player 1 plays player 2, player 3 plays player 4, and so on. In the next round, the winner of 1 versus 2 plays the winner of 3 versus 4, the winner of 5 versus 6 plays the winner of 7 versus 8, and so on. Subsequent rounds follow the same pattern. If the better player always wins each match in the tournament, what is the probability of the best and second-best players meeting in the final round?

SOLUTIONS TO PRACTICE PROBLEMS 5

1. $\frac{2}{3}$. Let F be the event that the up side is red and E the event that the down side is red. $\Pr(F) = \frac{1}{2}$, since half the faces are red. $F \cap E$ is the event that both sides of the card are red—that is, that the card which is red on both sides was selected, an event with probability $\frac{1}{3}$. By (2),

$$\Pr(E|F) = \frac{\Pr(E \cap F)}{\Pr(F)} = \frac{\frac{1}{3}}{\frac{1}{2}} = \frac{2}{3}.$$

(This result may seem more intuitively evident when you realize that two-thirds of the time the card will have the same color on the bottom as on the top.)

2. By the hint,

$$\Pr(E \cap F') = \Pr(E) - \Pr(E \cap F)$$
$$= \Pr(E) - \Pr(E) \cdot \Pr(F) \qquad \text{(since } E \text{ and } F \text{ are independent)}$$
$$= \Pr(E)[1 - \Pr(F)]$$
$$= \Pr(E) \cdot \Pr(F'). \qquad \text{(by the complement rule)}.$$

Therefore, E and F' are independent events.

6.6. Tree Diagrams

In solving many probability problems, it is helpful to represent the various events and their associated probabilities by a *tree diagram*. To explain this useful notion, suppose that we wish to compute the probability of an event which results from performing a sequence of experiments. The various outcomes of each experiment

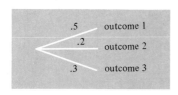

FIGURE 1

are represented as branches emanating from a point. For example, Fig. 1 represents an experiment with three outcomes. Notice that each branch has been labeled with the probability of the associated outcome. For example, the probability of outcome 1 is .5.

We represent experiments performed one after another by stringing together diagrams of the above sort, proceeding from left to right. For example, the diagram in Fig. 2 indicates that first we perform experiment A, having three outcomes, labeled 1–3. If the outcome is 1 or 2, we then perform experiment B. If the outcome is 3, we perform experiment C. The probabilities on the right are conditional probabilities. For example, the top probability is the probability of outcome a (of B) given outcome 1 (of A). The probability of a sequence of outcomes may then be computed by multiplying the probabilities along a path. For example, to calculate the probability of outcome 2 followed by outcome b, we must calculate $\Pr(2 \text{ and } b) = \Pr(2) \cdot \Pr(b|2)$. To carry out this calculation, trace out the sequence of outcomes. Multiplying the probabilities along the path gives $(.2)(.6) = .12$—the probability of outcome 2 followed by outcome b.

The next example illustrates the use of tree diagrams in calculating probabilities.

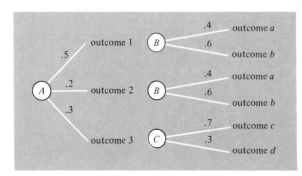

FIGURE 2

EXAMPLE 1 A pollster is hired by a presidential candidate to determine his support among the voters of Pennsylvania's two big cities: Philadelphia and Pittsburgh. The pollster designs the following sampling technique: Select one of the cities at random and then poll a voter selected at random from that city. Suppose that in Philadelphia two-fifths of the voters favor the Republican candidate and three-fifths favor the Democratic candidate. Suppose that in Pittsburgh two-thirds of the voters favor the Republican candidate and one-third favor the Democratic candidate.

(a) Draw a tree diagram describing the survey.

(b) Find the probability that the voter polled is from Philadelphia and favors the Republican candidate.

(c) Find the probability that the voter favors the Republican candidate.

(d) Find the probability that the voter is from Philadelphia, given that he favors the Republican candidate.

Solution (a) The survey proceeds in two steps: first, select a city, and second, poll a voter. Figure 3(a) shows the possible outcomes of the first step and the associated probabilities. For each outcome of the first step there are two possibilities for the second step: the person selected could favor the Republican or the Demo-

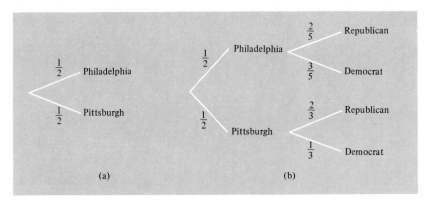

(a)   (b)

FIGURE 3

crat. In Fig. 3(b) we have represented these possibilities by drawing branches emanating from each of the outcomes of the first step. The probabilities on the new branches are actually conditional probabilities. For instance, $\frac{2}{5} = \Pr(\text{Rep}\mid\text{Phila})$, the probability that the voter favors the Republican candidate, given that the voter is from Philadelphia.

(b) $\Pr(\text{Phila} \cap \text{Rep}) = \Pr(\text{Phila}) \cdot \Pr(\text{Rep}\mid\text{Phila}) = \frac{1}{2} \cdot \frac{2}{5} = \frac{1}{5}.$

That is, the probability is $\frac{1}{5}$ that the combined outcome corresponds to the path highlighted in Fig. 4(a). We have written the probability $\frac{1}{5}$ at the end of the path to which it corresponds.

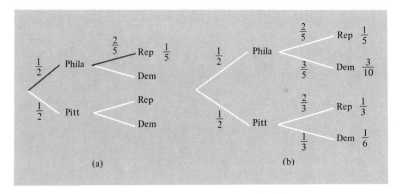

(a)   (b)

FIGURE 4

(c) In Fig. 4(b) we have computed the probabilities for each path of the tree as in part (b) above. Namely, the probability for a given path is the product of the probabilities for each of its segments. We are asked for $\Pr(\text{Rep})$. There are two paths through the tree leading to Republican, namely Philadelphia-Republican and Pittsburgh-Republican. The probabilities of these two paths are $\frac{1}{5}$ and $\frac{1}{3}$, respectively. So the probability that the Republican is favored equals $\frac{1}{5} + \frac{1}{3} = \frac{8}{15}$.

(d) Here we are asked for $Pr(Phila\,|\,Rep)$. By the definition of conditional probability

$$Pr(Phila\,|\,Rep) = \frac{Pr(Phila \cap Rep)}{Pr(Rep)} = \frac{\frac{1}{5}}{\frac{8}{15}} = \frac{3}{8}.$$

Note that from part (c) we might be led to conclude that the Republican candidate is leading, with $\frac{8}{15}$ of the vote. However, we must always be careful when interpreting surveys. The results depend heavily on the survey design. For example, the survey above drew half of its sample from each of the cities. However, Philadelphia is a much larger city and is leaning toward the Democratic candidate—so much so, in fact, that in terms of popular vote the Democratic candidate would win, contrary to our expectations drawn from (c). A pollster must be very careful in designing the procedure for selecting people.

We now finally solve the medical diagnosis problem mentioned in Section 1 and in the Introduction as Problem 2.

EXAMPLE 2 Suppose that the reliability of a skin test for active pulmonary tuberculosis (TB) is specified as follows: Of people with TB, 98% have a positive reaction and 2% have a negative reaction; of people free of TB, 99% have a negative reaction and 1% have a positive reaction. From a large population of which 2 per 10,000 persons have TB, a person is selected at random and given a skin test, which turns out to be positive. What is the probability that the person has active pulmonary tuberculosis?

Solution The given data are organized in Fig. 5. The procedure called for is as follows: First select a person at random from the population. There are two possible outcomes: The person has TB [$Pr(TB) = 2/10{,}000 = .0002$] or the person does not have TB [$Pr(\text{not TB}) = .9998$]. For each of these two possibilities the possible test results and conditional probabilities are given. Multiplying the probabilities along each of the paths through the tree gives the probabilities of the different outcomes. The resulting probabilities are written on the right in Fig. 5. The problem asks for the conditional probability that a person has TB, given that the test is positive. By definition

FIGURE 5

$$Pr(TB\,|\,POS) = \frac{Pr(TB \cap POS)}{Pr(POS)} = \frac{.000196}{.000196 + .009998} = \frac{.000196}{.010194} \approx .02.$$

Therefore, the probability is .02 that a person with a positive skin test has TB. In other words, although the skin test is quite reliable, only about 2% of those with a positive test turn out to have active TB. This result must be taken into account when large-scale medical diagnostic tests are planned. Because the group of people without TB is so much larger than the group with TB, the small error in the former group is magnified to the point where it dominates the calculation.

Note The numerical data presented in Example 2 are only approximate. Variations in air quality for different localities within the United States cause variations in the incidence of TB and the reliability of skin tests.

Tree diagrams come in all shapes and sizes. Three or more branches might emanate from a single point, for example, and some trees may not have the symmetry of those in Examples 1 and 2. Tree diagrams arise whenever an activity can be thought of as a sequence of simpler activities.

EXAMPLE 3   A box contains five good light bulbs and two defective ones. Bulbs are selected one at a time (without replacement) until a good bulb is found. Find the probability that the number of bulbs selected is (i) one, (ii) two, (iii) three.

Solution   The initial situation in the box is shown in Fig. 6(a). A bulb selected at random will be good (G) with probability $\frac{5}{7}$ and defective (D) with probability $\frac{2}{7}$. If a good bulb is selected, the activity stops. Otherwise, the situation is as shown in Fig. 6(b), and a bulb selected at random has probability $\frac{5}{6}$ of being good and probability $\frac{1}{6}$ of

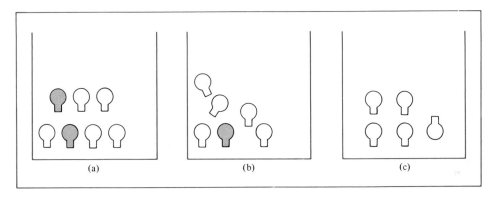

(a)   (b)   (c)

FIGURE 6

FIGURE 7

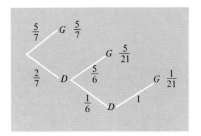

being defective. If the second bulb is good, the activity stops. If the second bulb is defective, then the situation is as shown in Fig. 6(c). At this point a bulb has probability 1 of being good.

The tree diagram corresponding to the sequence of activities is given in Fig. 7. Each of the three paths has a different length. The probability associated with each path has been computed by multiplying the probabilities for its branches. The first path corresponds to the situation where only one bulb is selected, the second path corresponds to two bulbs, and the third path to three bulbs. Therefore,

(i) $\Pr(1) = \frac{5}{7}$ (ii) $\Pr(2) = \frac{5}{21}$ (iii) $\Pr(3) = \frac{1}{21}$.

PRACTICE PROBLEMS 6

Fifty percent of the students enrolled in a business statistics course had previously taken a finite math course. Thirty percent of these students received an A for the statistics course, whereas 20% of the other students received an A for the statistics course.

1. Draw a tree diagram and label it with the appropriate probabilities.

2. What is the probability that a student selected at random previously took a finite math course and did not receive an A in the statistics course?

3. What is the probability that a student selected at random from among the business statistics students received an A in the statistics course?

4. What is the conditional probability that a student previously took a finite math course, given that he received an A in the statistics course?

EXERCISES 6

In Exercises 1–4, draw trees representing the sequence of experiments.

1. Experiment I is performed. Outcome a occurs with probability .3 and outcome b occurs with probability .7. Then experiment II is performed. Its outcome c occurs with probability .6 and its outcome d occurs with probability .4.

2. Experiment I is performed twice. The three outcomes of experiment I are equally likely.

3. A stereo repair shop uses a two-step diagnostic procedure to repair amplifiers. Step I locates the problem in an amplifier with probability .8. Step II (which is executed only if step I fails to locate the problem) locates the problem with probability .6.

4. A training program is used by a corporation to direct hirees to appropriate jobs. The program consists of two steps. Step I identifies 30% as management trainees, 60% as nonmanagerial workers, and 10% as to be fired. In step II, 75% of the management trainees are assigned to managerial positions, 20% to nonmanagerial positions, and 5% are fired. In step II, 60% of the nonmanagerial workers are kept in the same category, 10% are assigned to management positions, and 30% are fired.

5. Refer to Exercise 3. What is the probability that the procedure will fail to locate the problem?

6. Refer to Exercise 4. What is the probability that a randomly chosen hiree will be assigned to a management position at the end of the training period?

7. Refer to Exercise 4. What is the probability that a randomly chosen hiree will be fired by the end of the training period?

8. Refer to Exercise 4. What is the probability that a randomly chosen hiree will be designated a management trainee but *not* be appointed to a management position?

9. Suppose that we have a white urn containing two white and one red balls and we have a red urn containing one white and three red balls. An experiment consists of selecting at random a ball from the white urn and then (without replacing the first ball) selecting at random a ball from the urn having the color of the first ball. Find the probability that the second ball is red.

10. Color blindness is a sex-linked, inherited condition that is much more common among men than women. Suppose that 5% of all men and .4% of all women are color-blind. A person is chosen at random and found to be color-blind. What is the probability that the person is male? (You may assume that 50% of the population are men and 50% are women.)

11. A mouse is put into a T-maze (a maze shaped like a "T"). In this maze he has the choice of turning to the left and being rewarded with cheese or going to the right and receiving a mild shock. Before any conditioning takes place (i.e., on trial 1), the mouse is equally likely to go to the left or to the right. After the first trial his decision is influenced by what happened on the previous trial. If he receives cheese on any trial, the probabilities of his going to the left or right become .9 and .1, respectively, on the following trial. If he receives the electric shock on any trial, the probabilities of his going to the left or right on the next trial become .7 and .3, respectively. What is the probability that the mouse will turn left on the second trial?

12. Refer to Exercise 11. What is the probability that the mouse will turn left on the third trial?

13. A factory has two machines that produce bolts. Machine I produces 60% of the daily output of bolts, and 3% of its bolts are defective. Machine II produces 40% of the daily output, and 2% of its bolts are defective.

    (a) What is the probability that a bolt selected at random will be defective?

    (b) If a bolt is selected at random and found to be defective, what is the probability that it was produced by machine I?

14. Three ordinary quarters and a fake quarter with two heads are placed in a hat. One quarter is selected at random and tossed twice. If the outcome is "HH," what is the probability that the fake quarter was selected?

15. Suppose that the reliability of a test for hepatitis is specified as follows: of people with hepatitis, 95% have a positive reaction and 5% have a negative reaction; of people free of hepatitis, 90% have a negative reaction and 10% have a positive reaction. From a large population, of which .05% of the people have hepatitis, a person is selected at random and given the test. If the test is positive, what is the probability that the person actually has hepatitis?

16. A small citrus farmer has two groves. This season, grove I produced 3000 crates of oranges and 1000 crates of grapefruit. Grove II produced 5000 crates of oranges and 1000 crates of grapefruit. A crate is selected at random and is found to contain grapefruit. What is the probability that the crate is from grove I?

17. Suppose that during any year the probability of an accidental nuclear war is .0001 (provided, of course, that there hasn't been one in a previous year). Draw a tree diagram representing the possibilities for the next three years. What is the probability that there will be an accidental nuclear war during the next three years?

18. Refer to Exercise 17. What is the probability that there will be an accidental nuclear war during the next n years?

19. A coin is to be tossed at most five times. The tosser wins as soon as the number of heads exceeds the number of tails and loses as soon as three tails have been tossed. Draw a tree diagram for this game and calculate the probability of winning.

20. Suppose that, instead of tossing a coin, the player in Exercise 19 draws up to five cards from a deck consisting only of three red and three black cards. The drawer wins as

soon as the number of red cards exceeds the number of black cards and loses as soon as three black cards have been drawn. Does the tree diagram for the card game have the same shape as the tree diagram for the coin game? Is there any difference in the probability of winning? If so, which game has the greater probability of winning?

21. A man has been guessing the colors of cards drawn from a standard deck. During the first 50 draws he kept track of the number of cards of each color. What is the probability of guessing the color of the fifty-first card?

SOLUTIONS TO PRACTICE PROBLEMS 6

1.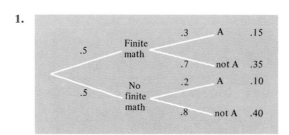

2. The event "finite math and not A" corresponds to the second path of the tree diagram, which has probability .35.

3. This event is satisfied by the first and third paths and therefore has probability .15 + .10 = .25.

4. $\text{Pr(finite math}|A) = \dfrac{\text{Pr(finite math and A)}}{\text{Pr(A)}} = \dfrac{.15}{.25} = .6.$

6.7. Bayes' Theorem

FIGURE 1

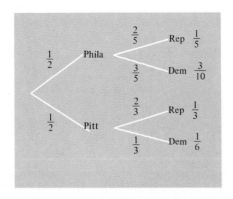

Let us reconsider the polling survey in Example 1 of Section 6.6. See Fig. 1. We obtained $\text{Pr(Phila}|\text{Rep})$ by finding $\dfrac{\text{Pr(Phila} \cap \text{Rep})}{\text{Pr(Rep)}}$. Let us analyze the components of this calculation. First,

$$\text{Pr(Phila} \cap \text{Rep}) = \text{Pr(Phila)} \cdot \text{Pr(Rep}|\text{Phila)}.$$

Second,

$$\text{Pr(Rep)} = \text{Pr(Phila} \cap \text{Rep}) + \text{Pr(Pitt} \cap \text{Rep})$$
$$= \text{Pr(Phila)} \cdot \text{Pr(Rep}|\text{Phila)} + \text{Pr(Pitt)} \cdot \text{Pr(Rep}|\text{Pitt)}.$$

Denote the events "Phila," "Pitt," "Rep," and "Dem" by the letters A, B, R, and D, respectively. Then,

$$\text{Pr}(\text{Phila}|\text{Rep}) = \text{Pr}(A|R) = \frac{\text{Pr}(A) \cdot \text{Pr}(R|A)}{\text{Pr}(A) \cdot \text{Pr}(R|A) + \text{Pr}(B) \cdot \text{Pr}(R|B)}.$$

This is a special case of Bayes' theorem.

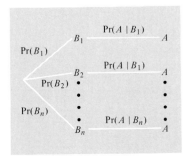

FIGURE 2

To derive Bayes' theorem in general, we consider a two-stage tree. Suppose that at the first stage there are the events $B_1, B_2, \ldots, B_n$, which are mutually exclusive and exhaust all possibilities. Let us examine only the paths of the tree leading to event A at the second stage of the experiment. (See Fig. 2.) Suppose we are given that the event A occurs. What is $\text{Pr}(B_1|A)$? First, consider $\text{Pr}(B_1 \cap A)$. This can be seen from Fig. 2 to be $\text{Pr}(B_1) \cdot \text{Pr}(A|B_1)$. Next we calculate $\text{Pr}(A)$. Recall that A occurs at stage 2, preceded at stage 1 by either event $B_1, B_2, \ldots$, or B_n. Since $B_1, B_2, \ldots, B_n$ are mutually exclusive,

$$\text{Pr}(A) = \text{Pr}(B_1 \cap A) + \text{Pr}(B_2 \cap A) + \cdots + \text{Pr}(B_n \cap A).$$

Each of the elements in the sum can be calculated by the product rule, or directly from Fig. 2:

$$\text{Pr}(B_1 \cap A) = \text{Pr}(B_1) \cdot \text{Pr}(A|B_1)$$

$$\text{Pr}(B_2 \cap A) = \text{Pr}(B_2) \cdot \text{Pr}(A|B_2)$$

$$\vdots$$

$$\text{Pr}(B_n \cap A) = \text{Pr}(B_n) \cdot \text{Pr}(A|B_n).$$

The result is the following:

Bayes' Theorem If $B_1, B_2, \ldots, B_n$ are mutually exclusive events and $B_1 \cup B_2 \cup \cdots \cup B_n = S$, then for any event A in S,

$$\text{Pr}(B_1|A) = \frac{\text{Pr}(B_1) \cdot \text{Pr}(A|B_1)}{\text{Pr}(B_1) \cdot \text{Pr}(A|B_1) + \text{Pr}(B_2) \cdot \text{Pr}(A|B_2) + \cdots + \text{Pr}(B_n) \cdot \text{Pr}(A|B_n)},$$

$$\text{Pr}(B_2|A) = \frac{\text{Pr}(B_2) \cdot \text{Pr}(A|B_2)}{\text{Pr}(B_1) \cdot \text{Pr}(A|B_1) + \text{Pr}(B_2) \cdot \text{Pr}(A|B_2) + \cdots + \text{Pr}(B_n) \cdot \text{Pr}(A|B_n)},$$

and so forth.

EXAMPLE 1 Solve the tuberculosis skin test problem of Example 2 of the previous section by using Bayes' theorem.

Solution The observed event A is "positive skin test result." There are two possible events leading to A—namely

$$B_1 = \text{"person has tuberculosis"}$$

$$B_2 = \text{"person does not have tuberculosis."}$$

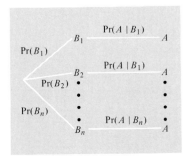

We wish to calculate $\Pr(B_1 | A)$. From the data given we have

$$\Pr(B_1) = \frac{2}{10{,}000} = .0002$$

$$\Pr(B_2) = .9998$$

$$\Pr(A|B_1) = \Pr(\text{POS}|\text{TB}) = .98$$

$$\Pr(A|B_2) = \Pr(\text{POS}|\text{not TB}) = .01.$$

Therefore, by Bayes' theorem,

$$\Pr(B_1|A) = \frac{\Pr(B_1)\,\Pr(A|B_1)}{\Pr(B_1)\,\Pr(A|B_1) + \Pr(B_2)\,\Pr(A|B_2)}$$

$$= \frac{(.0002)(.98)}{(.0002)(.98) + (.9998)(.01)} \approx .02,$$

in agreement with our calculation of Example 2 of the previous section.

The advantages of Bayes' theorem over the use of tree diagrams are: (1) we do not need to draw the tree diagram to calculate the desired probability, and (2) we need not compute extraneous probabilities. These advantages become very significant in dealing with experiments with many outcomes.

EXAMPLE 2   A printer has seven book-binding machines. For each machine Table 1 gives the proportion of the total book production that it binds and the probability that the machine produces a defective binding.

TABLE 1

Machine	Proportion of books bound	Probability of defective binding
1	.10	.03
2	.05	.03
3	.20	.02
4	.15	.02
5	.25	.01
6	.15	.02
7	.10	.03

For instance, machine 1 binds 10% of the books and produces a defective binding with probability .03. Suppose that a book is selected at random and found to have a defective binding. What is the probability that it was bound by machine 1?

Solution Let B_i $(i = 1, 2, \ldots, 7)$ be the event that the book was bound by machine i and let A be the event that the book has a defective binding. Then, for example, $\Pr(B_1) = .10$ and $\Pr(A \mid B_1) = .03$. The problem asks for the reversed conditional probability, $\Pr(B_1 \mid A)$. By Bayes' theorem

$$\Pr(B_1 \mid A)$$

$$= \frac{\Pr(B_1) \Pr(A \mid B_1)}{\Pr(B_1) \Pr(A \mid B_1) + \Pr(B_2) \Pr(A \mid B_2) + \cdots + \Pr(B_7) \Pr(A \mid B_7)}$$

$$= \frac{(.10)(.03)}{(.10)(.03) + (.05)(.03) + (.20)(.02) + (.15)(.02) + (.25)(.01) + (.15)(.02) + (.10)(.03)}$$

$$= \frac{.003}{.02} = .15.$$

PRACTICE PROBLEMS 7

Refer to Example 2. Suppose that a book is selected at random and found to have a defective binding.

1. What is the probability that the book was bound by machine 2?

2. By what machine is the book most likely to have been bound?

EXERCISES 7

1. An automobile insurance company has determined the accident rate (probability of having at least one accident during a year) for various age groups. (See Table 2.) Suppose that a policyholder calls in to report an accident. What is the probability that he is over 60?

TABLE 2

Age group	Proportion of total insured	Accident rate
Under 21	.05	.06
21–30	.10	.04
31–40	.25	.02
41–50	.20	.015
51–60	.30	.025
Over 60	.10	.04

2. An electronic device has six different types of transistors. For each type of transistor, Table 3 gives the proportion of the total number of transistors of that type and the failure rate (probability of failing within one year). If a transistor fails, what is the probability that it is type 1?

TABLE 3

Type	Proportion of total	Failure rate
1	.30	.0002
2	.25	.0004
3	.20	.0005
4	.10	.001
5	.05	.002
6	.10	.004

3. The enrollment in a certain course is 10% freshmen, 30% sophomores, 40% juniors, and 20% seniors. Past experience has shown that the likelihood of receiving an A in the course is .2 for freshmen, .4 for sophomores, .3 for juniors, and .1 for seniors. Find the probability that a student who receives an A is a sophomore.

4. A metropolitan police department maintains statistics of larcenies reported in the various precincts of the city. It records the proportion of the city population in each precinct and the precinct larceny rate (= the proportion of the precinct population reporting a larceny within the past year). These statistics are summarized in Table 4. A larceny victim is randomly chosen from the city population. What is the probability he comes from Precinct 3?

TABLE 4

Precinct	Proportion of population	Larceny rate
1	.20	.01
2	.10	.02
3	.40	.05
4	.30	.04

TABLE 5

Annual family income	Proportion of people	Proportion having two or more cars
< $10,000	.10	.2
$10,000–$14,999	.20	.5
$15,000–$19,999	.35	.6
$20,000–$24,999	.30	.75
≥ $25,000	.05	.9

5. Table 5 gives the distribution of incomes and that of two-car families by income level for a certain suburban county. Suppose that a randomly chosen family has two or more cars. What is the probability that its income is at least $25,000 per year?

6. Table 6 gives the distribution of voter registration and voter turnouts for a certain city. A randomly chosen person is questioned at the polls. What is the probability that the person is an Independent?

TABLE 6

	Proportion registered	Proportion turnout
Democrat	.50	.4
Republican	.20	.5
Independent	.30	.7

7. A crime investigator feels 70% certain that the suspect being held for a theft is guilty. She then discovers that the thief was left-handed. Twenty percent of the population are left-handed, and the suspect is also left-handed. In light of this new evidence, what probability should the investigator now assign to the guilt of the suspect?

8. Table 7 shows the percentages of various portions of the U.S. population based on age and sex. Suppose that a person is chosen at random from the entire population.

   (a) What is the probability that the person chosen is male?

   (b) Given that the person chosen is male, find the probability that he is between 5 and 19 years old.

TABLE 7

	1980 U.S. Population				
Age group	Under 5 yrs.	5–19 yrs.	20–44 yrs.	45–64 yrs.	Over 65 yrs.
% of pop.	7	25	37	20	11
% male	51	51	49	41	40

9. A multinational company has five divisions A, B, C, D, and E. The percentage of employees from each division who speak at least two languages fluently is shown in Table 8.

   (a) Find the probability that an employee selected at random is bilingual.

(b) Find the probability that a bilingual employee selected at random works for division C.

TABLE 8

Division	Number of employees	Percent of employees that are bilingual
A	20,000	20
B	15,000	15
C	25,000	12
D	30,000	10
E	10,000	10
Total	100,000	

10. A specially made pair of dice has only one- and two-spots on the faces. One of the dice has three faces with a one-spot and three faces with a two-spot. The other die has two faces with a one-spot and four faces with a two-spot. One of the dice is selected at random and then rolled six times. If a two-spot shows up only once, what is the probability that it is the die with four two-spots?

11. Al and Bob are the same age, 20. They enter a lottery in which a person wins if he guesses a number between 0 and 99 correctly. Both Al and Bob independently have a .2 probability of entering their age on the lottery ticket and a .8 probability of entering some other number selected at random. If Bob wins the lottery, what is the probability that Al also won?

12. A drug-testing laboratory produces false negative results 2% of the time and false positive results 5% of the time. Suppose that the laboratory has been hired by a company in which 10% of the employees use drugs.

(a) If an employee tests positive for drug use, what is the probability that he or she actually uses drugs?

(b) What is the probability that a nondrug user will test positive for drug use twice in a row?

(c) What is the probability that someone who tests positive twice in a row is not a drug user?

13. Thirteen cards are dealt from a deck of 52 cards.

(a) What is the probability that the ace of spades is one of the 13 cards?

(b) Suppose one of the 13 cards is chosen at random and found *not* to be the ace of spades. What is the probability that *none* of the 13 cards is the ace of spades?

(c) Suppose the experiment in part (b) is repeated a total of 10 times (replacing the card looked at each time), and the ace of spaces is not seen. What is the probability that the ace of spaces actually *is* one of the 13 cards?

SOLUTIONS TO PRACTICE PROBLEMS 7

1. The problem asks for $\Pr(B_2|A)$. Bayes' theorem gives this probability as a quotient with numerator $\Pr(B_2)\Pr(A|B_2)$ and the same denominator as in the solution to Example 2. Therefore,

$$\Pr(B_2|A) = \frac{\Pr(B_2)\,\Pr(A|B_2)}{.02} = \frac{(.05)(.03)}{.02} = .075.$$

2. To solve this problem we must compute the seven conditional probabilities $\Pr(B_1|A)$, $\Pr(B_2|A), \ldots, \Pr(B_7|A)$ and see which one is the largest. The first two have already been computed. Using the method of the preceding problem we find that $\Pr(B_3|A) = .20$, $\Pr(B_4|A) = .15$, $\Pr(B_5|A) = .125$, $\Pr(B_6|A) = .15$, and $\Pr(B_7|A) = .15$. Therefore, the book was most likely bound by machine 3.

Chapter 6: CHECKLIST

- ☐ Experiment
- ☐ Outcome
- ☐ Sample space
- ☐ Trial
- ☐ Event
- ☐ Impossible event
- ☐ Certain event
- ☐ Intersection, union, and complement of events
- ☐ Mutually exclusive events
- ☐ Elementary event
- ☐ Probability of an event
- ☐ Addition principle
- ☐ Inclusion-exclusion principle
- ☐ Odds
- ☐ Equally likely outcomes
- ☐ Complement rule
- ☐ $\Pr(E|F)$
- ☐ Independent events
- ☐ Tree diagram
- ☐ Bayes' theorem

Chapter 6: SUPPLEMENTARY EXERCISES

1. A coin is to be tossed five times. What is the probability of obtaining at least one head?

2. Suppose that we toss a coin three times and observe the sequence of heads and tails. Let E be the event that "the first toss lands heads" and F the event that "there are more heads than tails." Are E and F independent?

3. Each box of a certain brand of candy contains either a toy airplane or a toy gun. If one-third of the boxes contain an airplane and two-thirds contain a gun, what is the probability that a person who buys two boxes of candy will receive both an airplane and a gun?

4. A committee consists of five men and five women. If three people are selected at random from the committee, what is the probability that they will all be men?

5. Out of the 50 colleges in a certain state, 25 are private, 15 offer engineering majors, and 5 are private colleges offering engineering majors. If a college is selected at random, what is the conditional probability that it offers an engineering major given that it is a public college?

6. An auditing procedure for income tax returns has the following characteristics: If the return is incorrect, the probability is 90% that it will be rejected, and if the return is correct, the probability is 95% that it will be accepted. Suppose that 80% of all income tax returns are correct. If a return is audited and rejected, what is the probability that the return was actually correct?

7. Two archers shoot at a moving target. One can hit the target with probability $\frac{1}{4}$ and the other with probability $\frac{1}{3}$. Assuming that their efforts are independent events, what is the probability that

   (a) both will hit the target? (b) at least one will hit the target?

8. If the odds in favor of an event are 7 to 5, what is the probability that the event will occur?

9. In an Olympic swimming event, two of the seven contestants are American. The contestants are randomly assigned to lanes 1 through 7. What is the probability that the Americans are assigned to the first two lanes?

10. An urn contains three balls numbered 1, 2, and 3. Balls are drawn one at a time without replacement until the sum of the numbers drawn is four or more. Find the probability of stopping after exactly two balls are drawn.

11. A red die and a green die are tossed as a pair. Let E be the event that "the red die shows a 2" and let F be the event that "the sum of the numbers is 8." Are the events E and F independent?

12. Let E and F be events with $\Pr(E) = .4$, $\Pr(F) = .3$, and $\Pr(E \cup F) = .5$. Find $\Pr(E|F)$.

13. A supermarket has three employees who package and weigh produce. Employee A records the correct weight 98% of the time. Employees B and C record the correct weight 97% and 95% of the time, respectively. Employees A, B, and C handle 40%, 40%, and 20% of the packaging, respectively. A customer complains about the incorrect weight recorded on a package he has purchased. What is the probability that the package was weighed by employee C?

14. Three people are chosen at random. What is the probability that at least two of them were born on the same day of the week?

15. Let B and A be independent events for which the probability that at least one of them occurs is $\frac{1}{2}$ and the probability that B occurs but A does not occur is $\frac{1}{3}$. Find $\Pr(A)$.

16. An urn contains 10 balls numbered 1 through 10. Seven balls are drawn one at a time at random without replacement. Find the probability that exactly three odd numbered balls are drawn and they occur on odd-numbered draws from the urn.

17. Each of three sealed opaque envelopes contains two bills. One envelope contains two $1 bills, another contains two $5 bills, and the third contains a $1 bill and a $5 bill. An envelope is selected at random and a bill is taken from the envelope at random. If the bill is a $5 bill, what is the probability that the other bill in the envelope is also a $5 bill?

18. A carnival huckster has placed a coin under one of three cups and asks you to guess which cup contains the coin. After you select a cup, he removes one of the unselected cups, which he guarantees does not contain the coin. You may now either stay with your original choice or switch to the other remaining cup. What decision will give you the greatest probability of winning?

19. The odds of an American worker living within 20 minutes of work is 13 to 12. What is the probability that a worker selected at random lives within 20 minutes of work?

20. Twenty-six percent of all Americans are under 18 years old. What are the odds that a person selected at random is under 18?

21. If the nine letters A, C, D, E, I, N, O, T, and U are arranged to form a word, what is the probability that it will be one of the meaningful words EDUCATION, AUCTIONED, or CAUTIONED?

22. An island contains an equal number of one-headed, two-headed, and three-headed dragons. If a dragon head is picked at random, what is the likelihood of its belonging to a one-headed dragon?

7

PROBABILITY AND STATISTICS

7.1 Frequency and Probability Distributions

7.2 The Mean

7.3 The Variance and Standard Deviation

7.4 Binomial Trials

7.5 The Normal Distribution

7.6 Normal Approximation to the Binomial Distribution

Statistics is the branch of mathematics that deals with data—their collection, description, analysis, and use in prediction. In this chapter we present some topics in statistics which can be used as a springboard to further study. Since we are presenting a series of topics, rather than a comprehensive survey, we shall bypass large areas of statistics without saying anything about them. However, the discussion should give you some feeling for the subject. In Section 1 we discuss the problem of describing data by means of a distribution and a histogram. We also introduce the concept of random variables. In Sections 2 and 3 we introduce the two most frequently used descriptive statistics: the mean and standard deviation. In Section 3 we illustrate how Chebychev's inequality can be used to make estimations. In Sections 4 to 6 we discuss the probability distributions most frequently employed in statistical work: the binomial and the normal distributions.

7.1. Frequency and Probability Distributions

Our goal in this section is to describe a given set of data in terms that allow for interpretation and comparison. As we shall see, both graphical and tabular displays of data can be useful for this purpose.

Our modern technological society has a fetish about gathering statistical data. It is hardly possible to glance at a newspaper or a magazine and not be confronted with massive arrays of statistics gathered from studies of schools, churches, the economy, and so forth. One of the chief tasks confronting us is to interpret in a meaningful way the data collected and to make decisions based on our interpretations. The mathematical tools for doing this belong to that part of mathematics called *statistics*.

To get an idea of the problems considered in statistics, let us consider a concrete example. Mr. Jones, a businessman, is interested in purchasing a car dealership. Two dealerships are for sale, and each dealer has provided him with data describing past sales. Dealership A provided 1 year's worth of data, dealership B, 2 years' worth. The data are summarized in Table 1. The problem confronting Mr. Jones is that of analyzing the data to determine which car dealership to buy.

TABLE 1

Dealership A		Dealership B	
Weekly sales	*Number of occurrences*	*Weekly sales*	*Number of occurrences*
5	2	5	20
6	2	6	0
7	13	7	0
8	20	8	10
9	10	9	12
10	4	10	50
11	1	11	12

TABLE 2

Dealership A	
Weekly sales	Proportion of occurrences
5	$\frac{2}{52} \approx .04$
6	$\frac{2}{52} \approx .04$
7	$\frac{13}{52} = .25$
8	$\frac{20}{52} \approx .38$
9	$\frac{10}{52} \approx .19$
10	$\frac{4}{52} \approx .08$
11	$\frac{1}{52} \approx .02$

These data are presented in a form often used in statistical surveys. For each possible value of a statistical variable (in this case the number of cars sold weekly) we have tabulated the number of occurrences. Such a tabulation is called a *frequency distribution*. Although a frequency distribution is a very useful way of displaying and summarizing survey data, it is by no means the most efficient form in which to analyze such data. For example, it is difficult to compare dealership A with dealership B using only Table 1.

Comparisons are much more easily made if we use proportions rather than actual numbers of occurrences. For example, instead of recording that dealership A had weekly sales of 5 cars during two of the weeks of the year, let us record that the proportion of the observed weeks in which dealership A had weekly sales of 5 was $\frac{2}{52} \approx .04$. Similarly, by dividing each of the entries in the right column by 52, we obtain a new table describing the sales of dealership A (Table 2).* We similarly can construct a new table for dealership B (Table 3).

These tables are called *relative frequency distributions*. In general, consider an experiment with the numerical outcomes $x_1, x_2, \ldots, x_r$. Suppose that the number

TABLE 3

Dealership B	
Weekly sales	Proportion of occurrences
5	$\frac{20}{104} \approx .19$
6	0
7	0
8	$\frac{10}{104} \approx .10$
9	$\frac{12}{104} \approx .12$
10	$\frac{50}{104} \approx .48$
11	$\frac{12}{104} \approx .12$

* For simplification we shall round off the data of this example to two decimal places.

of occurrences of x_1 is f_1, the number of occurrences of x_2 is f_2, and so forth. The frequency distribution lists all the outcomes of the experiment and the number of times each occurred. (For the sake of simplicity, we usually arrange $x_1, x_2, \ldots, x_r$ in increasing order.)

Outcome	Frequency	Relative frequency
x_1	f_1	f_1/n
x_2	f_2	f_2/n
$\vdots$	$\vdots$	
x_r	f_r	
Total	n	1

Suppose that the total number of occurrences is n. Then the relative frequency of the outcome x_1 is f_1/n, the relative frequency of the outcome x_2 is f_2/n, and so forth. The relative frequency distribution pairs each outcome with its relative frequency. The sum of the frequencies in a frequency distribution is n. The sum of the relative frequencies in a relative frequency distribution is 1.

The frequency or relative frequency distribution is obtained directly from the performance of an experiment and the collection of data observed at each trial of the experiment. For example, we might imagine a coin-tossing experiment in which a coin is tossed five times and the number of occurrences of heads is observed. On each performance of the experiment we might observe 0, 1, 2, 3, 4, or 5 heads. We could repeat the experiment, say 90 times, and record the outcomes. We have collected data like that in Table 4. While doing the experiment we would record the frequencies $f_1, f_2, \ldots, f_5$ of the various outcomes and then divide by 90 to obtain the relative frequencies $f_1/90, f_2/90, \ldots, f_5/90$. The sum of the frequency column is the number of trials of the experiment (here, 90), and the sum of the relative frequency column is 1.

TABLE 4

Number of heads	Frequency	Relative frequency
0	3	$\frac{3}{90} \approx .03$
1	14	$\frac{14}{90} \approx .16$
2	23	$\frac{23}{90} \approx .26$
3	27	$\frac{27}{90} \approx .30$
4	17	$\frac{17}{90} \approx .18$
5	6	$\frac{6}{90} \approx .07$
Total	90	$\frac{90}{90} = 1.00$

It is often possible to gain useful insight into an experiment by representing its relative frequency distribution in graphical form. For instance, let us graph the relative frequency distribution for car dealership A. Begin by drawing a number line (Fig. 1).

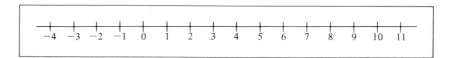

FIGURE 1

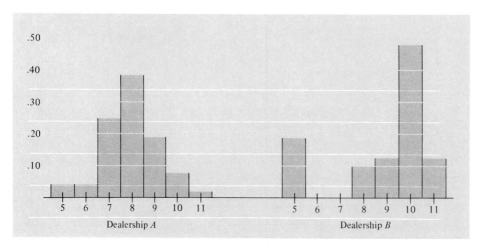

Dealership *A*

Dealership *B*

FIGURE 2

FIGURE 3 Relative frequency distribution for number of heads in 5 tosses—Experimental results

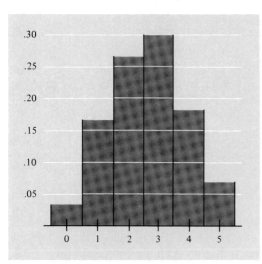

The numbers that represent possible outcomes of the experiment (weekly car sales) are 5, 6, 7, 8, 9, 10, 11. Locate each of these numbers on the number line. Above each number erect a rectangle whose base is one unit wide and whose height is the relative frequency of that number. Above the number 5, for example, we draw a rectangle of height .04. The completed graph is shown in Fig. 2. For the sake of comparison we have also drawn a graph of the relative frequency distribution for dealership B.

Graphs of the type just drawn are called *histograms*. They vividly illustrate the data being considered. For example, comparison of the histograms of Fig. 2 reveals significant differences between the two dealerships. On the one hand, dealership A is very consistent. Most weeks its sales are in the middle range of 7, 8, or 9. On the other hand, dealership B can often achieve very high sales (it had sales of 10 in 48% of the weeks) at the expense of a significant number of weeks of low sales (sales of 5 in 19% of the weeks).

The histogram for the coin-tossing experiment recorded above appears in Fig. 3.

It is possible to use histograms to represent the relative frequency of events as areas. To illustrate the procedure, consider the histogram corresponding to dealership A. Each rectangle has width 1 and height equal to the relative frequency of a particular outcome of the experiment. For instance, the highest rectangle

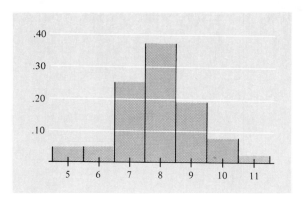

FIGURE 4

is centered over the number 8 and has height .38, the relative frequency of the outcome 8. Note that the area of the rectangle is

$$\text{area} = (\text{height})(\text{width}) = (.38)(1) = .38.$$

In other words, the area of the rectangle equals the relative frequency of the corresponding outcome. We have verified this in the case of the outcome 8, but it is true in general. In a similar fashion, we may represent the relative frequency of more complicated events as areas. Consider, for example, the event $E =$ "sales between 7 and 10 inclusive." This event consists of the set of outcomes $\{7, 8, 9, 10\}$, and so its relative frequency is the sum of the respective relative frequencies of the outcomes 7, 8, 9, and 10. Therefore, the relative frequency of the event E is the area of the shaded region of the histogram in Fig. 4.

An important thing to notice is that so far we have made the tables and the histograms for actual, as opposed to theoretical, experiments. That is, the tables and the histograms were produced from collections of sample data that were obtained by actually recording the outcomes of experiments. We shall now look at theoretical experiments and introduce the important notion of *probability distribution*. In many cases, data of an actual experiment are best interpreted when we can construct a theoretical model for the experiment.

Let us reconsider the coin-tossing experiment in which a coin was tossed five times and the number of occurrences of heads recorded. If the coins are fair coins, then we can set up a model for the experiment by noting once again that the possible outcomes are 0, 1, 2, 3, 4, or 5 heads. We will construct the *probability distribution* for this experiment by listing the outcomes in the sample space with their probabilities. The probabilities of 0, 1, 2, 3, 4, 5 heads can be obtained with the methods of Chapter 6. The number of distinct sequences of 5 tosses is 2^5 or 32. The number of sequences having k heads (and $5 - k$ tails) is $C(5, k) = \binom{5}{k}$. Thus

$$\Pr(k \text{ heads}) = \frac{\binom{5}{k}}{2^5} \qquad (k = 0, 1, 2, 3, 4, 5).$$

The probability distribution for the experiment is, therefore:

Number of heads	Probability
0	$\dfrac{\binom{5}{0}}{2^5} = \dfrac{1}{32}$
1	$\dfrac{\binom{5}{1}}{2^5} = \dfrac{5}{32}$
2	$\dfrac{\binom{5}{2}}{2^5} = \dfrac{10}{32}$
3	$\dfrac{\binom{5}{3}}{2^5} = \dfrac{10}{32}$
4	$\dfrac{\binom{5}{4}}{2^5} = \dfrac{5}{32}$
5	$\dfrac{\binom{5}{5}}{2^5} = \dfrac{1}{32}$

The histogram for a probability distribution is constructed in the same way as the histogram for a relative frequency distribution. Each outcome is represented on the number line, and above each outcome we erect a rectangle of width 1 and of height equal to the probability corresponding to that outcome (see Fig. 5).

We note that the histogram in Fig. 5 is based on a theoretical model of coin tossing, whereas the histogram in Fig. 3 was drawn from experimental results only available after the experiment is actually performed and observed.

Just as we used the histogram for the relative frequency distribution to picture the relative frequencies of an event, we may also use the histogram of a probability distribution to picture the probability of an event. For instance, to find the probability of at least 3 heads on 5 tosses of a fair coin, we need only add the probabilities of the outcomes: 3 heads, 4 heads, 5 heads. That is,

$$\text{Pr(at least 3 heads)} = \text{Pr(3 heads)} + \text{Pr(4 heads)} + \text{Pr(5 heads)}.$$

Since each of these probabilities is equal to the area of a rectangle in the histogram of Fig. 5, the area of the shaded region in Fig. 6 equals the probability of the

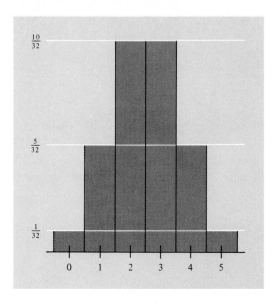

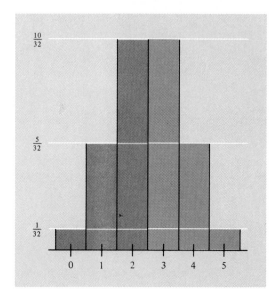

FIGURE 5   Probability distribution for number of
heads in 5 tosses—Theoretical results

FIGURE 6

event. This result is a special case of the following fact:

> Suppose that the rectangles of a histogram of a probability distribution all
> have width 1. Then the probability of an event E is the sum of the areas of
> the rectangles corresponding to the outcomes in E.

EXAMPLE 1   Suppose that the histogram of a probability distribution is as given in Fig. 7.
Shade in the portion of the histogram whose area is the probability of the event
"more than 50."

FIGURE 7

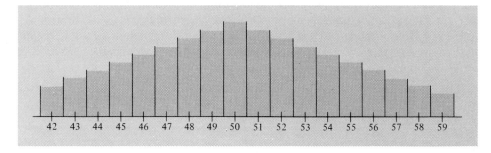

Solution The event "more than 50" is the set of outcomes {51, 52, 53, 54, 55, 56, 57, 58, 59}. So we shade in the portion of the histogram corresponding to these outcomes (Fig. 8). The area of the shaded region is the probability of the event "more than 50."

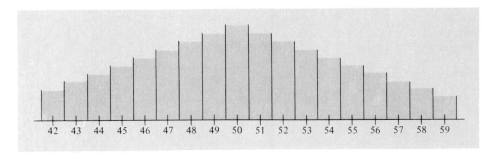

FIGURE 8

Random Variables Consider a theoretical experiment with numerical outcomes. Denote the outcome of the experiment by the letter X. Thus, for example, if the experiment consists of observing the result of tossing a single die, then X assumes one of the six values 1, 2, 3, 4, 5, 6. Since the values of the variable X are determined by the unpredictable random outcomes of the experiment, X is called a *random variable* or, more specifically, the *random variable associated with the experiment*. The random variable notation is often convenient and is in common use in probability and statistics texts. In considering several different experiments, it is sometimes necessary to use letters other than X to stand for random variables. It is customary, however, to use only capital letters, such as X, Y, Z, W, U, V, for random variables.

If k is one of the possible outcomes of the experiment with associated random variable X, then we denote the probability of the outcome k by

$$\Pr(X = k).$$

For example, if we consider the die-tossing experiment described above, then

$$\Pr(X = 3)$$

denotes the probability of throwing a 3.

Rather than speak of the probability distribution associated with the model of an experiment, we can speak of the *probability distribution associated with the corresponding random variable*. Such a probability distribution is a table listing the various values of X (i.e., outcomes of the experiment) and their associated probabilities with $p_1 + p_2 + \cdots + p_r = 1$:

k	$\Pr(X = k)$
x_1	p_1
x_2	p_2
$\vdots$	$\vdots$
x_r	p_r

The advantage of the random-variable notation is that the variable X can be treated algebraically and we can consider expressions such as X^2. This is just the random variable corresponding to the experiment whose outcomes are the squares of the outcomes of the original experiment. Similarly, we can consider random variables such as $X + 3$ and $(X - 2)^2$. One important example of algebraic manipulation of random variables appears in Section 7.3. There are many others.

EXAMPLE 2 Suppose that a random variable X has probability distribution given by the following table:

k	$\Pr(X = k)$
-1	.2
0	.3
1	.1
2	.4

Determine the probability distribution of the random variable X^2.

Solution The outcomes of X^2 are the squares of the outcomes of X. The probabilities of the outcomes of X^2 are determined by the probabilities of the outcomes of X. The possible outcomes of X^2 are $k = 1$ (which results from the case $X = -1$ or from the case $X = 1$), $k = 0$, and $k = 4$. Since $\Pr(X^2 = 1) = \Pr(X = -1) + \Pr(X = 1)$, the probability distribution of X^2 is

k	$\Pr(X^2 = k)$
0	.3
1	.3
4	.4

EXAMPLE 3 Let X denote the random variable defined as the sum of the upper faces appearing when two dice are thrown. Determine the probability distribution of X and draw its histogram.

Solution The experiment of throwing two dice leads to 36 possibilities, each having probability $\frac{1}{36}$.

(1, 1)	(1, 2)	(1, 3)	(1, 4)	(1, 5)	(1, 6)
(2, 1)	(2, 2)	(2, 3)	(2, 4)	(2, 5)	(2, 6)
(3, 1)	(3, 2)	(3, 3)	(3, 4)	(3, 5)	(3, 6)
(4, 1)	(4, 2)	(4, 3)	(4, 4)	(4, 5)	(4, 6)
(5, 1)	(5, 2)	(5, 3)	(5, 4)	(5, 5)	(5, 6)
(6, 1)	(6, 2)	(6, 3)	(6, 4)	(6, 5)	(6, 6)

The sum of the numbers in each pair gives the value of X. For example, the pair (3, 1) corresponds to $X = 4$. Note that the pairs corresponding to a given value of

X lie on a diagonal, as shown above, where we have indicated all pairs corresponding to $X = 4$. It is now easy to calculate the number of pairs corresponding to a given value of X and from it the probability $\Pr(X = k)$. For example, there are three pairs adding to four, so

$$\Pr(X = 4) = \tfrac{3}{36} = \tfrac{1}{12}.$$

Performing this calculation for all k from 2 to 12 gives the following probability distribution and histogram:

k	$\Pr(X = k)$	k	$\Pr(X = k)$
2	$\tfrac{1}{36}$	8	$\tfrac{5}{36}$
3	$\tfrac{1}{18}$	9	$\tfrac{1}{9}$
4	$\tfrac{1}{12}$	10	$\tfrac{1}{12}$
5	$\tfrac{1}{9}$	11	$\tfrac{1}{18}$
6	$\tfrac{5}{36}$	12	$\tfrac{1}{36}$
7	$\tfrac{1}{6}$		

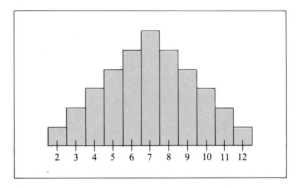

FIGURE 9

PRACTICE PROBLEMS 1

1. In a certain carnival game a wheel is divided into five equal parts, of which two are red and three are white. The player spins the wheel until the marker lands on "red" or until three spins have occurred. The number of spins is observed. Determine the probability distribution for this experiment.

2. Refer to the carnival game of Problem 1. Suppose that the player pays $1 to play this game and receives 50 cents for each spin. Determine the probability distribution for the experiment of playing the game and observing the player's earnings.

EXERCISES 1

1. Table 5(a) gives the frequency distribution for the final grades in a course. (Here $A = 4$, $B = 3$, $C = 2$, $D = 1$, $F = 0$). Determine the relative frequency distribution associated with these data and draw the associated histogram.

TABLE 5

Grade	Number of occurrences
0	2
1	3
2	10
3	6
4	4

(a)

Number of cars waiting	Number of occurrences
0	0
1	9
2	21
3	15
4	12
5	3

(b)

2. The number of cars waiting to be served at a gas station was counted at the beginning of every minute during the morning rush hour. The frequency distribution is given in Table 5(b). Determine the relative frequency distribution associated with these data and draw the histogram.

3. The telephone company counted the number of people dialing the weather each minute on a rainy morning from 5 A.M. to 6 A.M. The frequency distribution is given in Table 6(a). Determine the relative frequency distribution associated with these data.

TABLE 6

Number of calls during minute	Number of occurrences
20	3
21	3
22	0
23	6
24	18
25	12
26	0
27	9
28	6
29	3

(a)

Number produced during hour	Number of occurrences
50	2
51	0
52	4
53	6
54	14
55	8
56	4
57	0
58	0
59	2

(b)

4. A production manager counted the number of items produced each hour during a 40-hour workweek. The frequency distribution is given in Table 6(b). Determine the relative frequency distribution associated with these data.

5. A fair coin is tossed three times and the number of heads is observed. Determine the probability distribution for this experiment and draw its histogram.

6. An urn contains three red balls and four white balls. A sample of three balls is selected at random and the number of red balls observed. Determine the probability distribution for this experiment and draw its histogram.

7. An archer can hit the bull's-eye of the target with probability $\frac{1}{3}$. She shoots until she hits the bull's-eye or until four shots have been taken. The number of shots is observed. Determine the probability distribution for this experiment.

8. A die is rolled and the number on the top face is observed. Determine the probability distribution for this experiment and draw its histogram.

9. In a certain carnival game the player selects two balls at random from an urn containing two red balls and four white balls. The player receives $5 if he draws two red balls and $1 if he draws one red ball. He loses $1 if no red balls are in the sample. Determine the probability distribution for the experiment of playing the game and observing the player's earnings.

10. In a certain carnival game a player pays a dollar and then tosses a fair coin until either a "head" occurs or he has tossed the coin four times. He receives 50 cents for each toss. Determine the probability distribution for the experiment of playing the game and observing the player's earnings.

11. Figure 9(a) is the histogram for a probability distribution. What is the probability that the outcome is between 5 and 7 inclusive?

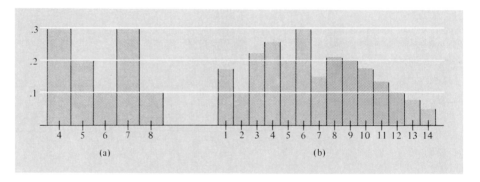

FIGURE 9

12. Figure 9(b) is the histogram for a probability distribution. To what event do the shaded rectangles correspond?

Let the random variables X and Y have the following probability distributions:

k	$Pr(X = k)$	k	$Pr(Y = k)$
0	.1	5	.3
1	.2	10	.4
2	.3	15	.1
3	.2	20	.1
4	.2	25	.1

Determine the probability distributions of the following random variables:

13. X^2 14. Y^2 15. $X - 1$ 16. $Y - 15$

17. $\frac{1}{5} Y$ 18. $2X^2$ 19. $(X + 1)^2$ 20. $(\frac{1}{5} Y + 1)^2$

SOLUTIONS TO PRACTICE PROBLEMS 1

1. Since the outcomes are the number of spins, there are three possible outcomes: one, two, and three spins. The probabilities for each of these outcomes can be computed from a tree diagram. For instance, the outcome two (spins) occurs if the first spin lands on white and the second spin on red. The probability of this outcome is $\frac{3}{5} \cdot \frac{2}{5} = \frac{6}{25}$.

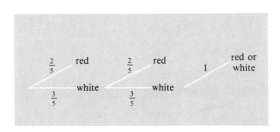

Outcome	Probability
1	$\frac{2}{5}$
2	$\frac{6}{25}$
3	$\frac{9}{25}$

2. The same game is being played as in Problem 1, except that now the outcome we are concentrating on is the player's financial situation at the end of the game. The player's earnings depend on the number of spins as follows: one spin results in $-\$.50$ earnings (i.e., a loss of 50 cents), two spins result in \$0 earnings (i.e., breaking even) and 3 spins result in \$.50 earnings (i.e., the player ends up ahead by 50 cents). The probabilities for these three situations are the same as before.

Earnings	Probability
$-\$.50$	$\frac{2}{5}$
0	$\frac{6}{25}$
\$.50	$\frac{9}{25}$

7.2. The Mean

It is important at this point to recognize the difference between a *population* and a *sample*. A population is a set of all elements about which information is desired. A sample is a subset of a population that is analyzed in an attempt to estimate certain properties of the entire population. For instance, suppose that we are interested in finding out some characteristics of the automobile industry but do not have access to the entire population of dealerships in the United States. We must choose a random and representative sample of these dealerships and concentrate on gathering information from it.

A numerical descriptive measurement made on a sample is called a *statistic*. Such a measurement made on a population is called a *parameter* of the population. Usually, since we cannot have access to entire populations, we rely on our experimental results to obtain statistics, and we attempt to use the statistics to estimate the parameters of the population. One of the measurements we all have used is the arithmetic average. For instance, we could choose a representative random sample of $n = 200$ car dealerships across the country, determine the average weekly sales for these dealerships, and then use this value to estimate the average weekly sales of all car dealerships in the United States.

It is a familiar notion to find the *average* (also called the *mean*) of a set of numbers. For example, to obtain the average of 11, 17, 18, and 10, we add these numbers and divide by 4:

$$[\text{average}] = \frac{11 + 17 + 18 + 10}{4} = 14$$

If we have gathered a sample of n numbers $x_1, x_2, \ldots, x_n$, the *sample mean* is

$$\bar{x} = \frac{x_1 + x_2 + \cdots + x_n}{n}. \tag{1}$$

EXAMPLE 1   Compute the sample mean of the weekly sales of dealership A of the preceding section.

Solution   Recall that the weekly sales of dealership A are given in the table:

Weekly sales	Number of occurrences
5	2
6	2
7	13
8	20
9	10
10	4
11	1

Thus we may form the sample mean of the weekly sales figures as follows:

$$\tfrac{1}{52}\big[(5+5)+(6+6)+\underbrace{(7+\cdots+7)}_{13\ \text{times}}+\underbrace{(8+\cdots+8)}_{20\ \text{times}}+\underbrace{(9+\cdots+9)}_{10\ \text{times}}+\underbrace{(10+\cdots+10)}_{4\ \text{times}}+11\big]$$

$$= \frac{5 \cdot 2 + 6 \cdot 2 + 7 \cdot 13 + 8 \cdot 20 + 9 \cdot 10 + 10 \cdot 4 + 11}{52}$$

$$= 5 \cdot \tfrac{2}{52} + 6 \cdot \tfrac{2}{52} + 7 \cdot \tfrac{13}{52} + 8 \cdot \tfrac{20}{52} + 9 \cdot \tfrac{10}{52} + 10 \cdot \tfrac{4}{52} + 11 \cdot \tfrac{1}{52} \tag{2}$$

$$\approx 7.96.$$

Thus the sample mean of the weekly sales is approximately 7.96 cars.

Note: We considered the data from dealership A to be a sample since it was only one year's data and we are interested in making comparisons and using them to predict the future sales of the two dealerships.

Let us reexamine the calculation above. The sample mean is given by the expression (2). Another way of writing this expression is:

[mean weekly sales] $=$ 5 · [rel. freq. that weekly sales $=$ 5]

$+$ 6 · [rel. freq. that weekly sales $=$ 6]

$+$ 7 · [rel. freq. that weekly sales $=$ 7]

$+$ 8 · [rel. freq. that weekly sales $=$ 8]

$+$ 9 · [rel. freq. that weekly sales $=$ 9]

$+$ 10 · [rel. freq. that weekly sales $=$ 10]

$+$ 11 · [rel. freq. that weekly sales $=$ 11].

That is, to compute the mean weekly sales we need only add up the products [number of sales] · [relative frequency of occurrence] over all possible sales figures. If the data for any sample are displayed in a frequency table or a relative frequency table, the sample mean can be calculated in a similar fashion. Since the calculation of $\bar{x}$ depends on sample values, $\bar{x}$ is a *statistic*.

Sample Mean   Suppose that an experiment has as outcomes the numbers $x_1, x_2, \ldots, x_r$. Suppose the frequency of x_1 is f_1, the frequency of x_2 is f_2, and so forth, and that

$$f_1 + f_2 + \cdots + f_r = n.$$

Then

$$\bar{x} = \frac{x_1 f_1 + x_2 f_2 + \cdots + x_r f_r}{n},$$

or

$$\bar{x} = x_1 \left(\frac{f_1}{n} \right) + x_2 \left(\frac{f_2}{n} \right) + \cdots + x_r \left(\frac{f_r}{n} \right).$$

Assume the population size is N and that $x_1, x_2, \ldots, x_N$ are the population values. The *population mean*, denoted by the Greek letter μ (mu), is given by the formula

$$\mu = \frac{x_1 + x_2 + \cdots + x_N}{N}$$

If the data have been grouped into a frequency or relative frequency table, formulas analogous to the ones for samples are used:

$$\mu = x_1 \left(\frac{f_1}{N} \right) + x_2 \left(\frac{f_2}{N} \right) + \cdots + x_r \left(\frac{f_r}{N} \right).$$

To help distinguish the parameter from the statistic, all parameters are denoted by Greek letters and all statistics by English letters. We use the common convention of denoting a sample size by lowercase n and a population size by uppercase N.

EXAMPLE 2 An ecologist observes the life expectancy of a certain species of deer held in captivity. Based on a population of 1000 deer, he observes the following data:

Age at death (years)	Number observed
1	0
2	60
3	180
4	250
5	200
6	120
7	50
8	120
9	20

What is the mean life expectancy of this population of deer?

Solution We convert the given data into a relative frequency distribution by replacing observed frequencies by relative frequencies [= (observed frequency)/1000].

Age at death	Relative frequency
1	0
2	.06
3	.18
4	.25
5	.20
6	.12
7	.05
8	.12
9	.02

The mean of these data is

$$\mu = 1 \cdot 0 + 2 \cdot (.06) + 3 \cdot (.18) + 4 \cdot (.25) + 5 \cdot (.20) + 6 \cdot (.12)$$
$$+ 7 \cdot (.05) + 8 \cdot (.12) + 9 \cdot (.02) = 4.87.$$

So the mean life expectancy of this population of deer is 4.87 years.

EXAMPLE 3 Which car dealership should Mr. Jones buy if he wants the one which will, on the average, sell the most cars?

Solution   We have seen in Example 1 that the mean of the sample data for dealership A is $\bar{x}_A = 7.96$. On the other hand, associated to the data for dealership B—namely,

Weekly sales	Relative frequency
5	$\frac{20}{104}$
6	0
7	0
8	$\frac{10}{104}$
9	$\frac{12}{104}$
10	$\frac{50}{104}$
11	$\frac{12}{104}$

we find the sample mean:

$$\bar{x}_B = 5 \cdot \left(\tfrac{20}{104}\right) + 6 \cdot 0 + 7 \cdot 0 + 8 \cdot \left(\tfrac{10}{104}\right) + 9 \cdot \left(\tfrac{12}{104}\right) + 10 \cdot \left(\tfrac{50}{104}\right) + 11 \cdot \left(\tfrac{12}{104}\right)$$
$$= 8.85.$$

Thus, the average sales of dealership A are 7.96 cars per week, whereas those of dealership B are 8.85 cars per week. If we make the assumption that past sales history predicts future sales, Mr. Jones should buy dealership B.

Expected Value   The sample and population mean have analogs in the theoretical setting of random variables. Suppose that X is a random variable with the following probability distribution:

x_i	$\Pr(X = x_i)$
x_1	p_1
x_2	p_2
$\vdots$	$\vdots$
x_N	p_N

Then the values of X (namely, $x_1, x_2, \ldots, x_N$) are the possible outcomes of an experiment. The *expected value* of X, denoted $E(X)$, is defined as follows:

> **The Expected Value of the Random Variable X**
>
> $$E(X) = x_1 p_1 + x_2 p_2 + \cdots + x_N p_N.$$

Since the value of p_i represents the theoretical relative frequency of the outcome x_i, the formula for $E(X)$ is similar to the formula for the mean of a relative frequency distribution. Thus, the expected value of the random variable X is also called the *mean* of the probability distribution of X and may be denoted either by $E(X)$ or μ_X. Frequently $E(X)$ is used interchangeably with the Greek letter μ when the context is clear.

TABLE 1 X = Number of Heads in 5
Tosses of a Fair Coin

k	$\Pr(X = k)$	$k \cdot \Pr(X = k)$
0	$\frac{1}{32}$	0
1	$\frac{5}{32}$	$\frac{5}{32}$
2	$\frac{10}{32}$	$\frac{20}{32}$
3	$\frac{10}{32}$	$\frac{30}{32}$
4	$\frac{5}{32}$	$\frac{20}{32}$
5	$\frac{1}{32}$	$\frac{5}{32}$
Totals	1	$\mu = \frac{80}{32} = 2.5$

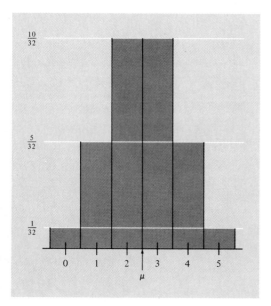

The expected value of a random variable is the center of the probability distribution in the sense that it is the balance point of the histogram. For example, let X = the number of heads in 5 tosses of a fair coin. The probability distribution appears in Table 1 and the histogram in Fig. 1. We can calculate the mean μ of X:

$$\mu_X = 0(\tfrac{1}{32}) + 1(\tfrac{5}{32}) + 2(\tfrac{10}{32}) + 3(\tfrac{10}{32}) + 4(\tfrac{5}{32}) + 5(\tfrac{1}{32})$$

$$= \tfrac{80}{32} = 2.5.$$

The mean is shown at the bottom of the histogram (Fig. 1). In contrast, we note that the sample mean, $\bar{x}$, for the coin-tossing experiment tabulated in Section 7.1 was 2.67 (see Table 2). We rarely find that the sample mean $\bar{x}$ is exactly the theoretical value μ_X.

FIGURE 1

TABLE 2

# Heads x_i	Frequency f_i	Relative frequency (f_i/n)	$x_i \cdot (f_i/n)$
0	3	.03	0
1	14	.16	.16
2	23	.26	.54
3	27	.30	.90
4	17	.18	.72
5	6	.07	.35
Totals	$n = 90$	1	$2.67 = \bar{x}$

EXAMPLE 4   Let the random variable X denote the sum of the faces appearing after throwing two dice. Determine $E(X)$.

Solution   We determined the probability distribution of X in Example 3 of Section 1:

k	$\Pr(X = k)$	k	$\Pr(X = k)$
2	$\frac{1}{36}$	8	$\frac{5}{36}$
3	$\frac{1}{18}$	9	$\frac{1}{9}$
4	$\frac{1}{12}$	10	$\frac{1}{12}$
5	$\frac{1}{9}$	11	$\frac{1}{18}$
6	$\frac{5}{36}$	12	$\frac{1}{36}$
7	$\frac{1}{6}$		

Therefore,

$$E(X) = 2 \cdot \tfrac{1}{36} + 3 \cdot \tfrac{1}{18} + 4 \cdot \tfrac{1}{12} + 5 \cdot \tfrac{1}{9} + 6 \cdot \tfrac{5}{36} + 7 \cdot \tfrac{1}{6}$$
$$+ 8 \cdot \tfrac{5}{36} + 9 \cdot \tfrac{1}{9} + 10 \cdot \tfrac{1}{12} + 11 \cdot \tfrac{1}{18} + 12 \cdot \tfrac{1}{36} = 7.$$

Clearly, 7 is the balance point of the histogram shown in Section 1.

The expected value of a random variable may be used to analyze games of chance, as the next two examples show.

EXAMPLE 5   Two people play a dice game. A single die is thrown. If the outcome is 1 or 2, then A pays B \$2. If the outcome is 3, 4, 5, or 6, then B pays A \$4. What are the long-run expected winnings for A?

Solution   Let X be the random variable representing the payoff to A. Then X assumes the possible values -2 and 4. Moreover, since the probability of 1 or 2 on the die is $\frac{1}{3}$, we have

$$\Pr(X = -2) = \tfrac{1}{3}.$$

Similarly,

$$\Pr(X = 4) = \tfrac{2}{3}.$$

Therefore,

$$E(X) = (-2) \cdot \tfrac{1}{3} + 4 \cdot \tfrac{2}{3} = 2.$$

In other words, the expected payoff to A is \$2 per play. If the game is repeated a large number of times, then on the average A should profit \$2 per play. For example, in 1000 games, we expect A will profit \$2000.

In evaluating a game of chance we use the expected value of the winnings to determine how fair the game is. The expected value of a completely fair game is zero. Let us compute the expected value of the winnings for two variations of the game roulette. American and European roulette games differ in both the nature of the wheel and the rules for playing.

American roulette wheels have 38 numbers (1 through 36 plus 0 and 00), of which 18 are red, 18 are black, and 2 are green. Many different types of bets are possible. We shall consider the "red" bet. When you bet \$1 on red, you win \$1 if a red number appears and lose \$1 otherwise.

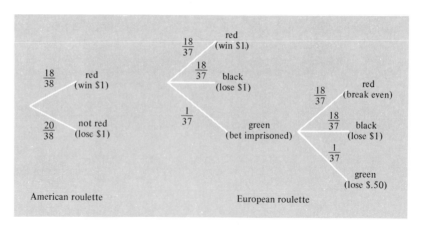

FIGURE 2

European roulette wheels have 37 numbers (1 through 36 plus 0). The rules of European roulette differ from the American rules. One variation is as follows: When you bet $1 on "red," then you win $1 if the ball lands on a red number and lose $1 if the ball lands on a black number. However, if the ball lands on the green number (0), then your bet stays on the table (the bet is said to be "imprisoned") and the payoff is determined by the result of the next spin. If a red number appears, you receive your $1 bet back, and if a black number appears, you lose your $1 bet. However, if the green number (0) appears, you get back half of your bet, $.50. The tree diagrams for American and European roulette are given in Fig. 2.

EXAMPLE 6   (a) Set up the probability distribution tables for the earnings in American roulette and European roulette for the $1 bet on red.

(b) Compute the expected values for the probability distributions in part (a).

Solution   (a) For American roulette there are only two possibilities: earnings of $1 or of $-\$1$. These occur with probabilities $\frac{18}{38}$ and $\frac{20}{38}$, respectively.

For European roulette the possible earnings are $1, $0, $-\$.50$, or $-\$1$. There are two ways in which to lose $1, one with probability $\frac{18}{37}$ and the other with probability $\frac{1}{37} \cdot \frac{18}{37} = \frac{18}{1369}$. Therefore, $\Pr(\text{lose } \$1) = \frac{18}{37} + \frac{18}{1369} = \frac{684}{1369}$. These probability distributions are tabulated in Table 3.

(b) American roulette:

$$\mu = 1 \cdot \tfrac{18}{38} + (-1) \cdot \tfrac{20}{38} = -\tfrac{2}{38} \approx -.0526.$$

European roulette:

$$\mu = 1 \cdot \tfrac{18}{37} + 0 \cdot \tfrac{18}{1369} + (-\tfrac{1}{2})\tfrac{1}{1369} + (-1)\tfrac{684}{1369}$$

$$= -\tfrac{1}{74} \approx -.0135.$$

The secrets of Nicholas Dandolas, one of the famous gamblers of the twentieth century, are revealed by Ted Thackrey, Jr., in *Gambling Secrets of Nick the Greek* (Rand McNally, 1968). The chapter entitled "Roulette" is subtitled "For Europeans Only." Looking at the probabilities of winning does not reveal any

TABLE 3

American roulette		European roulette	
Earnings	Probability	Earnings	Probability
1	$\frac{18}{38}$	1	$\frac{18}{37}$
-1	$\frac{20}{38}$	0	$\frac{18}{1369}$
		$-\frac{1}{2}$	$\frac{1}{1369}$
		-1	$\frac{684}{1369}$

significant advantage of European roulette over American roulette: $\frac{18}{37}$ is not much bigger than $\frac{18}{38}$. Also, the chance in European roulette to break even is very small. The real difference between the two games is revealed by the expected values. Someone playing American roulette will lose, on the average, about $5\frac{1}{4}$ cents per \$1 bet, whereas for European roulette the average loss is about $1\frac{1}{3}$ cents. In both cases you expect to lose money in the long run, but in American roulette you lose nearly four times as much.

In summary, there are three things to which the mean applies: samples, populations, and probability distributions.

PRACTICE PROBLEMS 2

1. A 74-year-old man pays \$100 for a 1-year life insurance policy, which pays \$2000 in the event that he dies during the next year. According to life insurance tables, the probability of a 74-year-old man's living one additional year is .95. Write down the probability distribution for the possible financial outcome and determine its expected value.

2. According to life insurance tables, the probability that a 74-year-old man will live an additional 5 years is .7. How much should a 74-year-old be willing to pay for a policy that pays \$2000 in the event of death any time within the next 5 years?

EXERCISES 2

1. Find the expected value for the probability distribution in Table 4(a).

2. Find the expected value for the probability distribution in Table 4(b).

TABLE 4

Values	Probability	Values	Probability
0	.15	-1	.1
1	.2	$-\frac{1}{2}$	.4
2	.1	0	.25
3	.25	$\frac{1}{2}$	.2
4	.3	1	.05
(a)		(b)	

3. A college student received the following course grades for 10 (three-credit) courses during his freshman year: 4, 4, 4, 3, 3, 3, 3, 2, 2, 1.

   (a) Find his grade point average by adding the grades and dividing by 10.

   (b) Write down the relative frequency table.

   (c) Find the mean of the relative frequency distribution in part (b).

4. An Olympic gymnast received the following scores from six judges: 9.8, 9.8, 9.4, 9.2, 9.2, 9.0.

   (a) Find the average score by adding the scores and dividing by 6.

   (b) Write down the relative frequency table.

   (c) Find the mean of the relative frequency distribution in part (b).

5. Table 5 gives the relative frequency of the number of cavities for two groups of children trying different brands of toothpaste. Calculate the sample means to determine which group had the fewer cavities.

TABLE 5

Group A		Group B	
Number of cavities	Relative frequency	Number of cavities	Relative frequency
0	.3	0	.2
1	.3	1	.3
2	.2	2	.3
3	.1	3	.1
4	.1	4	.1
5	.1	5	.0

6. Table 6 gives the possible returns of two different investments and their probabilities. Calculate the means of the probability distributions to determine which investment has the greater expected return.

TABLE 6

Investment A		Investment B	
Return	Probability	Return	Probability
$1000	.2	−$3000	.1
$2000	.5	0	.3
$3000	.3	$4000	.6

7. In American roulette, a bettor may place a $1 bet on any one of the 38 numbers on the roulette wheel. He wins $35 (plus the return of his bet) if the ball lands on his number; otherwise, he loses his bet. Write down the probability distribution for the earnings from this type of bet and find the expected value.

8. In American roulette, a dollar may be bet on a pair of numbers. The expected earnings for this type of bet is $-\$\frac{1}{19}$. How much money does the bettor receive if the ball lands on one of the two numbers?

9. In a carnival game, the player selects balls one at a time, without replacement, from an urn containing two red and four white balls. The game proceeds until a red ball is drawn. The player pays $1 to play the game and receives $\$\frac{1}{2}$ for each ball drawn. Write down the probability distribution for the player's earnings and find its expected value.

10. In a carnival game, the player selects two coins from a bag containing two silver dollars and six slugs. Write down the probability distribution for the winnings and determine how much the player would have to pay so that he would break even, on average, over many repetitions of the game.

11. Using life insurance tables, a retired man determines that the probability of living 5 more years is .9. He decides to take out a life insurance policy which will pay $10,000 in the event that he dies during the next 5 years. How much should he be willing to pay for this policy? (Do not take account of interest rates or inflation.)

12. Using life insurance tables, a retired couple determines that the probability of living 5 more years is .9 for the man and .95 for the woman. They decide to take out a life insurance policy which will pay $10,000 if either one dies during the next 5 years and $15,000 if both die during that time. How much should they be willing to pay for this policy? (Assume that their life spans are independent events.)

13. A pair of dice is tossed and the larger of the two numbers showing is recorded. Find the expected value of this experiment.

SOLUTIONS TO PRACTICE PROBLEMS 2

1. There are two possibilities. If the man lives until the end of the year, he loses $100. If he dies during the year, his estate gains by $1900 (the $2000 settlement minus the $100 premium).

Outcome	Probability
$-\$100$	.95
$\$1900$	.05

$$\mu = (-100)(.95) + (1900)(.05) = 0.$$

(Thus, if the insurance company insures a large number of people, it should break even. Its profits will result from the interest that it earns on the money being held.)

2. Let x denote the cost of the policy. The probability distribution is given below.

Outcome	Probability
$-x$	.7
$2000 - x$	.3

$$\mu = (-x)(.7) + (2000 - x)(.3)$$

$$= -.7x + 600 - .3x$$

$$= 600 - x.$$

The expected value will be zero if $x = 600$. Therefore, the man should be willing to pay $600 for his policy.

7.3. The Variance and Standard Deviation

In the preceding section we introduced three analogous concepts: the mean of a sample ($\bar{x}$), the mean of a population (μ), and the mean or expected value of a random variable ($E(X)$). The mean is probably the single most important number that can be used to describe a sample, a population, or a probability distribution of a random variable. The next most important number is the variance.

Roughly speaking, the variance measures the dispersal or spread of a distribution about its mean. The more closely concentrated the distribution about its mean, the smaller the variance; the more spread out, the larger the variance. Thus, for example, the probability distribution whose histogram is drawn in Fig. 1(a) has a smaller variance than that in Fig. 1(b).

Let us now define the variance of a probability distribution of a random variable. Suppose X is a random variable with values $x_1, x_2, \ldots, x_N$ and respective probabilities $p_1, p_2, \ldots, p_N$. Suppose that the mean is μ. Then the deviations of the various outcomes from the mean are given by the N differences

$$x_1 - \mu, x_2 - \mu, \ldots, x_N - \mu.$$

Since we want to give weight to the various deviations according to their likelihood of occurrence, it is tempting to multiply each deviation by its probability of occurrence. However, this will not lead to a very satisfactory measure of deviation from the mean. This is because some of the differences will be positive and others negative. In the process of addition, deviations from the mean (both positive and negative deviations) will combine to yield a zero total deviation. To correct this, we consider instead the squares of the differences:

$$(x_1 - \mu)^2, (x_2 - \mu)^2, \ldots, (x_N - \mu)^2,$$

FIGURE 1

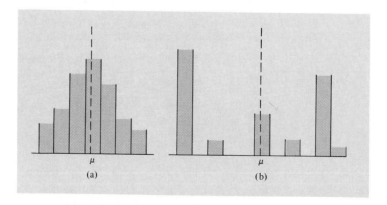

(a)   (b)

which are all ≥ 0. To obtain a measure of deviation from the mean, we multiply each of these expressions by the probability of the corresponding outcome. The number thus obtained is called the *variance of the probability distribution* (or of the associated random variable). That is, the variance of a probability distribution is given by the formula

$$[\text{variance}] = (x_1 - \mu)^2 p_1 + (x_2 - \mu)^2 p_2 + \cdots + (x_N - \mu)^2 p_N. \qquad (1)$$

EXAMPLE 1 Compute the variance of the following probability distribution:

Outcome	Probability
0	.1
1	.3
2	.5
3	.1

Solution The mean is given by

$$\mu = 0 \cdot (.1) + 1 \cdot (.3) + 2 \cdot (.5) + 3 \cdot (.1) = 1.6.$$

In the notation of random variables, the calculation of the variance may be summarized as follows:

k	$\Pr(X = k)$	$k - \mu$	$(k - \mu)^2$	$(k - \mu)^2 \Pr(X = k)$
0	.1	$0 - 1.6 = -1.6$	2.56	.256
1	.3	$1 - 1.6 = -.6$	.36	.108
2	.5	$2 - 1.6 = .4$	.16	.080
3	.1	$3 - 1.6 = 1.4$	1.96	.196

$$[\text{variance}] = .256 + .108 + .080 + .196 = .640.$$

Actually, a much more commonly used measure of dispersal about the mean is the *standard deviation*, which is just the square root of the variance:

$$[\text{standard deviation}] = \sqrt{[\text{variance}]}.$$

The most commonly used notation for standard deviation is the Greek letter σ (sigma). Thus, for example, for the probability distribution of Example 1 we have

$$\sigma = \sqrt{[\text{variance}]} = \sqrt{.64} = .8.$$

We denote the variance by σ^2 (sigma squared). The reason for using the standard deviation as opposed to the variance is that the former is expressed in the same units of measurement as X, whereas the latter is not.

In a similar way, we can define the variance and standard deviation for a collection of data—either for an entire population or for a sample drawn from a population. If we have collected data for an entire population under study, with values $x_1, x_2, \ldots, x_N$, then the variance can be found by first finding the mean, μ, and then finding σ^2:

$$\sigma^2 = \frac{1}{N}[(x_1 - \mu)^2 + (x_2 - \mu)^2 + \cdots + (x_N - \mu)^2].$$

If the data have been grouped into a frequency table or relative frequency table, the appropriate formulas would be:

$$\sigma^2 = \frac{1}{N}[(x_1 - \mu)^2(f_1) + (x_2 - \mu)^2(f_2) + \cdots + (x_r - \mu)^2(f_r)]$$

or

$$\sigma^2 = (x_1 - \mu)^2\left(\frac{f_1}{N}\right) + (x_2 - \mu)^2\left(\frac{f_2}{N}\right) + \cdots + (x_r - \mu)^2\left(\frac{f_r}{N}\right),$$

where the value x_1 occurs with frequency f_1, the value x_2 occurs with frequency f_2, and so forth. Recall that N denotes the population size.

EXAMPLE 2   Compute the variance and the standard deviation for the population of scores on a five-question quiz as tabulated below:

Scores	Frequency
0	4
1	9
2	6
3	14
4	18
5	9
Total	60

Solution   We first find μ.

$$\mu = \tfrac{1}{60}[0(4) + 1(9) + 2(6) + 3(14) + 4(18) + 5(9)] = \tfrac{180}{60} = 3.$$

We find σ^2 by subtracting 3 from each of the test scores, squaring the differences, weighting each with its frequency, and dividing the resulting sum by $N = 60$. The computation is shown in the chart below:

x_i	f_i	$(x_i - \mu)$	$(x_i - \mu)^2$	$(x_i - \mu)^2(f_i)$
0	4	-3	9	36
1	9	-2	4	36
2	6	-1	1	6
3	14	0	0	0
4	18	1	1	18
5	9	2	4	36
Totals	60			132

Therefore, $\sigma^2 = \frac{132}{60} = 2.2$. The standard deviation, which is found by taking the square root of the variance, is $\sigma \approx 1.48$.

As we discussed in the previous section, we sometimes have only sample values at our disposal. If so, we must use sample statistics to estimate the population parameters. For example, we can use the sample mean $\bar{x}$ as an estimate of the population mean μ, and the sample variance as an estimate of the population variance, σ^2. In fact, if samples of size n were chosen repeatedly from a population and the sample mean were computed for each sample, then the average of these means should be close to the value of the population mean, μ. This is a desirable property for any statistic used to estimate a parameter—the averages of the statistic get arbitrarily close to the actual value of the parameter as the number of samples increases. Such an estimate is said to be *unbiased*. Thus $\bar{x}$ is an unbiased estimate of μ.

The situation with the variance is a little trickier. In order to have an unbiased estimate of the population variance σ^2, we must define the *sample variance*, s^2, in a slightly peculiar way. Suppose that $x_1, x_2, \ldots, x_n$ are the n sample values. Then

$$s^2 = \frac{1}{n-1} \left[(x_1 - \bar{x})^2 + (x_2 - \bar{x})^2 + \cdots + (x_n - \bar{x})^2 \right]$$

and the sample standard deviation is $s = \sqrt{s^2}$. This is the usual definition and the way in which most statistical calculators do the computation of the sample variance. Note that the divisor is one less than the sample size. With this definition, s^2 is an unbiased estimate of σ^2. The formula for the sample standard deviation when the data are presented by a frequency table is analogous. For example

$$s^2 = \frac{1}{n-1} \left[(x_1 - \bar{x})^2 (f_1) + (x_2 - \bar{x})^2 (f_2) + \cdots + (x_r - \bar{x})^2 (f_r) \right]$$

> The sample mean, $\bar{x}$, is an unbiased estimate of the population mean μ. The sample variance, s^2, is an unbiased estimate of the population variance σ^2.

EXAMPLE 3 Compute the sample standard deviations for the frequency distributions of sales in car dealerships A and B.

Solution The frequency distribution for dealership A is given by

Weekly Sales	Frequency
5	2
6	2
7	13
8	20
9	10
10	4
11	1
Total	$n = 52$

In Example 1 of Section 7.2, we found the sample mean of the weekly sales to be $\bar{x}_A = 7.96$. Recall that we are treating the data collected from these dealerships as samples (one year's data from dealership A and two year's data from dealership B). Therefore, the sample variance for dealership A is given by

$$s_A^2 = \frac{1}{51}[(5 - 7.96)^2 \cdot 2 + (6 - 7.96)^2 \cdot 2 + (7 - 7.96)^2 \cdot 13$$
$$+ (8 - 7.96)^2 \cdot 20 + (9 - 7.96)^2 \cdot 10 + (10 - 7.96)^2 \cdot 4$$
$$+ (11 - 7.96)^2 \cdot 1] = 1.45.$$

The sample standard deviation, s_A, for dealership A is given by

$$s_A = \sqrt{s_A^2} = \sqrt{1.45} \approx 1.20 \text{ cars.}$$

In a similar way, we find that the sample standard deviation of dealership B is $s_B \approx 2.18$ cars. Since s_A is smaller than s_B, dealership B exhibited greater variation than dealership A during the time the sales were observed. On the average, dealership B had higher weekly sales, but those of dealership A showed greater consistency. The sample statistics might help Mr. Jones decide which dealership to buy. He will have to decide if consistency is more important than the size of long-run average sales per week.

There is an alternate formula for σ^2 which can simplify its calculation:

$$\sigma^2 = E(X^2) - \mu^2$$

EXAMPLE 4   Use the alternate formula for the variance to find σ^2 for X, the number of heads in 5 tosses of a coin.

Solution   We tabulate the essential values:

k	$\Pr(X = k)$	k^2	$k \Pr(X = k)$	$k^2 \Pr(X = k)$
0	$\frac{1}{32}$	0	0	0
1	$\frac{5}{32}$	1	$\frac{5}{32}$	$\frac{5}{32}$
2	$\frac{10}{32}$	4	$\frac{20}{32}$	$\frac{40}{32}$
3	$\frac{10}{32}$	9	$\frac{30}{32}$	$\frac{90}{32}$
4	$\frac{5}{32}$	16	$\frac{20}{32}$	$\frac{80}{32}$
5	$\frac{1}{32}$	25	$\frac{5}{32}$	$\frac{25}{32}$
Totals			$\frac{80}{32}$	$\frac{240}{32}$

From the chart above we can see that $E(X^2) = \frac{240}{32} = \frac{15}{2}$, and $\mu = \frac{80}{32} = \frac{5}{2}$. Thus,

$$\sigma^2 = E(X^2) - \mu^2$$
$$= \frac{15}{2} - \left(\frac{5}{2}\right)^2 = \frac{15}{2} - \frac{25}{4} = \frac{5}{4}.$$

One of the big advantages of this alternate formula is that μ and σ^2 can be calculated at the same time; this means that we need only make one pass through the probability distribution. In addition, if the mean is an unwieldly number, we are spared the tedious task of subtracting it from the values of the variable, which sometimes produces even more unwieldly numbers that must be squared and summed.

The variance and standard deviation are used in many sophisticated statistical analyses, which are beyond the scope of this book. For example, we can use the value of the sample mean to estimate the population mean. The sample standard deviation helps us to determine the degree of accuracy of our estimate. If the sample standard deviation is small, indicating that the population is not widely dispersed about its mean, the estimated value of μ is likely to be close to the actual value of μ.

What does the standard deviation tell us about the dispersal of the data about the mean? *Chebychev's inequality* helps us to see that the larger the standard deviation, the more likely it is that we find extreme values in the data. The probability that an outcome falls more than c units away from the mean is at most σ^2/c^2.

Chebychev Inequality Suppose that a probability distribution with numerical outcomes has expected value μ and standard deviation σ. Then the probability that a randomly chosen outcome lies between $\mu - c$ and $\mu + c$ is at least $1 - (\sigma^2/c^2)$.

A verification of the Chebychev inequality can be found in most elementary statistics texts.

EXAMPLE 5 Suppose that a probability distribution has mean 5 and standard deviation 1. Use the Chebychev inequality to estimate the probability that an outcome lies between 3 and 7.

Solution Here $\mu = 5$, $\sigma = 1$. Since we wish to estimate the probability of an outcome lying between 3 and 7, we set $\mu - c = 3$ and $\mu + c = 7$. Thus, $c = 2$. Then by the Chebychev inequality the desired probability is at least $1 - (\sigma^2/c^2) = 1 - \frac{1}{4} = .75$. That is, if the experiment is repeated a large number of times, we expect at least 75% of the outcomes to be between 3 and 7. Also, at most 25% of the outcomes fall below 3 or above 7.

The Chebychev inequality has many practical applications, one of which is illustrated in the next example.

EXAMPLE 6 Apex Drug Supply Company sells bottles containing 100 capsules of penicillin. Owing to the bottling procedure, not every bottle contains exactly 100 capsules. Assume that the average number of capsules in a bottle is indeed 100 ($\mu = 100$) and the standard deviation is 2 ($\sigma = 2$). If the company ships 5000 bottles, estimate the number having between 95 and 105 capsules inclusive.

Solution By Chebychev's inequality, the proportion of bottles having between $100 - 5$ and $100 + 5$ capsules should be at least $1 - (2^2/5^2) = \frac{21}{25} = .84$. That is, we expect at least 84% of the 5000 bottles, or 4200 bottles, to have a number of capsules in the desired range.

Note that the estimate provided by Chebychev's inequality is a crude one. In more advanced statistics books, you can find sharper estimates. Also, in the case of the normal distribution, we shall provide a much more precise way of estimating the probability of falling within c units of the mean, based on use of a table of areas under the normal curve. (See Section 5.)

PRACTICE PROBLEMS 3

1. (a) Compute the variance of the probability distribution in Table 1.

TABLE 1

Outcome	Probability
21	$\frac{1}{16}$
22	$\frac{1}{8}$
23	$\frac{5}{8}$
24	$\frac{1}{8}$
25	$\frac{1}{16}$

(b) Using Table 1, find the probability that the outcome is between 22 and 24 inclusive.

2. Refer to the probability distribution of Problem 1. Use the Chebychev inequality to approximate the probability that the outcome is between 22 and 24.

EXERCISES 3

1. Compute the variance of the probability distribution in Table 2(a).

TABLE 2

Outcome	Probability
70	.5
71	.2
72	.1
73	.2

(a)

Outcome	Probability
-1	$\frac{1}{8}$
$-\frac{1}{2}$	$\frac{3}{8}$
0	$\frac{1}{8}$
$\frac{1}{2}$	$\frac{1}{8}$
1	$\frac{2}{8}$

(b)

Compute the variance of the probability distribution in Table 2(b).

Determine by inspection which one of the probability distributions, A or B in Fig. 2, has the greater variance.

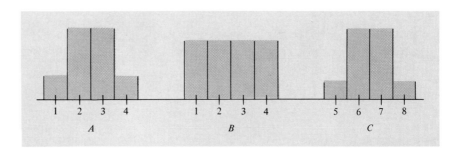

4. Determine by inspection which one of the probability distributions, B or C in Fig. 2 has the greater variance.

5. Table 3 gives the probability distribution for the possible returns from two different investments.

   (a) Compute the mean and variance for each investment.

   (b) Which investment has the higher expected return (i.e., mean)?

   (c) Which investment is the less risky (i.e., has lesser variance)?

   TABLE 3

Investment A		Investment B	
Return ($ millions)	Probability	Return ($ millions)	Probability
-10	$\frac{1}{5}$	0	.3
20	$\frac{3}{5}$	10	.4
25	$\frac{1}{5}$	30	.3

6. Two golfers recorded their scores for 20 nine-hole rounds of golf. Golfer A's scores were 39, 39, 40, 40, 40, 40, 40, 40, 41, 41, 41, 41, 41, 41, 41, 42, 43, 43, 43, 44. Golfer B's scores were 40, 40, 40, 41, 41, 41, 41, 42, 42, 42, 42, 42, 43, 43, 43, 43, 43, 43, 44, 44.

   (a) Compute the sample mean and variance of the scores of each golfer.

   (b) Who is the better golfer?

   (c) Who is the more consistent golfer?

7. Table 4 gives the relative frequency distribution for the weekly sales of two businesses.

   (a) Compute the population mean and variance for each business.

   TABLE 4

Business A			Business B	
Sales	Relative frequency		Sales	Relative frequency
100	.1		100	0
101	.2		101	.2
102	.3		102	0
103	0		103	.2
104	0		104	.1
105	.2		105	.2
106	.2		106	.3

   (b) Which business has the better sales record?

   (c) Which business has the more consistent sales record?

8. Student A received the following course grades during her freshman year: 4, 4, 4, 4, 3, 3, 2, 2, 2, 0. Student B received the following course grades during her freshman year: 4, 4, 4, 4, 4, 4, 3, 1, 1, 1.

   (a) Write down the relative frequency distribution tables for each student and compute the population means and variances.

   (b) Which student had the better grade point average?

   (c) Which student was the more consistent?

9. Suppose that a probability distribution has mean 35 and standard deviation 5. Use the Chebychev inequality to estimate the probability that an outcome will lie between

   (a) 25 and 45.   (b) 20 and 50.   (c) 29 and 41.

10. Suppose that a probability distribution has mean 8 and standard deviation .4. Use the Chebychev inequality to estimate the probability that an outcome will lie between

   (a) 6 and 10.   (b) 7.2 and 8.8.   (c) 7.5 and 8.5.

11. For certain types of fluorescent lights the number of hours a bulb will burn before requiring replacement has a mean of 3000 hours and a standard deviation of 250 hours. Suppose that 5000 such bulbs are installed in an office building. Estimate the number that will require replacement between 2000 and 4000 hours from the time of installation.

12. An electronics firm determines that the number of defective transistors in each batch averages 15 with standard deviation 10. Suppose that 100 batches are produced. Estimate the number of batches having between 0 and 30 defective transistors.

13. Suppose that a probability distribution has mean 75 and standard deviation 6. Use the Chebychev inequality to find the value of c for which the probability that the outcome lies between $75 - c$ and $75 + c$ is at least $\frac{7}{16}$.

14. Suppose that a probability distribution has mean 17 and standard deviation .2. Use the Chebychev inequality to find the value of c for which the probability that the outcome lies between $17 - c$ and $17 + c$ is at least $\frac{15}{16}$.

15. The probability distribution for the sum of numbers obtained from tossing a pair of dice is given in Table 5(a).

    (a) Compute the mean and variance of this probability distribution.

    (b) Using the table, give the probability that the number is between 4 and 10 inclusive.

    (c) Use the Chebychev inequality to estimate the probability that the number is between 4 and 10 inclusive.

TABLE 5

Number	Probability
2	$\frac{1}{36}$
3	$\frac{2}{36}$
4	$\frac{3}{36}$
5	$\frac{4}{36}$
6	$\frac{5}{36}$
7	$\frac{6}{36}$
8	$\frac{5}{36}$
9	$\frac{4}{36}$
10	$\frac{3}{36}$
11	$\frac{2}{36}$
12	$\frac{1}{36}$

(a)

Number of "ones"	Probability
0	.112
1	.269
2	.296
3	.197
4	.089
5	.028
6	.007
7	.001
8	.000
9	.000
10	.000
11	.000
12	.000

(b)

16. The probability distribution for the number of "ones" obtained from tossing 12 dice is given in Table 5(b). This probability distribution has mean 2 and standard deviation 1.291 ($\sigma^2 = \frac{5}{3}$).

    (a) Using the table, give the probability that the number of "ones" is between 0 and 4 inclusive.

    (b) Use the Chebychev inequality to estimate the probability that the number of "ones" is between 0 and 4 inclusive.

17. Redo Example 1 using the alternate formula for the variance.

18. If X is a random variable, then the variance of X equals the variance of $X - a$ for any number a. Redo Exercise 1 using this result with $a = 70$.

19. If X is a random variable, then the variance of aX equals a^2 times the variance of X. Verify this result for the random variable in Exercise 2 with $a = 2$.

20. If X is a random variable, then $E(X - a) = E(X) - a$, and $E(aX) = aE(X)$ for any number a. Give intuitive justifications of these results.

7.4. Binomial Trials

In this section we fix our attention on the simplest experiments: those with just two outcomes. These experiments, called *binomial trials* (or *Bernoulli trials*), occur in many applications. Here are some examples of binomial trials.

1. Flip a coin and observe the outcome, heads or tails.
2. Administer a drug to a sick individual and classify the reaction as "effective" or "ineffective."
3. Manufacture a light bulb and classify it as "nondefective" or "defective."

The outcomes of a binomial trial are usually called "success" and "failure." Of course, the labels "success" and "failure" need have no connection with the usual meanings of these words. For example, in experiment 2 above we might label the outcome "ineffective" as "success" and "effective" as "failure." Throughout our discussion of binomial trials we will always denote the probability of "success" by p and probability of "failure" by q. Since a binomial experiment has only two outcomes, we have $p + q = 1$ or

$$q = 1 - p.$$
(1)

Suppose that we consider a particular binomial trial. Let us perform the following experiment. Repeat the binomial trial n times and observe the number of successes which occur. Assume that the n successive trials are independent of one another. The fundamental problem of the theory of binomial trials is to calculate the probabilities of the various outcomes of this experiment. Let X denote the random variable associated with the experiment. Then X denotes the number of successes which occur. For example, if $n = 20$ and $X = 3$, then 3 of the 20 binomial trials yield success and 17 failure. The possible values of X (the possible outcomes of the experiment) are $0, 1, 2, \ldots, n$. We shall let $\Pr(X = k)$ denote the probability that $X = k$, namely, the probability that k of the n trials result in success. Moreover, let μ denote the expected value of X and σ^2 the variance. Then we have the following results:

For $k = 0, 1, \ldots, n$,

$$\Pr(X = k) = \binom{n}{k} p^k q^{n-k}$$
(2)

$$\mu = np$$
(3)

$$\sigma^2 = npq$$
(4)

Note that the right-hand side of (2) is one of the terms in the binomial expansion of $(p + q)^n$ (see Section 5.7). We say that X is a binomial random variable with parameters n and p. The derivations of (2) and (3) are given at the end of this section.

Let X be the number of heads in 5 tosses of a fair coin. In Example 4 of Section 7.3, we carried out a lengthy calculation to find the mean and variance of X. Since X is a binomial random variable with $n = 5$ and $p = \frac{1}{2}$, formulas (3) and (4) can be used to compute the same results with much less effort: $\mu = 5 \cdot \frac{1}{2} = \frac{5}{2}$ and $\sigma^2 = 5 \cdot \frac{1}{2} \cdot \frac{1}{2} = \frac{5}{4}$. Let us now consider some further applications of formulas (2), (3), and (4).

EXAMPLE 1 Each time at bat the probability that a baseball player gets a hit is .300. He comes up to bat four times in a game. Assume that his times at bat are independent trials. Find the probability that he gets (a) exactly two hits and (b) at least two hits.

Solution Each at-bat is considered an independent binomial trial. A "success" is a hit. So $p = .300$, $q = 1 - p = .700$, and $n = 4$. Therefore X is the number of hits in four at-bats.

(a) We need to determine $\Pr(X = 2)$. From formula (2) with $k = 2$ we have

$$\Pr(X = 2) = \binom{4}{2}(.300)^2(.700)^{4-2} = 6(.09)(.49) = .2646.$$

(b) "At least two hits" means $X \geq 2$. Applying formula (2) with $k = 2$, 3, and 4, we have

$$\Pr(X \geq 2) = \Pr(X = 2) + \Pr(X = 3) + \Pr(X = 4).$$

$$= \binom{4}{2}(.300)^2(.700)^2 + \binom{4}{3}(.300)^3(.700)^1 + \binom{4}{4}(.300)^4(.700)^0$$

$$= 6(.09)(.49) + 4(.027)(.700) + 1(.0081)(1)$$

$$= .2646 + .0756 + .0081 = .3483.$$

So the batter can be expected to get at least two hits out of four at-bats in about 35% of the games.

EXAMPLE 2 Statistics* show that 52% of all married women in the U.S. are in the labor force. Five married U.S. women are randomly selected. Find the probability that at least one of them is in the labor force. Assume that each selection is an independent binomial trial.

Solution Let "success" be "in the labor force." Then $p = .52$, $q = 1 - p = .48$, and $n = 5$. Therefore X is the number of women (out of the five selected) that are in the labor

---

* *1986 Almanac* (Boston: Houghton Mifflin Company, 1986), p. 749.

force. Then

$$\Pr(X \geq 1) = 1 - \Pr(X = 0) = 1 - \binom{5}{0}(.52)^0(.48)^5$$

$$\approx 1 - .0255 = .9745.$$

Thus, in a group of five randomly selected married U.S. women, there is a 97.45% chance that at least one is in the labor force.

EXAMPLE 3   A plumbing supplies manufacturer produces faucet washers which are packaged in boxes of 300. Quality control studies have shown that 2% of the washers are defective.

(a) What is the probability that a box of washers contains exactly 9 defective washers?

(b) What is the average number of defective washers per box?

Solution   (a) Deciding whether a single washer is or is not defective is a binomial trial. Since we wish to consider the number of defective washers in a box, let "success" be the outcome "defective." Then $p = .02$, $q = 1 - .02 = .98$, $n = 300$. The probability that 9 out of 300 washers are defective equals

$$\Pr(X = 9) = \binom{300}{9}(.02)^9(.98)^{291} \approx .07.$$

(b) The average number of defective washers per box is given by the expected value $\mu = np = 300 \cdot (.02) = 6$.

EXAMPLE 4   The recovery rate for a certain cattle disease is 25%. If 40 cattle are afflicted with the disease, what is the probability that exactly 10 will recover?

Solution   In this example the binomial trial consists of observing a single cow, with recovery as success. Then $p = .25$, $q = 1 - .25 = .75$, $n = 40$. The probability of 10 successes is

$$\Pr(X = 10) = \binom{40}{10}(.25)^{10}(.75)^{30} \approx .14.$$

The numerical work in Examples 3 and 4 required an electronic calculator with an x^y key [to compute, for example, $(.98)^{291}$]. In other problems the calculations can be stickier. Suppose, for example, that we return to Example 4. What is the probability that 16 or more cattle recover? Using formula (2) to compute the probabilities that 16, 17, . . . , 40 cattle recover, the desired probability is

$$\Pr(X = 16) + \Pr(X = 17) + \cdots + \Pr(X = 40)$$

$$= \binom{40}{16}(.25)^{16}(.75)^{24} + \binom{40}{17}(.25)^{17}(.75)^{23} + \cdots + \binom{40}{40}(.25)^{40}(.75)^0.$$

Each of the terms in this sum is difficult to compute. The thought of computing all of them should be sufficient motivation to seek an alternate approach. Fortunately, there exists a reasonably simple method of approximating a sum of this type. Let us illustrate the technique with an example.

Toss a coin 20 times and observe the number of heads. By using formula (2) with $n = 20$ and $p = .5$ we can calculate the probability of k heads. The results are displayed in Table 1. The data of Table 1 can be displayed in histogram form, as in Fig. 1.

TABLE 1 Probability of k Heads (to four places)

k	0	1	2	3	4	5	6	7	8	9	10
Probability of k heads	.0000	.0000	.0002	.0011	.0046	.0148	.0370	.0739	.1201	.1602	.1762

k	11	12	13	14	15	16	17	18	19	20
Probability of k heads	.1602	.1201	.0739	.0370	.0148	.0046	.0011	.0002	.0000	.0000

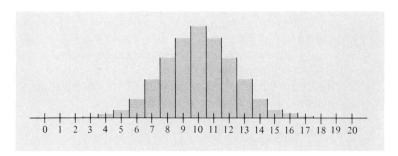

FIGURE 1

As we have seen, various probabilities may be interpreted as areas. For example, the probability that at most 9 heads occur is equal to the sum of the areas of all the rectangles to the left of the central one (Fig. 2). The shape of the graph in Fig. 1 and Fig. 2 suggests that we might be able to approximate such

FIGURE 2

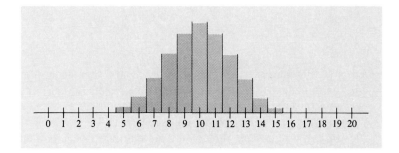

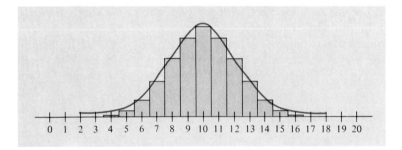

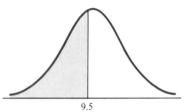

FIGURE 4

FIGURE 3

areas by using a smooth bell-shaped curve. The curve shown in Fig. 3 is a good candidate. It is called a *normal curve* and plays an important role in statistics and probability. (Tables giving the areas under normal curves have been constructed and can be used to approximate binomial probabilities.) For instance, the area of the shaded rectangles in Fig. 2 is approximately the same as the area under the normal curve to the left of 9.5 shown in Fig. 4. As another example, the probability of obtaining exactly 10 heads in 20 tosses of a coin is the area of the shaded rectangle in Fig. 5(a), which is approximated by the area under the normal curve shown in Fig. 5(b).

In order to be able to use normal curves in our computations, we need to study them more closely. In Section 7.5 we discuss how to find the areas of regions under normal curves, and in Section 7.6 we apply this skill to computing probabilities arising in binomial distributions.

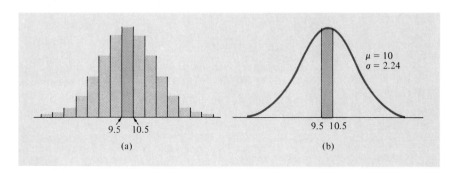

FIGURE 5

Verification of Formula (2)   We first consider the case where $n = 3$ and then generalize. Consider a three-trial binomial experiment with two possible results (S or F) on each trial. Assume that the trials are independent and the probability of a "S" on each trial is p. It follows that the probability of "F" on each trial is $1 - p$, which we denote by q. The tree in Fig. 6 represents the possible outcomes of the experiment.

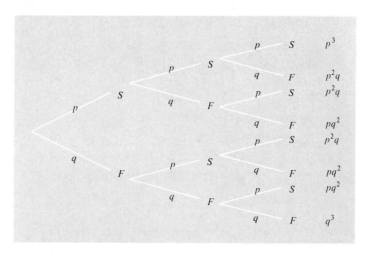

FIGURE 6

The probability of each individual branch is obtained by multiplying the probabilities along the branch. Then, the probability of two successes in three trials, for example, is the sum of the probabilities of each branch representing two S's and one F. There are three such branches, each with probability p^2q. For the general case, suppose we want the probability of k successes in n trials. Each branch having k S's and $(n-k)$ F's has the associated probability p^kq^{n-k}. How many such branches are there in the tree? We can use the methods of Chapter 5 to count the number of branches with k S's and $(n-k)$ F's. What we want to know is the number of ways in which we can arrange k S's and $(n-k)$ F's. This is just $\binom{n}{k}$. So,

$$\Pr(X = k) = \binom{n}{k}p^kq^{n-k}.$$

Verification of Formula (3)

Note that

$$\mu = 0 \cdot \Pr(X = 0) + 1 \cdot \Pr(X = 1) + \cdots + n \cdot \Pr(X = n)$$

$$= 0 \cdot \binom{n}{0}p^0q^{n-0} + 1 \cdot \binom{n}{1}p^1q^{n-1} + \cdots + n \cdot \binom{n}{n}p^nq^{n-n}.$$

Moreover, note that

$$k\binom{n}{k} = k \cdot \frac{n(n-1)\cdot\ldots\cdot(n-k+1)}{k(k-1)\cdot\ldots\cdot 2\cdot 1} = n \cdot \frac{(n-1)\cdot\ldots\cdot(n-k+1)}{(k-1)\cdot\ldots\cdot 2\cdot 1}$$

$$= n \cdot \binom{n-1}{k-1} \qquad (k = 1, 2, 3, \ldots, n).$$

Therefore,

$$\mu = 1 \cdot \binom{n}{1} p^1 q^{n-1} + 2 \cdot \binom{n}{2} p^2 q^{n-2} + \cdots + n \cdot \binom{n}{n} p^n q^0$$

$$= n \binom{n-1}{0} p^1 q^{n-1} + n \binom{n-1}{1} p^2 q^{n-2} + \cdots + n \binom{n-1}{n-1} p^n q^0$$

$$= np \left[\binom{n-1}{0} p^0 q^{n-1} + \binom{n-1}{1} pq^{n-2} + \cdots + \binom{n-1}{n-1} p^{n-1} q^0 \right]$$

$$= np(p + q)^{n-1} \qquad \text{(by the binomial theorem)}$$

$$= np \qquad \qquad \text{(since } p + q = 1 \text{).}$$

PRACTICE PROBLEMS 4

1. A number is selected at random from the numbers 0 through 9999. What is the probability that the number is a multiple of 5?

2. If the experiment in Problem 1 is repeated 200 times, with replacement, what is the expected number of numbers that will be multiples of 5?

EXERCISES 4

1. A single die is tossed four times. Find the probability that exactly two of the tosses show a "one."

2. Find the probability of obtaining exactly three heads when tossing a fair coin six times.

3. A salesman determines that the probability of making a sale to a customer is $\frac{1}{4}$. What is the probability of making a sale to three of the next four customers?

4. A basketball player makes free throws with probability .7. What is the probability of making exactly two out of five free throws?

5. Suppose that 60% of the voters in a state intend to vote for a certain candidate. What is the probability that a survey polling five people reveals that two or less intend to vote for that candidate?

6. An exam consists of six "true or false" questions. What is the probability that a person can get five or more correct by just guessing?

7. Sixty percent of all students at a university are female. A committee of five students is selected at random. Only one is a woman. Find the probability that no more than one woman is selected. What might be your conclusion about the way the committee was chosen?

8. Ten percent of all undergraduates at a university are chemistry majors. In a random sample of eight students, find the probability that exactly two are chemistry majors.

9. Thirty percent of all cars crossing a toll bridge have a commuter sticker. What is the probability that among ten randomly selected cars waiting to cross the bridge at least two have commuter stickers?

10. (a) Show that

$$\binom{n}{k} = \binom{n}{n-k}$$

for $k = 0, 1, 2, \ldots, n$.

(b) Let X be the random variable associated with binomial trials with $n = 10$ and $p = \frac{1}{2}$. Show that

$$\Pr(X = k) = \Pr(X = 10 - k)$$

for $k = 0, 1, 2, \ldots, n$.

11. A jury has 12 jurors. A vote of at least 10 of 12 for "guilty" is necessary for a defendant to be convicted of a crime. Assume that each juror acts independently of the others and that the probability that any one juror makes the correct decision on a defendant is .80. If the defendant is guilty, what is the probability that the jury makes the correct decision?

12. Every offspring inherits a gene for hair color from each parent. We denote the dominant gene by A and the recessive gene by a. If a person has AA or Aa, then the person exhibits the dominant characteristic. We call a person with the genes Aa a *hybrid*. A person with aa exhibits the recessive characteristic. Two hybrid parents have three children. Find the probability that at least one child exhibits the recessive characteristic.

13. A single die is rolled 10 times and the number of sixes observed. What is the probability that a six appears 9 times, given that it appears at least 9 times?

14. A coin is tossed until 4 heads occur. What is the probability that the fourth head occurs on the tenth toss?

SOLUTIONS TO PRACTICE PROBLEMS 4

1. The number of multiples of 5 that occur in the numbers from 0 to 9999 is 2000. Since each of the 10,000 choices is as likely as any other, the probability is $\dfrac{2000}{10,000} = .2$.

2. Let "success" be "the selected number is a multiple of 5." Then the selection of a number at random is a binomial trial. So $p = .2$ and $n = 200$. The expected number of successes is $np = 200 \cdot (.2) = 40$.

7.5. The Normal Distribution

In Section 7.4 we saw that the histogram for a binomial random variable is approximated by the region under a smooth curve called a *normal curve*. Actually, these curves have value in their own right in that they represent random variables associated with experiments having infinitely many possible outcomes. Let us now take a glimpse into the realm of so-called continuous probability by studying such experiments—namely, experiments with *normally distributed outcomes*. For such experiments, the probabilities of events are computed as areas under normal curves. It is no exaggeration to say that experiments with normally distributed outcomes are among the most significant in probability theory. Such experiments abound in the world around us. Here are a few examples:

1. Choose an individual at random and observe his or her IQ.
2. Choose a 1-day-old infant and observe its weight.
3. Choose an 8-year-old male at random and observe his height.
4. Choose a leaf at random from a particular tree and observe its length.
5. A lumber mill is cutting planks which are supposed to be 8 feet long; choose a plank at random and observe its actual length.

Associated to each of the foregoing experiments is a bell-shaped curve, as shown in Fig. 1. Such a curve is called a *normal curve*. The curve is symmetric about a vertical line drawn through its highest point. This line of symmetry indicates the mean value of the corresponding experiment. The mean value is denoted as usual by the Greek letter μ. Thus, for example, if in experiment 4 above the average length of the leaves on the tree is 5 inches, then $\mu = 5$ and the corresponding bell-shaped curve is symmetric about the line $x = 5$.

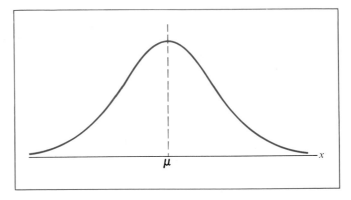

FIGURE 1

The connection between an experiment with normally distributed outcomes and its associated normal curve is this:

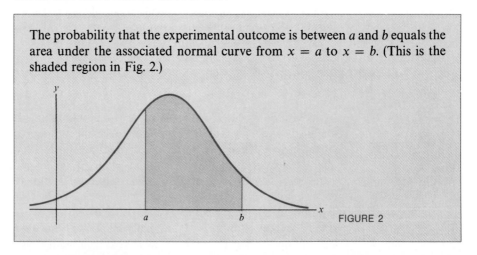

The probability that the experimental outcome is between a and b equals the area under the associated normal curve from $x = a$ to $x = b$. (This is the shaded region in Fig. 2.)

FIGURE 2

The total area under a normal curve is always 1. This is due to the fact that the probability that the variable X corresponding to the distribution takes on some numerical value on the x-axis is 1.

EXAMPLE 1 A certain experiment has normally distributed outcomes with mean $\mu = 1$. Shade the region corresponding to the probabilities of the following outcomes.

(a) The outcome lies between 1 and 3.

(b) The outcome lies between 0 and 2.

(c) The outcome is less than .5.

(d) The outcome is greater than 2.

Solution The outcomes are plotted along the x-axis. We then shade the appropriate area under the curve.

(a)
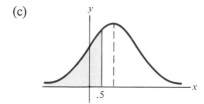

(b)
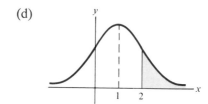

(c)
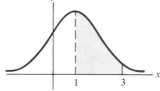

(d)
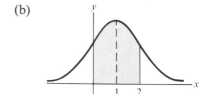

There are many different normal curves with the same mean. For instance, in Fig. 3 we have drawn three normal curves, all with $\mu = 0$. Roughly speaking, the difference between these normal curves is in the width of the center "hump." A sharper hump indicates that the outcomes are more likely to be close to the mean. A flatter hump indicates a greater likelihood for the outcomes to be spread out.

FIGURE 3

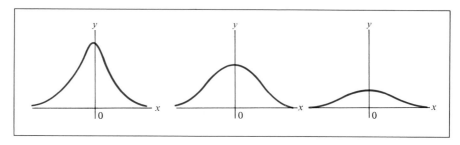

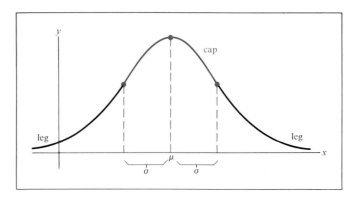

FIGURE 4

As we have seen, the spread of the outcomes about the mean is described by the standard deviation, denoted by the Greek letter σ. In the case of a normal curve, the standard deviation has a simple geometric meaning: The normal curve "twists" (or, in calculus terminology, "inflects") at a distance σ on either side of the mean (Fig. 4). More specifically, a normal curve may be thought of as made up of two pieces, a "cap," which looks like an upside-down bowl, and a pair of legs which curve in the opposite direction. The places at which the cap and legs are joined are at a distance σ from the mean. Thus, it is clear that the size of σ controls the sharpness of the hump.

A normal curve is completely described by its mean μ and standard deviation σ. In fact, given μ and σ, we may write down the equation of the associated normal curve:

$$y = \frac{1}{\sigma\sqrt{2\pi}}\, e^{-(1/2)[(x-\mu)/\sigma]^2},$$

where $\pi \approx 3.1416$ and $e \approx 2.7183$. Fortunately, we will not need this rather complicated formula in what follows. But it is only fair to say that all theoretical work on the normal curve ultimately rests on this equation.

For our purposes we will compute areas under normal curves by consulting a table. One might expect that a separate table would be needed for each normal curve, but such is not the case. Only one table is needed: the table corresponding to the so-called *standard normal curve*, which is the one for which $\mu = 0$, $\sigma = 1$. So let us begin our discussion of areas under normal curves by considering the standard normal curve.

We usually use the letter Z to denote a random variable having the standard normal distribution. Let z be any number and let $A(z)$ denote the area under the standard normal curve to the left of z (Fig. 5). Table 1 gives $A(z)$ for various values of z, with the values of $A(z)$ rounded to four decimal places. Thus, $A(z) = \Pr(Z \leq z)$.* A more extensive table can be found in Table 1 of the Appendix. The efficient

---

* We could have said that $A(z) = \Pr(Z < z)$. However, since the region strictly to the left of z and that region with the line segment at z adjoined have the same area, $\Pr(Z < z)$ and $\Pr(Z \leq z)$ are the same. We always use the $\leq$ symbol.

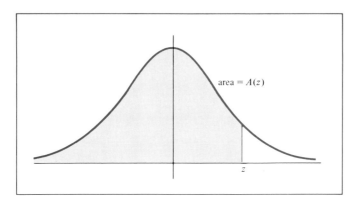

FIGURE 5

TABLE 1

z	$A(z)$	z	$A(z)$	z	$A(z)$
−4.00	.0000	−1.25	.1056	1.50	.9332
−3.75	.0001	−1.00	.1587	1.75	.9599
−3.50	.0002	−.75	.2266	2.00	.9772
−3.25	.0006	−.50	.3085	2.25	.9878
−3.00	.0013	−.25	.4013	2.50	.9938
−2.75	.0030	0	.5000	2.75	.9970
−2.50	.0062	.25	.5987	3.00	.9987
−2.25	.0122	.50	.6915	3.25	.9994
−2.00	.0228	.75	.7734	3.50	.9998
−1.75	.0401	1.00	.8413	3.75	.9999
−1.50	.0668	1.25	.8944	4.00	1.0000

use of these tables depends on the following three facts:

1. The standard normal curve is symmetric about $z = 0$.
2. The total area under the standard normal curve is 1.
3. The probability that the standard normal variable Z lies to the left of the number z is the area $A(z)$ in Fig. 5.

These facts allow us to use the tables to find the areas of various types of regions.

EXAMPLE 2   Use Table 1 to determine the areas of the regions under the standard normal curve pictured in Fig. 6.

Solution   (a) This region is just the portion of the curve to the left of −.5. So its area is $A(-.5)$. Looking down the middle pair of columns of the table, we find that $A(-.5) = .3085$. This means that $\Pr(Z \leq -.5) = .3085$.

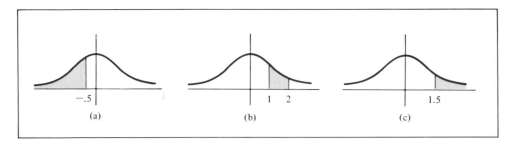

FIGURE 6

(b) This region results from beginning with the region to the left of 2 and subtracting the region to the left of 1. We obtain an area of $A(2) - A(1) = .9772 - .8413 = .1359$. Thus, $\Pr(1 \leq Z \leq 2) = .1359$.

(c) This region can be thought of as the entire region under the curve, with the region to the left of 1.5 removed. Therefore, the area is $1 - A(1.5) = 1 - .9332 = .0668$. So, $\Pr(Z \geq 1.5) = .0668$.

EXAMPLE 3   Find the value of z for which $\Pr(Z \geq z) = .1056$ (See Fig. 7).

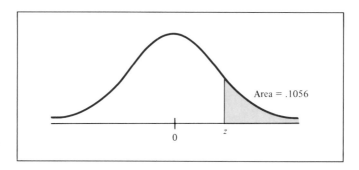

FIGURE 7

Solution   Since the area under the standard normal curve is 1 and the curve is symmetric about $z = 0$, the area of the portion to the right of 0 must be .5. We draw a sketch of the standard normal curve, placing z on the axis to the right of 0. (This way, the area to the right of z will be less than .5.) Table 1 gives the values of $A(z)$, which are left tail areas. The area to the left of our z is

$$A(z) = 1 - .1056 = .8944.$$

From Table 1, we find that the value of z for which $A(z) = .8944$ is 1.25.

Percentiles   In large-scale testing, scores are frequently reported as percentiles rather than as raw scores. What does it mean to say that our score is at the 90th percentile? It means that 90% of the population attained scores that were below ours and 10% had scores that were above ours.

> If a score S is the pth percentile, then $p\%$ of all scores fall below S, and $(100 - p)\%$ of all scores fall above S.

EXAMPLE 4 What is the 50th percentile of the standard normal distribution?

Solution The standard normal curve is symmetric about $z = 0$, and the total area under the curve is 1. Thus, 50% of the values of the standard normal variable fall below 0, and $(100 - 50)\% = 50\%$ of its values fall above 0. So 0 is the 50th percentile of the standard normal distribution.

EXAMPLE 5 What is the 95th percentile of the standard normal distribution?

Solution We shall call the value that we seek z_{95} to remind us that it is a score and that the probability that an outcome is to the left of it is 95%. See Fig. 8.

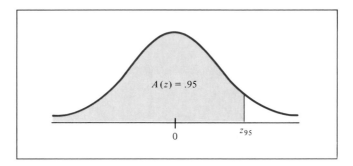

$A(z) = .95$

0 z_{95}

FIGURE 8

Since Table 1 of the Appendix gives areas to the left of values of z, we should search the column marked $A(z)$ for the area we need—.95. We find that the closest value to .95 is .9505, and $A(1.65) = .9505$. Hence $z_{95} \approx 1.65$. This means that 95% of the time the standard normal variable falls below 1.65. Since $\mu = 0$ and $\sigma = 1$, another way of stating the result is that in the standard normal distribution, 95% of the values are less than 1.65 standard deviations above the mean.

The problem of finding the area of a region under any normal curve can be reduced to finding the area of a region under the standard normal curve. To illustrate the computation procedure, let us consider a numerical example.

EXAMPLE 6 Find the area under the normal curve with $\mu = 3$, $\sigma = 2$ from $x = 1$ to $x = 5$. This represents $\Pr(1 \leq X \leq 5)$ for a random variable X having the given normal distribution.

Solution We have sketched the described region in Fig. 9. It extends from one standard deviation below the mean to one standard deviation above. Draw the corresponding region under the standard normal curve. That is, draw the region from one standard deviation below to one standard deviation above the mean (Fig. 10). It is a theorem that this new region has the same area as the original one. But the area in Fig. 10 may be computed from Table 1 as $A(1) - A(-1)$. So our desired area is

$$A(1) - A(-1) = .8413 - .1587 = .6826.$$

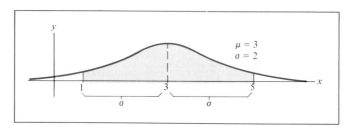

FIGURE 9

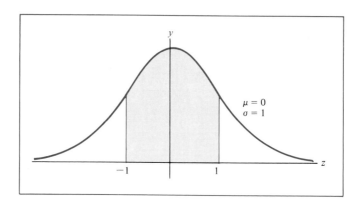

FIGURE 10

EXAMPLE 7 Consider the normal curve with $\mu = 12$, $\sigma = 1.5$. Find the area of the region under the curve between $x = 11.25$ and $x = 15$ (Fig. 11).

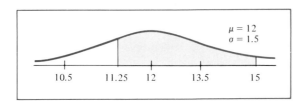

$\mu = 12$
$\sigma = 1.5$

10.5   11.25   12   13.5   15

FIGURE 11

Solution Expressed as a probability, we want to find $\Pr(11.25 \le X \le 15)$ for a random variable X having a normal distribution with $\mu = 12$ and $\sigma = 15$. The number 11.25 is .75 below the mean 12. And .75 is $.75/1.5 = .5$ standard deviations. The number 15 is 3 above the mean. And 3 is $3/1.5 = 2$ standard deviations. Therefore, the region has the same area as the region under the standard normal curve from $-.5$ to 2, which is $A(2) - A(-.5) = .9772 - .3085 = .6687$.

Suppose that a normal curve has mean μ and standard deviation σ. Then the area under the curve from $x = a$ to $x = b$ is

$$A\left(\frac{b - \mu}{\sigma}\right) - A\left(\frac{a - \mu}{\sigma}\right).$$

The numbers $b - \mu$ and $a - \mu$, respectively, measure the distances of b and a from the mean. The numbers $(b - \mu)/\sigma$ and $(a - \mu)/\sigma$ express these distances as multiples of the standard deviation σ. So the area under the normal curve from $x = a$ to $x = b$ is computed by expressing x in terms of standard deviations from the mean and then treating the curve as if it were the standard normal curve.

We summarize the procedure:

If X is a random variable having a normal distribution with mean μ and standard deviation σ, then

$$\Pr(a \le X \le b) = \Pr\left(\frac{a - \mu}{\sigma} \le Z \le \frac{b - \mu}{\sigma}\right) = A\left(\frac{b - \mu}{\sigma}\right) - A\left(\frac{a - \mu}{\sigma}\right)$$

and

$$\Pr(X \le x) = \Pr\left(Z \le \frac{x - \mu}{\sigma}\right) = A\left(\frac{x - \mu}{\sigma}\right).$$

where Z has the standard normal distribution and $A(z)$ is the area under that distribution to the left of z.

Let us now use our knowledge of areas under normal curves to calculate probabilities arising in some applied problems.

EXAMPLE 8 Suppose that for a certain population the birth weights of infants in pounds is normally distributed with $\mu = 7.75$ and $\sigma = 1.25$. Find the probability that an infant's birth weight is more than 9 pounds 10 ounces. (*Note:* 9 pounds 10 ounces $= 9\frac{5}{8}$ pounds.)

Solution Let $X =$ infant's birth weight. Then X is a random variable having a normal distribution with $\mu = 7.75$ and $\sigma = 1.25$ pounds. $\Pr(X \geq 9\frac{5}{8})$ is given by the area

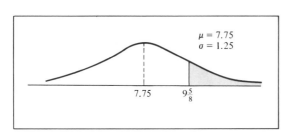

$\mu = 7.75$
$\sigma = 1.25$

7.75 $9\frac{5}{8}$

under the appropriate normal curve to the right of $9\frac{5}{8}$—that is, the area shaded in Fig. 12. Since $9\frac{5}{8} = 9.625$, the number $9\frac{5}{8}$ lies $9.625 - 7.75 = 1.875$ units above the mean. In turn, this is $1.875/1.25 = 1.5$ standard deviations. We can find the corresponding z value in one calculation by finding

$$z = \frac{x - \mu}{\sigma} = \frac{9.625 - 7.75}{1.25} = 1.5$$

FIGURE 12

Thus 9.625 is 1.5 standard deviations above the mean. The area we seek is sketched under the standard normal curve in Fig. 13 and is

$$1 - A(1.5) = 1 - .9332 = .0668.$$

So the probability that an infant weighs more than 9 pounds 10 ounces is .0668.

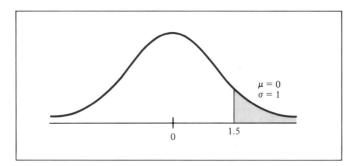

$\mu = 0$
$\sigma = 1$

0   1.5

FIGURE 13

EXAMPLE 9 A wholesale produce dealer finds that the number of boxes of bananas sold each day is normally distributed with $\mu = 1200$ and $\sigma = 100$. Find the probability that the number sold on a particular day is less than 1000.

Solution Let $X =$ the number of boxes of bananas sold each day. Since daily sales are normally distributed, the desired probability, $\Pr(X \leq 1000)$, is the area to the left of 1000 in the normal curve drawn in Fig. 14. The number 1000 is 2 standard devia-

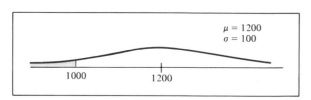

$\mu = 1200$
$\sigma = 100$

1000   1200

FIGURE 14

tions below the mean; that is, $x = 1000$ corresponds to a z value of

$$z = \frac{x - \mu}{\sigma} = \frac{1000 - 1200}{100} = -2.$$

Therefore, the area we seek is $A(-2) = .0228$ shown in Fig. 15. The probability that less than 1000 boxes will be sold in a day is .0228.

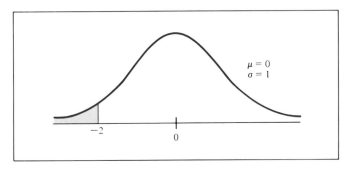

FIGURE 15

EXAMPLE 10   Find the 95th percentile of infant birth weights if infant birth weights are normally distributed with $\mu = 7.75$ and $\sigma = 1.25$ pounds.

Solution   Let us denote the 95th percentile of the infant birth weights by x_{95}. This means that the area to the left of x_{95} under the normal curve with $\mu = 7.75$ and $\sigma = 1.25$ is 95% (Fig. 16).

   The corresponding value for the standard normal random variable would be z_{95}, which we already know to be 1.65. Thus x_{95} is 1.65 standard deviations above μ. That is, $x_{95} = 7.75 + (1.65)(1.25) = 9.81$ pounds. Therefore, 95% of all infants have birth weights below 9.81 pounds (9 lbs 13 oz).

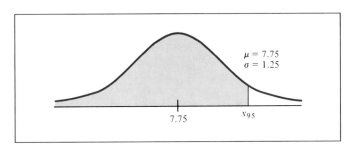

FIGURE 16

EXAMPLE 11   The wholesale produce dealer of Example 9 wants to be 99% sure that she has enough boxes of bananas on hand each day to meet the demand. How many should she stock each day?

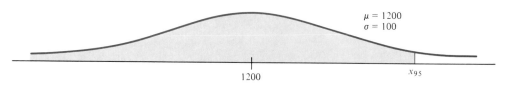

$\mu = 1200$
$\sigma = 100$

1200

x_{95}

FIGURE 17

Solution Let x be the number of boxes of bananas that the produce dealer should stock. Since she wants to be 99% sure that the demand for boxes of bananas does not exceed x, we must find the 99th percentile of a normal distribution with $\mu = 1200$ and $\sigma = 100$. To help us remember what x really is, we will rename it x_{99}. The corresponding value for the standard normal random variable is z_{99}. Figures 17 and 18 show the appropriate areas—first under the given normal curve and then under the standard normal curve.

The area, $A(z)$, that we seek under the standard normal curve is .99. Referring to Table 1 of the Appendix, we get closest with $A(z) = .9906$, corresponding to $z_{99} = 2.35$. The value z_{99} is 2.35 standard deviations above the mean of its distribution. We conclude that x_{99} is also 2.35 standard deviations above its mean. Hence, $x_{99} = 1200 + (2.35)(100) = 1435$ boxes. Therefore, we expect that on 99% of the days, 1435 boxes of bananas will meet the demand. More than 1435 boxes should be needed 1% of the time (one day out of 100).

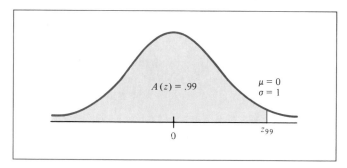

$A(z) = .99$

$\mu = 0$
$\sigma = 1$

0

z_{99}

FIGURE 18

We summarize the technique in finding percentiles of normal distributions:

> If x_p is the pth percentile of a normal distribution with mean μ and standard deviation σ, then
>
> $$x_p = \mu + (z_p)(\sigma)$$
>
> where z_p is the pth percentile of the standard normal distribution.

PRACTICE PROBLEMS 5

1. Refer to Fig. 19(a). Find the value of z for which the area of the shaded region is .0802.

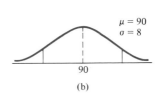

$$\mu = 0$$
$$\sigma = 1$$

$$\mu = 90$$
$$\sigma = 8$$

(a)   (b)   FIGURE 19

2. Refer to the normal curve in Fig. 19(b). Express the following numbers in terms of standard deviations from the mean.

(a) 90   (b) 82   (c) 94   (d) 104

EXERCISES 5

Use the table for $A(z)$ to find the areas of the shaded regions under the standard normal curve.

1.
1.25

2.
−.75   1

3.
.25

4.
−1   1

5.
.5 1.5

6.
−1

7.
−.5 .5

8.
−1.25

In Exercises 9–12, find the value of z for which the area of the shaded region under the standard normal curve is as specified.

9. Area is .0401.

10. Area is .0456.

z

$-z$ z

11. Area is .5468.

12. Area is .6915.

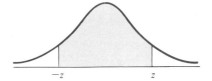
$-z$ z

$-z$

In Exercises 13–16 determine μ and σ by inspection.

13.

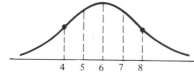

14.

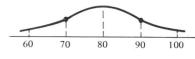

15.

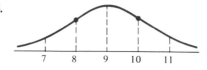

16.

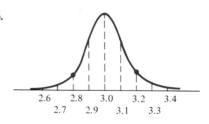

Exercises 17–20 refer to the normal curve with $\mu = 8$, $\sigma = \frac{3}{4}$.

17. Convert 6 into standard deviations from the mean.

18. Convert $9\frac{1}{4}$ into standard deviations from the mean.

19. What value of x corresponds to 10 standard deviations above the mean?

20. What value of x corresponds to 2 standard deviations below the mean?

In Exercises 21–24, find the areas of the shaded regions under the given normal curves.

21.

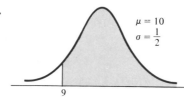

$\mu = 10$
$\sigma = \frac{1}{2}$

22.

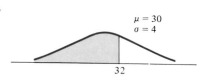

$\mu = 30$
$\sigma = 4$

23.

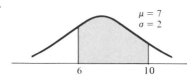

$\mu = 7$
$\sigma = 2$

24.

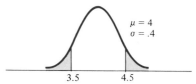

$\mu = 4$
$\sigma = .4$

25. Suppose that the height (at the shoulder) of adult African bull bush elephants is normally distributed with $\mu = 3.3$ meters and $\sigma = .2$ meter. The elephant on display at the Smithsonian Institution has height 4 meters and is the largest elephant on record. What is the probability that an adult African bull bush elephant has height 4 meters or more?

26. At a soft-drink bottling plant, the amount of cola put into the bottles is normally distributed with $\mu = 16\frac{3}{4}$ ounces and $\sigma = \frac{1}{2}$. What is the probability that a bottle will contain less than 16 ounces?

27. Bolts produced by a machine are acceptable provided that their length is within the range 5.95 to 6.05 centimeters. Suppose that the lengths of the bolts produced are normally distributed with $\mu = 6$ centimeters and $\sigma = .02$. What is the probability that a bolt will be of an acceptable length?

28. Suppose that IQ scores are normally distributed with $\mu = 100$ and $\sigma = 10$. What percent of the population have IQ scores of 125 or more?

29. The amount of gas sold weekly by a certain gas station is normally distributed with $\mu = 30{,}000$ gallons and $\sigma = 4000$. If the station has 39,000 gallons on hand at the beginning of the week, what is the probability of its running out of gas before the end of the week?

30. Suppose that the lifetimes of a certain type of light bulb are normally distributed with $\mu = 1200$ hours and $\sigma = 160$. Find the probability that a light bulb will burn out in less than 1000 hours.

31. Assume that SAT verbal scores for a freshman class at a university are normally distributed with mean 520 and standard deviation 75.

   (a) The top 10% of the students are placed into the honors program for English. What is the lowest score for admittance into the honors program?

   (b) What is the range of the middle 90% of the SAT verbal scores at this university?

32. A mail-order house uses an average of 300 mailing bags per day. The number of bags needed each day is approximately normally distributed with $\sigma = 50$. How many bags must the company have on hand at the beginning of a day to be 99% certain that all orders can be filled?

33. The lifetime of a certain brand of tires is normally distributed with mean $\mu = 30{,}000$ miles and standard deviation $\sigma = 5000$ miles. The company has decided to issue a warranty for the tires but does not want to replace more than 2% of the tires that it sells. At what mileage should the warranty take effect?

34. Let X be a random variable with $\mu = 4$ and $\sigma = .5$.

   (a) Use the Chebyshev inequality to estimate $\Pr(3 \le X \le 5)$.

   (b) If X were normally distributed, what would be the exact probability that X is between 3 and 5 inclusive?

   (c) Reconcile the differences between the answers to (a) and (b).

35. Let X be the amount of soda released by a soft-drink dispensing machine into a 6-ounce cup. Assume that X is normally distributed with $\sigma = .25$ ounces, and that the average "fill" can be set by the vendor.

   (a) At what quantity should the average "fill" be set so that no more than .5% of the releases overflow the cup?

   (b) Using the average "fill" found in (a), determine the minimal amount that will be dispensed in 99% of the cases.

SOLUTIONS TO PRACTICE PROBLEMS 5

1. 1.75. Owing to the symmetry of normal curves, each piece of the shaded region has area $\frac{1}{2}(.0802) = .0401$. Therefore, $A(-z) = .0401$ and so by Table 1, $-z = -1.75$. Thus, $z = 1.75$.

2. (a) 0. Since 90 *is* the mean, it is 0 standard deviations from the mean.

(b) −1. Since 82 is 90 − 8, it is 8 units or 1 standard deviation below the mean.

(c) .5. Here 94 = 90 + 4 is 4 units or .5 standard deviation above the mean.

(d) 1.75. Here 104 = 90 + 14 is 14 units or $\frac{14}{8}$ = 1.75 standard deviations above the mean.

7.6. Normal Approximation to the Binomial Distribution

In Section 7.4, we saw that complicated and tedious calculations can arise from the binomial probability distribution. For instance, determining the probability of getting at least 20 threes in 100 tosses of a die requires the computation

$$\binom{100}{20}(\tfrac{1}{6})^{20}(\tfrac{5}{6})^{80} + \binom{100}{21}(\tfrac{1}{6})^{21}(\tfrac{5}{6})^{79} + \cdots + \binom{100}{100}(\tfrac{1}{6})^{100}(\tfrac{5}{6})^{0}$$

Mathematicians have shown that such probabilities can be closely approximated by using normal curves.

Consider the histograms of the binomial distributions in Figs. 1 through 4. The number of trials, n, increases from 5 to 25, with p fixed at .3. As n increases, the shape of the histogram more closely conforms to the shape of the region under a normal curve.

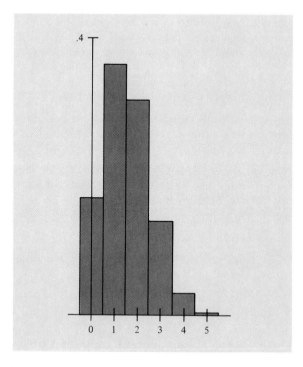

k	$\Pr(X = k)$
0	.168070
1	.360150
2	.308700
3	.132300
4	.028350
5	.002430

FIGURE 1   Binomial Distribution: $n = 5, p = .3$

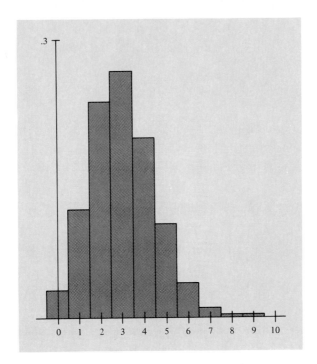

k	Pr($X = k$)
0	.028248
1	.121061
2	.233474
3	.266828
4	.200121
5	.102919
6	.036757
7	.009002
8	.001447
9	.000138
10	.000006

FIGURE 2   Binomial Distribution: $n = 10, p = .3$

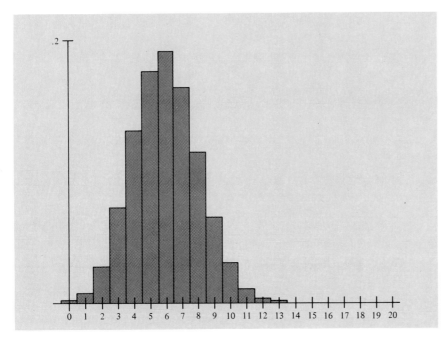

k	Pr($X = k$)
0	.000798
1	.006839
2	.027846
3	.071604
4	.130421
5	.178863
6	.191639
7	.164262
8	.114397
9	.065370
10	.030817
11	.012007
12	.003859
	.001018
14	.000218
15	.000037
16	.000005
17	.000001
18	.000000
19	.000000
20	.000000

FIGURE 3   Binomial Distribution: $n = 20, p = .3$

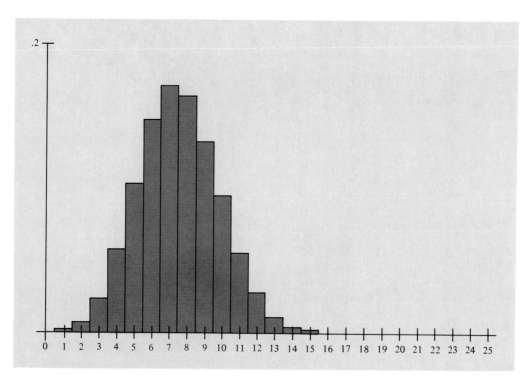

k	$Pr(X = k)$	k	$Pr(X = k)$
0	.000134	13	.011476
1	.001437	14	.004216
2	.007390	15	.001325
3	.024280	16	.000355
4	.057231	17	.000081
5	.103016	18	.000015
6	.147166	19	.000002
7	.171194	20	.000000
8	.165080	21	.000000
9	.133636	22	.000000
10	.091636	23	.000000
11	.053554	24	.000000
12	.026777	25	.000000

FIGURE 4   Binomial Distribution: $n = 25$, $p = .3$

We have the following result:

> Suppose that we perform a sequence of n binomial trials with probability of success p and probability of failure q and observe the number of successes. Then the histogram for the resulting probability distribution may be approximated by the normal curve with $\mu = np$ and $\sigma = \sqrt{npq}$.

Note: This approximation is very accurate when both $n \cdot p > 5$ and $n \cdot q > 5$.

EXAMPLE 1 Refer to Example 3, Section 7.4. A plumbing supplies manufacturer produces faucet washers which are packaged in boxes of 300. Quality control studies have shown that 2% of the washers are defective. What is the probability that more than 10 of the washers in a single box are defective?

Solution Let X = the number of defective washers in a box. Then X is a binomial random variable with $n = 300$ and $p = .02$. We will use the approximating normal curve with

$$\mu = np = 300(.02) = 6$$

$$\sigma = \sqrt{npq} = \sqrt{300(.02)(.98)} \approx 2.425.$$

The probability that more than 10 of the washers in a single box are defective is the sum of the areas of the rectangles centered at 11, 12, ..., 300 in the histogram for the random variable X. See Fig. 5. The corresponding region under the approximating normal curve is shaded in Fig. 6. This is the area under the standard normal curve to the right of

$$z = \frac{10.5 - \mu}{\sigma} = \frac{10.5 - 6}{2.425} \approx 1.85.$$

The area of the region is $1 - A(1.85) = 1 - .9678 = .0322$. Therefore, approximately 3.22% of the boxes should contain more than 10 defective washers.

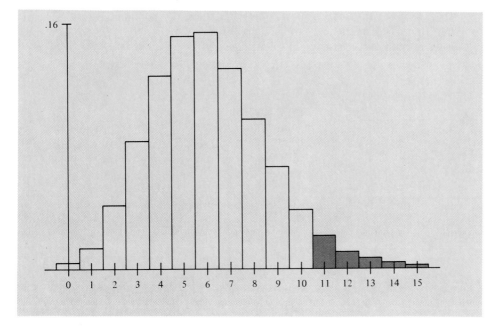

FIGURE 5

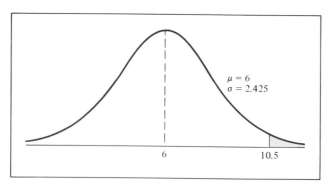

FIGURE 6

Note: In Fig. 6 we shaded the region to the right of 10.5 rather than to the right of 10. This gives a better approximation to the corresponding area under the histogram, since the rectangle corresponding to 11 "successes" has its left endpoint at 10.5.

Let us consider an application to medical research.

EXAMPLE 2   Consider the cattle disease of Example 4, of Section 7.4, from which 25% of the cattle recover. A veterinarian discovers a serum to combat the disease. In a test of the serum she observes that 16 of a herd of 40 recover. Suppose that the serum had not been used. What is the likelihood that at least 16 cattle would have recovered?

Solution   Let X be the number of cattle that recover. Then X is a binomial random variable with $n = 40$ independent trials. If the serum is not used, $p = .25$. The approximating normal curve has

$$\mu = np = 40(.25) = 10$$
$$\sigma = \sqrt{npq} = \sqrt{40(.25)(.75)} \approx 2.74.$$

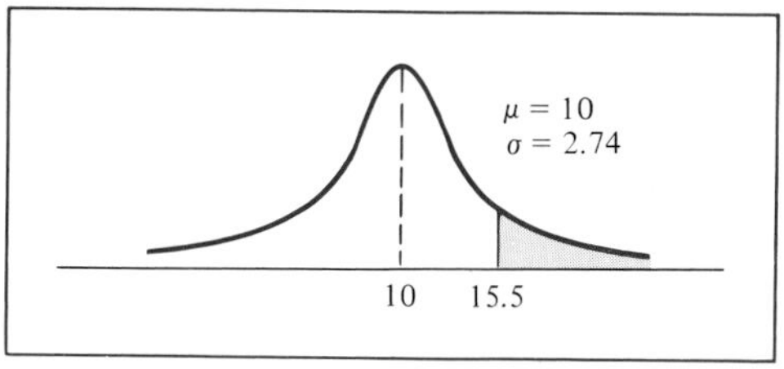

The likelihood that at least 16 cattle would have recovered is $\Pr(X \geq 16)$. This corresponds to the area under the normal curve to the right of 15.5 (Fig. 7). The area to the right of 15.5 under a normal curve with $\mu = 10$ and $\sigma = 2.74$ is the same as the area under the standard normal curve to the right of

$$z = \frac{15.5 - \mu}{\sigma} = \frac{15.5 - 10}{2.74} \approx 2.01.$$

FIGURE 7

We find $1 - A(2.01) \approx 1 - A(2.00) = 1 - .9772 = .0228$. Thus, if the serum were not used, the veterinarian would expect about a 2% chance that 16 or more cattle recover. Thus, the 16 observed recoveries probably did not occur by chance. The veterinarian can reasonably conclude that the serum is effective against the disease.

EXAMPLE 3   Assume that a fair coin is tossed 100 times. Find the probability of observing exactly 50 heads.

Solution   Let X be the number of heads on $n = 100$ binomial trials with $p = .5$ probability of "success" on each trial. We are to find the area of the rectangle extending from 49.5 to 50.5 on the x-axis in the histogram for X. The actual probability is

$$\binom{100}{50}(.5)^{50}(.5)^{50} = \binom{100}{50}(.5)^{100}.$$

We approximate the probability with an area under the normal curve with $\mu = np = 100(.5) = 50$ and $\sigma = \sqrt{npq} = \sqrt{100(.5)(.5)} = 5$. The area we seek is sketched in Fig. 8.

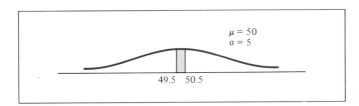

FIGURE 8

It is the area under the standard normal curve from

$$z = \frac{49.5 - 50}{5} = -.10 \quad \text{to} \quad z = \frac{50.5 - 50}{5} = .10.$$

Using Table 1 in the Appendix, we find

$$A(.10) - A(-.10) = .5398 - .4602 = .0796.$$

So the likelihood of getting exactly 50 heads in 100 tosses of a fair coin is fairly small—about 8%.

EXAMPLE 4   Find the probability that in 100 tosses of a fair die we observe at least 20 threes.

Solution   Let X be the number of threes observed in $n = 100$ trials, with $p = \frac{1}{6}$ the probability of a three on each trial. Then we need to find the area to the right of 19.5 under the normal curve with

$$\mu = np = 100(\tfrac{1}{6}) \approx 16.7$$

and

$$\sigma = \sqrt{npq} = \sqrt{100(\tfrac{1}{6})(\tfrac{5}{6})} \approx 3.73.$$

This area is

$$1 - A\left(\frac{19.5 - 16.7}{3.73}\right) = 1 - A(.75) = 1 - .7734 = .2266.$$

Therefore, the probability of observing at least 20 threes is about .2266.

PRACTICE PROBLEMS 6

1. A new drug is being tested on laboratory mice. The mice have been given a disease for which the recovery rate is $\frac{1}{2}$.

   (a) In the first experiment the drug is given to 5 of the mice and all 5 recover. Find the probability that the success of this experiment was due to luck. That is, find the probability that 5 out of 5 mice recover in the event that the drug has no effect on the illness.

   (b) In a second experiment the drug is given to 25 mice and 18 recover. Find the probability that 18 or more recover in the event that the drug has no effect on the illness.

2. What conclusions can be drawn from the results in Problem 1?

EXERCISES 6

In Exercises 1–14, use the normal curve to approximate the probability.

1. An experiment consists of 25 binomial trials, each having probability $\frac{1}{5}$ of success. Use an approximating normal curve to estimate the probability of

   (a) Exactly 5 successes.

   (b) Between 3 and 7 successes inclusive.

   (c) Less than 10 successes.

   An experiment consists of 18 binomial trials, each having probability $\frac{2}{3}$ of success. Use an approximating normal curve to estimate the probability of

   (a) Exactly 10 successes.

   (b) Between 8 and 16 successes inclusive.

   (c) More than 12 successes.

3. Laboratory mice are given an illness for which the usual recovery rate is $\frac{1}{6}$. A new drug is tested on 20 of the mice, and 8 of them recover. What is the probability that 8 or more would have recovered if the 20 mice had not been given the drug?

4. A person claims to have ESP (extrasensory perception). A coin is tossed 16 times, and each time he is asked to predict in advance whether the coin will land "heads" or "tails." He predicts correctly 75% of the time (i.e., on 12 tosses). What is the probability of being correct 12 or more times by pure guessing?

5. A wine-taster claims that she can usually distinguish between domestic and imported wines. As a test, she is given 100 wines to test and correctly identifies 63 of them. What is the probability that she accomplished that good a record by pure guessing? That is, what is the probability of being correct 63 or more times out of 100 by pure guessing?

6. In American roulette, the probability of winning when betting "red" is $\frac{9}{19}$. What is the probability of being ahead after betting the same amount 90 times?

7. A basketball player makes each free throw with probability $\frac{3}{4}$. What is the probability of making 68 or more shots out of 75 trials?

8. A bookstore determines that two-fifths of the people who come into the store make a purchase. What is the probability that of the 54 people who come into the store during a certain hour, less than 14 make a purchase?

9. A baseball player has a lifetime batting average of .310. Find the probability that he gets at least 6 hits in 20 times at bat.

10. An advertising agency, which reached 25% of its target audience with its old campaign, has devised a new advertising campaign. In a sample of 1000 people, it finds that 290 have been reached by the new advertising campaign. What is the probability that at least 290 people would have been reached by the old campaign? Does the new campaign seem to be more effective?

11. A washing machine manufacturer finds that 2% of his washing machines break down within the first year. Find the probability that less than 15 out of a lot of 1000 washers break down within 1 year.

12. The incidence of colorblindness among the men in a certain country is 20%. Find the expected number of colorblind men in a random sample of 70 men. What is the probability of finding exactly that number of colorblind men in a sample of size 70?

13. In a random sample of 250 college students, 50 of them own a compact-disc player. Estimate the probability that a college student chosen at random owns a compact-disc player. If actually 25% of all college students own compact-disc players, what is the probability that in a random survey of 250 students, at most 50 of them own the devices?

14. The probabilities of failure for each of three independent components in a device are .01, .02, and .01, respectively. The device fails only if all three components fail. Out of a lot of 1 million devices, how many would be expected to fail? Find the probability that more than three devices in the lot fail.

SOLUTIONS TO PRACTICE PROBLEMS 6

1. (a) Giving the drug to a single mouse is a binomial trial with "recovery" as "success" and "death" as "failure." If the drug has no effect, then the probability of success is $\frac{1}{2}$. The probability of five successes in five trials is given by formula (2), with $n = 5$, $p = \frac{1}{2}$, $q = \frac{1}{2}, k = 5$.

$$\Pr(X = 5) = \binom{5}{5}(\tfrac{1}{2})^5(\tfrac{1}{2})^0 = (\tfrac{1}{2})^5 = \tfrac{1}{32} = .03125.$$

(b) As in part (a), this experiment is a binomial experiment with $p = \frac{1}{2}$. However, now $n = 25$. The probability that 18 or more mice recover is

$$\Pr(X = 18) + \Pr(X = 19) + \cdots + \Pr(X = 25).$$

This probability is the area of the shaded portion of the histogram in Fig.   The histogram can be approximated by the normal curve with $\mu = 25 \cdot \frac{1}{2} = 12.5$ and $\sigma = \sqrt{25 \cdot \frac{1}{2} \cdot \frac{1}{2}} = \sqrt{\frac{25}{4}} = \frac{5}{2} = 2.5$ [Fig. 9(b)]. Since the shaded portion of the histogram begins at the point 17.5, the desired probability is approximately the area of the shaded region under the normal curve. The number 17.5 is $(17.5 - 12.5)/2.5 = 5/2.5 = 2$ standard deviations to the right of the mean. Hence, the area under the curve is

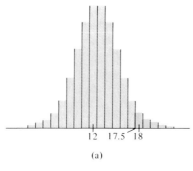
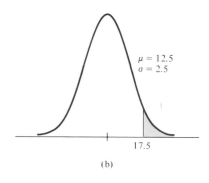

(a)   (b)

FIGURE 9

$1 - A(2) = 1 - .9772 = .0228$. Therefore, the probability that 18 or more mice recover is approximately .0228.

2. Both experiments offer convincing evidence that the drug is helpful in treating the illness. The likelihood of obtaining the results by pure chance is slim. The second experiment might be considered more conclusive than the first, since the result has a lower probability of being due to chance.

Chapter 7: CHECKLIST

☐ Relative frequency distribution
☐ Probability distribution
☐ Histogram
☐ Random variable
☐ Average value, mean, expected value
☐ Variance
☐ Standard deviation
☐ Chebychev inequality
☐ Normal distribution
☐ Standard normal curve
☐ Binomial random variable
☐ Use of normal curve to approximate binomial probabilities
☐ Percentiles

Chapter 7: SUPPLEMENTARY EXERCISES

1. An experiment consists of three binomial trials, each having probability $\frac{1}{3}$ of success.

    (a) Determine the probability distribution table for the number of successes.

    (b) Use the table to compute the mean and variance of the probability distribution.

2. Find the area of the shaded region under the standard normal curve shown in Fig. 1(a).

3. Find the area of the shaded region under the normal curve with $\mu = 5, \sigma = 3$ shown in Fig. 1(b).

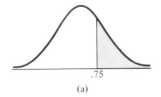
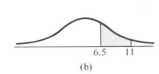

.75   6.5   11

(a)   (b)   FIGURE 1

4. An archer has probability .3 of hitting a certain target. What is the probability of hitting the target exactly two times in four attempts?

5. Suppose that a probability distribution has mean 10 and standard deviation $\frac{1}{3}$. Use the Chebychev inequality to estimate the probability that an outcome will lie between 9 and 11.

6. Table 1 gives the probability distribution of the random variable X. Compute the mean and the variance of the random variable.

TABLE 1

k	$\Pr(X = k)$
0	.2
1	.3
5	.1
10	.4

7. The height of adult males in the United States is normally distributed with $\mu = 5.75$ feet and $\sigma = .2$ feet. What percent of the adult male population has height of 6 feet or greater?

8. An urn contains four red balls and four white balls. An experiment consists of selecting at random a sample of four balls and recording the number of red balls in the sample. Set up the probability distribution and compute its mean and variance.

9. In a certain city two-fifths of the registered voters are women. Out of a group of 54 voters allegedly selected at random for jury duty, 13 are women. A local civil liberties group has charged that the selection procedure discriminated against women. Use the normal curve to estimate the probability of 13 or less women being selected in a truly random selection process.

10. In a complicated production process, $\frac{1}{4}$ of the items produced have to be readjusted. Find the probability that out of a batch of 75 items, between 8 and 22 (inclusive) of the items require readjustment.

11. Figure 2(a) is a normal curve with $\mu = 80$ and $\sigma = 15$. Find the value of h for which the area of the shaded region is .8664.

12. Figure 2(b) is a standard normal curve. Find the value of z for which the area of the shaded region is .7734.

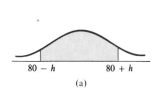 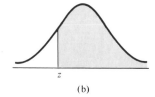

80 − h	80 + h	z
(a)		(b)

FIGURE 2

13. The World Series in baseball consists of a sequence of games which terminates when one team (the winner) wins its fourth game. If the two teams are equally likely to win any one game, what is the probability that the series will last exactly four games? Five? Six? Seven?

14. A true-false exam consists of ten 10-point questions. The instructor informs the students that six of the answers are *true* and four are *false*. An unprepared student decides to guess the answer to each question with the use of the spinner in Fig. 3 that gives *true* 60% of the time. Determine the student's expected score. Can you think of a better strategy?

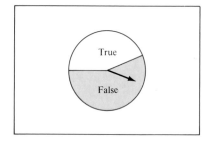

FIGURE 3

MARKOV PROCESSES

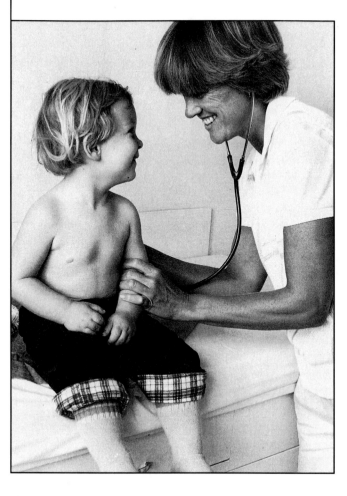

8.1 The Transition Matrix

8.2 Regular Stochastic Matrices

8.3 Absorbing Stochastic Matrices

Suppose that we perform, one after the other, a sequence of experiments which have the same set of outcomes. The probabilities of the various outcomes of a particular experiment of the sequence may depend in some way on the outcomes of preceding experiments. The nature of such a dependency may be very complicated. In the extreme, the outcome of the current experiment may depend on the entire history of the outcomes of preceding experiments. However, there is a simple type of dependency which occurs frequently in applications and which we can analyze with fair ease. Namely, we suppose that the probabilities of the various outcomes of the current experiment depend (at most) on the outcome of the preceding experiment. In this case the sequence of experiments is called a *Markov process*. In this chapter we present some of the most elementary ideas concerning Markov processes and their applications.

8.1. The Transition Matrix

Here are some Markov processes that arise in applications.

EXAMPLE 1 (*Investment*) A particular utility stock is very stable and, in the short run, the probability that it increases or decreases in price depends only on the result of the preceding day's trading. The price of the stock is observed at 4 P.M. each day and is recorded as "increased," "decreased," or "unchanged." The sequence of observations forms a Markov process.

EXAMPLE 2 (*Medicine*) A doctor tests the effect of a new drug on high blood pressure. Based on the effects of metabolism, a given dose is eliminated from the body in 24 hours. Blood pressure is measured once a day and is recorded as "high," "low," or "normal." The sequence of measurements forms a Markov process.

EXAMPLE 3 (*Sociology*) A sociologist postulates that the likelihood that, in certain countries, a woman will enter the labor force depends primarily on whether the woman's mother worked. He designs an experiment to test this hypothesis by viewing the sequence of career choices of a woman, her daughters, her granddaughters, her great-granddaughters, and so on as a Markov process.

Let us now introduce some vocabulary and mathematical machinery with which to study Markov processes. The experiments are performed at regular time intervals and have the same set of outcomes. These outcomes are called *states* and the outcome of the current experiment is referred to as the *current state* of the process. After each time interval the process changes its state. The transition from state to state can be described by tree diagrams, as is shown in the next example.

EXAMPLE 4 Refer to the utility stock of Example 1. Suppose that if the stock increases one day, the probability that on the next day it increases is .3, remains unchanged .2, decreases .5. On the other hand, if the stock is unchanged one day, the probability that on the next day it increases is .6, remains unchanged .1, decreases .3. If the stock decreases one day, the probability that it increases the next day is .3, is unchanged .4, decreases .3. Represent the possible transitions between states and their probabilities by tree diagrams.

Solution The Markov process has three states: "increases," "unchanged," and "decreases." The transitions from the first state ("increase") to the other states are:

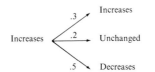

Note that each branch of the tree has been labeled with the probability of the corresponding transition. Similarly, the tree diagrams corresponding to the other two states are:

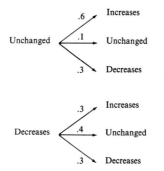

The three tree diagrams of Example 4 may be summarized in a single matrix. We insert the probabilities from a given tree down a column of the matrix, so that each column of the matrix records the information about transitions from one particular state. So the first column of the matrix is $\begin{bmatrix} .3 \\ .2 \\ .5 \end{bmatrix}$, corresponding to transitions from the state "increases." The complete matrix is:

		Current state		
		Increases	Unchanged	Decreases
Next state	Increases	.3	.6	.3
	Unchanged	.2	.1	.4
	Decreases	.5	.3	.3

This matrix, which records all data about transitions from one state to another, is called the *transition matrix of the Markov process*.

EXAMPLE 5 (*Women in the Labor Force*)   Census studies in Holland reveal that 60% of the daughters of working women also work and that 20% of the daughters of non-working women work. Assume that this trend remains unchanged from one generation to the next. Determine the corresponding transition matrix.

Solution There are two states, which we label "work" and don't work." The first column corresponds to transitions from the first state—that is, from "work." The probability that the daughter of a working woman also works is .6. So the probability that the daughter of a working woman chooses *not* to work is $1 - .6 = .4$. Therefore, the first column is

$$\begin{bmatrix} .6 \\ .4 \end{bmatrix}.$$

In similar fashion, the second column is

$$\begin{bmatrix} .2 \\ .8 \end{bmatrix}.$$

The transition matrix is, therefore,

		Current generation	
		Work	Don't work
Next generation	Work	.6	.2
	Don't work	.4	.8

Here is the form of a general transition matrix for a Markov process:

		Current state				
		State 1	$\cdots$	State j	$\cdots$	State n
Next state	State 1					
	$\vdots$					
	State i			$\Pr(\text{next } i \mid \text{current } j)$		
	$\vdots$					
	State n					

Note that this matrix satisfies the following two properties:

1. All entries are ≥ 0.
2. The sum of the entries in each column is 1.

Any square matrix satisfying properties 1 and 2 is called a *stochastic matrix*. (The word "stochastic" derives from the Greek word "stochastices," which means a person who predicts the future.)

Let us examine further the Markov process of Example 5. Currently, 25% of Dutch women work and 75% do not work. This distribution is described by the column matrix

$$\begin{bmatrix} .25 \\ .75 \end{bmatrix}_0,$$

which is called a *distribution matrix*. The subscript 0 is added to denote that this matrix describes generation 0, the current generation. Their daughters constitute generation 1, and their granddaughters generation 2. There is a distribution matrix for each generation: the distribution matrix for generation n is

$$\begin{bmatrix} p_W \\ p_{DW} \end{bmatrix}_n,$$

where p_W is the percentage of women in generation n who work and p_{DW} is the percentage who don't work. Of course, the numbers p_W and p_{DW} are also probabilities. The number p_W is the probability that a woman selected at random from generation n works. Similarly, p_{DW} is the probability that a woman selected at random from generation n does not work. Shortly we shall give a method for calculating the distribution matrix for generation n.

In general, whenever a Markov process applies to a group with members in r possible states, a distribution matrix of the form

$$\begin{bmatrix} p_1 \\ p_2 \\ \vdots \\ p_r \end{bmatrix}_0$$

gives the initial percentages of members in each of the r states. Similarly, a matrix of the same type (with the subscript n) gives the percentages of members in each of the r states after n time periods. Note that each of the percentages in the distribution matrix is also a probability—the probability that a randomly selected member will be in the corresponding state.

EXAMPLE 6 Currently, 25% of Dutch women work. Use the stochastic matrix of Example 5 to determine the percentages of working women in each of the next two generations.

Solution Let us denote by A the stochastic matrix of Example 5:

Current generation

		Work	Don't work	
Next generation	Work	.6	.2	
	Don't work	.4	.8	$= A.$

The initial distribution matrix is

$$\begin{bmatrix} .25 \\ .75 \end{bmatrix}_0.$$

Our goal is to compute the distribution matrices for generations 1 and 2. With an eye toward performing similar calculations for other countries, let us use letters rather than specific numbers in these distribution matrices. Let

$$\left[\quad \right]_0 = \begin{bmatrix} x_0 \\ y_0 \end{bmatrix} \quad \left[\quad \right]_1 = \begin{bmatrix} x_1 \\ y_1 \end{bmatrix} \quad \left[\quad \right]_2 = \begin{bmatrix} x_2 \\ y_2 \end{bmatrix}.$$

The state transitions from generation 0 to generation 1 may be displayed in a tree diagram.

Generation 0	Generation 1	Probabilities
	work	$.6x_0$
work	don't work	$.4x_0$
	work	$.2y_0$
don't work	don't work	$.8y_0$

Adding together the probabilities for the two paths leading to "work" in generation 1 gives

$$x_1 = .6x_0 + .2y_0. \tag{1}$$

Similarly, adding together the probabilities for the two paths leading to "don't work" in generation 1 gives

$$y_1 = .4x_0 + .8y_0. \tag{2}$$

Thus (1) and (2) show that x_1 and y_1 can be computed from this pair of equations.

$$\begin{cases} .6x_0 + .2y_0 = x_1 \\ .4x_0 + .8y_0 = y_1. \end{cases}$$

This system of equations is equivalent to the one matrix equation

$$\begin{bmatrix} .6 & .2 \\ .4 & .8 \end{bmatrix} \begin{bmatrix} x_0 \\ y_0 \end{bmatrix} = \begin{bmatrix} x_1 \\ y_1 \end{bmatrix} \tag{3}$$

Or, to write equation (3) symbolically,

$$A \left[\quad \right]_0 = \left[\quad \right]_1. \tag{4}$$

Now it is easy to do the arithmetic to compute the distribution matrix for generation 1 of Dutch women:

$$A \left[\quad \right]_0 = \begin{bmatrix} .6 & .2 \\ .4 & .8 \end{bmatrix} \begin{bmatrix} .25 \\ .75 \end{bmatrix}_0 = \begin{bmatrix} .30 \\ .70 \end{bmatrix}_1.$$

That is, 30% of Dutch women in generation 1 will work, 70% will not.

To compute the distribution matrix for generation 2 we use the same reasoning as above. There we showed that to get $\begin{bmatrix} \\ \end{bmatrix}_1$ from $\begin{bmatrix} \\ \end{bmatrix}_0$, just multiply by A.

Similarly,

$$A\begin{bmatrix} \\ \end{bmatrix}_1 = \begin{bmatrix} \\ \end{bmatrix}_2.$$

However, by (4), we have a formula for $\begin{bmatrix} \\ \end{bmatrix}_1$ which we can insert into the last equation, getting*

$$\begin{bmatrix} \\ \end{bmatrix}_2 = A\begin{bmatrix} \\ \end{bmatrix}_1 = A\left(A\begin{bmatrix} \\ \end{bmatrix}_0 \right) = A^2\begin{bmatrix} \\ \end{bmatrix}_0.$$

In other words,

$$A^2\begin{bmatrix} \\ \end{bmatrix}_0 = \begin{bmatrix} \\ \end{bmatrix}_2. \tag{5}$$

A simple calculation gives

$$A^2 = \begin{bmatrix} .44 & .28 \\ .56 & .72 \end{bmatrix},$$

so that now we can compute the distribution for generation 2 of Dutch women:

$$\begin{bmatrix} .44 & .28 \\ .56 & .72 \end{bmatrix} \begin{bmatrix} .25 \\ .75 \end{bmatrix}_0 = \begin{bmatrix} .32 \\ .68 \end{bmatrix}_2.$$

That is, after two generations 32% of Dutch women work and 68% do not.

An argument similar to that used to derive equation (5) can be used to show that

$$A^3\begin{bmatrix} \\ \end{bmatrix}_0 = \begin{bmatrix} \\ \end{bmatrix}_3, \quad A^4\begin{bmatrix} \\ \end{bmatrix}_0 = \begin{bmatrix} \\ \end{bmatrix}_4, \quad A^5\begin{bmatrix} \\ \end{bmatrix}_0 = \begin{bmatrix} \\ \end{bmatrix}_5, \ldots.$$

A condensed notation is

$$A^n\begin{bmatrix} \\ \end{bmatrix}_0 = \begin{bmatrix} \\ \end{bmatrix}_n \quad (n = 1, 2, 3, \ldots). \tag{6}$$

That is, to compute the distribution matrix for generation n merely compute the product of A^n times the distribution matrix for generation 0. Equation (6) can be used to predict the distribution matrix for any number of generations into the

---

* Just as in elementary algebra, we define the powers $A^2, A^3, \ldots$ of the matrix A by $A^2 = A \cdot A$, $A^3 = A \cdot A \cdot A, \ldots$.

future, starting from any given distribution matrix. Let us now look at an entirely different type of situation which can be described by a Markov process.

EXAMPLE 7 Suppose that taxis pick up and deliver passengers in a city which is divided into three zones. Records kept by the drivers show that of the passengers picked up in zone I, 50% are taken to a destination in zone I, 40% to zone II, and 10% to zone III. Of the passengers picked up in zone II, 40% go to zone I, 30% to zone II, and 30% to zone III. Of the passengers picked up in zone III, 20% go to zone I, 60% to zone II, and 20% to zone III. Suppose that at the beginning of the day 60% of the taxis are in zone I, 10% in zone II, and 30% in zone III. What is the distribution of taxis in the various zones after all have had one rider? two riders?

Solution This situation is an example of a Markov process. The states are the zones. The initial distribution of the taxis gives the zeroth distribution matrix:

$$\begin{bmatrix} .6 \\ .1 \\ .3 \end{bmatrix}_0 .$$

The stochastic matrix associated to the process is the one giving the probabilities of taxis starting in any one zone and ending up in any other. There is one column for each zone:

$$\text{From zone}$$

$$\begin{array}{cc} & \begin{array}{ccc} \text{I} & \text{II} & \text{III} \end{array} \\ \text{To zone} \quad \begin{array}{c} \text{I} \\ \text{II} \\ \text{III} \end{array} & \begin{bmatrix} .5 & .4 & .2 \\ .4 & .3 & .6 \\ .1 & .3 & .2 \end{bmatrix} = A. \end{array}$$

After all taxis have had one passenger, the distribution matrix is just

$$A \begin{bmatrix} .6 \\ .1 \\ .3 \end{bmatrix}_0 = \begin{bmatrix} .5 & .4 & .2 \\ .4 & .3 & .6 \\ .1 & .3 & .2 \end{bmatrix} \begin{bmatrix} .6 \\ .1 \\ .3 \end{bmatrix}_0 = \begin{bmatrix} .40 \\ .45 \\ .15 \end{bmatrix}_1 .$$

That is, 40% of the taxis are in zone I, 45% in zone II, and 15% in zone III. After all taxis have had two passengers, the distribution matrix is

$$A^2 \begin{bmatrix} .6 \\ .1 \\ .3 \end{bmatrix}_0 ,$$

which, after some arithmetic, can be shown to be

$$\begin{bmatrix} .410 \\ .385 \\ .205 \end{bmatrix}_2 .$$

That is, after two passengers, 41% of the taxis are in zone I, 38.5% are in zone II, and 20.5% in zone III.

The crucial formula used in both Examples 6 and 7 is

$$A \begin{bmatrix} \quad \\ \quad \end{bmatrix}_0 = \begin{bmatrix} \quad \\ \quad \end{bmatrix}_1. \tag{7}$$

From this one follows the more general formula

$$A^n \begin{bmatrix} \quad \\ \quad \end{bmatrix}_0 = \begin{bmatrix} \quad \\ \quad \end{bmatrix}_n.$$

We carefully proved (7) in the special case of Example 6. Let us do the same for Example 7. Suppose that the initial distribution of taxis is

$$\begin{bmatrix} x_0 \\ y_0 \\ z_0 \end{bmatrix}.$$

How many taxis end up in zone I after one passenger? The taxis in zone I come from three sources—zones I, II, and III. And the first row of the stochastic matrix A gives the percentages of taxis starting out in each of the zones and ending up in zone I:

[percent of taxis going to zone I] = [.5] · [percent of taxis in zone I]

$$+ [.4] \cdot [\text{percent of taxis in zone II}]$$

$$+ [.2] \cdot [\text{percent of taxis in zone III}]$$

$$= .5x_0 + .4y_0 + .2z_0.$$

Indeed, the first entry, x_1, in the product

$$\begin{bmatrix} .5 & .4 & .2 \\ .4 & .3 & .6 \\ .1 & .3 & .2 \end{bmatrix} \begin{bmatrix} x_0 \\ y_0 \\ z_0 \end{bmatrix} = \begin{bmatrix} x_1 \\ \\ \end{bmatrix}$$

is just $.5x_0 + .4y_0 + .2z_0$. Similarly, the proportions of taxis in zones II and III coincide with the other two entries in the matrix product. So equation (7) really holds.

PRACTICE PROBLEMS 1

1. Is $\begin{bmatrix} \frac{2}{5} & 1 \\ \frac{3}{5} & .2 \\ \frac{2}{5} & -.3 \end{bmatrix}$ a stochastic matrix?

2. An elementary learning process consists of subjects participating in a sequence of events. Experiment shows that of the subjects not conditioned to make the correct response at

the beginning of any event, 40% will be conditioned to make the correct response at the end of the event. Once a subject is conditioned to make the correct response, he stays conditioned.

(a) Set up the 2 × 2 stochastic matrix with columns and rows labeled N (not-conditioned) and C (conditioned) which describes this situation.

(b) Compute A^3.

(c) If initially all the subjects are not conditioned, what percent of them will be conditioned after three events?

EXERCISES 1

In Exercises 1–6, determine whether or not the matrix is stochastic.

1. $\begin{bmatrix} 1 & .8 \\ 0 & .2 \end{bmatrix}$

2. $\begin{bmatrix} \frac{1}{3} & \frac{1}{3} \\ \frac{2}{3} & \frac{2}{3} \end{bmatrix}$

3. $\begin{bmatrix} .4 & .3 & .2 \\ .6 & .7 & .8 \end{bmatrix}$

4. $\begin{bmatrix} .4 & .5 & .1 \\ .3 & .4 & 0 \\ .3 & .2 & .9 \end{bmatrix}$

5. $\begin{bmatrix} \frac{1}{6} & \frac{5}{12} & 0 \\ \frac{1}{2} & \frac{1}{4} & .5 \\ \frac{1}{3} & \frac{1}{3} & .5 \end{bmatrix}$

6. $\begin{bmatrix} 1 & 0 & 0 \\ 0 & 1 & 0 \\ 0 & 0 & 1 \end{bmatrix}$

7. Referring to Example 5 (women in the labor force), consider a typical group of Danish women, of whom 40% currently work. Assume that the same percentage of daughters follow in their mothers' footsteps as with the Dutch women—that is, those given by the matrix $A = \begin{bmatrix} .6 & .2 \\ .4 & .8 \end{bmatrix}$. Use A and A^2 to determine the proportion of working women in the next two generations.

8. Repeat Exercise 7 for the women of Belgium, of whom 33% currently work. (Round off the percentage of women to the nearest whole percent.)

9. Refer to Example 7 (taxi zones). If originally 40% of the taxis start in zone I, 40% in zone II, and 20% in zone III, how will the taxis be distributed after each has taken one passenger?

10. (*T-maze*) Each day mice are put into a T-maze (a maze shaped like a "T"). In this maze they have the choice of turning to the left (rewarded with cheese) or to the right (receive cheese along with mild shock). After the first day their decision whether to turn left or right is influenced by what happened on the previous day. Of those that go to the left on a certain day, 90% go to the left on the next day and 10% go to the right. Of those that go to the right on a certain day, 70% go to the left on the next day and 30% go to the right.

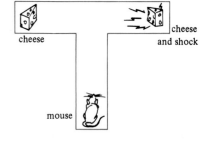

(a) Set up the 2×2 stochastic matrix with columns and rows labeled L, R which describes this situation.

(b) Compute the square of the matrix in part (a).

(c) Suppose that on the first day (day 0) 50% go to the left and 50% go to the right. So, the initial distribution is given by the column matrix $\begin{bmatrix} .5 \\ .5 \end{bmatrix}_0$. Using the matrices in parts (a) and (b), find the distribution matrices for the next two days, days 1 and 2.

(d) Make a guess as to the percentage of mice that will go to the left after 50 days. (Do not compute.)

11. (*Gym*)   A group of physical fitness devotees work out in the gym every day. The workouts vary from strenuous to moderate to light. When their exercise routine was recorded, the following observation was made: Of the people who work out strenuously on a particular day, 40% will work out strenuously on the next day and 60% will work out moderately. Of the people who work out moderately on a particular day, 50% will work out strenuously and 50% will work out lightly on the next day. Of the people working out lightly on a particular day, 30% will work out strenuously the next day, 20% moderately, and 50% lightly.

(a) Set up the 3×3 stochastic matrix with columns and rows labeled S, M, and L which describes these transitions.

(b) Suppose that on a particular Monday 80% have a strenuous, 10% a moderate, and 10% a light workout. What percent will have a strenuous workout on Wednesday?

12. (*Bird Migrations*)   The matrix below describes the migration patterns of a species of bird from year to year among three habitats, I, II, and III.

$$\begin{array}{cc} & \text{From habitat} \\ & \begin{array}{ccc} \text{I} & \text{II} & \text{III} \end{array} \\ \text{To habitat} \quad \begin{array}{c} \text{I} \\ \text{II} \\ \text{III} \end{array} & \begin{bmatrix} .78 & .07 & .15 \\ .12 & .85 & .05 \\ .10 & .08 & .80 \end{bmatrix} \end{array}$$

(a) What percentage of the birds that begin the year in habitat III migrate to habitat I during a year?

(b) Explain the meaning of the percentage .85 appearing in the center of the matrix.

(c) If there are 1000 birds in each habitat at the beginning of a year, how many will be in each habitat at the end of the year? At the end of two years?

13. (*Voter Patterns*)   For a certain group of states, it was observed that 70% of the Democratic governors were succeeded by Democrats and 30% by Republicans. Also, 40% of the Republican governors were succeeded by Democrats and 60% by Republicans.

(a) Set up the 2×2 stochastic matrix with columns and rows labeled D and R which displays these transitions.

(b) Compute A^2 and A^3.

(c) Suppose that all the current governors are Democrats. Assuming that the current trend holds for three elections, what percent of the governors will then be Democrats?

14. Suppose the matrices $\begin{bmatrix} .5 & .4 \\ .5 & .6 \end{bmatrix}$ and $\begin{bmatrix} .3 \\ .7 \end{bmatrix}$ are the stochastic and initial distribution matrices of a Markov process. Find the third and fourth distribution matrices. (Round entries to two decimal places.)

15. $\begin{bmatrix} \frac{1}{3} & \frac{1}{3} \\ \frac{2}{3} & \frac{2}{3} \end{bmatrix}$

16. $\begin{bmatrix} 1 & \frac{1}{2} \\ 0 & \frac{1}{2} \end{bmatrix}$

17. $\begin{bmatrix} .1 & .3 \\ .9 & .7 \end{bmatrix}$ (round off to two decimal places)

18. $\begin{bmatrix} 0 & 1 \\ 1 & 0 \end{bmatrix}$

19. $\begin{bmatrix} .2 & .2 & .2 \\ .3 & .3 & .3 \\ .5 & .5 & .5 \end{bmatrix}$

Stochastic matrices for which some power contains no zero entries are called *regular* matrices. In Exercises 20 and 21, determine whether or not the given matrix is regular by looking at A and A^2.

20. $\begin{bmatrix} \frac{2}{3} & 0 \\ \frac{1}{3} & 1 \end{bmatrix}$

21. $\begin{bmatrix} 0 & .4 \\ 1 & .6 \end{bmatrix}$

SOLUTIONS TO PRACTICE PROBLEMS 1

1. No. It fails on all three conditions. The matrix is not square, the entry $-.3$ is not ≥ 0, and the sum of the entries in the first (and second) column is not equal to 1.

2.

$\quad\quad N \quad C$

(a) $\begin{matrix} N \\ C \end{matrix} \begin{bmatrix} .6 & 0 \\ .4 & 1 \end{bmatrix}$. Since there are only two possibilities and 40% of those not-conditioned become conditioned, the remainder or 60% stay not-conditioned. After each event 100% of the conditioned stay conditioned, and therefore 0% become not-conditioned.

(b) $A^2 = \begin{bmatrix} .6 & 0 \\ .4 & 1 \end{bmatrix}\begin{bmatrix} .6 & 0 \\ .4 & 1 \end{bmatrix} = \begin{bmatrix} .36 & 0 \\ .64 & 1 \end{bmatrix}$,

$A^3 = A^2 \cdot A = \begin{bmatrix} .36 & 0 \\ .64 & 1 \end{bmatrix}\begin{bmatrix} .6 & 0 \\ .4 & 1 \end{bmatrix} = \begin{bmatrix} .216 & 0 \\ .784 & 1 \end{bmatrix}$.

(c) Here $\begin{bmatrix} \\ \end{bmatrix}_0 = \begin{bmatrix} 1 \\ 0 \end{bmatrix}$. Therefore,

$$\begin{bmatrix} \\ \end{bmatrix}_3 = A^3 \begin{bmatrix} 1 \\ 0 \end{bmatrix} = \begin{bmatrix} .216 & 0 \\ .784 & 1 \end{bmatrix}\begin{bmatrix} 1 \\ 0 \end{bmatrix} = \begin{bmatrix} .216 \\ .784 \end{bmatrix}.$$

So, 78.4% will be conditioned after three events.

8.2. Regular Stochastic Matrices

In the preceding section we studied the percentages of working women in various generations in Holland. We showed that if $\begin{bmatrix} & \\ & \end{bmatrix}_0$ is the initial distribution matrix, then the distribution matrix $\begin{bmatrix} & \\ & \end{bmatrix}_n$ for the nth generation is given by

$$\begin{bmatrix} & \\ & \end{bmatrix}_n = A^n \begin{bmatrix} & \\ & \end{bmatrix}_0, \qquad (1)$$

where A is the stochastic matrix

$$A = \begin{bmatrix} .6 & .2 \\ .4 & .8 \end{bmatrix}.$$

In this section we are interested in determining long-term trends in Markov processes. To get an idea of what is meant, consider the example of the women in the labor force.

EXAMPLE 1 In Holland 25% of the women currently work.

(a) How many women work after $1, 2, 3, \ldots, 11$ generations?

(b) Estimate the long-term trend.

(c) Answer the same questions (a) and (b) for Denmark, assuming that 40% of all Danish women currently work and that the effect of maternal influence is the same as for Holland.

Solution (a) The percentages of women working in generation n (for $n = 1, 2, 3, \ldots$) can be determined from equation (1):

$$\begin{bmatrix} & \\ & \end{bmatrix}_n = \begin{bmatrix} .6 & .2 \\ .4 & .8 \end{bmatrix}^n \begin{bmatrix} .25 \\ .75 \end{bmatrix}_0.$$

After the mildly tedious job of raising the stochastic matrix to various powers, we obtain the following results:

Generation	Percent of women working	Generation	Percent of women working
0	25	6	33.30
1	30	7	33.32
2	32	8	33.33
3	32.8	9	33.33
4	33.12	10	33.33
5	33.25	11	33.33

(b) It appears from the accompanying table that the long-term trend is for one-third or $33\frac{1}{3}\%$ of all Dutch women to work.

(c) The corresponding results for Denmark can be computed by replacing the initial distribution matrix $\begin{bmatrix} .25 \\ .75 \end{bmatrix}_0$ by $\begin{bmatrix} .40 \\ .60 \end{bmatrix}_0$, reflecting that initially 40% of all Danish women work. The results of the calculations are:

Generation	Percent of women working	Generation	Percent of women working
0	40	6	33.36
1	36	7	33.34
2	34.4	8	33.34
3	33.76	9	33.34
4	33.50	10	33.33
5	33.40	11	33.33

Again, the long-term trend is for one-third of the women to work.

From the example above one might begin to suspect the following: The long-term trend is always for one-third of the women to work, independent of the initial distribution.

Verification   To see why this rather surprising fact should hold, it is useful to examine the powers of A:

$$A^2 = \begin{bmatrix} .44 & .28 \\ .56 & .72 \end{bmatrix} \qquad A^3 = \begin{bmatrix} .376 & .312 \\ .624 & .688 \end{bmatrix} \qquad A^4 = \begin{bmatrix} .3504 & .3248 \\ .6496 & .6752 \end{bmatrix}$$

$$A^5 = \begin{bmatrix} .3401 & .3299 \\ .6598 & .6701 \end{bmatrix} \qquad A^6 = \begin{bmatrix} .3361 & .3320 \\ .6639 & .6680 \end{bmatrix} \qquad A^7 = \begin{bmatrix} .3344 & .3328 \\ .6656 & .6672 \end{bmatrix}$$

$$A^8 = \begin{bmatrix} .3338 & .3331 \\ .6662 & .6669 \end{bmatrix} \qquad A^9 = \begin{bmatrix} .3335 & .3332 \\ .6665 & .6668 \end{bmatrix} \qquad A^{10} = \begin{bmatrix} .3334 & .3333 \\ .6666 & .6667 \end{bmatrix}.$$

As A is raised to further powers, the matrices approach

$$\begin{bmatrix} \frac{1}{3} & \frac{1}{3} \\ \frac{2}{3} & \frac{2}{3} \end{bmatrix}. \tag{2}$$

Now suppose that initially the proportion of women working is x_0. That is, the initial distribution matrix is

$$\begin{bmatrix} x_0 \\ 1 - x_0 \end{bmatrix}.$$

Then, after n generations, the distribution matrix is

$$\begin{bmatrix} \\ \end{bmatrix}_n = A^n \begin{bmatrix} \\ \end{bmatrix}_0 = A^n \begin{bmatrix} x_0 \\ 1 - x_0 \end{bmatrix}.$$

But after many generations n is large, so that A^n is approximately the matrix (2). And thus,

$$\begin{bmatrix} \ \\ \ \end{bmatrix}_n \approx \begin{bmatrix} \frac{1}{3} & \frac{1}{3} \\ \frac{2}{3} & \frac{2}{3} \end{bmatrix} \begin{bmatrix} x_0 \\ 1 - x_0 \end{bmatrix} = \begin{bmatrix} \frac{1}{3}x_0 + \frac{1}{3}(1 - x_0) \\ \frac{2}{3}x_0 + \frac{2}{3}(1 - x_0) \end{bmatrix} = \begin{bmatrix} \frac{1}{3} \\ \frac{2}{3} \end{bmatrix}.$$

In other words, after n generations approximately one-third of the women work and two-thirds do not.

From the calculations above we see that the stochastic matrix A possesses a number of very special properties. First, as n gets large, A^n approaches the matrix $\begin{bmatrix} \frac{1}{3} & \frac{1}{3} \\ \frac{2}{3} & \frac{2}{3} \end{bmatrix}$. Second, any initial distribution approaches the distribution $\begin{bmatrix} \frac{1}{3} \\ \frac{2}{3} \end{bmatrix}$ after many generations. The limiting matrix $\begin{bmatrix} \frac{1}{3} & \frac{1}{3} \\ \frac{2}{3} & \frac{2}{3} \end{bmatrix}$ is called the *stable matrix* of A, and the limiting distribution $\begin{bmatrix} \frac{1}{3} \\ \frac{2}{3} \end{bmatrix}$ is called the *stable distribution* of A. Finally, note that all the columns of the stable matrix are the same and are equal to the stable distribution.

The matrices that share the properties above with A are very important. For these matrices one can predict a long-term trend, and this trend is independent of the initial distribution. An important class of matrices with these properties is the class of *regular stochastic matrices*.

A stochastic matrix is said to be *regular* if some power has all positive entries.

EXAMPLE 2   Which of the following stochastic matrices are regular?

$$(a) \ \begin{bmatrix} .6 & .2 \\ .4 & .8 \end{bmatrix} \qquad\qquad (b) \ \begin{bmatrix} 0 & .5 \\ 1 & .5 \end{bmatrix} \qquad\qquad (c) \ \begin{bmatrix} 0 & 1 \\ 1 & 0 \end{bmatrix}$$

Solution   (a) All entries are positive, so the matrix is regular.

(b) Here a zero occurs in the first power. However,

$$\begin{bmatrix} 0 & .5 \\ 1 & .5 \end{bmatrix}^2 = \begin{bmatrix} .5 & .25 \\ .5 & .75 \end{bmatrix},$$

which has all positive entries. So the original matrix is regular.

(c) Note that

$$\begin{bmatrix} 0 & 1 \\ 1 & 0 \end{bmatrix}^2 = \begin{bmatrix} 1 & 0 \\ 0 & 1 \end{bmatrix} \qquad \begin{bmatrix} 0 & 1 \\ 1 & 0 \end{bmatrix}^3 = \begin{bmatrix} 0 & 1 \\ 1 & 0 \end{bmatrix}$$

$$\begin{bmatrix} 0 & 1 \\ 1 & 0 \end{bmatrix}^4 = \begin{bmatrix} 1 & 0 \\ 0 & 1 \end{bmatrix} \qquad \begin{bmatrix} 0 & 1 \\ 1 & 0 \end{bmatrix}^5 = \begin{bmatrix} 0 & 1 \\ 1 & 0 \end{bmatrix}.$$

The even powers of the matrix are the 2×2 identity matrix, and the odd powers are the original matrix. Every power has a zero in it, so the matrix is not regular.

Regular matrices share all the properties observed in the special case. Moreover, there is a simple technique for computing the stable distribution (see property 4 below) which spares us from having to multiply matrices.

Let A be a regular stochastic matrix.

1. The powers A^n approach a certain matrix as n gets large. This limiting matrix is called the *stable matrix* of A.

2. For any initial distribution $\begin{bmatrix} \\ \end{bmatrix}_0$, $A^n \begin{bmatrix} \\ \end{bmatrix}_0$ approaches a certain distribution $\begin{bmatrix} \\ \end{bmatrix}$. This limiting distribution is called the *stable distribution* of A.

3. All columns of the stable matrix are the same; they equal the stable distribution.

4. The stable distribution X can be determined by solving the system of linear equations
$$\begin{cases} \text{sum of the entries of } X = 1, \\ AX = X. \end{cases}$$

EXAMPLE 3 Use property 4 to determine the stable distribution of the regular stochastic matrix

$$A = \begin{bmatrix} .6 & .2 \\ .4 & .8 \end{bmatrix}.$$

Solution Let $X = \begin{bmatrix} x \\ y \end{bmatrix}$ be the stable distribution. The condition "sum of the entries of $X = 1$" yields the equation

$$x + y = 1.$$

The condition $AX = X$ gives the equations

$$\begin{bmatrix} .6 & .2 \\ .4 & .8 \end{bmatrix} \begin{bmatrix} x \\ y \end{bmatrix} = \begin{bmatrix} x \\ y \end{bmatrix}$$

or

$$\begin{cases} .6x + .2y = x \\ .4x + .8y = y. \end{cases}$$

So we have the system

$$\begin{cases} x + y = 1 \\ .6x + .2y = x \\ .4x + .8y = y. \end{cases}$$

Combining terms in the second and third equations and eliminating the decimals by multiplying by 10, we have

$$\begin{cases} x + y = 1 \\ -4x + 2y = 0 \\ 4x - 2y = 0. \end{cases}$$

Note that the second and third equations are the same, except for a factor -1, so the last equation may be omitted. Now the system reads

$$\begin{cases} x + y = 1 \\ -4x + 2y = 0. \end{cases}$$

The diagonal form is obtained by the Gaussian elimination method:

$$\begin{bmatrix} 1 & 1 & | & 1 \\ -4 & 2 & | & 0 \end{bmatrix} \xrightarrow{\ [2] + 4[1]\ } \begin{bmatrix} 1 & 1 & | & 1 \\ 0 & 6 & | & 4 \end{bmatrix}$$

$$\xrightarrow{\ \frac{1}{6}[2]\ } \begin{bmatrix} 1 & 1 & | & 1 \\ 0 & 1 & | & \frac{2}{3} \end{bmatrix}$$

$$\xrightarrow{\ [1] + (-1)[2]\ } \begin{bmatrix} 1 & 0 & | & \frac{1}{3} \\ 0 & 1 & | & \frac{2}{3} \end{bmatrix}.$$

Thus

$$x = \tfrac{1}{3}, \qquad y = \tfrac{2}{3}.$$

So the stable distribution is $\begin{bmatrix} x \\ y \end{bmatrix} = \begin{bmatrix} \frac{1}{3} \\ \frac{2}{3} \end{bmatrix}$, as we observed before. Note that once the stable distribution is determined, the stable matrix is easy to find. Just place the stable distribution in every column:

$$\begin{bmatrix} \frac{1}{3} & \frac{1}{3} \\ \frac{2}{3} & \frac{2}{3} \end{bmatrix}.$$

EXAMPLE 4 In Section 1 we studied the distribution of taxis in three zones of a city. The movement of taxis from zone to zone was described by the regular stochastic matrix

$$A = \begin{bmatrix} .5 & .4 & .2 \\ .4 & .3 & .6 \\ .1 & .3 & .2 \end{bmatrix}.$$

In the long run, what percentage of taxis will be in each of the zones?

Solution Let $X = \begin{bmatrix} x \\ y \\ z \end{bmatrix}$ be the stable distribution of A. Then x is the long-term percentage of taxis in zone I, y the percentage in zone II, and z the percentage in zone III.

X is determined by the equations

$$\begin{cases} x + y + z = 1 \\ AX = X, \end{cases}$$

or, equivalently,

$$\begin{cases} x + y + z = 1 \\ .5x + .4y + .2z = x \\ .4x + .3y + .6z = y \\ .1x + .3y + .2z = z. \end{cases}$$

Rewriting the equations with all the terms involving the variables on the left, we get

$$\begin{cases} x + y + z = 1 \\ -.5x + .4y + .2z = 0 \\ .4x - .7y + .6z = 0 \\ .1x + .3y - .8z = 0. \end{cases}$$

Applying the Gaussian elimination method to this system, we get the solution $x = .4, y = .4, z = .2$. Thus, after many trips, approximately 40% of the taxis are in zone I, 40% in zone II, and 20% in zone III.

We have come full circle. We began Chapter 2 by solving systems of linear equations. Matrices were developed as a tool for solving such systems. Then we found matrices to be interesting in their own right. Finally, in the current chapter, we have used systems of linear equations to answer questions about matrices.

APPENDIX Verification of Method for Obtaining the Stable Distribution

In this appendix we verify that the stable distribution X can be obtained by solving the system of equations

$$\begin{cases} \text{sum of the entries of } X = 1 \\ \phantom{\text{sum of the entries of } X} AX = X. \end{cases}$$

Let A be a regular stochastic matrix. Suppose that we take as the initial distribution its stable distribution X. Then the nth distribution matrix, or A^nX, approaches the stable distribution (by property 2), so that

$$X \approx A^nX \qquad \text{for large } n.$$

Therefore,

$$AX \approx A \cdot A^nX = A^{n+1}X.$$

But $A^{n+1}X$ is the $(n + 1)$st distribution matrix, which is also approximately X. Thus AX is approximately X. But the approximation can be made closer and closer by taking n large. Therefore, AX is arbitrarily close to X or

$$AX = X. \tag{3}$$

This is a matrix equation in X. There is one other condition on X: X is a distribution matrix. Therefore,

$$\text{sum of the entries of } X = 1. \tag{4}$$

And so, by (3) and (4), we find this system of linear equations for the entries of X:

$$\begin{cases} \text{sum of the entries of } X = 1 \\ \qquad\qquad\qquad AX = X. \end{cases}$$

PRACTICE PROBLEMS 2

1. Is $\begin{bmatrix} 0 & .2 & .5 \\ .5 & 0 & .5 \\ .5 & .8 & 0 \end{bmatrix}$ a regular stochastic matrix? Explain your answer.

2. Find the stable matrix for the regular stochastic matrix in Problem 1.

3. In a study of cigarette smokers it was determined that of the people who smoked menthol cigarettes on a particular day, 10% smoked menthol the next day and 90% smoked nonmenthol. Of the people who smoked nonmenthol cigarettes on a particular day, the next day 30% smoked menthol and 70% smoked nonmenthol. In the long run, what percent of the people will be smoking nonmenthol cigarettes on a particular day?

EXERCISES 2

In Exercises 1–6, determine whether or not the matrix is a regular stochastic matrix.

1. $\begin{bmatrix} \frac{1}{4} & \frac{2}{7} \\ \frac{3}{4} & \frac{5}{7} \end{bmatrix}$

2. $\begin{bmatrix} .6 & 0 \\ .4 & 1 \end{bmatrix}$

3. $\begin{bmatrix} .3 & 1 \\ .7 & 0 \end{bmatrix}$

4. $\begin{bmatrix} 1 & 0 & .7 \\ 0 & 1 & .2 \\ 0 & 0 & .1 \end{bmatrix}$

5. $\begin{bmatrix} 0 & .8 & 0 \\ 1 & .1 & .5 \\ 0 & .1 & .5 \end{bmatrix}$

6. $\begin{bmatrix} .6 & .6 & .6 \\ .3 & .3 & .3 \\ .1 & .1 & .1 \end{bmatrix}$

In Exercises 7–11 find the stable distribution for the given regular stochastic matrix.

7. $\begin{bmatrix} .5 & .1 \\ .5 & .9 \end{bmatrix}$

8. $\begin{bmatrix} .4 & 1 \\ .6 & 0 \end{bmatrix}$

9. $\begin{bmatrix} .8 & .3 \\ .2 & .7 \end{bmatrix}$

10. $\begin{bmatrix} .3 & .1 & .2 \\ .4 & .8 & .6 \\ .3 & .1 & .2 \end{bmatrix}$

11. $\begin{bmatrix} .1 & .4 & .7 \\ .6 & .4 & .2 \\ .3 & .2 & .1 \end{bmatrix}$

12. Refer to Exercise 10 of Exercises 1 of the previous section (T-maze). What percentage of the mice will be going to the left after many days?

13. Refer to Exercise 11 of Exercises 1 of the previous section (gym). In the long run what percentage of the people will have a strenuous workout on a particular day?

14. Refer to Exercise 12 of Exercises 1 of the previous section (voter patterns). In the long run what proportion of the governors will be Democrats?

15. (*Genetics*) With respect to a certain gene, geneticists classify individuals as dominant, recessive, or hybrid. In an experiment, individuals are crossed with hybrids, then their offspring are crossed with hybrids, and so on. For dominant individuals 50% of their offspring will be dominant and 50% will be hybrid. For the recessive individuals 50% of their offspring will be recessive and 50% hybrid. For hybrid individuals (to be crossed with hybrids) their offspring will be 25% dominant, 25% recessive, and 50% hybrid. In the long run what percent of the individuals in a generation will be dominant?

16. Commuters can get into town by car or bus. Surveys have shown that for those taking their car on a particular day, 20% take their car the next day and 80% take a bus. Also, for those taking a bus on a particular day, 50% take their car the next day and 50% take a bus. In the long run what percentage of the people take a bus on a particular day?

17. As shown in Example 2, $\begin{bmatrix} 0 & 1 \\ 1 & 0 \end{bmatrix}$ is not a regular stochastic matrix. Show that the matrix has $\begin{bmatrix} .5 \\ .5 \end{bmatrix}$ as a stable distribution and explain why this fact does not contradict the main premise of this section.

SOLUTIONS TO PRACTICE PROBLEMS 2

1. Yes. It is easily seen to be stochastic. Although it has some zero entries, there are no zero entries in

$$A^2 = \begin{bmatrix} .35 & .40 & .10 \\ .25 & .50 & .25 \\ .40 & .10 & .65 \end{bmatrix}.$$

2. $\begin{cases} x + y + z = 1 \\ \begin{bmatrix} 0 & .2 & .5 \\ .5 & 0 & .5 \\ .5 & .8 & 0 \end{bmatrix} \begin{bmatrix} x \\ y \\ z \end{bmatrix} = \begin{bmatrix} x \\ y \\ z \end{bmatrix}, \end{cases}$ or $\begin{cases} x + y + z = 1 \\ \quad .2y + .5z = x \\ .5x \quad\quad + .5z = y \\ .5x + .8y \quad\quad = z, \end{cases}$

or $\begin{cases} x + y + z = 1 \\ -x + .2y + .5z = 0 \\ .5x - y + .5z = 0 \\ .5x + .8y - z = 0. \end{cases}$

To simplify the arithmetic, multiply each of the last three equations of the system by 10 to eliminate the decimals. Then apply the Gaussian elimination method.

$\begin{cases} x + y + z = 1 \\ -10x + 2y + 5z = 0 \\ 5x - 10y + 5z = 0 \\ 5x + 8y - 10z = 0, \end{cases}$ $\begin{cases} x + y + z = 1 \\ 12y + 15z = 10 \\ -15y \quad\quad = -5 \\ 3y - 15z = -5. \end{cases}$

Next interchange the second and third equations and pivot about $-15y$.

$$\begin{cases} x + y + z = 1 \\ -15y = 5 \\ 12y + 15z = 10 \\ 3y - 15z = -5, \end{cases} \qquad \begin{cases} x + z = \tfrac{2}{3} \\ y = \tfrac{1}{3} \\ 15z = 6 \\ -15z = -6. \end{cases}$$

Pivoting about $15z$ yields $x = \tfrac{4}{15}$, $y = \tfrac{5}{15}$, $z = \tfrac{6}{15}$, so the stable matrix is

$$\begin{bmatrix} \tfrac{4}{15} & \tfrac{4}{15} & \tfrac{4}{15} \\ \tfrac{5}{15} & \tfrac{5}{15} & \tfrac{5}{15} \\ \tfrac{6}{15} & \tfrac{6}{15} & \tfrac{6}{15} \end{bmatrix}.$$

3. The regular stochastic matrix describing this daily transition is

$$\begin{array}{cc} & \begin{array}{cc} M & N \end{array} \\ \begin{array}{c} M \\ N \end{array} & \begin{bmatrix} .1 & .3 \\ .9 & .7 \end{bmatrix}. \end{array}$$

The stable distribution is found by solving

$$\begin{cases} x + y = 1 \\ \begin{bmatrix} .1 & .3 \\ .9 & .7 \end{bmatrix}\begin{bmatrix} x \\ y \end{bmatrix} = \begin{bmatrix} x \\ y \end{bmatrix} \end{cases} \text{ or } \begin{cases} x + y = 1 \\ .1x + .3y = x \\ .9x + .7y = y \end{cases} \text{ or } \begin{cases} x + y = 1 \\ -.9x + .3y = 0 \\ .9x - .3y = 0. \end{cases}$$

Since the last two equations are essentially the same, we need only solve the system consisting of the first two equations. Multiply the second equation by 10:

$$\begin{cases} x + y = 1 \\ -9x + 3y = 0 \end{cases} \rightarrow \begin{cases} x + y = 1 \\ 12y = 9 \end{cases} \rightarrow \begin{cases} x = \tfrac{1}{4} \\ y = \tfrac{3}{4}. \end{cases}$$

So the stable distribution is $\begin{bmatrix} \tfrac{1}{4} \\ \tfrac{3}{4} \end{bmatrix}$. The stable distribution tells us that in the long run, 25% smoke menthol and 75% smoke nonmenthol cigarettes on any particular day.

8.3. Absorbing Stochastic Matrices

In this section we study long-term trends for a certain class of matrices which are not regular—the absorbing stochastic matrices. By way of introduction, recall some general facts about stochastic matrices.

Stochastic matrices, such as

$$\begin{bmatrix} .3 & .5 & .1 & 0 \\ .2 & .2 & .8 & 0 \\ .1 & .3 & 0 & 0 \\ .4 & 0 & .1 & 1 \end{bmatrix},$$

describe state-to-state changes in certain processes. Each column of a stochastic matrix describes the transitions (or movements) from one specific state. For example, the first column of the stochastic matrix above indicates that at the end of one time period the probability is .3 that an object in state 1 stays in state 1,

.2 that it goes to state 2, .1 that it goes to state 3, and .4 that it goes to state 4. Similarly, the second column indicates the probabilities for transitions from state 2, the third column from state 3, and the fourth from state 4.

If A is a stochastic matrix, then the columns of A^2 describe the transitions from the various states *after two time periods*. For instance, the third column of A^2 indicates the probabilities of an object starting out in state 3 and ending up in each of the states after two time periods. Similarly, the columns of A^n indicate the transitions from the various states after n periods.

Consider the stochastic matrix described above. Its fourth column illustrates a curious phenomenon. It indicates that if an object starts out in state 4, after one time period the probabilities of going to states 1, 2, or 3 is 0 and the probability of going to state 4 is 1. In other words, all of the objects in state 4 stay in state 4. A state with this property is called an *absorbing state*. More precisely, an absorbing state is a state that always leads back to itself.

EXAMPLE 1   Find all absorbing states of the stochastic matrix

$$\begin{bmatrix} 1 & 0 & .3 & 0 \\ 0 & 1 & .1 & 1 \\ 0 & 0 & .5 & 0 \\ 0 & 0 & .1 & 0 \end{bmatrix}.$$

Solution   To determine which states are absorbing one must look at the columns. The first column describes the transitions from state 1. It says that state 1 leads to state 1 100% of the time and to the other states 0% of the time. So state 1 is absorbing. Column 2 describes transitions from state 2. It says that state 2 leads to state 2 100% of the time. So state 2 is absorbing. Clearly, the third column says that state 3 is not absorbing. For example, state 3 leads to state 1 with probability .3. At first glance, column 4 seems to say that state 4 is absorbing. But it is not, because column 4 says that state 4 leads to state 2 100% of the time.

Based on the example above, we can easily determine the absorbing states of any stochastic matrix: First, the corresponding column has a single 1 and the remaining entries 0. Second, the lone 1 must be located on the main diagonal of the matrix. That is, its row and column number must be the same. So, for example, state i is an absorbing state if and only if the ith entry in the ith column is 1 and all the remaining entries in that column are 0.

An *absorbing stochastic matrix* is a stochastic matrix in which (1) there is at least one absorbing state and (2) from any state it is possible to get to at least one absorbing state, either directly or through one or more intermediate states.

EXAMPLE 2   Is the matrix

$$\begin{bmatrix} 1 & 0 & .3 & 0 \\ 0 & 1 & .1 & 1 \\ 0 & 0 & .5 & 0 \\ 0 & 0 & .1 & 0 \end{bmatrix}$$

an absorbing stochastic matrix?

Solution In Example 1 we showed that states 1 and 2 were absorbing. From column 3 we see that state 3 can lead to both state 1 and state 2: state 1 with probability .3 and state 2 with probability .1. From column 4 we see that state 4 does not lead to state 1, but it does lead to state 2. Thus states 3 and 4 both lead to absorbing states, and so the matrix is an absorbing stochastic matrix.

In general, processes described by stochastic matrices can oscillate indefinitely from state to state in such a way that they exhibit no long-term trend. An example was given in Section 2 using the matrix

$$\begin{bmatrix} 0 & 1 \\ 1 & 0 \end{bmatrix}.$$

The idea of introducing absorbing states is to reduce the degree of oscillation. For when an absorbing state is reached, the process no longer changes. The main result of this section is that absorbing stochastic matrices exhibit a long-term trend. Further, we can determine this trend using a simple computational procedure.

When considering an absorbing stochastic matrix, we will always arrange the states so that the absorbing states come first, then the nonabsorbing states.

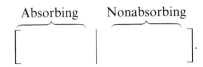

When the states are ordered in this manner, an absorbing stochastic matrix can be partitioned, or subdivided, into four submatrices.

$$\begin{array}{cc} \overbrace{\text{Absorbing}} & \overbrace{\text{Nonabsorbing}} \\ \left[\begin{array}{c|c} I & S \\ \hline 0 & R \end{array}\right]. \end{array}$$

The matrix I is an identity matrix and 0 denotes a matrix having all entries 0. The matrices S and R are the two pieces corresponding to the nonabsorbing states. For example, in the case of the absorbing stochastic matrix of Example 2, this partition is given by

$$\left[\begin{array}{cc|cc} 1 & 0 & .3 & 0 \\ 0 & 1 & .1 & 1 \\ \hline 0 & 0 & .5 & 0 \\ 0 & 0 & .1 & 0 \end{array}\right].$$

EXAMPLE 3 The victims of a certain disease are classified into three states: cured, dead from the disease, or sick. Once a person is cured, he is permanently immune. Each year 70% of those sick are cured, 10% die from the disease, and 20% remain ill.

(a) Determine a stochastic matrix describing the progression of the disease.

(b) Determine the absorbing states.

Solution (a) There is one column for each state. Transferring the given data to the matrix gives the result:

$$
\begin{array}{c}
 \\
\text{Cured} \\
\text{Dead} \\
\text{Sick}
\end{array}
\begin{array}{ccc}
\text{Cured} & \text{Dead} & \text{Sick} \\
\left[\begin{array}{ccc}
1 & 0 & .7 \\
0 & 1 & .1 \\
0 & 0 & .2
\end{array}\right].
\end{array}
$$

(b) The absorbing states are "cured" and "dead," since these are the states that always lead back to themselves. The matrix is an absorbing stochastic matrix, since there is at least one absorbing state and the other state, "sick," can lead to an absorbing state. (In fact, in this case it leads to either of the absorbing states.)

EXAMPLE 4 Find the long-term trend of the absorbing stochastic matrix

$$
A = \left[\begin{array}{cc|c}
1 & 0 & .2 \\
0 & 1 & .4 \\
\hline
0 & 0 & .4
\end{array}\right].
$$

Solution The long-term trend is found by raising A to various powers. Here are the results, accurate to three decimal places:

$$
A^2 = \left[\begin{array}{cc|c}
1 & 0 & .28 \\
0 & 1 & .56 \\
\hline
0 & 0 & .16
\end{array}\right]
\qquad
A^3 = \left[\begin{array}{cc|c}
1 & 0 & .312 \\
0 & 1 & .624 \\
\hline
0 & 0 & .064
\end{array}\right]
$$

$$
A^4 = \left[\begin{array}{cc|c}
1 & 0 & .325 \\
0 & 1 & .650 \\
\hline
0 & 0 & .025
\end{array}\right]
\qquad
A^5 = \left[\begin{array}{cc|c}
1 & 0 & .330 \\
0 & 1 & .660 \\
\hline
0 & 0 & .010
\end{array}\right]
$$

$$
A^6 = \left[\begin{array}{cc|c}
1 & 0 & .332 \\
0 & 1 & .664 \\
\hline
0 & 0 & .004
\end{array}\right]
\qquad
A^7 = \left[\begin{array}{cc|c}
1 & 0 & .333 \\
0 & 1 & .666 \\
\hline
0 & 0 & .001
\end{array}\right]
$$

Based on this numerical evidence, it appears as if the powers of A approach the matrix

$$
\left[\begin{array}{cc|c}
1 & 0 & \frac{1}{3} \\
0 & 1 & \frac{2}{3} \\
\hline
0 & 0 & 0
\end{array}\right].
$$

In other words, after a large number of time intervals, 100% of those originally in state 1 remain in state 1 (column 1), 100% of those originally in state 2 remain in state 2 (column 2), and those originally in state 3 end up in state 1 with probability one-third and in state 2 with probability two-thirds.

The preceding example exhibits three important features of the long-term behavior exhibited by all absorbing stochastic matrices. First, as in the case of regular stochastic matrices, the powers approach a particular matrix. This limiting matrix is called a *stable matrix*. Second, for absorbing stochastic matrices the long-term trend depends on the initial state. For example, the stable matrix just computed gives different results, depending on the state in which you start. This is reflected by the fact that the three columns are different. (In the case of regular stochastic matrices, the long-term trend does not depend on the initial distribution. All of the columns of the stable matrix are the same. They all equal the stable distribution.) The third important point to notice is that, no matter what the initial state, all objects eventually go to absorbing states. The absorbing states act like magnets and attract all objects to themselves in the long run.

EXAMPLE 5 Find the long-term trend of the disease described in Example 3.

Solution   The stochastic matrix here is

$$A = \left[\begin{array}{cc|c} 1 & 0 & .7 \\ 0 & 1 & .1 \\ \hline 0 & 0 & .2 \end{array}\right].$$

Its first few powers are given by

$$A^2 = \left[\begin{array}{cc|c} 1 & 0 & .84 \\ 0 & 1 & .12 \\ \hline 0 & 0 & .04 \end{array}\right] \qquad A^3 = \left[\begin{array}{cc|c} 1 & 0 & .868 \\ 0 & 1 & .124 \\ \hline 0 & 0 & .008 \end{array}\right].$$

$$A^4 = \left[\begin{array}{cc|c} 1 & 0 & .873 \\ 0 & 1 & .125 \\ \hline 0 & 0 & .002 \end{array}\right] \qquad A^5 = \left[\begin{array}{cc|c} 1 & 0 & .875 \\ 0 & 1 & .125 \\ \hline 0 & 0 & .000 \end{array}\right].$$

It appears that the powers approach the matrix

$$\left[\begin{array}{cc|c} 1 & 0 & \frac{7}{8} \\ 0 & 1 & \frac{1}{8} \\ \hline 0 & 0 & 0 \end{array}\right].$$

This is the stable matrix. In other words, reading from column 3, of those initially in state 3, "sick," the probability is seven-eighths of eventually being cured, and one-eighth of eventually dying of the disease.

In Examples 4 and 5 we have computed the stable matrix by raising the given stochastic matrix to various powers. Actually, there is a formal computational procedure for determining the stable matrix. Suppose that we partition an absorbing stochastic matrix into submatrices

$$A = \left[\begin{array}{c|c} I & S \\ \hline 0 & R \end{array}\right].$$

The stable matrix of A is

$$\left[\begin{array}{c|c} I & S(I-R)^{-1} \\ \hline 0 & 0 \end{array}\right].$$

[*Note* The identity matrix I in $(I-R)^{-1}$ is chosen to be the same size as R in order to make the matrix subtraction permissible.]

EXAMPLE 6 Use the formula above to determine the stable matrices of

(a) $\left[\begin{array}{cc|c} 1 & 0 & .2 \\ 0 & 1 & .4 \\ 0 & 0 & .4 \end{array}\right]$
(b) $\left[\begin{array}{cc|c} 1 & 0 & .7 \\ 0 & 1 & .1 \\ 0 & 0 & .2 \end{array}\right]$

Solution (a) In this case $S = \left[\begin{array}{c} .2 \\ .4 \end{array}\right]$, $R = [.4]$. The stable matrix is

$$\left[\begin{array}{c|c} I & S(I-R)^{-1} \\ \hline 0 & 0 \end{array}\right] = \left[\begin{array}{cc|c} 1 & 0 & \\ 0 & 1 & S(I-R)^{-1} \\ 0 & 0 & 0 \end{array}\right].$$

Since R is 1×1, to compute $I - R$ we let I be the 1×1 identity matrix $[1]$. Then $I - R = [1] - [.4] = [1 - .4] = [.6]$. Now $(I - R)^{-1}$ is a matrix which, when multiplied by $I - R$ or $[.6]$, gives the 1×1 identity matrix $[1]$. Thus $(I - R)^{-1} = [.6]^{-1} = [1/.6]$ and

$$S(I-R)^{-1} = \left[\begin{array}{c} .2 \\ .4 \end{array}\right][1/.6] = \left[\begin{array}{c} .2/.6 \\ .4/.6 \end{array}\right] = \left[\begin{array}{c} \frac{1}{3} \\ \frac{2}{3} \end{array}\right].$$

Therefore, the stable matrix is given by

$$\left[\begin{array}{cc|c} 1 & 0 & \frac{1}{3} \\ 0 & 1 & \frac{2}{3} \\ 0 & 0 & 0 \end{array}\right].$$

(b) In this case $S = \left[\begin{array}{c} .7 \\ .1 \end{array}\right]$, $R = [.2]$, $I - R = [1] - [.2] = [.8]$, $(I - R)^{-1} = [1/.8]$. Therefore,

$$S(I-R)^{-1} = \left[\begin{array}{c} .7 \\ .1 \end{array}\right][1/.8] = \left[\begin{array}{c} .7/.8 \\ .1/.8 \end{array}\right] = \left[\begin{array}{c} \frac{7}{8} \\ \frac{1}{8} \end{array}\right].$$

So the stable matrix is

$$\left[\begin{array}{c|c} I & S(I-R)^{-1} \\ \hline 0 & 0 \end{array}\right] = \left[\begin{array}{cc|c} 1 & 0 & \frac{7}{8} \\ 0 & 1 & \frac{1}{8} \\ 0 & 0 & 0 \end{array}\right].$$

EXAMPLE 7 (*Gambler's Ruin*) Consider a game of chance with the following characteristics. A person repeatedly bets $1 each play. If he wins, he receives $1.* If he goes broke, he stops playing. Also, if he accumulates $3, he stops playing. On each play the probability of winning is .4 and of losing .6. What is the probability of eventually accumulating $3 if he starts with $1? $2?

Solution There are four states, corresponding to having $0, $3, $1, or $2. The first two are absorbing states. The stochastic matrix is

$$
\begin{array}{c}
 \\
\$0 \\
\$3 \\
\$1 \\
\$2
\end{array}
\begin{array}{cc}
\begin{array}{cccc}
\$0 & \$3 & \$1 & \$2
\end{array} \\
\left[
\begin{array}{cc|cc}
1 & 0 & .6 & 0 \\
0 & 1 & 0 & .4 \\
\hline
0 & 0 & 0 & .6 \\
0 & 0 & .4 & 0
\end{array}
\right].
\end{array}
$$

The third column is derived in this way: If he has $1, there is a .6 probability of losing $1, which would mean going to state $0. Thus the first entry in the third column is .6. There is no way to get from $1 to $3 or from $1 to $1 after one play. So the second and third entries are 0. There is a .4 probability of winning $1—that is, of going from $1 to $2. So the last entry is .4. The fourth column is derived similarly.

   In this example

$$
S = \begin{bmatrix} .6 & 0 \\ 0 & .4 \end{bmatrix} \qquad R = \begin{bmatrix} 0 & .6 \\ .4 & 0 \end{bmatrix}.
$$

To compute $S(I - R)^{-1}$ observe that

$$
I - R = \begin{bmatrix} 1 & 0 \\ 0 & 1 \end{bmatrix} - \begin{bmatrix} 0 & .6 \\ .4 & 0 \end{bmatrix} = \begin{bmatrix} 1 & -.6 \\ -.4 & 1 \end{bmatrix}.
$$

To compute $(I - R)^{-1}$ recall that

$$
\begin{bmatrix} a & b \\ c & d \end{bmatrix}^{-1} = \begin{bmatrix} d/\Delta & -b/\Delta \\ -c/\Delta & a/\Delta \end{bmatrix}, \qquad \Delta = ad - bc \neq 0.
$$

So, in this example, $\Delta = 1 \cdot 1 - (-.6)(-.4) = 1 - .24 = .76$ and

$$
(I - R)^{-1} = \begin{bmatrix} 1 & -.6 \\ -.4 & 1 \end{bmatrix}^{-1} \approx \begin{bmatrix} 1.32 & .79 \\ .53 & 1.32 \end{bmatrix}
$$

$$
S(I - R)^{-1} = \begin{bmatrix} .6 & 0 \\ 0 & .4 \end{bmatrix} \begin{bmatrix} 1.32 & .79 \\ .53 & 1.32 \end{bmatrix} = \begin{bmatrix} .79 & .47 \\ .21 & .53 \end{bmatrix}.
$$

---

* That is, he receives his bet of $1 plus winnings of $1.

Thus the stable matrix is

$$\left[\begin{array}{c|c} I & S(I-R)^{-1} \\ \hline 0 & 0 \end{array}\right] = \left[\begin{array}{cc|cc} 1 & 0 & .79 & .47 \\ 0 & 1 & .21 & .53 \\ \hline 0 & 0 & 0 & 0 \\ 0 & 0 & 0 & 0 \end{array}\right].$$

We are interested in the probability that he ends up with $3. The percentage is different for each of the two starting amounts $1 and $2. Recall the meaning of the rows and columns:

$$\begin{array}{c} \\ \$0 \\ \$3 \\ \$1 \\ \$2 \end{array} \begin{array}{cc} \$0 \quad \$3 & \$1 \quad \$2 \\ \left[\begin{array}{cc|cc} 1 & 0 & .79 & .47 \\ 0 & 1 & .21 & .53 \\ \hline 0 & 0 & 0 & 0 \\ 0 & 0 & 0 & 0 \end{array}\right]. \end{array}$$

Looking at the $1 column, we see that if he starts with $1, the probability is .21 that he ends up with $3. Looking at the $2 column, the probability is .53 that he ends up with $3.

PRACTICE PROBLEMS 3

1. When an absorbing stochastic matrix is partitioned, the submatrix in the upper left is an identity matrix, denoted by I. Also, when finding the stable matrix, we subtract the submatrix R from an identity matrix, also denoted by I. Are these two identity matrices the same?

2. Let A be an absorbing stochastic matrix. Interpret the entries of A^2.

3. Is $\begin{bmatrix} 1 & .4 & 0 \\ 0 & .2 & .1 \\ 0 & .4 & .9 \end{bmatrix}$ an absorbing stochastic matrix?

EXERCISES 3

In Exercises 1–4, determine whether the given matrix is an absorbing stochastic matrix.

1. $\begin{bmatrix} 1 & 0 & 0 & 0 \\ 0 & 1 & 0 & 0 \\ 0 & 0 & .8 & .1 \\ 0 & 0 & .2 & .9 \end{bmatrix}$

2. $\begin{bmatrix} 1 & 0 & 0 & .3 \\ 0 & 1 & 0 & .2 \\ 0 & 0 & 1 & .2 \\ 0 & 0 & 0 & .3 \end{bmatrix}$

3. $\begin{bmatrix} 1 & 0 & .4 \\ 0 & .5 & .3 \\ 0 & .5 & .3 \end{bmatrix}$

4. $\begin{bmatrix} 1 & 0 & 0 & 0 \\ 0 & 1 & 0 & 0 \\ 0 & 0 & 0 & 1 \\ 0 & 0 & 1 & 0 \end{bmatrix}$

The matrices in Exercises 5–10 are absorbing stochastic matrices. In each, identify R and S and compute the stable matrix.

5. $\begin{bmatrix} 1 & 0 & .3 \\ 0 & 1 & .2 \\ 0 & 0 & .5 \end{bmatrix}$

6. $\begin{bmatrix} 1 & 0 & \frac{1}{2} \\ 0 & 1 & \frac{1}{6} \\ 0 & 0 & \frac{1}{3} \end{bmatrix}$

7. $\begin{bmatrix} 1 & 0 & .1 & 0 \\ 0 & 1 & .5 & .2 \\ 0 & 0 & .3 & .6 \\ 0 & 0 & .1 & .2 \end{bmatrix}$

8. $\begin{bmatrix} 1 & 0 & \frac{1}{4} & \frac{1}{6} \\ 0 & 1 & \frac{1}{6} & 0 \\ 0 & 0 & \frac{1}{4} & \frac{1}{2} \\ 0 & 0 & \frac{1}{3} & \frac{1}{3} \end{bmatrix}$

9. $\begin{bmatrix} 1 & 0 & 0 & .1 & .2 \\ 0 & 1 & 0 & .3 & 0 \\ 0 & 0 & 1 & 0 & .2 \\ 0 & 0 & 0 & .5 & 0 \\ 0 & 0 & 0 & .1 & .6 \end{bmatrix}$

10. $\begin{bmatrix} 1 & 0 & \frac{1}{4} & 0 & \frac{1}{3} \\ 0 & 1 & \frac{1}{4} & 0 & 0 \\ 0 & 0 & 0 & \frac{1}{3} & \frac{1}{6} \\ 0 & 0 & \frac{1}{2} & \frac{1}{3} & 0 \\ 0 & 0 & 0 & \frac{1}{3} & \frac{1}{2} \end{bmatrix}$ $Note:$ $\begin{bmatrix} 1 & -\frac{1}{3} & -\frac{1}{6} \\ -\frac{1}{2} & \frac{2}{3} & 0 \\ 0 & -\frac{1}{3} & \frac{1}{2} \end{bmatrix}^{-1} = \begin{bmatrix} \frac{3}{2} & 1 & \frac{1}{2} \\ \frac{9}{8} & \frac{9}{4} & \frac{3}{8} \\ \frac{3}{4} & \frac{3}{2} & \frac{9}{4} \end{bmatrix}$

11. Suppose that the following data were obtained from the records of a certain 2-year college. Of those who were freshmen (F) during a particular year, 80% became sophomores (S) the next year and 20% dropped out (D). Of those who were sophomores during a particular year, 90% graduated (G) by the next year and 10% dropped out.

    (a) Set up the absorbing stochastic matrix with states D, G, F, S which describes this transition.

    (b) Find the stable matrix.

    (c) Determine the probability that an entering freshman will eventually graduate.

12. A research article* on the application of stochastic matrices to mental illness considers a person to be in one of four states: state I—chronically insane and hospitalized; state II—dead, with death occurring while unhospitalized; state III—sane; state IV—mildly insane and unhospitalized. States I and II are absorbing states. Suppose that of the people in state III, after one year 1.994% will be in state II, 98% in state III, and .006% in state IV. Also, suppose that of the people in state IV, after one year 2% will be in state I, 3% in state II, and 95% in state IV.

    (a) Set up the absorbing stochastic matrix that describes this transition.

    (b) Find the stable matrix.

    (c) Determine the probability that a person who is currently well will eventually be chronically insane.

---

* A. W. Marshall and H. Goldhamer, "An Application of Markov Processes to the Study of the Epidemiology of Mental Diseases," *American Statistical Association Journal*, March 1955, pp. 99–129.

SOLUTIONS TO PRACTICE PROBLEMS 3

1. Sometimes they are, but in general they are not. The size of the first identity matrix equals the number of absorbing states, whereas the size of the second identity matrix is the same as that of the matrix R, which equals the number of nonabsorbing states.

2. Consider the entry a_{ij}, in the ith row, jth column. Suppose that an object begins in state j. Then the entry gives the probability that it ends up in state i after two time periods.

3. Yes. (i) Its first state is absorbing. (ii) From each state it is possible to get to the first state. Note that it is possible to go directly from state 2 to state 1, since $a_{12} = .4 \neq 0$. Since $a_{13} = 0$, it is not possible to go directly from state 3 to state 1. However, this can be accomplished indirectly by going from state 3 to state 2 (possible since $a_{23} = .1 \neq 0$) and then from state 2 to state 1.

Chapter 8: CHECKLIST

☐ Markov process
☐ Transition matrix
☐ Stochastic matrix
☐ Distribution matrix
☐ A^n
☐ $A^n \begin{bmatrix} \end{bmatrix}_0 = \begin{bmatrix} \end{bmatrix}_n$

☐ Regular stochastic matrix
☐ Stable matrix of a regular stochastic matrix
☐ Stable distribution of a regular stochastic matrix
☐ Absorbing state
☐ Absorbing stochastic matrix
☐ Stable matrix of an absorbing stochastic matrix

Chapter 8: SUPPLEMENTARY EXERCISES

In Exercises 1–6, determine whether or not the given matrix is stochastic. If so, determine if it is regular, absorbing, or neither.

1. $\begin{bmatrix} 1 & .3 & 0 & 0 \\ 0 & .1 & 0 & 0 \\ 0 & .4 & .7 & .4 \\ 0 & .2 & .3 & .6 \end{bmatrix}$

2. $\begin{bmatrix} .1 & .1 & .1 & .1 \\ .2 & .2 & .2 & .2 \\ .3 & .3 & .3 & .3 \\ .4 & .4 & .4 & .4 \end{bmatrix}$

3. $\begin{bmatrix} 0 & .3 \\ 1 & .7 \end{bmatrix}$

4. $\begin{bmatrix} 1 & 0 & 0 \\ 0 & 1 & \frac{1}{3} \\ 0 & 0 & \frac{2}{3} \end{bmatrix}$

5. $\begin{bmatrix} 1 & \frac{1}{2} & 0 \\ 0 & \frac{1}{2} & 0 \\ 0 & \frac{1}{2} & 1 \end{bmatrix}$

6. $\begin{bmatrix} 1 & 0 & 0 & .3 \\ 0 & 1 & 0 & .3 \\ 0 & 0 & .5 & .3 \\ 0 & 0 & .5 & .1 \end{bmatrix}$

7. Find the stable distribution for the regular stochastic matrix $\begin{bmatrix} .6 & .5 \\ .4 & .5 \end{bmatrix}$.

8. Find the stable matrix for the absorbing stochastic matrix

$$\begin{bmatrix} 1 & 0 & 0 & \frac{1}{8} & \frac{1}{4} \\ 0 & 1 & 0 & \frac{1}{8} & 0 \\ 0 & 0 & 1 & 0 & \frac{1}{4} \\ 0 & 0 & 0 & \frac{1}{4} & \frac{1}{2} \\ 0 & 0 & 0 & \frac{1}{2} & 0 \end{bmatrix}.$$

9. In a certain community currently 10% of the people are H (high income), 60% are M (medium income), and 30% are L (low income). Studies show that for the children of H parents, 50% also become H, 40% become M, and 10% become L. Of the children of M parents, 40% become H, 30% become M, and 30% become L. Of the children of L parents, 30% become H, 50% become M, and 20% become L.

   (a) Set up the 3×3 stochastic matrix that describes this situation.

   (b) What percent of the children of the current generation will have high incomes?

   (c) In the long run what proportion of the population will have low incomes?

10. In a certain factory some machines are properly adjusted and some need adjusting. Technicians randomly inspect machines and make adjustments. Suppose that of the machines that are properly adjusted on a particular day, 80% will also be properly adjusted the following day and 20% will need adjusting. Also, of the machines that need adjusting on a particular day, 30% will be properly adjusted the next day and 70% will still need adjusting.

   (a) Set up the 2×2 stochastic matrix with columns labeled P (properly adjusted) and N (need adjusting) which describes this situation.

   (b) If initially all the machines are properly adjusted, what percent will need adjusting after 2 days?

   (c) In the long run, what percent will be properly adjusted each day?

11. Find the stable matrix for the absorbing stochastic matrix

$$\begin{bmatrix} 1 & 0 & \frac{1}{6} & \frac{1}{2} & \frac{2}{5} \\ 0 & 1 & 0 & 0 & \frac{2}{5} \\ 0 & 0 & 0 & 0 & 0 \\ 0 & 0 & \frac{2}{3} & \frac{1}{2} & 0 \\ 0 & 0 & \frac{1}{6} & 0 & \frac{1}{5} \end{bmatrix}.$$

12. Figure 1 gives the layout of a house with four rooms connected by doors. Room I contains a mousetrap and room II contains cheese. A mouse, after being placed in one of the rooms, will search for cheese and if unsuccessful after one minute, will exit to another room by selecting one of the doors at random. (For instance, if the mouse is in room III, after one minute, he will go to room II with probability $\frac{1}{3}$ and to room IV with probability $\frac{2}{3}$.) A mouse entering room I will be trapped and therefore no longer move. Also, a mouse entering room II will remain in that room.

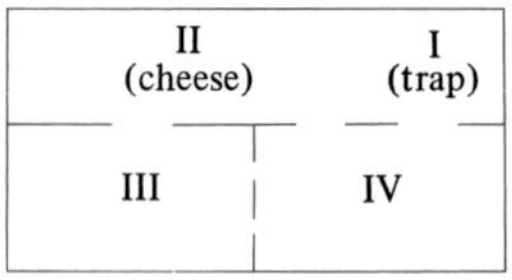

II (cheese)	I (trap)
III	IV

FIGURE 1

(a) Set up the 4 × 4 absorbing stochastic matrix that describes this situation.

(b) If a mouse begins in room IV, what is the probability that he will find the cheese after 2 minutes?

(c) If a mouse begins in room IV, what is the probability that he will find the cheese in the long run?

13. Which of the following is the stable distribution for the regular stochastic matrix?

$$\begin{bmatrix} .4 & .4 & .2 \\ .1 & .1 & .3 \\ .5 & .5 & .5 \end{bmatrix}?$$

(a) $\begin{bmatrix} .6 \\ .4 \\ 1 \end{bmatrix}$ (b) $\begin{bmatrix} .2 \\ .3 \\ .5 \end{bmatrix}$ (c) $\begin{bmatrix} .3 \\ .2 \\ .5 \end{bmatrix}$

14. A city has two competing news programs. From a survey of regular listeners it was determined that of those who listen to station A on a particular day, 90% listen to station A the next day and 10% listen to station B. Of those who listen to station B on a particular day, 20% listen to station A the next day and 80% listen to station B. If today 50% of the regular listeners listen to each station, what percentage of them would you expect to listen to station A 2 days from now?

9

THE THEORY OF GAMES

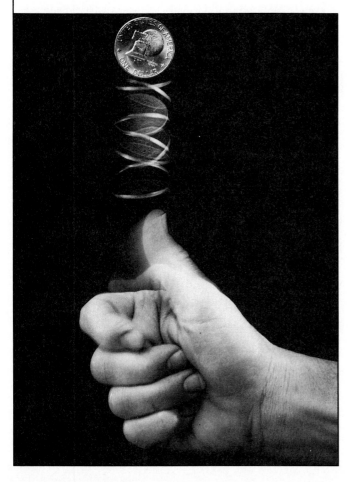

9.1 Games and Strategies

9.2 Mixed Strategies

9.3 Determining Optimal Mixed Strategies

One of the more interesting developments of twentieth-century mathematics has been the theory of games, a branch of mathematics used to analyze competitive phenomena. This theory has been applied extensively in many fields, including business, economics, psychology, and sociology. Mathematically, the theory of games blends the theory of matrices with probability theory. Although an extensive discussion is well beyond the scope of this book, we hope to give the flavor of the subject and some indication of the wide range of its applications.

9.1. Games and Strategies

Let us begin our study of game theory by analyzing a typical competitive situation. Suppose that in a certain town there are two furniture stores, Reliable Furniture Company and Cut-Rate Furniture Company, which compete for all furniture sales in the town. Each of the stores is planning a Labor Day sale and each has the option of marking its furniture down by 10% or 20%. The results of their decisions affect the total percentage of the market that each captures. On the basis of an analysis of past consumer tendencies, it is estimated that if Reliable chooses a 10% discount and so does Cut-Rate, then Reliable will capture 60% of the sales. If Reliable chooses a 10% discount but Cut-Rate chooses 20%, then Reliable will capture only 35% of the sales. On the other hand, if Reliable chooses a 20% discount but Cut-Rate chooses 10%, then Reliable will get 80% of the sales. If Reliable chooses a 20% discount and Cut-Rate also chooses 20%, then Reliable will get 50% of the sales. Each store is able to determine the other store's discount prior to the start of the sale and adjust its own discount accordingly. If you were a consultant to Reliable, what discount would you choose to obtain as large a share of the sales as possible?

In order to analyze the various possibilities, let us summarize the given data in a matrix. For the sake of brevity, denote Reliable by R and Cut-Rate by C. Then the data can be summarized as follows:

$$
\begin{array}{cc}
 & C \text{ discount} \\
R \text{ discount} &
\begin{array}{c c}
 & \begin{array}{cc} 10\% & 20\% \end{array} \\
\begin{array}{c} 10\% \\ 20\% \end{array} &
\left[\begin{array}{cc} .6 & .35 \\ .8 & .5 \end{array} \right]
\end{array}
\end{array}.
$$

For example, the number in the second row, first column corresponds to an R discount of 20% and a C discount of 10%. In this case R will capture 80% or .8 of the sales.

We may view R's choice of discount as choosing one of the rows of the matrix. Similarly, C's choice of discount amounts to choosing one of the columns of the matrix. Suppose that R and C both act rationally. What will be the result? Let us view things from R's perspective first. In scanning his options he sees a .8 in the second row. So his first reaction might be to take a 20% discount and try for 80% of the sales. However, this route is very risky. As soon as C learns that R has chosen a 20% discount, C will set a 20% discount and lower R's share of the sales to 50%. So the result of choosing row 2 will be for R to capture only 50% of the sales. On the other hand, if R chooses row 1, then C will naturally choose a 20% discount to give R a 35% share of the sales. Of the options open to R, the 50% share is clearly the most desirable. So R will choose a 20% discount.

What about C? In setting his discount C must choose a column of the matrix. Since the entries represent R's share of the sales, C wishes to make a choice resulting in as *small* a number as possible. If C chooses column 1, then R will immediately respond with a 20% discount in order to acquire 80% of the sales, a disaster for C. On the other hand, if C chooses column 2, then R will choose a 20% discount to obtain 50% of the sales. The best option open to C is to choose a 20% discount.

Thus we see that if both stores act rationally, they will each choose 20% discounts and each will capture 50% of the sales.

The competitive situation above is an example of a (mathematical) *game*. In such a game there are two or more players. In the example above the players were R and C. Each player is allowed to make a move. In the example above the moves are the choices of discount. As the result of a move by each player, there is a payoff to each player. In the example above the payoff to each player (store) is the percentage of total sales he captures. In our example, then, we have solved a problem which can be posed for any game:

Fundamental Problem of Game Theory   How should each player decide his move in order to maximize his gain?

Indeed, in our example, R and C chose moves such that each maximized his own share of sales.

Throughout this chapter we consider only games with two players, whom we shall denote by R and C. (R and C stand for row and column, respectively.) Suppose that R can make moves $R_1, R_2, \ldots, R_m$ and that C can make moves $C_1, C_2, \ldots, C_n$. Further suppose that a move R_i by R and C_j by C results in a payoff of a_{ij} to R. Then the game can be represented by the *payoff matrix*:

$$
\begin{array}{c}
\qquad\qquad\qquad C \text{ moves} \\[4pt]
\begin{array}{c c}
 & \begin{array}{cccc} C_1 & C_2 & \cdots & C_n \end{array} \\
R \text{ moves} \quad
\begin{array}{c} R_1 \\ R_2 \\ \vdots \\ R_m \end{array}
&
\left[
\begin{array}{cccc}
a_{11} & a_{12} & \cdots & a_{1n} \\
a_{21} & a_{22} & \cdots & a_{2n} \\
\vdots & \vdots & & \vdots \\
a_{m1} & a_{m2} & \cdots & a_{mn}
\end{array}
\right].
\end{array}
\end{array}
$$

Note that the payoff matrix is an $m \times n$ matrix with a_{ij} as the entry of the ith row, jth column. In our furniture store example the payoff matrix was just the matrix we used in our analysis. Note that a move by R corresponds to a choice of a *row* of the payoff matrix, whereas a move by C corresponds to a choice of a *column*.

Suppose that a given game is played repeatedly. The players can adopt various strategies to attempt to maximize their respective gains (or minimize their losses). In what follows we shall discuss the problem of determining strategies. The simplest type of strategy is one in which a player, on consecutive plays, consistently chooses the same row (or column). Such strategies are called *pure strategies* and will be discussed below. Strategies involving varied moves are called *mixed strategies* and will be discussed in Section 2.

In most examples of games the payoffs to C are related in a simple way to the corresponding payoffs to R. For example, for the furniture stores the payoff to C is 100% minus the payoff to R. In another common type of game a payoff to R of a given amount results in a loss to C of the same amount, and vice versa; for such games the sum of the gains on each play is zero, hence they are called *zero-sum games*. An illustration of such a game is provided in the next example.

EXAMPLE 1 Suppose that R and C play a coin-matching game. Each player can show either heads or tails. If R and C both show heads, then C pays R $5. If R shows heads and C shows tails, then R pays C $8. If R shows tails and C shows heads, then C pays R $3. If R shows tails and so does C, then C pays R $1.

(a) Determine the payoff matrix of this game.

(b) Suppose that R and C play the game repeatedly. Determine optimum pure strategies for R and C.

Solution (a) The payoff matrix is given by

$$\begin{array}{cc} \text{Heads} & \text{Tails} \end{array}$$
$$\begin{bmatrix} 5 & -8 \\ 3 & 1 \end{bmatrix}.$$

Each entry specifies a payoff from C to R. The entry "-8" denotes a negative payoff to R, that is, a gain of $8 to C.

(b) R would clearly like to choose heads so as to gain $5. However, if R consistently chooses heads, C will retaliate by choosing tails, causing R to lose $8. If R chooses tails, however, then the best C can do is choose tails to give a gain of $1 to R. So R should clearly choose tails. Now we look at the game from C's point of view. Clearly, C's objective is to minimize the payment to R. If C consistently chooses heads, then R will notice the pattern and choose heads, at a cost to C of $5. However, if C chooses tails, then the best that R can do is choose tails, thereby costing C $1. So clearly the optimum move for C is to choose tails.

The reasoning described above is rather cumbersome. There is, however, an easy way to summarize what we have done. Let us first describe R's reasoning. R seeks to choose a row of the matrix which will maximize the payoff to himself. However, once a row is chosen consistently, R can expect C to counter by choosing the least element of that row. Thus R should choose his move as follows:

Optimum Pure Strategy for R

1. For each row of the payoff matrix, determine the least element.
2. Choose the row for which this element is as large as possible.

Thus, for example, in the game of Example 1, we have circled the least element of each row:

$$\begin{bmatrix} 5 & \boxed{-8} \\ 3 & \boxed{1} \end{bmatrix}.$$

The largest circled element is 1. So R should choose the second row—that is, tails.

In a similar way we may describe the optimum strategy for C. C wishes to choose a column of the payoff matrix so as to minimize the payoff to R. However, C can expect R to adjust his choice to the maximum element of the column. Therefore, we can summarize the optimum strategy for C as follows:

Optimum Pure Strategy for C

1. For each column of the payoff matrix, determine the largest element.
2. Choose the column for which this element is as small as possible.

For example, in the game of Example 1 the largest element in each column has been circled:

$$\begin{bmatrix} \boxed{5} & -8 \\ 3 & \boxed{1} \end{bmatrix}.$$

The smallest circled element is 1, so C should choose the second column, or tails.

EXAMPLE 2   Determine optimum pure strategies for R and C for the game whose payoff matrix is

$$\begin{bmatrix} -1 & 5 \\ 1 & 4 \\ 0 & -1 \end{bmatrix}.$$

Solution   To determine the strategy for R we first circle the smallest element in each row:

$$\begin{bmatrix} \text{\textcircled{-1}} & 5 \\ \text{\textcircled{1}} & 4 \\ 0 & \text{\textcircled{-1}} \end{bmatrix}.$$

The largest of these is 1, so R should play the second row. To determine the strategy for C we circle the largest element in each column:

$$\begin{bmatrix} -1 & \text{\textcircled{5}} \\ \text{\textcircled{1}} & 4 \\ 0 & -1 \end{bmatrix}.$$

The smallest of these is 1, so C should play the first column.

All the games considered so far have an important characteristic in common: there is an entry in the payoff matrix which is *simultaneously* the minimum element in its row and the maximum element in its column. Such an entry is called a *saddle point* for the game. As we have seen in the examples considered above, if a game possesses a saddle point, then an optimum strategy is for R to choose the row containing the saddle point and for C to choose the column containing the saddle point.

A game need not have a saddle point. For example, the matrix

$$\begin{bmatrix} 2 & -2 \\ 0 & 1 \end{bmatrix}$$

is the payoff matrix of a game with no saddle point. The optimum pure strategy for R is to choose the row with the maximum of the circled elements in

$$\begin{bmatrix} 2 & \text{\textcircled{-2}} \\ \text{\textcircled{0}} & 1 \end{bmatrix}.$$

Thus, R chooses row 2. The optimum pure strategy for C is to choose the column with the minimum of the circled elements in

$$\begin{bmatrix} \text{\textcircled{2}} & -2 \\ 0 & \text{\textcircled{1}} \end{bmatrix}.$$

So C chooses column 2. No element is simultaneously the minimum element in its row and the maximum element in its column.

A game that has a saddle point is called a *strictly determined game*. If v is a saddle point for a strictly determined game, then if each player plays the optimum pure strategy, each repetition of the game will result in a payment of v to player R. The number v is called the *value* of the game.

EXAMPLE 3 Find the saddle point and value of the strictly determined game given by the payoff matrix:

$$\begin{bmatrix} -1 & -10 & 10 \\ 0 & 7 & 6 \\ 3 & 4 & 11 \\ 2 & 5 & 7 \end{bmatrix}.$$

Solution The least elements in the various rows are

$$\begin{bmatrix} -1 & \boxed{-10} & 10 \\ \boxed{0} & 7 & 6 \\ \boxed{3} & 4 & 11 \\ \boxed{2} & 5 & 7 \end{bmatrix}.$$

The maximum elements in the columns are

$$\begin{bmatrix} -1 & -10 & 10 \\ 0 & \boxed{7} & 6 \\ \boxed{3} & 4 & \boxed{11} \\ 2 & 5 & 7 \end{bmatrix}.$$

The element 3 in the third row, first column is a minimum in its row and a maximum in its column and so is a saddle point of the game. The value of the game is therefore 3: each repetition of the game, assuming optimum strategies, results in a payoff of 3 to R.

A game may have more than one saddle point. Consider the game with payoff matrix

$$\begin{bmatrix} \boxed{1} & 2 & \boxed{1} \\ \boxed{1} & 5 & \boxed{1} \\ 0 & -7 & -1 \end{bmatrix}.$$

Each circled element is both the minimum element in its row and the maximum element in its column. There are four saddle points representing four optimal strategies. The value of the game, regardless of strategy, is 1. If a game has more than one saddle point, then the value of the game is the same at each of them.

EXAMPLE 4 R and C play a game in which they show 1 or 2 fingers simultaneously. It is agreed that C pays R an amount equal to the total number of fingers shown less 3 cents. Find the optimal strategy for each player and the value of the game.

Solution The payoff matrix is given by

$$\begin{matrix} & 1 & 2 & & & 1 & 2 \\ 1 & \begin{bmatrix} 2-3 & 3-3 \\ 3-3 & 4-3 \end{bmatrix} & = & \begin{matrix} 1 \\ 2 \end{matrix} & \begin{bmatrix} -1 & 0 \\ 0 & 1 \end{bmatrix}. \end{matrix}$$

The saddle point is the element that is simultaneously the minimum of its row and the maximum of its column—so an optimal strategy is for R to show 2 fingers and for C to show 1 finger. The value of the game is 0.

PRACTICE PROBLEMS 1

Which of the following matrices are the payoff matrices of strictly determined games? For those which are, determine the saddle point and optimum pure strategy for each of the players.

1. $\begin{bmatrix} 1 & -1 & -3 \\ 0 & -2 & 3 \end{bmatrix}$
2. $\begin{bmatrix} 1 & -1 & 0 \\ 0 & -4 & 5 \end{bmatrix}$
3. $\begin{bmatrix} 1 & -2 & 1 \\ -2 & 1 & 1 \\ 1 & 1 & -2 \end{bmatrix}$

EXERCISES 1

Each of the following matrices is the payoff matrix for a strictly determined game. Determine optimum pure strategies for R and C.

1. $\begin{bmatrix} -1 & -2 \\ 0 & 3 \end{bmatrix}$
2. $\begin{bmatrix} -4 & 0 \\ 2 & 1 \end{bmatrix}$
3. $\begin{bmatrix} -2 & 4 & 1 \\ -1 & 3 & 5 \\ -3 & 5 & 2 \end{bmatrix}$

4. $\begin{bmatrix} 1 & -1 & 0 \\ 6 & 3 & 2 \\ 2 & -2 & 1 \end{bmatrix}$
5. $\begin{bmatrix} 0 & 3 \\ -1 & 1 \\ -2 & -4 \end{bmatrix}$
6. $\begin{bmatrix} 0 & -4 & 0 \\ -1 & -2 & 1 \end{bmatrix}$

Each of the following matrices is the payoff matrix for a strictly determined game. (a) Find a saddle point. (b) Determine the value of the game.

7. $\begin{bmatrix} 1 & 0 \\ 0 & -1 \end{bmatrix}$
8. $\begin{bmatrix} 2 & 3 \\ 4 & 5 \end{bmatrix}$

For each of the following games, give the payoff matrix and decide if the game is strictly determined. If so, determine the optimum strategies for R and C.

9. Suppose that R and C play a game by matching coins. On each play, C pays R the number of heads shown (0, 1, or 2) minus twice the number of tails shown.

10. In the child's game "scissor, paper, stone," each of two children call out one of the three words. If they both call out the same word then the game is a tie. Otherwise, "scissors" beats "paper" (since scissors can cut paper), "paper" beats "stone" (since paper can cover stone), and "stone" beats "scissors" (since stone can break scissors). Suppose that the loser pays a penny to the winner.

11. Two candidates for political office must decide to be for, against, or neutral on a certain referendum. Pollsters have determined that if candidate R comes out for the referendum, then he will gain 8000 votes if candidate C also comes out for the referendum, lose 1000 votes if candidate C comes out against, and gain 1000 votes if candidate C comes out

neutral. If candidate R comes out against, then he will lose 7000 votes (respectively gain 4000 votes, lose 2000 votes) if candidate C comes out for (respectively against, neutral on) the referendum. If candidate R is neutral, then he will gain 3000 votes if C is for or against, and gain 2000 votes if C is neutral.

12. TV stations R and C each have a quiz show and a situation comedy to schedule for their 1 and 2 o'clock time slots. If they both schedule their quiz shows at 1 o'clock, then station R will take $3000 in advertising revenue away from station C. If they both schedule their quiz shows at 2 o'clock, then station C will take $2000 in advertising revenue from R. If they choose different hours for the quiz show, then R will take $5000 in advertising from C by scheduling it at 2 o'clock, and $2000 by scheduling it at 1 o'clock.

13. Player R has two cards, a red 5 and a black 10. Player C has three cards, a red 6, a black 7, and a black 8. They each place one of their cards on the table. If the cards are the same color, R receives the difference of the two numbers. If the cards are of different colors, C receives the minimum of the two numbers.

SOLUTIONS TO PRACTICE PROBLEMS 1

1. Not strictly determined. The minimum elements of the rows are

$$\begin{bmatrix} 1 & -1 & \boxed{-3} \\ 0 & \boxed{-2} & 3 \end{bmatrix};$$

the maximums of the columns are

$$\begin{bmatrix} \boxed{1} & \boxed{-1} & -3 \\ 0 & -2 & \boxed{3} \end{bmatrix}.$$

No element is simultaneously the minimum in its row and the maximum in its column.

2. Strictly determined. The least elements of the rows are

$$\begin{bmatrix} 1 & \boxed{-1} & 0 \\ 0 & \boxed{-4} & 5 \end{bmatrix}.$$

The largest elements of the columns are

$$\begin{bmatrix} \boxed{1} & \boxed{-1} & 0 \\ 0 & -4 & \boxed{5} \end{bmatrix}.$$

Thus -1 is a saddle point. The optimum strategy for R is to choose row 1; the optimum strategy for C is to choose column 2.

3. Not strictly determined. The minimums of the rows are

$$\begin{bmatrix} 1 & \boxed{-2} & 1 \\ \boxed{-2} & 1 & 1 \\ 1 & 1 & \boxed{-2} \end{bmatrix}.$$

The largest element for each column is 1. (Note that there are two choices for each largest element.) But none of the largest column elements is a least row element.

9.2. Mixed Strategies

In the preceding section we introduced strictly determined games and gave a method for determining optimum strategies for each player. However, not all games are strictly determined. For example, consider the game with payoff matrix

$$\begin{bmatrix} -1 & 5 \\ 2 & -3 \end{bmatrix}.$$

The minimum entries of the rows are

$$\begin{bmatrix} \boxed{-1} & 5 \\ 2 & \boxed{-3} \end{bmatrix},$$

whereas the maximum entries of the columns are

$$\begin{bmatrix} -1 & \boxed{5} \\ \boxed{2} & -3 \end{bmatrix}.$$

Note that no matrix entry is simultaneously the minimum in its row and the maximum in its column. Note also that no simple strategy of the type considered in Section 1 is sufficient to both maximize R's winnings and minimize C's losses. To see this consider the game from R's point of view. Suppose that R repeatedly plays the strategy "first row," thereby attempting to win 5. After a few plays C will catch on to R's strategy and choose column 1, giving R a loss of 1. Similarly, if R consistently plays the strategy "second row," attempting to win 2, then C can thwart R by choosing the second column to give R a loss of 3. It is clear that, in order to maximize the payoff to himself, R should sometimes choose row 1 and sometimes row 2. One might expect that by choosing the rows on a probabilistic basis R can prevent C from anticipating his moves and amass enough positive payoffs to counteract the occasional negative ones. Thus we should investigate strategies of the type

A: $\begin{cases} \text{Choose row 1 with probability .5} \\ \text{Choose row 2 with probability .5} \end{cases}$

and

B: $\begin{cases} \text{Choose row 1 with probability .9} \\ \text{Choose row 2 with probability .1.} \end{cases}$

Such strategies are called *mixed strategies*. Either of the players can pursue such a strategy.

One way that R can carry out strategy A is to alternate between row 1 and row 2 on successive plays of the game. However, if C is at all clever he will recognize this pattern and determine his play accordingly. What R should do is toss a coin and play row 1 whenever it lands heads and row 2 whenever it lands tails. Then there is no way that C can anticipate R's choice.

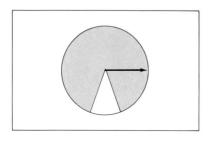

FIGURE 1

To carry out strategy B, R might use a card with a spinner attached at the center of a circle, 90% of the circle being blue and 10% white. R would then determine his play by spinning the spinner and choosing row 1 if the spinner landed on the blue part of the circle and row 2 if it landed on the white part. (See Fig. 1.)

It will be convenient to write mixed strategies in matrix form. Mixed strategies for R will be row matrices, and mixed strategies for C will be column matrices. Thus, for example, mixed strategy A above (for R) corresponds to the matrix

$$A: \quad [.5 \quad .5],$$

whereas mixed strategy B (for R) corresponds to

$$B: \quad [.9 \quad .1].$$

The mixed strategy in which C chooses column 1 with probability .6 and column 2 with probability .4 corresponds to the column matrix

$$\begin{bmatrix} .6 \\ .4 \end{bmatrix}.$$

In comparing different mixed strategies we use a number called their *expected value*. This number is just the average amount per game which is paid to R if the players pursue the given mixed strategies. The next example illustrates the computation of the expected value in a special case.

EXAMPLE 1 Suppose that a game has payoff matrix

$$\begin{bmatrix} -1 & 5 \\ 2 & -3 \end{bmatrix}.$$

Further suppose that R pursues the mixed strategy $[.9 \quad .1]$ and that C pursues the mixed strategy $\begin{bmatrix} .6 \\ .4 \end{bmatrix}$. Calculate the expected value of this game.

Solution Let us view each repetition of the game as an experiment. There are four possible outcomes:

(row 1, column 1)   (row 1, column 2)

(row 2, column 1)   (row 2, column 2).

Let us compute the probability (relative frequency) with which each of the outcomes occurs. For example, consider the outcome (row 1, column 1). According to our assumptions about the strategies of R and C, R chooses row 1 with probability .9 and C chooses column 1 with probability .6. Since R and C make their respective choices independently of one another, the events "row 1" and "column 1" are independent. Therefore, we can compute the probability of the outcome (row 1, column 1) as follows:

$$\text{Pr(row 1, column 1)} = \text{Pr(row 1)} \cdot \text{Pr(column 1)}$$

$$= (.9)(.6) = .54.$$

That is, the outcome "row 1, column 1" will occur with probability .54. Similarly, we compute the probabilities of the other three outcomes. Table 1 gives the probability and the amount won by R for each outcome. The average amount that R wins per play is then given by

$$(-1)(.54) + (5)(.36) + (2)(.06) + (-3)(.04) = 1.26.$$

Hence the expected value of the given strategies is 1.26.

TABLE 1

Outcome	R wins	Probability
Row 1, column 1	-1	$(.9)(.6) = .54$
Row 1, column 2	5	$(.9)(.4) = .36$
Row 2, column 1	2	$(.1)(.6) = .06$
Row 2, column 2	-3	$(.1)(.4) = .04$

We may generalize the computations above. To see the pattern let us first stick to 2×2 games. Suppose that a game has payoff matrix

$$\begin{bmatrix} a_{11} & a_{12} \\ a_{21} & a_{22} \end{bmatrix}.$$

Further suppose that R pursues a strategy $\begin{bmatrix} r_1 & r_2 \end{bmatrix}$. That is, R randomly chooses row 1 with probability r_1 and row 2 with probability r_2. Similarly, suppose that C pursues a strategy $\begin{bmatrix} c_1 \\ c_2 \end{bmatrix}$. Then the probabilities of the various outcomes can be tabulated as follows:

Outcome	Payoff to R	Probability
Row 1, column 1	a_{11}	$r_1 c_1$
Row 1, column 2	a_{12}	$r_1 c_2$
Row 2, column 1	a_{21}	$r_2 c_1$
Row 2, column 2	a_{22}	$r_2 c_2$

Thus, by following the reasoning used in the special case above, we see that the expected value of the strategies is the sum of the products of the payoffs times the corresponding probabilities:

$$a_{11}(r_1 c_1) + a_{12}(r_1 c_2) + a_{21}(r_2 c_1) + a_{22}(r_2 c_2).$$

On the average R gains this amount for each play. A somewhat tedious (but easy) calculation shows that

$$\begin{bmatrix} r_1 & r_2 \end{bmatrix} \begin{bmatrix} a_{11} & a_{12} \\ a_{21} & a_{22} \end{bmatrix} \begin{bmatrix} c_1 \\ c_2 \end{bmatrix} = \begin{bmatrix} r_1 & r_2 \end{bmatrix} \begin{bmatrix} a_{11}c_1 + a_{12}c_2 \\ a_{21}c_1 + a_{22}c_2 \end{bmatrix}$$

$$= [a_{11}(r_1c_1) + a_{12}(r_1c_2) + a_{21}(r_2c_1) + a_{22}(r_2c_2)]. \quad (1)$$

Formula (1) is a special case of the following general fact:

Expected Value of a Pair of Strategies Suppose that a game has payoff matrix

$$\begin{bmatrix} a_{11} & a_{12} & \cdots & a_{1n} \\ a_{21} & a_{22} & \cdots & a_{2n} \\ & \vdots & & \\ a_{m1} & a_{m2} & \cdots & a_{mn} \end{bmatrix}.$$

Suppose that R plays the strategy $\begin{bmatrix} r_1 & r_2 & \cdots & r_m \end{bmatrix}$ and that C plays the strategy

$$\begin{bmatrix} c_1 \\ c_2 \\ \vdots \\ c_n \end{bmatrix}.$$

Let e be the expected value of the pair of strategies. Then

$$\begin{bmatrix} r_1 & r_2 & \cdots & r_m \end{bmatrix} \begin{bmatrix} a_{11} & a_{12} & \cdots & a_{1n} \\ a_{21} & a_{22} & \cdots & a_{2n} \\ & \vdots & & \\ a_{m1} & a_{m2} & \cdots & a_{mn} \end{bmatrix} \begin{bmatrix} c_1 \\ c_2 \\ \vdots \\ c_n \end{bmatrix} = [e].$$

EXAMPLE 2 Suppose that a game has payoff matrix

$$\begin{bmatrix} 2 & 0 & -1 \\ -1 & 3 & 4 \end{bmatrix}$$

and that R plays the strategy $[.5 \quad .5]$. Which of the following strategies is more advantageous for C?

$$A = \begin{bmatrix} .6 \\ .3 \\ .1 \end{bmatrix} \quad \text{or} \quad B = \begin{bmatrix} .3 \\ .3 \\ .4 \end{bmatrix}.$$

Solution We compare the expected value with C using strategy A to that with C using strategy B. With strategy A we have

$$\begin{bmatrix} .5 & .5 \end{bmatrix} \begin{bmatrix} 2 & 0 & -1 \\ -1 & 3 & 4 \end{bmatrix} \begin{bmatrix} .6 \\ .3 \\ .1 \end{bmatrix} = [.90].$$

Using strategy B we have

$$[.5 \quad .5]\begin{bmatrix} 2 & 0 & -1 \\ -1 & 3 & 4 \end{bmatrix}\begin{bmatrix} .3 \\ .3 \\ .4 \end{bmatrix} = [1.20].$$

Thus strategy A will yield an average payment per play of .90 to R, whereas strategy B will yield an average payment per play of 1.20. Clearly, it is to C's advantage to choose strategy A.

EXAMPLE 3 The Acme Chemical Corporation has two plants, each situated on the banks of the Blue River, 10 miles from one another. A single inspector is assigned to check that the plants do not dump waste into the river. If he discovers plant A dumping waste, Acme is fined $20,000. If he discovers plant B dumping waste, Acme is fined $50,000. Suppose that the inspector visits one of the plants each day and that he chooses, on a random basis, to visit plant B 60% of the time. Acme schedules dumping from its two plants on a random basis, one plant per day, with plant B dumping waste on 70% of the days. How much is Acme's average fine per day?

Solution The competition between Acme and the inspector can be viewed as a game, whose matrix is

$$
\begin{array}{cc}
 & \text{Inspect A} \quad \text{Inspect B} \\
\begin{array}{c} \text{Plant A dumps} \\ \text{Plant B dumps} \end{array} &
\begin{bmatrix} -20{,}000 & 0 \\ 0 & -50{,}000 \end{bmatrix}
\end{array}
$$

The strategy of Acme is given by the row matrix $[.3 \quad .7]$. The strategy of the inspector is given by the column matrix $\begin{bmatrix} .4 \\ .6 \end{bmatrix}$. The expected value of the strategies is the matrix product

$$[.3 \quad .7]\begin{bmatrix} -20{,}000 & 0 \\ 0 & -50{,}000 \end{bmatrix}\begin{bmatrix} .4 \\ .6 \end{bmatrix} = [-6{,}000 \quad -35{,}000]\begin{bmatrix} .4 \\ .6 \end{bmatrix} = [-23{,}400].$$

In other words, Acme will be fined an average of $23,400 per day for its pollution of the river.

In the next section we will alter payoff matrices by adding a fixed constant to each entry so that all entries become positive numbers. This does not alter the essential character of the game, in that good strategies for the original matrix will also be good strategies for the new matrix. The only difference is that the expected value is increased by the constant added. This procedure enables us to apply the methods of linear programming to the determination of optimal mixed strategies for zero-sum games that are not strictly determined.

EXAMPLE 4 In Example 1 we saw that for the game with payoff matrix

$$\begin{bmatrix} -1 & 5 \\ 2 & -3 \end{bmatrix}$$

and strategies $[.9 \quad .1]$, $\begin{bmatrix} .6 \\ .4 \end{bmatrix}$, the expected value was 1.26. Compute the expected value of those strategies for the matrix obtained by adding 4 to each entry.

Solution The new matrix is

$$\begin{bmatrix} -1+4 & 5+4 \\ 2+4 & -3+4 \end{bmatrix} \quad \text{or} \quad \begin{bmatrix} 3 & 9 \\ 6 & 1 \end{bmatrix}.$$

The expected value of the strategies for the new matrix is

$$[.9 \quad .1]\begin{bmatrix} 3 & 9 \\ 6 & 1 \end{bmatrix}\begin{bmatrix} .6 \\ .4 \end{bmatrix} = [3.3 \quad 8.2]\begin{bmatrix} .6 \\ .4 \end{bmatrix} = 5.26.$$

As it should, the expected value has also increased by 4.

Suppose that the payoff matrix was

$$\begin{bmatrix} -\frac{1}{2} & \frac{5}{2} \\ 1 & -\frac{3}{2} \end{bmatrix}.$$

What is the expected value of the strategies $[.9 \quad .1]$ and $\begin{bmatrix} .6 \\ .4 \end{bmatrix}$? We see that

$$[.9 \quad .1]\begin{bmatrix} -\frac{1}{2} & \frac{5}{2} \\ 1 & -\frac{3}{2} \end{bmatrix}\begin{bmatrix} .6 \\ .4 \end{bmatrix} = .63.$$

We note that multiplying each element in the payoff matrix by 2 and adding 4 gives the matrix

$$\begin{bmatrix} 3 & 9 \\ 6 & 1 \end{bmatrix},$$

which with the strategies above gives expected value 5.26 (see the solution to Example 4). Multiplying each element of the payoff matrix by 2 and adding 4 to each element produces the same effect on the expected value:

$$5.26 = 2(.63) + 4.$$

PRACTICE PROBLEMS 2

1. Suppose that the payoff matrix of a game is

$$\begin{bmatrix} 4 & -2 \\ -3 & 1 \end{bmatrix}.$$

Suppose that R plays the strategy $[.6 \quad .4]$. Which of the two strategies $\begin{bmatrix} .5 \\ .5 \end{bmatrix}$ or $\begin{bmatrix} .7 \\ .3 \end{bmatrix}$ is better for C?

2. Answer the question in Problem 1 for the game whose payoff matrix is

$$\begin{bmatrix} 9 & 3 \\ 2 & 6 \end{bmatrix}.$$

(This matrix is obtained by adding 5 to each entry of the matrix in Problem 1.)

EXERCISES 2

1. Suppose that a game has payoff matrix

$$\begin{bmatrix} 3 & -1 \\ -7 & 5 \end{bmatrix}.$$

Calculate the expected values for the following strategies and determine which of the following situations is most advantageous to R.

(a) R plays $[.5 \quad .5]$, C plays $\begin{bmatrix} .5 \\ .5 \end{bmatrix}$.

(b) R plays $[1 \quad 0]$, C plays $\begin{bmatrix} .5 \\ .5 \end{bmatrix}$.

(c) R plays $[.3 \quad .7]$, C plays $\begin{bmatrix} .6 \\ .4 \end{bmatrix}$.

(d) R plays $[.75 \quad .25]$, C plays $\begin{bmatrix} .2 \\ .8 \end{bmatrix}$.

2. Suppose that a game has payoff matrix

$$\begin{bmatrix} 1 & 0 & 2 \\ -1 & 2 & 0 \\ 0 & -1 & -1 \end{bmatrix}.$$

Calculate the expected values for the following strategies and determine which of the following situations is most advantageous to C.

(a) R plays $[1 \quad 0 \quad 0]$, C plays $\begin{bmatrix} .5 \\ .4 \\ .1 \end{bmatrix}$.

(b) R plays $[.3 \quad .3 \quad .4]$, C plays $\begin{bmatrix} .4 \\ .4 \\ .2 \end{bmatrix}$.

(c) R plays $[0 \quad .5 \quad .5]$, C plays $\begin{bmatrix} .4 \\ 0 \\ .6 \end{bmatrix}$.

(d) R plays $[.1 \quad .1 \quad .8]$, C plays $\begin{bmatrix} .2 \\ .2 \\ .6 \end{bmatrix}$.

3. Refer to Example 3. Suppose that the inspector changes his strategy and visits plant B 80% of the time. How much is Acme's average fine per day?

4. Refer to Example 3. Suppose that the inspector visits plant B 30% of the time. How much is Acme's average fine per day?

5. Suppose that two players, R and C, write down letters of the alphabet. If both write down vowels or both consonants, then there is no payment to either player. If R writes a vowel

and C writes a consonant, then C pays R \$2. If R writes a consonant and C writes a vowel, then R pays C \$1. Suppose that R chooses a consonant 75% of the plays and C chooses a vowel 40% of the plays. What is the average loss (or gain) of R per play?

6. A businessman must decide whether to carry flood insurance. He may insure himself for \$2 million for \$100,000, \$1 million for \$50,000, or \$.5 million for \$30,000. His business is worth \$2 million. There is a flood serious enough to destroy his business an average of once every 10 years. In order to save insurance premiums, he decides each year on a probabilistic basis how much insurance to carry. He chooses \$2 million 20% of the time, \$1 million 20% of the time, \$.5 million 20% of the time, and no insurance 40% of the time. What is the average annual loss to the businessman?

SOLUTIONS TO PRACTICE PROBLEMS 2

1. If C plays $\begin{bmatrix} .5 \\ .5 \end{bmatrix}$, the expected value (to R) is

$$[.6 \quad .4]\begin{bmatrix} 4 & -2 \\ -3 & 1 \end{bmatrix}\begin{bmatrix} .5 \\ .5 \end{bmatrix} = [1.2 \quad -.8]\begin{bmatrix} .5 \\ .5 \end{bmatrix} = [.2].$$

If C plays $\begin{bmatrix} .7 \\ .3 \end{bmatrix}$, the expected value (to R) is

$$[.6 \quad .4]\begin{bmatrix} 4 & -2 \\ -3 & 1 \end{bmatrix}\begin{bmatrix} .7 \\ .3 \end{bmatrix} = [1.2 \quad -.8]\begin{bmatrix} .7 \\ .3 \end{bmatrix} = [.6].$$

Thus in the first case R gains an average of .2 per play, whereas in the second R gains .6. Since C wishes to minimize R's winnings, C should clearly play the first strategy.

2. The answer is the same as in Problem 1, since the new expected values will be 5 more than the original expected values and the first strategy will still be the best for C.

9.3. Determining Optimal Mixed Strategies

As we have seen, each choice of strategies by R and C results in an expected value, representing the average payoff to R per play. In this section we shall give a method for choosing the best strategies. Let us begin by clarifying our notion of optimality.

> *Optimal Strategy for R* To every choice of a strategy for R there is a best counterstrategy—that is, a strategy for C which results in the least expected value e. An *optimal mixed strategy for R* is one for which the expected value against C's best counterstrategy is as large as possible.

In a similar way we can define the optimal strategy for C.

> *Optimal Strategy for C* To every choice of a strategy for C there is a best counterstrategy—that is, a strategy for R which results in the largest expected value e. An *optimal mixed strategy for C* is one for which the expected value against R's best counterstrategy is as small as possible.

It is most surprising that the optimal strategies for R and C may be determined using linear programming. To see how this is done let us consider a particular problem.

EXAMPLE 1 Suppose that a game has payoff matrix

$$\begin{bmatrix} 5 & 3 \\ 1 & 4 \end{bmatrix}.$$

Reduce the determination of an optimal strategy for R to a linear programming problem.

Solution Suppose that R plays the strategy $[r_1 \quad r_2]$. What is C's best counterstrategy? If C plays $\begin{bmatrix} c_1 \\ c_2 \end{bmatrix}$, then the expected value of the game is

$$[r_1 \quad r_2]\begin{bmatrix} 5 & 3 \\ 1 & 4 \end{bmatrix}\begin{bmatrix} c_1 \\ c_2 \end{bmatrix} = [5r_1 + r_2 \quad 3r_1 + 4r_2]\begin{bmatrix} c_1 \\ c_2 \end{bmatrix}$$

$$= [(5r_1 + r_2)c_1 + (3r_1 + 4r_2)c_2].$$

If C pursues his best counterstrategy, then he will try to minimize the expected value of the game. That is, C will try to minimize

$$(5r_1 + r_2)c_1 + (3r_1 + 4r_2)c_2.$$

Since $c_1 \geq 0, c_2 \geq 0, c_1 + c_2 = 1$, this expression has as its minimum value the smaller of the terms $5r_1 + r_2$ or $3r_1 + 4r_2$. That is, if $5r_1 + r_2$ is the smaller, then C should choose the strategy $c_1 = 1, c_2 = 0$, whereas if $3r_1 + 4r_2$ is the smaller, then C should choose the strategy $c_1 = 0, c_2 = 1$. In any case the expected value of the game if C adopts his best counterstrategy is the smaller of $5r_1 + r_2$ and $3r_1 + 4r_2$. The goal of R is to maximize this expected value. In other words, the mathematical problem R faces is this:

Maximize the minimum of $5r_1 + r_2$ and $3r_1 + 4r_2$,

where $r_1 \geq 0, r_2 \geq 0, r_1 + r_2 = 1$.

Let v denote the minimum of $5r_1 + r_2$ and $3r_1 + 4r_2$. Clearly $v > 0$. Then

$$5r_1 + r_2 \geq v$$
$$3r_1 + 4r_2 \geq v. \tag{*}$$

Maximizing v is the same as minimizing $1/v$. Moreover, the inequalities (*) may be rewritten in the form

$$5\frac{r_1}{v} + \frac{r_2}{v} \geq 1$$
$$3\frac{r_1}{v} + 4\frac{r_2}{v} \geq 1. \tag{**}$$

Moreover, since $r_1 \geq 0, r_2 \geq 0, r_1 + r_2 = 1$, we see that

$$\frac{r_1}{v} \geq 0 \qquad \frac{r_2}{v} \geq 0 \qquad \frac{r_1}{v} + \frac{r_2}{v} = \frac{1}{v}. \tag{***}$$

This suggests that we introduce new variables:

$$y_1 = \frac{r_1}{v} \qquad y_2 = \frac{r_2}{v}.$$

Then (***) and (**) may be rewritten as

$$y_1 + y_2 = \frac{1}{v}$$
$$5y_1 + y_2 \geq 1$$
$$3y_1 + 4y_2 \geq 1$$
$$y_1 \geq 0 \qquad y_2 \geq 0.$$

We wish to minimize $1/v$, so we may finally state our original question in terms of a linear programming problem:

Minimize $y_1 + y_2$ subject to the constraints

$$\begin{cases} 5y_1 + y_2 \geq 1 \\ 3y_1 + 4y_2 \geq 1 \\ y_1 \geq 0 \quad y_2 \geq 0. \end{cases}$$

In terms of the solution to this linear programming problem we may calculate R's optimal strategy as follows:

$$r_1 = vy_1 \qquad r_2 = vy_2, \qquad \text{where } v = \frac{1}{y_1 + y_2}.$$

In the derivation above it was essential that the entries of the matrix were positive numbers, for this is how we derived that $v > 0$. The same reasoning used in Example 1 can be used in general to convert the determination of R's optimal strategy to a linear programming problem, *provided the payoff matrix has positive entries.* If the payoff matrix does not have positive entries, then just add a large positive constant to each of the entries so as to give a matrix with positive entries. The new matrix will have the same optimal strategy as the original one. However, since all its entries are positive, we may use the reasoning above to reduce determination of the optimal strategy to a linear programming problem.

Optimal Strategy for R   Let the payoff matrix of a game be

$$\begin{bmatrix} a_{11} & a_{12} & \cdots & a_{1n} \\ a_{21} & a_{22} & & a_{2n} \\ \vdots & \vdots & & \vdots \\ a_{m1} & a_{m2} & \cdots & a_{mn} \end{bmatrix},$$

where all entries of the matrix are positive numbers. Let $y_1, y_2, \ldots, y_m$ be chosen so as to minimize

$$y_1 + y_2 + \cdots + y_m$$

subject to the constraints

$$\begin{cases} y_1 \geq 0, y_2 \geq 0, \ldots, y_m \geq 0 \\ a_{11}y_1 + a_{21}y_2 + \cdots + a_{m1}y_m \geq 1 \\ a_{12}y_1 + a_{22}y_2 + \cdots + a_{m2}y_m \geq 1 \\ \qquad\qquad \vdots \\ a_{1n}y_1 + a_{2n}y_2 + \cdots + a_{mn}y_m \geq 1. \end{cases}$$

Let

$$v = \frac{1}{y_1 + y_2 + \cdots + y_m}.$$

Then an optimal strategy for R is $\begin{bmatrix} r_1 & r_2 \cdots & r_m \end{bmatrix}$, where $r_1 = vy_1, r_2 = vy_2, \ldots, r_m = vy_m$. Furthermore, if C adopts the best counterstrategy, then the expected value is v.

Note that the determination of $y_1, y_2, \ldots, y_m$ is a linear programming problem, whose solution can be obtained using either the method of Chapter 3 (if $m = 2$) or the simplex method of Chapter 4 (any m). The next example illustrates the result above.

EXAMPLE 2   Suppose that a game has payoff matrix

$$\begin{bmatrix} 5 & 3 \\ 1 & 4 \end{bmatrix}.$$

(a) Determine an optimal strategy for R.

(b) Determine the expected payoff to R if C uses the best counterstrategy.

Solution (a) The associated linear programming problem asks us to minimize $y_1 + y_2$ subject to the constraints

$$\begin{cases} y_1 \geq 0, \quad y_2 \geq 0 \\ 5y_1 + y_2 \geq 1 \\ 3y_1 + 4y_2 \geq 1. \end{cases}$$

In Fig. 1 we have sketched the feasible set for this problem and evaluated the objective function at each vertex.

The minimum value of $y_1 + y_2$ is $\frac{5}{17}$ and occurs when $y_1 = \frac{3}{17}$ and $y_2 = \frac{2}{17}$. Further,

$$v = \frac{1}{y_1 + y_2} = \frac{1}{\frac{5}{17}} = \frac{17}{5}$$

$$r_1 = vy_1 = \tfrac{17}{5} \cdot \tfrac{3}{17} = \tfrac{3}{5}$$

$$r_2 = vy_2 = \tfrac{17}{5} \cdot \tfrac{2}{17} = \tfrac{2}{5}.$$

Thus the optimal strategy for R is $\begin{bmatrix} \frac{3}{5} & \frac{2}{5} \end{bmatrix}$.

(b) The expected value against the best counterstrategy is $v = \frac{17}{5}$.

There is a similar linear programming technique for determining the optimal strategy for C.

FIGURE 1

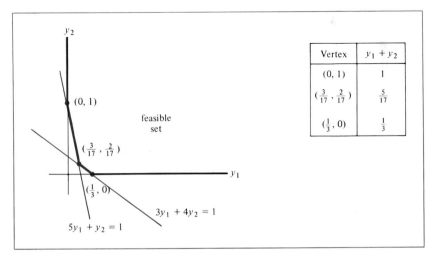

Optimal Strategy for C Let the payoff matrix of a game be

$$\begin{bmatrix} a_{11} & a_{12} & \cdots & a_{1n} \\ a_{21} & a_{22} & \cdots & a_{2n} \\ \vdots & & & \\ a_{m1} & a_{m2} & \cdots & a_{mn} \end{bmatrix},$$

where all entries of the matrix are positive numbers. Let $z_1, z_2, \ldots, z_n$ be chosen so as to maximize

$$z_1 + z_2 + \cdots + z_n$$

subject to the constraints

$$\begin{cases} z_1 \geq 0, z_2 \geq 0, \ldots, z_n \geq 0 \\ a_{11}z_1 + a_{12}z_2 + \cdots + a_{1n}z_n \leq 1 \\ a_{21}z_1 + a_{22}z_2 + \cdots + a_{2n}z_n \leq 1 \\ \vdots \\ a_{m1}z_1 + a_{m2}z_2 + \cdots + a_{mn}z_n \leq 1. \end{cases}$$

Let $v = 1/(z_1 + z_2 + \cdots + z_n)$. Then an optimal strategy for C is

$$\begin{bmatrix} c_1 \\ c_2 \\ \vdots \\ c_n \end{bmatrix},$$

where $c_1 = vz_1, c_2 = vz_2, \ldots, c_n = vz_n$.

EXAMPLE 3 Determine the optimal strategy for C for the game with payoff matrix

$$\begin{bmatrix} 5 & 3 \\ 1 & 4 \end{bmatrix}.$$

Solution We must maximize $z_1 + z_2$ subject to the constraints

$$\begin{cases} z_1 \geq 0, \quad z_2 \geq 0 \\ 5z_1 + 3z_2 \leq 1 \\ z_1 + 4z_2 \leq 1. \end{cases}$$

The solution is: The maximum value is $\frac{5}{17}$ and it occurs when $z_1 = \frac{1}{17}, z_2 = \frac{4}{17}$.

Therefore, $v = \frac{17}{5}$, and the optimal strategy for C is $\begin{bmatrix} c_1 \\ c_2 \end{bmatrix}$, where

$$c_1 = vz_1 = \frac{17}{5} \cdot \frac{1}{17} = \frac{1}{5}$$

$$c_2 = vz_2 = \frac{17}{5} \cdot \frac{4}{17} = \frac{4}{5}.$$

Notice that in both Examples 2 and 3 we obtained $v = \frac{17}{5}$. This was not just coincidence. This phenomenon always occurs, and the number v is called the *value* of the game. An easy computation shows that for the matrix of Examples 2 and 3, when R and C each use their optimum strategies, the expected value is $v = \frac{17}{5}$. That is,

$$\begin{bmatrix} \frac{3}{5} & \frac{2}{5} \end{bmatrix} \begin{bmatrix} 5 & 3 \\ 1 & 4 \end{bmatrix} \begin{bmatrix} \frac{1}{5} \\ \frac{4}{5} \end{bmatrix} = \begin{bmatrix} \frac{17}{5} \end{bmatrix}.$$

Let us briefly reconsider the calculations of optimal strategies for R and C. We begin in every case with the matrix for the game, A. Let us assume that A is an $m \times n$ matrix and that each entry of A is a positive number. To find the optimal strategy for C, we find the matrix Z that maximizes the objective function EZ subject to the constraints $AZ \leq B$ and $Z \geq \mathbf{0}$, where

$$E = \underbrace{\begin{bmatrix} 1 & 1 & 1 & \cdots & 1 \end{bmatrix}}_{n \text{ entries}}, \quad Z = \begin{bmatrix} z_1 \\ z_2 \\ \vdots \\ z_n \end{bmatrix}, \quad \text{and} \quad B = \left.\begin{bmatrix} 1 \\ 1 \\ \vdots \\ 1 \end{bmatrix}\right\} m \text{ entries.}$$

The dual of the linear programming problem associated with finding an optimal strategy for C is to find the matrix Y that minimizes the objective function $B^T Y$ subject to the constraints $A^T Y \geq E^T$ and $Y \geq \mathbf{0}$, where

$$Y = \begin{bmatrix} y_1 \\ y_2 \\ \vdots \\ y_m \end{bmatrix}.$$

But this is exactly the problem of finding an optimal strategy for R.

> The problem of finding the optimal strategy for R is a linear programming problem whose dual is the problem of finding an optimal strategy for C, and vice versa.

Our previous work with duality leads us to conclude:

1. If there exists an optimal strategy for R, then there exists an optimal strategy for C, and vice versa.
2. The minimum of the objective function $y_1 + y_2 + \cdots + y_m$ and the maximum of the objective function $z_1 + z_2 + \cdots + z_n$ are equal—since the linear programming problems are duals of each other. Hence, the value of the optimal strategy for C is the same as the value of the optimal strategy for R, that is,

$$\frac{1}{z_1 + z_2 + \cdots + z_n} = \frac{1}{y_1 + y_2 + \cdots + y_m}.$$

EXAMPLE 4   Determine the optimal strategies for R and C for the game with payoff matrix

$$\begin{bmatrix} 1 & -1 \\ -3 & 0 \end{bmatrix}.$$

Solution   We cannot apply our technique directly, since only one of the entries of the given matrix is a positive number. However, if we add 4 to each entry, then the new matrix will be

$$\begin{bmatrix} 5 & 3 \\ 1 & 4 \end{bmatrix},$$

which does have all positive entries. These two payoff matrices have the same optimal strategies. The only difference is that the new matrix has a value that is 4 more than the given matrix. Now, the optimal strategies and value of the new matrix were found in Examples 2 and 3 to be $\begin{bmatrix} \frac{3}{5} & \frac{2}{5} \end{bmatrix}, \begin{bmatrix} \frac{1}{5} \\ \frac{4}{5} \end{bmatrix}$, and $\frac{17}{5}$. Therefore, the optimal strategies for the given matrix are $\begin{bmatrix} \frac{3}{5} & \frac{2}{5} \end{bmatrix}$ and $\begin{bmatrix} \frac{1}{5} \\ \frac{4}{5} \end{bmatrix}$, and the value is $\frac{17}{5} - 4 = -\frac{3}{5}$.

EXAMPLE 5   Use the simplex method and resulting tableau to determine the optimal strategies for the game of Example 4.

Solution   As in Example 4, add 4 to each entry to get a matrix with positive entries, and set up the tableau for finding the optimal strategy for C. [We choose this linear programming problem because it is a maximization problem.] The transformed matrix A is

$$\begin{bmatrix} 5 & 3 \\ 1 & 4 \end{bmatrix}.$$

To find the optimal strategy for C we need to find the values of z_1 and z_2 that maximize $z_1 + z_2$ subject to the constraints

$$\begin{cases} 5z_1 + 3z_2 \le 1 \\ z_1 + 4z_2 \le 1 \\ z_1 \ge 0, \quad z_2 \ge 0. \end{cases}$$

We set up the tableau using slack variables t and u and display the initial and final tableaux:

	z_1	z_2	t	u	M	
t	5	3	1	0	0	1
u	1	4	0	1	0	1
M	-1	-1	0	0	1	0

	z_1	z_2	t	u	M	
z_1	1	0	$\frac{4}{17}$	$-\frac{3}{17}$	0	$\frac{1}{17}$
z_2	0	1	$-\frac{1}{17}$	$\frac{5}{17}$	0	$\frac{4}{17}$
M	0	0	$\frac{3}{17}$	$\frac{2}{17}$	1	$\frac{5}{17}$

Thus, the solution is $z_1 = \frac{1}{17}$, $z_2 = \frac{4}{17}$ with $M = z_1 + z_2 = \frac{5}{17}$. Then $v = \frac{17}{5}$, and the optimal strategy for C is

$$\begin{bmatrix} vz_1 \\ vz_2 \end{bmatrix} = \begin{bmatrix} \frac{1}{5} \\ \frac{4}{5} \end{bmatrix}.$$

This agrees with previous solutions and the value of the matrix is $\frac{17}{5}$, which is 4 more than the original matrix. Thus, the value of the game is $\frac{17}{5} - 4 = -\frac{3}{5}$.

But the optimal strategy for R can be read from the tableau since y_1 and y_2 are the values of the variables in the dual of the problem we solved. So $y_1 = t = \frac{3}{17}$, $y_2 = u = \frac{2}{17}$, and $M = \frac{5}{17}$. Then $v = \frac{17}{5}$ and the optimal strategy for R is $\begin{bmatrix} vy_1 & vy_2 \end{bmatrix} = \begin{bmatrix} \frac{3}{5} & \frac{2}{5} \end{bmatrix}$.

We actually have a very useful fact.

> *Fundamental Theorem of Game Theory*
>
> Every two-person zero-sum game has a solution.

Verification of the Fundamental Theorem of Game Theory

If the given two-person game has a saddle point, then the game is strictly determined, and optimal strategies for R and C are given by the position of the saddle point.

If the game is not strictly determined, then let us assume that the $m \times n$ payoff matrix A has only positive entries. We let B be an $m \times 1$ column matrix in which each entry is 1, and let E be a $1 \times n$ row matrix of 1's. Then:

1. There is an optimal feasible solution to the problem:

$$\text{Maximize } M = EZ \text{ subject to } AZ \leq B \text{ and } Z \geq 0. \qquad \text{(P)}$$

2. There is an optimal feasible solution to the problem:

$$\text{Minimize } M = B^T Y \text{ subject to } A^T Y \geq E^T \text{ and } Y \geq 0. \qquad \text{(D)}$$

3. The solutions to (P) and (D) give a solution to the game.

To see that (1) and (2) hold, we note that there is a feasible solution for the inequalities of the primal problem (P). The $n \times 1$ matrix of zeros, $Z = 0$, satisfies $AZ \leq B$. Also, there is a feasible solution for the inequalities of the dual problem (D). This can be seen by noting that since every element of the matrix A is positive, we can find an $m \times 1$ matrix $Y \geq 0$ with sufficiently large entries to guarantee that $A^T Y \geq E^T$. Since the inequalities of both (P) and (D) have a feasible solution, the Fundamental Theorem of Duality (Chapter 4) tells us that both (P) and (D) have optimal feasible solutions.

Let Z^* and Y^* be optimal feasible solutions of (P) and (D), respectively. Say that

$$Z^* = \begin{bmatrix} z_1^* \\ z_2^* \\ \vdots \\ z_n^* \end{bmatrix} \quad \text{and} \quad Y^* = \begin{bmatrix} y_1^* \\ y_2^* \\ \vdots \\ y_m^* \end{bmatrix}.$$

Then the maximum for (P)

$$M = z_1^* + z_2^* + \cdots + z_n^*$$

equals the minimum for (D)

$$M = y_1^* + y_2^* + \cdots + y_m^*$$

and

$$AZ^* \leq B \quad \text{and} \quad A^T Y^* \geq E^T.$$

Recall that B is an $m \times 1$ matrix of 1's and E^T is an $n \times 1$ matrix of 1's.

M must be strictly greater than zero since at least one of the y_i^* must be >0 in order for $A^T Y^* \geq E^T$ to hold. Therefore $\dfrac{1}{M}$ is defined. We let

$$C = \begin{bmatrix} \dfrac{1}{M} z_1^* \\ \dfrac{1}{M} z_2^* \\ \vdots \\ \dfrac{1}{M} z_n^* \end{bmatrix} = \begin{bmatrix} c_1 \\ c_2 \\ \vdots \\ c_n \end{bmatrix}$$

and

$$R = \begin{bmatrix} \dfrac{1}{M} y_1^* & \dfrac{1}{M} y_2^* & \cdots & \dfrac{1}{M} y_m^* \end{bmatrix} = \begin{bmatrix} r_1 & r_2 & \cdots & r_m \end{bmatrix}$$

Furthermore, C and R represent optimal strategies for players C and R, respectively. To verify that C and R are legitimate strategies, we note that since $M > 0$, $Z^* \geq \mathbf{0}$, and $Y^* \geq \mathbf{0}$, every entry in C and R is nonnegative. We only need to check that

$$c_1 + c_2 + \cdots + c_n = 1, \qquad r_1 + r_2 + \cdots + r_m = 1.$$

This follows directly from the definitions of M, C, and R.

PRACTICE PROBLEMS 3

1. Determine the optimal strategy for C for the game with payoff matrix

$$\begin{bmatrix} 2 & 14 \\ 6 & 12 \\ 8 & 6 \end{bmatrix}.$$

2. Determine by inspection the optimal strategies for C for the games whose payoff matrices are:

(a) $\begin{bmatrix} 0 & 12 \\ 4 & 10 \\ 6 & 4 \end{bmatrix}$ (b) $\begin{bmatrix} 6 & 0 \\ 4 & 3 \\ 8 & -1 \end{bmatrix}$

EXERCISES 3

Determine optimal strategies for R and for C for the games whose payoff matrices are:

1. $\begin{bmatrix} 2 & 4 \\ 5 & 3 \end{bmatrix}$ 2. $\begin{bmatrix} 2 & 3 \\ 3 & 2 \end{bmatrix}$ 3. $\begin{bmatrix} 3 & -6 \\ -5 & 4 \end{bmatrix}$

4. $\begin{bmatrix} 5 & 2 \\ 7 & 1 \end{bmatrix}$ 5. $\begin{bmatrix} 4 & 1 \\ 2 & 4 \end{bmatrix}$ 6. $\begin{bmatrix} 5 & -8 \\ 3 & 6 \end{bmatrix}$

Determine optimal strategies for R for the games whose payoff matrices are:

7. $\begin{bmatrix} 3 & 5 & -1 \\ 4 & -1 & 6 \end{bmatrix}$ 8. $\begin{bmatrix} -2 & 1 & 0 \\ 2 & 0 & 1 \end{bmatrix}$

Determine optimal strategies for C for the games whose payoff matrices are:

9. $\begin{bmatrix} -3 & 1 \\ 4 & 2 \\ 1 & 0 \end{bmatrix}$ 10. $\begin{bmatrix} 0 & 2 \\ 2 & -1 \\ 1 & 0 \end{bmatrix}$

11. A rumrunner attempts to smuggle rum into a country having two ports. Each day the coast guard is able to patrol only one of the ports. If the rumrunner enters via an unpatrolled port, he will be able to sell his rum for a profit of $7000. If he enters the first port and it is patrolled that day, he is certain to be caught and will have his rum (worth $1000) confiscated and be fined $1000. If he enters the second port (which is big and crowded) and it is patrolled that day, he will have time to jettison his cargo and thereby escape a fine.

   (a) What is the optimal strategy for the rumrunner?

   (b) What is the optimal strategy for the coast guard?

   (c) How profitable is rumrunning? That is, what is the value of the game?

12. (*Which hand?*)   Ralph puts a coin in one of his hands and Carl tries to guess which hand holds the coin. If Carl guesses incorrectly, he must pay Ralph $2. If Carl guesses correctly, then Ralph must pay him $3 if the coin was in the left hand and $1 if it was in the right.

(a) What is the optimal strategy for Ralph?

(b) What is the optimal strategy for Carl?

(c) Whom does this game favor?

13.   The Carter Company can choose between two advertising strategies (I and II). Its most important competitor, Rosedale Associates, has a choice of three advertising strategies (*a*, *b*, *c*). The estimated payoff to Rosedale Associates away from the Carter Company is given by the payoff matrix

$$
\begin{array}{c}
 \\
a \\
b \\
c
\end{array}
\begin{array}{cc}
\text{I} & \text{II}
\end{array}
\begin{bmatrix}
-2 & 1 \\
2 & -3 \\
1 & -2
\end{bmatrix}
$$

where the entries represent thousands of dollars per week. Determine the optimal strategies for each company.

SOLUTIONS TO PRACTICE PROBLEMS 3

1. The associated linear programming problem is: Maximize $z_1 + z_2$ subject to the constraints

$$
\begin{cases}
z_1 \geq 0, \quad z_2 \geq 0 \\
2z_1 + 14z_2 \leq 1 \\
6z_1 + 12z_2 \leq 1 \\
8z_1 + 6z_2 \leq 1.
\end{cases}
$$

In Fig. 2 we have sketched the feasible set and evaluated the objective function at each vertex. The maximum value of $z_1 + z_2$ is $\frac{4}{30}$, which is achieved at the vertex $(\frac{3}{30}, \frac{1}{30})$. Therefore,

$$
v = \frac{1}{z_1 + z_2} = \frac{1}{\frac{4}{30}} = \frac{30}{4},
$$

$$
c_1 = v \cdot z_1 = \frac{30}{4} \cdot \frac{3}{30} = \frac{3}{4},
$$

$$
c_2 = v \cdot z_2 = \frac{30}{4} \cdot \frac{1}{30} = \frac{1}{4}.
$$

That is, the optimal strategy for C is $\begin{bmatrix} \frac{3}{4} \\ \frac{1}{4} \end{bmatrix}$.

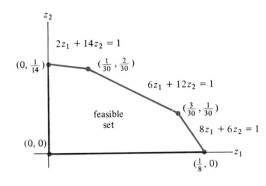

$2z_1 + 14z_2 = 1$

$(0, \frac{1}{14})$

$(\frac{1}{30}, \frac{2}{30})$

$6z_1 + 12z_2 = 1$

$(\frac{3}{30}, \frac{1}{30})$

feasible set

$8z_1 + 6z_2 = 1$

$(0, 0)$

$(\frac{1}{8}, 0)$

Vertex	$z_1 + z_2$
$(0, 0)$	0
$(0, \frac{1}{14})$	$\frac{1}{14}$
$(\frac{1}{30}, \frac{2}{30})$	$\frac{3}{30}$
$(\frac{3}{30}, \frac{1}{30})$	$\frac{4}{30}$
$(\frac{1}{8}, 0)$	$\frac{1}{8}$

FIGURE 2

2. (a) $\begin{bmatrix} \frac{3}{4} \\ \frac{1}{4} \end{bmatrix}$. If we add 2 to each entry, we obtain the payoff matrix of Problem 1, so these two games have the same optimal strategies.

(b) Always play column 2. This game is strictly determined and has the entry 3 as saddle point. It is a good idea always to check for a saddle point before looking for a mixed strategy.

Chapter 9: CHECKLIST

☐ Payoff matrix
☐ Zero-sum game
☐ Optimum pure strategy for R
☐ Optimum pure strategy for C
☐ Saddle point
☐ Strictly determined game
☐ Value of a strictly determined game
☐ Expected value of a mixed strategy
☐ Optimal mixed strategy for R and C
☐ Use of linear programming to compute optimal mixed strategies
☐ Value of a game

Chapter 9: SUPPLEMENTARY EXERCISES

State whether or not the games having the following payoff matrices are strictly determined. If so, give the optimal pure strategies and the values of the strategies.

1. $\begin{bmatrix} 5 & -1 & 1 \\ -3 & 5 & 1 \\ 4 & 3 & 2 \end{bmatrix}$

2. $\begin{bmatrix} 1 & 2 & 3 \\ 3 & 2 & 1 \end{bmatrix}$

3. $\begin{bmatrix} 0 & 1 \\ 1 & 0 \\ 2 & -1 \end{bmatrix}$

4. $\begin{bmatrix} 2 & 1 & 2 \\ -1 & 0 & 3 \\ 4 & 1 & -4 \end{bmatrix}$

Determine the expected value of each given pair of mixed strategies for the given payoff matrix.

5. $\begin{bmatrix} \frac{3}{4} & \frac{1}{4} \end{bmatrix}$, $\begin{bmatrix} \frac{1}{3} \\ \frac{2}{3} \end{bmatrix}$; $\begin{bmatrix} 0 & 24 \\ 12 & -36 \end{bmatrix}$

6. $\begin{bmatrix} \frac{1}{2} & \frac{1}{2} \end{bmatrix}$, $\begin{bmatrix} \frac{1}{3} \\ \frac{1}{3} \\ \frac{1}{3} \end{bmatrix}$; $\begin{bmatrix} -6 & 6 & 0 \\ 0 & -12 & 24 \end{bmatrix}$

7. $\begin{bmatrix} .2 & .3 & .5 \end{bmatrix}$, $\begin{bmatrix} .4 \\ .6 \end{bmatrix}$; $\begin{bmatrix} 1 & 0 \\ -3 & 1 \\ 0 & 5 \end{bmatrix}$

8. $\begin{bmatrix} .1 & .1 & .8 \end{bmatrix}$, $\begin{bmatrix} .4 \\ .3 \\ .3 \end{bmatrix}$; $\begin{bmatrix} 0 & 1 & 3 \\ -1 & 0 & 2 \\ -3 & -2 & 0 \end{bmatrix}$

Determine the optimal strategies for R and C for the games with the following payoff matrices:

9. $\begin{bmatrix} -3 & 4 \\ 2 & -2 \end{bmatrix}$

10. $\begin{bmatrix} 3 & -6 \\ -4 & 4 \end{bmatrix}$

11. Determine the optimal strategy for R for the game with payoff matrix

$$\begin{bmatrix} 5 & -2 & 0 \\ 1 & 4 & 1 \end{bmatrix}.$$

12. Determine the optimal strategy for C for the game with payoff matrix

$$\begin{bmatrix} 1 & 3 \\ 3 & 1 \\ 4 & 2 \end{bmatrix}.$$

13. Ruth and Carol play the following game. Both have two cards, a two and a six. Each puts one of her cards on the table. If both put down the same denomination, Ruth pays Carol $3. Otherwise, Carol pays Ruth as many dollars as the denomination of Carol's card.

   (a) Find the optimal strategies for Ruth and Carol.

   (b) Whom does this game favor?

14. An investor is considering purchasing one of three stocks. Stock A is regarded as conservative, stock B as speculative, and stock C as highly risky. If the economic growth during the coming year is strong, then stock A should increase in value by $3000, stock B by $6000, and stock C by $15,000. If the economic growth during the next year is average, then stock A should increase in value by $2000, stock B by $2000, and stock C by $1000. If the economic growth is weak, then stock A should increase in value by $1000 and stocks B and C decrease in value by $3000 and $10,000, respectively.

   (a) Set up the 3 × 3 payoff matrix showing the investing gains for the possible stock purchases and levels of economic growth.

   (b) What is the investor's optimal strategy?

10

THE MATHEMATICS
OF FINANCE

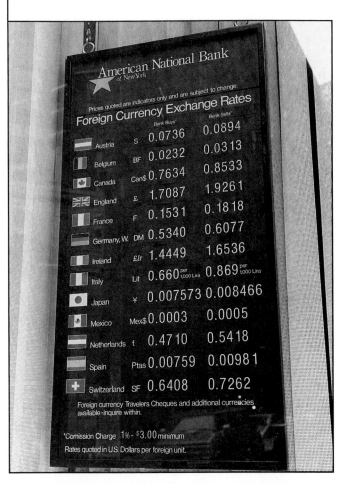

10.1 Interest

10.2 Annuities

10.3 Amortization of Loans

10.1. Interest

When you deposit money into a savings account, the bank pays you a fee for the use of your money. This fee is called *interest* and is determined by the amount deposited, the duration of the deposit, and the quoted interest rate. The amount deposited is called the *principal* and the amount to which the principal grows (after the addition of interest) is called the *compound amount*.

Here are the entries in a hypothetical bank passbook:

Date	Deposits	Withdrawals	Interest	Balance
1/1/80	$100.00			$100.00
4/1/80			$2.00	102.00
7/1/80			2.04	104.04
10/1/80			2.08	106.12
1/1/81			2.12	108.24

Note the following facts about this passbook:

1. The principal is $100.00. The compound amount after 1 year is $108.24.

2. Interest is being paid four times per year (or in financial language, *quarterly*).

3. Each quarter, the interest is 2% of the previous balance That is, $2.00 is 2% of $100.00, $2.04 is 2% of $102.00, and so on. Since $4 \times 2\%$ is 8%, we say that the passbook is earning 8% *annual interest compounded quarterly*.

As in the passbook above, interest rates are usually stated as annual percentage rates, with the interest to be *compounded* (i.e., computed) a certain number of times per year. Some common frequencies for compounding are:

Number of interest periods per year	Length of each interest period	Interest called:
1	One year	Compounded annually
2	Six months	Compounded semiannually
4	Three months	Compounded quarterly
12	One month	Compounded monthly
52	One week	Compounded weekly
365	One day	Compounded daily

Of special importance is the *interest per period*, denoted i, which is calculated by dividing the annual percentage by the number of interest periods per year. For example, in our passbook above, the annual percentage rate was 8%, the interest was compounded quarterly, and the interest per period was $8\%/4 = 2\%$.

EXAMPLE 1 Determine the interest per period for each of the following interest rates.

    (a) 10% interest compounded semiannually

    (b) 6% interest compounded monthly

Solution (a) The annual percentage rate is 10% and the number of interest periods is 2. Therefore,

$$i = \frac{10\%}{2} = 5\%.$$

Note In decimal form, $i = .05$.

    (b) The annual percentage rate is 6% and the number of interest periods is 12. Therefore,

$$i = \frac{6\%}{12} = \frac{1}{2}\%.$$

Note In decimal form, $i = .005$.

    Consider a savings account in which the interest per period is i. Then the interest earned for the next period is i times the current balance. The next balance, B_{next}, is then computed by adding this interest to the current balance, B_{current}. That is,

$$B_{\text{next}} = B_{\text{current}} + i \cdot B_{\text{current}}. \tag{1}$$

EXAMPLE 2 Compute the interest and the balance for the first two interest periods for a deposit of $1000 at 10% interest compounded semiannually.

Solution Here $i = 5\%$ or $.05$. The interest for the first period is 5% of $1000, or $50. By formula (1), the balance after the first interest period is

$$B_{\text{first}} = 1000 + .05(1000)$$
$$= 1000 + 50$$
$$= \$1050.$$

Similarly, since 5% of 1050 is 52.50, the interest for the second period is $52.50 and the balance is

$$B_{\text{second}} = B_{\text{first}} + .05B_{\text{first}}$$
$$= 1050 + .05(1050)$$
$$= \$1050 + 52.50$$
$$= \$1102.50.$$

Let us compute B_{second} using another method.

$$B_{\text{second}} = B_{\text{first}} + .05B_{\text{first}}$$
$$= 1 \cdot B_{\text{first}} + .05B_{\text{first}}$$
$$= (1 + .05)B_{\text{first}}$$
$$= (1.05)B_{\text{first}}$$
$$= (1.05)1050$$
$$= \$1102.50.$$

The alternative method for computing B_{second} just presented can be generalized. Namely, we always have

$$B_{\text{next}} = B_{\text{current}} + i \cdot B_{\text{current}}$$
$$= 1 \cdot B_{\text{current}} + i \cdot B_{\text{current}}.$$

Therefore,

$$B_{\text{next}} = (1 + i)B_{\text{current}}.$$

This last result says that balances for successive time periods are computed by multiplying by $1 + i$.

The formula for the balance after any number of interest periods is now easily derived:

Principal	P
Balance after 1 interest period	$(1 + i)P$
Balance after 2 interest periods	$(1 + i) \cdot (1 + i)P$ or $(1 + i)^2 P$
Balance after 3 interest periods	$(1 + i) \cdot (1 + i)^2 P$ or $(1 + i)^3 P$
Balance after 4 interest periods	$(1 + i)^4 P$
$\vdots$	
Balance after n interest periods	$(1 + i)^n P$

Denote the compound amount by the letter F (suggestive of "future value"). Then, the compound amount after n interest periods is given by the formula

$$F = (1 + i)^n P, \tag{2}$$

where i is the interest rate per period and P is the principal.

Values of $(1 + i)^n$ for specific values of i and n are easily determined using either a calculator or tables. A brief table of useful values has been included as Table 2 of the Appendix.

EXAMPLE 3 Apply formula (2) to the savings account passbook discussed at the beginning of this section and calculate the compound amount after 1 year and after 5 years.

Solution The principal P is \$100. Since the interest rate is 8% compounded quarterly, we have $i = 2\%$ or .02. One year consists of four interest periods, so $n = 4$. Therefore, the compound amount after 1 year is

$$F = (1 + .02)^4 \cdot 100$$

$$= (1.02)^4 \cdot 100$$

$$= (1.08243216) \cdot 100 \qquad \text{(from Table 2)}$$

$$= 108.243216$$

$$= \$108.24 \qquad \text{(after rounding off to the nearest cent).}$$

Five years consists of $n = 5 \times 4 = 20$ interest periods. Therefore, the compound amount after 5 years is

$$F = (1.02)^{20} \cdot 100$$

$$= (1.48594740) \cdot 100$$

$$= \$148.59.$$

The next example is a variation on the previous one and introduces a new concept, present value.

EXAMPLE 4 How much money must be deposited now in order to have \$1000 after 5 years if interest is paid at an 8% annual rate compounded quarterly?

Solution As in Example 3, we have $i = .02$ and $n = 20$. However, now we are given F and are asked to solve for P.

$$F = (1 + i)^n P$$

$$1000 = (1.02)^{20} P$$

$$P = \frac{1000}{(1.02)^{20}}.$$

From Table 2, $(1.02)^{20} = 1.48594740$. However, the cumbersome arithmetic can be avoided by using Table 3 of the Appendix which tabulates values of $1/(1 + i)^n$ for various values of i and n.

$$P = \frac{1}{(1.02)^{20}} \cdot 1000$$

$$= (.67297133) \ 1000$$

$$= \$672.97.$$

We say that \$672.97 is the present value of \$1000 5 years from now at 8% interest compounded quarterly.

In general, the present value of F dollars at a given interest rate and given length of time is the amount of money P that must be deposited now in order for the compound amount to grow to F dollars in the given length of time. From formula (2), we see that the present value P may be computed from the following formula:

$$P = \frac{1}{(1 + i)^n} \cdot F. \tag{3}$$

EXAMPLE 5 Determine the present value of a $10,000 payment to be received on January 1, 2000, if it is now May 1, 1991, and money can be invested at 6% interest compounded monthly.

Solution Here $F = 10,000$, $n = 104$ (the number of months between the two given dates), and $i = \frac{1}{2}\%$. By formula (3),

$$P = \frac{1}{(1 + i)^n} \cdot F$$

$$= (.59529136)\ 10,000$$

$$= \$5952.91$$

Therefore, $5952.91 invested on May 1, 1991, will grow to $10,000 by January 1, 2000.

The interest that we have been discussing so far is the most prevalent type of interest and is known as *compound interest*. There is another type of interest, called *simple interest*, which is used in some financial circumstances. Let us now discuss this type of interest.

Interest rates for simple interest are given as an annual percentage rate r. Interest is earned *only* on the principal P and the interest is rP for each year. Therefore, the interest earned in n years is nrP. So the amount A after n years is the original amount plus the interest earned. That is,

$$A = P + nrP$$

$$= 1 \cdot P + nrP$$

$$= (1 + nr)P. \tag{4}$$

EXAMPLE 6 Calculate the amount after 4 years if $1000 is invested at 7% simple interest.

Solution Apply formula (4) with $P = \$1000$, $n = 4$, and $r = 7\%$ or .07.

$$A = (1 + nr)P$$

$$= [1 + 4(.07)]1000$$

$$= (1.28)1000$$

$$= \$1280.$$

In Example 6, had the money been invested at 7% compound interest, with annual compounding, then the compound amount would have been $1310.80.

Money invested at simple interest is earning interest only on the principal amount. However, with compound interest, after the first interest period, the interest is also earning interest. Thus, if we compare two savings accounts, each earning interest at the same stated annual rate, but with one earning simple interest and one earning compound interest, the latter will grow at a faster rate.

Let us close this section with a summary of the key formulas we have developed so far.

Compound Interest

$$\text{Compound amount}: \quad F = (1 + i)^n P$$

$$\text{Present value}: \quad P = \left[\frac{1}{(1 + i)^n} \right] F$$

where i is the interest rate per period and n is the number of interest periods.

Simple Interest

$$\text{Amount}: \quad A = (1 + nr)P$$

PRACTICE PROBLEMS 1

1. (a) In Table 3 of the Appendix, look up the value of $1/(1 + i)^n$ for $i = 6\%$ and $n = 10$.

   (b) Calculate the present value of $1000 to be received 10 years in the future at 6% interest compounded annually.

2. Calculate the compound amount after 2 years of $1 at 26% interest compounded weekly.

3. Calculate the amount of $2000 after 6 months if invested at 10% simple interest.

EXERCISES 1

1. Determine i and n for the following situations.

   (a) 12% interest compounded monthly for 2 years.

   (b) 8% interest compounded quarterly for 5 years.

   (c) 10% interest compounded semiannually for 20 years.

2. Determine i and n for the following situations.

   (a) 6% interest compounded annually for 3 years.

   (b) 6% interest compounded monthly from January 1, 1990, to July, 1, 1991.

   (c) 9% interest compounded quarterly from January 1, 1990, to October 1, 1990.

3. Determine i, n, P, and F for the following situations.

   (a) $500 invested at 6% interest compounded annually grows to $631.24 in 4 years.

   (b) $800 invested on January 1, 1991, at 24% interest compounded monthly will grow to $8612.13 by January 1, 2001.

   (c) $2974.62 is deposited on January 1, 1991. The balance on July 1, 2000, is $9000 and the interest is 12% interest compounded semiannually.

4. Determine i, n, P, and F for each of the following situations.

   (a) The amount of money that must be deposited now at 26% interest compounded weekly in order to have $7500 in 1 year is $5786.63.

   (b) $3000 deposited at 12% interest compounded monthly will grow to $107,848.92 in 30 years.

   (c) In 1626, Peter Minuit, the first director-general of New Netherlands province, purchased Manhattan Island for trinkets and cloth valued at about $24. Had this money been invested at 8% interest compounded quarterly, it would have amounted to $39,125,072,000,000 by 1981.

5. Calculate the compound amount of $1000 after 2 years if deposited at 6% interest compounded monthly.

6. Calculate the present value of $10,000 payable in 5 years at 12% interest compounded semiannually.

7. Calculate the present value of $100,000 payable in 25 years at 12% interest compounded monthly.

8. Calculate the compound amount of $1000 after 1 year if deposited at 7.3% interest compounded daily.

9. Six thousand dollars is deposited in a savings account at 12% interest compounded monthly. Find the balance after 3 years and the amount of interest earned during that time.

10. Two thousand dollars is deposited in a savings account at 12% interest compounded semiannually. Find the balance after 7 years and the amount of interest earned during that time.

11. (a) How much money would have to be deposited now at 6% interest compounded monthly to accumulate to $4000 in 2 years?

    (b) How much of the $4000 would be interest?

    (c) Prepare a table showing the growth of the account for the first 3 months.

12. (a) How much money would have to be deposited now at 8% interest compounded quarterly to accumulate to $10,000 in 15 years?

    (b) How much of the $10,000 would be interest?

    (c) Prepare a table showing the growth of the account for the first 3 quarters.

13. If you had invested $10,000 on January 1, 1990, at 24% interest compounded quarterly, how much would you have on April 1, 1996?

14. In order to have $10,000 on his twenty-fifth birthday, how much would a person just turned 21 have to invest if the money will earn 24% interest compounded monthly?

15. Mr. Smith wishes to purchase a $10,000 sailboat upon his retirement in 3 years. He has just won the state lottery and would like to set aside enough cash in a savings account paying 4% interest compounded quarterly to buy the boat upon retirement. How much should he deposit?

16. Ms. Jones has just invested $100,000 at 6% interest compounded annually. How much money will she have in 20 years?

17. Consider the following savings account passbook:

Date	Deposits	Withdrawals	Interest	Balance
1/1/91	$1000.00			$1000.00
2/1/91			$10.00	1010.00
3/1/91			10.10	1020.10

(a) What annual interest rate is this bank paying?

(b) Give the interest and balance on 4/1/91.

(c) Give the interest and balance on 1/1/93.

18. Consider the following saving account passbook:

Date	Deposits	Withdrawals	Interest	Balance
1/1/90	$10,000.00			$10,000.00
7/1/90			$600.00	10,600.00
1/1/91			636.00	11,236.00

(a) Give the interest and balance on 7/1/91.

(b) Give the interest and balance on 1/1/99.

19. Is it more profitable to receive $1000 now or $1700 in 9 years? Assume that money can earn 6% interest compounded annually.

20. Is it more profitable to receive $7000 now or $10,000 in 9 years? Assume that money can earn 4% interest compounded quarterly.

21. Would you rather earn 30% interest compounded annually or 26% interest compounded weekly?

22. Would you rather earn 8% interest compounded annually or 7.3% interest compounded daily?

23. On January 1, 1988, a deposit was made into a savings account paying interest compounded quarterly. The balance on January 1, 1991, was $10,000 and the balance on April 1, 1991, was $10,200. How large was the deposit?

24. During the 1970s, a deposit was made into a savings account paying 4% interest compounded quarterly. On January 1, 1991, the balance was $2020. What was the balance on October 1, 1990?

Exercises 25–36 concern simple interest.

25. Determine r, n, P, and A for each of the following situations.

    (a) $500 invested at 7% simple interest grows to $517.50 in 6 months.

    (b) In order to have $580 after 2 years at 8% simple interest, $500 must be deposited.

26. Determine r, n, P, and A for each of the following situations.

    (a) At 15% simple interest, $1000 deposited on January 1, 1991, will be worth $1100 on September 1, 1991.

    (b) At 20% simple interest, in order to have $6000 in 5 years, $3000 must be deposited now.

27. Calculate the amount after 3 years if $1000 is deposited at 9% simple interest.

28. Calculate the amount after 18 months if $2000 is deposited at 8% simple interest.

29. Find the present value of $3000 in 2 years at 25% simple interest.

30. Find the present value of $2000 in 4 years at 15% simple interest.

31. Determine the (simple) interest rate at which $800 grows to $1000 in 6 months.

32. How many years are required for $1000 to grow to $1210 at 14% simple interest?

33. Determine the amount of time required for money to double at 10% simple interest.

34. Derive the formula for the (simple) interest rate at which P dollars grows to F dollars in n years. That is, express r in terms of P, A, and n.

35. Derive the formula for the present value of A dollars in n years at simple interest rate r.

36. Derive the formula for the number of years required for P dollars to grow to A dollars at simple interest rate r.

37. Compute the compound amount after 1 year for $100 invested at 24% interest compounded quarterly. What simple interest rate will yield the same amount in 1 year?

38. Compute the compound amount after 1 year for $100 invested at 12% interest compounded monthly. What simple interest rate will yield the same amount in 1 year?

Suppose that interest is compounded. Then the annual percentage rate is called the *nominal rate*. The simple interest rate that yields the same amount after 1 year is called the *effective rate*. For example, refer to Exercise 37. The nominal rate there is 24% and corresponds to an effective rate of 26.25%.

39. Calculate the effective rate for 12% interest compounded semiannually.

40. Calculate the effective rate for 26% interest compounded weekly.

41. Calculate the effective rate for 6% interest compounded monthly.

42. Calculate the effective rate for 9% interest compounded quarterly.

43. Suppose that an investment earns a nominal rate of a (expressed as a decimal) compounded twice a year. What is the effective rate?

44. Suppose that an investment of P dollars earns a nominal rate of a compounded n times per year. What is the compound amount after 1 year?

45. Suppose that an investment of P dollars earns a nominal rate of a compounded n times per year. What is the effective rate?

SOLUTIONS TO PRACTICE PROBLEMS 1

1. (a) Look up the entry in Table 3 of the Appendix in the row labeled "10" and the column labeled 6%. That entry is .55839478.

(b) Here we are given the value in the future $F = \$1000$ and are asked to find the present value P. Interest compounded annually has just one interest period per year, so $n = 10/1 = 10$ and $i = 6\%/1 = 6\%$. By formula (3),

$$P = \frac{1}{(1+i)^n} F$$

$$= \frac{1}{(1+i)^n} \cdot 1000 \qquad \text{(with } i = 6\% \text{ and } n = 10\text{)}$$

$$= (.55839478) \cdot 1000 \qquad [\text{from part (a)}]$$

$$= 558.39478$$

$$= \$558.39 \qquad \text{(rounding off to the nearest cent).}$$

2. Here we are given the present value, $P = 1$, and are asked to find the value F at a future time. Interest compounded weekly has 52 interest periods each year, so $n = 2 \times 52 = 104$ and $i = 26\%/52 = \frac{1}{2}\%$. By formula (2), we have

$$F = (1+i)^n P$$

$$= (1+i)^n \cdot 1 \qquad \text{(with } i = \frac{1}{2}\%, \quad n = 104\text{)}$$

$$= (1.67984969) \cdot 1, \qquad \text{(using Table 2)}$$

$$= \$1.68.$$

[*Note:* In general, whenever $P = \$1$, then $F = (1+i)^n \cdot 1 = (1+i)^n$. This explains why $(1+i)^n$ is often referred to as the *compound amount of* $\$1$.]

3. In simple interest problems, time should be expressed in terms of years. Since 6 months is $\frac{1}{2}$ year, formula (4) gives

$$A = (1 + nr)P$$

$$= [1 + \tfrac{1}{2}(.10)]2000$$

$$= (1.05)2000$$

$$= \$2100.$$

10.2. Annuities

An annuity is a sequence of equal payments made at regular intervals of time. Here are two illustrations.

1. As the proud parent of a newborn daughter, you decide to save for her college education by depositing $30 at the end of each month into a savings account paying 6% interest compounded monthly. Eighteen years from now after you make the last of 216 payments, the account will contain $11,620.60.

2. Having just won the state lottery, you decide not to work for the next 5 years. You want to deposit enough money into the bank so that you can withdraw $900 at the end of each month for 60 months. If the bank pays 6% interest compounded monthly, you must deposit $46,553.

The payments in the foregoing financial transactions are called *rent*. The amount of a typical rent payment is denoted by the letter R. Thus, in the examples above, we have $R = \$30$ and $R = \$900$, respectively.

In illustration 1, you make equal payments to a bank in order to generate a large sum of money in the future. This sum, namely $11,620.60, is called the *future value of the annuity*.

In illustration 2, the bank will make equal payments to you in order to pay back the sum of money that you currently deposit. The value of the current deposit, namely $46,553, is called the *present value of the annuity*.

In this section we will derive formulas for future values and present values of annuities. However, as a mathematical preliminary, we must first discuss sums of geometric progressions.

Let a be any number. Then the *geometric progression* determined by a is the sequence of nonnegative powers of a:

$$1, a, a^2, a^3, a^4, \ldots$$

Let us assume that $a \neq 1$. Then in this case, the sum of the first n terms of the geometric progression determined by a is given by the formula

$$1 + a + a^2 + \cdots + a^{n-1} = \frac{a^n - 1}{a - 1} \tag{1}$$

For instance, with $a = 2$ and $n = 3$, formula (1) states that

$$1 + 2 + 2^2 = \frac{2^3 - 1}{2 - 1}.$$

Doing the arithmetic shows that this result is correct since both sides equal 7.

Formula (1) is derived by multiplying $a^{n-1} + \cdots + a^2 + a + 1$ by $a - 1$.

$$
\begin{array}{r}
a^{n-1} + \cdots + a^2 + a + 1 \\
a - 1 \\
\hline
-a^{n-1} - \cdots - a^2 - a - 1 \\
a^n + a^{n-1} + \cdots + a^2 + a \\
\hline
a^n \phantom{+ a^{n-1} + \cdots + a^2 + a} - 1
\end{array}
$$

That is,

$$(a - 1)(1 + a + a^2 + \cdots + a^{n-1}) = a^n - 1.$$

Divide both sides of this last equation by $a - 1$ (which is possible since a is not equal to 1) and formula (1) results.

Let us now apply formula (1) to annuities. Suppose that an annuity consists of a sequence of n equal payments, each of R dollars. Suppose that the annuity payments are deposited into an account paying compound interest at the rate of i per interest period. We will further suppose that there is a single annuity payment per interest period and that the payment is made at the end of the interest period. Let us derive a formula for the future value of the annuity: that is, a formula for the balance of the account immediately after the last payment.

Each payment accumulates interest for a different number of interest periods, so let us calculate the balance in the account as the sum of n compound amounts, one corresponding to each payment.

Payment	Amount	Number of Interest Periods on Deposit	Compound Amount
1	R	$n - 1$	$(1 + i)^{n-1}R$
2	R	$n - 2$	$(1 + i)^{n-2}R$
$\vdots$			
$n - 2$	R	2	$(1 + i)^2 R$
$n - 1$	R	1	$(1 + i)R$
n	R	0	R

Denote the future value of the annuity by F. Then F is the sum of the numbers in the right-hand column:

$$F = R + (1 + i)R + (1 + i)^2 R + \cdots + (1 + i)^{n-1}R$$

$$= [1 + (1 + i) + (1 + i)^2 + \cdots + (1 + i)^{n-1}]R$$

$$= \frac{(1 + i)^n - 1}{(1 + i) - 1} R,$$

where we have applied formula (1) with $a = 1 + i$. Therefore, we have the result

$$F = \frac{(1 + i)^n - 1}{i} \cdot R.$$

The expression $[(1 + i)^n - 1]/i$ occurs often in financial analysis and is denoted by the special symbol $s_{\overline{n}|i}$ (read "s sub n angle i"). Values of $s_{\overline{n}|i}$ may be computed using a financial calculator or by consulting an appropriate table. We have included a brief table as Table 4 of the Appendix. We may summarize our calculation above as follows.

> Suppose that an annuity consists of n payments of R dollars each, deposited at the ends of consecutive interest periods into an account with interest compounded at a rate i per period. Then the future value F of the annuity is given by the formula
>
> $$F = s_{\overline{n}|i}R.$$

(2)

EXAMPLE 1 Calculate the future value of an annuity of $100 per month for 5 years at 6% interest compounded monthly.

Solution Here $R = 100$ and $i = \frac{1}{2}\%$. Since a payment is made at the end of each month for 5 years, there will be $5 \times 12 = 60$ payments. So $n = 60$. Therefore,

$$F = s_{\overline{n}|i}R$$

$$= s_{\overline{60}|1/2\%} \cdot 100$$

$$= (69.77003051) \cdot 100 \qquad \text{(by Table 4)}$$

$$= \$6977.00$$

Formula (2) can also be used to determine the rent necessary to achieve a certain future value.

$$F = s_{\overline{n}|i}R$$

$$R = \frac{F}{s_{\overline{n}|i}} = \frac{1}{s_{\overline{n}|i}}F.$$

Thus we have established the following result.

> Suppose that an annuity of n payments has future value F and has interest compounded at the rate i per period. Then the rent R is given by
>
> $$R = \frac{1}{s_{\overline{n}|i}} \cdot F. \qquad (3)$$

Table 5 of the Appendix gives values of $1/s_{\overline{n}|i}$ for various n and i.

EXAMPLE 2 Ms. Adams would like to buy a $30,000 airplane when she retires in 8 years. How much should she deposit at the end of each half-year into an account paying 12% interest compounded semiannually so that she will have enough money to purchase the airplane?

Solution Here $n = 16$, $i = 6\%$ and $F = 30,000$. Therefore,

$$R = \frac{1}{s_{\overline{n}|i}} \cdot F$$

$$= \frac{1}{s_{\overline{16}|6\%}} \cdot 30,000$$

$$= (.03895214) \cdot 30,000 \qquad \text{(by Table 5)}$$

$$= \$1168.56.$$

She should deposit $1168.56 at the end of each half-year period.

The *present value* of an annuity is the amount of money it would take to finance the sequence of annuity payments. More specifically, the present value of an annuity is the amount you would need to deposit in order to provide the desired sequence of annuity payments and leave a balance of zero at the end of the term. Let us now find a formula for the present value of an annuity. There are two ways to proceed. On the one hand, the desired present value could be computed as the sum of the present values of the various annuity payments. This computation would make use of the formula for the sum of the first n terms of a geometric progression. However, there is a much "cleaner" derivation which proceeds indirectly to obtain a formula relating the present value P and the rent R.

As before, let us assume that our annuity consists of n payments made at the ends of interest periods, with interest compounded at a rate i per interest period.

Situation 1 : Suppose that the P dollars were just left in the account and that the annuity payments were not withdrawn. At the end of the n interest periods there would be $(1 + i)^n P$ dollars in the account.

Situation 2 : Suppose that the payments are withdrawn but immediately are redeposited into another account having the same rate of interest. At the end of the n interest periods, there would be $s_{\overline{n}|i} R$ dollars in the new account.

In both of these situations, P dollars is deposited and it, together with all the interest generated, is earning income at the same interest rate for the same amount of time. Therefore, the final amounts of money in the accounts should be the same. That is,

$$(1 + i)^n P = s_{\overline{n}|i} R$$

$$P = \frac{s_{\overline{n}|i}}{(1 + i)^n} \cdot R.$$

Let us now substitute the value of $s_{\overline{n}|i}$, namely $[(1 + i)^n - 1]/i$, into the formula above. If we denote by $a_{\overline{n}|i}$ the expression

$$a_{\overline{n}|i} = \frac{(1 + i)^n - 1}{i(1 + i)^n},$$

then the formula above may be written in the simple form

$$P = a_{\overline{n}|i} R; \qquad R = \frac{1}{a_{\overline{n}|i}} P$$

Tables 6 and 7 of the Appendix give, respectively, the values of $a_{\overline{n}|i}$ and $1/a_{\overline{n}|i}$ for various values of n and i. Let us record the main result of our discussion above.

The present value P and the rent R of an annuity of n payments with interest compounded at a rate i per interest period are related by the formulas

$$P = a_{\overline{n}|i} R; \qquad R = \frac{1}{a_{\overline{n}|i}} P$$

EXAMPLE 3   How much money must you deposit now at 6% interest compounded quarterly in order to be able to withdraw $3000 at the end of each quarter year for 2 years?

Solution   Here $R = 3000$, $i = 1.5\%$, and $n = 8$. We are asked to calculate the present value of the sequence of payments.

$$P = a_{n|i}R$$

$$= a_{\overline{8}|1.5\%} \cdot 3000$$

$$= (7.48592508) \cdot 3000 \qquad \text{(by Table 6)}$$

$$= \$22{,}457.78.$$

EXAMPLE 4   If you deposit $1000 into a fund paying 26% interest compounded weekly, how much can you withdraw at the end of each week for 1 year?

Solution   Here $P = 1000$, $i = \frac{1}{2}\%$, and $n = 52$. We are asked to calculate the periodic payment, that is, the rent.

$$R = \frac{1}{a_{n|i}} P$$

$$= \frac{1}{a_{\overline{52}|1/2\%}} \cdot 1000$$

$$= (.02188675) \cdot 1000 \quad \text{(by Table 7)}$$

$$= \$21.89.$$

Remark   In this section we have considered only annuities with payments made at the end of each interest period. Such annuities are called *ordinary annuities*. Annuities that have payments at the beginning of the interest period are called *annuities due*. Annuities whose payment period is different from the interest period are called *general annuities*.

PRACTICE PROBLEMS 2

Decide whether or not each of the following annuities are ordinary annuities, that is, the type of annuities considered in this section. If so, identify n, i, and R and calculate the present value or the future value, whichever is appropriate.

1.   You make a deposit at 24% interest compounded monthly into a fund that pays you $1 at the end of each month for 5 years.

2.   At the end of each week for 2 years you deposit $10 into a savings account earning 18% interest compounded monthly.

3.   At the end of each month for 2 years, you deposit $10 into a savings account earning 18% interest compounded monthly.

EXERCISES 2

1. Determine the following sums by using formula (1).

   (a) $1 + 2 + 4 + 8 + 16$ (b) $1 - 2 + 4 - 8 + 16$

   (c) $1 + \frac{1}{2} + \frac{1}{4} + \frac{1}{8} + \frac{1}{16}$ (d) $3 + \frac{3}{2} + \frac{3}{4} + \frac{3}{8} + \frac{3}{16}$

2. Determine the following sums by using formula (1).

   (a) $1 + 11 + 121 + 1331 + 14{,}641$ $[Note: 11^5 = 161{,}051.]$

   (b) $1 - 9 + 81 - 729 + 6561$ $[Note: 9^5 = 59{,}049.]$

   (c) $1 + .5 + .25 + .125 + .0625$ $[Note: (.5)^5 = .03125.]$

   (d) $4 + 2 + 1 + \frac{1}{2} + \frac{1}{4}$

3. For each of the following annuities, specify i, n, R, and F.

   (a) If at the end of each month, \$50 is deposited into a savings account paying 6% interest compounded monthly, the balance after 10 years will be \$8193.97.

   (b) Mr. Smith is saving to buy a \$65,000 yacht in 2000. Since 1990, he has been depositing \$1767 at the end of each half-year into a fund paying 12% interest compounded semiannually.

4. For each of the following annuities, specify i, n, R, and P.

   (a) A retiree deposits \$72,582 into a bank paying 6% interest compounded monthly and withdraws \$520 at the end of each month for 20 years.

   (b) In order to receive \$700 at the end of each quarter-year from 1981 until 1986, Ms. Jones deposited \$11,446 into a savings account paying 8% interest compounded quarterly.

5. Calculate the future value of an annuity of \$100 per month for 5 years at 12% interest compounded monthly.

6. Calculate the rent of an annuity at 12% interest compounded semiannually if payments are made every half-year and the future value after 7 years is \$10,000.

7. Calculate the rent of an annuity at 8% interest compounded quarterly with payments made every quarter-year for 7 years and present value \$100,000.

8. Calculate the present value of an annuity of \$1000 per year for 10 years at 6% interest compounded annually.

9. A person deposits \$500 into a savings account at the end of every month for 4 years at 12% interest compounded monthly.

   (a) Find the balance at the end of 4 years.

   (b) How much interest will be earned during the 4 years?

   (c) Prepare a table showing the deposits, balance, and interest for the first 3 months.

10. A person deposits \$800 into a savings account at the end of every quarter year for 5 years at 6% interest compounded quarterly.

    (a) Find the balance at the end of 5 years.

(b) How much interest will be earned during the 5 years?

(c) Prepare a table showing the deposits, balance, and interest for the first three quarters.

11. How much must Jim save each month in order to have $12,000 to buy a new car in 3 years if the interest rate is 18% compounded monthly? How much of the $12,000 does Jim actually deposit and how much is interest?

12. Valerie needs $5000 three years from now in order to pay off a loan. How much must she save each quarter for the next 3 years if interest rates are 8% compounded quarterly? How much of the $5000 will be interest?

13. When Bridget takes a new job, she is offered a $2000 bonus now or the option of an extra $200 a month for the next year. If interest rates are 6% compounded monthly, which choice is better and by how much?

14. When Michael started college, his father gave him $5000 deposited in a savings account earning 18% interest compounded monthly. Michael withdraws $500 a month starting at the beginning of the second month. Prepare a schedule showing the monthly balance in the account for the first four months.

15. A city has a debt of $1,000,000 falling due in 15 years. How much money must it deposit at the end of each half-year into a savings fund at 12% interest compounded semiannually in order to raise this amount?

16. On January 1, 1990, Tom decided to save for exactly 1 year for a 10-speed bike by depositing $10 at the end of each month into a savings account paying 6% interest compounded monthly. How much did he accumulate?

17. During Jack's freshman year at college, his father had been sending him $100 per month for incidental expenses. For the sophomore year, his father decided instead to make a deposit into a savings account on August 1 and have his son withdraw $100 on the first of each month until May 1. If the bank pays 6% interest compounded monthly, how much should Jack's father deposit?

18. Suppose that a magazine subscription costs $9 per year and that you receive a magazine at the end of each month. At an interest rate of 12% compounded monthly, how much are you actually paying for each issue?

19. Is it more profitable to receive $1000 at the end of each month for 10 years or to receive a lump sum of $230,000 at the end of 10 years? Assume that money can earn 12% interest compounded monthly.

20. Is it more profitable to receive a lump sum of $10,000 at the end of 3 years or to receive $750 at the end of each quarter-year for 3 years? Assume that money can earn 8% interest compounded quarterly.

21. Suppose that you deposited $1000 into a savings account on January 1, 1990, and deposited an additional $100 into the account at the end of each quarter-year. If the bank pays 8% interest compounded quarterly, how much will be in the account on January 1, 1999?

22. Suppose that you opened a savings account on January 1, 1990, and made a deposit of $100. In 1991, you began depositing $10 into the account at the end of each month. If the bank pays 6% interest compounded monthly, how much money will be in the account on January 1, 1995?

23. Ms. Jones deposited $100 at the end of each month for 10 years into a savings account earning 6% interest compounded monthly. However, she deposited an additional $1000 at the end of the seventh year. How much money was in the account at the end of the tenth year?

24. Redo Exercise 23 for the situation where Ms. Jones withdrew $1000 at the end of the seventh year instead of depositing it.

A *perpetuity* is an annuity whose payments are to continue forever. Exercises 25 and 26 concern such annuities.

25. A grateful alumnus decides to donate a permanent scholarship of $1200 per year. How much money should be deposited in the bank at 12% interest compounded annually in order to be able to supply the money for the scholarship at the end of each year?

26. Show that to establish a perpetuity paying R dollars at the end of each interest period, it requires a deposit of R/i dollars, where i is the interest rate per interest period.

A *deferred annuity* is an annuity whose term is to start to some future date. Exercises 27–30 concern such annuities.

27. On his tenth birthday, a boy inherits $10,000, which is to be used for his college education. The money is deposited into a trust fund which will pay him R dollars on his 18th, 19th, 20th, and 21st birthdays. Find R if the money earns 6% interest compounded annually.

28. Refer to Exercise 27. Find the size of inheritance that would result in $10,000 per year during the college years (ages 18–21, inclusive).

29. On December 1, 1990, a philanthropist set up a permanent trust fund to buy Christmas presents for needy children. The fund will provide $6000 each year beginning on December 1, 2000. How much must have been set aside in 1990 if the money earns 6% interest compounded annually?

30. Show that the rent paid by a deferred annuity of n payments which are deferred by m interest periods is given by the formula

$$R = \frac{i(1 + i)^{n+m}}{(1 + i)^n - 1} \cdot P.$$

31. One dollar is deposited in a savings account with an interest rate of i per interest period. At the end of each interest period, the earned interest is withdrawn from the account and deposited into a second account earning the same rate of interest.

   (a) How much money will be in the first account after n interest periods?

   (b) How much money will be in the second account after n interest periods?

   (c) Since both the original deposit and the interest are all on deposit and earning interest throughout the entire n interest periods, the amounts in parts (a) and (b) must add up to $(1 + i)^n$. Use this fact to derive $s_{n|i} = [(1 + i)^n - 1]/i$.

32. Show that $\dfrac{1}{a_{\overline{n}|i}} - \dfrac{1}{s_{\overline{n}|i}} = i$.

33. Show that $s_{\overline{n+1}|i} = (1 + i)s_{\overline{n}|i} + 1$.

34. One hundred dollars is deposited in the bank at the end of each year for 20 years. During the first 5 years, the bank paid 6% interest compounded annually and after that paid 7% interest compounded annually. Show that the account balance after 20 years is $(1.07)^{15} \cdot 100s_{\overline{5}|6\%} + 100s_{\overline{15}|7\%}$.

35. A municipal bond pays 8% interest compounded semiannually on its face value of $5000. The interest is paid at the end of every half-year period. Fifteen years from now the face value of $5000 will be returned. The current market interest rate is 12% compounded semiannually. How much should you pay for the bond?

36. A businessman wishes to lend money for a second mortgage on some local real estate. Suppose that the mortgage pays $500 per month for a 5-year period. Suppose that you can invest your money in certificates of deposit paying 12% interest compounded monthly. How much should you offer for the mortgage?

37. Suppose that a business note for $50,000 carries an interest rate of 18% compounded monthly. Suppose that the business pays only interest for the first 5 years and then repays the loan amount plus interest in equal monthly installments for the next 5 years.

(a) Calculate the monthly payments during the second 5-year period.

(b) Assume that the current market interest rate for such loans is 12%. How much should you be willing to pay for such a note?

38. A lottery winner is to receive $1000 a month for the next 5 years. How much is this sequence of payments worth today if interest rates are 12% compounded monthly? How is the difference between this amount and the $60,000 paid out beneficial to the agency running the lottery?

SOLUTIONS TO PRACTICE PROBLEMS 2

1. An ordinary annuity with $n = 60$, $i = 2\%$, and $R = 1$. You will make a deposit now, in the present, and then withdraw money each month. The amount of this deposit is the present value of the annuity.

$$P = a_{\overline{n}|i}R$$

$$= a_{\overline{60}|2\%} \cdot 1$$

$$= (34.76088668) \cdot 1 \qquad \text{(by Table 6)}$$

$$= \$34.76.$$

[*Notes*: (1) When $R = 1$, we have $P = a_{\overline{n}|i}$. This explains why $a_{\overline{n}|i}$ is often called the *present value of $1*. (2) For this transaction, the future value of the annuity has no significance. At the end of 5 years, the fund will have a balance of 0.]

2. Not an ordinary annuity since the payment period (1 week) is different from the interest period (1 month).

3. An ordinary annuity with $n = 24$, $i = 1.5\%$, and $R = \$10$. There is no money in the account now, in the present. However, in 2 years, in the future, money will have accumulated. So for this annuity, only the future value has significance.

$$F = s_{\overline{n}|i}R$$

$$= s_{\overline{24}|1.5\%} \cdot 10$$

$$= (28.63352080) \cdot 10 \quad \text{(by Table 4)}$$

$$= \$286.34.$$

10.3. Amortization of Loans

In this section we analyze the mathematics of paying off loans. The loans we shall consider will all be repaid in a sequence of equal payments at regular time intervals, with the payment intervals coinciding with the interest periods. The process of paying off such a loan is called *amortization*. In order to obtain a feeling for the amortization process, let us consider a particular case, the amortization of a $563 loan to buy a color television set. Suppose that this loan charges interest at a 12% rate with interest compounded monthly and that the monthly payments are $116 for 5 months. The repayment process is summarized in the following chart.

Payment number	Amount	Interest	Applied to Principal	Unpaid balance
1	$116	$5.63	$110.37	$452.63
2	116	4.53	111.47	341.16
3	116	3.41	112.59	228.57
4	116	2.29	113.71	114.85
5	116	1.15	114.85	0.00

Note the following facts about the financial transactions above.

1. Payments are made at the end of each month. The payments have been carefully calculated to pay off the debt, with interest in the specified time interval.
2. Since $i = 1\%$, the interest to be paid each month is 1% of the unpaid balance at the end of the previous month. That is, 5.63 is 1% of 563, 4.53 is 1% of 452.63, and so on.
3. Although we write just one check each month for $116, we regard part of the check as being applied to payment of that month's interest. The remainder, namely 116 − [interest], is regarded as being applied to repayment of part of the principal amount.
4. The unpaid balance at the end of each month is the previous unpaid balance minus the portion of the payment applied to the principal. A loan can be paid off early by just paying the current unpaid balance.

The four factors that describe the amortization process above are:

the principal	$563
the interest rate	12% compounded monthly
the term	5 months
the monthly payment	$116

The important fact to recognize is that the sequence of payments in the amortization above constitute an annuity, with the person taking out the loan paying the interest. Therefore, the mathematical tools developed in Section 2 suffice to analyze the amortization. In particular, we could determine the monthly payment or the principal once the other three factors have been specified.

EXAMPLE 1 Suppose that a loan has an interest rate of 12% compounded monthly and a term of 5 months.

(a) Given that the principal is $563, calculate the monthly payment.

(b) Given that the monthly payment is $116, calculate the principal.

(c) Given that the monthly payment is $116, calculate the unpaid balance after 3 months.

Solution The sequence of payments constitute an annuity with the monthly payments as rent and the principal as the present value. Also, $n = 5$ and $i = 1\%$.

(a) Since $R = (1/a_{\overline{n}|i})P$, we see that

$$R = \frac{1}{a_{\overline{5}|1\%}} \cdot 563$$

$$= (.20603980) \cdot 563 \qquad \text{(by Table 7)}$$

$$= \$116.$$

(b) Since $P = a_{\overline{n}|i}R$, we see that

$$P = a_{\overline{5}|1\%} \cdot 116$$

$$= (4.85343124) \cdot 116$$

$$= \$563.00.$$

(c) The unpaid balance is most easily calculated by regarding it as the amount necessary to retire the debt. Therefore, it must be sufficient to generate, with interest, the sequence of two remaining payments. That is, the unpaid balance is the present value of an annuity of two payments of $116. So we see that

$$[\text{unpaid balance after 3 months}] = a_{\overline{2}|1\%} \cdot 116$$

$$= (1.97039506) \cdot 116$$

$$= \$228.57.$$

A mortgage is a long-term loan used to purchase real estate. The real estate is used as collateral to guarantee the loan.

EXAMPLE 2 On December 31, 1980, a house was purchased with the buyer taking out a 30-year $60,050 mortgage at 18% interest, compounded monthly. The mortgage payments are made at the end of each month.

(a) Calculate the amount of the monthly payment.

(b) Calculate the unpaid balance of the loan on December 31, 2006, just after the 312th payment.

(c) How much interest will be paid during the month of January 2007?

(d) How much of the principal will be paid off during the year 2006?

(e) How much interest will be paid during the year 2006?

Solution (a) Denote the monthly payment by R. Then \$60,050 is the present value of an annuity of $n = 360$ payments, with $i = 1.5\%$. Therefore,

$$R = \frac{1}{a_{\overline{n}|i}} P$$

$$= \frac{1}{a_{\overline{360}|1.5\%}} \cdot 60{,}050$$

$$= (.01507085) \cdot 60{,}050 \qquad \text{(by Table 7)}$$

$$= \$905.00.$$

(b) The remaining payments constitute an annuity of 48 payments. Therefore, the unpaid balance is the present value of that annuity.

$$[\text{unpaid balance}] = a_{\overline{n}|i} R$$

$$= a_{\overline{48}|1.5\%} \cdot 905$$

$$= (34.04255365) \cdot 905 \qquad \text{(by Table 6)}$$

$$= \$30{,}808.51$$

(c) The interest paid during 1 month is i, the interest rate per month, times the unpaid balance at the end of the preceding month. Therefore,

$$[\text{interest for January 2007}] = 1.5\% \text{ of } \$30{,}808.51$$

$$= (.015) \cdot 30{,}808.51$$

$$= \$462.13.$$

(d) Since the portions of the monthly payments applied to repay the principal have the effect of reducing the unpaid balance, this question may be answered by calculating how much the unpaid balance will be reduced during 2006. Reasoning as in part (b), we determine that the unpaid balance on December 31, 2005 (just after the 300th payment), is equal to \$35,639.14. Therefore,

$$[\text{amount of principal repaid in 2006}]$$

$$= [\text{unpaid balance Dec. 31, 2005}] - [\text{unpaid balance Dec. 31, 2006}]$$

$$= \$35{,}639.14 - \$30{,}808.51 = \$4830.63.$$

(e) During the year 2006, the total amount paid is $12 \times 905 = 10{,}860$ dollars. But by part (d), \$4830.63 is applied to repayment of principal, the remainder being applied to interest.

$$[\text{interest in 2006}] = [\text{total amount paid}] - [\text{principal repaid in 2006}]$$

$$= 10{,}860 - 4830.63$$

$$= \$6029.37.$$

Note that in the early years of a mortgage, most of each monthly payment is applied to interest. For the mortgage above, the interest portion will exceed the principal portion until the twenty-sixth year.

We can easily derive a formula which illuminates exactly how the unpaid balance of a mortgage changes from period to period. Denote the current unpaid balance by B_{cur}. Then

$$B_{next} = B_{cur} - (R - [\text{interest}]_{next})$$
$$= B_{cur} - (R - iB_{cur})$$
$$= B_{cur} + iB_{cur} - R$$
$$= (1 + i)B_{cur} - R. \qquad (1)$$

Successive unpaid balances are computed by multiplying by $1 + i$ and subtracting R.

EXAMPLE 3   Refer to Example 2. Compute the unpaid balance of the loan on January 31, 2007, just after the 313th payment.

Solution   By formula (1),

$$B_{Jan} = (1 + i)B_{Dec} - R$$
$$= (1.015) \cdot 30{,}808.51 - 905$$
$$= \$30{,}365.64$$

Sometimes amortized loans stipulate a *balloon payment* at the end of the term. For instance, you might pay \$200 at the end of each quarter year for 3 years and an additional \$1000 at the end of the third year. The \$1000 is a balloon payment.

EXAMPLE 4   How much money can you borrow at 24% interest compounded quarterly if you agree to pay \$200 at the end of each quarter-year for 3 years and in addition a balloon payment of \$1000 at the end of the third year?

Solution   Here you are borrowing in the present and repaying in the future. The amount of the loan will be the present value of *all* the future payments. The future payments consist of an annuity and a lump-sum payment. Let us calculate the present values of each of these separately. Now, $i = 6\%$ and $n = 12$.

$$[\text{present value of annuity}] = a_{\overline{n}|i}R$$
$$= a_{\overline{12}|6\%} \cdot 200$$
$$= (8.38384394) \cdot 200 \qquad \text{(by Table 6)}$$
$$= \$1676.77.$$

$$[\text{present value of balloon payment}] = \frac{1}{(1+i)^n} \cdot F$$

$$= \frac{1}{(1+i)^n} \cdot 1000 \qquad (\text{for } i = 6\%, n = 12)$$

$$= .49696936 \cdot (1000)$$

$$= \$496.97.$$

Therefore, the amount you can borrow is $\$1676.77 + \$496.97 = \$2173.74$.

PRACTICE PROBLEMS 3

1. The word "amortization" comes from the French "a mort" meaning "at the point of death." Justify the word.

2. Explain why only present values and not future values arise in amortization problems.

EXERCISES 3

1. A loan of $10,000 is to be repaid with monthly payments for 5 years at 24% interest compounded monthly. Calculate the monthly payment.

2. Find the monthly payment on a $100,000 25-year mortgage at 12% interest compounded monthly.

3. How much money can you borrow at 12% interest compounded semiannually if the loan is to be repaid at half-year intervals for 10 years and you can afford to pay $1000 per half-year?

4. You buy a car with a down payment of $500 and $100 per month for 3 years. If the interest rate is 18% compounded monthly, how much did the car cost?

5. Consider a $58,331 30-year mortgage at interest rate 12% compounded monthly with a $600 monthly payment.

    (a) How much interest is paid the first month?

    (b) How much of the first month's payment is applied to paying off the principal?

    (c) What is the unpaid balance after 1 month?

    (d) What is the unpaid balance at the end of 25 years?

    (e) How much of the principal is repaid during the twenty-sixth year?

    (f) How much interest is paid during the 301st month?

6. Consider a $13,406.16 loan for 7 years at 24% interest compounded quarterly and a payment of $1000 per quarter-year.

    (a) Compute the unpaid balance after 5 years.

(b) How much interest is paid during the fifth year?

(c) How much principal is repaid in the first payment?

(d) What is the total amount of interest paid on the loan?

7. Susie takes out a car loan for $8000 for a term of 3 years at 12% interest compounded monthly.

(a) Find her monthly payments.

(b) Find the total amount she pays for the car.

(c) Find the total amount of interest she pays.

(d) Find the amount she still owes after 1 year.

(e) Find the amount she still owes after 2 years.

(f) Find the total interest she pays in year 2.

(g) Prepare an amortization schedule for the first 4 months.

8. James buys a house for $90,000. He puts $10,000 down and then finances the rest at 9% interest compounded monthly for 25 years.

(a) Find his monthly payments.

(b) Find the total amount he pays for the house.

(c) Find the total amount of interest he pays.

(d) Find the amount he still owes after 23 years.

(e) Find the amount he still owes after 24 years.

(f) Find the total amount of interest he pays in year 23.

(g) Prepare the amortization schedule for the first 3 months.

9. A mortgage at 15% interest compounded monthly with a monthly payment of $1125 has an unpaid balance of $10,000 after 350 months. Find the unpaid balance after 351 months.

10. A loan with a quarterly payment of $1500 has an unpaid balance of $10,000 after 30 quarters and an unpaid balance of $9000 after 31 quarters. If interest is compounded quarterly, find the interest rate.

11. A loan is to be amortized over an 8-year term at 12% interest compounded semiannually, payments of $1000 every 6 months, and a balloon payment of $10,000 at the end of the term. Calculate the amount of the loan.

12. A loan of $105,504.50 is to be amortized over a 5-year term at 12% interest compounded monthly with monthly payments and a $10,000 balloon payment at the end of the term. Calculate the monthly payment.

13. Write out a complete amortization schedule (as in the table at the beginning of this section) for the amortization of a $1000 loan with monthly payments at 12% interest compounded monthly for 4 months.

14. Write out a complete schedule (as on page 471) for the amortization of a $10,000 loan with payments every 6 months at 12% interest compounded semiannually for 1 year.

15. You purchase a $120,000 house, pay $20,000 down, and take out a 30-year mortgage with monthly payments, at an interest rate of 24% compounded monthly. How much money will you be paying each month?

16. In 1980, you purchased a house and took out a 25-year $50,000 mortgage at 6% interest compounded monthly. In 1990, you sold the house for $150,000. How much money did you have left after you paid the bank the unpaid balance on the mortgage?

17. You are considering the purchase of a condominium to use as a rental property. You estimate that you can rent the condominium for $1200 per month and that taxes, insurance, and maintenance costs will run about $200 per month. If interest rates are 18% compounded monthly, how large a 25-year mortgage can you assume and still have the rental income cover the monthly expenses?

18. Consider formula (1). Derive an analogous formula for the balance of an annuity for which regular payments are deposited into a savings account and accrue with interest.

19. A car is purchased for $5548.88 with $2000 down and a loan to be repaid at $100 a month for 3 years followed by a balloon payment. If the interest rate is 24% compounded monthly, how large will the balloon payment be?

20. A real estate speculator purchases a tract of land for $1 million and assumes a 25-year mortgage at 18% interest compounded monthly.

    (a) What is his monthly payment?

    (b) Suppose that at the end of 5 years the mortgage is changed to a 10-year term for the remaining balance. What is the new monthly payment?

    (c) Suppose that after 5 more years, the mortgage is required to be repaid in full. How much will then be due?

21. Suppose that you make annual payments of $5000 for 20 years into an annuity paying 6% interest compounded annually.

    (a) What is the value of the annuity at the end of the twentieth year?

    (b) Suppose that you elect to have your annuity repaid to you over a 10-year period in annual installments. What is the annual payment you will receive?

    (c) Suppose that after you receive payments for 5 years you elect to have the remainder of your annuity paid to you in a lump sum. How much will you receive?

A *sinking fund* is a pool of money accumulated by a corporation or government to repay a specific debt at some future date.

22. A corporation wishes to deposit money into a sinking fund at the end of each half year in order to repay $50 million in bonds in 10 years. It can expect to receive a 12% (compounded semiannually) return on its deposits to the sinking fund. How much should the deposits be?

23. The Federal National Mortgage Association ("Fannie Mae") puts $30 million dollars at the end of each month into a sinking fund paying 12% interest compounded monthly. The sinking fund is to be used to repay debentures that mature 15 years from the creation of the fund. How large is the face amount of the debentures assuming the sinking fund will exactly pay them off?

24. A corporation borrows $5 million to erect a new headquarters. The financing is arranged using industrial development bonds, to be repaid in 15 years. How much should it deposit into a sinking fund at the end of each quarter if the sinking fund earns 8% interest compounded quarterly?

25. A corporation sets up a sinking fund to replace some aging machinery. It deposits $100,000 into the fund at the end of each month for 10 years. The sinking fund earns 12% interest compounded monthly. The equipment originally cost $6 million. However, the cost of the equipment is rising 6% each year. Will the sinking fund be adequate to replace the equipment? If not, how much additional money is needed?

26. A corporation sets up a sinking fund to replace an aging warehouse. The cost of the warehouse today would be $8 million. However, the corporation plans to replace the warehouse in 5 years. It estimates that the cost of the warehouse will increase 10% annually. The sinking fund will earn 12% interest compounded monthly. What should be the monthly payments to the sinking fund?

SOLUTIONS TO PRACTICE PROBLEMS 3

1. A portion of each payment is applied to reducing the debt and by the end of the term, the debt is totally annihilated.

2. The debt is formed when the creditor gives you a lump sum of money now, in the present. The lump sum of money is gradually repaid by you, with interest, thereby generating the annuity. At the end of the term, in the future, the loan is totally paid off, so there is no more debt. That is, the future value is always zero!

Remark You are actually functioning like a savings bank, since you are paying the interest. Think of the creditor as depositing the lump sum with you and then making regular withdrawals until the balance is 0.

Chapter 10: CHECKLIST

- ☐ Interest
- ☐ Principal
- ☐ Compound amount
- ☐ Compound interest
- ☐ Interest per period
- ☐ Present value
- ☐ Simple interest
- ☐ Annuity
- ☐ $s_{\overline{n}|i}$
- ☐ $a_{\overline{n}|i}$
- ☐ Future value of an annuity
- ☐ Present value of an annuity
- ☐ Geometric progression

□ Rent of an annuity
□ Amortization
□ Mortgage
□ Balloon payment

Chapter 10: SUPPLEMENTARY EXERCISES

1. Mr. West wishes to purchase a condominium for $80,000 cash upon his retirement 10 years from now. How much should he deposit at the end of each month into an annuity paying 12% interest compounded monthly in order to accumulate the required savings?

2. What is the monthly payment on a $150,000 30-year mortgage at 18% interest?

3. The income of a typical family in a certain city is currently $19,200 per year. Family finance experts recommend that mortgage payments not exceed 25% of a family's income. Assuming a current mortgage interest rate of 18% for a 30-year mortgage with monthly payments, how large a mortgage can the typical family in that city afford?

4. Calculate the compound amount of $50 after a year if deposited at 7.3% compounded daily.

5. Which is a better investment: 10% compounded annually or $9\frac{1}{8}$% compounded daily? [*Useful fact:* $(1.00025)^{365} = 1.09553$.]

6. Ms. Smith deposits $200 per month into a bond fund yielding 12% interest compounded monthly. How much are her holdings worth after 5 years?

7. A real estate investor takes out a $200,000 mortgage subject to the following terms: For the first 5 years, the payments will be the same as the monthly payments on a 15-year mortgage at 12% interest compounded monthly. The balance will then be payable in full.

   (a) What are the monthly payments for the first 5 years?

   (b) What balance will be owed after 5 years?

8. College expenses at a private college currently average $12,000 per year. It is estimated that these expenses are increasing at the rate of 1% per month. What is the estimated cost of a year of college 10 years from now?

9. What is the present value of $50,000 in 10 years at 12% interest compounded monthly?

10. An investment will pay $10,000 in 2 years and $5,000 in 3 years. If the current market interest rate is 12% compounded monthly, what should a rational person be willing to pay for the investment?

11. A woman purchases a car for $12,000. She pays $3,000 as a down payment and finances the remaining amount at 24% interest compounded monthly for 4 years. What is her monthly car payment?

12. A businessman buys a $100,000 piece of manufacturing equipment on the following terms: Interest will be charged at a rate of 12% compounded semiannually, but no payments will be made until 2 years after purchase. Beginning at that time equal semi-annual payments will be made for 5 years. Determine the semiannual payment.

13. A retired person has set aside a fund of $105,003.50 for his retirement. This fund is in a bank account paying 12% interest compounded monthly. How much can he draw out of the account at the end of each month so that there is a balance of $30,000 at the end of 15 years? [*Hint*: First compute the present value of the $30,000.]

14. A business loan of $509,289.22 is to be paid off in monthly payments for 10 years with a $100,000 balloon payment at the end of the tenth year. The interest rate on the loan is 24% compounded monthly. Calculate the monthly payment.

15. Ms. Jones saves $100 per month for 30 years at 12% interest compounded monthly. How much are her accumulated savings worth?

16. An apartment building is currently generating an income of $2000 per month. Its owners are considering a 10-year second mortgage at 18% interest compounded monthly in order to pay for repairs. How large a second mortgage can the income of the apartment house support?

17. Investment A generates $1000 at the end of each year for 10 years. Investment B generates $5000 at the end of the fifth year and $5000 at the end of the tenth year. Assume a market rate of interest of 6% compounded annually. Which is the better investment?

18. A 5-year bond has a face value of $1000 and is currently selling for $800. The bond pays $10 interest at the end of each month and, in addition, will repay the $1000 face value at the end of the fifth year. The market rate of interest is currently 18% compounded monthly. Is the bond a bargain? Why or why not? [*Note*: $1.015^{-60} = .40930$.]

19. Calculate the effective rate for 10% interest compounded semiannually.

20. Calculate the effective rate for 18% interest compounded monthly.

21. A person makes an initial deposit of $10,000 into a savings account and then deposits $1000 at the end of each quarter-year for 15 years. If the interest rate is 8% compounded quarterly, how much money will be in the account after 15 years?

22. A $7000 car loan at 24% interest compounded monthly is to be repaid with 36 equal monthly payments. Write out an amortization schedule for the first 6 months of the loan.

23. A person pays $200 at the end of each month for 10 years into an account paying 1% interest per month compounded monthly. At the end of the tenth year, the payments cease, but the balance continues to earn interest. What is the value of the balance at the end of the twentieth year?

24. A savings fund currently contains $300,000. It is decided to pay out this amount with 12% interest compounded monthly over a 5-year period. What are the monthly payments?

25. Calculate the following sum:

$$1 + 3 + 3^2 + 3^3 + 3^4 + \cdots + 3^{11}.$$

[*Hint*: $3^{12} = 531{,}441$.]

11

DIFFERENCE EQUATIONS AND MATHEMATICAL MODELS

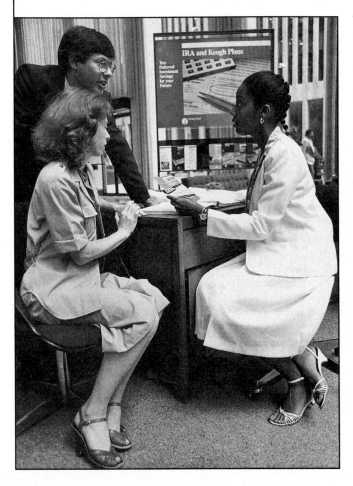

11.1   Introduction of Difference Equations, I

11.2   Introduction of Difference Equations, II

11.3   Graphing Difference Equations

11.4   Mathematics of Personal Finance

11.5   Modeling with Difference Equations

In this chapter we discuss a number of topics from the mathematics of finance—compound interest, mortgages, and annuities. As we shall see, all such financial transactions can be described by a single type of equation, called a *difference equation*. Furthermore, the same type of difference equation can be used to model many other phenomena, such as the spread of information, radioactive decay, and population growth, to mention a few.

11.1. Introduction to Difference Equations, I

In order to understand what difference equations are and how they arise, consider two examples concerned with financial transactions in a savings account.

EXAMPLE 1 Suppose that a savings account initially contains $40 and earns 6% interest, compounded annually. Determine a formula that describes how to compute each year's balance from the previous year's balance.

Solution The balances in the account for the first few years can be given as in the chart below, where y_0 is the initial balance (balance after zero years), y_1 is the balance after one year, and so on.

Year	Balance	Interest for year
0	$y_0 = \$40$	$(.06)40 = 2.40$
1	$y_1 = 42.40$	$(.06)42.40 = 2.54$
2	$y_2 = 44.94$	$(.06)44.94 = 2.70$
3	$y_3 = 47.64$	

Once the balance is known for a particular year, the balance at the end of the next year is computed as follows:

[balance at end of next year] = [balance at end of this year]

$$+ \text{[interest on balance at end of this year].}$$

That is,

$$y_1 = y_0 + .06y_0$$
$$y_2 = y_1 + .06y_1$$
$$y_3 = y_2 + .06y_2.$$

(Notice that since $3 = 2 + 1$, the last equation is $y_{2+1} = y_2 + .06y_2$.) In general, if y_n is the balance after n years, then y_{n+1} is the balance at the end of the next year, so

$$y_{n+1} = y_n + .06y_n \qquad \text{for } n = 0, 1, 2, \ldots.$$

This equation can be simplified:

$$y_{n+1} = y_n + .06y_n$$
$$= 1 \cdot y_n + .06y_n$$
$$= (1 + .06)y_n$$
$$= 1.06y_n. \qquad (1)$$

In other words, the balance after $n + 1$ years is 1.06 times the balance after n years. This formula describes how the balance is computed in successive years. For instance, using this formula, we can compute y_4. Indeed, setting $n = 3$ and using the value of y_3 from the chart above,

$$y_4 = y_{3+1} = 1.06y_3 = 1.06(47.64) = \$50.50.$$

In a similar way we can compute all of the year-end balances, one after another, by using formula (1).

EXAMPLE 2   Suppose that a savings account contains \$40 and earns 6% interest, compounded annually. At the end of each year a \$3 withdrawal is made. Determine a formula which describes how to compute each year's balance from the previous year's balance.

Solution   As in the preceding example, we compute the first few balances in a straightforward way and organize the data in a chart.

Balance	+ Interest for year	− Withdrawal
$y_0 = \$40$	$(.06)40 = 2.40$	\$3
$y_1 = 39.40$	$(.06)39.40 = 2.36$	3
$y_2 = 38.76$		

Reasoning as in the preceding example, let $y_n =$ the balance after n years. Then, by analyzing the calculations of the above chart, we see that

$$y_{n+1} \quad = \quad y_n \quad + \quad .06y_n \quad - \quad 3$$
$$[\text{new balance}] = [\text{old balance}] + [\text{interest on old balance}] - [\text{withdrawal}]$$

or

$$y_{n+1} = 1.06y_n - 3. \qquad (2)$$

This formula allows us to compute the values of y_n successively. For example, the chart above gives $y_2 = 38.76$. Therefore, from equation (2) with $n = 2$,

$$y_3 = (1.06)y_2 - 3 = (1.06)(38.76) - 3 = 38.09.$$

Equations of the form (1) and (2) are examples of what are called difference equations. More precisely, a difference equation is an equation of the form

$$y_{n+1} = ay_n + b,$$

where a, b are specific numbers. For example, for the difference equation (1) we have $a = 1.06$, $b = 0$. For the difference equation (2) we have $a = 1.06$, $b = -3$. A difference equation gives a procedure for calculating the term y_{n+1} from the preceding term y_n, thereby allowing one to compute all the terms—provided, of course, that a place to start is given. For this purpose one is usually given a specific value for y_0. Such a value is called an *initial value*. In both of the previous examples the initial value was 40.

Whenever we are given a difference equation, our goal is to determine as much information as possible about the terms y_0, y_1, y_2, and so on. To this end there are three things we can do:

1. *Generate the first few terms.* This is useful in giving us a feeling for how successive terms are generated.

2. *Graph the terms.* The terms that have been generated can be graphed by plotting the points $(0, y_0)$, $(1, y_1)$, $(2, y_2)$, and so on. Corresponding to the term y_n, we plot the point (n, y_n). The resulting graph (Fig. 1) depicts how the terms increase or decrease as n increases. In Figs. 2 and 3 we have drawn the graphs corresponding to the difference equations of Examples 1 and 2.

FIGURE 1

FIGURE 2 Graph of
$y_{n+1} = 1.06y_n$, $y_0 = 40$.

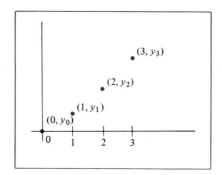

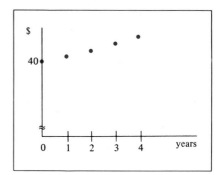

3. *Solve the difference equation.* **By** a *solution* of a difference equation we mean a general formula from which we can directly calculate any term without first having to calculate all of the terms preceding it. One can always write down a solution using the values of a, b, and y_0. Assume for now that $a \neq 1$. Then a solution of $y_{n+1} = ay_n + b$ is given by

$$ y_n = \frac{b}{1-a} + \left(y_0 - \frac{b}{1-a}\right)a^n, \qquad a \neq 1. \qquad (3) $$

(This formula will be derived in the next section.)

EXAMPLE 3   Solve the difference equation $y_{n+1} = 1.06y_n$, $y_0 = 40$.

Solution   Here $a = 1.06$, $b = 0$, $y_0 = 40$. So $b/(1-a) = 0$, and from equation (3),

$$ y_n = 0 + (40 - 0)(1.06)^n $$
$$ y_n = 40(1.06)^n. $$

EXAMPLE 4   Solve the difference equation $y_{n+1} = 1.06y_n - 3$, $y_0 = 40$. Determine y_{21}.

Solution   Here $a = 1.06$, $b = -3$, $y_0 = 40$. So $b/(1-a) = -3/(1-1.06) = -3/-.06 = 50$. Therefore,

$$ y_n = 50 + (40 - 50)(1.06)^n $$
$$ y_n = 50 - 10(1.06)^n. $$

In particular,

$$ y_{21} = 50 - 10(1.06)^{21}. $$

Using a calculator, we find that $(1.06)^{21} \approx 3.4$, so that

$$ y_{21} = 50 - 10(3.4) = 50 - 34 = 16. $$

So, after 21 years, the bank account of Example 2 will contain about $16. [If we did not have the solution $y_n = 50 - 10(1.06)^n$, we would need to perform 21 successive calculations in order to determine y_{21}.]

In the next two examples we apply our entire three-step procedure to analyze some specific difference equations.

EXAMPLE 5   Apply the three-step procedure to study the difference equation $y_{n+1} = .2y_n + 4.8$, $y_0 = 1$.

Solution First we compute a few terms:

$$y_0 = 1$$

$$y_1 = .2(1) + 4.8 = 5$$

$$y_2 = .2(5) + 4.8 = 5.8$$

$$y_3 = .2(5.8) + 4.8 = 5.96$$

$$y_4 = .2(5.96) + 4.8 = 5.992.$$

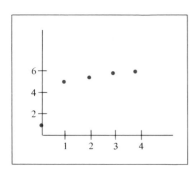

FIGURE 4 Graph of
$y_{n+1} = .2y_n + 4.8,$
$y_0 = 1.$

Next, we graph these terms (Fig. 4). That is, we plot $(0, 1)$, $(1, 5)$, $(2, 5.8)$, $(3, 5.96)$, $(4, 5.992)$. Note that the values of y_n increase. Finally, we solve the difference equation. Here $a = .2$, $b = 4.8$, $y_0 = 1$. Thus

$$\frac{b}{1 - a} = \frac{4.8}{1 - .2} = \frac{4.8}{.8} = 6$$

$$y_n = 6 + (1 - 6)(.2)^n = 6 - 5(.2)^n.$$

EXAMPLE 6 Apply the three-step procedure to study the difference equation $y_{n+1} = -.8y_n + 9$, $y_0 = 10$.

Solution The first few terms are

$$y_0 = 10$$

$$y_1 = -.8(10) + 9 = -8 + 9 = 1$$

$$y_2 = -.8(1) + 9 = -.8 + 9 = 8.2$$

$$y_3 = -.8(8.2) + 9 = 2.44$$

$$y_4 = -.8(2.44) + 9 = 7.048.$$

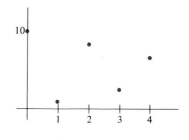

The accompanying graph corresponds to these terms. Notice that in this case the points oscillate up and down. To solve the difference equation note that

$$\frac{b}{1 - a} = \frac{9}{1 - (-.8)} = \frac{9}{1.8} = 5.$$

Thus

$$y_n = 5 + (10 - 5)(-.8)^n = 5 + 5(-.8)^n.$$

Difference equations can be used to describe many real-life situations. The solution of a particular difference equation in such instances yields a mathematical model of the situation. For example, consider the banking problem of Example 2. We described the activity in the account by a difference equation, where y_n represents the amount of money in the account after n years. Solving the difference equation, we found that

$$y_n = 50 - 10(1.06)^n.$$

And this formula gives a mathematical model of the bank account.

Here is another example of a mathematical model derived using difference equations.

EXAMPLE 7 Suppose that the population of a certain country is currently 6 million. The growth of this population attributable to an excess of births over deaths is 2% per year. Further, the country is experiencing immigration at the rate of 40,000 people per year.

(a) Find a mathematical model for the population of the country.

(b) What will be the population of the country after 35 years?

Solution (a) Let y_n denote the population (in millions) of the country after n years. Then $y_0 = 6$. The growth in the population in year $n + 1$ due to an excess of births over deaths is $.02y_n$. There are $.04$ (million) immigrants each year. Therefore,

$$y_{n+1} = y_n + .02y_n + .04 = 1.02y_n + .04.$$

So the terms satisfy the difference equation

$$y_{n+1} = 1.02y_n + .04, \qquad y_0 = 6.$$

Here

$$\frac{b}{1 - a} = \frac{.04}{1 - 1.02} = \frac{.04}{-.02} = -2$$

$$y_n = -2 + (6 - (-2))(1.02)^n$$

or

$$y_n = -2 + 8(1.02)^n.$$

This last formula is our desired mathematical model for the population.

(b) To determine the population after 35 years merely compute y_{35}:

$$y_{35} = -2 + 8(1.02)^{35} \approx -2 + 8(2) = 14,$$

since $(1.02)^{35} \approx 2$. So the population after 35 years will be about 14 million.

PRACTICE PROBLEMS 1

1. Consider the difference equation $y_{n+1} = -2y_n + 21$, $y_0 = 7.5$.

   (a) Generate y_0, y_1, y_2, y_3, y_4 from the difference equation.

   (b) Graph these first few terms.

   (c) Solve the difference equation.

2. Use the solution in Problem 1(c) to obtain the terms y_0, y_1, y_2.

EXERCISES 1

For each of the difference equations in Exercises 1–6, identify a and b and compute $b/(1-a)$.

1. $y_{n+1} = 4y_n - 6$
2. $y_{n+1} = -3y_n + 16$
3. $y_{n+1} = -\frac{1}{2}y_n$

4. $y_{n+1} = \frac{1}{3}y_n + 4$
5. $y_{n+1} = -\frac{2}{3}y_n + 15$
6. $y_{n+1} = .5y_n - 4$

In Exercises 7–14: (a) Generate y_0, y_1, y_2, y_3, y_4 from the difference equation. (b) Graph these first few terms. (c) Solve the difference equation.

7. $y_{n+1} = \frac{1}{2}y_n - 1, y_0 = 10$
8. $y_{n+1} = .5y_n + 5, y_0 = 2$

9. $y_{n+1} = 2y_n - 3, y_0 = 3.5$
10. $y_{n+1} = 5y_n - 32, y_0 = 8$

11. $y_{n+1} = -.4y_n + 7, y_0 = 17.5$
12. $y_{n+1} = -2y_n, y_0 = \frac{1}{2}$

13. $y_{n+1} = 2y_n - 16, y_0 = 15$
14. $y_{n+1} = 2y_n + 3, y_0 = -2$

15. The solution to $y_{n+1} = .2y_n + 4.8$, $y_0 = 1$ is $y_n = 6 - 5(.2)^n$. Use the solution to compute y_0, y_1, y_2, y_3, y_4.

16. The solution to $y_{n+1} = -.8y_n + 9$, $y_0 = 10$ is $y_n = 5 + 5(-.8)^n$. Use the solution to compute y_0, y_1, y_2.

17. One thousand dollars is deposited into a savings account paying 5% interest compounded annually. Let y_n be the amount after n years. What is the difference equation showing how to compute y_{n+1} from y_n?

18. The population of a certain country is currently 70 million but is declining at the rate of 1% each year. Let y_n be the population after n years. Find a difference equation showing how to compute y_{n+1} from y_n.

19. One thousand dollars is deposited into an account paying 5% interest compounded annually. At the end of each year $100 is added to the account. Let y_n be the amount in the account after n years. Find a difference equation satisfied by y_n.

20. The population of a certain country is currently 70 million but is declining at the rate of 1% each year, owing to an excess of deaths over births. In addition, the country is losing a million people each year due to emigration. Let y_n be the population after n years. Find a difference equation satisfied by y_n.

21. Consider the difference equation $y_{n+1} = y_n + 2, y_0 = 1$.

    (a) Generate y_0, y_1, y_2, y_3, y_4.

    (b) Sketch the graph.

    (c) Why cannot formula (3) be used to obtain the solution?

22. Multiply: $(1 + a + a^2)(1 - a)$. (This result is needed in Section 11.2.)

23. Suppose that you take a consumer loan for $55 at 20% annual interest and pay off $36 at the end of each year for two years. Compute the balance on the loan immediately after you make the first payment (i.e., the amount you would pay if you wanted to settle the account at the beginning of the second year).

24. Refer to Exercise 23. Set up the difference equation for y_n, the balance after n years.

SOLUTIONS TO PRACTICE PROBLEMS 1

1. (a) $y_0 = 7.5$.

   $y_1 = -2(7.5) + 21 = -15 + 21 = 6$.
   $y_2 = -2(6) + 21 = -12 + 21 = 9$.
   $y_3 = -2(9) + 21 = -18 + 21 = 3$.
   $y_4 = -2(3) + 21 = -6 + 21 = 15$.

   (b) We graph the points $(0, 7.5)$, $(1, 6)$, $(2, 9)$, $(3, 3)$, $(4, 15)$. (In order to accommodate the points, we use a different scale for each axis.)

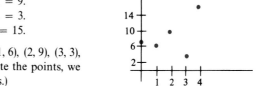

   (c) $y_n = b/(1 - a) + (y_0 - b/(1 - a))a^n$.

   Since $a = -2$ and $b = 21$, $\dfrac{b}{1 - a} = \dfrac{21}{1 - (-2)} = \dfrac{21}{3} = 7$.

   $y_n = 7 + (7.5 - 7)(-2)^n = 7 + \frac{1}{2}(-2)^n$.

2. $y_n = 7 + \frac{1}{2}(-2)^n$.

   $y_0 = 7 + \frac{1}{2}(-2)^0 = 7 + \frac{1}{2} = 7.5$ (Here we have used the fact that any number raised to the zeroth power is 1.)

   $y_1 = 7 + \frac{1}{2}(-2)^1 = 7 + \frac{1}{2}(-2) = 7 - 1 = 6$.
   $y_2 = 7 + \frac{1}{2}(-2)^2 = 7 + \frac{1}{2}(4) = 7 + 2 = 9$.

11.2. Introduction to Difference Equations, II

The main result of the preceding section is:

> The difference equation $y_{n+1} = ay_n + b$, with $a \neq 1$, has solution
> $$y_n = \frac{b}{1 - a} + \left(y_0 - \frac{b}{1 - a}\right)a^n. \tag{1}$$

When $a = 1$, the difference equation $y_{n+1} = a \cdot y_n + b$ is $y_{n+1} = 1 \cdot y_n + b$ or just $y_{n+1} = y_n + b$. In this case the solution is given by a formula different from (1):

> The difference equation $y_{n+1} = y_n + b$ has the solution
> $$y_n = y_0 + bn. \tag{2}$$

EXAMPLE 1 (a) Solve the difference equation $y_{n+1} = y_n + 2$, $y_0 = 3$.

(b) Find y_{100}.

Solution (a) Since $a = 1$, the solution is given by (2). Here $b = 2$, $y_0 = 3$. Therefore,

$$y_n = 3 + 2n.$$

(b) $y_{100} = 3 + 2(100) = 203$.

Note Without (2), determining y_{100} would require 100 computations.

Both (1) and (2) are derived at the end of this section. Before deriving them, however, let us see what they tell us about simple interest, compound interest, and consumer loans.

Simple Interest Suppose that a certain amount of money is deposited in a savings account. If interest is paid only on the initial deposit (and not on accumulated interest), then the interest is called *simple*. For example, if $40 is deposited at 6% simple interest, then each year the account earns .06(40), or $2.40. So the bank balance accumulates as follows:

Year	Amount	Interest
0	$40	$2.40
1	42.40	2.40
2	44.80	2.40
3	47.20	

EXAMPLE 2 (a) Find a formula for y_n, the amount in the account above at the end of n years.

(b) Find the amount at the end of 10 years.

Solution (a) Let $y_n =$ the amount at the end of n years. So $y_0 = 40$. Moreover,

[amount at end of $n + 1$ years] = [amount at end of n years] + [interest].

$$y_{n+1} \qquad = \qquad y_n \qquad + \qquad 2.40$$

This difference equation has $a = 1$, $b = 2.40$, so from formula (2),

$$y_n = y_0 + bn$$
$$= 40 + (2.40)n.$$

This is the desired formula.

(b) $y_{10} = 40 + (2.40)10 = 40 + 24.00 = \$64$.

Compound Interest When interest is calculated on the current amount in the account (instead of on the amount initially deposited), the interest is called *compound*. The interest discussed in Section 11.1 was compound, being computed on the balance at the end of each year. Such interest is called *annual* compound interest. Often interest is compounded more than once a year. For example, interest might be stated as 6% compounded *semiannually*. This means that interest is computed every 6 months, with 3% given for each 6-month period. At the end of each such period the interest is added to the balance, which is then used to compute the interest for the next 6-month period. Similarly, 6% *interest compounded* *quarterly* means .06/4 or .015 interest four times a year. And 6% interest compounded six times a year means .06/6 or .01 interest each interest period of 2 months. Or, in general, if 6% interest is compounded k times a year, then .06/k interest is earned k times a year. This illustrates the following general principle:

If interest is at a yearly rate r and is compounded k times per year, then the interest rate per period (denoted i) is $i = r/k$.

EXAMPLE 3 Suppose that the interest rate is 6% compounded monthly. Find a formula for the amount after n months.

Solution Here $r = .06$, $k = 12$, so that the monthly interest rate is $i = .005$. Let y_n denote the amount after n months. Then, reasoning as in the preceding section,

$$y_{n+1} = y_n + .005y_n$$
$$= 1.005y_n.$$

The solution of this difference equation is obtained as follows:

$$\frac{b}{1-a} = 0$$

$$y_n = 0 + (y_0 - 0)(1.005)^n$$
$$= y_0(1.005)^n.$$

EXAMPLE 4 Suppose that interest is computed at the rate i per interest period. Find a general formula for the balance y_n after n periods under (a) simple interest, (b) compound interest.

Solution (a) With simple interest, the interest each period is just i times the initial amount—that is, $i \cdot y_0$. Therefore,

$$y_{n+1} = y_n + iy_0.$$

Apply (2) with $b = iy_0$. Then

$$y_n = y_0 + (iy_0)n.$$

(b) With compound interest, the interest each period is i times the current balance or $i \cdot y_n$. Therefore,

$$y_{n+1} = y_n + iy_n$$
$$= (1 + i)y_n.$$

Here $b/(1 - a) = 0$, so that

$$y_n = 0 + (y_0 - 0)(1 + i)^n$$
$$= y_0(1 + i)^n.$$

Summarizing:

> If y_0 dollars is deposited at interest rate i per period, then the amount after n interest periods is:
>
> Simple interest: $\quad y_n = y_0 + (iy_0)n.$ $\qquad\qquad$ (3)
>
> Compound interest: $\quad y_n = y_0(1 + i)^n.$ $\qquad\qquad$ (4)

EXAMPLE 5 How much money will you have after seven years if you deposit $40 at 8% interest compounded quarterly?

Solution Apply formula (4). Here a period is a quarter or 3 months. In 7 years there are $7 \cdot 4 = 28$ periods. The interest rate per period is

$$i = \frac{.08}{4} = .02.$$

The amount after 28 periods is

$$y_{28} = 40(1.02)^{28}$$
$$= \$69.64.$$

[To compute $(1.02)^{28}$ we used a calculator. The answer could have been left in the form $40(1.02)^{28}$.]

Consumer Loans   It is common for people to buy cars or appliances "on time." Basically, they are borrowing money from the dealer and repaying it (with interest) with several equal payments until the loan is paid off. Each time period, part of the payment goes toward paying off the interest and part toward reducing the balance of the loan. A consumer loan used to purchase a house is called a *mortgage*.

EXAMPLE 6 Suppose that a consumer loan of $2400 carries an interest rate of 12% compounded annually and a yearly payment of $1000.

(a) Write down a difference equation for y_n, the balance owed after n years.

(b) Compute the balances after 1, 2, and 3 years.

Solution (a) At the end of each year the new balance is computed as follows:

$$[\text{new balance}] = [\text{previous balance}] + [\text{interest}] - [\text{payment}].$$

Since the interest is compound, it is computed on the previous balance:

$$y_{n+1} = y_n + .12y_n - 1000.$$

Thus

$$y_{n+1} = 1.12y_n - 1000, \qquad y_0 = 2400,$$

the desired difference equation.

(b) $y_1 = 1.12y_0 - 1000 = (1.12)(2400) - 1000 = \$1688$
$y_2 = 1.12y_1 - 1000 = (1.12)(1688) - 1000 = 891$
$y_3 = 1.12y_2 - 1000 = (1.12)(891) - 1000 \approx 0.$

Although savings accounts using simple interest are practically unheard of, consumer loans are often computed using simple interest. Under simple interest, interest is paid on the entire amount of the initial loan, not just on the outstanding balance.

EXAMPLE 7 Rework the preceding example, except assume that the interest is now simple.

Solution (a) As before,

[new balance] = [previous balance] + [interest] − [payment].

Now, however, since the interest is simple, it is computed on the original loan y_0. So

$$y_{n+1} = y_n + .12y_0 - 1000, \qquad y_0 = 2400.$$

(b) $y_1 = y_0 + .12y_0 - 1000 = \$1688$
$y_2 = y_1 + .12y_0 - 1000 = 976$
$y_3 = y_2 + .12y_0 - 1000 = 264.$

Remark Note that after three years the loan of Example 7 is not yet paid off, whereas the loan of Example 6 is. Therefore, the loan at 12% simple interest is more expensive than the one at 12% compound interest. Actually, a 12% simple interest loan is equivalent to a 16% compound interest loan. It was at one time a common practice to advertise loans in terms of simple interest to make the interest rate seem cheaper. But the Federal Truth in Lending Law now requires that all loans be stated in terms of their equivalent compound interest rate.

Verification of Formulas (1) and (2) The derivation of formula (1) depends on the formula for the sum of the powers of a number. If $a \neq 1$, then

$$1 + a = \frac{1 - a^2}{1 - a}$$

$$1 + a + a^2 = \frac{1 - a^3}{1 - a}$$

$$1 + a + a^2 + a^3 = \frac{1 - a^4}{1 - a}.$$

In general, for any positive integer r the sum of the first r powers of a is given by $1 - a^{r+1}$ divided by $1 - a$. Symbolically:

If $a \neq 1$, then

$$1 + a + a^2 + \cdots + a^r = \frac{1 - a^{r+1}}{1 - a},$$
(5)

where r is any positive integer.

Note that the condition $a \neq 1$ is essential to avoid dividing by zero on the right.

Illustration of Formula (5)

When $a = .9$ and $r = 2$, then

$$1 + a + a^2 = 1 + .9 + .9^2 = 1 + .9 + .81 = 2.71$$

$$\frac{1 - a^3}{1 - a} = \frac{1 - .9^3}{1 - .9} = \frac{1 - .729}{1 - .9} = \frac{.271}{.1} = 2.71.$$

Verification of Formula (5)

Form the product

$$(1 + a + a^2 + \cdots + a^r)(1 - a).$$

Multiplying the product out, we get

$$1 + a + a^2 + \cdots + a^r$$

$$\frac{1 - a}{1 + a + a^2 + \cdots + a^r}$$

$$\frac{-a - a^2 - a^3 + \cdots - a^{r+1}}{1 + 0 + 0 + \cdots + 0 - a^{r+1}}.$$

Thus

$$(1 + a + a^2 + \cdots + a^r)(1 - a) = 1 - a^{r+1}.$$

If $a \neq 1$, then $1 - a \neq 0$, and we may divide both sides by $1 - a$ to get formula (5).

The form in which formula (5) will be needed is as follows. Let n be any integer > 1. Then

$$1 + a + a^2 + \cdots + a^{n-1} = \frac{1 - a^n}{1 - a}.$$
(6)

This formula is obtained by replacing r in formula (5) by $n - 1$.

From the difference equation $y_{n+1} = ay_n + b$ we get

$$y_1 = ay_0 + b$$

$$y_2 = ay_1 + b = a(ay_0 + b) + b$$

$$= a^2 y_0 + ab + b$$

$$= a^2 y_0 + b(1 + a)$$

$$y_3 = ay_2 + b = a(a^2 y_0 + ab + b) + b$$

$$= a^3 y_0 + a^2 b + ab + b$$

$$= a^3 y_0 + b(1 + a + a^2).$$

The pattern that clearly develops is

$$y_n = a^n y_0 + b(1 + a + a^2 + \cdots + a^{n-1}). \tag{7}$$

By formula (6)

$$y_n = a^n y_0 + b \cdot \frac{1 - a^n}{1 - a}$$

$$= a^n y_0 + b \cdot \frac{1}{1 - a} - b \cdot \frac{a^n}{1 - a}$$

$$= \frac{b}{1 - a} + \left(y_0 - \frac{b}{1 - a} \right) a^n,$$

which is formula (1).

Formula (1) gives the solution of the difference equation $y_{n+1} = ay_n + b$ in the case $a \neq 1$. The reasoning above also gives the solution in case $a = 1$—namely equation (7) holds for any value of a. In particular, for $a = 1$ equation (7) reads

$$y_n = a^n y_0 + b(1 + a + a^2 + \cdots + a^{n-1})$$

$$= 1^n y_0 + b(1 + 1 + 1^2 + \cdots + 1^{n-1})$$

$$= y_0 + b\underbrace{(1 + 1 + 1 + \cdots + 1)}_{n \text{ ones}}$$

$$= y_0 + bn.$$

Thus we have derived formula (2).

PRACTICE PROBLEMS 2

1. Solve the following difference equations.

   (a) $y_{n+1} = -.2y_n + 24$, $y_0 = 25$.   (b) $y_{n+1} = y_n - 3$, $y_0 = 7$.

2. In 1626, Peter Minuit, the first director-general of New Netherlands province, purchased Manhattan Island for trinkets and cloth valued at about $24. Suppose that this money had been invested at 7% interest compounded quarterly. How much would it have been worth by the U.S. bicentennial year, 1976?

EXERCISES 2

1. Solve the difference equation $y_{n+1} = y_n + 5$, $y_0 = 1$.

2. Solve the difference equation $y_{n+1} = y_n - 2$, $y_0 = 50$.

In Exercises 3–8, find an expression for the amount of money in the bank after five years where the initial deposit is $80 and the annual interest rate is as given.

3. 9% compounded monthly

4. 4% simple interest

5. 100% compounded daily

6. 8% compounded quarterly

7. 7% simple interest

8. 12% compounded semiannually

9. Find the general formula for the amount of money accumulated after t years when A dollars is invested at annual interest rate r compounded k times per year.

10. Determine the amount of money accumulated after one year when $1 is deposited at 40% interest compounded

    (a) annually   (b) semiannually   (c) quarterly

11. For the difference equation $y_{n+1} = 2y_n - 10$ generate y_0, y_1, y_2, y_3 and draw the graph corresponding to the initial condition:

    (a) $y_0 = 10$ (b) $y_0 = 11$ (c) $y_0 = 9$

12. For the difference equation $y_{n+1} = \frac{1}{2}y_n + 5$ generate y_0, y_1, y_2, y_3 and draw the graph corresponding to the initial condition:

    (a) $y_0 = 10$ (b) $y_0 = 18$ (c) $y_0 = 2$

In Exercises 13–16, solve the difference equation and, by inspection, determine the long-run behavior of the terms (i.e., the behavior of the terms as n gets large).

13. $y_{n+1} = .4y_n + 3$, $y_0 = 7$

14. $y_{n+1} = 3y_n - 12$, $y_0 = 10$

15. $y_{n+1} = -5y_n$, $y_0 = 2$

16. $y_{n+1} = -.7y_n + 3.4$, $y_0 = 3$

17. A bank loan of $38,900 at 9% interest compounded monthly is made in order to buy a house and is paid off at the rate of $350 per month for 20 years. (Such a loan is called a *mortgage*.) The balance at any time is the amount still owed on the loan, that is, the amount which would have to be paid out to repay the loan all at once at that time. Find the difference equation for y_n, the balance after n months.

18. Refer to Exercise 17. Express in mathematical notation the fact that the loan is paid off after 20 years.

19. A house is purchased for $50,000 and depreciated over a 25-year period. Let y_n be the (undepreciated) value of the house after n years. Determine and solve the difference

equation for y_n assuming straight-line depreciation (i.e., each year the house depreciates by one twenty-fifth of its original value).

20. Refer to Exercise 19. Determine and solve the difference equation for y_n assuming the double-declining balance method of depreciation (i.e., each year the house depreciates by two twenty-fifths of its value at the beginning of that year).

SOLUTIONS TO PRACTICE PROBLEMS 2

1. (a) Since $a = -.2 \neq 1$, use formula (1).

$$\frac{b}{1 - a} = \frac{24}{1 - (-.2)} = \frac{24}{1.2} = \frac{240}{12} = 20,$$

$$y_n = 20 + (25 - 20)(-.2)^n = 20 + 5(-.2)^n.$$

(b) Since $a = 1$, use formula (2).

$$y_n = 7 + (-3)n = 7 - 3n.$$

2. Since interest is compounded quarterly, the interest per period is $.07/4 = .0175$. Three hundred and fifty years consists of $4(350) = 1400$ interest periods. Therefore by (4), the amount accumulated is

$$y_{1400} = 24(1.0175)^{1400}.$$

(This amount is approximately 850 billion dollars, which is more than Manhattan Island was worth in 1976.)

11.3. Graphing Difference Equations

In this section we introduce a method for sketching the graph of the difference equation $y_{n+1} = ay_n + b$ (with initial value y_0) directly from the three numbers a, b, and y_0. As we shall see, the graphs arising from difference equations can be completely described by two characteristics—vertical direction and long-run behavior. To begin we introduce some vocabulary to describe graphs.

The *vertical direction* of a graph refers to the up-and-down motion of successive terms. A graph *increases* if it rises when read from left to right—that is,

FIGURE 1
Increasing graphs.

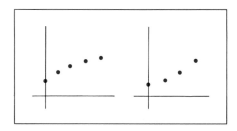

if the terms get successively larger. A graph *decreases* if it falls when read from left to right—that is, if the terms get successively smaller. Figure 1 shows the graphs of two difference equations. Both graphs increase. Figure 2 shows two examples of decreasing graphs. A graph that is either increasing or decreasing is called *monotonic*. That is, a graph is monotonic if it always heads in one direction—up or down.

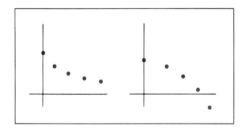

FIGURE 2 Decreasing graphs.

The extreme opposite of a monotonic graph is one that changes its direction with every term. Such a graph is called *oscillating*. Figure 3 shows two examples of oscillating graphs. A difference equation having an oscillating graph has terms y_n that alternately increase and decrease.

In addition to the monotonic and oscillating graphs, there are the *constant* graphs, which always remain at the same height. That is, all the terms y_n are the same. A constant graph is illustrated in Fig. 4.

One of the main results of this section is that the graph of a difference equation $y_{n+1} = ay_n + b$ is always either monotonic, oscillating, or constant. This threefold classification gives us all the possibilities for the vertical direction of the graph.

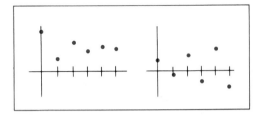

FIGURE 3 Oscillating graphs.

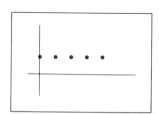

FIGURE 4 Constant graph

Long-run behavior refers to the eventual behavior of the graph. Most graphs of difference equations exhibit one of two types of long-run behavior. Some approach a horizontal line and are said to be *asymptotic* to the line. Some go indefinitely high or indefinitely low and are said to be *unbounded*. These phenomena are illustrated in Figs. 5 and 6.

The dashed horizontal lines in Figs. 5 and 6 each have the equation $y = b/(1 - a)$. Note that in Fig. 5 the terms move steadily closer to the dashed line and in Fig. 6 they move steadily farther away from the dashed line. In the first case we say that the terms are *attracted* to the dashed line and in the second case we say that they are *repelled* by it.

FIGURE 5 Asymptotic graphs.

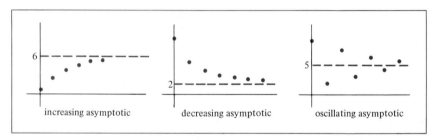

increasing asymptotic   decreasing asymptotic   oscillating asymptotic

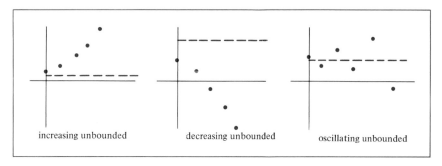

FIGURE 6   Unbounded graphs.

Constant Graphs   The general formula for the nth term of a difference equation where $a \neq 1$ is

$$y_n = \frac{b}{1-a} + \left(y_0 - \frac{b}{1-a}\right)a^n.$$

It is clear that as n varies so does a^n, and this makes y_n vary with n—unless, of course, the coefficient of a^n is 0, in which case the graph is constant. Thus we have the following result:

> The graph of $y_{n+1} = ay_n + b$ $(a \neq 1)$ is constant if $y_0 - b/(1-a) = 0$; that is, if $y_0 = b/(1-a)$.

Thus, when the graph starts out on the line $y = b/(1-a)$, it stays on the line.

EXAMPLE 1   Sketch the graph of the difference equation $y_{n+1} = 2y_n - 1$, $y_0 = 1$.

Solution   First compute $b/(1-a)$:

$$\frac{b}{1-a} = \frac{-1}{1-2} = 1.$$

But $y_0 = 1$. So $b/(1-a)$ and y_0 are the same and the graph is constant, always equal to 1 (Fig. 7).

FIGURE 7

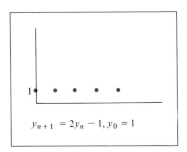

$y_{n+1} = 2y_n - 1, y_0 = 1$

Throughout this section assume that $a \neq \pm 1$. Then the graph of a given difference equation either is constant or is one of the types shown in Figs. 5 and 6. We can determine the nature of the graph by looking at the coefficients a and b and the initial value y_0. We have seen how constant graphs are handled. Thus, assume that we are dealing with a nonconstant graph—that is, the graph of a difference equation for which a^n actually affects the formula for y_n. Monotonic graphs may be differentiated from oscillating graphs by the following test.

> *Test 1* If $a > 0$, then the graph of $y_{n+1} = ay_n + b$ is monotonic. If $a < 0$, then the graph is oscillating.

The next two examples provide a convincing argument for Test 1.

EXAMPLE 2 Discuss the vertical direction of the graph of $y_{n+1} = -.8y_n + 9$, $y_0 = 50$.

Solution The formula for y_n yields

$$y_n = 5 + 45(-.8)^n.$$

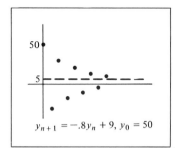

$y_{n+1} = -.8y_n + 9, y_0 = 50$

Note that the term $(-.8)^n$ is alternately positive and negative, since any negative number to an even power is positive and any negative number to an odd power is negative. Therefore, the expression $45(-.8)^n$ is alternately positive and negative. So $y_n = 5 + 45(-.8)^n$ is computed by alternately adding and subtracting something from 5. Thus y_n oscillates around 5. In this example $a = -.8$ (a negative number), so that the behavior just observed is consistent with that predicted by Test 1. The graph is sketched in Fig. 8. Note that in this case 5 is just $b/(1 - a)$, so the graph oscillates about the line $y = b/(1 - a)$.

FIGURE 8

The reasoning of the preceding example works whenever $a < 0$. The oscillation of a^n from positive to negative and back forces the term $(y_0 - [b/(1 - a)])a^n$ to swing up and back from positive to negative. So the value of

$$y_n = \frac{b}{1 - a} + \left(y_0 - \frac{b}{1 - a} \right)a^n$$

swings up and back, above and below $b/(1 - a)$. In other words, the graph oscillates about the line $y = b/(1 - a)$.

EXAMPLE 3 Discuss the vertical direction of the graph of $y_{n+1} = .8y_n + 9$, $y_0 = 50$.

Solution In this case $b/(1 - a) = 45$, and the formula for y_n gives

FIGURE 9

$$y_n = 45 + 5(.8)^n.$$

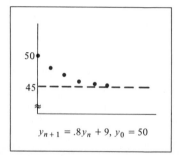

$y_{n+1} = .8y_n + 9, y_0 = 50$

The expression $5(.8)^n$ is always positive. As n gets larger, $5(.8)^n$ gets smaller, so that y_n decreases to 45. That is, the graph is steadily decreasing to $y = 45$ (Fig. 9). Here $a = .8 > 0$, so Test 1 correctly predicts that the graph is monotonic.

It is possible to determine whether a graph is asymptotic or unbounded, using the following result.

> *Test 2* If $|a| < 1$, then the graph of $y_{n+1} = ay_n + b$ is asymptotic to the line $y = b/(1 - a)$. If $|a| > 1$, then the graph is unbounded and moves away from the line $y = b/(1 - a)$.

Let us examine some difference equations in light of Test 2.

EXAMPLE 4 Discuss the graphs of the difference equation $y_{n+1} = .2y_n + 4.8$, with $y_0 = 1$ and with $y_0 = 11$.

Solution The graphs are shown in Fig. 10. Since $a = .2$, which has absolute value less than 1, Test 2 correctly predicts that each graph is asymptotic to the line $y = b/(1 - a) = 4.8/(1 - .2) = 6$. When the initial value is less than 6, the graph increases and moves toward the line $y = 6$. When the initial value is greater than 6, the graph decreases toward the line $y = 6$. In each case the graph is *attracted* to the line $y = 6$.

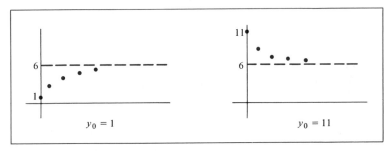

FIGURE 10

EXAMPLE 5 Apply Test 2 to the difference equation $y_{n+1} = -.8y_n + 9$, $y_0 = 50$.

Solution $|a| = |-.8| = .8 < 1$. So by Test 2 the graph is asymptotic to the line $y = b/(1 - a) = 5$. This agrees with the graph as drawn in Fig. 8.

EXAMPLE 6 Discuss the graphs of $y_{n+1} = 1.4y_n - 8$, with $y_0 > 20$ and with $y_0 < 20$.

Solution The graphs are shown in Fig. 11. Since $a = 1.4 > 0$, Test 1 predicts that the graphs are monotonic. Since $|a| = |1.4| = 1.4 > 1$, Test 2 predicts that the graphs are

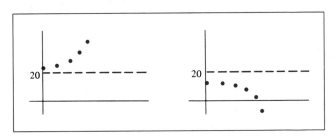

FIGURE 11

unbounded. Here

$$\frac{b}{1-a} = \frac{-8}{1-(1.4)} = \frac{-8}{-.4} = 20.$$

Both graphs move away from the line $y = 20$ as if being *repelled* by a force.

EXAMPLE 7 Discuss the graph of the difference equation $y_{n+1} = -2y_n + 60$, where $y_0 > 20$.

Solution A graph is drawn in Fig. 12. Since $a = -2 < 0$, Test 1 predicts that the graph is oscillating. Since $|a| = |-2| = 2 > 1$, Test 2 predicts that the graph is unbounded. Here

$$\frac{b}{1-a} = \frac{60}{1-(-2)} = \frac{60}{3} = 20.$$

Notice that successive points move farther away from $y = 20$—that is, they are repelled by the line.

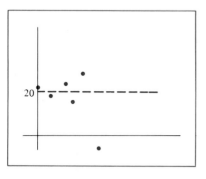

FIGURE 12

Verification of Test 2 The key to verifying Test 2 is to make the following two observations:

1. If $|a| < 1$, then the powers $a, a^2, a^3, a^4, \ldots$, become successively smaller and approach 0. (For example, if $a = .4$, then this sequence of powers is $.4, .16, .064, .0256, \ldots$.)
2. If $|a| > 1$, then the powers $a, a^2, a^3, a^4, \ldots$, become unbounded; that is, they become arbitrarily large. (For example, if $a = 3$, then this sequence is $3, 9, 27, 81, \ldots$.)

To verify Test 2 look at the formula for y_n:

$$y_n = \frac{b}{1-a} + \boxed{\left(y_0 - \frac{b}{1-a}\right)a^n}.$$

If $|a| < 1$, then the powers of a get smaller and smaller; so the boxed term approaches 0. That is, y_n approaches $b/(1-a)$ and the graph is asymptotic to $y = b/(1-a)$. If $|a| > 1$, the powers of a are unbounded, so that the boxed term becomes arbitrarily large in magnitude. Thus y_n is unbounded, and so is the graph.

In the case where $|a| < 1$ the graphs (whether oscillating or monotone) are *attracted* steadily to the line $y = b/(1-a)$. In the case where $|a| > 1$ the graphs are *repelled* by the line $y = b/(1-a)$. Thinking of graphs as being attracted or repelled helps us in making a rough sketch.

Sign of a:	$\begin{cases} \text{Positive} \\ \text{Negative} \end{cases}$	Monotonic Oscillating
Size of a:	$\begin{cases} \|a\| < 1 \\ \|a\| > 1 \end{cases}$	Attract Repel

Figure 13 shows some of the graphs already examined with the appropriate descriptive words labeling the line $y = b/(1 - a)$.

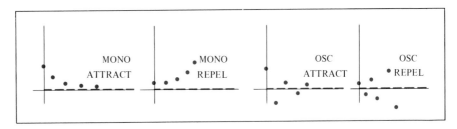

MONO
ATTRACT MONO
REPEL OSC
ATTRACT OSC
REPEL

FIGURE 13

Based on Tests 1 and 2 and the discussion of constant graphs, we can state a procedure for making a rough sketch of a graph without solving or generating terms.

To sketch the graph of $y_{n+1} = ay_n + b, a \neq 0, \pm 1$:

1. Draw the line $y = b/(1 - a)$ as a dashed line.
2. Plot y_0. If y_0 is on the line $y = b/(1 - a)$, the graph is constant and the procedure terminates.
3. If a is positive, write **MONO**, since the graph is then monotonic. If a is negative, write **OSC**, since the graph is then oscillating.
4. If $|a| < 1$, write **ATTRACT**, since the graph is attracted to the line $y = b/(1 - a)$. If $|a| > 1$, write **REPEL**, since the graph is repelled from the line.
5. Use all the information to sketch the graph.

EXAMPLE 8 Sketch the graph of $y_{n+1} = .6y_n + 8, y_0 = 50$.

Solution 1. $\dfrac{b}{1 - a} = \dfrac{8}{1 - (.6)} = \dfrac{8}{.4} = 20$. So draw the line

$y = 20$.

2. $y_0 = 50$, which is not on the line $y = 20$. So the graph is not constant.

3. $a = .6$, which is positive. Write MONO above the dashed line.

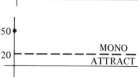

4. $|a| = |6| = .6 < 1$. Write ATTRACT below the dashed line.

5. The information above tells us to start at 50 and move monotonically toward the line $y = 20$.

EXAMPLE 9   Sketch the graph of $y_{n+1} = -1.5y_n + 5$, $y_0 = 2.6$.

Solution   1. $\dfrac{b}{1-a} = \dfrac{5}{1-(-1.5)} = \dfrac{5}{2.5} = 2$. So draw the line $y = 2$.

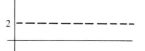

2. $y_0 = 2.6$, which is not on the line $y = 2$. So the graph is not constant.

3. $a = -1.5$, which is negative. Write OSC above the dashed line.

4. $|a| = |-1.5| = 1.5 > 1$. Write REPEL below the dashed line.

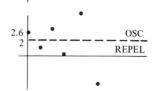

5. The above information says that the graph begins 2.6, oscillates about and is steadily repelled by the line $y = 2$.

The procedure above does not give an exact graph, but it shows the nature of the graph. This is exactly what is needed in many applications.

EXAMPLE 10   Suppose that the yearly interest rate on a mortgage is 9% compounded monthly and that you can afford to make payments of $300 per month. How much can you afford to borrow?

Solution   Let i = the monthly interest rate, R = the monthly payment, and y_n = the balance after n months. y_0 = the initial amount of the loan. Then the balance after $n + 1$ months equals the balance after n months plus the interest on that balance minus the monthly payment. That is,

$$y_{n+1} = y_n + iy_n - R$$
$$= (1 + i)y_n - R.$$

In this particular example $i = .09/12 = .0075$ and $R = 300$, so the difference equation reads

$$y_{n+1} = 1.0075y_n - 300.$$

Apply our graph-sketching technique to this difference equation. Here

$$\frac{b}{1 - a} = \frac{-300}{1 - (1.0075)} = \frac{-300}{-.0075} = 40,000.$$

Since $a = 1.0075$ is positive and $|a| = 1.0075 > 1$, the words MONO and REPEL describe the graphs.

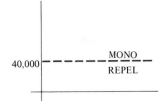

Let us see what happens for various initial values. If $y_0 > 40,000$, then the balance increases indefinitely.

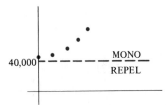

If $y_0 = 40,000$, then the balance will always be 40,000.

If $y_0 < 40,000$, then the balance decreases steadily and eventually reaches 0, at which time the mortgage is paid off.

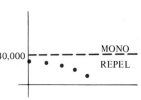

Therefore, the loan must be less than $40,000 in order for it to be paid off eventually.

PRACTICE PROBLEMS 3

1. A parachutist opens her parachute after reaching a speed of 100 feet per second. Suppose that y_n, the speed n seconds after opening the parachute, satisfies $y_{n+1} = .1y_n + 14.4$. Sketch the graph of y_n.

2. Rework Problem 1 if the parachute opens when the speed is 10 feet per second.

3. Upon retirement, a person deposits a certain amount of money into the bank at 6% interest compounded monthly and withdraws $100 at the end of each month.

   (a) Set up the difference equation for y_n, the amount in the bank after n months.

   (b) How large must the initial deposit be so that the money will never run out?

EXERCISES 3

Each of the graphs pictured below comes from a difference equation of the form $y_{n+1} = ay_n + b$. In Exercises 1–8, list all graphs that have the stated property.

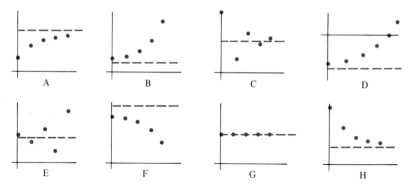

1. Monotonic   2. Increasing   3. Unbounded   4. Constant

5. Repelled from $y = b/(1 - a)$ 6. Decreasing

7. $|a| < 1$ 8. $a < 0$

In Exercises 9–14, sketch a graph having the given characteristics.

9. $y_0 = 4$, monotonic, repelled from $y = 2$

10. $y_0 = 5$, monotonic, attracted to $y = -3$

11. $y_0 = 2$, oscillating, attracted to $y = 6$

12. $y_0 = 7$, monotonic, repelled from $y = 10$

13. $y_0 = -2$, monotonic, attracted to $y = 5$

14. $y_0 = 1$, oscillating, repelled from $y = 2$

Exercises 15–20, make a rough sketch of the graph of the difference equation without generating terms or solving the difference equation.

15. $y_{n+1} = 3y_n + 4, y_0 = 0$

16. $y_{n+1} = .3y_n + 3.5, y_0 = 2$

17. $y_{n+1} = .5y_n + 3, y_0 = 6$

18. $y_{n+1} = 4y_n - 18, y_0 = 5$

19. $y_{n+1} = -2y_n + 12, y_0 = 5$

20. $y_{n+1} = -.6y_n + 6.4, y_0 = 1$

21. A particular news item was broadcast regularly on radio and TV. Let y_n be the number of people who had heard the news within n hours after broadcasting began. Sketch the graph of y_n, assuming that it satisfies the difference equation, $y_{n+1} = .7y_n + 3000, y_0 = 0$.

22. The radioactive element strontium 90 emits particles and slowly decays. Let y_n be the amount left after n years. Then y_n satisfies the difference equation $y_{n+1} = .98y_n$. Sketch the graph of y_n if initially there is 10 milligrams of strontium 90.

23. Laws of supply and demand cause the price of oats to fluctuate from year to year. Suppose that the current price is $1.25 per bushel and that the price n years from now, p_n, satisfies the difference equation $p_{n+1} = -.6p_n + 1.6$. (Prices are assumed to have been adjusted for inflation.) Sketch the graph of p_n.

24. Under ideal conditions a bacteria population satisfies the difference equation $y_{n+1} = 1.4y_n, y_0 = 1$, where y_n is the size of the population (in millions) after n hours. Sketch a graph which shows the growth of the population.

25. Suppose that the interest rate on a mortgage is 9% compounded monthly. If you can afford to pay $450 per month, how much money can you borrow?

26. A municipal government can take out a long-term construction loan at 8% interest compounded quarterly. Assuming that it can pay back $100,000 per quarter, how much money can it borrow?

27. A person makes an initial deposit into a savings account paying 6% interest compounded annually. He plans to withdraw $120 at the end of each year.

(a) Find the difference equation for y_n, the amount after n years.

(b) How large must y_0 be such that the money will not run out?

28. Suppose that a loan of $10,000 is to be repaid at $120 per month and that the annual interest rate is 12% compounded monthly. Then the interest for the first month is .01 · (10,000) or $100. The $120 paid at the end of the first month can be thought of as paying the $100 interest and paying $20 toward the reduction of the loan. Therefore, the balance after 1 month is $10,000 − $20 = $9980. How much of the $120 paid at the end of the second month goes for interest and how much is used to reduce the loan? What is the balance after 2 months?

SOLUTIONS TO PRACTICE PROBLEMS 3

1. $\dfrac{b}{1-a} = \dfrac{14.4}{1-(.1)} = \dfrac{14.4}{.9} = \dfrac{144}{9} = 16$. $y_0 = 100$, which is greater than $b/(1-a)$. Since $a = .1$ is positive, the terms y_n are monotonic. Since $|a| = .1 < 1$, the terms are attracted to the line $y = 16$. Now, plot $y_0 = 100$, draw the line $y = 16$, and write in the words MONO and ATTRACT [Fig. 14(a)]. Since the terms are attracted to the line monotonically they must move downward and asymptotically approach the line [Fig. 14(b)].

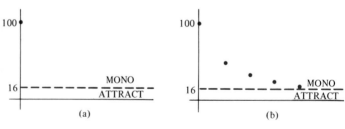

FIGURE 14

(Notice that the terminal speed, 16 feet per second, does not depend on the speed of the parachutist when the parachute is opened.)

2. Everything is the same as above except that now $y_0 < 16$. The speed increases to a terminal speed of 16 feet per second.

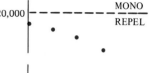

3. (a) $i = .06/12 = .005$.

$$y_{n+1} = y_n + \text{(interest)} - \text{(withdrawal)}$$

$$= y_n + .005y_n - 100$$

$$= (1.005)y_n - 100.$$

(b) $\dfrac{b}{1-a} = \dfrac{-100}{1-(1.005)} = \dfrac{-100}{-.005} = 20{,}000$. Since $a = 1.005 > 0$, the graph is monotonic. Since $|a| = 1.005 > 1$, the graph is repelled from the line $y = 20{,}000$.

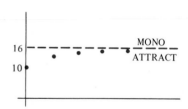

If $y_0 < 20{,}000$, then the amount of money in the account decreases and eventually runs out.

If $y_0 = 20{,}000$, then the amount of money in the account stays constant at 20,000.

If $y_0 > 20{,}000$, then the amount of money in the account grows steadily and is unbounded.

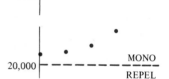

Answer: $y_0 \geq 20{,}000$.

11.4. Mathematics of Personal Finance

In this section we apply the theory of difference equations to the study of mortgages and annuities.

Mortgages   Most families take out bank loans to pay for a new house. Such a loan, called a *mortgage*, is used to purchase the house and then is repaid with interest in monthly installments over a number of years, usually 25 or 30. The monthly payments are computed so that after exactly the correct length of time the unpaid balance* is 0 and the loan is thereby paid off.

Mortgages can be described by difference equations, as follows. Let y_n be the unpaid balance on the mortgage after n months. In particular, y_0 is the initial amount borrowed. Let i denote the monthly interest rate and R the monthly mortgage payment. Then

$$\begin{bmatrix} \text{balance after} \\ n+1 \text{ months} \end{bmatrix} = \begin{bmatrix} \text{balance after} \\ n \text{ months} \end{bmatrix} + \begin{bmatrix} \text{interest for} \\ \text{month} \end{bmatrix} - [\text{payment}]$$

$$y_{n+1} \qquad = \qquad y_n \qquad + \qquad iy_n \qquad - \qquad R$$

$$y_{n+1} = (1+i)y_n - R. \tag{1}$$

EXAMPLE 1   Suppose that you can afford to pay $300 per month and the yearly interest rate is 9% compounded monthly. Exactly how much can you borrow if the mortgage is to be paid off in 30 years?

Solution   The monthly interest rate is $.09/12 = .0075$, so equation (1) becomes in this case

$$y_{n+1} = 1.0075y_n - 300.$$

Further, we are given that the mortgage runs for 30 years, or 360 months. Thus

$$y_{360} = 0. \tag{2}$$

Our problem is to determine y_0, the amount of the loan. From our general theory we first compute $b/(1-a)$:

$$\frac{b}{1-a} = \frac{-300}{1-(1.0075)} = 40,000.$$

The formula for y_{360} is then given by

$$y_{360} = \frac{b}{1-a} + \left(y_0 - \frac{b}{1-a} \right) a^{360}$$

$$= 40,000 + (y_0 - 40,000)(1.0075)^{360}.$$

---

* The balance after n months is the amount that would have to be paid at that time in order to retire the debt.

Using a calculator, we find that

$$(1.0075)^{360} \approx 14.73.$$

Therefore,

$$y_{360} = 40{,}000 + (y_0 - 40{,}000)(14.73).$$

However, by equation (2), $y_{360} = 0$, so that

$$0 = 40{,}000 + (y_0 - 40{,}000)(14.73)$$

$$= 14.73 y_0 - 549{,}200$$

$$y_0 = \frac{549{,}200}{14.73} = 37{,}284.45.$$

Thus the initial amount—that is, the amount which can be borrowed—is $37,284.45.

In the early months almost all of the monthly payment goes toward interest. With each passing month, however, the amount of interest declines and the amount applied to reducing the debt increases. This phenomenon is illustrated for selected months in Table 1.

TABLE 1

	$n = 0$	$n = 1$	$n = 2$	$n = 120$	$n = 240$	$n = 348$
Balance on loan	37,284.45	37,264.08	37,243.56	33,343.22	23,681.85	3429.00
Interest	279.63	279.48	279.33	250.07	177.61	25.72
Reduction of debt	20.37	20.52	20.67	49.93	122.39	274.28

EXAMPLE 2 Suppose that we have a 30-year mortgage for $7000 at 12% interest, compounded monthly. Find the monthly payment.

Solution Here the monthly interest rate i is just $.12/12 = .01$ and $y_0 = 7000$. Since the mortgage is for 30 years, we have

$$y_{360} = 0.$$

The difference equation for the mortgage is

$$y_{n+1} = 1.01 y_n - R, \qquad y_0 = 7000.$$

Now $b/(1 - a) = -R/(1 - 1.01) = R/.01 = 100R$. The formula for y_n is

$$y_n = 100R + (7000 - 100R)(1.01)^n.$$

Thus

$$0 = y_{360} = 100R + (7000 - 100R)(1.01)^{360}.$$

To complete the problem solve this equation for R. Using a calculator, $(1.01)^{360} \approx$ 35.95. Thus

$$100R + (7000 - 100R)(35.95) = 0$$

$$3495R = 251{,}650$$

$$R = \$72.00.$$

Annuities   The term *annuity* has several meanings. For our purposes an annuity is a bank account into which equal sums are deposited at regular intervals, either weekly, monthly, quarterly, or annually. The money draws interest and accumulates for a certain number of years, after which it becomes available to the investor. Annuities are often used to save for a child's college education or to generate funds for retirement.

The growth of money in an annuity can be described by a difference equation. Let $y_n = $ [the amount of money in the annuity after n time periods], $i = $ [the interest rate per time period], $D = $ [the deposit per time period]. Then $y_0 = $ [the amount after 0 time periods] $ = 0$.* Moreover,

$$y_{n+1} = \text{[previous amount]} + \text{[interest]} + \text{[deposit]}$$

$$= \qquad y_n \qquad + \quad iy_n \quad + \quad D$$

$$= (1 + i)y_n + D.$$

This last equation is the difference equation of the annuity.

EXAMPLE 3   Suppose that $20 is deposited into an annuity at the end of every quarter-year and that interest is earned at the annual rate of 8%, compounded quarterly. How much money is in the annuity after 10 years?

Solution   Since 10 years $= 40$ quarters, the problem is to compute y_{40}. Now $D = 20$ and $i = .08/4 = .02$. So the difference equation reads

$$y_{n+1} = 1.02y_n + 20.$$

In this case

$$\frac{b}{1 - a} = \frac{20}{1 - 1.02} = -1000.$$

Thus

$$y_n = -1000 + [0 - (-1000)](1.02)^n$$

$$= -1000 + 1000(1.02)^n.$$

In particular, setting $n = 40$,

$$y_{40} = -1000 + 1000(1.02)^{40}.$$

---

* One model for the situation $y_0 = 0$ is a payroll savings plan, where a person signs up at time zero and has the first deduction made at the end of the next pay period.

Using a calculator, $(1.02)^{40} \approx 2.21$. Thus

$$y_{40} = -1000 + 1000(2.21)$$
$$= -1000 + 2210$$
$$= 1210.$$

Thus after 10 years the account contains $1210.

EXAMPLE 4 How much money must be deposited at the end of each quarter into an annuity at 8% interest compounded quarterly in order to have $10,000 after 15 years?

Solution Here D is unknown. But $i = .02$. Also, 15 years $= 60$ quarters, so that $y_{60} = 10,000$. The difference equation in this case reads

$$y_{n+1} = 1.02y_n + D \qquad y_0 = 0.$$

Then

$$\frac{b}{1-a} = \frac{D}{1-1.02} = \frac{D}{-.02} = -50D$$

and

$$y_n = -50D + (0 - (-50D))(1.02)^n$$
$$= -50D + 50D(1.02)^n.$$

Set $n = 60$. Then

$$y_{60} = -50D + 50D(1.02)^{60}.$$

However, from the statement of the problem $y_{60} = 10,000$. Thus we have the equation

$$10,000 = -50D + 50D(1.02)^{60},$$

to be solved for D. But $(1.02)^{60} \approx 3.28$, so that

$$-50D + 50(3.28)D = 10,000$$
$$-50D + 164D = 10,000$$
$$114D = 10,000$$
$$D = 10,000/114$$
$$= \$87.72.$$

PRACTICE PROBLEMS 4

1. Suppose that you deposit $650,000 into a bank account paying 5% interest compounded annually and you withdraw $50,000 at the end of each year. Find a difference equation for y_n, the amount in the account after n years.

2. Refer to Problem 1. How much money will be in the account after 20 years? [*Note:* $(1.05)^{20} \approx 2.65.$]

3. Refer to Problem 1. Assume that the money is tax free and that you could earn 5% interest compounded annually. Would you rather have $650,000 now or $50,000 a year for 20 years?

EXERCISES 4

In Exercises 1–4, give the difference equation for y_n, the amount (or balance) after n interest periods.

1. A mortgage loan of $32,500 at 9% interest compounded monthly and having monthly payments of $261.50.

2. A bank deposit of $1000 at 6% interest compounded semiannually.

3. An annuity for which $4000 is deposited into an account at 6% interest compounded quarterly and $200 is added to the account at the end of each quarter.

4. A bank account into which $20,000 is deposited at 6% interest compounded monthly and $100 is withdrawn at the end of each month.

5. How much money can you borrow at 12% interest compounded monthly if the loan is to be paid off in monthly installments for 10 years and you can afford to pay $660 per month? [*Note:* $(1.01)^{120} \approx 3.3.$]

6. Find the monthly payment on a $38,000, 25-year mortgage at 12% interest compounded monthly. [*Note:* $(1.01)^{300} \approx 20.$]

7. Find the amount accumulated after 20 years if, at the end of each year, $300 is deposited into an account paying 6% interest compounded annually. [*Note:* $(1.06)^{20} \approx 3.2.$]

8. How much money would you have to deposit at the end of each month into an annuity paying 6% interest compounded monthly in order to have $6000 after 12 years? [*Note:* $(1.005)^{144} \approx 2.$]

9. How much money would you have to put into an account initially at 8% interest compounded quarterly in order to have $6000 after 14 years? [*Note:* $(1.02)^{56} \approx 3.$]

10. How much money would you have to put into a bank account paying 6% interest compounded monthly in order to be able to withdraw $150 each month for 30 years? [*Note:* $(1.005)^{360} \approx 6.$]

11. In order to buy a car, a person borrows $4000 from the bank at 12% interest compounded monthly. The loan is to be paid off in 3 years with equal monthly payments. What will the monthly payments be? [*Note:* $(1.01)^{36} \approx 1.43.$]

12. How much money would you have to deposit at the end of each month into an annuity paying 6% interest compounded monthly in order to have $1620 after 4 years? [*Note:* $(1.005)^{48} \approx 1.27.$]

1. [Amount after $n + 1$ years] = [amount after n years] + [interest for $(n + 1)$st year] − [withdrawal at end of year].

$$y_{n+1} = y_n + .05y_n - 50{,}000$$

$$= (1.05)y_n - 50{,}000, \qquad y_0 = 650{,}000.$$

2. Solve the difference equation and set $n = 20$.

$$\frac{b}{1 - a} = \frac{-50{,}000}{1 - (1.05)} = \frac{-50{,}000}{-.05} = 1{,}000{,}000,$$

$$y_n = 1{,}000{,}000 + (650{,}000 - 1{,}000{,}000)(1.05)^n$$

$$= 1{,}000{,}000 - 350{,}000(1.05)^n,$$

$$y_{20} \approx 1{,}000{,}000 - 350{,}000(2.65) = 72{,}500.$$

3. \$650,000. According to Problem 2, this money could be deposited into the bank. You could take out \$50,000 per year and still have \$72,500 left over after 20 years.

11.5. Modeling with Difference Equations

In this section we show how difference equations may be used to build mathematical models of a number of phenomena. Since the difference equation describing a situation contains as much data as the formula for y_n, we shall regard the difference equation itself as a mathematical model. We demonstrate in this section that the difference equation and a sketch of its graph can often be of more use than an explicit solution.

The concept of proportionality will be needed to develop some of these models.

Proportionality To say that two quantities are proportional is the same as saying that one quantity is equal to a constant times the other quantity. For instance, in a state having a 4% sales tax the sales tax on an item is proportional to the price of the item, since

$$[\text{sales tax}] = .04[\text{price}].$$

Here the constant of proportionality is .04. In general, for two proportional quantities we have

$$[\text{first quantity}] = k[\text{second quantity}],$$

where k is some fixed constant of proportionality. Note that if both quantities are positive and if the first quantity is known to always be smaller than the second, then k is a positive number less than 1; that is, $0 < k < 1$.

EXAMPLE 1   (*Radioactive Decay*)   Certain forms of natural elements, such as uranium 238, strontium 90, and carbon 14, are radioactive. That is, they decay, or dissipate, over a period of time. Physicists have found that this process of decay obeys the following law: Each year the amount which decays is proportional to the amount present at the start of the year. Construct a mathematical model describing radioactive decay.

Solution   Let y_n = the amount left after n years. The physical law states that

$$[\text{amount that decays in year } n + 1] = k \cdot y_n,$$

where $k \cdot y_n$ represents a constant of proportionality times the amount present at the start of that year. Therefore,

$$y_{n+1} \quad = \quad y_n \quad - \quad ky_n$$

$$\begin{bmatrix} \text{amount after} \\ \text{year } n + 1 \end{bmatrix} = \begin{bmatrix} \text{amount after} \\ \text{year } n \end{bmatrix} - \begin{bmatrix} \text{amount that decays} \\ \text{in year } n + 1 \end{bmatrix}$$

or

$$y_{n+1} = (1 - k)y_n,$$

where k is a constant between 0 and 1. This difference equation is the mathematical model for radioactive decay.

EXAMPLE 2   Experiment shows that for cobalt 60 (a radioactive form of cobalt used in cancer therapy) the constant k above is given by $k = .12$.

(a) Write the difference equation for y_n in this case.

(b) Sketch the graph of the difference equation.

Solution   (a) Setting $k = .12$ in the result of the preceding example, we get

$$y_{n+1} = (1 - .12)y_n$$

or

$$y_{n+1} = .88y_n.$$

(b) Since .88 is positive and less than 1, the graph is monotonic and attracted to the line $y = b/(1 - a) = 0/(1 - .88) = 0$. The graph is sketched in Fig. 1.

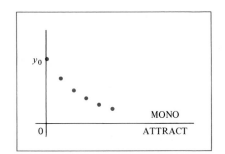

MONO

ATTRACT

FIGURE 1

EXAMPLE 3   (*Growth of Bacteria*)   A bacteria culture grows in such a way that each hour the increase in the number of bacteria in the culture is proportional to the total number present at the beginning of the hour. Sketch a graph depicting the growth of the culture.

Solution Let y_n = the number of bacteria present after n hours. Then the increase during the next hour is ky_n, where k is a positive constant of proportionality. Therefore,

$$y_{n+1} \quad = \quad y_n \quad + \quad ky_n$$

$$\begin{bmatrix} \text{number after} \\ n+1 \text{ hours} \end{bmatrix} = \begin{bmatrix} \text{number after} \\ n \text{ hours} \end{bmatrix} + \begin{bmatrix} \text{increase during} \\ (n+1)\text{st hour} \end{bmatrix},$$

so

$$y_{n+1} = (1+k)y_n.$$

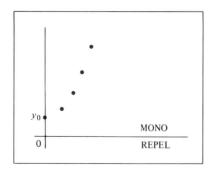

MONO

REPEL

FIGURE 2

This last equation gives a mathematical model of the growth of the culture. Since $a = 1 + k > 0$ and $|a| > 1$, the graph is monotonic and repelled from $b/(1 - a) = 0$. Thus the graph is as drawn in Fig. 2.

EXAMPLE 4 (*Spread of Information*) Suppose that at 8 A.M. on a Saturday the local radio and TV stations in a town start broadcasting a certain piece of news. The number of people learning the news each hour is proportional to the number who had not yet heard it by the end of the preceding hour.

(a) Write down a difference equation describing the spread of the news through the population of the town.

(b) Sketch the graph of this difference equation for the case where the constant of proportionality is .3 and the population of the town is 50,000.

Solution The example states that two quantities are proportional. The first quantity is the number of people learning the news each hour and the second is the number who have not yet heard it by the end of the preceding hour.

(a) Let y_n = the number of people who have heard the news after n hours. The terms $y_0, y_1, y_2, \ldots$ are increasing.

Let P = the total population of the town. Then the number of people who have not yet heard the news after n hours is $P - y_n$. Thus the assumption can be stated in the mathematical form

$$y_{n+1} \quad = \quad y_n \quad + \quad k(P - y_n)$$

$$\begin{bmatrix} \text{number who know} \\ \text{after } n+1 \text{ hours} \end{bmatrix} = \begin{bmatrix} \text{number who know} \\ \text{after } n \text{ hours} \end{bmatrix} + \begin{bmatrix} \text{number who learn} \\ \text{during } (n+1)\text{st hour} \end{bmatrix},$$

where k is a constant of proportionality. The constant k tells how fast the news is traveling. It measures the percentage of the uninformed population which hears the news each hour. It is clear that $0 < k < 1$.

(b) For part (b) we assume that $k = .3$ and that the town has a population of 50,000. Further, measure the number of people in thousands. Then $P = 50$

and the mathematical model of the spread of the news is

$$y_{n+1} = y_n + .3(50 - y_n)$$
$$= y_n + 15 - .3y_n$$
$$= .7y_n + 15.$$

Suppose that initially no one had heard the news, so that

$$y_0 = 0.$$

Now we may sketch the graph:

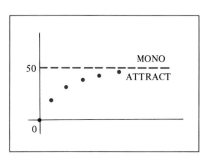

FIGURE 3

$$\frac{b}{1 - a} = \frac{15}{1 - .7} = \frac{15}{.3} = 50.$$

Since $a = .7$, the graph is monotonic and attracted to the line $y = 50$ (Fig. 3). Note that the number of people who hear the news increases and approaches the population of the entire town. The increases between consecutive terms are large at first, then become smaller. This coincides with the intuitive impression that the news spreads rapidly at first, then progressively slower.

EXAMPLE 5 *(Supply and Demand)* This year's level of production and price for most agricultural products greatly affects the level of production and price next year. Suppose that the current crop of soybeans in a certain country is 80 million bushels. Let q_n denote the quantity* of soybeans grown n years from now, and let p_n denote the market price* in n years. Suppose experience has shown that q_n and p_n are related by the following equations:

$$p_n = 20 - .1q_n$$
$$q_{n+1} = 5p_n - 10.$$

Draw a graph depicting the changes in production from year to year.

Solution What we seek is the graph of a difference equation for q_n.

FIGURE 4

$$q_{n+1} = 5p_n - 10$$
$$= 5(20 - .1q_n) - 10$$
$$= 100 - .5q_n - 10$$
$$= -.5q_n + 90.$$

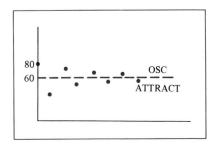

This is a difference equation for q_n with $a = -.5$, $b = 90$, $b/(1 - a) = 90/(1 - (-.5)) = 60$. Since a is negative, the graph is oscillating. Since $|a| = .5 < 1$, the graph is attracted to the line $y = b/(1 - a) = 60$. The initial condition is $q_0 = 80$. So the graph is as drawn in Fig. 4. Note that the graph oscillates

---

* q_n in terms of millions of bushels, p_n in terms of dollars per bushel.

and approaches 60. That is, the crop size fluctuates from year to year and eventually gets close to 60 million bushels.

PRACTICE PROBLEMS 5

1. (*Glucose Infusion*) Glucose is being given to a patient intravenously at the rate of 100 milligrams per minute. Let y_n be the amount of glucose in the blood after n minutes, measured in milligrams. Suppose that each minute the body takes from the blood 2% of the amount of glucose present at the beginning of that minute. Find a difference equation for y_n and sketch its graph. [*Hint:* Each minute the amount of glucose in the blood is increased by the intravenous infusion and decreased by the absorption into the body.]

2. (*Light at Ocean Depths*) Sunlight is absorbed by water and so, as one descends into the ocean, the intensity of light diminishes. Suppose that at each depth, going down one more meter causes a 20% decrease in the intensity of the sunlight. Find a difference equation for y_n, the intensity of the light at a depth of n meters, and sketch its graph.

EXERCISES 5

1. In a certain country with current population 100 million, each year the number of births is 3% and the number of deaths 1% of the population at the beginning of the year. Find the difference equation for y_n, the population after n years. Sketch its graph.

2. A small city with current population 50,000 is experiencing an emigration of 600 people each year. Assuming that each year the increase in population due to natural causes is 1% of the population at the start of that year, find the difference equation for y_n, the population after n years. Sketch its graph.

3. After a certain drug is injected, each hour the amount removed from the bloodstream by the body is 25% of the amount in the bloodstream at the beginning of the hour. Find the difference equation for y_n, the amount in the bloodstream after n hours, and sketch the graph.

4. The atmospheric pressure at sea level is 14.7 pounds per square inch. Suppose that at any elevation an increase of 1 mile results in a decrease of 20% of the atmospheric pressure at that elevation. Find the difference equation for y_n, the atmospheric pressure at elevation n miles, and sketch its graph.

5. A sociological study* was made to examine the process by which doctors decide to adopt a new drug. Certain doctors who had little interaction with other physicians were called "isolated." Out of 100 isolated doctors, each month the number who adopted the new drug that month was 8% of those who had not yet adopted the drug at the beginning of the month. Find a difference equation for y_n, the number of isolated physicians using the drug after n months, and sketch its graph.

6. A cell is put into a fluid containing an 8 milligrams/liter concentration of a solute. (This concentration stays constant throughout.) Initially, the concentration of the solute in the cell is 3 milligrams/liter. The solute passes through the cell membrane at such a rate

---

* James S. Coleman, Elihu Katz, and Herbert Menzel, "The Diffusion of an Innovation Among Physicians," *Sociometry*, 20 (1957), 253–270.

that each minute the increase in concentration in the cell is 40% of the difference between the outside concentration and the inside concentration. Find the difference equation for y_n, the concentration of the solute in the cell after n minutes, and sketch its graph.

7. Psychologists have found that in certain learning situations in which there is a maximum amount which can be learned, the additional amount learned each minute is proportional to the amount yet to be learned at the beginning of that minute. Let 12 units of information be the maximum amount that can be learned and let the constant of proportionality be 30%. Find a difference equation for y_n, the amount learned after n minutes, and sketch its graph.

8. Consider two genes A and a in a population, where A is a dominant gene and a is a recessive gene controlling the same genetic trait. (That is, A and a belong to the same locus.) Suppose that initially 80% of the genes are A and 20% are a. Suppose that in each generation .003% of genes A mutate to gene a. Find a difference equation for y_n, the percentage of genes a after n generations, and sketch its graph. [*Note:* The percentage of genes A $= 1 -$ (the percentage of genes a).]

9. Thirty thousand dollars is deposited in a savings account paying 5% interest compounded annually, and $1000 is withdrawn from the account at the end of each year. Find the difference equation for y_n, the amount in the account after n years, and sketch its graph.

10. Rework Exercise 9 where $15,000 is deposited initially.

11. When a cold object is placed in a warm room, each minute its increase in temperature is proportional to the difference between the room temperature and the temperature of the object at the beginning of the minute. Suppose that the room temperature is 70°F, the initial temperature of the object is 40°F, and the constant of proportionality is 20%. Find the difference equation for y_n, the temperature of the object after n minutes, and sketch its graph.

12. Suppose that the annual amount of electricity used in the United States will increase at a rate of 7% each year and that this year 2.6 trillion kilowatt-hours are being used. Find a difference equation for y_n, the number of kilowatt-hours to be used during the year that is n years from now, and sketch its graph.

13. Suppose that in Example 5 the current price of soybeans is $10 per bushel. Find the difference equation for p_n and sketch its graph. [*Note:* Since $p_n = 20 - .1q_n$ holds for each year, it holds for the $(n + 1)$st year. That is, $p_{n+1} = 20 - .1q_{n+1}$.]

SOLUTIONS TO PRACTICE PROBLEMS 5

1. The amount of glucose in the blood is affected by two factors. It is being increased by the steady infusion of glucose; each minute the amount is increased by 100 milligrams. On the other hand, the amount is being decreased each minute by $.02y_n$. Therefore,

$$y_{n+1} \quad = \quad y_n \quad + \quad 100 \quad - \quad .02y_n$$

$$\begin{bmatrix} \text{amount after} \\ n + 1 \text{ minutes} \end{bmatrix} = \begin{bmatrix} \text{amount after} \\ n \text{ minutes} \end{bmatrix} + \begin{bmatrix} \text{increase due} \\ \text{to infusion} \end{bmatrix} - \begin{bmatrix} \text{amount taken} \\ \text{by the body} \end{bmatrix}$$

or

$$y_{n+1} = .98y_n + 100.$$

In order to sketch the graph, first compute $b/(1 - a)$.

$$\frac{b}{1 - a} = \frac{100}{1 - .98} = \frac{100}{.02} = 5000.$$

The amount of glucose in the blood rises and approaches 5000 milligrams.

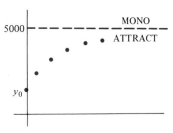

2. The change in intensity in going from depth n meters to $n + 1$ meters is 20% of the intensity at n meters. Therefore,

$$y_{n+1} = y_n - .20y_n$$

$$\begin{bmatrix} \text{intensity at} \\ n + 1 \text{ meters} \end{bmatrix} = \begin{bmatrix} \text{intensity at} \\ n \text{ meters} \end{bmatrix} - \begin{bmatrix} \text{change in} \\ \text{intensity} \end{bmatrix}$$

$$y_{n+1} = .8y_n.$$

Thus $b/(1 - a) = 0$, and the graph is as follows.

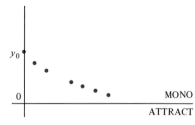

Chapter 11: CHECKLIST

☐ Difference equation
☐ Initial value
☐ Graph of difference equation
☐ Solution of difference equation for $a \neq 1$
☐ Solution of difference equation for $a = 1$
☐ Simple interest
☐ Compound interest
☐ Increasing and decreasing graphs
☐ Monotonic and oscillating graphs
☐ Constant graph
☐ Technique for sketching graph
☐ Mortgage
☐ Annuity

Chapter 11: SUPPLEMENTARY EXERCISES

1. Consider the difference equation $y_{n+1} = -3y_n + 8, y_0 = 1$.
   (a) Generate y_1, y_2, y_3 from the difference equation.
   (b) Solve the difference equation.
   (c) Use the solution in part (b) to obtain y_4.

2. Consider the difference equation $y_{n+1} = y_n - \frac{3}{2}, y_0 = 10$.
   (a) Generate y_1, y_2, y_3 from the difference equation.
   (b) Solve the difference equation.
   (c) Use the solution in part (b) to obtain y_6.

3. How much money would you have to put into a savings account initially at 8% interest compounded quarterly in order to have $6600 after 10 years? [*Note:* $(1.08)^{10} \approx 2.16$; $(1.08)^{40} \approx 21.7$; $(1.02)^{10} \approx 1.2$; $(1.02)^{40} \approx 2.2$.]

4. How much money would you have in the bank after 2 years if you deposited $1000 at 5.2% interest compounded weekly? [*Note:* $(1.052)^2 \approx 1.107$; $(1.001)^{104} \approx 1.110$; $(1.12)^2 \approx 1.254$; $(1.02)^{52} \approx 2.800$.]

5. Make a rough sketch of the graph of the difference equation $y_{n+1} = -\frac{1}{3}y_n + 8$, $y_0 = 10$ without generating terms.

6. Make a rough sketch of the graph of the difference equation $y_{n+1} = 1.5y_n - 2$, $y_0 = 5$ without generating terms.

7. The population of a certain city is currently 120,000. The growth rate of the population due to an excess of births to deaths is 3% per year. Furthermore, each year 600 people are moving out of the city. Let y_n be the population after n years.

   (a) Find the difference equation for the population growth.

   (b) What will be the population after 20 years? [*Note:* $(1.03)^{20} \approx 1.8$.]

8. The monthly payment on a $35,000 30-year mortgage at 12% interest compounded monthly is $360. Let y_n be the unpaid balance of the mortgage after n months.

   (a) Give the difference equation expressing y_{n+1} in terms of y_n. Also, give y_0.

   (b) What is the unpaid balance after seven years? [*Note:* $(1.12)^7 \approx 2.21$; $(1.12)^{84} \approx 13,624$; $(1.01)^7 \approx 1.07$; $(1.01)^{84} \approx 2.3$.]

9. How much money must be deposited at the end of each week into an annuity at $5\frac{1}{5}\%$ ($= .052$) interest compounded weekly in order to have $40,000 after 21 years? [*Note:* $(1.001)^{21} \approx 1.02$; $(1.001)^{1092} \approx 3$.]

10. Find the monthly payment on a $33,100 20-year mortgage at 6% interest compounded monthly. [*Note:* $(1.005)^{240} \approx 3.31$.]

11. How much money can you borrow at 8% interest compounded annually if the loan is to be paid off in yearly installments for 18 years and you can afford to pay $2400 per year? [*Note:* $(1.08)^{18} \approx 4$.]

12. A college alumnus pledges to give his alma mater $5 at the end of each month for 4 years. If the college puts the money in a savings account earning 6% interest compounded monthly, how much money will be in the account at the end of 4 years? [*Note:* $(1.06)^{48} \approx 16.39$; $(1.06)^4 \approx 1.26$; $(1.005)^{48} \approx 1.27$; $(1.005)^4 \approx 1.02$.]

13. An unknown candidate for governor in a state having 1,000,000 voters mounts an extensive media campaign. Each day, 10% of the voters who do not yet know about him become aware of his candidacy. Let y_n be the number of voters who are aware of his candidacy after n days. Find a difference equation for y_n and sketch its graph.

14. Suppose that 100 people were just elected to a certain state legislature and that after each term 8% of those still remaining from this original group will either retire or not be reelected. Let y_n be the number of legislators from the original group of 100 who are still serving after n terms. Find a difference equation for y_n and sketch its graph.

12

LOGIC

12.1 Introduction to Logic

12.2 Truth Tables

12.3 Implication

12.4 Logical Implication and Equivalence

12.5 Valid Argument

12.6 Predicate Calculus

12.1. Introduction to Logic

In this chapter we introduce logic: the foundation for mathematics, coherent argument, and computing. The idea of logical argument was introduced in writing by Aristotle. For centuries his principles have formed the basis for systematic thought, communication, debate, law, mathematics, and science. More recently, the concept of a computing machine and the essence of programming are applications of the ancient ideas we present here. Also, new developments in mathematical logic have helped to promote significant advances in artificial intelligence.

The building blocks of logic are statements, connectives, and the rules for calculating the truth or falsity of compound statements. We begin with statements.

A *statement* is a declarative sentence whose truth can be determined.

Here are some examples of statements:

1. George Washington was the first president of the United States.
2. The New York Knicks won the NBA basketball championship in 1989.
3. The number of atoms in the universe is 10^{75}.
4. Ronald Reagan knew what Ollie North was doing.
5. If x is a positive real number and $x^2 = 9$, then $x = 3$.

We know that statement 1 is true. Statement 2 is false. Statements 3 and 4 are either true or false, but we do not know which. Neither is both true *and* false. Statement 5 is true.

To better understand what a statement is, we list below some nonstatements.

6. He is a real nice guy.
7. Do your homework!
8. If $x^2 = 9$, then $x = 3$.
9. Is this course fun?

Items 6 and 8 are not statements because the person "he" and the variable x are not specified. There are circumstances under which we would claim that item 6 is true and others under which we would claim that item 6 is false; clearly, it depends on who "he" is. Likewise, if x were a positive real number, then item 8 would be true. If x were -3, then item 8 would be false. Items 7 and 9 are not declarative sentences.

The *truth value* of a statement is either *true* or *false*. The problem of deciding the truth value of a declarative sentence will be dealt with in a simple manner, although it is far from simple. The assignment of a truth value to a statement may be obvious, as in the statement "George Washington was the first president of the United States." The truth value of the statement "The number of the atoms in the universe is 10^{75}," on the other hand, is much more problematic. Logic cannot be used to determine the truth value of a simple statement. However, if there is agreement on the truth value of simple statements that are components of a compound statement, the rules of mathematical logic determine the truth value of the compound statement.

Logic forms the basis for analyzing legal briefs, political rhetoric, and family discussions. It allows us to understand another's point of view as well as to expose weaknesses in an argument that lead to unfounded conclusions. We combine statements naturally when we speak, using connective words like "and" and "or." We state implications with the words "if" and "then." Frequently, we negate statements with the word "not." A *compound statement* is formed by combining statements using the words "and," "or," "not," or "if, then." A *simple statement* is a statement that is not a compound statement. A compound statement is analyzed as a combination of simple statements. Its truth value is determined according to mathematical rules given in the next sections.

EXAMPLE 1   Give the simple statements in each of the following compound statements.

(a) The number 6 is even and the number 5 is odd.

(b) Tom Jones does a term paper or takes the final exam.

(c) If England is in the Common Market, then the British eat Spanish oranges and Italian melons.

Solution   (a) The simple statements are: "The number 6 is even" and "The number 5 is odd."

(b) The simple statements are: "Tom Jones does a term paper" and "Tom Jones takes the final exam."

(c) The simple statements are: "England is in the Common Market," "The British eat Spanish oranges," and "The British eat Italian melons."

It is sometimes helpful to add English words to clarify the implied meaning in simple statements.

To be able to develop the rules of logic and logical argument, we need to deal with any logical statement, rather than specific examples. We use the letters p, q, r, and so on to represent simple statements. They are not really the statements themselves, but variables for which a statement may be substituted. For example, we can let p represent a statement in general. Then, if we wish, we can specify that for the moment p will represent the statement "The number 5 is odd." In this particular case, p has the truth value *true*. We might decide to let p represent the statement "There are 13 states in the United States today." Then the truth value of the statement p is *false*.

The use of these logical variables is similar to the use of variables x, y, z, and so on, to represent unspecified numbers in algebra. In algebra, we manipulate the symbols (such as x, y, $+$, and the implied multiplication and exponentiation) in expressions according to specified rules to establish identities such as

$$(x + y)^2 = x^2 + 2xy + y^2,$$

which are true no matter which numbers are substituted for x and y. In logic, we will manipulate the symbols in compound statements (p, q, r, "and," "or," "not," and "if, then") to look for expressions that are true no matter which statements are substituted for p, q, r, and so on.

It is useful to be able to write a compound statement in terms of its component parts and to use symbols to represent the connectives. We use $\wedge$ to represent the word "and" and $\vee$ to represent the word "or."

EXAMPLE 2   Write the compound statement "Ina likes popcorn and Fred likes peanuts" in symbolic form.

Solution   We first let p represent the statement "Ina likes popcorn" and we let q represent the statement "Fred likes peanuts." We use the symbol $\wedge$ to represent the word "and." Thus, we represent our compound statement symbolically as $p \wedge q$.

EXAMPLE 3   Write the compound statement "The United States trades with Japan or Germany trades with Japan" in symbolic form.

Solution   We use the letter p to represent "The United States trades with Japan" and the letter q to represent the statement "Germany trades with Japan." We use the symbol $\vee$ to represent the word "or." Then the compound sentence is of the form $p \vee q$.

The symbol $\sim$ represents "not" so that, with p as defined in the previous example, we would use $\sim p$ to represent the statement "The United States does not trade with Japan."

EXAMPLE 4   Let p represent "Fred likes Cindy" and let q represent "Cindy likes Fred." Use connectives $\wedge$, $\vee$, and $\sim$ to represent the compound sentences given below.

(a) Fred and Cindy like each other.

(b) Fred likes Cindy but Cindy does not like Fred.

(c) Fred and Cindy dislike each other.

(d) Fred likes Cindy or Cindy likes Fred.

Solution   (a) "Fred and Cindy like each other" should be rewritten as "Fred likes Cindy and Cindy likes Fred." This is expressed symbolically as $p \wedge q$. Statement (b) can be written $p \wedge \sim q$. Note that the word "but" here means "and." Statement (c) can be rewritten as "Fred dislikes Cindy and Cindy dislikes Fred." This is written symbolically as $\sim p \wedge \sim q$. The last statement (d) is simply $p \vee q$.

EXAMPLE 5   Let p represent the statement "The interest rate is 10%" and let q be the statement "The Dow Jones average is over 2000." Write the English statements corresponding to each of the following.

(a) $p \vee q$ (b) $p \wedge q$ (c) $p \wedge \sim q$ (d) $\sim p \vee \sim q$

Solution   (a) The statement $p \vee q$ can be written, "The interest rate is 10% or the Dow Jones average is over 2000."

(b) We write $p \wedge q$ as "The interest rate is 10% and the Dow Jones average is over 2000."

(c) The statement $p \wedge q$ becomes "The interest rate is 10% and the Dow Jones average is less than or equal to 2000."

(d) The statement $\sim p \vee \sim q$ can be written as "The interest rate is not 10% or the Dow Jones average is less than or equal to 2000."

The symbol $\rightarrow$ represents an implication. The statement $p \rightarrow q$ is read "p implies q" or "if p, then q." Thus, using the representations of Example 5, the English sentence "If the interest rate is 10%, then the Dow Jones average is over 2000" can be written as $p \rightarrow q$.

EXAMPLE 6   Let p denote the statement "The train stops in Washington" and let q denote the statement "The train stops in New York." Write the following statements in symbolic form.

(a) The train stops in New York and Washington.

(b) The train stops in Washington but not in New York.

(c) The train does not stop in New York.

(d) The train stops in New York or Washington.

(e) The train stops in New York or Washington but not in both.

(f) If the train stops in New York, it does not stop in Washington.

Solution   (a) $p \wedge q$, (b) $p \wedge \sim q$, (c) Negate q to get $\sim q$, (d) $p \vee q$, and (e) $(p \vee q) \wedge \sim(p \wedge q)$. The parentheses make the statement clear. We discuss the need for them in a later section. (f) $q \rightarrow \sim p$.

How are truth values assigned to compound statements? The assignment depends on the truth values of the simple statements and the connectives in the compound form. The rules are discussed in the next section.

PRACTICE PROBLEMS 1

1. Determine which of the following sentences are statements.

(a) The earnings of IBM went up from 1989 to 1990.

(b) The national debt of the United States is $6 trillion.

(c) What an exam that was!

(d) Abraham Lincoln was the sixteenth president of the United States.

(e) Lexington is the capital of Kentucky or Albany is the capital of the United States.

(f) When was the Civil War?

2. Let p denote the statement "Sally is the class president" and let q denote "Sally is an accounting major." Translate the symbolic statements into proper English.

(a) $p \wedge q$ (b) $\sim p$ (c) $p \vee q$

(d) $(\sim p) \vee q$ (e) $p \wedge \sim q$ (f) $\sim p \vee \sim q$

EXERCISES 1

In Exercises 1–15, determine which sentences are statements.

1. The number 3 is even.

2. The 1939 World's Fair was held in Miami.

3. The price of coffee depends on the rainfall in Brazil.

4. The Nile River flows through Asia.

5. What a way to go!

6. If snow falls on the Rockies, people are skiing in Aspen.

7. Why is the sky blue?

8. Moisture in the atmosphere determines the type of cloud formation.

9. No aircraft carrier is assigned to the Indian Ocean.

10. The number of stars in the universe is 10^{60}.

11. $x + 3 \geq 0$.

12. He is a brave fellow.

13. Let us pray.

14. The Louvre and the Metropolitan Museum of Art contain paintings by Leonardo da Vinci.

15. If a United States coin is fair, the chance of getting a head is $\frac{1}{3}$.

16. Let p denote the statement "Arizona has the largest U.S. Indian population" and let q denote the statement "Arizona is the site of the O.K. Corral." Write out the following statements in proper English sentences.

(a) $\sim p$ (b) $\sim p \vee q$ (c) $\sim q \wedge p$

(d) $p \vee q$ (e) $\sim p \wedge \sim q$ (f) $\sim (p \vee q)$

17. Let p denote the statement "Ozone is opaque to ultraviolet light" and let q denote the statement "Life on earth requires ozone." Write out the following statements in proper English.

(a) $p \wedge q$ (b) $\sim p \vee q$ (c) $\sim p \vee \sim q$ (d) $\sim (\sim q)$

18. Let p denote the statement "Papyrus is the earliest form of paper" and let q denote "The papyrus reed is found in Africa." Put the following statements into symbolic form.

(a) Papyrus is not the earliest form of paper.

(b) The papyrus reed is not found in Africa or papyrus is not the earliest form of paper.

(c) Papyrus is the earliest form of paper and the papyrus reed is not found in Africa.

19. Let p denote the statement "Florida borders Alabama" and let q denote the statement "Florida borders Mississippi." Put the following into symbolic form.

(a) Florida borders Alabama or Mississippi.

(b) Florida borders Alabama but not Mississippi.

(c) Florida borders Mississippi but not Alabama.

(d) Florida borders neither Alabama nor Mississippi.

SOLUTIONS TO PRACTICE PROBLEMS 1

1. Statements appear in (a), (b), (d), and (e). Both (c) and (f) are not statements.

2. (a) "Sally is the class president and Sally is an accounting major."

(b) "Sally is not the class president."

(c) "Sally is the class president or Sally is an accounting major."

(d) "Sally is not the class president or Sally is an accounting major."

(e) "Sally is the class president and Sally is not an accounting major."

(f) "Sally is not the class president or Sally is not an accounting major."

12.2. Truth Tables

In this section, we discuss how the truth values of the statements $p \land q$, $p \lor q$, and $\sim p$ depend on the truth values of p and q.

The simple statements will be denoted p, q, r, and so on. A *statement form* is an expression formed from simple statements and connectives according to the following rules.

A simple statement is a statement form.
If p is a statement form, $\sim p$ is a statement form.
If p and q are statement forms, then so are $p \land q$, $p \lor q$, and $p \to q$.

EXAMPLE 1   Show that each of the following is a statement form according to the definition. Assume that p, q, and r are simple statements.

(a) $(p \land \sim q) \to r$ (b) $\sim (p \to (q \lor \sim r))$

Solution   (a) Since p, q, and r are simple statements, they are statement forms. Since q is a statement form, so is $\sim q$. Then $p \land \sim q$ is again a statement form. Since r is a statement form, $(p \land \sim q) \to r$ is a statement form.

(b) Since r is a statement form, so is $\sim r$. Since both q and $\sim r$ are statement forms, $q \vee \sim r$ is a statement form. However, p is also a statement form; thus $p \rightarrow (q \vee \sim r)$ is a statement form. The negation of a statement form is a statement form, so $\sim(p \rightarrow (q \vee \sim r))$ is also.

These definitions allow us to study the statement calculus: the manipulation, verification, and simplification of logical statement forms. It will help us to see when two different statement forms are logically equivalent in the sense that they have the same truth values.

The main mechanism for determining the truth values of statement forms is known as a *truth table*. It is also possible to use tree diagrams (see Chapter 6) to determine the truth values of statement forms. We begin with truth tables.

Consider any simple statement p. Then p has one of the two truth values T (TRUE) or F (FALSE). We can list the possible values of p in a table:

p
T
F

Clearly, $\sim p$ (not p) is a form derived from p and has two possible values also. When p has the truth value T, $\sim p$ has the truth value F, and vice versa.

We represent the truth value of $\sim p$ in a *truth table*.

p	$\sim p$
T	F
F	T

The statement form $p \wedge q$ (p and q) is made up of two statements represented by p and q and the connective $\wedge$. The statement p can have one of the truth values T or F. Similarly, q can have one of the truth values T or F. Hence, by the multiplication principle, there are four possible pairs of truth values for p and q.

Their *conjunction*, $p \wedge q$, is true if and only if both p is true and q is true. The truth table for conjunction is

p	q	$p \wedge q$
T	T	T
T	F	F
F	T	F
F	F	F

The *disjunction* of p and q, p ∨ q (p or q), is true if either p is true or q is true or both p and q are true. In English, the word "or" is ambiguous; we have to distinguish between the exclusive and inclusive "or." For example, in the sentence "Ira either will go to Princeton or Stanford," the word "or" is assumed to be exclusive since Ira will choose one of the schools, but not both. On the other hand, in the sentence, "Diana is smart or she is rich," the "or" probably is inclusive since either Diana is smart, Diana is rich, or Diana is both smart and rich. The mathematical statement form p ∨ q uses the inclusive "or" and is unambiguous, as the truth table makes clear.

p	q	p ∨ q
T	T	T
T	F	T
F	T	T
F	F	F

Note that T stands for TRUE and F stands for FALSE. Each line of the truth table corresponds to a combination of truth values for the components p and q. There are only two lines in the table for ∼p, while for the others there are four.

What we have described so far is a system of calculating the truth value of a statement form based on the truth of its component statement forms. The beauty of truth tables is that they can be used to determine the truth values of more elaborate statement forms by reapplying the basic rules.

EXAMPLE 2   Construct a truth table for the statement form

$$\sim(p \vee q).$$

Solution   We write all the components of the statement form so that they can be evaluated for the four possible pairs of values for p and q. Note that we enter the truth values for p and q in the same order as in the previous tables. It is a good idea to use this order all the time; that is,

p	q
T	T
T	F
F	T
F	F

The truth table for the form ∼(p ∨ q) contains a column for each calculation.

p	q	p ∨ q	∼(p ∨ q)
T	T	T	F
T	F	T	F
F	T	T	F
F	F	F	T

The third column represents the truth values of the disjunction of the first two columns and the fourth column represents the negation of the third column. The statement form $\sim(p \vee q)$ is TRUE in one case: when both p and q are FALSE. Otherwise, $\sim(p \vee q)$ is FALSE.

EXAMPLE 3 Construct a truth table for the statement form

$$\sim(p \wedge \sim q) \vee p.$$

Solution

p	q	$\sim q$	$p \wedge \sim q$	$\sim(p \wedge \sim q)$	$\sim(p \wedge \sim q) \vee p$
T	T	F	F	T	T
T	F	T	T	F	T
F	T	F	F	T	T
F	F	T	F	T	T
(1)	(2)	(3)	(4)	(5)	(6)

We fill in each column by using the rules already established. Column 3 is the negation of the truth values in column 2. Column 4 represents the conjunction of columns 1 and 3. Column 5 is the negation of the statement form whose values appear in column 4. Finally, column 6 is the disjunction of columns 5 and 1. We see that this statement form has truth value TRUE no matter what the truth values of the statements p and q. Such a statement is called a tautology.

> A statement form that has truth value TRUE regardless of the truth values of the individual statement variables it contains is called a *tautology*.

> A statement form that has truth value FALSE regardless of the truth values of the individual statement variables it contains is called a *contradiction*.

There is a more efficient way to prepare the truth table of Example 3. Use the same order for entering possible truth values of p and q. Put the statement form at the top of the table. Fill in columns under each operation as you need them, working from the inside out. We label the columns in the order in which we fill them in.

p	q	$\sim$	$(p$	$\wedge$	$\sim q)$	$\vee$	p
T	T	T		F	F		T
T	F	F		T	T		T
F	T	T		F	F		T
F	F	T		F	T		T
(1)	(2)	(5)		(4)	(3)		(6)

We entered the values of p and q in columns 1 and 2, and then entered the values for $\sim q$ in column 3. We used the conjunction of columns 1 and 3 to fill in column 4, applied the negation to column 4 to get column 5, and combined columns 5 and 1 to get the values in column 6.

EXAMPLE 4   Construct the truth table for the statement form

$$(\sim p \vee \sim q) \wedge (p \wedge \sim q).$$

Solution   Again, we will use the short form of the table, putting the statement form at the top of the table, and labeling each column in the order in which it was completed.

p	q	$(\sim p$	$\vee$	$\sim q)$	$\wedge$	$(p$	$\wedge$	$\sim q)$
T	T	F	F	F	F	T	F	F
T	F	F	T	T	T	T	T	T
F	T	T	T	F	F	F	F	F
F	F	T	T	T	F	F	F	T
(1)	(2)	(3)	(5)	(4)	(7)		(6)	

There are some unlabeled columns. These were just recopied for convenience and clarity. It is not necessary to include them. Column 7, the conjunction of columns 5 and 6, tells us that the statement form is true if and only if p is TRUE and q is FALSE.

EXAMPLE 5   Find the truth table for the statement form

$$(p \vee q) \wedge \sim (p \wedge q).$$

Solution

p	q	$(p \vee q)$	$\wedge$	$\sim$	$(p \wedge q)$
T	T	T	F	F	T
T	F	T	T	T	F
F	T	T	T	T	F
F	F	F	F	T	F
(1)	(2)	(3)	(6)	(5)	(4)

Column 6, the conjunction of columns 3 and 5, gives the truth values of the statement form.

We note that the statement form in Example 5 can be read as "p or q and not both p and q." This is the exclusive p "or" q. We will find it a useful connective and denote it by $\oplus$. The statement form $p \oplus q$ is true exactly when p is true or q is true but not both p and q are true. The truth table of $p \oplus q$ is

p	q	$p \oplus q$
T	T	F
T	F	T
F	T	T
F	F	F

If there are three simple statements in a statement form and we denote them p, q, and r, then each could take on the truth values TRUE or FALSE. Hence, by the multiplication principle, there are $2 \times 2 \times 2 = 8$ different assignments to the three variables together. There are eight lines in a truth table for such a statement form. Again, in the truth table we will list all the T's and then all the F's for the first variable. Then we alternate TT followed by FF in the second column.

Finally, for the third variable we alternate T with F. This assures that we have listed all the possibilities and facilitates comparison of the final results.

EXAMPLE 6 Construct a truth table for $(p \vee q) \wedge [(p \vee r) \wedge \sim r]$.

Solution Note the order in which the T's and F's are listed in the first three columns. The columns are numbered in the order in which they were filled.

p	q	r	\multicolumn{6}{c}{$(p \vee q) \wedge [(p \vee r) \wedge \sim r]$}
T	T	T	T	F	T	F	F	
T	T	F	T	T	T	T	T	
T	F	T	T	F	T	F	F	
T	F	F	T	T	T	T	T	
F	T	T	T	F	T	F	F	
F	T	F	T	F	F	F	T	
F	F	T	F	F	T	F	F	
F	F	F	F	F	F	F	T	
(1)	(2)	(3)	(4)	(8)	(5)	(7)	(6)	

The statement form has truth value TRUE if and only if p is TRUE and r is FALSE (regardless of the truth value of q).

The formal evaluation of the truth table of a statement form can help in deciding the truth value when particular sentences are substituted for the statement variables.

EXAMPLE 7 Let p denote the statement "London is the capital of England" and let q denote the statement "Venice is the capital of Italy." Determine the truth value of each of the following statements.

(a) $p \wedge q$

(b) $p \vee q$

(c) $\sim p \wedge q$

(d) $\sim p \vee \sim q$

(e) $\sim (p \wedge q)$

(f) $p \oplus q$

Solution First, we note that p has truth value TRUE and q has truth value FALSE. (Rome is the capital of Italy.) Thus, the conjunction in (a) is FALSE. The disjunction in (b) is TRUE because p is TRUE. In (c), both $\sim p$ and q are FALSE, so their conjunction is FALSE. Since $\sim q$ is TRUE, statement (d) is TRUE. Since (a) is FALSE, and the statement in (e) is its negation, (e) is TRUE. Since p is TRUE and q is FALSE, $p \oplus q$ is TRUE.

Logic and Computer Languages In many computer languages, such as BASIC, the logical connectives are incorporated into programs that depend on logical decision making. The symbols are replaced by their original English words in the BASIC language; for example;

AND	$\wedge$
OR	$\vee$
NOT	$\sim$
XOR	$\oplus$ (XOR stands for exclusive or)

If the statements p and q are assigned truth values T and F, then the truth tables for the connectives AND, OR, XOR, and NOT are as shown in Fig. 1.

FIGURE 1

p	q	p AND q	p OR q	p XOR q		p	NOT p
T	T	T	T	F		T	F
T	F	F	T	T		F	T
F	T	F	T	T			
F	F	F	F	F			

Use of a Tree to Represent a Statement Form

Another method of describing a statement form is by using a mathematical structure called a *tree*. We have already seen tree diagrams in Chapter 6. The statement form $(p \wedge \sim q) \vee r$ is diagrammed in Fig. 2. We begin by writing the statement form at the top of the tree. Two branches are then required, one for each of the operands: $(p \wedge \sim q)$ and r. We continue until we get to forms involving only a simple statement alone. From $(p \wedge \sim q)$, we require two branches, one for each of the operands p and $\sim q$. Finally, one branch is needed to perform the negation of q.

If we have specific truth values for the statement variables, the tree can be used to determine the truth value of the statement form by entering the values for all the variables and filling in truth values for the forms, working from the bottom to the top of the tree. Suppose that we are interested in the value of the statement form $(p \wedge \sim q) \vee r$ when p is true, q is false, and r is false. Figure 3 shows how we fill in values at the nodes and proceed up the tree to its top for the truth value of the statement form.

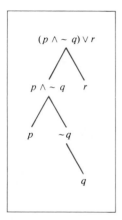

FIGURE 2

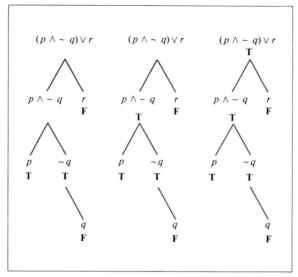

FIGURE 3

534   CHAP. 12: *Logic*

PRACTICE PROBLEMS 2

1. Construct the truth table for $(p \vee \sim r) \wedge q$.

2. Construct the truth table for $p \oplus \sim q$.

3. Let p denote "May follows April" and let q denote "June follows May." Determine the truth values of the following.

   (a) $p \wedge \sim q$ (b) $\sim(p \vee \sim q)$

   (c) $p \oplus q$ (d) $\sim[(p \wedge q) \oplus \sim q]$

EXERCISES 2

In Exercises 1–18, construct truth tables for the given statement forms.

1. $p \wedge \sim q$	2. $(p \oplus q) \wedge r$	3. $(p \wedge \sim r) \vee q$
4. $\sim(p \wedge r) \vee q$	5. $\sim[(p \wedge r) \vee q]$	6. $p \wedge \sim p$
7. $p \vee \sim p$	8. $(p \vee q) \wedge \sim r$	9. $p \oplus (q \vee r)$
10. $p \vee (q \wedge r)$	11. $(p \vee q) \wedge (p \vee r)$	12. $(p \vee q) \wedge (p \vee \sim r)$
13. $(p \vee q) \wedge \sim(p \vee q)$	14. $(\sim p \vee q) \wedge r$	15. $\sim(p \vee q) \wedge r$
16. $\sim[(p \vee q) \wedge r]$	17. $\sim p \vee (q \wedge r)$	18. $(p \oplus q) \oplus \sim(p \oplus q)$

19. Compare the truth tables of $\sim p \vee \sim q$ and $\sim(p \wedge q)$.

20. Compare the truth tables of $p \wedge (q \vee r)$ and $(p \wedge q) \vee (p \wedge r)$.

21. Compare the truth tables of $p \oplus q$ and $(p \vee q) \wedge \sim(p \wedge q)$.

22. How many possible truth tables can you construct for statement forms involving two variables, p and q? (*Hint:* Consider the number of ways to complete the last column of a truth table for each of the four possible pairs of values for p and q.)

23. Compare the truth tables of $(p \wedge q) \vee r$ and $p \wedge (q \vee r)$.

24. Define the connective $\ominus$ by the truth table

p	q	$p \ominus q$
T	T	T
T	F	F
F	T	F
F	F	T

Construct the truth table of the following.

   (a) $p \ominus \sim q$ (b) $(p \ominus q) \ominus r$

   (c) $p \ominus (q \ominus r)$ (d) $\sim(p \ominus q) \wedge (p \oplus q)$

25. Define $p|q$ by the truth table.

p	q	$p|q$
T	T	F
T	F	T
F	T	T
F	F	T

This connective is called the *Sheffer stroke* (or NAND). Construct truth tables for the following.

(a) $p|p$ (b) $(p|p)|(q|q)$ (c) $(p|q)|(p|q)$ (d) $p|((p|q)|q)$

26. Compare the truth tables of the connectives $\sim$, $\vee$, and $\wedge$ with those constructed in Exercise 25. Do you see any similarities?

27. Let p denote "John Lennon was a member of the Beatles" and let q be the statement "The Beatles came from Spain." Determine the truth values of the following.

(a) $p \vee \sim q$ (b) $\sim p \wedge q$ (c) $p \oplus q$

(d) $\sim p \oplus q$ (e) $\sim(p \oplus q)$ (f) $(p \vee q) \oplus \sim q$

28. Let p be the statement "There were 14 original American colonies" and let q denote "Utah was one of the original colonies." Determine the truth value of each of the following.

(a) $p \vee q$ (b) $\sim p \vee q$ (c) $\sim p \wedge \sim q$ (d) $p \oplus \sim q$

29. Let p be the statement "A rectangle has three sides" and q be the statement "A right angle has 90 degrees." Determine the truth value of each of the following.

(a) $p \wedge \sim q$ (b) $\sim(p \oplus q)$ (c) $p \wedge q$ (d) $\sim p \wedge \sim q$

30. Draw a tree for the statement form given in Exercise 1. Assuming p has truth value T and q has value F, use the tree to find the truth value of the statement form.

31. Draw a tree for the statement form given in Exercise 2. Assuming that p has truth value T, q has truth value F, and r has truth value F, use the tree to find the truth value of the statement form.

32. Repeat Exercise 31 for the statement in Exercise 9.

33. Repeat Exercise 31 for the statement in Exercise 12.

SOLUTIONS TO PRACTICE PROBLEMS 2

1.

p	q	r	$(p \vee \sim r) \wedge q$		
T	T	T	T	F	T
T	T	F	T	T	T
T	F	T	T	F	F
T	F	F	T	T	F
F	T	T	F	F	F
F	T	F	T	T	T
F	F	T	F	F	F
F	F	F	T	T	F
(1)	(2)	(3)	(5)	(4)	(6)

2.

p	q	$p \oplus \sim q$	
T	T	T	F
T	F	F	T
F	T	F	F
F	F	T	T
(1)	(2)	(4)	(3)

3. Both p and q are TRUE. Thus, (a) is FALSE. Item (b) is the negation of the statement $(p \vee \sim q)$, which is TRUE since p is TRUE. Hence, the statement in (b) is FALSE. Since both p and q are TRUE, $p \oplus q$ is FALSE. To analyze (d), we consider that $(p \wedge q)$ is TRUE and $\sim q$ is FALSE, so $[(p \wedge q) \oplus \sim q]$ is TRUE. Its negation is FALSE.

12.3. Implication

We are quite familiar with implications. For example, "If Jay is caught smoking in the rest room, he is suspended from school" is an implication. Note that, although Jay may avoid being caught smoking in the rest room, he still might be suspended for some other infraction. Symbolically, we represent the statement "Jay is caught smoking in the rest room" by p, while we represent "Jay is suspended from school" by q. The implication is represented by the *conditional* connective $\rightarrow$ and we write $p \rightarrow q$. The truth table for the statement form using the conditional is given below.

p	q	$p \rightarrow q$
T	T	T
T	F	F
F	T	T
F	F	T

The *conditional* statement form $p \rightarrow q$ has truth value FALSE only when p has truth value TRUE and q has truth value FALSE.

In the case given above, if p is TRUE, then "Jay is caught smoking in the rest room" is TRUE, while if q were FALSE, then "Jay is not suspended" is TRUE. The implication $p \rightarrow q$ is FALSE in this case.

We call p the *hypothesis* and q the *conclusion*. Note that if the hypothesis is FALSE then the implication $p \rightarrow q$ is TRUE. There are several ways to read $p \rightarrow q$ in English:

1. p implies q
2. If p, then q
3. p only if q
4. q, if p

5. p is sufficient for q

6. q is a necessary condition for p

Note that $p \rightarrow q$ means that p is sufficient for q, so if you have p, then you get q. On the other hand, $q \rightarrow p$ means that p is necessary for q, so if you want q, then you must have p.

EXAMPLE 1   In each of the following statements, determine the hypothesis and the conclusion.

(a) Bill goes to the party only if Greta goes to the party.

(b) Sue goes to the party if Craig goes to the party.

(c) For 6 to be even, it is sufficient that its square, 36, be even.

Solution   (a) The statement is in the form of p only if q. Thus, the hypothesis is "Bill goes to the party" and the conclusion is "Greta goes to the party." We could rewrite the statement as "If Bill goes to the party, then Greta goes to the party." The statement "If Bill goes to the party, then Greta goes to the party" seems to mean that Greta is following Bill around, whereas the statement "Bill goes to the party only if Greta goes to the party" makes the romance seem quite the opposite. However, the second statement should be interpreted as follows: if "Bill goes to the party" is true, then that means Greta must also have gone— for that was the ONLY reason he would go. The unemotional interpretation of the logical form shows the equivalence of the two statements.

(b) The statement is of the form q, if p. It has hypothesis "Craig goes to the party" and conclusion "Sue goes to the party."

(c) This is of the form p is sufficient for q, with the hypothesis p of the form "The square of the integer 6 is even" and the conclusion q of the form "The integer 6 is even."

EXAMPLE 2   Determine whether each of the following statements is true or false:

(a) If Paris is in France, then the Louvre is in Paris.

(b) If the Louvre is in Paris, then $2 + 3 = 7$.

(c) If $2 + 3 = 7$, then the Louvre is in Paris.

(d) If $2 + 3 = 7$, then Paris is in Spain.

(e) Paris is in Spain only if $2 + 3 = 7$.

Solution   The statement (a) is TRUE because both the hypothesis and the conclusion are TRUE. The statement (b) is FALSE because the hypothesis is TRUE but the conclusion is FALSE. The last three statements are TRUE because the hypothesis in each is FALSE. Note that the last statement has hypothesis "Paris is in Spain" and conclusion "$2 + 3 = 7$."

EXAMPLE 3   Construct the truth table of the statement "If the president dies or becomes incapacitated, then the vice-president becomes president."

Solution   We let p be the statement "The president dies"; we let q be the statement "The president becomes incapacitated." We let r be the statement "The vice-president becomes president." The statement form is $(p \vee q) \rightarrow r$. A truth table for the statement form appears below.

p	q	r	$(p \vee q)$	$\rightarrow r$
T	T	T	T	T
T	T	F	T	F
T	F	T	T	T
T	F	F	T	F
F	T	T	T	T
F	T	F	T	F
F	F	T	F	T
F	F	F	F	T
(1)	(2)	(3)	(4)	(5)

The statement is false only in the case where the president dies or becomes incapacitated but the vice-president does not become president.

It is important to note that $p \rightarrow q$ is not the same as $q \rightarrow p$. We should not confuse the hypothesis and the conclusion in an implication. An example will demonstrate the difference between $p \rightarrow q$ and $q \rightarrow p$. Consider the implication "If the truck carries ice cream, then it is refrigerated." If the implication is in the form $p \rightarrow q$, then $q \rightarrow p$ is the implication "If the truck is refrigerated, then it carries ice cream." These implications need not have the same truth values. The first statement form is probably true; the second need not be. The truth tables of $p \rightarrow q$ and $q \rightarrow p$ show the differences.

p	q	$p \rightarrow q$	$q \rightarrow p$
T	T	T	T
T	F	F	T
F	T	T	F
F	F	T	T

The implication $q \rightarrow p$ is called the *converse* of the statement $p \rightarrow q$. Thus, given the statement "If the Los Angeles Dodgers won the pennant, some games of the World Series are in California," its converse is "If some games of the World Series are in California, then the Los Angeles Dodgers won the pennant." While the original statement is a tautology, the converse is not. There are situations in which the hypothesis of the converse is true, but the conclusion is false. Some games of the World Series may be in California because the San Francisco Giants, the Oakland Athletics, or the California Angels, rather than the Los Angeles Dodgers, won the pennant.

EXAMPLE 4   Construct the truth table for the conjunction of $p \to q$ and its converse. That is, find the truth table of the statement form $(p \to q) \land (q \to p)$.

Solution

p	q	$(p$	$\to$	$q)$	$\land$	$(q$	$\to$	$p)$
T	T		T		T		T	
T	F		F		F		T	
F	T		T		F		F	
F	F		T		T		T	
(1)	(2)		(3)		(5)		(4)	

We note that the statement form has truth value TRUE whenever p and q have the same truth value: either both TRUE or both FALSE.

The statement form $(p \to q) \land (q \to p)$ is referred to as the *biconditional*, which we write as $p \leftrightarrow q$. The statement $p \leftrightarrow q$ is read as "p, if and only if q" or "p is necessary and sufficient for q." The truth table for the biconditional is shown below.

p	q	$p \leftrightarrow q$
T	T	T
T	F	F
F	T	F
F	F	T

The biconditional is a statement form because it can be expressed as $(p \to q) \land (q \to p)$. In some sense, which will be discussed further in the next section, the biconditional expresses a kind of equivalence of p and q. That is, the biconditional statement form is TRUE whenever p and q are both TRUE or both FALSE. The biconditional is FALSE whenever p and q have different truth values.

Using connectives, we have seen how we can string together several simple statements into compound and fairly complex statement forms. Such forms should not be confusing: what is the meaning of $\sim p \lor q$? Do we mean $\sim(p \lor q)$ or do we mean $(\sim p) \lor q$? To clarify matters and to avoid the use of too many parentheses, we define an order of *precedence* for the connectives. This dictates which of the connectives should be applied first.

> The order of precedence for logical connectives is
>
> $$\sim, \quad \land, \quad \lor, \quad \to, \quad \leftrightarrow$$

in which one applies $\sim$ first, then $\land$, and so on. If there is any doubt, insert parentheses to clarify the statement form. Using the order of precedence, we see that $\sim p \lor q$ is $(\sim p) \lor q$.

EXAMPLE 5 Compare the truth tables of $(\sim p) \vee q$ and $\sim(p \vee q)$.

Solution

p	q	$(\sim p) \vee q$		$\sim(p \vee q)$	
T	T	F	T	F	T
T	F	F	F	F	T
F	T	T	T	F	T
F	F	T	T	T	F
(1)	(2)	(3)	(4)	(6)	(5)

Note that columns 4 and 6 are different.

EXAMPLE 6 Insert parentheses in the statement to show the proper order for the application of the connectives.

$$p \wedge q \vee r \rightarrow \sim s \wedge r$$

Solution Reading from the left, we apply the $\sim$ first, so $\sim s \wedge r$ becomes $(\sim s) \wedge r$. We then scan for the connective $\wedge$, since that is next in the precedence list. So $p \wedge q \vee r$ becomes $(p \wedge q) \vee r$. We apply the $\rightarrow$ last, and the statement can be written

$$[(p \wedge q) \vee r] \rightarrow [(\sim s) \wedge r]$$

Parentheses have highest priority in the precedence. Clearly, $p \wedge (q \vee (r \rightarrow \sim s)) \wedge r$ has the same symbols as the statement of Example 6, but the parentheses have defined a different statement form.

That the statement form $p \vee q \vee r$ may be written either as $(p \vee q) \vee r$ or as $p \vee (q \vee r)$ is shown in another section. The statement form $p \wedge q \wedge r$ can be written as $(p \wedge q) \wedge r$ or as $p \wedge (q \wedge r)$, as is shown later.

Implications and Computer Languages The logical connective $\rightarrow$ is used in writing computer programs, although in some programming languages it is expressed with the words IF, THEN. Thus, we might see an instruction of the form "IF $a = 3$ THEN LET $s = 0$." Strictly speaking, this is not a statement form because "LET $s = 0$" is not a statement. However, that command means "the computer will set $s = 0$." Thus, if the value of a is 3, the computer sets s to 0. If $a \neq 3$, the program just continues. The instruction "IF ... THEN ... ELSE" allows for a branch in the program. For example, "IF $a = 3$ THEN LET $s = 0$ ELSE LET $s = 1$" will assign value 0 to s if $a = 3$ and assign value 1 to s if $a \neq 3$.

EXAMPLE 7 For the given input values of A and B, use the program to determine the value of C. The asterisk ($*$) denotes multiplication.

$$\text{IF } (A * B) + 6 \geq 10$$

$$\text{THEN LET } C = A * B$$

$$\text{ELSE LET } C = 10$$

(a) A $= -2$, B $= -7$ (b) A $= -2$, B $= 3$ (c) A $= 2$, B $= 2$

Solution   (a) We determine that $(A * B) + 6 = 14 + 6 = 20$. Hence, we set $C = 14$.

(b) In this case, $(A * B) + 6 = 0$. So we set $C = 10$.

(c) $(A * B) + 6 = 10$. Hence $C = 4$.

Implication and Common Language   The mathematical uses of the conditional and the biconditional are very precise. However, colloquial speech is rarely as precise. For example, consider the statement, "Peter gets dessert only if he eats his broccoli." Technically, this means that if Peter gets dessert, he eats his broccoli. However, Peter and his parents probably interpret the statement to mean that Peter gets dessert if and only if he eats his broccoli. This imprecision in language is sometimes confusing; we make every effort to avoid confusion in mathematicas by adhering to strict rules for using the conditional and biconditional.

PRACTICE PROBLEMS 3

1. Using p: "A square is a rectangle" and q: "A rectangle has four sides," write out the statements in symbolic form. Name the hypothesis and the conclusion in each statement.

    (a) A square is a rectangle if a rectangle has four sides.

    (b) A rectangle has four sides if a square is a rectangle.

    (c) A rectangle has four sides only if a square is a rectangle.

    (d) A square is a rectangle is sufficient to show that a rectangle has four sides.

    (e) For a rectangle to have four sides, it is necessary for a square to be a rectangle.

2. Let p denote the statement "There are 48 states in the United States," let q denote the statement "The American flag is red, white, and blue," and let r be the statement "Maine is on the east coast." Determine the truth value of each of the following statement forms.

    (a) $p \wedge q \to r$ (b) $p \vee q \to r$

    (c) $\sim p \wedge q \leftrightarrow r$ (d) $p \vee \sim q \to \sim r$

3. Use the program to assign a value to D in each of the cases listed below.

$$\text{LET } C = A + B$$

$$\text{IF } (C > 0) \text{ OR } (A > 0)$$

$$\text{THEN LET } D = 1$$

$$\text{ELSE LET } D = -1$$

    (a) $A = 4, B = 6$ (b) $A = 4, B = -6$

    (c) $A = 4, B = -4$ (d) $A = -2, B = 4$

    (e) $A = -6, B = 4$ (f) $A = -2, B = -2$

EXERCISES 3

Construct a truth table for each of the statement forms in Exercises 1–15.

1. $p \to \sim q$ 2. $p \vee (q \to \sim r)$

3. $(p \oplus q) \to q$ 4. $(p \oplus q) \to r$

5. $(\sim p \wedge q) \to r$ 6. $\sim(p \to q)$

7. $(p \to q) \leftrightarrow (\sim p \vee q)$ 8. $p \oplus (q \to r)$

9. $(p \to q) \to r$ 10. $p \to (q \to r)$

11. $\sim(p \vee q) \to (\sim p \wedge r)$ 12. $p \to (p \oplus q)$

13. $(p \vee q) \leftrightarrow (p \wedge q)$ 14. $[(\sim p \wedge q) \oplus q] \to p$

15. $[p \wedge (q \vee r)] \leftrightarrow [(p \wedge q) \vee (p \wedge r)]$

Let p be the statement "Abraham Lincoln was the sixteenth president of the United States" and let q be the statement "The battle of Gettysburg took place at the O.K. Corral." Determine the truth value of each of the statement forms in Exercises 16–25.

16. $p \to q$ 17. $\sim p \to q$

18. $p \to \sim q$ 19. $q \to p$

20. $p \leftrightarrow q$ 21. $(p \oplus q) \to p$

22. $(p \vee q) \to q$ 23. $(p \wedge \sim q) \to (\sim p \oplus q)$

24. $[p \wedge (\sim q \to \sim p)] \oplus q$ 25. $p \to [p \wedge (p \oplus q)]$

In Exercises 26–33, write the statement forms in symbols using the conditional ($\to$) or the biconditional ($\leftrightarrow$) connective. Name the hypothesis and the conclusion in each conditional form. We let p be the statement "Sally studied" and q be the statement "Sally passes."

26. If Sally studied, then she passes.

27. Sally studied if and only if Sally passes.

28. If Sally passes, then Sally studied.

29. Sally passes only if Sally studied.

30. That Sally studied is sufficient for Sally to pass.

31. Sally's studying is necessary for Sally to pass.

32. Sally passes implies that Sally studied.

33. If Sally did not study, then Sally does not pass.

In Exercises 34–38, let p denote the TRUE statement "The die is fair" and let q be the TRUE statement "The probability of a 2 is $\frac{1}{6}$." Write each of the statement forms in symbols, name the hypothesis and the conclusion in each, and determine whether the statement is true or false.

34. If the die is fair, the probability of a 2 is $\frac{1}{6}$.

35. If the die is not fair, the probability of a 2 is not $\frac{1}{6}$.

36. The probability of a 2 is $\frac{1}{6}$ only if the die is fair.

37. The die is not fair if the probability of a 2 is not $\frac{1}{6}$.

38. The die is fair implies that the probability of a 2 is $\frac{1}{6}$.

39. Give the hypothesis and the conclusion in each statement.

   (a) Healthy people live a long life.

   (b) The train stops at the station only if a passenger requests it.

   (c) For a plant to grow, it is necessary that it be exposed to sunlight.

   (d) Only if Jane goes to the store, I will go to the store.

40. State the converse of each of the following statements.

   (a) If Jane runs 20 miles, Jane is tired.

   (b) Cindy loves Fred only if Fred loves Cindy.

   (c) Jon cashes a check if the bank is open.

   (d) Errors are clear only if the documentation is complete.

   (e) Sally's eating the vegetables is a necessary condition for Sally's getting dessert.

   (f) Sally's eating the vegetables is sufficient for Sally's getting dessert.

41. State the converse of each of the following statements.

   (a) City Sanitation collects the garbage if the mayor calls.

   (b) The price of beans goes down only if there is no drought.

   (c) If goldfish swim in Lake Erie, Lake Erie is fresh water.

   (d) If tap water is not salted, then it boils slowly.

42. State the hypothesis and conclusion in each statement form.

   (a) Copa beach is crowded if the weather is hot and sunny.

   (b) Our team wins a game only if I carry a rabbit's foot.

   (c) If I carry a rabbit's foot, our team wins a game.

   (d) Ivy is green is sufficient for it to be healthy.

43. Determine the output value of A in the program when input values of X and Y are given.

   $$\text{LET } Z = X + Y$$
   $$\text{IF } (Z \neq 0) \text{ AND } (X > 0)$$
   $$\text{THEN LET } A = 6$$
   $$\text{ELSE LET } A = 4$$

   (a) X = 0, Y = 0 (b) X = 8, Y = −8

   (c) X = −3, Y = 3 (d) X = −3, Y = 8

   (e) X = 8, Y = −3 (f) X = 3, Y = −8

44. For the given input values of A and B, use the program to find the output values of X.

$$\text{LET } C = A + B$$

$$\text{IF } ((A > 0) \text{ OR } (B > 0)) \text{ AND } (C > 0)$$

$$\text{THEN LET } X = 100$$

$$\text{ELSE LET } X = -100$$

(a) $A = 2, B = 2$ (b) $A = 2, B = -2$

(c) $A = 2, B = -5$ (d) $A = -2, B = -5$

(e) $A = -5, B = 3$ (f) $A = -5, B = 8$

45. For the input values of A and B, use the program to determine the value of Y.

$$\text{LET } C = A * B$$

$$\text{IF } ((C \geq 10) \text{ AND } (A < 0)) \text{ OR } (B < 0)$$

$$\text{THEN LET } Y = 7$$

$$\text{ELSE LET } Y = 0$$

(a) $A = -2, B = -6$ (b) $A = -1, B = -6$

(c) $A = -2, B = 6$ (d) $A = 6, B = -1$

(e) $A = 4, B = 3$ (f) $A = 3, B = 1$

46. For the given input values of X and Y, use the program to determine the output value of Z.

$$\text{IF } Y \neq 0$$

$$\text{THEN LET } Z = X/Y$$

$$\text{ELSE LET } Z = -1{,}000{,}000$$

(a) $X = 6, Y = 2$ (b) $X = 10, Y = 5$

(c) $X = 5, Y = 10$ (d) $X = 0, Y = 10$

(e) $X = 10, Y = 0$ (f) $X = -1{,}000{,}000, Y = 1$

47. For the given input values of A and B, use the program to determine the value of C.

$$\text{IF } ((A < 0) \text{ AND } (B < 0)) \text{ OR } (B \geq 6)$$

$$\text{THEN LET } C = (A * B) + 4$$

$$\text{ELSE LET } C = 0$$

(a) $A = -1, B = -2$ (b) $A = -2, B = 8$

(c) $A = -2, B = 3$ (d) $A = 3, B = -2$

(e) $A = 3, B = 8$ (f) $A = 3, B = -3$

48. For the given input values for A and B, use the program to determine the output value of C.

$$\text{IF } ((A \geq 5) \text{ OR } (B \geq 5)) \text{ AND } (B \leq 10)$$

$$\text{THEN LET } C = A - B$$

$$\text{ELSE LET } C = -30$$

(a) A = 7, B = 7   (b) A = 7, B = 3

(c) A = 7, B = 12   (d) A = 4, B = 7

(e) A = 4, B = 3   (f) A = 4, B = 12

49. For the given input values of A and B, find the value of X in the program.

$$\text{LET } C = A - B$$

$$\text{IF } (C < 0) \text{ OR } (B < 0)$$

$$\text{THEN LET } D = 5 * C$$

$$\text{ELSE LET } D = 0$$

$$\text{LET } X = D + 3$$

(a) A = 0, B = 0   (b) A = 6, B = 3

(c) A = −5, B = 3   (d) A = 3, B = 5

(e) A = 5, B = −3   (f) A = −5, B = −3

50. We let S be the weekly salary for a consultant who works H hours at a rate of R dollars per hour. We take a deduction of 4% on salaries up to $1000 per week and deduct the maximum of $40 from salaries above $1000. Find the pay P for each of the consultants given the rate and the number of hours worked in a week.

$$\text{LET } S = R * H$$

$$\text{IF } S < 1000$$

$$\text{THEN LET } D = .04 * S$$

$$\text{ELSE LET } D = 40$$

$$\text{LET } P = S - D$$

(a) R = 10, H = 40   (b) R = 20, H = 40

(c) R = 25, H = 20   (d) R = 25, H = 40

(e) R = 30, H = 40   (f) R = 35, H = 40

SOLUTIONS TO PRACTICE PROBLEMS 3

1. (a) This is of the form $q \to p$ with hypothesis q and conclusion p. Statement (b) is of the form $p \to q$, where p is the hypothesis and q is the conclusion. Statement (c) is of the form $q \to p$, since it states that "If a rectangle has four sides, then a square is a rectangle." Note that "only if" signals the clause that is the conclusion. In (c), q is the hypothesis and p is the conclusion. Statement (d) is of the form $p \rightarrow q$ with hypothesis p and conclusion q. Statement (e) is the converse of (d) and is of the form $q \rightarrow p$ with hypothesis q and conclusion p.

2. First, we note that p, q, and r have truth values FALSE, TRUE, and TRUE, respectively. Use the order of precedence to decide which connectives are applied first.

(a) TRUE. Since $p \wedge q$ is FALSE, the implication $(p \wedge q) \rightarrow r$ is TRUE.

(b) TRUE. Here $p \vee q$ is TRUE and r is TRUE, so $(p \vee q) \rightarrow r$ is TRUE.

(c) TRUE. Since p is FALSE and q is TRUE, $(\sim p) \wedge q$ is TRUE. Since r is also TRUE, $(\sim p \wedge q) \leftrightarrow r$ is TRUE.

(d) TRUE. Since p is FALSE and $\sim q$ is also FALSE, $p \wedge (\sim q)$ is FALSE. Thus, $(p \wedge \sim q) \rightarrow r$ is TRUE.

3.

	A	B	C	D
(a)	4	6	10	1
(b)	4	−6	−2	1
(c)	4	−4	0	1
(d)	−2	4	2	1
(e)	−6	4	−2	−1
(f)	−2	−2	−4	−1

For example, in (b) we check first to see if $(C > 0)$ is TRUE. It is not. We check to see if $(A > 0)$ is TRUE. It is. So we let $D = 1$. In (e), however both $(C > 0)$ and $(A > 0)$ are FALSE, so we let $D = -1$.

12.4. Logical Implication and Equivalence

Different statement forms may have the same truth tables. Suppose, for example, that p and q are simple statements and we construct the truth tables of both $p \rightarrow q$ and $\sim p \vee q$.

p	q	$p \rightarrow q$	$\sim p \vee q$
T	T	T	F T
T	F	F	F F
F	T	T	T T
F	F	T	T T
(1)	(2)	(3)	(4) (5)

We may compare columns 3 and 5 to see that, for any given pair of truth values assigned to p and q, the statement forms $p \rightarrow q$ and $\sim p \vee q$ have the same truth values.

> Two statement forms that have the same truth tables are called *logically equivalent*.

Recall that we defined a *tautology* to be a statement form that has truth value TRUE regardless of the truth values of its component statements. We denote a tautology with the letter t. A statement form that always has truth value FALSE regardless of the truth values of its component statements is called a *contradiction*. We use the letter c to denote a contradiction.

EXAMPLE 1 Show that $p \lor \sim p$ is a tautology and that $p \land \sim p$ is a contradiction.

Solution We construct the truth tables for the statement forms.

p	$\sim p$	$p \lor \sim p$	$p \land \sim p$
T	F	T	F
F	T	T	F

No matter what the value of p, $p \lor \sim p$ is TRUE and $p \land \sim p$ is FALSE.

EXAMPLE 2 Show that $\sim(p \land q)$ is logically equivalent to $\sim p \lor \sim q$.

Solution We construct the truth table to demonstrate that $\sim(p \land q)$ is logically equivalent to $\sim p \lor \sim q$.

p	q	$\sim(p \land q)$		$\sim p \lor \sim q$		
T	T	F	T	F	F	F
T	F	T	F	F	T	T
F	T	T	F	T	T	F
F	F	T	F	T	T	T
(1)	(2)	(4)	(3)	(5)	(7)	(6)

Since columns 4 and 7 are the same, $\sim(p \land q)$ and $\sim p \lor \sim q$ have the same truth tables and therefore are logically equivalent.

For convenience, we denote compound statement forms by capital letters. We will use P, Q, R, and so on, to denote statement forms such as $p \to q$, $p \land (q \to \sim r)$, $r \lor (p \land \sim(r \lor q))$, and so on. As we saw in the last section, $P \leftrightarrow Q$ is TRUE whenever P and Q have the same truth values (either both TRUE or both FALSE). And $P \leftrightarrow Q$ is FALSE if the truth values of P and Q differ. Thus, P is logically equivalent to Q if and only if $P \leftrightarrow Q$ is a tautology. We write $P \Leftrightarrow Q$ when P and Q are logically equivalent. In other words,

---

P and Q are logically equivalent; we write $P \Leftrightarrow Q$, whenever $P \leftrightarrow Q$ is a tautology.

We note that in comparing columns 4 and 7 in Example 2 we find that $\sim(p \wedge q)$ has the same truth table as $\sim p \vee \sim q$. This is reflected in the entries in column 8.

$$
\begin{array}{cccccccc}
p & q & \sim(p \wedge q) & \leftrightarrow & \sim p & \vee & \sim q \\
\text{T} & \text{T} & \text{F} & \text{T} & \text{T} & \text{F} & \text{F} & \text{F} \\
\text{T} & \text{F} & \text{T} & \text{F} & \text{T} & \text{F} & \text{T} & \text{T} \\
\text{F} & \text{T} & \text{T} & \text{F} & \text{T} & \text{T} & \text{T} & \text{F} \\
\text{F} & \text{F} & \text{T} & \text{F} & \text{T} & \text{T} & \text{T} & \text{T} \\
(1) & (2) & (4) & (3) & (8) & (5) & (7) & (6)
\end{array}
$$

This logical equivalency is important and is one of *De Morgan's laws*. We summarize some important logical equivalences in Table 1.

TABLE 1   Logical Equivalences

1.	$\sim \sim p \Leftrightarrow p$	Double negation
2a.	$(p \vee q) \Leftrightarrow (q \vee p)$	Commutative laws
b.	$(p \wedge q) \Leftrightarrow (q \wedge p)$	
c.	$(p \leftrightarrow q) \Leftrightarrow (q \leftrightarrow p)$	
3a.	$[(p \vee q) \vee r] \Leftrightarrow [p \vee (q \vee r)]$	Associative laws
b.	$[(p \wedge q) \wedge r] \Leftrightarrow [p \wedge (q \wedge r)]$	
4a.	$[p \vee (q \wedge r)] \Leftrightarrow [(p \vee q) \wedge (p \vee r)]$	Distributive laws
b.	$[p \wedge (q \vee r)] \Leftrightarrow [(p \wedge q) \vee (p \wedge r)]$	
5a.	$(p \vee p) \Leftrightarrow p$	Idempotent laws
b.	$(p \wedge p) \Leftrightarrow p$	
6a.	$(p \vee c) \Leftrightarrow p$	Identity laws
b.	$(p \vee t) \Leftrightarrow t$	
c.	$(p \wedge c) \Leftrightarrow c$	
d.	$(p \wedge t) \Leftrightarrow p$	
7a.	$(p \vee \sim p) \Leftrightarrow t$	
b.	$(p \wedge \sim p) \Leftrightarrow c$	
8a.	$\sim(p \vee q) \Leftrightarrow (\sim p \wedge \sim q)$	De Morgan laws
b.	$\sim(p \wedge q) \Leftrightarrow (\sim p \vee \sim q)$	
9.	$(p \rightarrow q) \Leftrightarrow (\sim q \rightarrow \sim p)$	Contrapositive
10a.	$(p \rightarrow q) \Leftrightarrow (\sim p \vee q)$	Implication
b.	$(p \rightarrow q) \Leftrightarrow \sim(p \wedge \sim q)$	
11.	$(p \leftrightarrow q) \Leftrightarrow [(p \rightarrow q) \wedge (q \rightarrow p)]$	Equivalence
12.	$(p \rightarrow q) \Leftrightarrow [(p \wedge \sim q) \rightarrow c]$	Reductio ad absurdum

The logical equivalences are important for understanding common English as well as for more formal analysis of mathematical statements and applications to computing. For example, the *contrapositive rule* is used frequently in speech. It states the equivalence of $p \rightarrow q$ with $\sim q \rightarrow \sim p$.

The *contrapositive* of the statement form $p \rightarrow q$
is the statement form $\sim q \rightarrow \sim p$.

EXAMPLE 3   State the contrapositive of "If Jay does not play Lotto, he will not win the Lotto jackpot."

Solution   We let p be the statement "Jay does not play Lotto" and q be the statement "He will not win the Lotto jackpot." We use the *double negation* rule from Table 1 to see that the negation of p is the statement "Jay plays Lotto," while the negation of q is the statement "He will win the Lotto jackpot." The contrapositive is the statement "If Jay will win the Lotto jackpot, then Jay plays Lotto." It is logically equivalent to the original statement form.

To prove the *contrapositive* rule, we use the truth table to show that $(p \rightarrow q) \Leftrightarrow (\sim q \rightarrow \sim p)$.

p	q	$(p \rightarrow q)$	$\leftrightarrow$	$(\sim q$	$\rightarrow$	$\sim p)$
T	T	T	T	F	T	F
T	F	F	T	T	F	F
F	T	T	T	F	T	T
F	F	T	T	T	T	T
(1)	(2)	(3)	(7)	(4)	(6)	(5)

A comparison of columns 3 and 6 shows that $p \rightarrow q$ and $\sim q \rightarrow \sim p$ have the same truth tables. Column 7 shows that the biconditional is a tautology.

$$p \rightarrow q$$

is equivalent to its contrapositive

$$\sim q \rightarrow \sim p.$$

EXAMPLE 4   A recent Associated Press article quoted a prominent official as having said, "If Mr. Jones is innocent of a crime, then he is not a suspect." State the contrapositive. Do you think the statement is TRUE or FALSE?

Solution   The contrapositive of the statement is "If Mr. Jones is a suspect, then he is guilty of a crime." That seems FALSE, so the original statement is also FALSE.

The *De Morgan laws* allow us to negate statements having the disjunctive or conjunctive connectives. It is important to recognize that when the negation is brought inside the parentheses it changes the connective from $\wedge$ to $\vee$, or vice versa.

EXAMPLE 5   Negate the statement "The earth's orbit is round and a year has 365 days."

Solution   We are asked to negate a statement of the form $p \wedge q$. By rule 8b, the negation can be written as "The earth's orbit is not round or a year does not have 365 days." This is true whenever the original statement is false.

EXAMPLE 6 Rewrite the statement "It is false that Jack and Jill went up the hill."

Solution We use the De Morgan laws to negate a statement of the form $(p \wedge q)$ to get "Jack did not go up the hill or Jill did not go up the hill."

EXAMPLE 7 Use the De Morgan laws to negate the statement form $(p \wedge q) \wedge r$.

Solution $\sim((p \wedge q) \wedge r) \Leftrightarrow (\sim(p \wedge q) \vee (\sim r)) \Leftrightarrow ((\sim p \vee \sim q) \vee \sim r)$. Note that each time the negation is applied to the conjunction or the disjunction the connective changes.

The rules in Table 1 allow us to simplify statement forms in a variety of ways. In place of any of the simple statements p, q, and r, we may substitute any compound statement. Provided we substitute consistently throughout, the equivalence holds. The rules are similar to the rules of algebra in that the need and the purpose for simplification depend on the situation. Frequently, however, we can simplify statement forms containing $\rightarrow$ by eliminating the implication. As the *implication* rule

$$(p \rightarrow q) \Leftrightarrow (\sim p \vee q)$$

shows, we can reduce a statement involving $\rightarrow$ to one involving other connectives. In fact, as the next examples demonstrate, we may express statements involving any of the connectives using only $\sim$ and $\vee$ or only $\sim$ and $\wedge$.

EXAMPLE 8 Eliminate the connective $\rightarrow$ in the statement form and simplify as much as possible.

$$(\sim p \wedge \sim q) \rightarrow (q \rightarrow r).$$

Solution Eliminate both implications and simplify by first replacing p by $(\sim p \wedge \sim q)$ and q by $(q \rightarrow r)$ in rule 10a. Then reapply 10a to the implication $(q \rightarrow r)$. We have

$$(\sim p \wedge \sim q) \rightarrow (q \rightarrow r)$$

$\Leftrightarrow \sim(\sim p \wedge \sim q) \vee (q \rightarrow r)$	Implication
$\Leftrightarrow \sim(\sim p \wedge \sim q) \vee (\sim q \vee r)$	Implication
$\Leftrightarrow (\sim \sim p \vee \sim \sim q) \vee (\sim q \vee r)$	De Morgan law
$\Leftrightarrow (p \vee q) \vee (\sim q \vee r)$	Double negation
$\Leftrightarrow [p \vee (q \vee \sim q) \vee r]$	Associative law
$\Leftrightarrow [(q \vee \sim q) \vee p \vee r]$	Commutative law
$\Leftrightarrow [t \vee (p \vee r)]$	Rule 7a
$\Leftrightarrow t$	Identity rule

Thus, the original statement is equivalent to a tautology.

We used the principle of *substitution* in the solution to the previous example. This allowed us to substitute statement forms for the simple statements in the rules of Table 1. We also substituted equivalent statement forms in interpreting the rules. The *substitution principles* are formally stated as follows:

Substitution Principles

Suppose P and R are statement forms and $P \Leftrightarrow R$. If R is substituted for P in a statement form Q, the resulting statement form Q' is logically equivalent to Q.

Suppose that P and Q are statement forms containing the simple statement p. Assume that $P \Leftrightarrow Q$. If R is any statement form and is substituted for every p in P to yield P' and for every p in Q to yield Q', then $P' \Leftrightarrow Q'$.

EXAMPLE 9 Rewrite the statement form

$$(p \to q) \wedge (q \vee r)$$

(a) using only the connectives $\sim$ and $\vee$

(b) using only the connectives $\sim$ and $\wedge$.

Solution (a) $(p \to q) \wedge (q \vee r)$

$\Leftrightarrow (\sim p \vee q) \wedge (q \vee r)$	Implication
$\Leftrightarrow (q \vee \sim p) \wedge (q \vee r)$	Commutative law
$\Leftrightarrow q \vee (\sim p \wedge r)$	Distributive law
$\Leftrightarrow q \vee (\sim p \wedge \sim \sim r)$	Double negation
$\Leftrightarrow q \vee \sim (p \vee \sim r)$	De Morgan law

(b) $(p \to q) \wedge (q \vee r)$

$\Leftrightarrow (\sim p \vee q) \wedge (q \vee r)$	Implication
$\Leftrightarrow (\sim p \vee \sim \sim q) \wedge (q \vee r)$	Double negation
$\Leftrightarrow \sim (p \wedge \sim q) \wedge (q \vee r)$	De Morgan law
$\Leftrightarrow \sim (p \wedge \sim q) \wedge \sim \sim (q \vee r)$	Double negation
$\Leftrightarrow \sim (p \wedge \sim q) \wedge \sim (\sim q \wedge \sim r)$	De Morgan law

---

EXAMPLE 10 Use the implication rule to show that

$$[(p \wedge \sim r) \to (q \to r)] \Leftrightarrow [\sim (p \wedge \sim r) \vee (q \to r)].$$

Solution   The implication rule states that $p \rightarrow q \Leftrightarrow \sim p \vee q$. Substitute the statement form $p \vee \sim r$ for each p on both sides of the logical equivalence. Substitute $q \rightarrow r$ for each q in the logical equivalence.

Recall that we have already seen the De Morgan laws in another form in Chapter 5. In that case, S and T were sets and

$$(S \cup T)' = S' \cap T' \quad \text{and} \quad (S \cap T)' = S' \cup T'.$$

The De Morgan laws for unions and intersections of sets are closely related to the De Morgan laws for the connectives $\vee$ and $\wedge$. In Section 6, we will explore this further.

We have seen in this section how statement forms that look quite different can be logically equivalent. Statement forms sometimes can be simplified by using the logical equivalencies of Table 1. Logical argument is based not only on logical equivalences, but also on those statements that logically imply others.

> Given statement forms P and Q, we say that P *logically implies* Q whenever $P \rightarrow Q$ is a tautology. We write this as $P \Rightarrow Q$. $P \Rightarrow Q$ means that $P \rightarrow Q$ is a tautology.

This means that $P \Rightarrow Q$ if, whenever P is true, then Q is true. To determine whether a statement form P logically implies a statement form Q, we only need to consider those rows of the truth table for which P is TRUE or those for which Q is FALSE.

EXAMPLE 11   Verify $[(p \vee q) \wedge \sim q] \Rightarrow p$.

Solution   Let statement form $P = (p \vee q) \wedge \sim q$ and statement form $Q = p$. We are interested in verifying that $P \rightarrow Q$ is a tautology. In a straightforward approach, we can construct the entire truth table.

p	q	$[(p \vee q)$	$\wedge$	$\sim q]$	$\rightarrow$	p
T	T	T	F	F	T	
T	F	T	T	T	T	
F	T	T	F	F	T	
F	F	F	F	T	T	
(1)	(2)	(3)	(5)	(4)	(6)	

Column 6 shows that the implication is a tautology. We could have shortened our work. The only situation in which an implication $P \rightarrow Q$ is FALSE is when P is TRUE and Q is FALSE. We check out these possibilities by first letting P be TRUE. This requires that both $(p \vee q)$ and $\sim q$ be TRUE. This occurs when q is FALSE and p is TRUE, and this is the only possibility for P to be TRUE. The row of the truth table we consider is

$$p \quad q \quad [(p \vee q) \wedge \sim q] \to p$$
$$\text{T} \quad \text{F} \quad \quad \text{T} \quad \text{T} \quad \text{T} \quad \text{T}$$

In this example it was simple to construct the entire truth table, but, in general, we can save a lot of work by considering only the relevant cases.

Table 2 summarizes some useful logical implications.

TABLE 2   Logical Implications

1.	$p \Rightarrow (p \vee q)$	Addition
2.	$(p \wedge q) \Rightarrow p$	Simplification
3.	$[p \wedge (p \to q)] \Rightarrow q$ (law of detachment)	Modus ponens
4.	$[(p \to q) \wedge \sim q] \Rightarrow \sim p$	Modus tollens
5.	$[(p \vee q) \wedge \sim p] \Rightarrow q$	Disjunctive syllogism
6.	$[(p \to q) \wedge (q \to r)] \Rightarrow (p \to r)$	Hypothetical syllogism
7a.	$[(p \to q) \vee (r \to s)] \Rightarrow [(p \vee r) \to (q \vee s)]$	Constructive
7b.	$[(p \to q) \wedge (r \to s)] \Rightarrow [(p \wedge r) \to (q \wedge s)]$	dilemmas

EXAMPLE 12   Verify logical implication 3 (modus ponens).

Solution   We must show that

$$[p \wedge (p \to q)] \to q$$

is a tautology. A truth table will suffice.

$$p \quad q \quad [p \wedge (p \to q)] \to q$$

p	q	$[p$	$\wedge$	$(p \to q)]$	$\to$	q
T	T		T	T		T
T	F		F	F		T
F	T		F	T		T
F	F		F	T		T
(1)	(2)		(4)	(3)		(5)

Looking at the last column, we see that the implication is a tautology.

PRACTICE PROBLEMS 4

1. Prove $(p \to q) \Leftrightarrow (\sim p \vee q)$.

2. Rewrite the statement

$$(p \to \sim q) \to (r \wedge p)$$

eliminating the connective $\to$ and simplifying if possible.

3. Assume the statement "Jim goes to the ballgame only if Ted gets tickets" is true.

    (a) Write the contrapositive and give its truth value.

    (b) Write the converse and give its truth value.

    (c) Write the negation and give its truth value.

EXERCISES 4

1. Show that $[(p \to q) \wedge q] \to p$ is not a tautology.

2. Show that the distributive laws hold:

    (a) $[p \vee (q \wedge r)] \Leftrightarrow [(p \vee q) \wedge (p \vee r)]$

    (b) $[p \wedge (q \vee r)] \Leftrightarrow [(p \wedge q) \vee (p \wedge r)]$

3. Show that the second law of implication holds:

$$(p \to q) \Leftrightarrow \sim(p \wedge \sim q).$$

4. Write a statement equivalent to

$$\sim(p \wedge \sim q) \to (r \wedge p)$$

    using only $\sim$ and $\wedge$.

5. The Sheffer stroke is defined by the truth table

p	q	$p|q$
T	T	F
T	F	T
F	T	T
F	F	T

    (a) Show that $\sim p \Leftrightarrow p|p$.

    (b) Show that $p \vee q \Leftrightarrow (p|p)|(q|q)$.

    (c) Show that $p \wedge q \Leftrightarrow (p|q)|(p|q)$.

    (d) Write $p \to q$ using only the Sheffer stroke.

    (e) Write $p|q$ using only $\sim$ and $\wedge$.

    (f) Write $p \oplus q$ using only the Sheffer stroke.

6. Without using truth tables, show that

$$[(p \vee \sim q) \wedge r] \to p \Leftrightarrow (p \vee q) \vee \sim r$$

7. Show that

$$(p \to q) \Leftrightarrow [(p \wedge \sim q) \to c]$$

    using the fact that c denotes a contradiction; that is, c is always FALSE. Read the statement aloud.

8. (a) Prove that

$$p \Rightarrow [q \to (p \wedge q)].$$

    (b) True or false?

$$p \Leftrightarrow [q \to (p \wedge q)].$$

9. True or false?
$$(p \lor q) \Rightarrow [q \to (p \land q)].$$

10. True or false?
$$[p \lor (p \land q)] \Leftrightarrow p.$$

11. Write an equivalent form of $p \oplus q$ using only $\sim$ and $\lor$.

12. Write each statement using only connectives $\sim$ and $\lor$.

   (a) $p \land \sim q \to p$ (b) $(p \to r) \land (q \lor r)$

   (c) $p \to [r \land (p \lor q)]$ (d) $(p \land q) \to (\sim q \lor r)$

13. Negate the following statements.

   (a) Arizona borders California and Arizona borders Nevada.

   (b) There are tickets available or the agency can get tickets.

   (c) The killer's hat was either white or gray.

14. Negate the following statements.

   (a) Montreal is a province and Ottawa is a province in Canada.

   (b) The salesman goes to the customer or the customer calls the salesman.

   (c) The hospital does not admit psychiatric patients or orthopedic patients.

15. Negate the following statements.

   (a) If I had a ticket to the theater, I spent a lot of money.

   (b) Basketball is played on an indoor court only if the players wear sneakers.

   (c) The stock market is going up implies that the interest rates are going down.

   (d) For man to stay healthy, it is sufficient that man have enough water.

16. For each statement, give the contrapositive, the converse, and the negation. Determine the truth value in each case.

   (a) If a rectangle has equal sides, it is a square.

   (b) An airplane flies faster than the speed of sound only if it is a Concorde.

   (c) If the intersection of two sets is not empty, then the union of the two sets is not empty.

   (d) If a coin is fair, the probability of a head is $\frac{1}{2}$.

17. Give the contrapositive and the converse of each statement and give the truth value of each.

   (a) If a bird is small, it is a hummingbird.

   (b) If two nonvertical lines have the same slope, they are parallel.

   (c) If we are in Paris, we must be in France.

   (d) If a road is one-way, you cannot legally make a U-turn.

18. Bill, Sue, and Alice are lined up facing forward with Bill first, then Sue, then Alice. From a collection which they know contains three blue and two red hats, hats are placed on their heads while they are blindfolded. Blindfolds are removed but they continue to face forward and see only those hats in front of them. Bill sees none, Sue sees Bill's, and Alice sees both Sue's and Bill's. Alice claims she does not know what color her hat is. Sue also claims that she does not know what color her hat is. Now Bill knows what color his hat is. What is it?

19. A prisoner is given one chance for freedom. He may ask one yes-or-no question of either of two guards. Each guard allows access to one of two unmarked doors. One is the door to freedom, the other the door to death. One guard always tells the truth, the other always lies. What question should the prisoner ask?

SOLUTIONS TO PRACTICE PROBLEMS 4

1. We show that $(p \rightarrow q) \leftrightarrow (\sim p \vee q)$ is a tautology.

p	q	$(p \rightarrow q)$	$\leftrightarrow$	$(\sim p$	$\vee$	$q)$
T	T	T	T	F	T
T	F	F	T	F	F
F	T	T	T	T	T
F	F	T	T	T	T
(1)	(2)	(3)	(6)	(4)	(5)

2. $(p \rightarrow \sim q) \rightarrow (r \wedge p)$

$\Leftrightarrow (\sim p \vee \sim q) \rightarrow (r \wedge p)$ Implication

$\Leftrightarrow [\sim(\sim p \vee \sim q)] \vee (r \wedge p)$ Implication

$\Leftrightarrow (p \wedge q) \vee (r \wedge p)$ De Morgan law

$\Leftrightarrow (p \wedge q) \vee (p \wedge r)$ Commutative law

$\Leftrightarrow p \wedge (q \vee r)$ Distributive law

3. (a) Contrapositive: "If Ted does not get the tickets, Jim does not go to the ballgame." TRUE.

(b) Converse: "If Ted gets the tickets, Jim goes to the ballgame." Truth value is unknown.

(c) We negate a statement of the form $p \rightarrow q$. We use the implication law $(p \rightarrow q) \Leftrightarrow (\sim p \vee q)$, form the negative, and apply a De Morgan law to get $[\sim(p \rightarrow q)] \Leftrightarrow (p \wedge \sim q)$. Negation: "Jim goes to the ballgame and Ted does not get tickets." FALSE.

12.5. Valid Argument

In this section, we study methods of valid argument. We rely heavily on the Table of Logical Implications presented in the previous section. Let us begin with an example.

EXAMPLE 1   Analyze the argument, given that the first two statements are true.

> If Marvin studies mathematics, then he is smart.
>
> Marvin is not smart.
>
> Therefore, Marvin does not study mathematics.

Solution   Let p denote "Marvin studies mathematics" and let q denote "Marvin is smart." The argument is of the form

$$\text{If } [(p \to q) \wedge \sim q], \text{ then } \sim p.$$

Note that

$$[(p \to q) \wedge \sim q] \Rightarrow \sim p$$

is the rule of modus tollens of Table 2 of the previous section. Thus, the implication

$$[(p \to q) \wedge \sim q] \to p$$

is a tautology and, therefore, always TRUE. The argument presented is valid.

The technique used in the example can be extended to more complex arguments.

An *argument* or a *proof* is a set of statements

$$H_1, H_2, \ldots, H_n$$

each of which is assumed to be true and a statement C that is claimed to have been deduced from them. The statements

$$H_1, H_2, \ldots, H_n$$

are called *hypotheses* and the statement C is called the *conclusion*. We say that the argument is *valid* if and only if

$$H_1 \wedge H_2 \wedge \cdots \wedge H_n \Rightarrow C.$$

The statement $H_1 \wedge H_2 \wedge \cdots \wedge H_n \to C$ is a tautology provided it is never false; if each hypothesis is true, the conclusion is also true.

In the example given above, the argument is valid because, for the hypotheses

$$H_1 = (p \to q) \quad \text{and} \quad H_2 = \sim q$$

and the conclusion

$$C = \sim p,$$

we have the logical implication, $H_1 \wedge H_2 \Rightarrow C$.

There are several important points to make here. First, although the conclusion may be a true statement, the argument presented may or may not be valid. Also, if one or more of the premises is false, it is possible for a valid argument to result in a conclusion that is false. The logical implications stated earlier in this chapter can be restated in the form of *rules of inference* as given in Table 1.

TABLE 1   Rules of Inference

	From	Conclude	
1.	P	$P \lor Q$	Addition
2.	$P \land Q$	P	Subtraction
3.	$P \land (P \to Q)$	Q	Modus ponens
4.	$(P \to Q) \land \sim Q$	$\sim P$	Modus tollens
5.	$(P \lor Q) \land \sim P$	Q	Disjunctive syllogism
6.	$(P \to Q) \land (Q \to R)$	$P \to R$	Hypothetical syllogism
7a.	$(P \to Q) \lor (R \to S)$	$(P \lor R) \to (Q \lor S)$	
b.	$(P \to Q) \land (R \to S)$	$(P \land Q) \to (Q \land S)$	Constructive dilemmas

EXAMPLE 2   Suppose that the following statements are true.

> Pat is going to the office or to dinner with Jan.
> If Pat is going to the office, then it is not sunny.
> It is sunny.

Prove: Pat is going to dinner with Jan.

Solution   Let s, p, and d denote the statements

$$s = \text{``It is sunny.''}$$

$$p = \text{``Pat is going to the office.''}$$

$$d = \text{``Pat is going to dinner with Jan.''}$$

We write the argument step by step. Numbers in parentheses refer to the previous steps used in the argument.

1.	$p \lor d$	Hypothesis
2.	$p \to \sim s$	Hypothesis
3.	s	Hypothesis
4.	$\sim p$	Modus tollens (2, 3)
5.	d	Disjunctive syllogism (1, 4)

Step 4 results from the implication $[(p \to \sim s) \land s] \Rightarrow \sim p$. And step 5 results from the implication $[(p \lor d) \land \sim p] \Rightarrow d$. We have made convenient substitutions in the rules in Table 1.

EXAMPLE 3   Show that the following argument is valid.

> Either I study mathematics or economics.
> If I have to take English, then I do not study economics.
> I do not study mathematics.
> Therefore, I do not have to take English.

Solution   Let p, q, and r represent the statements

$$p = \text{``I study mathematics.''}$$

$$q = \text{``I study economics.''}$$

$$r = \text{``I have to take English.''}$$

The hypotheses are

$$p \vee q, \quad r \rightarrow {\sim}q, \quad \text{and} \quad {\sim}p.$$

We write the argument step by step. Numbers in parentheses refer to the steps used in the argument.

1. $p \vee q$	Hypothesis
2. $r \rightarrow {\sim}q$	Hypothesis
3. ${\sim}p$	Hypothesis
4. q, since $[(p \vee q) \wedge {\sim}p] \Rightarrow q$	Disjunctive syllogism (1, 3)
5. ${\sim}r$, since $[(r \rightarrow {\sim}q) \wedge q] \Rightarrow {\sim}r$	Modus tollens (2, 4)

Thus, we have shown that r can be proven false by a valid argument from the given hypotheses.

EXAMPLE 4   Verify that the following argument is valid.

If it is raining, then my car stalls or will not start.
If my car stalls, then I arrive late for school.
If my car does not start, I do not go to school.
Therefore, if it is raining, either I arrive late for school or I do not go to school.

Solution   Let p, q, r, s and u be the following statements.

$$p = \text{``It is raining.''}$$

$$q = \text{``My car stalls.''}$$

$$r = \text{``My car starts.''}$$

$$s = \text{``I arrive late for school.''}$$

$$u = \text{``I go to school.''}$$

The argument proceeds as follows.

1. $p \rightarrow (q \vee {\sim}r)$	Hypothesis
2. $q \rightarrow s$	Hypothesis
3. ${\sim}r \rightarrow {\sim}u$	Hypothesis
4. $(q \vee {\sim}r) \rightarrow (s \vee {\sim}u)$	Constructive dilemma (2, 3)
5. $p \rightarrow (s \vee {\sim}u)$	Hypothetical syllogism (1, 4)

Therefore, a valid conclusion is "If it rains, either I arrive late for school or I do not go to school."

In trying to deduce that

$$H_1 \wedge H_2 \wedge \cdots \wedge H_n \Rightarrow C,$$

we sometimes find it easier to prove that the contrapositive of

$$H_1 \wedge H_2 \wedge \cdots \wedge H_n \to C$$

is a tautology. Thus, we would need to show that

$$\sim C \to \sim(H_1 \wedge H_2 \wedge \cdots \wedge H_n)$$

is a tautology. If we assume that $\sim C$ is TRUE, then, of course, the premise is that C is FALSE. The only case we need to consider to prove the tautology is the case where $\sim C$ is TRUE. In that case, we must show that $\sim(H_1 \wedge H_2 \wedge \cdots \wedge H_n)$ is TRUE also. This requires that the statement

$$(\sim H_1) \vee (\sim H_2) \vee \cdots \vee (\sim H_n)$$

be TRUE (use the De Morgan law). The disjunction is TRUE if and only if at least one of its components is TRUE. So we must show that $(\sim H_i)$ is TRUE for some subscript i. Hence, we are required to show that H_i is FALSE for at least one i. This is called an *indirect proof.*

We summarize the idea.

Indirect Proof   To prove

$$(H_1 \wedge H_2 \wedge \cdots \wedge H_n) \Rightarrow C,$$

we assume the conclusion C is FALSE and then prove that at least one of the hypotheses H_i must be FALSE.

EXAMPLE 5   Use an indirect proof to show that the following argument is valid.

> If I am happy, then I do not eat too much.
> I eat too much or I spend money.
> I do not spend money.
> Therefore, I am not happy.

Solution   We will begin symbolically. Let

$$p = \text{"I am happy."}$$

$$q = \text{"I eat too much."}$$

$$r = \text{"I spend money."}$$

We wish to show that $[(p \rightarrow \sim q) \wedge (q \vee r) \wedge \sim r] \Rightarrow \sim p$. Assume, by way of indirect proof, that the conclusion is FALSE.

1.	p	Negation of conclusion
2.	$p \rightarrow \sim q$	Hypothesis
3.	$\sim q$	Modus ponens (1, 2)
4.	$q \vee r$	Hypothesis
5.	r	Disjunctive syllogism (3, 4)

Therefore, the given hypothesis $\sim r$ is false. We express the proof in words. Let us assume the negative of the conclusion; that is, we assume "I am happy." Then "I do not eat too much" (modus ponens). We were to assume "I eat too much or I spend money"; hence (modus tollens), it is true that "I spend money." This means that the hypothesis "I do not spend money" cannot be true. Assumption of the negation of the conclusion leads us to claim one of the hypotheses is false. Thus, we have a valid indirect proof.

PRACTICE PROBLEMS 5

1. Show that the argument is valid.

> If goldenrod is yellow, then violets are blue.
> Either pine trees are not green or goldenrod is yellow.
> Pine trees are green.
> Therefore, violets are blue.

2. Show by indirect proof that the argument is valid.

> If I go to the beach, I cannot study.
> Either I study or I work as a waiter.
> I go to the beach.
> Therefore, I work as a waiter.

EXERCISES 5

In Exercises 1–10, show that the argument is valid.

1. Either Sue goes to the movies or she reads. Sue does not go to the movies. Therefore Sue reads.

2. If the class votes for an oral final, the teacher is glad. If the exam is not scheduled for a Monday, the teacher is sad. The exam is scheduled for a Friday. Therefore, the class doesn't vote for an oral final.

3. If my allowance comes this week and I pay the rent, then my bank account will be

in the black. If I do not pay the rent, I will be evicted. I am not evicted and my allowance comes. Therefore, my bank account is in the black.

4. Either Jane is in sixth grade implies that Jane understands fractions or Jane is in sixth grade implies that Jane is in a remedial math class. Jane is in sixth grade. Therefore, either Jane understands fractions or she is in a remedial math class.

5. If the price of oil increases, the OPEC countries are in agreement. If there is no U.N. debate, the price of oil increases. The OPEC countries are in disagreement. Therefore, there is a U.N. debate.

6. If Jill wins, then Jack loses. If Peter wins, then Paul loses. Either Jill wins or Peter wins. Therefore, either Jack loses or Paul loses.

7. If the germ is present, then the rash and the fever are present. The fever is present. The rash is not present. Therefore, the germ is not present.

8. If Hal is a politician, he is a liar or a fraud. Hal is not a liar. He is not a fraud. Therefore, Hal is not a politician.

9. If the material is cotton or rayon, it can be made into a dress. The material cannot be made into a dress. Therefore, it is not rayon.

10. If there is money in my account and I have a check, then I will pay the rent. If I do not have a check, then I am evicted. Therefore, if I am not evicted and if I do not pay the rent, then there is no money in my account.

In Exercises 11–20, test the validity of the arguments.

11. If the salaries go up, then more people apply. Either more people apply or the salaries go up. Therefore, the salaries go up.

12. If Rita studies, she gets good grades. Rita gets bad grades. Therefore, Rita does not study.

13. Either the balloon is yellow or the ribbon is pink. If the balloon is filled with helium, then the balloon is a green one. The balloon is filled with helium. Therefore, the ribbon is pink.

14. If the job offer is for at least $30,000 or has 5 weeks vacation, I will accept the position. If the offer is for less than $30,000, then I will not accept the job and I will owe rent money. I will not accept the job. Therefore, I will owe rent.

15. If the papa bear sits, then the mama bear stands. If the mama bear stands, the baby bear crawls on the floor. The baby bear is standing. Therefore, the papa bear is not sitting.

16. If it is snowing, I wear my boots. It is not snowing. Therefore, I am not wearing boots.

17. If wheat prices are steady, exports will increase or the GNP will be steady. Wheat prices are steady and the GNP is steady. Therefore, exports will increase.

18. If we eat out, either Mom or Dad will treat, I pay for dinner. Therefore, we do not eat out.

19. If Tim is industrious, then Tim is in line for a promotion. Either Tim is in line for a promotion or he is thinking of leaving. Therefore, if Tim is thinking of leaving, then Tim is industrious.

20. If I pass history, then I do not go to summer school. If I go to summer school, I will take a course in French. Therefore, if I go to summer school, then either I do not pass history or I take a course in French.

In Exercises 21–25, use indirect proof to show the argument is valid.

21. Sam goes to the store only if he needs milk. Sam does not need milk. Therefore, Sam does not go to the store.

22. If it rains hard, there will be no picnic. If Dave brings the Frisbee, the kids will be happy. The kids are not happy and there is a picnic. Therefore, it does not rain hard and Dave did not bring the Frisbee.

23. If the newspaper and television both report a crime, then it is a serious crime. If a person was killed, then the newspaper reports the crime. A person is killed. Television reports the crime. Therefore, the crime is serious.

24. If Linda feels ill, she takes aspirin. If she runs a fever, she does not take a bath. If Linda does not feel ill, she takes a bath. Linda runs a fever. Therefore, she takes an aspirin.

SOLUTIONS TO PRACTICE PROBLEMS 5

1. Let g = "Goldenrod is yellow."

   v = "Violets are blue."

   p = "Pine trees are green."

   The following steps show the argument is valid.

   (1) $g \rightarrow v$ Hypothesis

   (2) $\sim p \vee g$ Hypothesis

   (3) p Hypothesis

   (4) g Disjunctive syllogism (2, 3)

   (5) v Modus ponens (1, 4)

2. Let p = "I go to the beach."

   q = "I study."

   r = "I work as a waiter."

   Begin by assuming the negation of the conclusion.

   (1) $\sim r$ Negation of conclusion

   (2) $p \rightarrow \sim q$ Hypothesis

   (3) $q \vee r$ Hypothesis

   (4) q Disjunctive syllogism (1, 3)

   (5) $\sim p$ Modus tollens (2, 4)

   Since one of the hypotheses has been shown to be false, the argument is valid.

12.6. Predicate Calculus

In previous discussion, we noted that a sentence of the form

"The number x is even."

is not a statement. Although the sentence is a declarative one, we cannot determine whether the statement is true or false because we do not know to what x refers. For example, if $x = 3$, then the statement is FALSE. If $x = 6$, then the statement is TRUE.

> We define an *open sentence* $p(x)$ to be a declarative sentence that becomes a statement when x is given a particular value chosen from a universe of values.

Consider the open sentence $p(x) = $ "If x is even, $x - 6 > 8$." Let us consider as possible values for x all those integers greater than zero, so the universe $U = \{1, 2, 3, 4, \ldots\}$. Then the open sentence $p(x)$ represents many statements, one for each value of x chosen from U. Let us state $p(x)$ and record its truth value for several possible values of x. We note that the open sentence $p(x)$ becomes the statement $p(1)$ when we substitute the specific value 1 for the indeterminate letter x. Recall that an implication $p \to q$ is false if and only if p is true and q is false.

$p(1) = $ "If 1 is even, $-5 > 8$" is TRUE (since the hypothesis is FALSE).

$p(2) = $ "If 2 is even, $-4 > 8$" is FALSE.

$p(3) = $ "If 3 is even, $-3 > 8$" is TRUE.

$p(4) = $ "If 4 is even, $-2 > 8$" is FALSE.

$p(20) = $ "If 20 is even, $14 > 8$" is TRUE.

$p(21) = $ "If 21 is even, $15 > 8$" is TRUE.

In fact, $p(x)$ is TRUE for any odd integer x. And $p(x)$ is TRUE for all values of x such that x is even and $x > 14$. The only x for which $p(x)$ is FALSE is x both even and between 2 and 14, inclusive. That is, $p(2)$, $p(4)$, $p(6)$, $p(8)$, $p(10)$, $p(12)$, and $p(14)$ are FALSE. All other values of x make $p(x)$ a TRUE statement.

EXAMPLE 1 Let

$$p(x) = \text{"If } x > 4, \text{ then } x + 10 > 14\text{."}$$

be an open sentence. Let $x \in U$, where $U = \{1, 2, 3, 4, \ldots\}$. Find the truth value of each statement formed when these values are substitued for x in $p(x)$.

Solution $p(1)$ is TRUE because if $x = 1$, the hypothesis is FALSE.
$p(2)$ is TRUE because if $x = 2$, the hypothesis is FALSE.
$p(3)$ is TRUE because if $x = 3$, the hypothesis is FALSE.
$p(4)$ is TRUE because if $x = 4$, the hypothesis is FALSE.
$p(5)$ is TRUE because if $x = 5$, then $x + 10 > 14$.

In fact, $p(x)$ is TRUE for all values of U. We say that for all $x \in U$, if $x > 4$, then $x + 10 > 14$.

The statement

$$\text{For all } x \in U, \, p(x)$$

is symbolized by

$$\forall x \in U \; p(x).$$

> The statement $\forall x \in U \; p(x)$ is TRUE if and only if $p(x)$ is TRUE for every $x \in U$.

We call the symbol $\forall$ the *universal quantifier* and we read it as "for all," "for every," or "for each." The universal set must be known in order to decide if $\forall x \in U \; p(x)$ is TRUE or FALSE. We emphasize that, whereas $p(x)$ is an open sentence with no assignable truth value, $\forall x \in U \; p(x)$ is a legitimate logical statement. At times, the notation will be abbreviated to $\forall x \; p(x)$ when the universe U is clear, or to $\forall x \, [p(x)]$ for further clarity.

From the example, we can see that the statement "For all $x \in U$, if $x > 4$, then $x + 10 > 14$" is a TRUE statement for $U = \{1, 2, 3, \ldots\}$.

EXAMPLE 2 Let $U = \{1, 2, 3, 4, 5, 6\}$. Determine the truth value of the statement

$$\forall x, \, [(x - 4)(x - 8) > 0].$$

Solution We let $p(x)$ be the open statement "$(x - 4)(x - 8) > 0$." We consider the truth value of $p(1)$, $p(2)$, $p(3)$, $p(4)$, $p(5)$, and $p(6)$. We note that $p(1)$ is TRUE because $(1 - 4)(1 - 8) = (-3)(-7) = 21$ and $21 > 0$. We find that $p(2)$ and $p(3)$ are also TRUE. However, $p(4)$ is FALSE because $(4 - 4)(4 - 8) = 0$. We need not check any other values from U. Already, we know that

$$\forall x \in U, \, p(x) \text{ is FALSE.}$$

We note that there are values of x in U for which $p(x)$ is TRUE.

The statement

$$\text{There exists an } x \text{ in } U \text{ such that } p(x)$$

is symbolized by

$$\exists x \in U \; p(x).$$

> The statement $\exists x \in U \; p(x)$ is TRUE if and only if there is at least one element $x \in U$ such that $p(x)$ is TRUE.

The symbol $\exists$ is called the *existential quantifier*. We read the existential quantifier as "there exists x such that $p(x)$," "for some x, $p(x)$," or "there is some x for which

$p(x)$." We may write $\exists x\, p(x)$ when the universal set is clear. When the universal set does not appear explicitly in the statement, U must be made clear in order for $\exists x\, p(x)$ be a statement and to have a truth value.

EXAMPLE 3 Determine the truth value of the following statements where $U = \{1, 2, 3, 4, 5, 6, 7, 8\}$.

(a) $\forall x\, (x + 3 < 15)$ (b) $\exists x\, (x > 5)$

(c) $\exists x\, (x^2 = 0)$ (d) $\forall x\, (0 < x < 10)$

(e) $\forall x\, [(x + 2 > 5) \vee (x \leq 3)]$ (f) $\exists x\, [(x - 1)(x + 2) = 0]$

Solution (a) is TRUE because for every $x \in U$, $x + 3$ is less than 15. The statement in (b) is TRUE because, for example, $7 \in U$ and $7 > 5$. Statement (c) is FALSE because there is no $x \in U$ that satisfies the equation $x^2 = 0$. Statement (d) is TRUE. Statement (e) is TRUE because for each $x \in U$ either $x + 2 > 5$ or $x \leq 3$. Statement (f) is TRUE because $1 \in U$ and $(1 - 1)(1 + 2) = 0$.

Sometimes it is convenient to let U be a very large set, the real numbers for example, and to restrict U within the statement itself. Consider the statement

For every positive integer x, either x is even or $x + 1$ is even.

One way to write this is to let U be the set of positive integers and to write

$$\forall x \in U\, [(x \text{ is even}) \vee (x + 1 \text{ is even})]$$

An equivalent statement that more explicitly shows the universal set would be

$$\forall x \in U\, [(x \text{ is a positive integer}) \rightarrow ((x \text{ is even}) \vee (x + 1 \text{ is even}))].$$

The universal set can then be the set of all integers. The universe is restricted by using an implication with the hypothesis that x is a positive integer in the open sentence $p(x)$. We can allow U to be any set containing the positive integers and restrict U by using such an implication.

A statement with the existential quantifier can be rephrased to include the universe in the statement. For example, let U be the set of positive integers and consider the statement

$$\exists x \in U\, [(x - 3)(x + 6) = 0]$$

We can rewrite the statement as

$$\exists x \in U\, [x \text{ is a positive integer} \wedge ((x - 3)(x + 6) = 0)]$$

We could let U be the set of all integers or the reals, if we wish. Any set containing the positive integers would be an acceptable choice for U. The existential operator and the connective $\wedge$ can be used to restrict the universe to the positive integers.

We frequently have reason to negate a quantified statement. Of course, the negation changes the truth value from TRUE to FALSE or from FALSE to TRUE. The negations should be properly worded.

EXAMPLE 4   Negate the statement

"For every positive integer x, either x is even or $x + 1$ is even."

Solution   We could say, "It is not the case that for every positive integer x either x is even or $x + 1$ is even." That is not entirely satisfactory, however. What we mean is that there exists a positive integer for which it is not the case that x is even or $x + 1$ is even. Assuming that U is the set of positive integers, this is written symbolically as

$$\sim[\forall x \,((x \text{ is even}) \vee (x + 1 \text{ is even}))] \Leftrightarrow \exists x \,[\sim((x \text{ is even}) \vee (x + 1 \text{ is even}))]$$

$$\Leftrightarrow \exists x \,[(x \text{ is odd}) \wedge (x + 1 \text{ is odd})]$$

The last equivalence follows from the De Morgan laws, which allow us to rewrite a statement of the form $\sim(p \vee q)$ as $(\sim p \wedge \sim q)$. The negation of the original statement then is "There exists a positive integer x such that x is odd and $x + 1$ is odd."

EXAMPLE 5   Negate the statement "There exists a positive integer x such that $x^2 - x - 2 = 0$."

Solution   One way to negate the statement is to say, "It is not the case that there exists a positive integer x such that $x^2 - x - 2 = 0$." A better way is to say, "For all positive integers x, $x^2 - x - 2 \neq 0$." Note that

$$\sim[\exists x \,(x^2 - x - 2 = 0)] \Leftrightarrow [\forall x \,(x^2 - x - 2 \neq 0)].$$

We summarize the rules for the negation of quantified statements. The negation of a quantified statement can be rewritten by bringing the negation inside the open statement and replacing $\exists$ by $\forall$, or vice versa.

$$\sim[\exists x \, p(x)] \Leftrightarrow [\forall x \sim p(x)]$$
$$\sim[\forall x \, p(x)] \Leftrightarrow [\exists x \sim p(x)]$$

These rules are also called De Morgan laws.

EXAMPLE 6   Write the negation of each of the following statements.

(a) All university students like football.

(b) There is a mathematics textbook that is both short and clear.

(c) For every positive integer x, if x is even then $x + 1$ is odd.

Solution   (a) Here the universe is "university students" and $p(x)$ is the statement "Student x likes football." We negate $\forall x \, p(x)$ to get $\exists x \,[\sim p(x)]$, which in words is "There exists a university student who does not like football."

(b) The universe is the set of all mathematics textbooks. We have a statement of the form $\exists x \,[p(x) \wedge q(x)]$, where $p(x)$ denotes "The mathematics textbook is short" and $q(x)$ denotes "The mathematics textbook is clear." The negation of the statement is of the form $\sim [\exists x\, (p(x) \wedge q(x))] \Leftrightarrow [\forall x \sim (p(x) \wedge q(x))] \Leftrightarrow [\forall x\, (\sim p(x) \vee \sim q(x))]$ This translates into the English sentence "Every mathematics textbook is either not short or not clear."

(c) The universe is the set of positive integers. The statement is of the form

$$\forall x\, (p(x) \rightarrow q(x)),$$

where $p(x)$ is the statement "x is even" and $q(x)$ is "$x + 1$ is odd." We write the negation

$$\sim [\forall x\, (p(x) \rightarrow q(x))] \Leftrightarrow [\exists x \sim (p(x) \rightarrow q(x))] \Leftrightarrow [\exists x \sim (\sim p(x) \vee q(x))]$$
$$\Leftrightarrow [\exists x\, (p(x) \wedge \sim q(x))].$$

In English, this is "There exists a positive integer x such that x is even and $x + 1$ is also even."

We will have occasion to use sentences that have two open variables. For example, the sentence "For every child there exists an adult who cares for him or her" has two variables: the child and the adult. We can write the statement symbolically using two variables x and y. We let the universe for the x variable be the set of all children and the universe for the y variable be the set of all adults. Then we let $p(x, y) = $ "y cares for x." The statement can be written symbolically as

$$\forall x\, \exists y\, p(x, y).$$

Binding each of the variables with the universal or existential quantifier and a universal set, gives a legitimate logical statement.

EXAMPLE 7   Let the universe for each of the variables be the set of all Americans. Let

$$p(x, y) = x \text{ is taller than } y$$

$$q(x, y) = x \text{ is heavier than } y.$$

Write the English sentences from the symbolic statements.

(a) $\forall x\, \forall y\, [p(x, y) \rightarrow q(x, y)]$

(b) $\forall x\, \exists y\, [p(x, y) \wedge q(x, y)]$

(c) $\forall y\, \exists x\, [p(x, y) \vee q(x, y)]$

(d) $\exists x\, \exists y\, [p(x, y) \wedge q(x, y)]$

(e) $\exists x\, \exists y\, [p(x, y) \wedge q(y, x)]$

Solution   (a) For all Americans x and y, if x is taller than y, then x is heavier than y.

(b) For every American x, there is an American y such that x is taller and heavier than y.

(c) For every American y, there is an American x such that x is taller than y or x is heavier than y.

(d) There are Americans x and y such that x is taller than y and x is heavier than y.

(e) There are Americans x and y such that x is taller than y and y is heavier than x.

The rules for the negation of statements with more than one variable are just repeated applications of the De Morgan laws mentioned for single-variable statements. We have

$$\sim[\forall x\ \forall y\ p(x, y)] \Leftrightarrow \exists x\ [\sim \forall y\ p(x, y)] \Leftrightarrow \exists x\ \exists y\ [\sim p(x, y)]$$

and

$$\sim[\exists x\ \exists y\ p(x, y)] \Leftrightarrow \forall x\ [\sim \exists y\ p(x, y)] \Leftrightarrow \forall x\ \forall y\ [\sim p(x, y)]$$

We summarize these rules and several others in Table 1.

TABLE 1

1. $\sim[\forall x\ \forall y\ p(x, y)] \Leftrightarrow \exists x\ \exists y\ [\sim p(x, y)]$	
2. $\sim[\forall x\ \exists y\ p(x, y)] \Leftrightarrow \exists x\ \forall y\ [\sim p(x, y)]$	
3. $\sim[\exists x\ \forall y\ p(x, y)] \Leftrightarrow \forall x\ \exists y\ [\sim p(x, y)]$	De Morgan laws
4. $\sim[\exists x\ \exists y\ p(x, y)] \Leftrightarrow \forall x\ \forall y\ [\sim p(x, y)]$	
5. $\forall x\ \forall y\ p(x, y) \Leftrightarrow \forall y\ \forall x\ p(x, y)$	
6. $\exists x\ \exists y\ p(x, y) \Leftrightarrow \exists y\ \exists x\ p(x, y)$	
7. $\exists x\ \forall y\ p(x, y) \Rightarrow \forall y\ \exists x\ p(x, y)$	

Notice that rule 7 is a logical implication and not a logical equivalence. Any other exchanges of $\exists$ and $\forall$ should be handled very carefully. They are unlikely to give equivalent statements even in specific cases.

EXAMPLE 8   Let the universe for both variables be the nonnegative integers $0, 1, 2, 3, 4, \ldots$. Write out the statements in words and decide on the truth value of each.

(a) $\forall x\ \forall y\ [2x = y]$ (b) $\forall x\ \exists y\ [2x = y]$

(c) $\exists x\ \forall y\ [2x = y]$ (d) $\exists x\ \forall y\ [2y = x]$

(e) $\exists y\ \forall x\ [2x = y]$ (f) $\exists x\ \exists y\ [2x = y]$

(g) $\forall y\ \exists x\ [2x = y]$

Solution   (a) For every pair of nonnegative integers x and y, $2x = y$. This is FALSE because, for example, it is FALSE in the case $x = 3$, $y = 10$.

(b) For every nonnegative integer x, there is a nonnegative integer y such that $2x = y$. This is TRUE because, once having chosen any nonnegative number x, we can let y be the double of x.

(c) There exists a nonnegative integer x such that, for every nonnegative integer, y, $2x = y$. This statement is FALSE because, no matter what x we choose, $2x = y$ is TRUE only for y equal to twice x and for no other nonnegative number.

(d) There is a nonnegative number x such that, for all nonnegative numbers y, $2y = x$. This is FALSE for similar reasons as in (c).

(e) There exists a nonnegative integer y such that, for all nonnegative integers x, $2x = y$. The statement is FALSE. For it to be TRUE, we would need to find a specific value of y that can be fixed and for which, no matter what nonnegative integer x we choose, $2x = y$.

(f) There exist nonnegative integers x and y such that $2x = y$. This is TRUE. One choice that makes the statement $2x = y$ TRUE is $x = 4$ and $y = 8$.

(g) For every nonnegative integer y, there exists a nonnegative integer x such that $2x = y$. This is FALSE because if $y = 5$ there exists no nonnegative integer x such that $2x = 5$.

A statement of the form $\forall x\, p(x)$ is FALSE if $\exists x\, [\sim p(x)]$ is TRUE. If we want to show that the statement $\forall x\, p(x)$ is FALSE, we need only find a value for x for which $p(x)$ is FALSE. Such an example is called a *counterexample*. We used such a counterexample in parts (a) and (g) of Example 8.

Analogy between Sets and Statements Actually, the symbols and the truth values in the tables are suggestive of set theory definitions introduced in Chapter 5. Recall that we use U to denote the universe and the letters S and T to denote sets in that universe. Then the union of the sets S and T is defined as

$$S \cup T = \{x \in U : x \in S \text{ or } x \in T\}.$$

If we allow $p(x)$ to represent "$x \in S$" and $q(x)$ to represent "$x \in T$," then it is TRUE that $x \in S \cup T$ whenever the statement $p(x) \vee q(x)$ is TRUE. Thus, $\forall x\, [x \in S \cup T \leftrightarrow p(x) \vee q(x)]$. Similarly, we recall the definition of the intersection of two sets:

$$S \cap T = \{x \in U : x \in S \text{ and } x \in T\}.$$

It is TRUE that $x \in S \cap T$ whenever the statement $p(x) \wedge q(x)$ is TRUE, where $p(x)$ and $q(x)$ are as above. Thus, $\forall x\, [x \in S \cap T \leftrightarrow p(x) \wedge q(x)]$. In addition, we have

$$S' = \{x \in U : x \notin S\}.$$

Thus, $x \in S'$ whenever $\sim p(x)$ is TRUE. Hence, $\forall x\, [x \in S' \leftrightarrow \sim p(x)]$.

The exclusive "or" corresponds to the *symmetric difference of S and T* defined by $S \oplus T = \{x \in U : x \in S \cup T \text{ but } x \notin S \cap T\}$. That is, it is TRUE that $x \in S \oplus T$ whenever the statement $p(x) \oplus q(x)$ is TRUE or $\forall x\, [x \in S \oplus T \leftrightarrow p(x) \oplus s(x)]$.

As with the connectives $\vee$ and $\wedge$, the connectives $\rightarrow$ and $\leftrightarrow$ have counterparts in set theory. We will assume that the universe is U and that S and T are subsets of U. We let $p(x)$ be the statement "$x \in S$" and let $q(x)$ be the statement

"$x \in T$." Then we have seen that S is a subset of T if and only if every element of S is also an element of T. We say that

S is a subset of T if and only if $(x \in S) \to (x \in T)$ is TRUE for all x,

or, equivalently

S is a subset of T if and only if $\forall x \, [p(x) \to q(x)]$ is TRUE for all x.

Two sets are equal if they have the same elements. We claim that

$$S = T \text{ if and only if } (x \in S) \leftrightarrow (x \in T) \text{ is TRUE for all } x,$$

that is,

$$S = T \text{ if and only if } \forall x \, [p(x) \leftrightarrow q(x)] \text{ is TRUE for all } x.$$

EXAMPLE 9 Let $U = \{1, 2, 3, 4, 5, 6\}$, let $S = \{x \in U : x \leq 3\}$, and let $T = \{x \in U : x \text{ divides } 6\}$. Show $S \subseteq T$.

Solution We must show that $\forall x \, [(x \in S) \to (x \in T)]$ is TRUE. An implication is FALSE only in the case where the hypothesis is TRUE and the conclusion is FALSE. Let us suppose then that the hypothesis is TRUE and show that the conclusion must also be TRUE in that case. We assume that $x \in S$. By the definition of the set S, this means that x is 1 or 2 or 3. Then $x \in T$, since in each of the cases x divides 6 and is an element of T. Hence, $(x \in S) \to (x \in T)$ cannot be FALSE and must be TRUE for all $x \in U$.

Recall that we have already seen the De Morgan laws in another form in Chapter 5. In that case, we assumed that S and T were sets and showed that

$$(S \cup T)' = S' \cap T' \quad \text{and} \quad (S \cap T)' = S' \cup T'.$$

We know that $x \in S \cup T$ if and only if $(x \in S) \vee (x \in T)$ is TRUE. So $x \notin S \cup T$ if and only if the negation of the statement is TRUE. Thus $x \in (S \cup T)'$ if and only if $\sim[(x \in S) \vee (x \in T)]$ is TRUE. By the De Morgan law in Table 1 of Section 4, we know that $x \in (S \cup T)'$ if and only if $\sim(x \in S) \wedge \sim(x \in T)$ is TRUE. Thus, $x \in (S \cup T)'$ if and only if $(x \notin S) \wedge (x \notin T)$ is TRUE. This means that $x \in S' \cap T'$. The De Morgan laws for unions and intersections of sets are closely related to the De Morgan laws for the connectives $\vee$ and $\wedge$.

Colloquial Usage Again, we point out that colloquial usage is frequently not precise. For example, the statement "All students in finite mathematics courses do not fail" is strictly interpreted to mean that $\forall x \in U \, [x \text{ does not fail}]$ with $U =$ the set of students in finite mathematics courses. This means, of course, that no one fails. On the other hand, some might loosely interpret this to mean that not every student in finite mathematics courses fails. This is the statement $\exists x \in U \, [x \text{ does not fail}]$. We require precision of language in mathematics and do *not* accept the second interpretation as correct.

PRACTICE PROBLEMS 6

1. Write the following statement symbolically and write out its negation in English: "There exists a flower that can grow in sand and is not subject to mold."

2. For the universe $U = \{0, 1, 2, 3, 4, 5, 6, 7, 8\}$, determine the truth values of the following statements.

   (a) $\forall x \, (x > 2)$ (b) $\exists x \, (x > 2)$

   (c) $\forall x \, (x^2 < 100)$ (d) $\exists x \, [(x - 1 = 4) \wedge (3x + 5 = 20)]$

3. Consider the universe for both variables x and y to be the set of nonnegative integers $\{0, 1, 2, 3, 4, \ldots\}$. Write out the statements in English and determine the truth value of each.

   (a) $\forall x \, \forall y \, [x < y]$ (b) $\forall x \, \exists y \, [x < y]$

   (c) $\forall y \, \exists x \, [x < y]$ (d) $\exists x \, \forall y \, [x < y]$

   (e) $\exists x \, \exists y \, [x < y]$ (f) $\forall x \, \forall y \, [(x < y) \vee (y < x)]$

EXERCISES 6

1. Let $U = \{1, 2, 3, 4, 5, 6\}$. Determine the truth value of

   $$p(x) = [(x \text{ is even}) \text{ or } (x \text{ is divisible by } 3)]$$

   for the given values of x.

   (a) $x = 1$ (b) $x = 4$ (c) $x = 3$

   (d) $x = 6$ (e) $x = 5$

2. Determine the truth value of $p(x)$ for the values of x chosen from the universe of all letters of the alphabet where

   $p(x) = [(x \text{ is a vowel}) \text{ and } (x \text{ is in the word ABLE})]$.

   (a) $x = a$ (b) $x = d$ (c) $x = b$ (d) $x = i$

3. Consider the universe of all college students. Let $p(x)$ denote the open statement "x takes a writing course."

   (a) Write the statement "Every college student takes a writing course" in symbols.

   (b) Write the statement "Not all college students take a writing course" symbolically.

   (c) Write the statement "Every college student does not take a writing course" symbolically.

   (d) Do any of the statements in (a), (b), or (c) imply one another? Explain.

4. Recently, as the Amtrak train pulled into the Baltimore station, the conductor announced, "All doors do not open." Since passengers were permitted to get off at the station, what do you think the conductor meant to say?

5. An alert California teacher chided "Dear Abby" (*Baltimore Sun*, March 1, 1989) recently for her statement "Confidential to Eunice: All men do not cheat on their wives." Let $p(x)$ be the statement "x cheats on his wife" on the universe of all men. Write out Abby's statement symbolically. Rewrite the statement using the existential quantifier. Do you think the statement is TRUE or FALSE? What do you think Abby really meant to say?

6. Consider the universe of all orange juice. Write the following symbolically, using $p(x) =$ "x comes from Florida."

   (a) All orange juice does not come from Florida.

   (b) Some orange juice comes from Florida.

   (c) Not all orange juice comes from Florida.

   (d) Some orange juice does not come from Florida.

   (e) No orange juice comes from Florida.

   (f) Are any of the statements (a) through (e) equivalent to one another? Explain.

7. Let the universe be all university professors. Let $p(x)$ be the open statement "x likes poetry." Write the following statements symbolically.

   (a) All university professors like poetry.

   (b) Some university professors do not like poetry.

   (c) Some university professors like poetry.

   (d) Not all university professors like poetry.

   (e) All university professors do not like poetry.

   (f) No university professors like poetry.

   (g) Are any of the statements in (a) through (f) equivalent? Explain.

8. Let $U = \{0, 1, 2, 3, 4\}$. Let

$$p(x) = [x^2 > 9].$$

   Find the truth value of:

   (a) $\exists x\, p(x)$ (b) $\forall x\, p(x)$

9. Let $U = \{1, 2, 3, 4, 5, 6, 7, 8, 9\}$. Let

$$p(x) = [(x \text{ is prime}) \to (x^2 + 1 \text{ is even})]$$

   Find the truth value of:

   (a) $\exists x\, p(x)$ (b) $\forall x\, p(x)$

10. Let $U = \{3, 4, 5, 6, \ldots\}$. Let

$$p(x) = [(x \text{ is prime}) \to (x^2 + 1 \text{ is even})]$$

   Find the truth value of:

   (a) $\exists x\, p(x)$ (b) $\forall x\, p(x)$

11. Let the universe consist of all nonnegative integers. Let $p(x)$ be the statement "x is even." Let $q(x)$ be the statement "x is odd." Determine the truth value of:

(a) $\forall x \, [p(x) \vee q(x)]$

(b) $[\forall x \, p(x)] \vee [\forall x \, q(x)]$

(c) $\exists x \, [p(x) \vee q(x)]$

(d) $\exists x \, [p(x) \wedge q(x)]$

(e) $\forall x \, [p(x) \wedge q(x)]$

(f) $[\exists x \, p(x)] \wedge [\exists x \, q(x)]$

(g) $\forall x \, [p(x) \rightarrow q(x)]$

(h) $[\forall x \, p(x)] \rightarrow [\forall x \, q(x)]$

12. Let the universe consist of all real numbers. Let

$$p(x) = (x \text{ is positive}) \quad \text{and} \quad q(x) = (x \text{ is a perfect square})$$

Determine the truth value of

(a) $\forall x \, p(x)$

(b) $\forall x \, q(x)$

(c) $\exists x \, p(x)$

(d) $\exists x \, q(x)$

(e) $\exists x \, [p(x) \rightarrow q(x)]$

(f) $\exists x \, [q(x) \rightarrow p(x)]$

(g) $\forall x \, [p(x) \rightarrow q(x)]$

(h) $\forall x \, [q(x) \rightarrow p(x)]$

(i) $\forall x \, p(x) \rightarrow \exists x \, q(x)$

(j) $[\forall x \, p(x)] \rightarrow [\forall x \, q(x)]$

13. Negate the following statements.

(a) Every dog has its day.

(b) Some men fight wars.

(c) All mothers are married.

(d) For every pot, there is a cover.

(e) Not all children have pets.

(f) Not every month has 30 days.

14. Negate each statement by changing existential quantifiers to universal quantifiers, or vice-versa.

(a) Every stitch saves time.

(b) All books have hard covers.

(c) Some children are afraid of snakes.

(d) Not all computers have a hard disk.

(e) Some chairs do not have arms.

15. Consider the universe the set of nonnegative integers $= \{0, 1, 2, 3, 4, \ldots\}$. Write the English sentence for each symbolic statement. Determine the truth value of each statement. If the statement is FALSE, give a counterexample and write its negation out in words.

(a) $\forall x \, \forall y \, [x + y > 12]$

(b) $\forall x \, \exists y \, [x + y > 12]$

(c) $\exists x \, \forall y \, [x + y > 12]$

(d) $\exists x \, \exists y \, [x + y > 12]$

16. Consider the universe of all subsets of the set $A = \{a, b, c\}$. Let the variables x and y denote subsets of A. Find the truth value of each of the following statements and explain your answer. If the statement is FALSE, give a counterexample.

    (a) $\forall x \, \forall y \, [x \subseteq y]$

    (b) $\forall x \, \forall y \, [(x \subseteq y) \lor (y \subseteq x)]$

    (c) $\exists x \, \forall y \, [x \subseteq y]$

    (d) $\forall x \, \exists y \, [x \subseteq y]$

17. Let the universe for both variables x and y be the set $\{1, 2, 3, 4, 5, 6\}$. Let $p(x, y) =$ "x divides y." Give the truth values of each of the following statements; explain your answer, and give a counterexample in case the statement is FALSE.

    (a) $\forall x \, \forall y \, p(x, y)$

    (b) $\forall x \, \exists y \, p(x, y)$

    (c) $\exists x \, \forall y \, p(x, y)$

    (d) $\exists y \, \forall x \, p(x, y)$

    (e) $\forall y \, \exists x \, p(x, y)$

    (f) $\forall x \, p(x, x)$

18. If $p(x)$ is the statement "$x \in S$" and $q(x)$ denotes "$x \in T$," describe:

    (a) $x \in S' \cup T$

    (b) $x \in S \oplus T'$

    (c) $x \in S' \cap T'$

    (d) $x \in (S \cup T)'$

    Use logical statement forms using the statements $p(x)$ and $q(x)$.

19. Let $U = \{0, 1, 2, 3, 4, 5, 6, 7, 8, 9, 10, 11, 12, 13\}$, let $S = \{x \in U: x \geq 8\}$, and let $T = \{x \in U: x \leq 10\}$.

    (a) What implication must be TRUE if and only if S is a subset of T?

    (b) Is S a subset of T? Explain.

20. Is S is a subset of T? Let

$$U = \{1, 2, 3, 4, 5, 6, 7, 8, 9, 10, 11, 12\}$$

$$S = \{x \in U: x \text{ divides } 12 \text{ evenly}\}$$

$$T = \{x \in U: x \text{ is a multiple of } 2\}.$$

21. Prove that S is a subset of T, where

$$U = \{1, 2, 3, 4, 5, 6, 7, 8, 9\}$$

$$S = \{x \in U: x \text{ is a multiple of } 2\}$$

$$T = \{x \in U: x \text{ divides } 24 \text{ evenly}\}.$$

22. Prove that S is a subset of T, where

$$U = \{a, b, c, d, e, f, g, h\}$$

$$S = \{x \in U: x \text{ is a letter in } bad\}$$

$$T = \{x \in U: x \text{ is a letter in } badge\}.$$

23. Prove that $S = T$, where

$$U = \{1, 2, 3, 4, 5, 6\}$$

$$S = \{x \in U: x \text{ is a solution to } (x - 8)(x - 3) = 0\}$$

$$T = \{x \in U: x \text{ is a solution to } x^2 = 9\}.$$

1. Let the universe be the collection of all flowers. Let $p(x)$ be the open sentence "x can grow in sand" and let $q(x)$ be "x is subject to mold." The statement is then

$$\exists x \left[p(x) \wedge \sim q(x) \right].$$

The negation is

$$\forall x \left[\sim p(x) \vee q(x) \right].$$

This can be stated in English as

"Every flower cannot grow in sand or is subject to mold."

2. (a) FALSE. As a counterexample, consider $x = 2$.

   (b) TRUE. Consider $x = 3$.

   (c) TRUE. In fact, for every $x \in U$, $x^2 \leq 64$.

   (d) TRUE. Consider $x = 5$.

3. (a) For all nonnegative integers x and y, $x < y$. FALSE.

   (b) For every nonnegative integer x, there exists a nonnegative integer y such that $x < y$. TRUE. For any nonnegative integer x, we can let $y = x + 1$.

   (c) For every nonnegative integer y, there exists a nonnegative integer x such that $x < y$. FALSE. A counterexample is found by letting $y = 0$. Then there is no nonnegative integer x such that $x < y$.

   (d) There exists a nonnegative integer x such that, for all nonnegative integers y, $x < y$. FALSE. No matter what x is selected, letting $y = 0$ gives a counterexample.

   (e) There exist nonnegative integers x and y such that $x < y$. TRUE. Just let $x = 3$ and $y = 79$.

   (f) For every pair of nonnegative integers x and y, either $x < y$ or $y < x$. FALSE. Consider $x = y$; say, $x = 4$ and $y = 4$.

Chapter 12: CHECKLIST

☐ Logical statements
☐ Connectives and their truth tables ($\vee$, $\wedge$, $\sim$, $\rightarrow$, $\leftrightarrow$, $\oplus$)
☐ Logical equivalence and logical implication and the laws in Tables 1 and 2 in Section 4.
☐ De Morgan's laws
☐ Hypothesis, conclusion, contrapositive, converse, negation
☐ AND, OR, NOT, XOR, IF ... THEN, ELSE
☐ Logic trees
☐ Tautology and contradiction
☐ Valid argument and the rules for inference (Table 1, Section 5)
☐ Statements using the universal and existential quantifiers
☐ Predicate calculus of quantified statements

Chapter 12: SUPPLEMENTARY EXERCISES

1. Determine which of the following are statements.

   (a) The universe is 1 billion years old.

   (b) What a beautiful morning!

   (c) Mathematics is an important part of our culture.

   (d) He is a gentleman and a scholar.

   (e) All poets are men.

2. Write each of the statements below in "if ... , then ..." format.

   (a) The lines are perpendicular implies that their slopes are negative reciprocals of each other.

   (b) Goldfish can live in a fish bowl only if the water is aerated.

   (c) Jane uses her umbrella if it rains.

   (d) Only if Morris eats all his food does Sally give him a treat.

3. Write the contrapositive and the converse of each of the statements.

   (a) The Yankees are playing in Yankee Stadium if they are in New York City.

   (b) If the Richter scale indicates the earthquake is a 7, then the quake is considered major.

   (c) If a coat is made of fur, it is warm.

   (d) If Jane is in the USSR, she is in Moscow.

4. Negate the statements.

   (a) If two triangles are similar, their sides are equal.

   (b) There exists a real number x such that $x^2 = 5$.

   (c) For every positive integer n, if n is even, then n^2 is even also.

   (d) There exists a real number x such that $x^2 + 4 = 0$.

5. Determine which of the statements is a tautology.

   (a) $p \vee \sim p$ (b) $(p \to q) \leftrightarrow (\sim p \vee q)$

   (c) $(p \wedge \sim q) \leftrightarrow \sim(\sim p \wedge q)$ (d) $[p \to (q \to r)] \leftrightarrow [(p \to q) \to r]$

6. Construct a truth table for each of the following statements.

   (a) $p \to (\sim q \vee r)$ (b) $p \wedge (q \leftrightarrow (r \wedge p))$

7. Which of the logical implications is true?

   (a) $[p \wedge (\sim p \vee q)] \Rightarrow q$ (b) $[(p \to q) \wedge q] \Rightarrow p$

8. True or false?

   (a) $(\sim q \to \sim p) \Leftrightarrow (p \to q)$ (b) $[(p \to q) \wedge (r \to p)] \Leftrightarrow q$

9. True or false?

   (a) $[(p \to q) \land \sim p] \Rightarrow \sim q$ (b) $[(p \to q) \land \sim q] \Rightarrow \sim p$

10. For the given input for A and B, determine the output for Z.

$$\text{LET } C = 3 * A + B$$

$$\text{IF } ((C > 0) \text{ AND } (B > 3))$$

$$\text{THEN LET } Z = C$$

$$\text{ELSE LET } Z = 100$$

   (a) $A = 4, B = 5$ (b) $A = 10, B = 2$

   (c) $A = -4, B = 5$ (d) $A = -10, B = -2$

11. For the given inputs, determine the output for Z.

$$\text{IF } ((X > 0) \text{ AND } (Y > 0)) \text{ OR } (C \geq 10)$$

$$\text{THEN LET } Z = X * Y$$

$$\text{ELSE LET } Z = X + Y$$

   (a) $X = 5, Y = 10, C = 10$ (b) $X = 5, Y = -5, C = 10$

   (c) $X = -10, Y = -5, C = 2$ (d) $X = 2, Y = 5, C = 4$

12. The following statement is TRUE.

> If the deduction is allowed, then the tax law has been revised.

Give the truth value of each of the following statements if it can be determined directly from the truth of the original.

   (a) If the tax law has been revised, then the deduction is allowed.

   (b) If the tax law has not been revised, then the deduction is not allowed.

   (c) The deduction is allowed only if the tax law has been revised.

   (d) The deduction is allowed if the tax law has been revised.

   (e) The deduction is allowed or the tax law has been revised.

13. Assume the following statement is TRUE.

> If the voter is over the age of 21 and has a driver's license,
> the voter is eligible for free driver education.

Determine the truth value of each of the following directly from the original, if possible.

   (a) If the voter is eligible for free driver education, then the voter is over 21 and has a driver's license.

   (b) If the voter is not eligible for free driver education, then the voter is not over 21 and does not have a driver's license.

   (c) If the voter is not eligible for free driver education, then the voter is either not over 21 or does not hold a driver's license.

14. Assume the statement "All mathematicians like rap music" is TRUE. Determine the truth value of the following from this assumption, if possible.

   (a) Some mathematicians like rap music.

   (b) Some mathematicians do not like rap music.

   (c) There exists no mathematician who does not like rap music.

15. Assume the statement "Some apples are not rotten" is TRUE. Determine the truth value of each of the following using only that fact, if possible.

   (a) Some apples are rotten.

   (b) All apples are not rotten.

   (c) Not all apples are rotten.

16. Show that the argument is valid: If taxes go up, I sell the house and move to India. I do not move to India. Therefore, taxes do not go up.

17. Show that the argument is valid: I study mathematics and I study business. If I study business, then I cannot write poetry or I cannot study mathematics. Therefore, I cannot write poetry.

18. Show that the argument is valid: If I shop for a dress, then I wear high heels. If I have a sore foot, I do not wear high heels. I shop for a dress. Therefore, I do not have a sore foot.

19. Show that the argument is valid: Asters or dahlias grow in the garden. If it is spring, asters do not grow in the garden. It is spring. Therefore, dahlias grow in the garden.

20. Use indirect proof to show the argument is valid: If the professor gives a test, Nancy studies hard. If Nancy has a date, she takes a shower. If the professor does not give a test, she does not take a shower. Nancy has a date. Therefore, Nancy studies hard.

13

GRAPHS

13.1 Graphs as Models

13.2 Paths and Circuits

13.3 Hamiltonian Circuits and Spanning Trees

13.4 Directed Graphs

13.5 Matrices and Graphs

13.6 Trees

In this chapter we will discuss a special part of finite mathematics called graph theory. Many practical problems can be represented effectively with a mathematical structure called a graph. A detailed discussion will reveal the underlying theory and its usefulness in arriving at solutions to these common problems in business, sociology, geography, and management.

13.1 Graphs as Models

We begin with a few of the problems that lend themselves to solution using the pictorial representation that we will call a graph.

EXAMPLE 1 A field worker for a company that conducts polls is directed to visit all the houses in the two neighborhoods of the city pictured in Fig. 1. The streets are represented by line segments, the intersections of the streets by dots. The field worker wants to optimize her path through each neighborhood by parking her car at A, traversing each street exactly once, and returning to her car. What route should she choose in each case?

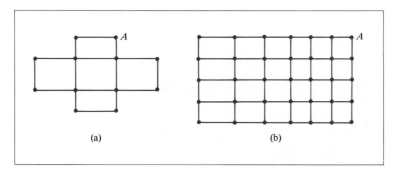

(a)   (b)

FIGURE 1

FIGURE 2

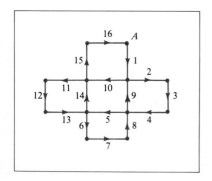

Solution With pencil and paper we try to solve the pollster's problem. We try to trace a path beginning and ending at A that covers each street exactly once. One such path is shown (Fig. 2) for the neighborhood in Fig. 1(a). There is no route that meets the pollster's requirements for the neighborhood of Fig. 1(b). The non-existence of such a path will be explained later. Could we have inspected the pictures quickly and known whether such a route could be found?

EXAMPLE 2   A complex component is to be manufactured by a microelectronics company. The company receives raw material, makes several parts from the material, installs the parts, sometimes in sequential order, and assembles the component. The chart below indicates each activity, its required time, and the immediate predecessors of the activity required before it can begin. What is the minimum amount of time needed for the entire job? On which days should specific jobs be done to attain this minimum?

	Activity	*Time (days)*	*Immediate Predecessors*
s.	Start	0	None
a.	Order material 1	3	None
b.	Order material 2	2	None
c.	Produce item 1	4	*a*
d.	Produce item 2	3	*a, b*
e.	Produce item 3	2	*a*
f.	Produce item 4	5	*c, e*
g.	Assemble final product	3	*d, f*
x.	End	0	*g*

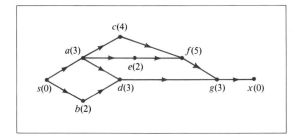

FIGURE 3

Solution   We draw a diagram (Fig. 3) of the manufacturing process by representing each activity with a dot. We draw a line connecting two dots if one of the activities represented is the immediate predecessor of the other activity, and we draw an arrow to indicate the direction of the process. Thus, since *a* is an immediate predecessor of *d*, we draw a line joining *a* to *d* with the arrow in the direction from *a* to *d*.

We can see that the minimum amount of time in which we can complete the manufacturing process of this component is the longest path (in time) through this graph from *s* to *x*. This will require 0 days for *s*, 3 days for *a*, 4 days for *c*, 5 days for *f* and 3 days for *g*. We would need at least $0 + 3 + 4 + 5 + 3 = 15$ days. Note that tasks such as *b* may be performed while others are being completed.

Assuming that we begin the process on day 1, we should perform *a* on day 1 and begin *b* by day 2. We may begin tasks *c*, *d*, and *e* on day 4. Task *f* begins on day 8, while task *g* begins on day 13. That task requires 3 days, so the process arrives at *x* on day 15.

EXAMPLE 3   The police force in a large city wants to set up a communications network so that various people in the group can communicate with each other. The question is how to design the network efficiently. We might also try to satisfy other criteria

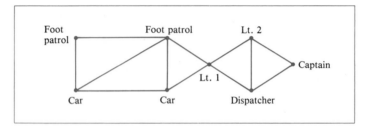

FIGURE 4

while we are at it, such as making the network invulnerable to sabotage via interruption. Figure 4 is an example of such a network, in which a line joining A to B indicates that A is able to communicate with B directly. Can this network be interrupted by a cut in a line? Can it be interrupted if there is a failure in someone's telephone?

Solution   Again, we can inspect the diagram and figure out what would happen if we removed any one line segment. Each person is able to communicate with everyone else no matter which line we might choose to cut. On the other hand, if we remove the telephone for Lieutenant 1, the communication network is disrupted.

The essential features of the objects we have been drawing are dots and the lines that connect them. We define a *graph* to be a finite collection of *vertices* (dots) and a finite collection of *edges* (line segments) where each edge connects two vertices or connects a vertex to itself. Some of the graphs we have drawn have directions indicated on the edges. Such graphs are called *directed graphs* or *digraphs*. We first concentrate our discussion on graphs; in Section 13.4 we discuss digraphs.

> A *graph* consists of two finite sets: the set V of vertices
> and the set E of edges such that each edge in E connects
> two vertices in V or connects one vertex in V to itself.

FIGURE 5

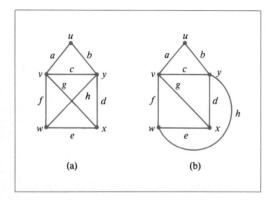

(a)   (b)

The graph in Fig. 5(a) has 5 vertices and 8 edges. Thus,

$$V = \{x, y, u, v, w\}$$

and

$$E = \{a, b, c, d, e, f, g, h\}.$$

Although the edge g joining x and v crosses the edge h joining w and y, the point of intersection of the two edges is not a vertex. In fact, we could have been a bit more creative to avoid that problem. Figure 5(b) is an equivalent graph drawn so that the edges intersect only at vertices of the graph. Note that the edge from *w* to *y* still appears in the pictorial representation. A graph for which it is possible to draw all edges so that two edges intersect only at vertices is called a *planar graph*. We will return to a discussion of equivalence and planar graphs later.

We turn our attention to Fig. 6. Note (see vertex *w*) that it is possible that a vertex has no edges identified with it. Such a vertex is called an *isolated vertex*. There are two edges connecting vertices *x* and *y*. These are called *parallel edges*. The edge that connects *z* to itself is called a *loop*. The graph has two loops.

We are naturally interested in those properties that characterize a graph. The *degree* of a vertex *v* is the number of edges with vertex *v*. If there is a loop at vertex *v*, that edge is counted twice in determining the degree of *v*. Let us record the degrees of each of the vertices in the graph of Fig. 6.

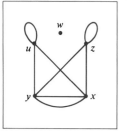

FIGURE 6

Vertex	Degree
u	4
w	0
x	4
y	4
z	4

We note that in this case the number of edges in the graph is 8 and the sum of the degrees of the vertices is $16 = 2 \times 8$. This is no accident. There is a formula that links the number of edges to the degrees of the vertices in any graph. Note that each edge has two vertices. Thus, each edge contributes twice to the sum of the degrees of the vertices. Even if the edge is a loop, we agreed to count the loop twice in calculating the degree of the vertex. Thus, the sum of the degrees of the vertices is twice the number of edges in the graph.

> **Graph Property 1** Let *G* be a graph. The sum of the degrees of the vertices of *G* is twice the number of edges of *G*.

We note that Graph Property 1 implies that the sum of the degrees of the vertices of a graph is an even integer. A vertex is called *even* if its degree is an even number. A vertex is called *odd* if its degree is an odd number.

FIGURE 7

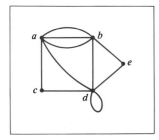

EXAMPLE 4 Find the degree of each of the vertices in the graph of Fig. 7. Verify Graph Property 1. How many of the vertices are even; how many are odd?

Solution

Vertex	Degree	Odd/Even
a	5	Odd
b	5	Odd
c	2	Even
d	6	Even
e	2	Even

The sum of the degrees is 20, which is twice 10, the number of edges. There are 2 vertices of odd degree and 3 of even degree.

It is helpful to recall several simple facts about integers.

1. If we add any number of even integers, the sum is even.
2. If we add an even integer to an odd integer, the sum is odd.
3. If we add an even number of odd integers, the sum is even.
4. If we add an odd number of odd integers, the sum is odd.

EXAMPLE 5 Use the graph drawn in Fig. 8 to help explain why the number of vertices of odd degree in any graph must be even.

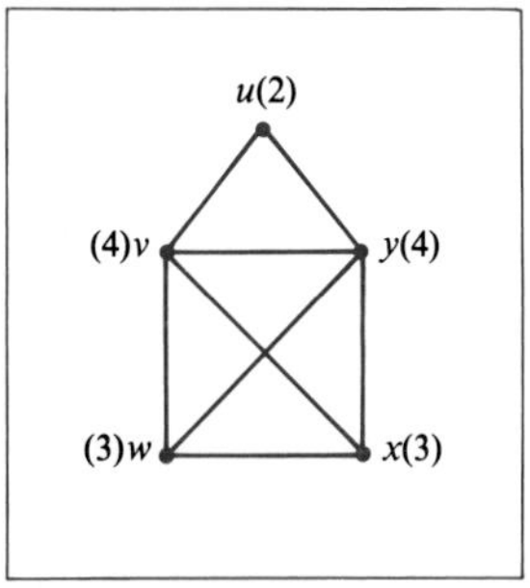

FIGURE 8

Solution The number in parentheses at each vertex is the degree of the vertex. The sum of the degrees of the vertices is even (here $5 + 5 + 2 + 6 + 2 = 20$). Vertices have either even or odd degree. We group vertices of the same type:

$$5 + 5 \quad + \quad 2 + 6 + 2 \quad = \quad 20$$

$$\left(\begin{array}{c}\text{sum of odd}\\\text{integers}\end{array}\right) + \left(\begin{array}{c}\text{sum of even}\\\text{integers}\end{array}\right) = \left(\begin{array}{c}\text{even}\\\text{integer}\end{array}\right)$$

$$N \quad + \quad \text{even} \quad = \quad \text{even}$$

Refer to the facts about integers listed above; we know that any sum of even integers is even. The sum of all the degrees of the vertices is also even. We have denoted the even sums on the equation. The other term in the sum, labeled N, is the sum of the odd degrees. N must be even. For if N were odd the sum on the right of the equation would also be odd. To summarize, N is even and also is the sum of odd integers. The sum of odd numbers is even only if there are an even number of terms in the sum, so N is the sum of an even number of odd numbers. (In this example, we have 2 odd vertices, giving $N = 10$.) This means the number of vertices of odd degree must be even.

> **Graph Property 2** Let G be a graph. The number of vertices of odd degree is always even.

EXAMPLE 6 Show that it is impossible to have a group of seven people such that each person knows exactly three other people in the group. Assume A has met B if and only if B has met A.

Solution If we drew a graph with seven vertices (one for each member of the group) and an edge joining two vertices if and only if the people represented by the vertices have met each other, we would have a graph with seven vertices, each of degree 3. But the number of vertices of odd degree must be even. So no such graph can be drawn. There can be no such group of seven people.

EXAMPLE 7   Can we construct a graph with six vertices of degrees 1, 3, 5, 4, 2, and 2, respectively?

Solution   The answer is no. The sum of the degrees of the vertices is $1 + 3 + 5 + 4 + 2 + 2 = 17$, which is not even, in violation of Graph Property 1. Furthermore, three vertices would have to be odd. That violates Graph Property 2.

We note that, if one has a graph G, the sum of the degrees of its vertices must be even. However, if we do not allow graphs that have more than one loop at a single vertex, it is possible to specify vertices and their degrees such that the sum of the degrees is even and yet be unable to draw the graph. Try, for example, to draw a graph (at most one loop at any vertex) with one vertex of degree 8 and two vertices of degree 2. The sum of the degrees of the vertices is 12, which is even, but no such graph exists.

EXAMPLE 8   How many vertices will a graph have if (a) it has 18 edges and each vertex has degree 3, (b) it has 18 edges, 4 vertices of degree 4, and the other vertices of degree 5?

Solution   (a) We know that the sum of the degrees of the vertices is twice the number of edges. A graph with 18 edges has a total degree of 36. Since each vertex has degree 3, the graph has $\frac{36}{3} = 12$ vertices.

(b) The sum of the degrees of the 4 vertices of degree 4 is 16. We let n be the number of vertices of degree 5 and note that the total of the degrees of all vertices in the graph must be

$$16 + 5n.$$

This is twice the number of edges in the graph; hence we have

$$16 + 5n = 2(18).$$

We find that $n = 4$. Thus, there are 4 vertices of degree 5 and the 4 given vertices of degree 4. The graph has a total of 8 vertices.

We have been using the term graph to include those with parallel edges. We will *now limit our discussion to simple graphs*—those without parallel edges.

Each branch of mathematics has a concept of equivalence, that is, being essentially the same. In geometry, two triangles are equivalent if they are congruent. In algebra, two equations are equivalent if they have the same solutions. We say that two simple graphs are *equivalent* if the vertices of the graphs can be paired so that, whenever two vertices in one graph are connected by an edge, then so are the corresponding vertices of the other graph.

Consider the simple graphs in Fig. 9(a) and (b). These graphs are equivalent. It is possible to see this easily by rotating the first graph through 180° and flipping

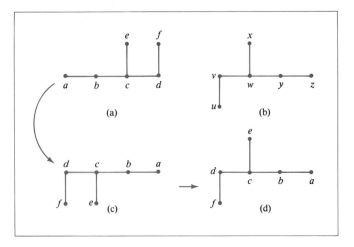

FIGURE 9

the edge from c to e. Of course, we also could go through systematic steps to verify that the graphs are equivalent by actually finding the pairing of the vertices in $\{a, b, c, d, e, f\}$ and $\{u, v, w, x, y, z\}$. We note that the two vertex sets have the same number of elements. This will be necessary for finding the pairing. Secondly, we note that the graphs each have 5 edges. If we mark the degree of each vertex, we have 3 vertices of degree 1, 2 vertices of degree 2, and 1 vertex of degree 3 in each of the graphs.

Vertex:	a	b	c	d	e	f	u	v	w	x	y	z
Degree:	1	2	3	2	1	1	1	2	3	1	2	1

Thus, the graphs have the same number of vertices of each degree. It is necessary to pair the vertex of degree 3 in one graph with the vertex of degree 3 in the other graph. So we pair c and w. Then we would pair e with x, since the edge from w to x can be paired to the edge from c to e. Note that x and e are each of degree 1. There are 2 vertices of degree 2 connected to each of c and w. A little experimentation reveals that d could be paired with v or y. We pair d with v. Then, clearly, we should pair f with u. The pairing of a with z and b with y follows by looking at their degrees. Here are the pairings:

$$a \to z \quad b \to y \quad c \to w \quad d \to v \quad e \to x \quad f \to u.$$

Check that the 5 edges pair up properly:

$$\text{edge from } a \text{ to } b \to \text{edge from } z \text{ to } y$$

$$\text{edge from } b \text{ to } c \to \text{edge from } y \text{ to } w$$

$$\text{edge from } c \text{ to } d \to \text{edge from } w \text{ to } v$$

$$\text{edge from } d \text{ to } f \to \text{edge from } v \text{ to } u$$

$$\text{edge from } c \text{ to } e \to \text{edge from } w \text{ to } x$$

Sometimes it is easy to show that two graphs are not equivalent.

EXAMPLE 9 (a) Decide if the simple graphs in Fig. 10(a) are equivalent.

(b) Do the same for the simple graphs of Fig. 10(b) and (c) Fig. 10(c).

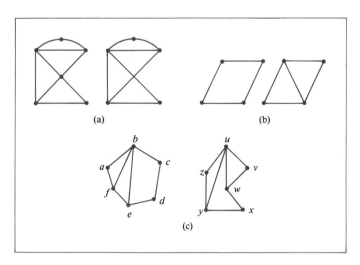

(a)

(b)

(c)

FIGURE 10

Solution (a) We count the number of vertices in each graph. There are 6 vertices in one and 5 in the other. The graphs cannot be equivalent.

(b) We count the number of edges in each graph. One has 4 edges; the other has 5 edges. The graphs cannot be equivalent.

(c) These graphs each have 6 vertices and 8 edges. We inspect the degrees of the vertices:

Vertices:	a	b	c	d	e	f		u	v	w	x	y	z
Degrees:	2	4	2	2	3	3		4	2	3	2	3	2.

They each have one vertex of degree 4, two of degree 3, and three of degree 2. However, if try to pair them up, we see that in the first graph there is an edge joining f and e (both of degree 3), but in the second graph there is no edge joining the vertices y and w (their counterparts of degree 3). The graphs are not equivalent.

To check for graph equivalence in simple graphs, we can use the following checklist. If the answer to any question is "no," the graphs are not equivalent. If the answer to every question is "yes," the graphs are equivalent.

1. Do the graphs have the same number of vertices?
2. Do the graphs have the same number of edges?

3. Do the graphs have the same number of vertices of each degree?

4. Can you find a pairing of vertices of the same degree so that, if there is an edge joining a pair in one graph, there is an edge in the other graph joining the corresponding pair?

For large graphs, the process of deciding whether two graphs are equivalent or not and verifying the answer may require a lot of work. For example, the graph G_1 with 15 vertices in Fig. 11 can be shown to be equivalent to G_2. You may use the table below to verify this.

v_1	v_2	v_3	v_4	v_5	v_6	v_7	v_8	v_9	v_{10}	v_{11}	v_{12}	v_{13}	v_{14}	v_{15}
u_7	u_6	u_5	u_4	u_3	u_2	u_{15}	u_{14}	u_1	u_{13}	u_{11}	u_{12}	u_{10}	u_9	u_8

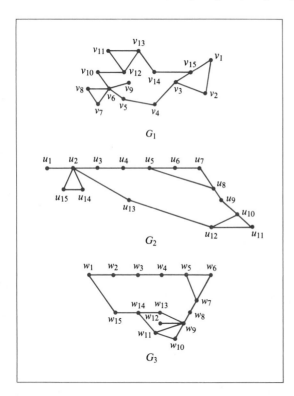

G_1

G_2

G_3

FIGURE 11

However, G_1 is not equivalent to G_3. In G_3, the vertex of degree 5 (w_9) is joined by an edge to a vertex of degree 3 but v_6 in G_1 is not joined to a vertex of degree 3.

We note that in each of the graphs there are 15 vertices, of which 1 is of degree 1, 9 are of degree 2, 4 are of degree 3, and 1 is of degree 5. If we simply try to match vertices without regard to their degrees, there are 15! ways to match the 15 vertices on any two graphs. Since 15! is 1,307,674,368,000, this can be a daunting procedure. We should match those vertices of the same degree. In any two of the graphs of Fig. 11, we have 1 way to match the vertices of degree 1, 9! ways of matching the vertices of degree 2, 4! ways of matching vertices of degree 3, and 1 way of matching the vertices of degree 1. Thus, there are $(1)(9!)(4!)(1) = 8{,}709{,}120$ ways to match vertices of the same degree. Of course, the matching of vertices is not sufficient. We must also make sure that the edges match as well. Later in this chapter, we will see that representing graphs by matrices may help somewhat.

It is not always possible to avoid having the intersection of edges occur at nonvertices. To see that it may be impossible, see Figs. 12(a) and (b). In Fig. 12(c), we show a graph equivalent to the graph of Fig. 12(a) redrawn so that intersections of edges occur only at vertices. However, no graph equivalent to the graph in Fig. 12(b) can be drawn in which edges intersect only at vertices. Try it! Graphs that can be drawn so that edges intersect only at vertices are called planar graphs.

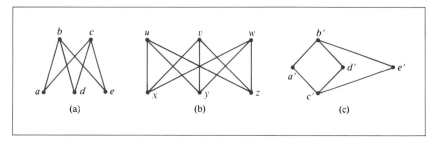

(a)   (b)   (c)

FIGURE 12

EXAMPLE 10   Show that the graphs in Fig. 13 are equivalent.

Solution   We go through the checklist. Both graphs have 4 vertices; both have 5 edges. We tabulate the degrees of the vertices.

$$\begin{array}{cccc|cccc}
\text{Vertices:} & u & v & w & x & a & b & c & d \\
\text{Degrees:} & 2 & 3 & 2 & 3 & 2 & 2 & 3 & 3.
\end{array}$$

Pair u with a and pair w with b. Try pairing v with c and x with d.

$$u \to a \quad w \to b \quad v \to c \quad x \to d.$$

Does it work? Check the edges.

$$\text{edge from } u \text{ to } v \;\to\; \text{edge from } a \text{ to } c$$

$$\text{edge from } v \text{ to } x \;\to\; \text{edge from } c \text{ to } d$$

$$\text{edge from } x \text{ to } w \;\to\; \text{edge from } d \text{ to } b$$

$$\text{edge from } w \text{ to } v \;\to\; \text{edge from } b \text{ to } c$$

$$\text{edge from } u \text{ to } x \;\to\; \text{edge from } a \text{ to } d$$

We have verified that the graphs are equivalent. We could also have used our geometric intuition to see that, if we could pick up the corner at w and rotate it 180° to the right, the two graphs would look more alike (Fig. 14).

FIGURE 13   FIGURE 14

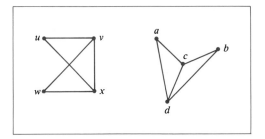

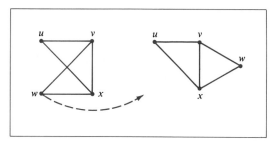

The appearance of differences in the drawing of graphs that are actually equivalent can cause some difficulty. We hope that understanding the idea of equivalence will help to avoid confusion.

PRACTICE PROBLEMS 1

1. Draw all simple graphs without loops that contain 4 vertices in which the vertices all have the same degree.

2. Determine whether the given situation contradicts Graph Property 1 or 2. If not, try to draw the graph.

   (a) Graph has three vertices of degree 1, two of degree 2, and one of degree 3.

   (b) Graph has two vertices of degree 1, three of degree 2, and one of degree 3.

   (c) Graph has 21 edges and 4 vertices of degree 4; all other vertices are of degree 5.

3. Determine whether the graphs in Fig. 15 are equivalent.

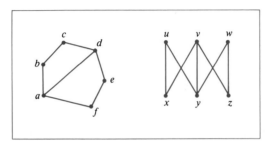

FIGURE 15

EXERCISES 1

Unless specifically stated, graphs in the following exercises may have parallel edges and loops.

1. For each of the graphs in Fig. 16:

   (a) Identify the loops and parallel edges.

FIGURE 16

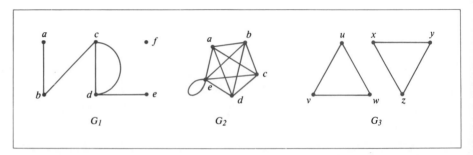

(b) Determine the degree of each of the vertices.

(c) Determine the number of vertices of odd degree.

(d) Verify that the sum of the degrees of the vertices is twice the number of edges in the graph.

2. For each of the graphs in Fig. 17, find the degree of each of the vertices. Verify that the number of vertices of odd degree is even, and show that the sum of the degrees of the vertices is twice the number of edges in the graph.

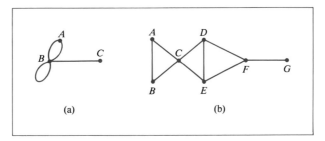

(a)   (b)

FIGURE 17

3. An *acquaintance* table is shown in Fig. 18. In it we place a ▌ in the row labeled x and the column labeled y if and only if x and y are acquainted with one another. If we represent each person with a vertex and join vertices if and only if the respective people are acquainted with one another, we can represent the table with a graph. Draw the graph for the given acquaintance table.

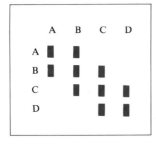

FIGURE 18

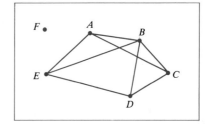

FIGURE 19

4. In the graph in Fig. 19, the vertices represent countries and an edge is drawn between two vertices if and only if a direct rail link exists between the countries they represent. Which country is connected by rail to the largest number of countries? Which country is isolated? In how many ways can we travel from A to D without returning to any country once it has been visited?

5. Show that it is impossible to have a group of 13 people each of whom is related to exactly 3 other people in the group.

6. Is it possible to draw a graph (may have parallel edges and loops) with 4 vertices where the degrees of the vertices are:

(a) 1, 1, 1, 4   (b) 1, 2, 3, 4   (c) 1, 2, 3, 3?

7. Draw graphs (they need not be simple) with 3 vertices having the given degrees if it is possible for such a graph to exist. If such a graph cannot be drawn, explain.

   (a) 1, 4, 5   (b) 2, 4, 4   (c) 3, 4, 4

8. Can we draw a graph having 5 vertices of degree 1, 3, 3, 4, 4? Why?

9. If a graph has 15 edges and each vertex has degree 3, how many vertices must it have?

10. A graph has 12 edges, 2 vertices of degree 5, and all other vertices of degree 2.

    (a) How many vertices does the graph have?

    (b) Draw such a graph.

11. Each of the graphs in Fig. 20 is planar; redraw each so that edges intersect only at vertices of the graph.

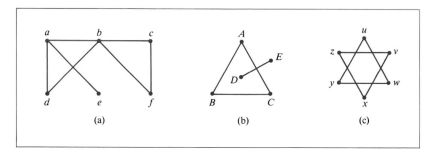

(a)   (b)   (c)

FIGURE 20

12. Each of the graphs in Fig. 21 is planar. Show that each can be redrawn so that edges intersect only at vertices of the graph.

FIGURE 21

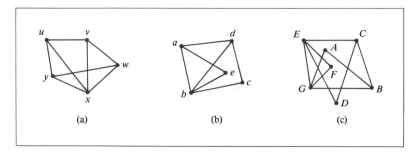

(a)   (b)   (c)

13. Show that the graphs in Fig. 22 are not equivalent.

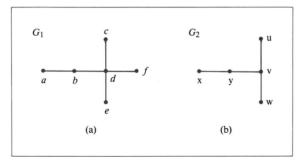

FIGURE 22

14. Show that the graphs in Fig. 23 are not equivalent.

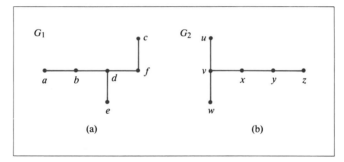

FIGURE 23

15. Show that the graphs in Fig. 24 are equivalent.

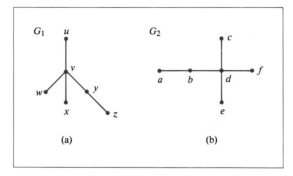

FIGURE 24

16. Show that the graphs in Fig. 25 are equivalent by giving a pairing of the vertices.

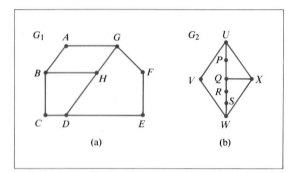

(a) (b)

FIGURE 25

In Exercises 17–21, decide if graphs G_1 and G_2 are equivalent. Explain by giving a property that holds in one and not the other or by giving a pairing of the vertices.

17.

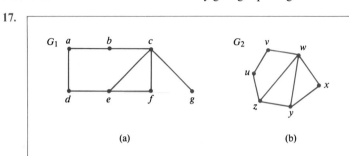

(a) (b)

FIGURE 26

18.

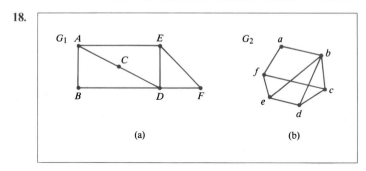

(a) (b)

FIGURE 27

19.

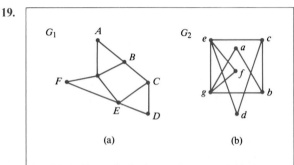

(a) (b)

FIGURE 28

20.

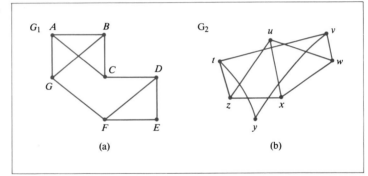

(a)   (b)

FIGURE 29

21.

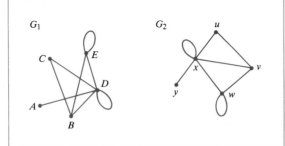

FIGURE 30

22. (a) If two graphs have the same number of vertices of the same degree, do they have the same number of edges?

    (b) Do they have to be equivalent? Why?

23. A snow plow must plow all the streets represented on the grid displayed in Figure 31. Construct a route so that the plow begins at A, exits at B, and goes down each street exactly once. If the garage is at A, can the plow begin and end at A, traversing each street exactly once?

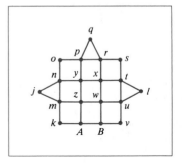

FIGURE 31

24. Draw all simple graphs (no parallel edges) with 2 vertices.

25. Draw all simple graphs having 3 vertices and 5 edges.

26. Draw all simple graphs with 4 vertices and 6 edges.

SOLUTIONS TO PRACTICE PROBLEMS 1

1. Simple graphs have no parallel edges. We are also excluding graphs with loops. The graphs on 4 vertices where each vertex has degree 0 through 3 are given in Fig. 32.

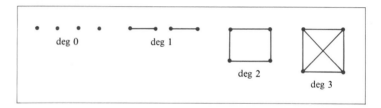

FIGURE 32

2. (a) $(3)(1) + (2)(2) + (1)(3) = 10$, which is even. There are an even number of vertices of odd degree. The graph appears in Fig. 33.

(b) There are an odd number of vertices of odd degree. Also, $(2)(1) + (3)(2) + (1)(3) = 11$, which is odd. Both Graph Properties are violated. No such graph exists.

(c) We would require n vertices of degree 5, where $(4)(4) + (n)(5) = (2)(21)$. This means we must find an integer solution to the equation $5n + 16 = 42$. There is no such solution. No such graph exists.

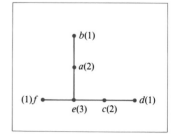

FIGURE 33

3. Use the pairings $a \rightarrow y, b \rightarrow u, c \rightarrow x, d \rightarrow v, e \rightarrow z, f \rightarrow w$.

13.2 Paths and Circuits

Let us consider the police communication system presented in Section 13.1. We reproduce the graph in Fig. 1, labeling the vertices with letters A through H and the edges with letters e_1 through e_{12}.

We consider several paths of communication from Foot Patrol 1 at vertex A to Lieutenant 1 at E. One route through which a message could be sent is $e_2 e_7$. Some others are $e_1 e_4 e_6$ and $e_1 e_3 e_5 e_6$. Each is a sequence of edges that links together the vertices A and E.

> A *path* in a graph G is a sequence of edges that links together two vertices.

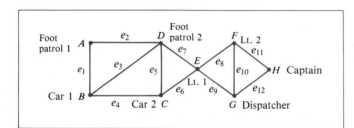

FIGURE 1

Paths might repeat edges; for example, $e_1e_3e_2e_1e_4e_6$ is a path from A to E in which edge e_1 is repeated. The *length of a path* is the number of edges in the path. The path $e_1e_3e_2e_1e_4e_6$ has length 6.

Associated with each path is a sequence of edges. Since adjacent edges have a vertex in common, a path also defines a sequence of vertices. For the graph of Fig. 1, the vertex sequence corresponding to the path $e_1e_2e_5e_6$ is *BADCE*. The number of vertices in the vertex sequence is one larger than the length of the path. In addition, if there are parallel edges, two different paths may have the same vertex sequence. Loops contribute the vertex twice to the sequence. We may always use the edge sequence to describe the path, but if there are parallel edges, the vertex sequence may not define a unique path. To simplify matters, *we limit discussion in this section to simple graphs*—those having no parallel edges. A sequence of vertices suffices to uniquely define a path in a simple graph.

Let us look at the simple graph in Fig. 2. Consider the vertex sequence *u y w x*. This defines a simple path from *u* to *x*. A *simple path* is a path in which no vertex is repeated. Of course, there are other simple paths from *u* to *x*, such as *u v x* or *u v w x*. The path *u v u y x* is not simple because the vertex *u* is repeated.

Suppose we want to trace a path from *u* back to itself. That is, we are seeking a *closed path* in the graph. The vertex sequence *u y x w v u* defines such a closed path. A *closed path* is a path in which the first and last vertices are the same. There are several closed paths starting and ending at *u*. Can you find some? We will have particular interest in closed paths that are *circuits* (sometimes called *cycles*)—closed paths in which no edge is repeated. The path *u y x v y u* is a closed path, but it is not a circuit because the edge *b* is used twice. We point out that the vertex *y* is also repeated, so the path is not simple.

A *simple circuit* is a circuit with no repeated vertices except for the first and the last. A simple circuit then is a closed path with no repeated edges or vertices. Question: In simple graphs, must a circuit be a simple circuit? Any circuit necessarily uses no edge twice. Is it possible to have a circuit that repeats a vertex? The answer is yes. Consider the circuit

$$u\ v\ w\ y\ v\ x\ y\ u$$

in the graph of Fig. 2. This is a circuit because no edges are repeated; however, clearly several vertices are repeated.

A graph is called *acyclic* if and only if it contains no circuits. The graphs of Figs. 1 and 2 are clearly not acyclic. We have already demonstrated several circuits. What does an acyclic graph look like?

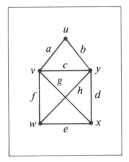

FIGURE 2

FIGURE 3

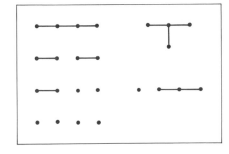

EXAMPLE 1   Draw all acyclic graphs having 4 vertices.

Solution   See the six graphs in Fig. 3.

EXAMPLE 2 Refer to the graph in Fig. 4 and determine whether each of the given paths is a simple path, a closed path, a circuit, or a simple circuit.

(a) $x\ y\ z\ u\ v\ x$

(b) $x\ y\ z\ u\ v\ z\ y\ x$

(c) $z\ w\ v\ z\ u\ y\ z$

(d) $u\ v\ z\ y$

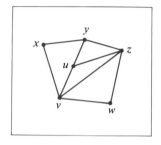

FIGURE 4

Solution (a) This is a closed path and a simple circuit. It begins and ends at x and no edges or vertices are repeated.

(b) This is a closed path because it begins and ends at x. However, it is not a circuit because the edge joining y to z is repeated.

(c) This is a circuit that begins and ends at the vertex z. However, although no edges are repeated, the vertex z appears three times in the sequence. Thus, this circuit is not a simple one.

(d) This is a simple path.

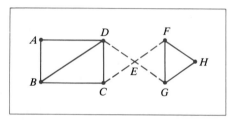

FIGURE 5

The graph given in Fig. 1 represents the communications network for a small police department. In Section 13.1, we asked whether the network is vulnerable to disruption. That is, will the removal of any person from the communication network disrupt the entire network? The removal of the vertex E corresponding to Lieutenant 1 will disrupt the entire network (Fig. 5). We say that the graph becomes disconnected.

A graph is *connected* if and only if given any pair of distinct vertices, there is a path between them. The graph in Fig. 1 is connected. Removal of the vertex E representing Lieutenant 1 produces a disconnected graph.

EXAMPLE 3 Which of the graphs in Fig. 6 are connected?

Solution G_1 is not connected. There is no path from vertex a to vertex b. G_2 is connected.

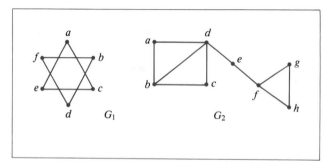

FIGURE 6

EXAMPLE 4 Name the vertices in graph G_2 of Fig. 6 whose removal would disconnect the graph.

Solution The removal of any one of the vertices d, e, or f would result in a disconnected graph.

Turning once again to G_2 of Fig. 6, we note that we can eliminate some **edges**, yet retain the connectedness of the network, and can define a path using each of the vertices of G_2. The network is connected as long as each person can communicate along some path to every other person. If we eliminate the edges from b to d, from c to d, and from f to g, we have a path through every vertex. The resulting graph has 7 edges. Is that the smallest number of edges in any path of G_2 through each vertex?

Given a connected graph G, our goal is to find a simple path using all the vertices of G and having a minimum number of edges. We call such a path a *path of minimal length* in G.

EXAMPLE 5 Find the path of minimal length needed to link any person in the network of Fig. 7 to any other person in the network. How many edges are in the path?

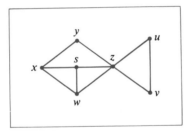

FIGURE 7

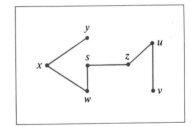

FIGURE 8

Solution You might start anywhere on the graph and try to trace a path that hits every vertex without lifting your pen. Since we are looking for a path with a minimal number of edges, we would want to avoid any duplications, so we avoid circuits. As we will see later, a path of minimal length must contain at least 6 edges. One example is the path y x w s z u v (see Fig. 8). If all other edges were removed, we would still have a connected graph.

The graph in Fig. 7 has 7 vertices. In order for us to have a connected graph using the 7 vertices, we must have at least 6 edges. Can you see why?

Let us try to see what is going on. Think about a connected graph with 2 vertices. Begin to trace a path from any vertex. Since the graph is connected, this vertex is not isolated so we can follow an edge to the other vertex. Thus, the connected graph with two vertices must have at least one edge. Now consider a connected graph with 3 vertices (refer to Fig. 9). Start at any vertex. It is not isolated so there is an edge connecting it to another of the vertices. There must be an edge joining the third vertex to one of the other two since the third vertex is not isolated

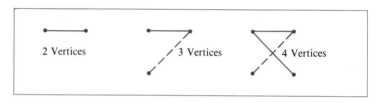

FIGURE 9

either. Thus, if we have a connected graph with 3 vertices, we must have at least 2 edges (see Fig. 9). Try to draw the picture for a connected graph with 4 vertices. One is drawn in Fig. 9. The smallest number of edges possible for that graph is 3. In fact, we have a general result.

> A connected graph with n vertices must have at least $n - 1$ edges.

EXAMPLE 6   Refer to the simple graphs in Figs. 10(a) and (b). Without lifting your pencil from the paper, try to trace a path in each graph from vertex u back to vertex u traversing each edge exactly once.

Solution   There are several correct solutions for Fig. 10(a). Here is one:

$$u \ y \ x \ z \ w \ x \ v \ w \ y \ v \ u.$$

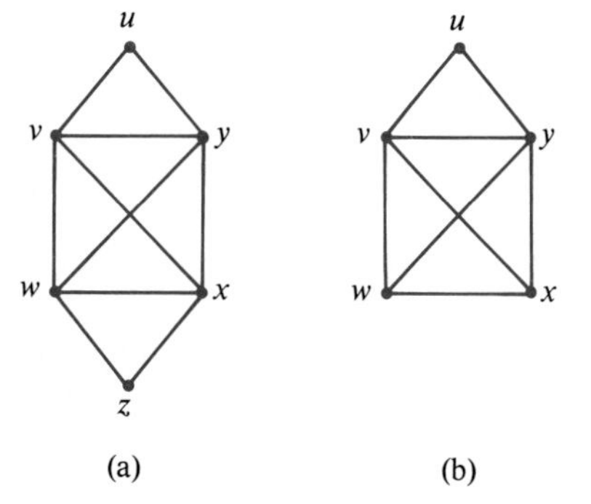

Try as you may on Fig. 10(b), there is no circuit starting at u and ending at u, using each edge exactly once. Let us analyze what goes wrong. The key is the fact that the graph has vertices of odd degree. For example, the vertex x has degree 3. Hence, if we enter on one edge at x and exit via another edge at x, we are left with still another edge with vertex x that eventually must be included in the circuit. But if we later enter toward x on that edge, there is no unused edge on which to exit. If we ever return to vertex x, we have no way out.

An *Euler path* is a path in which every edge of the graph is used exactly once. An *Euler circuit* is an Euler path that begins and ends at the same vertex. If a graph has an Euler circuit, then all its vertices are of even degree.

FIGURE 10

The easiest way to understand this property is to imagine that we begin at a vertex on the Euler (pronounced Oiler) circuit and begin to traverse the edges. At each stage, erase the edge as we progress. What has happened? At every vertex we have erased the edge leading in and the edge leading out. This means we have erased 2 edges at each vertex. Even if we have erased a loop, we have erased 2 edges at that vertex. As we progress, we reduce the degree of each vertex by 2 and we

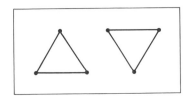

FIGURE 11

erase all the edges, so the degree of each of the vertices is eventually 0. Thus, each vertex must have had an even degree. The starting vertex must also have even degree because every time we leave the vertex (and erase an edge) we must later come back along another edge to eventually complete the circuit.

Do you think that the converse of the theorem is true? That is, if every vertex in a graph has even degree, can we find an Euler circuit? A trivial counterexample shows that this is not true (see Fig. 11). However, Euler proved a theorem that does give conditions that are both necessary and sufficient for the existence of an Euler circuit.

> **Euler's Theorem** Let G be a graph. Then G contains an Euler circuit if and only if
>
> 1. G is connected, and
> 2. Every vertex of G is of even degree.

We will not give a proof of Euler's theorem, but we will consider the practical matter of how to find an Euler circuit when we know one exists. We present a procedure (or *algorithm*) that accomplishes this.

EXAMPLE 7 (a) Explain why the graph in Fig. 12 has an Euler circuit, and (b) find an Euler circuit starting and ending at x.

Solution (a) The graph is connected and the degree of each vertex is even, so the graph has an Euler circuit.

(b) One Euler circuit starting and ending at x is

$$x \ y \ v \ u \ y \ z \ x.$$

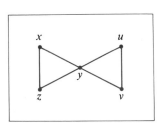

FIGURE 12

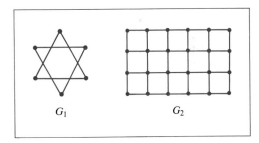

FIGURE 13

EXAMPLE 8 Explain why the graphs in Fig. 13 do not have Euler circuits.

Solution Graph G_1 is not connected so it cannot have an Euler circuit. Graph G_2 has several vertices of odd degree and therefore has no Euler circuit. Note that G_2 is similar to the rectangular street grids in many cities. A route beginning and ending at the same point, traversing every street exactly once, is not possible for such a city.

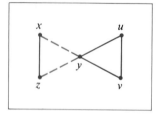

FIGURE 14

What procedure did we use to find the Euler circuit in Example 7? It seemed as if we just followed the edges around the graph. But consider what would have happened if we had begun with x, proceeded to y, and then to z (see Fig. 14). We see rather quickly that this is a bad choice, for we are forced to choose the edge from z to x and we are stranded at x; we cannot get to the other edges. An Euler circuit includes all the edges of the graph, so we want to be sure we can get to them. Notice (see Fig. 14) that, having chosen the edge from x to y and then the edge from y to z, we disconnected the graph. We want to avoid that. With this in mind, we state Fleury's algorithm for finding an Euler circuit whenever one exists.

Fleury's Algorithm for Finding an Euler Circuit Suppose G is a graph that contains an Euler circuit.

1. Select any vertex to begin.
2. From this vertex, select an edge whose removal will not disconnect the graph. Add the edge to the Euler circuit and erase the edge from the graph.
3. If an edge described in step 2 does not exist, there is only one choice of edge. Add the edge to the circuit, and erase it and the isolated vertex from the graph.
4. Continue from the next vertex.

EXAMPLE 9   Use Fleury's algorithm to find an Euler circuit for the graph of Fig. 15.

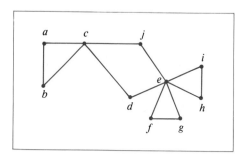
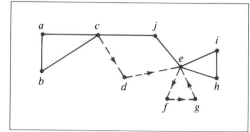

FIGURE 16

Solution   Let us begin at vertex c. Proceed to vertex d. Erase the edge from the graph; include it in the circuit. Continue on to vertex e. Erase the edge; include it in the circuit. At vertex e we have several choices. We note that proceeding to j disconnects the graph. We may choose to proceed to vertex f, then g, and return to e. We include the edges in the circuit and erase the edges from the graph. Figure 16 shows our progress so far. Dashed edges are in the circuit, but have been removed from the graph. As before, we cannot go from vertex e to vertex j since the removal of the edge from e to j would disconnect the graph. Fleury's algorithm forces us to reconsider. We proceed instead to vertex i (or h) and continue. Our Euler circuit is

$$c\ d\ e\ f\ g\ e\ i\ h\ e\ j\ c\ a\ b\ c.$$

EXAMPLE 10 (a) Is it possible to trace an Euler path from x to w on the graph of Fig. 17?

(b) From u to x?

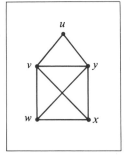

FIGURE 17

Solution An Euler path must use every edge in the graph exactly once. In (a), we are proceeding from vertex x to vertex w. One such path is given by

$$x\ y\ u\ v\ y\ w\ x\ v\ w.$$

In (b), after trying for a while, we wonder if it is possible to find an Euler path from u to x and see that it is not. We analyze the situation. There is a problem at w. Vertex w has odd degree. Thus, we would have to enter twice and exit once or enter once and exit twice. Since the path neither starts nor ends at w, this is impossible. If there are exactly two vertices of odd degree, there is an Euler path from one to the other.

We mention an additional theorem that can be used for determining if Euler paths exist.

Euler Path Property   Let G be a graph with 3 or more vertices. Then G contains an Euler path that is not a circuit if and only if

1. G is connected,
2. Two vertices are of odd degree, and
3. Every other vertex is of even degree.

We note that if there are 2 vertices of odd degree, then the path proceeds from one of the vertices to the other.

EXAMPLE 11 Consider the problem of plowing the snow from the streets of a small subdivision. Figure 18 is a map of the region; the vertices indicate where streets intersect. An edge indicates a street that must be plowed. Can we find a route that traverses each street exactly once?

Solution To assure that an Euler circuit exists, we have marked each vertex with its degree (Fig. 19).

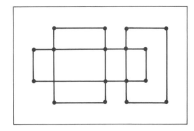

FIGURE 18

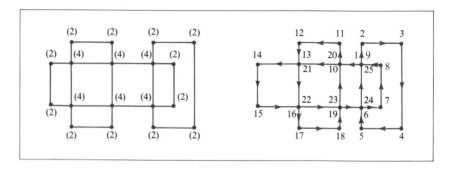

FIGURE 19

The graph is indeed connected and all vertices have even degree. We have marked the edges of the graph with arrows and the vertices sequentially to indicate one Euler circuit.

One of the most famous problems in graph theory involved the famous bridges of Königsberg across the Pregel River. The town, formerly part of Prussia, is now a part of the USSR and is called Kaliningrad. Part of the town is shown in Fig. 20(a). On Sundays, it is said, the people of Königsberg would stroll through the town and try to take a walk crossing each bridge exactly once and returning home. In 1736, the Swiss mathematician Leonhard Euler (1707–1783) showed that such a stroll was impossible (Euler's Theorem). We will represent the land masses by points (vertices) and the bridges by edges in a graph representation of Königsberg. Although the graph is not simple (it has parallel edges), the problem is of such historical importance that we temporarily lift our restriction to consider it.

We point out that the degrees of the vertices in the Königsberg bridge problem are all odd. Thus, it is not possible to cross every bridge exactly once and return to the starting point!

An interesting addendum to the Königsberg Bridge problem is that during World War II all seven bridges were destroyed. They were replaced by seven bridges placed roughly the way they were originally and one additional bridge. The graph representation has exactly 2 vertices of odd degree. Thus, although an Euler circuit does not exist, we could construct an Euler path that would begin at one of the vertices of odd degree and terminate at the other.

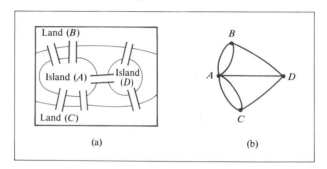

FIGURE 20

PRACTICE PROBLEMS 2

1. Use the graph in Fig. 21 to find the length of the paths in
   (a) through (d). Determine if the given path is a simple path,
   a closed path, a circuit, or a simple circuit.

   (a) *A B H G F*

   (b) *A B C H B A*

   (c) *A B C D F G C H A*

   (d) *B C D F G H B*

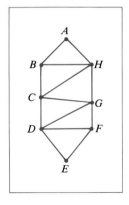

FIGURE 21

2. Determine if an Euler circuit exists in the graph in Fig. 22.

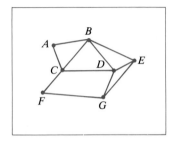

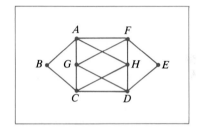

FIGURE 22   FIGURE 23

3. Show that an Euler circuit exists in the graph in Fig. 23. Find one.

EXERCISES 2

Assume that all graphs are simple (no parallel edges).

1. Use the graph in Fig. 24 to find the length of the given paths.

   (a) *a b c a*   (b) *b c d e f d*   (c) *a b c d f g h i*

   (d) *a b e d c a*   (e) *b c d e b*   (f) *e d f g h i g f*

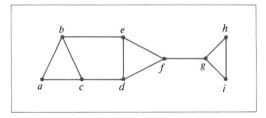

FIGURE 24

2. For each of the paths given in Exercise 1, determine if the path is:

> a simple path (sp)
> a closed path (cp)
> a circuit (c)
> a simple circuit (sc)

Justify your answers.

3. Find the lengths of each of the given paths in the graph in Fig. 25. Determine whether each is a simple path, a closed path, a circuit, or a simple circuit.

(a) *a f e d e*   (b) *a f c b*   (c) *a f c d d e*

(d) *a f e c b a*   (e) *a f c e f*   (f) *a b c d e f a*

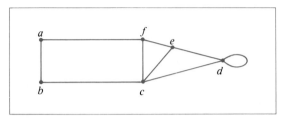

FIGURE 25

4. For the graph in Fig. 26, find:

(a) Four different simple paths

(b) Five different simple circuits

(c) Two different circuits that are not simple

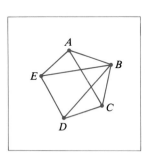

FIGURE 26

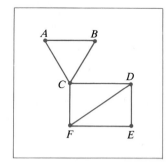

FIGURE 27

5. (a) Find four simple paths from *A* to *D* in the graph of Fig. 27.

(b) Find three simple circuits that begin and end at *C*. Are there any more?

6. Draw all acyclic connected graphs with 5 vertices.

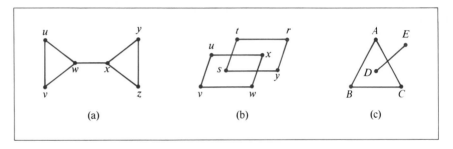

FIGURE 28

7. Which of the graphs in Fig. 28 is connected? Explain.

8. Suppose that *G* is a graph with 6 vertices and 4 edges. Can *G* be connected? Why?

9. Suppose *G* is a graph with no loops or parallel edges. It has 5 vertices and 7 edges. Can *G* be disconnected? Why? What if *G* has 5 vertices and 5 edges?

10. Why are there no Euler circuits in the graphs in Fig. 29?

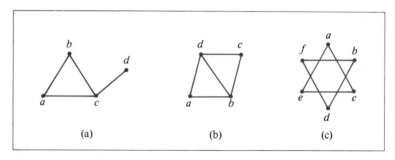

FIGURE 29

11. Which of the graphs shown in Fig. 30 are acyclic? Explain.

FIGURE 30

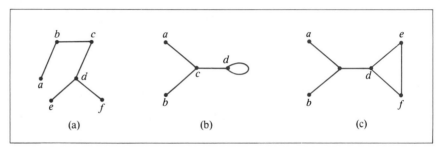

12. For which of the graphs in Fig. 31 can we find an Euler circuit? Explain.

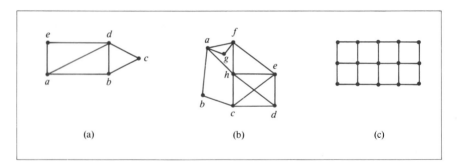

(a)   (b)   (c)

FIGURE 31

13. Use Fleury's algorithm, beginning with vertex v, to find an Euler circuit in the graphs in Fig. 32.

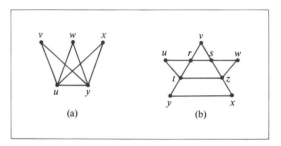

(a)   (b)

FIGURE 32

14. Use the floor plan of the house shown in Fig. 33 to draw a graph where each room and the area outside the house are represented by vertices and each door by an edge. Is it possible to begin outside the house and go through each doorway exactly once, returning to the outside? Find a route if it is possible.

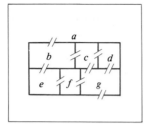   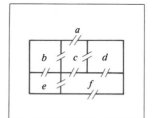

FIGURE 33   FIGURE 34

15. Use the floor plan in Fig. 34 to answer the question posed in Exercise 14.

16. A road inspector must investigate the roads in the area pictured in Fig. 35. Is there a route he can take that begins at city A, traverses each road exactly once and allows him to return to A? If so, find the route. If not, why not?

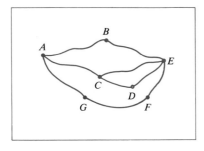

FIGURE 35

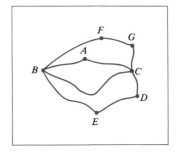

FIGURE 36

17. The road inspector of Exercise 16 needs to visit the area pictured in Fig. 36. Can he find a route that begins and ends at *A* and traverses each road exactly once? If so, find the route. If not, why not?

18. Can an Euler path be found in the graphs in Fig. 37? Find one, if you can. Otherwise, explain why one does not exist.

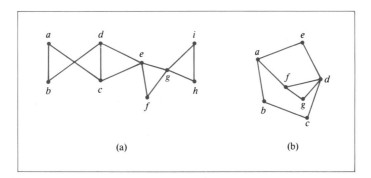

(a)   (b)

FIGURE 37

19. In Fig. 38, can you construct an Euler path so as to cross over each bridge exactly once? If one exists, where must it begin and end?

20. Draw a simple graph in which every vertex has even degree, but for which no Euler circuit exists.

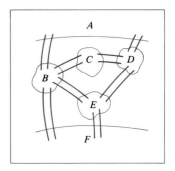

FIGURE 38

21. For the graph in Fig. 39, find the shortest path joining the given vertices and find its length.

(a) *b* to *d*   (b) *d* to *g*   (c) *c* to *g*

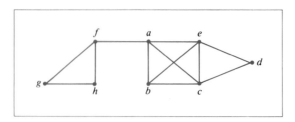

FIGURE 39

SOLUTIONS TO PRACTICE PROBLEMS 2

1. (a) *A B H G F* is a simple path of length 4, not closed and not a circuit.

(b) *A B C H B A* is a nonsimple closed path of length 5, but is not a circuit because the edge *BA* is repeated.

(c) *A B C D F G C H A* is a closed path of length 8, not a simple circuit, but a circuit nonetheless.

(d) B C D F G H B is a simple closed path of length 6 that is also a simple circuit.

2. Since vertices *E* and *G* have odd degree, no Euler circuit exists. However, there is an Euler path beginning at *E* and ending at *G*.

3. Since the graph is connected and all vertices are of even degree, an Euler circuit exists. One example is *A B C D E F A G C H D G F H A*.

13.3 Hamiltonian Circuits and Spanning Trees

EXAMPLE 1   Consider a school bus company. For the graph in Fig. 1, let us indicate by a vertex each of those corners at which the bus assigned to this neighborhood must pick up children. There is an edge between vertices if and only if there is a path the bus can take between the two corners. Is there a route that visits each corner exactly once and by which the bus begins and ends the route at school?

Solution   One circuit is indicated on the graph in Fig. 2.

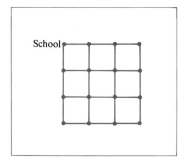

FIGURE 1

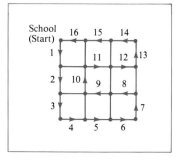

FIGURE 2

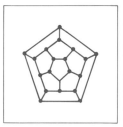

FIGURE 3

Example 1 shows a *Hamiltonian circuit*, that is, a circuit starting and ending at the same vertex that visits each other **vertex** exactly once. Sir William Rowan Hamilton (1805–1865) was born in Ireland. As a mathematician, he made significant contributions to algebra. His name is associated with circuits of this type because in 1857 he invented a game in which the player was to trace out the edges of a dodecahedron, while visiting each of the 20 vertices without a repetition, and return to the starting point. You might try Hamilton's puzzle on the dodecahedron drawn in Fig. 3. (Actually, in 1855, an English mathematician named Thomas R. Kirkman posed a similar problem to the Royal Society.)

Finding an Euler circuit and finding a Hamiltonian circuit is the difference in the two problems:

1. Find a route for a snow plow that traverses every street exactly once and returns to its starting point (Euler circuit).
2. Find a route for the mail truck that stops exactly once at each mail box to pick up outgoing mail and returns to the post office from which it began (Hamiltonian circuit).

In general, the problem of deciding whether a graph has a Hamiltonian circuit is more difficult than deciding on the existence of an Euler circuit. The traveling salesman problem is a problem of this type. It is concerned with a salesman who must visit each of several cities. The cost involved in traveling between any two cities is known, and the salesman wishes to construct a route that minimizes cost and visits each city exactly once before returning home. No satisfactory simple procedure is known for solving the general problem.

There are a few facts that can help us decide whether a graph has a Hamiltonian circuit. We mention a few of them without proof. First, a graph G with n vertices that has a Hamiltonian circuit must have at least n edges. This allows us to say that, if a graph with n vertices has fewer than n edges, it cannot have a Hamilton circuit. However, the fact that a graph with n vertices has at least n edges does not guarantee that the graph has a Hamiltonian circuit. We draw one such

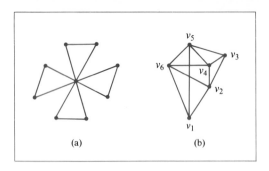

(a)   (b)

FIGURE 4

counterexample in Fig. 4(a). The graph has 9 vertices and 12 edges, but there is no way to trace a path that begins and ends at the same vertex and includes every other vertex exactly once.

> **Principle 1**   **If**
>
> 1. G is a simple graph
> 2. G has no loops
> 3. G has n vertices, where $n \geq 3$, and
> 4. the degree of each vertex of G is at least $n/2$,
>
> then G has a Hamiltonian circuit.

Again, we remind the reader that if the condition is satisfied the graph has a Hamiltonian circuit. However, we may have graphs with Hamiltonian circuits that do not satisfy the criteria. The graph in Fig. 4(b) satisfies the criteria of Principle 1 and therefore has a Hamiltonian circuit.

> **Principle 2**   **If**
>
> 1. G is a simple graph
> 2. G has no loops
> 3. G has n vertices
> 4. G has at least $\frac{1}{2}(n-1)(n-2) + 2$ edges,
>
> then G has a Hamiltonian circuit.

We note that the graph of Fig. 4(b) does not satisfy the criteria of Principle 2 because $n = 6$, but the number of edges is 11, while

$$\tfrac{1}{2}(n-1)(n-2) + 2 = \tfrac{1}{2}(5)(4) + 2 = 12.$$

However, as we have already noted, the graph does have a Hamiltonian circuit.

> **Principle 3**   Let G be a simple graph with no loops and n vertices, where $n \geq 3$. If, for any two vertices in G not connected by an edge, the sum of their degrees is at least n, then G has a Hamiltonian circuit.

Let us analyze the graph of Fig. 4(b) in light of Principle 3. We make a chart of all those vertex pairs that are not connected and calculate the sum of the degrees of their vertices:

Vertex Pairs That Are Not Connected	Sum of Degrees of Vertices
$v_1 \; v_3$	6
$v_1 \; v_4$	7
$v_2 \; v_5$	8
$v_3 \; v_6$	7

There are 6 vertices in the graph and the sums of the degrees of those pairs not connected with an edge is ≥ 6 for all such pairs. The graph is guaranteed by Principle 3 to have a Hamiltonian circuit.

Principle 3 tells us that, for example, if we have a simple graph G on 5 vertices with no loops, then in order for G to have a Hamiltonian circuit it is sufficient that the degrees of any 2 vertices not connected with an edge sum to at least 5.

EXAMPLE 2 Which of the graphs in Fig. 5 have Hamiltonian circuits? Use the principles if possible.

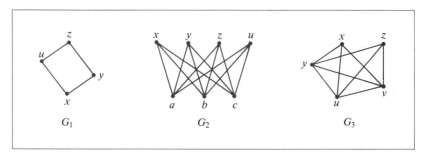

FIGURE 5

Solution We can see that the circuit $u \; x \; y \; z \; u$ is a Hamiltonian circuit in G_1. Graph G_1 has $n = 4$ vertices each with degree $n/2 = 2$. We could have come to the same conclusion by applying Principle 1. There is no way to draw a Hamiltonian circuit in G_2. We can draw a path using every vertex, but there is no way to close the path to form a circuit. We note that this is not a violation of Principle 1 because $n = 7$, and not every vertex has degree at least $7/2$. In particular, x, y, z, and u have degree 3. A Hamiltonian circuit exists in G_3 because it satisfies Principle 1. Note that G_3 also satisfies Principles 2 and 3. Since $n = 5$, Principle 2 requires that G_3 have at least $\frac{1}{2}(n - 1)(n - 2) + 2 = 9$ edges. Since the degree of each vertex is either 3 or 4, the sum of the degree of any two vertices not connected by an edge is 6, 7, or 8. In any case, the sum is at least $n = 5$. One Hamiltonian circuit in G_3 is $x \; y \; z \; u \; v \; x$.

There is a whole class of problems similar to the traveling salesman problem but much easier to solve. These involve connected graphs with weighted edges. The problem is to find a connected subgraph using all the vertices and having

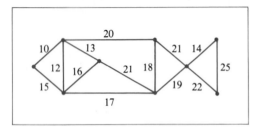

FIGURE 6

minimal total weight. Consider the weighted graph in Fig. 6. We could think of the graph as the representation of a pipeline through which we need to reach every customer (vertex). The weights might represent the unit cost of transmission of, say, natural gas through the pipeline. We try to find a pipeline system that will deliver gas to each vertex at the lowest possible unit cost. A *weighted graph* is a graph in which each edge has an associated positive number representing its weight.

A *subgraph* of a graph G is a graph whose vertex set is a subset of the vertex set of G and whose edge set is a subset of the edge set of G. There is an easy algorithm for finding a connected subgraph of a graph G with minimal total weight that includes each vertex of G. We first note that we are not looking for a path, only a subgraph of the original graph. Also, we would certainly not have any cycles (circuits) in our solution since we are trying to minimize the total weight while including every vertex of G. A graph with no cycles is called *acyclic*. Thus, our goal is to find a connected acyclic subgraph of G. A connected acyclic graph is called a *tree*. Since we wish to include every vertex of G, we say we want to *span* the graph G. In graph theory parlance, we are trying to find a *minimal spanning tree* of the graph G.

A connected acyclic graph is called a *tree* because the structure can always be redrawn to look like a natural tree with its branches and limbs. We study trees in more detail in section 13.6.

Here is the algorithm for finding the minimal spanning tree:

Minimal Spanning Tree Algorithm

1. Label the edges of the graph G: $e_1, e_2, \ldots, e_n$ in order of increasing weight. Thus,

$$\text{weight of } e_1 \leq \text{weight of } e_2 \leq \text{weight of } e_3 \leq \cdots .$$

2. Choose edge e_1 to begin the tree. Then choose the next edge in the sequence whose addition does not create a cycle.
3. Continue until no more edges can be added.

Why does the algorithm work? It operates in a greedy fashion, choosing the edges with smallest weight first. It ensures against cycles by stopping us from choosing edges which form them. It is possible that if several edges have the same weight the algorithm could result in different minimal spanning trees. All would have the same total minimal weight, however.

EXAMPLE 3   Find the minimal spanning tree for the graph in Fig. 6.

Solution   In Fig. 7 we show the graph labeled with edges in increasing order of weight. If two edges are tied, it does not matter which comes first (although different labeling might result in a different tree). We begin with e_1, which has weight 10. We then add

e_2, e_3, and e_4. However, we note that e_5 would complete a cycle ($e_1e_2e_5$), so we skip e_5 and check e_6. But we omit e_6 as well, since e_6 taken with e_2 and e_3 would form a cycle. We include e_7, e_8, and e_9, but omit e_{10}, e_{11}, and e_{12}. We conclude by including e_{13}. The algorithm gives the tree consisting of edges $e_1e_2e_3e_4e_7e_8e_9e_{13}$. The total weight of this tree is $10 + 12 + 13 + 14 + 17 + 18 + 19 + 22 = 125$. Natural gas sent along this pipeline will cost the minimal amount per unit and reach each of the vertices. Note that because the graph is connected we could pump gas from any vertex to any other. In Fig. 7(c) we have drawn the resulting graph to show its treelike structure.

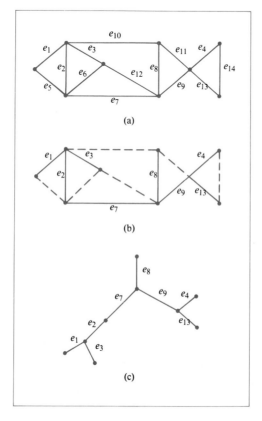

FIGURE 7

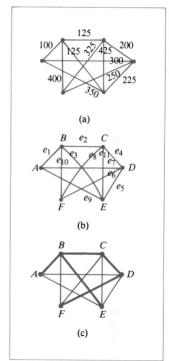

FIGURE 8

EXAMPLE 4 A technician for a cosmetics company must plan a telephone network linking distributors in cities A, B, C, D, E, and F. Tariffs are variable and she would like to plan a network for the lowest possible total cost. Using the map of Fig. 8(a) with the intercity tariffs shown (in \$1000's) on each edge, determine the minimal cost network.

Solution We have marked the edges in order of increasing weight by assigning e_1 to the edge of lowest weight, e_2 to the next weightier edge, and so on. Figure 8(b) is one such representation. Include e_1, e_2, e_3, and e_4, but since e_5 would complete a cycle we omit e_5. Of the other vertices, we include only e_6. The minimal spanning tree is drawn in Fig. 8(c). Its cost is $800,000.

There are a whole class of problems called *coloring problems*; these involve applications as diverse as coloring maps, describing relationships in social groups, and planning parties. One famous problem which was first proposed in 1852 is the *Four Color Problem*. To describe this problem, picture a section of an atlas map of the southeastern United States in Fig. 9. We assign a color to each state in such a way that any two adjacent states have different colors. What is the least number of colors required to color all the states in the map?

FIGURE 9

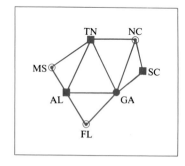

FIGURE 10

For a long time mathematicians believed that any map, no matter how complex, would require no more than four colors. This was finally proved in 1976 by Kenneth Appel and Wolfgang Haken of the University of Illinois using an enormous amount of computer time for the many thousands of cases they investigated. Interest in the problem persists as mathematicians seek a more elegant proof (independent of computers). We can see that the Four Color Problem is a graph theory problem by representing the map with a planar graph in which each state is represented by a vertex. An edge is drawn between two vertices if and only if their corresponding states are adjacent on the map. A representation of the map above with a coloring appears in Fig. 10. At least three colors are required. The coloring problem is then to find the least number of colors with which we can color the vertices so that no two vertices joined by an edge are the same color.

EXAMPLE 5 In each of the graphs in Fig. 11, determine whether you can color the vertices using exactly two colors in such a way that no pair of vertices joined by an edge are the same color. How many colors are necessary if two do not suffice?

Solution Two colors are sufficient to color the graphs in (a) and in (c). The graph in (b) requires three colors.

We add that from the graph in Fig. 11(b) we might conclude (try it) that any graph with a circuit consisting of an odd number of vertices cannot be colored

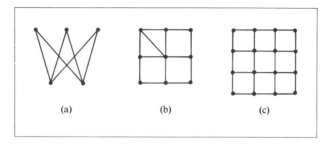

(a)   (b)   (c)

FIGURE 11

with only two colors. Rectangular grids, such as in the third example, can always be colored with two colors. (To see why, think of a checkerboard.) Each vertex in the first row of the grid can be marked alternatively with white and black, for example, starting with white. Then begin the next row with black, alternating white and black on the vertices of the second row. Since all the rows have the same number of vertices, adjacent vertices in the columns will have opposite colors. This can be repeated for any number of rows.

A graph G is called *bipartite* if and only if the vertices can be colored with two colors in such a way that no pair of vertices joined by an edge has the same color.

An interesting fact is the following: in a bipartite graph with a Hamiltonian circuit, the number of vertices of one color must be the same as the number of vertices of the other color. To see why, list the vertices in the Hamiltonian circuit. Except for the first and the last they must all be different. Furthermore, the color associated with any vertex differs from the color of its successor and the color of its predecessor in the circuit (since vertices connected by an edge have opposite colors). Thus, vertices alternate between the two colors, with no repeated vertices except the first and the last. There must be equal numbers of each color.

EXAMPLE 6   The chairperson of a department of eight faculty members wants to invite all eight to dinner, but there are ideological rifts in the department and some people do not speak to other people in the department. In the graph in Fig. 12(a), we have represented each of the department members by a vertex and have drawn an edge between two vertices if the people they represent cannot attend together. How many parties must the department chairperson give so as to assure that any two people

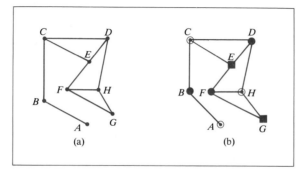

(a)   (b)

FIGURE 12

who will not attend together are invited to different dinners? Who should be invited to each party?

Solution   We note that if two vertices represent people who cannot attend together they are joined by an edge. Let's color those vertices different colors to indicate that they cannot attend the same party. Thus, we can find out how many parties we need by finding the fewest number of colors needed to color the graph. From Fig. 12(b), we see that three colors will color the graph. Vertices of the same color represent people who can be invited together. Possible guest lists are $\{A, C, H\}$, $\{E, G\}$, and $\{B, D, F\}$.

EXAMPLE 7   Suppose you are planning a meeting schedule for a day-long meeting. Several workshops are being planned for each time slot, and you want to schedule them so that you can maximize the satisfaction of the attendees. In the preregistration materials, those planning to attend indicate at most five workshops in which they would like to participate. What is the minimum number of time slots needed for the schedule if every person is to get his or her choice?

Solution   Here is a plan for the solution. Let each workshop be denoted by a vertex. Join two vertices with an edge if and only if at least one person wishes to take both workshops. The minimum number of time slots that can be used is the least number of colors needed to color the resulting graph. Since this could result in a large graph, we will leave the solution in theoretical form.

EXAMPLE 8   What is the minimal number of colors needed to color the diagram in Fig. 13(a) so that adjacent regions are different colors?

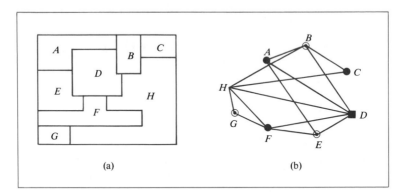

(a)   (b)

FIGURE 13

Solution   Draw a graph with a vertex for each area and an edge joining the vertices of adjacent areas (Fig. 13(b)). Color the graph so that no two vertices joined by an edge are the same color. Four colors are needed.

The *chromatic number* of a graph is the minimum number of colors needed to color the graph so that no two vertices joined by an edge are the same color. Clearly, a bipartite graph has chromatic number 2. The graph in Example 8 has

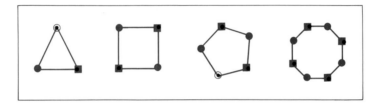

FIGURE 14

chromatic number 4. What is the chromatic number of a cycle? If the cycle contains no loops and no parallel edges, the chromatic number is 2 or 3. This can be seen in Fig. 14. If the cycle has an even number of vertices, two colors will suffice. Otherwise, three are required.

EXAMPLE 9   The floor plan of the first level of a house is shown in Fig. 15.

(a) Is it possible to enter the house at the front, exit from the rear, and travel through the house going through each doorway exactly once?

(b) Can one enter the house at the front, exit from the rear, going through each room exactly once?

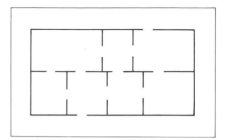

FIGURE 15

(c) If adjacent rooms are to be painted different colors, how few colors are necessary to paint the rooms of the house?

Solution   Figures 16(a) and (b) show the plan and the graph representation of it. We represent the rooms with vertices and the doorways with edges.

(a) Is there an Euler path from X to Y? There are 6 vertices of odd degree, so no such path exists.

FIGURE 16

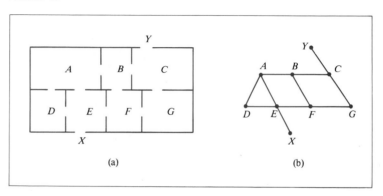

(b) Can we find a Hamiltonian path from X to Y? We construct such a path: $X\ E\ D\ A\ B\ F\ G\ C\ Y$.

(c) How many colors do we need to color the graph? Three colors are needed to color the graph; we need three colors of paint.

PRACTICE PROBLEMS 3

1. Find a Hamiltonian circuit for the graph in Fig. 17.

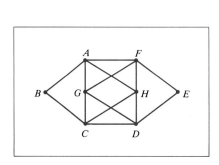

FIGURE 17   FIGURE 18

2. Find a minimal spanning tree for the graph in Fig. 18. The amount on each edge represents the cost of transportation of goods along the edge. What is the minimal cost?

3. Find the smallest number of colors needed to color the map in Fig. 19 so that no two adjacent regions are the same color.

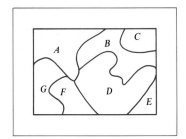

FIGURE 19

EXERCISES 3

1. Find a Hamiltonian circuit in the graphs of Fig. 20.

2. Find a Hamiltonian circuit in the graphs of Fig. 21.

3. For each graph in Fig. 22, tell whether an Euler circuit or a Hamiltonian circuit exists.

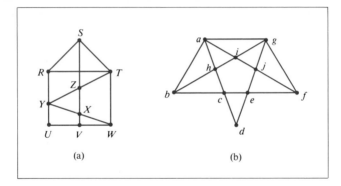

(a)

(b)

FIGURE 20

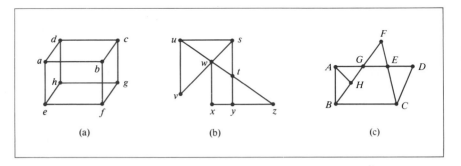

(a)

(b)

(c)

FIGURE 21

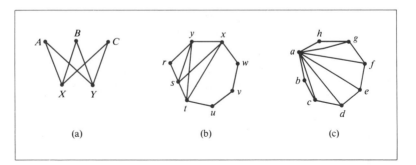

(a)

(b)

(c)

FIGURE 22

4. Let G be a simple graph on 5 vertices all of degree at least 3. Must G have a Hamiltonian circuit? Explain.

5. Let G be a simple graph on 6 vertices, one of which has degree 2. Must G have a Hamiltonian circuit? Explain.

6. Let G be a simple graph with 6 vertices and 13 edges. Must G have a Hamiltonian circuit? Explain.

7. Let G be a simple graph on 8 vertices. Find the smallest number of edges that guarantee that G has a Hamiltonian circuit.

8. Draw a simple graph on 6 vertices with 10 edges in which there is a Hamiltonian circuit.

9. Find the minimal spanning tree for the graph in Figure 23.

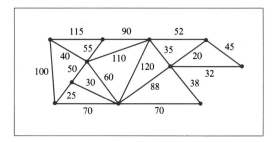

FIGURE 23

10. A communication line that must reach each of the locations represented by vertices is to be laid so as to require a minimal cost. How should this be done? The cost of each connection is marked on the graph in Figure 24.

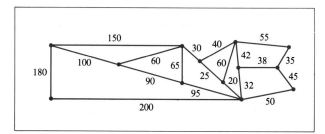

FIGURE 24

11. For each graph in Fig. 25, tell if the graph is bipartite. If not, what is the minimal number of colors necessary to color the graph?

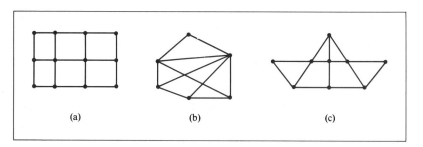

(a)   (b)   (c)

FIGURE 25

12. The table in Fig. 26 uses the symbol ▌ to indicate those people who do not like each other. Draw a graph with vertices representing people and draw an edge between two vertices if and only if the people represented do not like each other.

(a) How many colors are needed to color the graph?

(b) All vertices colored the same color represent those people who can be invited to the same party. How many parties are needed to entertain all the people and to be assured that any two people who do not like each other will not be invited to the same party?

FIGURE 26

13. Show that the graphs in Fig. 27 have no Hamiltonian circuits.

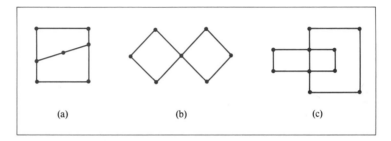

(a)   (b)   (c)

FIGURE 27

14. (a) Show that each of the graphs in Fig. 28 is bipartite.

(b) Show that each does not have a Hamiltonian circuit.

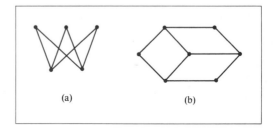

(a)   (b)

FIGURE 28

15. There are 12 senior students who need to complete their major in accounting this semester. The courses, $a, b, \ldots, g$, they need are tabulated below. How many time periods are needed so that they all can fulfill their requirements?

Allan: a, c, g Greg: f, g
Brad: a, b, d Helen: d, f
Carol: b, e Ira: a, d, g
David: a, c Jane: a, g
Evan: a, b Karen: b, c
Ferne: c, e Larry: e, f

1. A Hamiltonian circuit is
   $A\ B\ C\ H\ F\ E\ D\ G\ A$.

2. The minimal cost is 305 and the minimal spanning tree appears in Fig. 29.

3. The number of colors necessary to color the map is three.

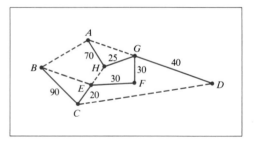

FIGURE 29

13.4 Directed Graphs

Let us consider the problem of scheduling a stop of an aircraft. The aircraft lands, passengers deplane, luggage and other cargo are unloaded, the plane is replenished with food, and it is cleaned and serviced. The new cargo is loaded, passengers board, and the plane takes off again. How much time must be allocated to this process? There are several stages in the operation, some of which are dependent on others. For example, incoming cargo must be unloaded before loading new cargo. Some jobs can be done simultaneously: passengers can deplane while cargo is unloaded. The table below shows which tasks must precede others and the time required for each task.

Task	Immediate Predecessors	Time Required for Task (min)
S (start)	None	0
A_1 (prepare to unload cargo and deplane passengers)	S	10
A_2 (deplane passengers)	A_1	20
A_3 (unload cargo)	A_1	30
A_4 (cleaning and food)	A_2	20
A_5 (maintenance and fuel)	A_1	30
A_6 (load cargo)	A_3	15
A_7 (board passengers)	A_4	20
E (end)	$A_5\ A_6\ A_7$	0

This process can be represented by a graph. We indicate the required order by directing the edges to obtain a *directed graph* or *digraph*. A directed graph D is a set of vertices and a set of edges for which each edge has an initial vertex and a terminal vertex. The digraph in Fig. 1 represents the aircraft problem. There is a directed edge from vertex A_i to vertex A_j if and only if activity A_i is an immediate predecessor of activity A_j. The times required for the activities appear in the large

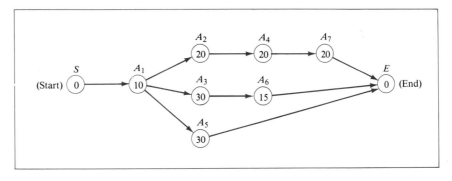

FIGURE 1

circles at the vertices. For convenience, we have a start (S) and an end (E) that require no time.

The time we need to allocate between arrival and departure is the length of the longest (time) path on Fig. 1 from start to end. There are three paths from start to end:

$SA_1A_2A_4A_7E$:   time $= 10 + 20 + 20 + 20 = 70$ minutes

$SA_1A_3A_6E$:   time $= 10 + 30 + 15 = 55$ minutes

SA_1A_5E: time $= 10 + 30 = 40$ minutes

The longest (time) path is 70 minutes, so we should allow 1 hour and 10 minutes between arrival and departure of the aircraft. The path $SA_1A_2A_4A_7E$ is called the *critical path*, and the directed graph used to represent the process is called the *activity digraph*. We should take note of the fact that, to reduce the amount of time necessary between arrival and departure of the aircraft, we first need to focus on reducing the time necessary for those activities on the critical path. For example, we can try to reduce the amount of time needed for A_4, cleaning the craft and loading the food. That would reduce the time length of the critical path. Reducing the time needed for maintenance and refueling of the aircraft (A_5) has no effect on the required time between arrival and departure because the activity A_5 is not on the critical path. Note, however, that if we were able to reduce the time needed for A_2 to 10 minutes and the time allocated to A_7 to 12 minutes the path $SA_1A_2A_4A_7E$ would be $10 + 10 + 20 + 12 = 52$ minutes. But that does not mean that we can turn the aircraft around in 52 minutes. We would need 55 minutes because the critical path would then become $SA_1A_3A_6E$. The digraph allows us not only to see where time savings can reduce the total time for a complicated process of inter-related activities, but also helps us to schedule those activities to minimize the necessary time.

From the activity digraph for the aircraft, we can see that times, in minutes, to begin each task are given by:

A_1	A_2	A_3	A_4	A_5	A_6	A_7	E
0	10	10	30	10	40	50	70

EXAMPLE 1   Use the table below to draw an activity digraph. Determine the critical path and give the beginning times for each activity to ensure the minimum time from start to end.

Activity	Immediate Predecessor	Time Required (days)
S	—	0
A_1	S	3
A_2	A_1	4
A_3	A_2	15
A_4	A_3	5
A_5	A_3	4
A_6	A_3	11
A_7	A_4, A_5	7
A_8	A_6	4
A_9	A_6, A_7	2
E	A_8, A_9	0

Solution   The activity digraph appears in Fig. 2(a). From the digraph we see that activities S, A_1, A_2, A_3, and E are common to all paths from S to E. The total amount of

FIGURE 2

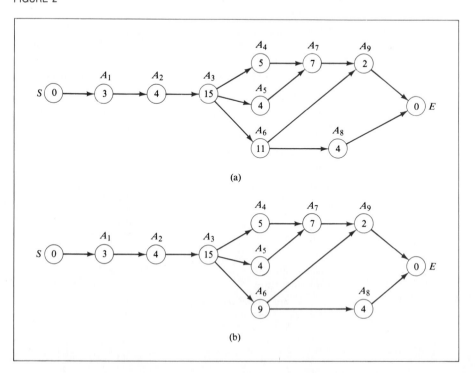

(a)

(b)

time for these activities is 22 days. The total time required on the paths from S to E is

$$SA_1A_2A_3A_4A_7A_9E: \quad 22 + 5 + 7 + 2 = 36 \text{ days}$$

$$SA_1A_2A_3A_5A_7A_9E: \quad 22 + 4 + 7 + 2 = 35 \text{ days}$$

$$SA_1A_2A_3A_6A_9E: \quad 22 + 11 + 2 = 35 \text{ days}$$

$$SA_1A_2A_3A_6A_8E: \quad 22 + 11 + 4 = 37 \text{ days}$$

The last of these paths is the critical path. The entire project will take a minimum of 37 days. We can schedule the work by following the digraph. We find the starting time of each activity by determining the length (in time) of the longest path from the start (S) to the activity in question.

Activity:	A_1	A_2	A_3	A_4	A_5	A_6	A_7	A_8	A_9	E
Starting Time (days):	0	3	7	22	22	22	27	33	34	37

We note that to find the time to begin A_7 we consider the two paths from start to a vertex preceding A_7. One is $SA_1A_2A_3A_4$, which requires 27 days. The other path is $SA_1A_2A_3A_5$, which requires 26 days. We use the longer of the two paths.

In fact, since it is frequently difficult to count all paths from start to finish, we can use this technique to calculate both the critical path and the starting time of each activity.

EXAMPLE 2 Suppose that in the project discussed above we are able to reduce the time required to do A_6 to 9 days. Determine the critical path then and find a schedule for the activities.

Solution The schedule is determined from Fig. 2(b).

Activity:	A_1	A_2	A_3	A_4	A_5	A_6	A_7	A_8	A_9	E
Starting Time (days):	0	3	7	22	22	22	27	31	34	36

The critical path is now $SA_1A_2A_3A_4A_7A_9E$, which has length 36 days.

In actuality, the schedule of very large and complicated projects, like the building of office buildings and shopping centers, requires far more complicated activity digraphs. To find the critical path may not be as simple as in our small example. An interesting feature of critical path analysis is that it can be set up in such a way that we can use the methods of linear programming to find the longest path.

Optimal Scheduling with Randomness The study of the scheduling of activities using a digraph can be enhanced for more sophisticated analysis using probability and statistics. We explore that notion at this point. Consider the digraph in Fig. 3. The activity table (Table 1) contains a bit more information than the type we considered earlier. In the table, we have added an optimistic time (a) and a pessimistic time (b) to the probable time (m) for completion of each task. This is a more realistic way of thinking of the entire project.

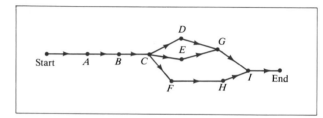

FIGURE 3

TABLE 1

Activity	Immediate Predecessor	a	m	b
A	Start	2	3	4
B	A	3	4	11
C	B	11	15	19
D	C	3.5	5	6.5
E	C	3	4	5
F	C	8	11	20
G	D, E	5	7	15
H	F	3	4	5
I	G, H	1.5	2	8.5
End	I	0	0	0

Studies have shown that the distribution of the probable time for completion of each task has what is called the *beta probability distribution*, and its mean and variance are given by the formulas

$$\text{mean time} = t = \frac{a + 4m + b}{6} \quad \text{and} \quad \text{variance} = \sigma^2 = \left(\frac{b - a}{6}\right)^2$$

The mean times (t) and the variances (σ^2) are calculated for each task, entered in Table 2, and displayed on the activity digraph (Fig. 4). For example, the mean time assigned to task F is $[8 + 4(11) + 20]/6 = 12$ while its variance is $[(20 - 8)/6]^2 = 4$.

TABLE 2

Activity	Mean Time t	σ^2
A	3	0.11
B	5	1.78
C	15	1.78
D	5	0.25
E	4	0.11
F	12	4.00
G	8	2.78
H	4	0.11
I	3	1.36

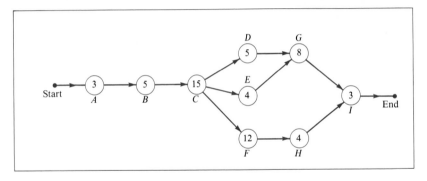

FIGURE 4

The usual calculations reveal that there are three paths from start to end with the longest and critical path ABCFHI for a total of 42 days. We continue the analysis with a look at the slack time for each activity and an estimate of the probability that we will complete the project within certain time constraints.

We work forward through the digraph in Fig. 4, using the expected times t at each stage to determine the earliest starting time (ES) and the *earliest finishing time* (EF) for each task. The earliest starting time for each activity is the largest value of all the earliest finishing times for all activities that enter that vertex in the digraph. For each activity

$$EF = ES + t.$$

That is, the earliest finishing time of any task is calculated by finding its earliest starting time and adding the mean time needed for the task.

For example, consider activity G. The earliest time it can start is the largest of the earliest finishing times of D and E since D and E are its immediate predecessors. These are tabulated in Table 3 for every task.

We can see the critical path by making a backward pass calculation as well. We use the last vertex first and let the earliest finishing time equal the latest

TABLE 3

Activity	ES	t	EF	Activity	LF	t	LS
A	0	3	3	I	42	3	39
B	3	5	8	H	39	4	35
C	8	15	23	G	39	8	31
D	23	5	28	F	35	12	23
E	23	4	27	E	35	4	31
F	23	12	35	D	35	5	30
G	28	8	36	C	23	15	8
H	35	4	39	B	8	5	3
I	39	3	42	A	3	3	0
End				Start			

finishing time at that vertex. The *latest starting time* (LS) and the *latest finishing time* (LF) are calculated for each activity as follows: the latest finishing time for an activity at a vertex is the smallest value of the latest starting time for all the activites that leave from that vertex. Then

$$LS = LF - t.$$

The *slack time* at each activity is defined as

$$slack = LS - ES = LF - EF.$$

For example, the earliest starting time for E is 23 since its immediate predecessor is C and the earliest finishing time for C is 23. The slack time for each activity is shown in Table 4.

TABLE 4

Activity	A	B	C	D	E	F	G	H	I
Slack	0	0	0	3	4	0	3	0	0
Var (σ^2)	.11	1.78	1.78			4.00		.11	1.36
t	3	5	15			12		4	3

The slack time for the activity E, for example, is LS $-$ ES for that activity. From the tables we find that LS is 27 while ES is 23. Thus, there is a leeway of $27 - 23 = 4$ days in the schedule. The activity E is not on a critical path. The critical activities are those with no slack time at all. Table 4 displays this information, showing the mean and the variance for the activities with zero slack time. Thus, we confirm that the critical activities are A, B, C, F, H, and I. The total time for these is 42 days.

We might assume that the time to actually complete the project is a normally distributed random variable. We have added the element of randomness by giving optimistic and pessimistic time estimates for each activity. We assume that the time T to completion is normally distributed with a mean of 42 days. If, in addition, we assume that the activities are independent with respect to the amount of time each takes, then the total variance of T is given approximately by the variance of the critical path: the sum of the variances of each task on the critical path.

$$\sigma^2 = .11 + 1.78 + 2.79 + 4.00 + .11 + 1.36 = 10.15$$

$$\sigma = \sqrt{10.15} = 3.19 \text{ days.}$$

Allowing 45 days for this project, what is the probability that we will finish in the allotted time? Assuming that T is normally distributed with $\mu = 42$ and $\sigma = 3.19$, we use the methods of Section 7.5 to find the probability.

$$\Pr(T \le 45) = \Pr\left(Z \le \frac{45 - 42}{3.19}\right) = \Pr(Z \le .94) \sim .8289.$$

Thus, if we allow 45 days for the project, we have an 83% chance of meeting the schedule.

How much time should we allocate if we wish to be 95% sure to complete the project as planned? We are seeking the 95th percentile of the normal distribution with mean $\mu = 42$ days and standard deviation $\sigma = 3.19$ days. The 95th percentile of the standard normal distribution is 1.65. We solve the equation below for t:

$$\frac{t - 42}{3.19} = 1.65.$$

We find $t = 47.26$ days. To be safe, we should allow 48 days.

Sophisticated activity digraphs are used internally in large computer systems. For a given number of processing units in the computer, and a list of priorities for the activities we expect the computer to accomplish, digraphs help the computer plan its own processing schedule. This technique is used in what is known as the list-processing algorithm and is used on computers with several processors.

Another example for which a digraph is useful is given by a street pattern in which direction of traffic is indicated. We can plan for one- and two-way streets by marking each edge (street) with appropriate arrows. We might consider the city pictured in Fig. 5(a) and ask whether a one-way street pattern exists that will allow us to get from A to B and back again. We would need to assign directions to the digraph so as to get a circuit without using any two-way streets. Can you find a solution? We have indicated one in Fig. 5(b).

A solution is not always possible. Look at the street pattern in Fig. 5(c). Try to find a one-way street pattern on which you can construct a circuit starting at A, passing through B, and returning to A.

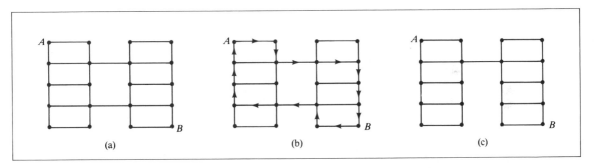

(a)   (b)   (c)

FIGURE 5

We will use the letter D to designate a digraph. A *path* in a digraph is a sequence of edges linking two vertices. A *closed path* is a path in which the first and the last vertices are the same. A closed path in which no edge is repeated is called a *circuit* (or a *cycle*). A digraph with no circuits is called *acyclic*. The digraph in Fig. 6(a) has a circuit ($w\ u\ x\ w$) and hence is not acyclic.

A digraph is said to be *connected* if the graph we get by omitting the directions on the edges is connected. We say that a digraph is *strongly connected* if,

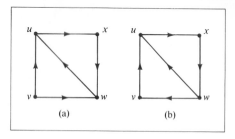

FIGURE 6

given any two vertices u and v, there is a path *from u to v*. We observe that the digraph in Fig. 6(a) is connected but it is not strongly connected. There is no path from vertex u to vertex v, for example. By changing the direction on the edge joining v and w, we have the strongly connected digraph shown in Fig. 6(b).

The *outdegree* of a vertex v (od v) is the number of edges in the digraph D having v as an initial point. The *indegree* of a vertex v (id v) is the number of edges in D having v as a terminal point. Then

$$\deg(v) = \text{od } v + \text{id } v$$

We observe in Fig. 6(a) that the indegree of vertex u is 2 and the outdegree of u is 1. Clearly, its degree is 3. The vertex v has indegree 0, which naturally means that there is no way to reach v from any other vertex in the graph.

Property 1 If D is a digraph with p vertices $v_1, v_2, \ldots, v_p$, then

$$\text{od } v_1 + \text{od } v_2 + \cdots + \text{od } v_p = \text{id } v_1 + \text{id } v_2 + \cdots + \text{id } v_p$$
$$= \text{number of edges in } D$$

EXAMPLE 3 A small subdivision of homes is being planned with a street pattern of one-way streets represented by the digraph D in Fig. 7.

(a) Is D acyclic? If not, give at least one circuit. That is, can we drive along at least one route beginning and ending at the same vertex?

(b) Is D connected? Is D strongly connected; that is, can we begin at any vertex and drive to any other vertex?

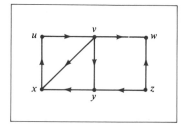

FIGURE 7

(c) Find the id and od of each vertex of D and verify that Property 1 holds.

Solution (a) The graph is not acyclic since u v x u is a circuit.

(b) The digraph is connected, but it is not strongly connected since there is no path from vertex w to any other vertex.

(c) We note that the number of edges in the digraph is 8. The id and od of each vertex are tabulated below. In each case the sum is 8.

Vertex	u	v	w	x	y	z	Sum
id	1	1	2	2	2	0	8
od	1	3	0	1	1	2	8

We might want to determine if an Euler circuit exists in a digraph. That is, given a connected digraph D, is it possible to find a path starting at any vertex, passing along each edge exactly once, and returning to our starting point? We recall that if an undirected connected graph G has an Euler circuit then every vertex has even degree. On the other hand, given a connected graph G, we only need to verify that each vertex has even degree to guarantee the existence of an Euler circuit in G. A similar argument can be used to prove the comparable theorem for digraphs.

> **Property 2** Let D be a connected digraph. Then D has an Euler circuit if and only if, for every vertex of D, od $v = $ id v.

Let us try to see why Property 2 holds. We assume that D is a digraph with an Euler circuit C. Choose an arbitrary vertex v of D. If v is not the initial or terminal vertex of the circuit C, then every time v occurs in the circuit it is entered and exited by different edges. We can imagine that as we traverse an edge we erase it from the graph (so it cannot be used again). Each time we encounter v we can imagine erasing the edge entering v and the edge exiting v, thus reducing both id v and od v by 1. Since all edges are eventually used, we eventually have id $v = 0$ and od $v = 0$. Thus, we must have had id $v = $ od v in D. If v is the initial vertex of the circuit C, then at the start of C we reduce od v by 1 and at the end of the circuit also reduce id v by 1.

Further inspection of Fig. 7 will reveal that there is no Euler circuit in the digraph. As we scan the vertices, we see that, except at vertex u, the indegree is unequal to the outdegree.

The proof of the converse requires that we assume D is connected and, for each v, od $v = $ id v. Then we need to show that an Euler circuit exists. We do not present the proof here.

The following statement is helpful in constructing an *Euler path* in D, that is, finding a path from a vertex u to a vertex v that traverses every edge of the digraph exactly once.

> **Property 3** Let D be a connected digraph. Then D has an Euler path if and only if D contains vertices u and v such that
>
> 1. od $u = $ id $u + 1$
> 2. id $v = $ od $v + 1$ and
> 3. id $w = $ od w for every other vertex in D.
>
> When an Euler path exists, the path begins at u and ends at v.

EXAMPLE 4   For the digraphs D_1 and D_2 given in Fig. 8, determine if an Euler circuit exists. Does an Euler path exist? If they do exist, find them.

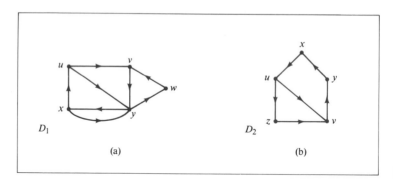

(a)   (b)

FIGURE 8

Solution   (a) We mark the id and od of each vertex. By Property 3, there are no Euler circuits and no Euler paths for the digraph in (a). There is no Euler circuit in the digraph in (b), but an Euler path indicated with numbers on the edges (in the order of traversal) can be traced from u to v because od $u =$ id $u + 1$ and id $v =$ od $v + 1$, while the outdegree and indegree of every other vertex are equal (see Fig. 9).

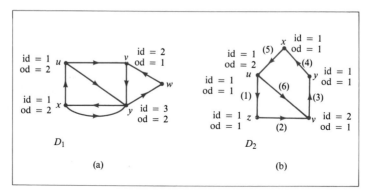

(a)   (b)

FIGURE 9

The theory of *Hamiltonian circuits* or paths for digraphs is similar to and as complex as that for graphs. The problem is to find a circuit beginning and ending at the same vertex and visiting every other vertex of the digraph exactly once. A *Hamiltonian path* is a nonclosed path that visits each vertex of the digraph exactly once. We will not mention any theorems here, but present an example for a trial and error solution. There is one application that arises naturally and for which we always have a Hamiltonian path—*tournaments*.

A *tournament* is a digraph in which, given any two vertices v_i and v_j, there is either a directed edge from v_i to v_j or a directed edge from v_j to v_i, but not both. Consider the following example.

EXAMPLE 5 Four tennis players are competing for the championship by playing a round-robin tournament. This means that each player plays every other player. We will not allow ties. Here is what happens in the tournament:

Allen defeats Calvin but is defeated by Boris and Dorian. Boris defeats Calvin and Dorian. Calvin defeats Dorian. Draw a digraph that describes the tournament and determine a winner, if you can.

Solution We draw a digraph with a vertex to represent each player (Fig. 10). An edge is directed from Allen to Calvin because Allen defeated Calvin. The digraph is a tournament because exactly one player wins each match. Since each edge directed out of a vertex represents a win for the player represented by the vertex, by calculating the outdegree of each vertex we can find the number of games won by each player. We see that

$$\text{od } A = \text{od } C = \text{od } D = 1, \text{ while od } B = 3.$$

Clearly, B is the winner if we choose to define the winner as the person with the most wins.

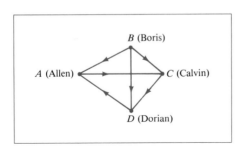

FIGURE 10   FIGURE 11

Can we find a Hamiltonian path in this digraph? We have marked it in Fig. 11. We visit the vertices in the order $B\ A\ C\ D$. This is not the only Hamiltonian path. Can you find another? What does the Hamiltonian path reveal about the tennis tournament? Clearly, in every path B is first. Each path gives a ranking of players. Unfortunately, in this tournament the players ranking second, third, and fourth are not uniquely determined.

EXAMPLE 6 A firm is planning to relocate its offices to another city. The management team makes a list of the five possibilities and determines pairwise preferences. These are shown in the list below. Use a digraph to establish where the firm should relocate.

Is the first choice unique? Find a priority list in case the first place choice is unavailable. Is the priority list unique?

Albany is preferred to Boston and Cleveland.
Dallas is preferred to Albany and Boston.
Eugene is preferred to Albany and Boston and Dallas.
Boston is preferred to Cleveland.
Cleveland is preferred over both Dallas and Eugene.

Solution From the digraph (Fig. 12) we see that Eugene has the highest outdegree. Thus, the firm should seriously consider relocation to Eugene. One Hamiltonian path is *E D A B C*. Another would be *E A C D B*. On the digraph representation in Fig. 12, we have marked each vertex v with the ordered pair (od v, id v). While the firm should relocate to Eugene, if that is not possible, there is no clear second choice.

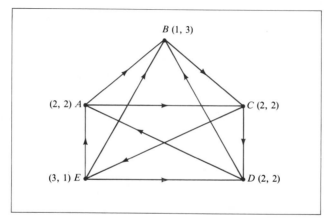

FIGURE 12

It would be preferable if we could ensure that a tournament be *transitive*. That is, if u is preferred to v and v is preferred to w, we would like to have u preferred to w. In the digraph representation, this means that, if there is an edge directed from u to v and an edge directed from v to w, then there must be an edge directed from u to w. In Fig. 13, we see what must occur if there are only three

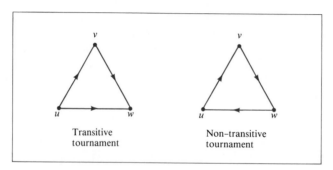

FIGURE 13

players. In sports, round-robin competitions frequently do not have this transitive property: player A may defeat player B, player B may defeat player C, and yet player C may defeat player A. However, preference listings very well may be of this form. We can characterize all transitive tournaments in a simple way.

Property 4   A tournament is transitive if and only if it is acyclic.

Suppose the tournament T has no cycles. Assume (u, v) and (v, w) are directed edges in T. Clearly, if T has no cycles, then (w, u) is not a directed edge of T. But then (u, w) must be in T. Thus, T is transitive.

Suppose T is transitive and suppose, by way of contradiction, that T contains a cycle C. Let the cycle be $v_1, v_2, \ldots, v_k, v_1$. Then (v_1, v_2) and (v_2, v_3) are edges in T. But T is transitive implies that (v_1, v_3) is an edge in T. Similarly, we can conclude that (v_1, v_k) is an edge of T. But the edge (v_k, v_1) is in the cycle C and therefore in T. This is a contradiction (there are no ties).

Since transitive tournaments are desirable, we wonder if, given any number of players, we can always construct a digraph that represents a transitive tournament. The next theorem guarantees that and more. We are assured that there is one and only one such tournament. Beware! This only allows us to write a preference list that is transitive. If we rename the vertices, we get a different list.

Property 5   For every positive integer p, there is a transitive tournament with p vertices. Moreover, any two such tournaments are equivalent.

The *length* of a path P in a digraph is the number of edges in P. The *distance $d(u, v)$ from vertex u to vertex v* is the length of the shortest path from u to v. [Note that $d(u, v)$ is not always defined.]

Property 6   Let T be a tournament. Let v be any vertex of T having maximum outdegree. Then the distance from v to any other vertex u of T is one or two.

How can we interpret this property? In a round-robin tournament (in which every team plays every other team and no ties are allowed), the winner (there may be more than one) has been defeated only by teams that have themselves been defeated by teams that have been defeated by the winner. For example, suppose that in a tournament with 16 teams the winner finishes with a record of 13 wins and 3 losses. Each of the 3 teams that beat the winner must have lost to at least one of the teams that lost to the winner. If this were not so, then one of the 3 teams would have defeated all 13 of the teams losing to the winner and would also have beaten the winner. Its record would be 14–2. And it would have had a better record than the winner!

The next result is useful in two ways. If a digraph represents a tournament (every vertex is connected by a directed edge to every other vertex and there are no loops or parallel edges), then the digraph has a Hamiltonian path. The property also allows teams to be ranked. The ranking gives a Hamiltonian path.

Property 7   Every tournament contains a Hamiltonian path.

Unfortunately, this guarantee of a ranking (a Hamiltonian path) need not yield a unique ranking. Fortunately, however, the following property ensures that in preference-type digraphs there is likely to be a unique Hamiltonian path.

Property 8   Every transitive tournament contains exactly one Hamiltonian path.

If we can show that a digraph D has no cycles, it will be transitive. Hence, if the digraph has no cycles, there is a unique Hamiltonian path. This path establishes a unique listing (or ranking) of our preferrences in such a digraph.

EXAMPLE 7   The ABC Construction Company has interviewed five candidates for chief foreman. Ullman is preferred to Victor, Xerxes, and Walters. Victor is preferred to Xerxes, while Walter is preferred to Victor and Xerxes. Yates is preferred to each of the other four. Find the person who should get the job and determine if it is a unique choice. If possible, obtain a unique ranking of the other candidates in case the winner declines the job offer.

Solution   The digraph representing the personnel preferences is given in Fig. 14. We have marked each vertex with its outdegree and indegree (od, id). By noting that there are no cycles, we are assured that the digraph is transitive and that there is a unique first choice and a unique ranking of the candidates. By listing the outdegree of each vertex, we can easily find that ranking. We find

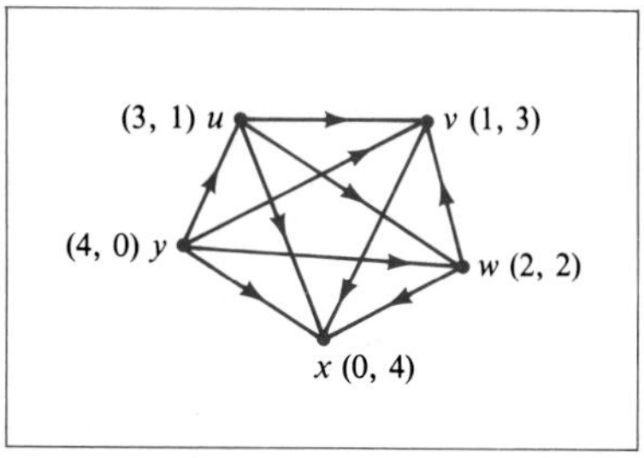

FIGURE 14

$$\text{Outdegree of:} \quad \begin{matrix} u & v & w & x & y \\ 3 & 1 & 2 & 0 & 4 \end{matrix}$$

Thus, the ranking of the candidates is Yates, Ullman, Walters, Victor, and, finally, Xerxes.

Although this example was fairly easy to do without knowing anything about graph theory, situations in which there are a large number of choices are effectively handled by a computer applying the appropriate graph theoretic procedure to obtain a unique ranking.

PRACTICE PROBLEMS 4

1. Determine the id and od of each vertex in the digraph in Fig. 15. Does an Euler circuit or an Euler path exist in this digraph? Explain.

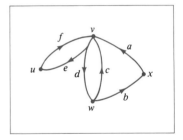

FIGURE 15

2. Draw the digraph and find the critical path for the writing of a grant proposal. How much time do we need? If the time required to do A_2 can be reduced to 16 days, what other task must be accomplished faster to assure we can get the grant proposal ready in 30 days?

	Tasks	Immediate Predecessors	Time for Task (days)
S	start	None	0
A_1	initial meeting	S	7
A_2	write narrative	A_1	21
A_3	solicit matching funds	A_1	14
A_4	invite consultants	A_1	12
A_5	get support letters	A_3, A_4	7
A_6	proofread narrative	A_2	3
A_7	write budget	A_3	5
A_8	write time table	A_2, A_5, A_7	3
A_9	type narrative	A_6	2
A_{10}	collate and submit	A_8, A_9	1
E	end	A_{10}	0

3. (a) Determine whether the digraphs in Fig. 16 are (i) acyclic, (ii) connected, or (iii) strongly connected.
   (b) Determine whether the digraphs have (i) an Euler circuit, (ii) an Euler path, (iii) a Hamiltonian circuit and/or (iv) a Hamiltonian path. Explain.

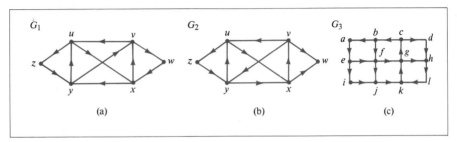

FIGURE 16

4. In a Ping-Pong round-robin tournament, there are six players, *A*, *B*, *C*, *D*, *E*, and *F*. The digraph in Fig. 17 shows the results of each match with an edge directed from the winner to the loser.

   (a) Is the tournament transitive?

   (b) Is there a unique winner? If so, who?

   (c) Is there a unique ranking of players determined from the tournament? If so, give it.

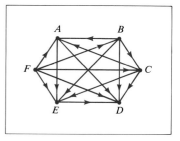

FIGURE 17

EXERCISES 4

1. Determine whether each of the digraphs in Fig. 18 is connected, strongly connected, neither, or both?

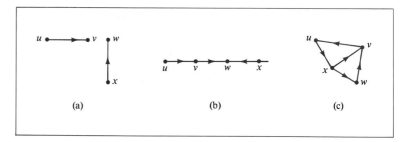

FIGURE 18

2. Determine whether each of the digraphs in Fig. 19 is connected, strongly connected, neither, or both?

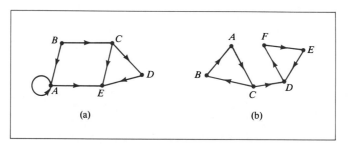

FIGURE 19

3. In a computer program, frequently one procedure calls another, which in turn calls other procedures. The digraph in Fig. 20 shows the relationship of six modules of a large program. The edge is directed from a procedure to one that calls it. Thus, (*D*, *E*) is a directed edge because procedure *D* is called by *E*.

(a) Find the indegree and the outdegree of each vertex.

(b) What do these numbers mean in terms of the computer program?

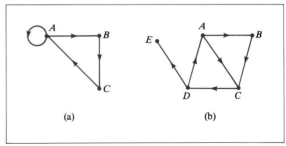

(a)

(b)

FIGURE 21

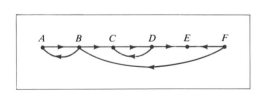

FIGURE 20

4. Which of the digraphs in Fig. 21 is acyclic? If the graph has a cycle, tell what it is.

5. Which of the digraphs in Fig. 22 is acyclic? If the graph is not acyclic, demonstrate some cycle.

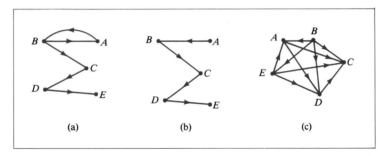

(a)   (b)   (c)

FIGURE 22

6. Which of the digraphs in Fig. 23 have Euler circuits? If one exists, give it. If none exists, why not? Is there an Euler path? Explain.

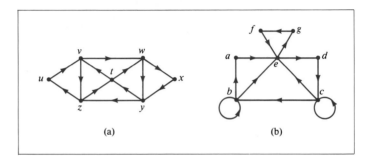

(a)   (b)

FIGURE 23

7. For the graphs in Fig. 24, tell if an Euler circuit exists. If one exists, tell what it is. If none exists, explain why. Is there an Euler path? Explain.

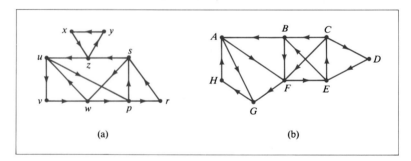

(a)   (b)

FIGURE 24

8. Decide if the digraphs in Fig. 25 are connected, strongly connected, acyclic, have an Euler circuit, and/or have an Euler path. Explain your answers.

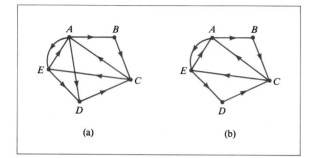

(a)   (b)

FIGURE 25

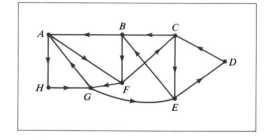

FIGURE 26

9. Find an Euler circuit in the digraph of Fig. 26 if such a circuit exists.

10. Which of the digraphs of Fig. 27 are connected, strongly connected, acyclic, have Euler paths, and/or have Euler circuits?

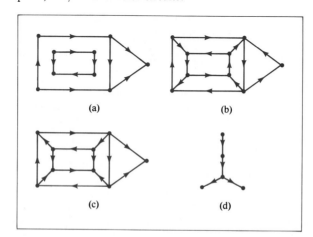

(a)   (b)

(c)   (d)

FIGURE 27

11. The process of packing to go away to college is sometimes a lengthy one. From the table below, determine the shortest possible time in which this person can be ready to leave. What is the critical path?

Task		Immediate Predecessors	Time to Do Task (days)
S	start	None	0
A_1	talk to roommate	S	2
A_2	make list	A_1	2
A_3	shop for clothes	A_2	8
A_4	shop for toiletries	A_2	2
A_5	buy trunk	A_3	1
A_6	wash clothes	A_3	3
A_7	mend and iron	A_6	4
A_8	pack trunk	A_4, A_5, A_7	3
A_9	send trunk	A_8	1
A_{10}	pack other things	A_8	2
A_{11}	load car	A_{10}	1
E	end	A_9, A_{11}	0

12. Given the digraph in Fig. 28 representing a scheduling problem, find the critical path. Numbers in circles represent the time (in minutes) necessary for the task. If the task A_3 can be done in 5 minutes, can we reduce the total time needed? If so, what is the new time? If A_8 can be done in 10 minutes, can we reduce the total time needed? If so, what is the new total time needed?

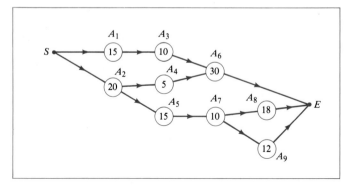

FIGURE 28

13. Draw a digraph representing the precedence table and determine the critical path.

Job	Immediate Predecessors	Completion Time (days)
Start		0
A	Start	15
B	Start	12
C	A	22
D	A, C	10
E	B, D	13
F	C, E	5
End	F	0

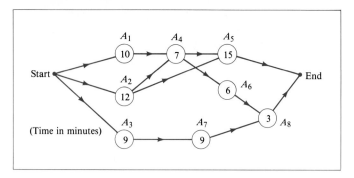

FIGURE 29

14. (a) Find the critical path for the digraph in Figure 29.

(b) How long does the entire job take?

(c) If A_5 could be reduced to 10 minutes, is the critical path the same?

(d) If not, give the new critical path and its time to completion.

(e) Suppose that each of the times given on the digraph in Fig. 29 represents the mean completion time (t) of the task given a = optimistic time, b = pessimistic time, and m = probable time with

$$t = (a + 4m + b)/6.$$

The values of a, b, and m are displayed on the precedence table. Find the probability that the job will be completed within 36 minutes.

Activity	S	A_1	A_2	A_3	A_4	A_5	A_6	A_7	A_8	E
Immed. Predec.	—	S	S	S	A_1, A_2	A_2, A_4	A_4	A_3	A_6, A_7	A_5, A_8
t	0	10	12	9	7	15	6	9	3	0
a	0	8	8	6	5	12	5	3	2	0
m	0	10	11	8	7	13	6	10	3	0
b	0	12	20	16	9	26	7	11	4	0

15. Let the vertices be labeled with the elements in the set $\{1, 2, 3, 4, 5, 6\}$. Draw the digraph in which there is a directed edge from vertex u to vertex v if and only if u divides v (with remainder 0).

16. Let the vertices be labeled with the elements in the set $\{2, 3, 4, 5, 6, 7, 8\}$. Draw the digraph in which there is a directed edge from u to v if and only if u divides v but $u \neq v$.

17. Label the vertices with the elements of the set $\{0, 1, 2, 3, 4, 5, 6, 7, 8\}$. Draw the digraph in which there is a directed edge from u to v if and only if $v - u$ is a multiple of 3 and $u < v$.

18. Label the vertices with the elements of the set $\{0, 1, 2, 3, 4\}$. Draw the digraph in which there is an edge from u to v if and only if $0 < u - v < 3$.

19. A *transition table* is a table that describes the states of a machine where certain inputs cause a change of state. Such a table appears below. Erroneous inputs have been left blank. Usually, when such tables control the operating system of a computer, an error message is generated when an incorrect input is used. We can draw a digraph to represent such a transition table by letting the vertices represent machine states and edges represent the inputs. An edge (labeled a) is directed from vertex A to vertex C since input of a into the machine in state A causes the transition of the machine to state C.

(a) Draw the digraph representing the transition table.

(b) If the process begins in state C and inputs are b, b, a, what is the final state?

(c) If the process begins in state B and the inputs are b, b, a, what is the final state?

States	Inputs a	b	c
A	C		
B	A	C	
C	C	A	
D	D	E	C
E			C

SOLUTIONS TO PRACTICE PROBLEMS 4

1.
Vertex	u	v	w	x
id	1	3	1	1
od	1	2	2	1

The connected digraph does not satisfy the requirements of Property 2. Hence, no Euler circuit exists. Since

$$\text{od } w = \text{id } w + 1,$$

$$\text{id } v = \text{od } v + 1,$$

$$\text{id } u = \text{od } u,$$

$$\text{and id } x = \text{od } x,$$

Property 3 states that an Euler path exists. It begins at w and ends at v.

2. The digraph is given in Fig. 30. The paths from S to E and the necessary time to complete them are:

$SA_1A_2A_6A_9A_{10}E$: time $= 7 + 21 + 3 + 2 + 1 = 34$ days

$SA_1A_2A_8A_{10}E$: time $= 7 + 21 + 3 + 1 = 32$ days

$SA_1A_3A_7A_8A_{10}E$: time $= 7 + 14 + 5 + 3 + 1 = 30$ days

$SA_1A_3A_5A_8A_{10}E$: time $= 7 + 14 + 7 + 3 + 1 = 32$ days

$SA_1A_4A_5A_8A_{10}E$: time $= 7 + 12 + 7 + 3 + 1 = 30$ days

Clearly, the critical path is the one taking 34 days. If A_2 can be reduced to 16 days duration, then the original critical path has duration 29 days. We would need to reduce

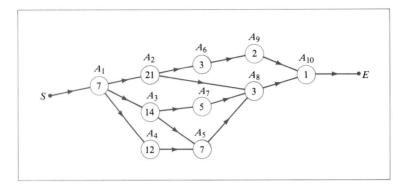

FIGURE 30

the time needed for A_1, A_3, A_5, or A_8 by 2 days to accomplish the entire process within 30 days.

3. (a) All the graphs have cycles. In both G_1 and G_2, $y\ u\ z\ y$ is a cycle. In G_3, $a\ e\ i\ j\ k\ g\ c\ b\ a$ is a cycle. All graphs are connected. But G_2 is not strongly connected. There is no path from w to y in G_2, for example.

   (b) In G_1, the indegree and the outdegree of each vertex are equal. Property 2 states there is an Euler circuit. Property 3 states there is no Euler path. There is no Hamiltonian circuit because the cycles in G_1 force us to repeat vertices in any circuit we choose. However, $u\ z\ y\ v\ w\ x$ is a Hamiltonian path in G_1.

   The graph G_2 has no Euler circuit or path. Properties 2 and 3 can be used to show this, since id $v = 1$ and od $v = 3$, and id $w = 2$ and od $w = 0$. There is a Hamiltonian path in G_2: $u\ z\ y\ x\ v\ w$. There is no Hamiltonian circuit because we can never leave vertex w.

   In G_3, there are neither Euler circuits nor Euler paths. A calculation of the id and od of each vertex shows that, for example, id $b \neq$ od b. Furthermore, many vertices have odd degree with $b\ c\ e\ h\ j$ and k having id and od differing by 1. There are no Hamiltonian paths and no Hamiltonian circuits in G_3.

4. (a) The digraph is acyclic; therefore, the tournament is transitive (Property 4).

   (b) F is the only winner.

   (c) The ranking is $FBCAED$. This can be read from the table below. We rank the players from the highest od to the lowest od, since the od counts the number of wins for the player.

Vertex	A	B	C	D	E	F
id	3	1	2	5	4	0
od	2	4	3	0	1	5

13.5 Matrices and Graphs

It would be helpful to have a bookkeeping device for storing the information provided by a graph. We can use a square matrix for the purpose. We list the vertices, in the same order, along the top and side margins of the matrix. The *adjacency matrix A(G) of a graph G* is the matrix obtained by letting the entry in the *i*th row, *j*th column be the number of edges joining the *i*th vertex with the *j*th vertex. A 0 denotes that the vertices are not joined by an edge. An adjacency matrix is not unique; but depends on the order of the vertices in the row and column headings.

EXAMPLE 1   Write out the adjacency matrix $A(G)$ of the graph G in Fig. 1.

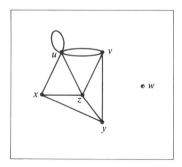

FIGURE 1

Solution   We note in the matrix below that the loop at *u* is recorded with a 1 in the first row, first column. No other vertex has a loop.

$$A(G) = \begin{array}{c c} & \begin{array}{c c c c c c} u & v & w & x & y & z \end{array} \\ \begin{array}{c} u \\ v \\ w \\ x \\ y \\ z \end{array} & \left[\begin{array}{c c c c c c} 1 & 2 & 0 & 1 & 0 & 1 \\ 2 & 0 & 0 & 0 & 1 & 1 \\ 0 & 0 & 0 & 0 & 0 & 0 \\ 1 & 0 & 0 & 0 & 1 & 1 \\ 0 & 1 & 0 & 1 & 0 & 1 \\ 1 & 1 & 0 & 1 & 1 & 0 \end{array} \right] \end{array}$$

We now consider only graphs with no parallel edges; that is, we restrict column is the same as the entry in the *j*th row, *i*th column. An edge that joins *u* and *v* also joins *v* and *u*.

We now consider only graphs with no parallel edges; that is, we restrict ourselves to simple graphs in all that follows in this section. The adjacency matrix of a simple graph has some special properties. Each entry a_{ij} must be either 0 or 1, since there is at most one edge from vertex v_i to vertex v_j. Since there is an edge from v_i to v_j if and only if there is an edge from v_j to v_i, the matrix $A(G)$ is symmetric; that is, $a_{ij} = a_{ji}$. If there is a loop at v_i, then $a_{ii} = 1$; otherwise, $a_{ii} = 0$.

If G is a simple graph, then $A(G)$ is a symmetric matrix of 0's and 1's.

What can we do with these matrices? First, we can store a graph in a computer in a form that can be handled mathematically. That is, instead of storing a picture of the graph, we can store all the relevant information from which the graph can be drawn.

FIGURE 2

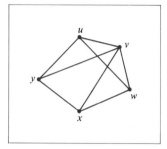

EXAMPLE 2   Draw the graph whose adjacency matrix is A.

$$A = \begin{array}{c c} & \begin{array}{c c c c c} u & v & w & x & y \end{array} \\ \begin{array}{c} u \\ v \\ w \\ x \\ y \end{array} & \left[\begin{array}{c c c c c} 0 & 1 & 1 & 0 & 1 \\ 1 & 0 & 1 & 1 & 1 \\ 1 & 1 & 0 & 1 & 0 \\ 0 & 1 & 1 & 0 & 1 \\ 1 & 1 & 0 & 1 & 0 \end{array} \right] \end{array}$$

Solution   The solution appears in Fig. 2.

We can read the degree of each vertex of a graph from an adjacency matrix. It is simply the sum of the entries in the row (or the column) corresponding to that vertex, if there is no loop there. If there is a loop, we need to add 1 to the row (or column) sum, since a loop adds 2 to the degree of a vertex.

Recall that two graphs are equivalent if the vertices of the graphs can be paired in such a way that, whenever two vertices of one graph are connected by an edge, then so are the corresponding vertices of the other graph. To determine if two graphs are equivalent, we can inspect the representations we have drawn and try to establish the identifications described in the definition. An alternative is to try to reorder the headings of the rows and columns in the adjacency matrices of the graphs so that the resulting matrices are identical.

EXAMPLE 3 Show that G_1 and G_2 (Fig. 3) are equivalent graphs by (a) establishing the identities described in the definition of equivalent graphs and (b) investigating their adjacency matrices $A(G_1)$ and $A(G_2)$.

Solution (a) There are 4 vertices in each of the vertex sets and there are 5 edges in each edge set. A little trial and error will show that we can identify vertices in the following way:

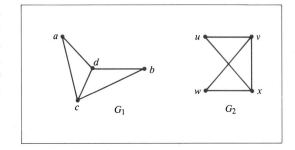

FIGURE 3

$$a \leftrightarrow w, \quad b \leftrightarrow u, \quad c \leftrightarrow v, \quad \text{and} \quad d \leftrightarrow x.$$

To show that the edge identification is correct, check the 6 edges of G_1 and the 6 edges of G_2.

Edge of G_1	Corresponding Edge of G_2	Check
From a to c	From w to v	✓
From a to d	From w to x	✓
From b to c	From u to v	✓
From b to d	From u to x	✓
From c to d	From v to x	✓

(b) The adjacency matrices of the graphs G_1 and G_2 are

$$A(G_1) = \begin{array}{c} \\ a \\ b \\ c \\ d \end{array} \begin{array}{cccc} a & b & c & d \\ \left[\begin{matrix} 0 & 0 & 1 & 1 \\ 0 & 0 & 1 & 1 \\ 1 & 1 & 0 & 1 \\ 1 & 1 & 1 & 0 \end{matrix}\right] \end{array} \qquad A(G_2) = \begin{array}{c} \\ u \\ v \\ w \\ x \end{array} \begin{array}{cccc} u & v & w & x \\ \left[\begin{matrix} 0 & 1 & 0 & 1 \\ 1 & 0 & 1 & 1 \\ 0 & 1 & 0 & 1 \\ 1 & 1 & 1 & 0 \end{matrix}\right] \end{array}$$

If the rows and columns of $A(G_2)$ are ordered w u v x, the resulting matrix will be identical to $A(G_1)$.

EXAMPLE 4 (a) Find the number of paths of length 2 from v_2 to v_4 in the graph G of Fig. 4.

(b) Write down an adjacency matrix of the graph of Fig. 4.

(c) Find the square of the adjacency matrix $A(G)$.

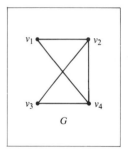

Solution (a) We can trace the paths of length 2 from v_2 to v_4. There are two such paths: $v_2v_1v_4$ and $v_2v_3v_4$.

(b) We saw (in the previous example) that an adjacency matrix of this graph is given by

$$A = \begin{array}{c} \\ v_1 \\ v_2 \\ v_3 \\ v_4 \end{array} \begin{array}{cccc} v_1 & v_2 & v_3 & v_4 \\ \begin{bmatrix} 0 & 1 & 0 & 1 \\ 1 & 0 & 1 & 1 \\ 0 & 1 & 0 & 1 \\ 1 & 1 & 1 & 0 \end{bmatrix} \end{array}$$

FIGURE 4

(c) We calculate the square of the matrix:

$$A^2 = \begin{array}{c} \\ v_1 \\ v_2 \\ v_3 \\ v_4 \end{array} \begin{array}{cccc} v_1 & v_2 & v_3 & v_4 \\ \begin{bmatrix} 2 & 1 & 2 & 1 \\ 1 & 3 & 1 & 2 \\ 2 & 1 & 2 & 1 \\ 1 & 2 & 1 & 3 \end{bmatrix} \end{array}$$

What does this matrix represent? The entry in the ith row, jth column of A^2 represents the number of paths of length 2 from v_i to v_j. Let us see why.

The entry in row 2, column 4 of A^2 is 2. Denote the entry by $a_{24}^{(2)}$. How was $a_{24}^{(2)}$ obtained? Recall from the definition of matrix multiplication that

$$a_{24}^{(2)} = a_{21}a_{14} + a_{22}a_{24} + a_{23}a_{34} + a_{24}a_{44}.$$

Thus,

$$a_{24}^{(2)} = 1 \cdot 1 + 0 \cdot 1 + 1 \cdot 1 + 1 \cdot 0 = 2.$$

FIGURE 5

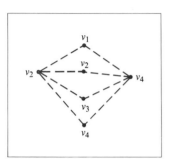

Each a_{ij} in the sum represents the number of edges between the ith and the jth vertex. The number a_{21} represents the number of edges from v_2 to v_1, and the number a_{14} represents the number of edges going from v_1 to v_4. The multiplication principle of Section 5.4 tells us that the product $a_{21}a_{14}$ is the number of two-edge paths from v_2 to v_4 with a stopover at v_1 (see Fig. 5).

Similarly, $a_{22}a_{24}$ is the number of two-edge paths from v_2 to v_4 with a stopover at v_2. Since there are no loops at v_2 in this graph ($a_{22} = 0$), there are no such paths. So $a_{22}a_{24}$ is zero. Again, $a_{23}a_{34}$ is the number of two-edge paths from v_2 to v_4 with a stopover at v_3, and $a_{24}a_{44}$ is the number of two-edge paths from v_2 to v_4 with a stopover at v_4 (why is this zero?).

The sum $a_{24}^{(2)}$ is the total number of two-edge paths from v_2 to v_4. We have found all such paths by looking at all possible stopovers. The two-edge paths from v_2 to v_4 were enumerated in (a): $v_2 v_1 v_4$ and $v_2 v_3 v_4$.

Assume that a graph has n vertices $v_1, v_2, \ldots, v_n$. The typical entry in the ith row, jth column of the square of the adjacency matrix is

$$a_{ij}^{(2)} = a_{i1} a_{1j} + a_{i2} a_{2j} + \cdots + a_{in} a_{nj}.$$

The sum $a_{ij}^{(2)}$ is 0 if and only if all the terms in the sum are zero, since each a_{ij} can take on only nonnegative values. Hence, if the sum is 0, there is no two-edge path from the ith vertex to the jth vertex. On the other hand, if the sum is not zero, then some term in the sum is not 0. Hence, for at least one of the values of $k = 1, 2, \ldots, n$, both a_{ik} and a_{kj} are not 0. Thus, there is a two-edge path from v_i to v_j with a stopover at v_k, giving at least one two-edge path from v_i to v_j.

If $A = A(G)$ is the adjacency graph of G, it represents the number of one-edge paths between vertices of G. Its square, A^2, represents the number of two-edge paths between vertices. If the i, j entry in A^2 is 0, there is no such path; otherwise, there is at least one such two-step path. In a similar way, the elements of A^3 can be interpreted as the number of three-step paths, and so on, for $A^4, A^5, \ldots$.

The i, j entry in A^m indicates the number of paths of length m between the ith vertex and the jth vertex.

Although we can tell by looking at a graph whether it is connected, if the graph information is stored in matrix form rather than as a picture, we need criteria related to the matrix representation of the graph for deciding if the graph is connected. The i, j entry in A^m is 0 if and only if no path of length m exists. This leads to the following theorems.

Theorem 1   If G is a graph with adjacency matrix A, and there exists a positive integer m such that A^m has only nonzero entries, then G is connected.

The converse of Theorem 1 is false; that is, although G may be connected, it is possible that there is no positive integer m such that A^m has only nonzero entries. Exercise 6 provides such an example.

Theorem 2   Let A be the adjacency matrix of a simple graph G. Then G is connected if and only if there exists an integer m such that every entry in $A + A^2 + \cdots + A^m$ is nonzero.

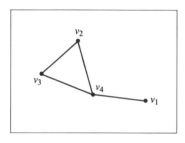

FIGURE 6

EXAMPLE 5   Find an adjacency matrix of the graph in Fig. 6. Find the number of paths of length 4 from v_4 to v_2. Give two of them. Use the adjacency matrix to show that the graph is connected.

Solution   The matrix $A = A(G)$ is given by

$$A(G) = A = \begin{bmatrix} 0 & 0 & 0 & 1 \\ 0 & 0 & 1 & 1 \\ 0 & 1 & 0 & 1 \\ 1 & 1 & 1 & 0 \end{bmatrix}$$

We find A^2 and then $A^4 = A^2 A^2$.

$$A^4 = \begin{bmatrix} 3 & 4 & 4 & 2 \\ 3 & 7 & 6 & 6 \\ 4 & 6 & 7 & 6 \\ 2 & 6 & 6 & 11 \end{bmatrix}$$

Since all the entries are nonzero, the graph is connected. Of course, when we have the picture in front of us, the connectedness is obvious. The entry in the 4th row, 2nd column is 6. Thus, there are 6 paths of length 4 from v_2 to v_4. Here are two of them:

$$v_4 v_1 v_4 v_3 v_2 \quad \text{and} \quad v_4 v_2 v_3 v_4 v_2.$$

Alternatively, we could have found $A + A^2$, where

$$A^2 = \begin{bmatrix} 1 & 1 & 1 & 0 \\ 1 & 2 & 1 & 1 \\ 1 & 1 & 2 & 1 \\ 0 & 1 & 1 & 3 \end{bmatrix}$$

and shown that $A + A^2$ has only nonzero entries. This is another way to verify that the graph is connected.

Although Theorems 1 and 2 seem to indicate that, theoretically, we might have to calculate A^m for large values of m to determine if a graph is connected, this is not quite so. If a graph has 4 vertices, for example, then any path of length at least 3 will either contain a cycle or contain every vertex. If we disregard the paths with cycles, then if there is a path between two distinct vertices there must be one of length less than or equal to 3. Thus, we need calculate at most A, A^2, and A^3. If a graph has n vertices, then to check its connectivity using Theorem 2 we need calculate at most $A, A^2, \ldots, A^{n-1}$. It is frequently the case that fewer powers of A will be needed.

We can use another matrix to record whether there exists a path between two distinct vertices in a graph. Inspection of the graph tells us whether such a path exists. The *reachability matrix of a graph G with n vertices* is the $n \times n$ matrix whose entry in the ith row, jth column is 1 if there is a path from v_i to v_j, and 0 if no such path exists. Denote the reachability matrix of the graph G as $R(G)$. Assume v_i is reachable from itself; the entries on the diagonal of $R(G)$ are all 1.

Given a graph such as that in Fig. 6, the reachability matrix is a 4×4 matrix of 1's since every vertex can be reached from any other along some path. We say that each vertex is clearly reachable from itself (use a path of length 0). The reachability matrix has a zero entry whenever there is no path between two vertices. In that case, the graph is disconnected.

It is possible to find the reachability matrix of a graph from an adjacency matrix. Let I denote the $n \times n$ identity matrix (it contains 1's on the main diagonal (upper left to lower right) and 0's elsewhere). The identity matrix represents the 0-edge paths between vertices (there is always one such path from any vertex to itself). We can obtain the reachability matrix $R(G)$ of a graph G with n vertices by first finding the matrix $M = I + A + A^2 + \cdots + A^{n-1}$ and then letting the entry in the ith row, jth column of $R(G)$, $r_{ij} = 1$ if M has a nonzero entry in the ith row, jth column and letting $r_{ij} = 0$ otherwise. The reachability matrix of a graph is a symmetric matrix of 0's and 1's with 1's along the main diagonal. Clearly, in Example 5, $I + A + A^2 + A^3$ has no zero entries, so the reachability matrix is a matrix of 1's.

> **Theorem 3** A simple graph is connected if and only if its reachability matrix is a matrix consisting only of 1's.

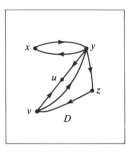

FIGURE 7

As we have already discussed, a *directed graph* or a *digraph* is a graph in which every edge is given a direction. We will use the letter D to denote a digraph. If D is a digraph, it has a set of vertices and a set of directed edges, actually denoted as ordered pairs of vertices, describing the direction of the edges. Thus, the edge (u, v) is the edge directed from u to v. It is possible to have edges directed between two vertices in opposite directions. For example, Fig. 7 shows a directed graph on 5 vertices. The directed edges are (x, y), (y, x), (y, z), (z, v), (v, u), (v, y), and (y, u). We assume that there is no ambiguity in giving the edge as (z, v). In our digraphs there are no parallel edges; so no other edge goes from z to v.

The *adjacency matrix of a digraph D*, $A(D)$, is formed by letting $a_{ij} = 1$ if there is a directed edge from the ith vertex to the jth vertex of D and $a_{ij} = 0$ if there is not.

EXAMPLE 6 Find the adjacency matrix $A(D)$ for the digraph D in Fig. 7.

Solution The adjacency matrix for D is

$$
\begin{array}{c c} & \begin{array}{c c c c c} u & v & x & y & z \end{array} \\ \begin{array}{c} u \\ v \\ x \\ y \\ z \end{array} & \left[\begin{array}{c c c c c} 0 & 0 & 0 & 0 & 0 \\ 1 & 0 & 0 & 1 & 0 \\ 0 & 0 & 0 & 1 & 0 \\ 1 & 0 & 1 & 0 & 1 \\ 0 & 1 & 0 & 0 & 0 \end{array}\right]. \end{array}
$$

Note that the adjacency matrix of a digraph need not be symmetric. The digraph in Fig. 7 contains an edge from v to y but does not contain an edge from y to v. Consequently, the matrix $A(D)$ contains a 1 in the 2nd row, 4th column, but contains a 0 in the 4th row, 2nd column.

The sum of the entries in the row corresponding to any vertex v is the outdegree of v; the sum of the entries in the column corresponding to any vertex v is the indegree of v. The sum of the entries in the matrix $A(D)$ is the total number of edges in the digraph.

As in the matrix representation of graphs, the matrix representation of digraphs provides a great deal of information about the structure of the digraph. If $A = A(D)$ is the adjacency matrix of a digraph, there is a path of length m from vertex v_i to vertex v_j whenever the entry in the ith row, jth column of A^m is nonzero. In working with digraphs, it can be extremely helpful to have a matrix representation of the graph, since it may be difficult to follow the paths to ascertain how many routes of length m lead from one vertex to another.

EXAMPLE 7 Find an adjacency matrix of the digraph representing the streets in one area of the city (Fig. 8). Note that some streets are one way and others are two way. Determine if there is a path of length 4 from the corner marked X to the corner marked Y. How many such paths are there?

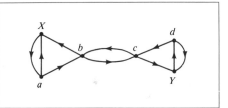

FIGURE 8

Solution An adjacency matrix of the digraph is

$$
\begin{array}{c c} & \begin{array}{c c c c c c} a & b & c & d & X & Y \end{array} \\ \begin{array}{c} a \\ b \\ c \\ d \\ X \\ Y \end{array} & \left[\begin{array}{c c c c c c} 0 & 1 & 0 & 0 & 1 & 0 \\ 0 & 0 & 1 & 0 & 1 & 0 \\ 0 & 1 & 0 & 0 & 0 & 1 \\ 0 & 0 & 1 & 0 & 0 & 1 \\ 1 & 0 & 0 & 0 & 0 & 0 \\ 0 & 0 & 0 & 1 & 0 & 0 \end{array}\right] \end{array}
$$

To determine if there is a path of length 4 from X to Y, find the matrix A^4 and check the entry in the 5th row, last column. To calculate A^4, first find A^2 and then find $A^2 \cdot A^2$.

$$A^2 = \begin{array}{c} \\ a \\ b \\ c \\ d \\ X \\ Y \end{array} \begin{array}{cccccc} a & b & c & d & X & Y \\ \begin{bmatrix} 1 & 0 & 1 & 0 & 1 & 0 \\ 1 & 1 & 0 & 0 & 0 & 1 \\ 0 & 0 & 1 & 1 & 1 & 0 \\ 0 & 1 & 0 & 1 & 0 & 1 \\ 0 & 1 & 0 & 0 & 1 & 0 \\ 0 & 0 & 1 & 0 & 0 & 1 \end{bmatrix} \end{array}, \qquad A^4 = \begin{array}{c} \\ a \\ b \\ c \\ d \\ X \\ Y \end{array} \begin{array}{cccccc} a & b & c & d & X & Y \\ \begin{bmatrix} 1 & 1 & 2 & 1 & 3 & 0 \\ 2 & 1 & 2 & 0 & 1 & 2 \\ 0 & 2 & 1 & 2 & 2 & 1 \\ 1 & 2 & 1 & 1 & 0 & 3 \\ 1 & 2 & 0 & 0 & 1 & 1 \\ 0 & 0 & 2 & 1 & 1 & 1 \end{bmatrix} \end{array}$$

Since the entry in the 5th row, last column is 1, there is one path of length 4 from X to Y.

The reachability matrix of a digraph is formed by inspecting the digraph. If there is a path from vertex v_i to vertex v_j, a 1 appears in the ith row, jth column. If there is no such path, a 0 appears. Assume that each vertex can be reached from itself (in a path of length 0), so the diagonal entries in the reachability matrix are all 1's.

From an adjacency matrix for the digraph, we can determine if one vertex is reachable from another by calculating the *reachability matrix R(D)* for the digraph D. If D has n vertices and adjacency matrix A, we first calculate the sum of the identity matrix with the first $n - 1$ powers of A:

$$I + A + A^2 + \cdots + A^{n-1}.$$

We form $R(D)$ by letting $r_{ij} = 1$ whenever the sum is nonzero in the ith row, jth column. We let $r_{ij} = 0$ otherwise.

EXAMPLE 8   For the digraph in Fig. 9(a), find an adjacency matrix and the corresponding reachability matrix.

FIGURE 9

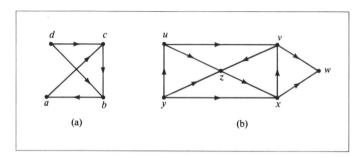

(a)   (b)

Solution An adjacency matrix A is given by

$$
\begin{array}{c@{}c}
 & \begin{array}{cccc} a & b & c & d \end{array} \\
\begin{array}{c} a \\ b \\ c \\ d \end{array} &
\left[
\begin{array}{cccc}
0 & 0 & 1 & 0 \\
1 & 0 & 0 & 0 \\
0 & 1 & 0 & 0 \\
0 & 1 & 1 & 0
\end{array}
\right]
\end{array}
$$

The powers of A are

$$
A^2 =
\begin{bmatrix}
0 & 1 & 0 & 0 \\
0 & 0 & 1 & 0 \\
1 & 0 & 0 & 0 \\
1 & 1 & 0 & 0
\end{bmatrix},
\qquad
A^3 =
\begin{bmatrix}
1 & 0 & 0 & 0 \\
0 & 1 & 0 & 0 \\
0 & 0 & 1 & 0 \\
1 & 0 & 1 & 0
\end{bmatrix}
$$

Finally, we add the 4×4 identity matrix to A, A^2, and A^3. The sum and the reachability matrix are

$$
\text{sum} =
\begin{bmatrix}
2 & 1 & 1 & 0 \\
1 & 2 & 1 & 0 \\
1 & 1 & 2 & 0 \\
2 & 2 & 2 & 1
\end{bmatrix}
\quad \text{and} \quad
R(D) =
\begin{bmatrix}
1 & 1 & 1 & 0 \\
1 & 1 & 1 & 0 \\
1 & 1 & 1 & 0 \\
1 & 1 & 1 & 1
\end{bmatrix}
$$

In the last column of the matrix, all entries except in the last row of $R(D)$ are zero; hence the vertex d cannot be reached from any other vertex. Once we exit from d there is no way to return, no matter how long the path.

A digraph is said to be *connected* if the graph obtained from it by ignoring the directions on the edges is a connected (undirected) graph. A digraph is *strongly connected* if, for any pair of vertices u and v, there is a path from u to v. We note that the digraph of Fig. 9(a) is connected but not strongly connected. There are no paths from vertex a to vertex d, for example.

> **Theorem 4** A digraph D is strongly connected if and only if its reachability matrix $R(D)$ has no zeros.

EXAMPLE 9 Determine if the digraph D given in Fig. 9(b) is strongly connected by using the reachability matrix. Are there any four-step paths from y to w? If so, how many?

Solution An adjacency matrix for the digraph is given by

$$
A = \begin{array}{c} \\ u \\ v \\ w \\ x \\ y \\ z \end{array}
\begin{array}{c} \begin{matrix} u & v & w & x & y & z \end{matrix} \\
\begin{bmatrix}
0 & 1 & 0 & 0 & 0 & 1 \\
0 & 0 & 1 & 0 & 0 & 1 \\
0 & 0 & 0 & 0 & 0 & 0 \\
0 & 1 & 1 & 0 & 0 & 0 \\
1 & 0 & 0 & 1 & 0 & 1 \\
0 & 0 & 0 & 1 & 0 & 0
\end{bmatrix}
\end{array}
$$

and from this we can caluate the powers of A and the reachability matrix. In the matrix A we note that the row corresponding to w contains only zeros. Thus, we can never leave the vertex w. Similarly, we can never enter the vertex y (the column corresponding to y contains only zeros). The graph cannot be strongly connected. Calculation of A^4 reveals that there are 2 four-step paths from y to w.

$$
A^4 = \begin{array}{c} \\ u \\ v \\ w \\ x \\ y \\ z \end{array}
\begin{array}{c} \begin{matrix} u & v & w & x & y & z \end{matrix} \\
\begin{bmatrix}
0 & 1 & 2 & 0 & 0 & 1 \\
0 & 0 & 1 & 0 & 0 & 1 \\
0 & 0 & 0 & 0 & 0 & 0 \\
0 & 1 & 1 & 0 & 0 & 0 \\
0 & 1 & 2 & 2 & 0 & 1 \\
0 & 0 & 0 & 1 & 0 & 0
\end{bmatrix}
\end{array}
$$

EXAMPLE 10 Show that a digraph with a directed Hamiltonian circuit must be strongly connected.

Solution Suppose that D is a digraph with a directed Hamiltonian circuit. Then there is a cycle that passes through each vertex of the graph exactly once. Thus, if there are n vertices, there is a circuit of $n - 1$ edges containing each of the vertices. This generates a path leading from any vertex to any other. Hence, the digraph is strongly connected.

PRACTICE PROBLEMS 5

1. Let A be the adjacency matrix of a simple graph.

$$
A = \begin{array}{c} \\ a \\ b \\ c \\ d \end{array}
\begin{array}{c} \begin{matrix} a & b & c & d \end{matrix} \\
\begin{bmatrix}
1 & 1 & 0 & 0 \\
1 & 0 & 1 & 0 \\
0 & 1 & 0 & 1 \\
0 & 0 & 1 & 0
\end{bmatrix}
\end{array}.
$$

(a) Determine the number of paths of length 3 from b to c by using the adjacency matrix.

(b) Determine from the adjacency matrix whether the graph is connected.

(c) Draw the graph for which A is the adjacency matrix.

(d) Find all the paths of length 3 from b to c.

2. For the adjacent digraph:

    (a) Find the adjacency matrix $A(D)$.

    (b) Find the number of paths of length 3 from v_3 to v_4 using the matrix $A(D)$.

    (c) Write out the paths of length 3 from v_3 to v_4.

    (d) Find the reachability matrix of the digraph.

    (e) Is D connected? Is D strongly connected? Explain.

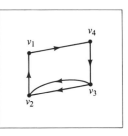

EXERCISES 5

1. (a) Find the adjacency matrix of the graph in Fig. 10.

    (b) Find the number of paths of length 2 from v_1 to v_4 in the graph of Fig. 10.

    (c) Find the reachability matrix of the graph in Fig. 10.

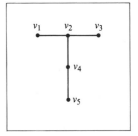

FIGURE 10

2. Let $A(G)$ be the adjacency matrix of a simple graph.

$$
\begin{array}{c@{\quad}ccccc}
 & v_1 & v_2 & v_3 & v_4 & v_5 \\
v_1 & 0 & 1 & 1 & 1 & 0 \\
v_2 & 1 & 0 & 1 & 0 & 1 \\
v_3 & 1 & 1 & 0 & 1 & 0 \\
v_4 & 1 & 0 & 1 & 0 & 0 \\
v_5 & 0 & 1 & 0 & 0 & 0
\end{array}
$$

(a) Draw the corresponding graph.

(b) Determine from $A(G)$ the number of paths of length 3 from v_1 to v_4.

(c) Determine from $A(G)$ whether G is connected. Explain.

(d) Find all the paths of length 3 from v_1 to v_4.

3. Let $A(G)$ be the adjacency matrix of a graph.

$$
\begin{array}{c@{\quad}ccccc}
 & v_1 & v_2 & v_3 & v_4 & v_5 \\
v_1 & 0 & 1 & 1 & 0 & 0 \\
v_2 & 1 & 0 & 1 & 0 & 0 \\
v_3 & 1 & 1 & 0 & 0 & 0 \\
v_4 & 0 & 0 & 0 & 0 & 1 \\
v_5 & 0 & 0 & 0 & 1 & 1
\end{array}
$$

(a) Determine from the adjacency matrix whether G is a connected graph.

(b) Draw the graph represented by $A(G)$.

4. (a) Find the adjacency matrix $A(G)$ for the graph G in Fig. 11.

(b) What is the *smallest* power of $A(G)$ that has no zero entries?

(c) Explain what your answer to (b) tells us about the graph.

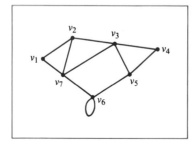

FIGURE 11   FIGURE 12

5. (a) Find the adjacency matrix of the graph in Fig. 12.

(b) Find the number of paths of length 3 from v_1 to v_4 by using the adjacency matrix.

(c) Find the reachability matrix of the graph.

6. Give an example of a graph on 3 vertices that is connected but for which A^3 contains at least one zero entry. Show A^3 for the graph you drew.

7. (a) Find the adjacency matrix A of the graph in Fig. 13.

(b) Find the *smallest* power of A that has no zero entries.

(c) How could you have determined the answer to (b) directly from the graph and without reference to the adjacency matrix? Explain.

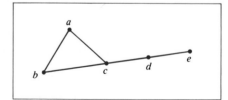

FIGURE 13

8. Draw a simple graph that has reachability matrix R.

$$R = \begin{bmatrix} 1 & 1 & 1 & 0 & 1 \\ 1 & 1 & 1 & 0 & 1 \\ 1 & 1 & 1 & 0 & 1 \\ 0 & 0 & 0 & 1 & 0 \\ 1 & 1 & 1 & 0 & 1 \end{bmatrix}$$

9. Draw a simple graph for which $A(G)^2$ is given by the matrix

$$\begin{bmatrix} 1 & 0 & 1 & 0 \\ 0 & 2 & 0 & 0 \\ 1 & 0 & 1 & 0 \\ 0 & 0 & 0 & 1 \end{bmatrix}$$

10. (a) Find the adjacency matrix for the digraph D in Fig. 14.

(b) Find the number of paths of length 2 from a to d.

(c) Write out those paths of length 2 from a to d.

(d) Find A^4.

(e) What can you say about paths of length 4 in this digraph?

(f) Find the reachability matrix of D.

(g) Is D connected? Is D strongly connected? Explain.

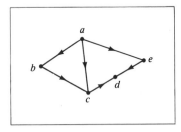

FIGURE 14   FIGURE 15

11. (a) Find the adjacency matrix for the digraph in Fig. 15.

(b) Find the number of paths of length 4 from a to b.

(c) Write out those paths.

(d) Find the reachability matrix for the digraph.

(e) Is the digraph connected? Is it strongly connected? Explain.

12. Let $A = A(D)$ be the adjacency matrix of a digraph D.

(a) From the matrix, determine the indegree and outdegree of each vertex.

(b) Draw the digraph.

(c) Is the digraph connected? Is it strongly connected?

(d) Is there an Euler circuit in this digraph?

$$A = \begin{bmatrix} 0 & 0 & 1 & 1 \\ 1 & 0 & 1 & 0 \\ 0 & 1 & 0 & 1 \\ 1 & 0 & 0 & 0 \end{bmatrix}$$

13. Let $A = A(D)$ be the adjacency matrix of a digraph D.

(a) For each of the vertices, find the indegree and the outdegree directly from the matrix.

(b) Draw the digraph.

(c) Determine the number of paths of length 3 from v_1 to v_4.

(d) Write out all the paths of length 3 from v_1 to v_4.

$$A = \begin{array}{c} \\ v_1 \\ v_2 \\ v_3 \\ v_4 \\ v_5 \end{array} \begin{array}{ccccc} v_1 & v_2 & v_3 & v_4 & v_5 \\ \begin{bmatrix} 1 & 1 & 0 & 0 & 0 \\ 0 & 0 & 1 & 1 & 0 \\ 1 & 0 & 0 & 0 & 1 \\ 1 & 1 & 0 & 1 & 0 \\ 0 & 0 & 1 & 1 & 0 \end{bmatrix} \end{array}$$

SOLUTIONS TO PRACTICE PROBLEMS

1. (a) The matrix A^3 appears below. The entry corresponding to the number of paths of length 3 from b to c is in row 2, column 3. Thus, there are three such paths.

   (b) The reachability matrix can be shown to have no zeros by finding $I + A + A^2 + A^3$. The graph is connected.

   (c) A graph represented by the matrix A is shown on the right.

   (d) The paths of length 3 from b to c are $b\ a\ b\ c$, $b\ c\ d\ c$, and $b\ c\ b\ c$.

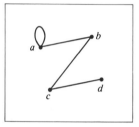

$$A^2 = \begin{bmatrix} 2 & 1 & 1 & 0 \\ 1 & 2 & 0 & 1 \\ 1 & 0 & 2 & 0 \\ 0 & 1 & 0 & 1 \end{bmatrix}, \quad A^3 = \begin{bmatrix} 3 & 3 & 1 & 1 \\ 3 & 1 & 3 & 0 \\ 1 & 3 & 0 & 2 \\ 1 & 0 & 2 & 0 \end{bmatrix}$$

2. (a) The adjacency matrix A is shown below.

   (b) We show A^3 below. There is one path of length 3 from v_3 to v_4. This is the entry in row 3, column 4 of A^3.

   (c) The only path of length 3 from v_3 to v_4 is $v_3 v_2 v_1 v_4$.

$$A = \begin{bmatrix} 0 & 0 & 0 & 1 \\ 1 & 0 & 1 & 0 \\ 0 & 1 & 0 & 0 \\ 0 & 0 & 1 & 0 \end{bmatrix}, \quad A^2 = \begin{bmatrix} 0 & 0 & 1 & 0 \\ 0 & 1 & 0 & 1 \\ 1 & 0 & 1 & 0 \\ 0 & 1 & 0 & 0 \end{bmatrix}, \quad A^3 = \begin{bmatrix} 0 & 1 & 0 & 0 \\ 1 & 0 & 2 & 0 \\ 0 & 1 & 0 & 1 \\ 1 & 0 & 1 & 0 \end{bmatrix}$$

   (d) To find the reachability matrix, we add $I + A + A^2 + A^3$ to find the sum matrix has no zeros. The reachability matrix is a matrix containing nothing but 1's.

$$I + A + A^2 + A^3 = \begin{bmatrix} 1 & 1 & 1 & 1 \\ 2 & 2 & 3 & 1 \\ 1 & 2 & 2 & 1 \\ 1 & 1 & 2 & 1 \end{bmatrix}$$

   (e) Since the reachability matrix has no zeros, the digraph is strongly connected. The digraph is connected because it is strongly connected.

13.6 Trees

In Section 13.3 we introduced the notion of a minimal spanning tree. In fact, trees have many applications.

EXAMPLE 1   Suppose your friend is thinking of an integer between 1 and 10 inclusive. Describe an efficient procedure for guessing the number if you can ask questions that can be answered only yes or no.

Solution   One procedure is to ask first if the number is less than or equal to 5. If the answer is yes, we know that the number is in the set {1, 2, 3, 4, 5}. If the answer is no, we know that the number is in the set {6, 7, 8, 9, 10}. In either case, we have reduced the search to a set that is half the size of the original. We continue in that way, dividing the set in half (or as close to half as we can) as we proceed. Figure 1 contains a schematic diagram of this process. Note that at most four questions are necessary to determine the chosen number.

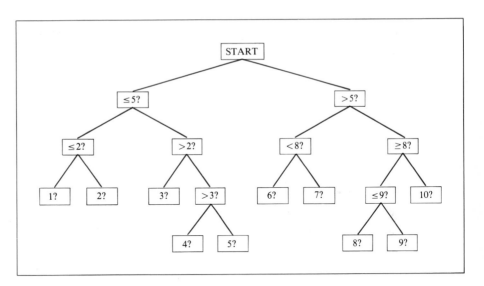

FIGURE 1

The diagram representing the procedure above is an example of a tree. It is called a tree because it has several branches, some branches have branches of their own, and so on, and the picture looks something like a tree (not necessarily growing in the standard direction). In the language of graph theory, a *tree* is a connected acyclic graph. A characterization of trees follows:

Tree Property 1 Let G be a simple graph with at least two vertices. The following are equivalent:

1. G is a tree.
2. Each pair of distinct vertices is joined by exactly one simple path.
3. G is connected, but would be disconnected if any edge were removed.
4. G is acyclic, but would contain a cycle if any edge were added.

EXAMPLE 2 Which of the graphs in Fig. 2 are trees? If the graph is not a tree, explain how it violates the definition.

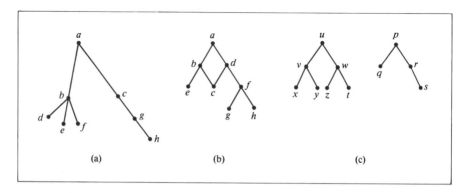

(a) (b) (c)

FIGURE 2

Solution The graph of Fig. 2(a) is a tree. The graph of Fig. 2(b) is not a tree because it contains a cycle. The graph of Fig. 2(c) is not connected; hence it is not a tree.

EXAMPLE 3 Sketch all trees having 6 vertices. There are 6 of them.

Solution Note that in each of the trees with 6 vertices (Fig. 3), there are exactly 5 edges. Also, in each tree there are some vertices of degree 1. Those vertices with degree 1 are

FIGURE 3

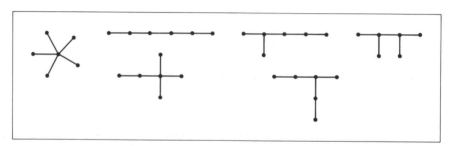

called *leaves*. Every tree with at least one edge has at least 2 leaves. To see this, just write down the longest acyclic path in the graph. The first and last vertices in this path cannot be equal (otherwise we would have a cycle). Each of those vertices is a leaf.

Tree Property 2   Let G be a simple graph with n vertices. The following statements are equivalent:

1. G is a tree.
2. G is acyclic and has $n - 1$ edges.
3. G is connected and has $n - 1$ edges.

EXAMPLE 4   A tree has 4 vertices of degree 3, and 3 vertices of degree 2. The remaining vertices have degree 1. Find the number of vertices in this tree.

Solution   Let n be the number of vertices of the tree.

Degree of Vertex	Number of Vertices	Sum of Degrees
3	4	12
2	3	6
1	$n - 7$	$n - 7$
Totals	n	$18 + (n - 7)$

Note that the sum of the degrees of the vertices in any graph is equal to twice the number of edges in the graph. Since we are working with a tree with n vertices, this graph has $n - 1$ edges. Thus, the sum of the degrees of the vertices is $2(n - 1) = 2n - 2$. We then get the following equation:

$$18 + (n - 7) = 2n - 2.$$

Solving this equation for n, we find that $n = 13$. Thus, there are 13 vertices in the tree, and 6 of them have degree 1. Exercise 20 asks for verification that such a tree exists.

The tree introduced in the solution to the first example is called a *rooted tree*. It is a tree in which a special vertex is designated to be the *root* of the tree. The root in the tree of Fig. 1 is START; it is assumed that edges are directed down from START to other vertices. In this way, a rooted tree is actually a directed graph that is connected and acyclic and for which there is a specified, unique vertex with indegree 0. The root may appear at the top, bottom, or side of the tree. We can omit the directional arrows on the edges when the direction is clear. A familiar example of a rooted tree is an organization chart of a company.

Suggestive terms are used to describe the parts of a tree. If there is a directed edge from u to v, we say that u is the *parent* of v and that v is the *child* of u. A leaf is a vertex that has no children. Every child in a tree has one parent, although a parent can have several children. A root has no parent and is the only parent in the tree with this property.

EXAMPLE 5   In the tree of Fig. 4, identify the root and the leaves. Find all the children of c.

Solution   The root is a. The leaves are all vertices with out-degree 0: d, e, f, g, and i. The children of c are g and h.

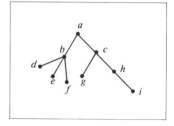

FIGURE 4

A tree in which every parent has at most two children is called a *binary tree*. The search procedure pictured in the graph of Fig. 1 is a binary tree. A search procedure whose graph representation is a binary tree is called a *binary search*. There are many opportunities to use binary trees.

EXAMPLE 6   Use a binary tree to store the following words in alphabetical order.

bad, even, guppy, apple, bear, dog, fish, whale.

Solution   Here is the algorithm we will use.

1. Start with any element in the list. (We will begin with *bad*.) That is the root. Label the root and draw two children [Fig. 5(a)].
2. Choose the next element in the list. If it precedes the root in alphabetical order, label the left child with its value. If it succeeds the root in alphabetical order, label the right child with its value. Draw left and right children for the newly labeled vertex. Continue.

FIGURE 5

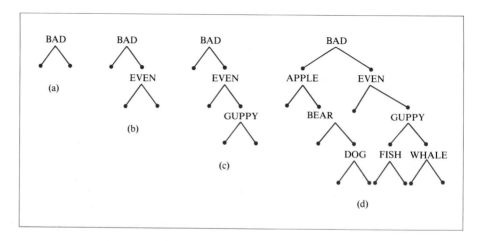

Say the root is *bad*. The next word is *even*. It succeeds *bad*, so it is the label on the right child [Fig. 5(b)]. We draw two children from the vertex at *even*. The next word is *guppy*. That succeeds *even*, so *guppy* is the label on the right child of *even* [Fig. 5(c)]. We draw two children from the vertex at *guppy*. Now, the word *apple* precedes *bad*, so *apple* is the label on the left child of *bad*. Add left and right children to that vertex. Continue. The tree that results from working through the entire list appears in Fig. 5(d).

The tree we formed to store our alphabetical list can be used to search for a word in our "dictionary." If the word is not in our dictionary, we can add it to the list in the proper place.

EXAMPLE 7 Use the tree of Fig. 5(d) (a) to determine if the word *elk* is in the stored list. If not, add it to the tree in its proper alphabetical order. (b) Do the same for the word *asp*.

Solution We compare *elk* with *bad*. Since *elk* succeeds *bad*, we proceed to the right child of *bad*. We compare *elk* with *even*. Since *elk* precedes *even*, we pass to the left child of *even*. The word *elk* succeeds *bear*, so we pass to the right child of *bear*, which is *dog*, and again, since *elk* succeeds *dog*, we pass the right child of *dog*. The word *elk* does not appear in the dictionary, so we add it as the right child of *dog*. Figure 6 shows the new tree. In answer to (b), the search reveals that we should add the word *asp* as the right child of *apple*, as shown in Fig. 7.

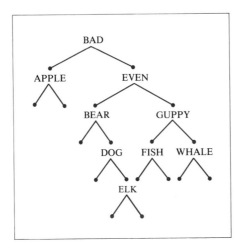

FIGURE 6

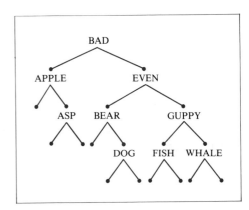

FIGURE 7

Another way of using trees is in showing the structure of algebraic expressions. The operators we will use here are + (add), − (subtract), ∗ (multiply), ÷ (divide), and ^ (exponentiation). An algebraic expression of the form $(x + y)^3$, commonly written as $(x + y)^3$, can be analyzed with a tree. The root of the tree representing the expression is the central operator, which indicates the last operation to be performed (in the above expression the root would be the exponentiation operator ^). Other vertices are labeled with operators, variables, or constants. The leaves cannot be operators. We do not put parentheses on the tree.

EXAMPLE 8 Find the algebraic expression represented by the tree in Fig. 8.

Solution We read from the bottom up, using the parent to get the operation to be performed on the children. From the left, we have $(a * b)$, which will be multiplied by the result of the operations on the right. Again, reading from the bottom, we have c, which is added to $d \div e$. The final expression is

$$(a * b) * [c + (d \div e)].$$

We assume in every algebraic expression that in order for the operations $+$, $-$, $*$, $\div$, and $^$ to be performed all of the arguments must have been evaluated. To find the value of $(a + b)$, a and b must be evaluated first. To find $(a + b) \div (b * c)$, we cannot use the operation $\div$ until both $a + b$ and $b * c$ have been evaluated. The root of the tree representing the expression is the central operation (the last operation that is to be performed). In labeling a tree to represent an algebraic expression we work from the central operator down. The root is the central operator. We then enter the arguments of the operator (variables or constants along with operators on them). Continue the process until the symbols (except parentheses) in the expression are all on the tree.

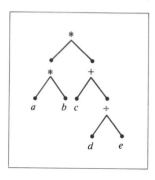

FIGURE 8

EXAMPLE 9 Use a tree to represent the algebraic expressions.

(a) $x + y$ (b) $(x * y) + z$

(c) $[x + (y \div z)]^{\wedge}(4 * y)$

Solution We use the operations $+$, $-$, $*$, $\div$, and $^$ (exponentiation) to label any vertices except leaves. Leaves are labeled with variables or constants. The operation denoted on a vertex operates on the left and right subtrees emanating from that vertex, reading from left to right.

FIGURE 9

(a) Note that the central operator is $+$. The left and right branches have vertices x and y, respectively. See Fig. 9(a).

(b) The central operator here is $+$, so it is the root. We must calculate both arguments in order to use the central operator. One is $x * y$; the other is z. We indicate this by having the left subtree show $x * y$, and a right branch show the variable z. See Fig. 9(b).

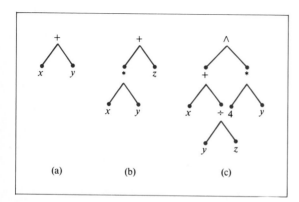

(a)   (b)   (c)

(c) Here the central operator is ^. To use this operation, we first must evaluate the expressions $x + (y \div z)$ and $4 * y$. The expression $x + (y \div z)$ has central operator $+$, so we have a subtree with root $+$. In turn, we must evaluate $y \div z$ before we can proceed. This requires an additional subtree. See Fig. 9(c).

PRACTICE PROBLEMS 6

1. Draw three distinct trees with 5 vertices. How many are there?

2. Is it possible to have a tree with 2 vertices of degree 3, 1 vertex of degree 2, and 6 vertices of degree 1 and 9 edges? Explain.

3. Find the tree that creates a dictionary by adding each of the words in the list: daily, milk, baby, silly, bib, crib, blanket, rattle.

EXERCISES 6

1. Draw all trees with 4 vertices.

2. Draw four distinct binary trees with 6 vertices.

3. Draw three distinct binary trees with 7 vertices.

4. Determine which of the graphs in Fig. 10 are trees. For those which are not trees, explain why not.

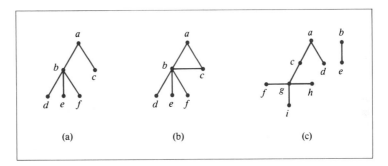

FIGURE 10

5. Name the leaves in each of the trees in Fig. 11.

6. Find (a) all the children of x and (b) all the descendants of x in each of the trees of Fig. 11.

7. Find the parent of x in each of the trees in Fig. 11.

8. Identify the root in each of the trees of Fig. 11

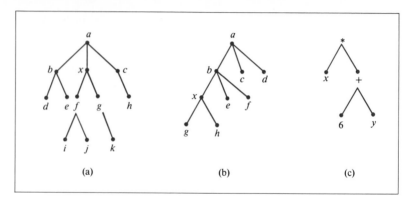

(a)   (b)   (c)

FIGURE 11

In Exercises 9–13, draw the graph with the given properties or tell why it cannot be drawn.

 9.   A tree that has 7 vertices and 12 edges.

10.   A tree that has vertices of degree 4, 3, 3, 3, 2, and 2.

11.   A connected graph with 3 edges and 4 vertices that is not a tree.

12.   A disconnected acyclic graph with 5 vertices and 4 edges.

13.   A tree with 3 vertices and 1 edge.

14.   Draw the binary tree that creates the dictionary from the following list: guppy, hamster, dog, cat, snake, canary, parakeet, pony.

15.   Make a binary tree showing the numerical ordering of 43, 67, 58, 2, 31, 91, and 34.

16.   Read the algebraic formula from the tree in Fig. 12.

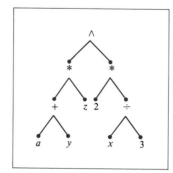

FIGURE 12

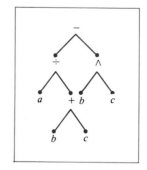

FIGURE 13

17.   Read the algebraic formula from the tree in Fig. 13.

18.   Construct the tree that represents the algebraic expression

$$[3 \div (a * c)] + (b - d).$$

19. Construct the tree that represents the algebraic expression

$$[(x * y)^3] \div (a + b).$$

20. Draw the tree described by Example 4.

21. Using the scheme described in Example 1, what is the maximum number of questions needed to determine the number if your friend can choose any integer between 1 and 100, inclusive? Between 1 and 1000, inclusive?

SOLUTIONS TO PRACTICE PROBLEMS 6

1. There are three different trees having 5 vertices.

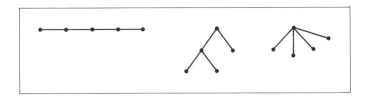

FIGURE 14

2. The sum of the degrees of the vertices is 14. The graph cannot have 9 edges; it must have 7 edges. A tree with 7 edges would have to have 8 vertices. There is no such tree.

3.
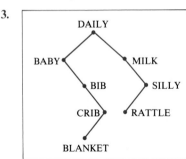

FIGURE 15

Chapter 13: CHECKLIST

☐ Graph, planar graph, parallel edges, loops
☐ Degree of a vertex; number of edges in graph, vertices of even and odd degree
☐ Simple graphs, graph equivalency
☐ Simple and closed paths, circuits, simple circuits, Euler and Hamiltonian paths and circuits
☐ Fleury's algorithm for finding an Euler circuit
☐ Acyclic and connected graphs, minimal spanning tree algorithm
☐ Graph coloring and bipartite graphs
☐ Digraphs; critical paths, indegree and outdegree, tournaments
☐ Euler and Hamiltonian circuits and paths in digraphs

□ Matrix representation of graphs; adjacency and reachability

□ Binary trees, rooted trees, tree representation of algebraic expressions and lists

Chapter 13: SUPPLEMENTARY EXERCISES

1. For each of the graphs in Fig. 1, give the degree of each vertex. Count the number of edges of each graph and verify the basic relationship between the sum of the degrees of the vertices and the number of edges in the graph.

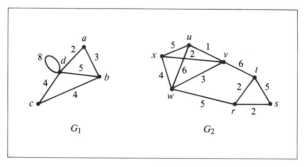

FIGURE 1

2. For each of the graphs in Fig. 1 give:

    (a) A simple path that is not closed.

    (b) A closed path that is not a circuit.

    (c) A circuit that is not a simple circuit.

3. Assume that each edge of the graphs in Fig. 1 is weighted with the value given on the edge. Find a minimal spanning tree for each graph and find its weight.

4. For each graph in Fig. 1, find an Euler circuit if one exists. If none exists, explain why.

5. For each of the graphs in Fig. 2, find the degree of each vertex. Verify that the number of vertices of odd degree is even in each case.

FIGURE 2

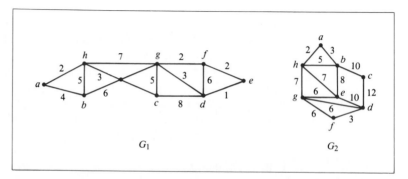

6. For each of the graphs in Fig. 2 determine:

   (a) A simple path that is not closed.

   (b) A closed path that is not simple.

   (c) A closed path that is not a circuit.

   (d) A circuit that is not a simple circuit.

7. Weights have been assigned to the edges of each of the graphs in Fig. 2. Use the algorithm to find a minimal spanning tree in each case and give its weight.

8. For each of the graphs in Fig. 2, determine if there is an Euler circuit. If one exists, describe it. If none exists, explain why.

9. For each of the graphs in Fig. 3, find an adjacency matrix. Find the number of paths of length 3 from v_1 to v_4 in each case.

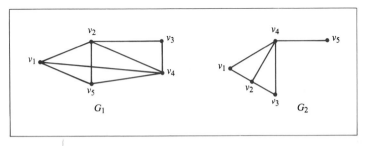

FIGURE 3

10. Determine which of the three graphs in Fig. 4 are equivalent to G_1.

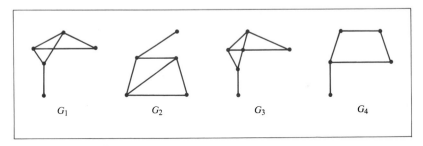

FIGURE 4

11. Draw the graph G with adjacency matrix $A(G)$ below. Determine if the graph G' in Fig. 5 is equivalent to G. Explain.

$$A(G) = \begin{bmatrix} 0 & 1 & 1 & 1 \\ 1 & 0 & 0 & 1 \\ 1 & 0 & 0 & 0 \\ 1 & 1 & 0 & 0 \end{bmatrix}$$

FIGURE 5

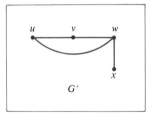

G'

12. For the floor plan of a house given in Fig. 6, determine if there is a path from the outside in which you can pass through each doorway exactly once and return outside. Show the path, if it exists. If it does not exist, explain why.

13. A house painter wishes to paint the interior of the house in Fig. 6 so that no two rooms joined by a doorway are the same color. What minimal number of colors does she need?

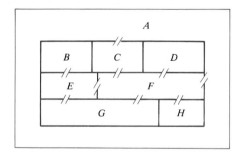

FIGURE 6

14. Determine the critical path for the publication of a magazine given the activity digraph in Fig. 7.

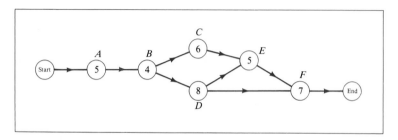

FIGURE 7

15. Construct a binary tree that shows the numerical ordering of the numbers 120, 187, 230, 102, 115, 134, and 190. Show how you would search the tree for the number 127. Insert it in the proper place on the tree.

16. Construct the binary trees that represent the algebraic expressions:

    (a) $[(4 * x) + y]\hat{\ }6$ (b) $(4 * x) + y\hat{\ }6$ (c) $[4 * (x + y)]\hat{\ }6$

17. Find the critical path for the activity digraph given in Fig. 8.

FIGURE 8

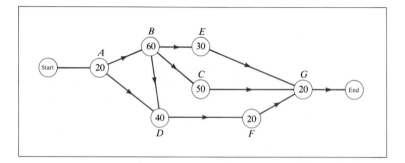

18. The graph in Fig. 9 represents the following: The vertices represent the people in a family; there is an edge between two members of the family if they need a car at the same time during the week. What is the minimal number of cars this family must own for everyone to have a car when it is needed?

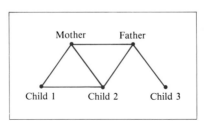

FIGURE 9

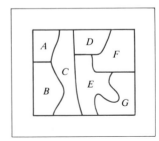

FIGURE 10

19. What is the minimal number of colors needed to color the map in Fig. 10? No two countries with a common border can be the same color.

20. A tournament is represented in the graph of Fig. 11. Determine if the tournament is transitive; find a winner and a ranking of players, if you can. Is your ranking unique?

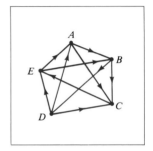

FIGURE 11

APPENDIX TABLES

TABLE 1 Areas under the Standard Curve

TABLE 2 $(1 + i)^n$ Compound Amount of $1 Invested for n Interest Periods at Interest i per Period

TABLE 3 $\dfrac{1}{(1 + i)^n}$ Present Value of $1. Principal that Will Accumulate to $1 in n Interest Periods at a Compound Rate of i per Period

TABLE 4 $s_{\overline{n}|i}$ Future Value of an Ordinary Annuity of n $1 Payments Each, Immediately after the Last Payment at Compound Interest Rate of i per Period

TABLE 5 $\dfrac{1}{s_{\overline{n}|i}}$ Rent per Period for an Ordinary Annuity of n Payments, Compound Interest Rate i per Period and Future Value $1

TABLE 6 $s_{\overline{n}|i}$ Present Value of an Ordinary Annuity of n Payments of $1 One Period before the First Payment with Interest Compounded at i per Period

TABLE 7 $\dfrac{1}{a_{\overline{n}|i}}$ Rent per Period for an Ordinary Annuity of n Payments Whose Present Value is $1 with Interest Compounded at i per Period

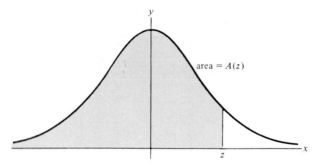

area = A(z)

TABLE 1   Areas under the Standard Normal Curve

z	A(z)	z	A(z)	z	A(z)	z	A(z)	z	A(z)
−3.50	.0002	−2.00	.0228	−.50	.3085	1.00	.8413	2.50	.9938
−3.45	.0003	−1.95	.0256	−.45	.3264	1.05	.8531	2.55	.9946
−3.40	.0003	−1.90	.0287	−.40	.3446	1.10	.8643	2.60	.9953
−3.35	.0004	−1.85	.0322	−.35	.3632	1.15	.8749	2.65	.9960
−3.30	.0005	−1.80	.0359	−.30	.3821	1.20	.8849	2.70	.9965
−3.25	.0006	−1.75	.0401	−.25	.4013	1.25	.8944	2.75	.9970
−3.20	.0007	−1.70	.0446	−.20	.4207	1.30	.9032	2.80	.9974
−3.15	.0008	−1.65	.0495	−.15	.4404	1.35	.9115	2.85	.9978
−3.10	.0010	−1.60	.0548	−.10	.4602	1.40	.9192	2.90	.9981
−3.05	.0011	−1.55	.0606	−.05	.4801	1.45	.9265	2.95	.9984
−3.00	.0013	−1.50	.0668	.00	.5000	1.50	.9332	3.00	.9987
−2.95	.0016	−1.45	.0735	.05	.5199	1.55	.9394	3.05	.9989
−2.90	.0019	−1.40	.0808	.10	.5398	1.60	.9452	3.10	.9990
−2.85	.0022	−1.35	.0885	.15	.5596	1.65	.9505	3.15	.9992
−2.80	.0026	−1.30	.0968	.20	.5793	1.70	.9554	3.20	.9993
−2.75	.0030	−1.25	.1056	.25	.5987	1.75	.9599	3.25	.9994
−2.70	.0035	−1.20	.1151	.30	.6179	1.80	.9641	3.30	.9995
−2.65	.0040	−1.15	.1251	.35	.6368	1.85	.9678	3.35	.9996
−2.60	.0047	−1.10	.1357	.40	.6554	1.90	.9713	3.40	.9997
−2.55	.0054	−1.05	.1469	.45	.6736	1.95	.9744	3.45	.9997
−2.50	.0062	−1.00	.1587	.50	.6915	2.00	.9772	3.50	.9998
−2.45	.0071	−.95	.1711	.55	.7088	2.05	.9798		
−2.40	.0082	−.90	.1841	.60	.7257	2.10	.9821		
−2.35	.0094	−.85	.1977	.65	.7422	2.15	.9842		
−2.30	.0107	−.80	.2119	.70	.7580	2.20	.9861		
−2.25	.0122	−.75	.2266	.75	.7734	2.25	.9878		
−2.20	.0139	−.70	.2420	.80	.7881	2.30	.9893		
−2.15	.0158	−.65	.2578	.85	.8023	2.35	.9906		
−2.10	.0179	−.60	.2743	.90	.8159	2.40	.9918		
−2.05	.0202	−.55	.2912	.95	.8289	2.45	.9929		

TABLE 2　$(1 + i)^n$　Compound Amount of $1 Invested for n Interest Periods at Interest Rate i per Period

n	$\frac{1}{50}\%$	$\frac{1}{2}\%$	1%	2%	6%
1	1.000200000	1.00500000	1.01000000	1.02000000	1.06000000
2	1.000400040	1.01002500	1.02010000	1.04040000	1.12360000
3	1.000600120	1.01507513	1.03030100	1.06120800	1.19101600
4	1.000800240	1.02015050	1.04060401	1.08243216	1.26247696
5	1.001000400	1.02525125	1.05101005	1.10408080	1.33822558
6	1.001200600	1.03037751	1.06152015	1.12616242	1.41851911
7	1.001400840	1.03552940	1.07213535	1.14868567	1.50363026
8	1.001601120	1.04070704	1.08285671	1.17165938	1.59384807
9	1.001601447	1.04591058	1.09368527	1.19509257	1.68947896
10	1.002001806	1.05114013	1.10462213	1.21899442	1.79084770
11	1.002202201	1.05639583	1.11566835	1.24337431	1.89829856
12	1.002402642	1.06167781	1.12682503	1.26824179	2.01219647
13	1.002603122	1.06698620	1.13809328	1.29360663	2.13292826
14	1.002802643	1.07232113	1.14947421	1.31947876	2.26090396
15	1.003004204	1.07768274	1.16096896	1.34586834	2.39655819
16	1.003204804	1.08307115	1.17257864	1.37278571	2.54035168
17	1.003405445	1.08848651	1.18430443	1.40024142	2.69277279
18	1.0035051	1.09392894	1.19614748	1.42824625	2.85433915
19	1.003806848	1.09939858	1.20810895	1.45681117	3.02559950
20	1.004007609	1.10489558	1.22019004	1.48594740	3.20713547
21	1.004208411	1.11042006	1.23239194	1.51566634	3.39956360
22	1.004409252	1.11597216	1.24471586	1.54597967	3.60353742
23	1.004610134	1.12155202	1.25716302	1.57689926	3.81974966
24	1.004811056	1.12715978	1.26973465	1.60843725	4.40893464
25	1.005012018	1.13279558	1.28243200	1.64060599	4.29187072
26	1.005213021	1.13845955	1.29525631	1.67341811	4.54938296
27	1.005414063	1.14415185	1.30820888	1.70688648	4.82234594
28	1.005615146	1.14987261	1.32129097	1.74102421	5.11168670
29	1.005816269	1.15562197	1.33450388	1.77584469	5.41838790
30	1.006017433	1.16140008	1.34784892	1.81136158	5.74349117
36	1.007225257	1.19668052	1.43076878	2.03988734	8.14725200
48	1.009645259	1.27048916	1.61222608	2.58707039	16.39387173
52	1.010453217	1.29609015	1.67768892	2.80032819	20.69688534
60	1.012071075	1.34885015	1.81669670	3.28103079	32.98769085
104	1.021015784	1.67984969	2.81464012	7.84183795	428.36106292
120	1.024287860	1.81939673	3.30038689	10.76516303	1,088.18774784
180	1.036652115	2.45409356	5.99580198	35.32083136	35,896.80101597
240	1.049165620	3.31020448	10.89255365	115.88873515	
300	1.061830136	4.46496981	19.78846626	380.23450806	
360	1.074647608	6.02257521	35.94964133	1,247.56112775	
365	1.075722685	6.17465278	37.78343433	1,377.40829197	

TABLE 3 $\dfrac{1}{(1 + i)^n}$ Present Value of $1. Principal that Will Accumulate to $1 in *n* Interest Periods at a Compound Rate of *i* per Period

n	\multicolumn{5}{c}{i}
	1/50%	1/2%	1%	2%	6%
1	.99980004	.99502488	.99009901	.98039216	.94339623
2	.99960012	.99007450	.98029605	.96116878	.88999644
3	.99940024	.98514876	.97059015	.94232233	.83961928
4	.99920040	.98024752	.96098034	.92384543	.79209366
5	.99900060	.97537067	.95146569	.90573081	.74725817
6	.99880084	.97051808	.94 04524	.88797138	.70496054
7	.99860112	.96568963	.93271805	.87056018	.66505711
8	.99840144	.96088520	.92348322	.85349037	.62741237
9	.99820180	.95610468	.91433982	.83675527	.59189846
10	.99800220	.95134794	.90528695	.82034830	.55839478
11	.99780264	.94661487	.89632372	.80426304	.52678753
12	.99760312	.94190534	.88744923	.78849318	.49696936
13	.99740364	.93721924	.87866260	.77303253	.46883902
14	.99720420	.93255646	.86996297	.75787502	.44230096
15	.99700479	.92791688	.86134947	.74301473	.41726506
16	.99680543	.92330037	.85282126	.72844581	.39364628
17	.99660611	.91870684	.84437749	.71416256	.37136442
18	.99640683	.91413616	.83601731	.70015937	.35034379
19	.99620759	.90958822	.82773992	.68643076	.33051301
20	.99600839	.90506290	.81954447	.67297133	.31180473
21	.99580923	.90056010	.81143017	.65977582	.29415540
22	.99561010	.89607971	.80339621	.64683904	.27750510
23	.99541102	.89162160	.79544179	.63415592	.26179726
24	.99521198	.88718567	.78756613	.62172149	.24697855
25	.99501298	.88277181	.77976844	.60953087	.23299863
26	.99481401	.87837991	.77204796	.59757928	.21981003
27	.99461509	.87400986	.76440392	.58586204	.20736795
28	.99441621	.86966155	.75683557	.57437455	.19563014
29	.99421736	.86533488	.74934215	.56311231	.18455674
30	.99401856	.86102973	.74192292	.55207089	.17411013
36	.99282657	.83564492	.69892495	.49022315	.12274077
48	.99044688	.78709841	.62026041	.38653761	.06099840
52	.98965492	.77155127	.59605806	.35710100	.04831645
60	.98807290	.74137220	.55044962	.30478227	.03031434
104	.97941686	.59529136	.35528521	.12752113	.00233448
120	.97628805	.54963273	.30299478	.09289223	.00091896
180	.96464377	.40748243	.16678336	.02831190	.00002786
240	.95313836	.30209614	.09180584	.00862897	.00000084
300	.94177018	.22396568	.05053449	.00262996	.00000003
360	.93053759	.16604193	.02781669	.00080156	.00000000
365	.92960762	.16195243	.02646663	.00072600	.00000000

TABLE 4 $s_{\overline{n}|i}$ Future Value of an Ordinary Annuity of n $1 Payments Each, Immediately after the Last Payment at Compound Interest Rate of i per Period

	i				
n	1/2%	1%	1.5%	2%	6%
1	1.00000000	1.00000000	1.00000000	1.00000000	1.00000000
2	2.00500000	2.01000000	2.01500000	2.02000000	2.06000000
3	3.01502500	3.03010000	3.04522500	3.06040000	3.18360000
4	4.03010013	4.06040100	4.09090337	4.12160800	4.37461600
5	5.05025063	5.10100501	5.15226693	5.20404016	5.63709296
6	6.07550188	6.15201506	6.22955093	6.30812096	6.97531854
7	7.10587939	7.21353521	7.32299419	7.43428338	8.39383765
8	8.14140879	8.28567056	8.43283911	8.58296905	9.89746791
9	9.18211583	9.36852727	9.55933169	9.75462843	11.49131598
10	10.22802641	10.46221254	10.70272167	10.94972100	13.18079494
11	11.27916654	11.56683467	11.86326249	12.16871542	14.97164264
12	12.33556237	12.68250301	13.04121143	13.41208973	16.86994120
13	13.39724018	13.80932804	14.23682960	14.68033152	18.88213767
14	14.46422639	14.94742132	15.45038205	15.97393815	21.01506593
15	15.53654752	16.09689554	16.68213778	17.29341692	23.27596988
16	16.61423026	17.25786449	17.93236984	18.63928525	25.67252808
17	17.69730141	18.43044314	19.20135539	20.01207096	28.21287976
18	18.78578791	19.61474757	20.48937572	21.41231238	30.90565255
19	19.87971685	20.81089504	21.79671636	22.84055863	33.75999170
20	20.97911544	22.01900399	23.12366710	24.29736980	36.78559120
21	22.08401101	23.23919403	24.47052211	25.78331719	39.99272668
22	23.19443107	24.47158598	25.83757994	27.29898354	43.39229028
23	24.31040322	25.71630183	27.22514364	28.84496321	46.99582769
24	25.43195524	26.97346485	28.63352080	30.42186247	50.81557735
25	26.55911502	28.24319950	30.06302361	32.03029972	54.86451200
26	27.69191059	29.52563150	31.51396896	33.67090572	59.15638272
27	28.83037015	30.82088781	32.98667850	35.34432383	63.70576568
28	29.97452200	32.12909669	34.48147867	37.05121031	68.52811162
29	31.12439461	33.45038766	35.99870085	38.79223451	73.63979832
30	32.28001658	34.78489153	37.53868137	40.56807921	79.05818622
36	39.33610496	43.07687836	47.27596921	51.99436719	119.12086666
48	54.09783222	61.22260777	69.56521929	79.35351927	256.56452882
52	59.21803075	67.76889215	77.92489152	90.01640927	328.28142239
60	69.77003051	81.66966986	96.21465171	114.05153942	533.12818089
104	135.96993732	181.46401172	246.93411381	342.09189731	7,122.68438195
120	163.87934681	230.03868946	331.28819149	488.25815171	18,119.79579725
180	290.81871245	499.58019754	905.62451261	1,716.04156785	
240	462.04089516	989.25536539	2,308.85437027	5,744.43675765	
300	692.99396243	1,878.84662619	5,737.25330834	18,961.72540308	
360	1,004.51504245	3,494.96413277	14,113.58539279	62,328.05638744	
365	1,034.93055669	3,678.34343329	15,209.49204803	68,820.41459830	

TABLE 5 $\dfrac{1}{s_{\overline{n}|i}}$ Rent per Period for an Ordinary Annuity of n Payments, Compound
Interest Rate i per Period, and Future Value $1

n	i				
	1/2%	1%	1.5%	2%	6%
1	1.00000000	1.00000000	1.00000000	1.00000000	1.00000000
2	.49875312	.49751244	.49627792	.49504950	.48543689
3	.33167221	.33002211	.32838296	.32675467	.31410981
4	.24813279	.24628109	.24444479	.24262375	.22859149
5	.19800997	.19603980	.19408932	.19215839	.17739640
6	.16459546	.16254837	.16052521	.15852581	.14336263
7	.14072854	.13862828	.13655616	.13451196	.11913502
8	.12282886	.12069029	.11858402	.11650980	.10103594
9	.10890736	.10674036	.10460982	.10251544	.08702224
10	.09777057	.09558208	.09343418	.09132653	.07586796
11	.08865903	.08645408	.08429384	.08217794	.06679294
12	.08106643	.07884879	.07667999	.07455960	.05927703
13	.07464224	.07241482	.07024036	.06811835	.05296011
14	.06913609	.06690117	.06472332	.06260197	.04758491
15	.06436436	.06212378	.05994436	.05782547	.04296276
16	.06018937	.05794460	.05576508	.05365013	.03895214
17	.05650579	.05425806	.05207966	.04996984	.03544480
18	.05323173	.05098205	.04880578	.04670210	.03235654
19	.05030253	.04805175	.04587847	.04378177	.02962086
20	.04766645	.04541531	.04324574	.04115672	.02718456
21	.04528163	.04303075	.04086550	.03878477	.02500455
22	.04311380	.04086372	.03870332	.03663140	.02304557
23	.04113465	.03888584	.03673075	.03466810	.02127848
24	.03932061	.03707347	.03492410	.03287110	.01967900
25	.03765186	.03540675	.03326345	.03122044	.01822672
26	.03611163	.03386888	.03173196	.02969923	.01690435
27	.03468565	.03244553	.03031527	.02829309	.01569717
28	.03336167	.03112444	.02900108	.02698967	.01459255
29	.03212914	.02989502	.02777878	.02577836	.01357961
30	.03097892	.02874811	.02663919	.02464992	.01264891
36	.02542194	.02321431	.02115240	.01923285	.00839483
48	.01848503	.01633384	.01437500	.01260184	.00389765
52	.01688675	.01475603	.01283287	.01110909	.00304617
60	.01433280	.01224445	.01039343	.00876797	.00187572
104	.00735457	.00551073	.00404966	.00292319	.00014040
120	.00610205	.00434709	.00301852	.00204810	.00005519
180	.00343857	.00200168	.00110421	.00058274	.00000167
240	.00216431	.00101086	.00043312	.00017408	.00000005
300	.00144301	.00053224	.00017430	.00005274	.00000000
360	.00099551	.00028613	.00007085	.00001604	.00000000
365	.00096625	.00027186	.00006575	.00001453	.00000000

TABLE 6 $a_{\overline{n}|i}$ Present Value of an Ordinary Annuity of n Payments of $1 One Period before the First Payment with Interest Compounded at i per Period

			i		
n	1/2%	1%	1.5%	2%	6%
1	.99502488	.99009901	.98522167	.98039216	.94339623
2	1.98509938	1.97039506	1.95588342	1.94156094	1.83339267
3	2.97024814	2.94098521	2.91220042	2.88388327	2.67301195
4	3.95049566	3.90196555	3.85438465	3.80772870	3.46510561
5	4.92586633	4.85343124	4.78264497	4.71345951	4.21236379
6	5.89638441	5.79547647	5.69718717	5.60143089	4.91732433
7	6.86207404	6.72819453	6.59821396	6.47199107	5.58238144
8	7.82295924	7.65167775	7.48592508	7.32548144	6.20979381
9	8.77906392	8.56601758	8.36051732	8.16223671	6.80169227
10	9.73041186	9.47130453	9.22218455	8.98258501	7.36008705
11	10.67702673	10.36762825	10.07111779	9.78684805	7.88687458
12	11.61893207	11.25507747	10.90750521	10.57534122	8.38384394
13	12.55615131	12.13374007	11.73153222	11.34837375	8.85268296
14	13.48870777	13.00370304	12.54338150	12.10624877	9.29498393
15	14.41662465	13.86505252	13.34323301	12.84926350	9.71224899
16	15.33992502	14.71787378	14.13126405	13.57770931	10.10589527
17	16.25863186	15.56225127	14.90764931	14.29187188	10.47725969
18	17.17276802	16.39826858	15.67256089	14.99203125	10.82760348
19	18.08235624	17.22600850	16.42616837	15.67846201	11.15811649
20	18.98741915	18.04555297	17.16863879	16.35143334	11.46992122
21	19.88797925	18.85698313	17.90013673	17.01120916	11.76407662
22	20.78405896	19.66037934	18.62082437	17.65804820	12.04158172
23	21.67568055	20.45582113	19.33086145	18.29220412	12.30337898
24	22.56286622	21.24338726	20.03040537	18.91392560	12.55035753
25	23.44563803	22.02315570	20.71961120	19.52345647	12.78335616
26	24.32401794	22.79520366	21.39863172	20.12103576	13.00316619
27	25.19802780	23.55960759	22.06761746	20.70689780	13.21053414
28	26.06768936	24.31644316	22.72671671	21.28127236	13.40616428
29	26.93302423	25.06578530	23.37607558	21.84438466	13.59072102
30	27.79405397	25.80770822	24.01583801	22.39645555	13.76483115
36	32.87101624	30.10750504	27.66068431	25.48884248	14.62098713
48	42.58031778	37.97395949	34.04255365	30.67311957	15.65002661
52	45.68974664	40.39419423	35.92874185	32.14494992	15.86139252
60	51.72556075	44.95503841	39.38026889	34.76088668	16.16142771
104	80.94172854	64.47147918	52.49436634	43.62394373	16.62775868
120	90.07345333	69.70052203	55.49845411	45.35538850	16.65135068
180	118.50351467	83.32166399	62.09556231	48.58440478	16.66620237
240	139.58077168	90.81941635	64.79573209	49.56855168	16.66665259
300	155.20686401	94.94655125	65.90090069	49.86850220	16.66666624
360	166.79161439	97.21833108	66.35324174	49.95992180	16.66666665
365	167.60951473	97.35333747	66.37572674	49.96369994	16.66666666

TABLE 7 $\dfrac{1}{a_{\overline{n}|i}}$ Rent per Period for an Ordinary Annuity of n Payments Whose Present Value is $1 with Interest Compounded at i per Period

			i		
n	1/2%	1%	1.5%	2%	6%
1	1.00500000	1.01000000	1.01500000	1.02000000	1.06000000
2	.50375312	.50751244	.51127792	.51504950	.54543689
3	.33667221	.34002211	.34338296	.34675467	.37410981
4	.25313279	.25628109	.25944479	.26262375	.28859149
5	.20300997	.20603980	.20908932	.21215839	.23739640
6	.16959546	.17254837	.17552521	.17852581	.20336263
7	.14572854	.14862828	.15155616	.15451196	.17913502
8	.12782886	.13069029	.13358402	.13650980	.16103594
9	.11390736	.11674036	.11960982	.12251544	.14702224
10	.10277057	.10558208	.10843418	.11132653	.13586796
11	.09365903	.09645408	.09929384	.10217794	.12679294
12	.08606643	.08884879	.09167999	.09455960	.11927703
13	.07964224	.08241482	.08524036	.08811835	.11296011
14	.07413609	.07690117	.07972332	.08260197	.10758491
15	.06936436	.07212378	.07494436	.07782547	.10296276
16	.06518937	.06794460	.07076508	.07365013	.09895214
17	.06150579	.06425806	.06707966	.06996984	.09544480
18	.05823173	.06098205	.06380578	.06670210	.09235654
19	.05530253	.05805175	.06087847	.06378177	.08962086
20	.05266645	.05541531	.05824574	.06115672	.08718456
21	.05028163	.05303075	.05586550	.05878477	.08500455
22	.04811380	.05086372	.05370332	.05663140	.08304557
23	.04613465	.04888584	.05173075	.05466810	.08127848
24	.04432061	.04707347	.04992410	.05287110	.07967900
25	.04265186	.04540675	.04826345	.05122044	.07822672
26	.04111163	.04386888	.04673196	.04969923	.07690435
27	.03968565	.04244553	.04531527	.04829309	.07569717
28	.03836167	.04112444	.04400108	.04698967	.07459255
29	.03712914	.03989502	.04277878	.04577836	.07357961
30	.03597892	.03874811	.04163919	.04464992	.07264891
36	.03042194	.03321431	.03615240	.03923285	.06839483
48	.02348503	.02633384	.02937500	.03260184	.06389765
52	.02188675	.02475603	.02783287	.03110909	.06304617
60	.01933280	.02224445	.02539343	.02876797	.06187572
104	.01235457	.01551073	.01904966	.02292319	.06014040
120	.01110205	.01434709	.01801852	.02204810	.06005519
180	.00843857	.01200168	.01610421	.02058274	.06000167
240	.00716431	.01101086	.01543312	.02017408	.06000005
300	.00644301	.01053224	.01517430	.02005274	.06000000
360	.00599551	.01028613	.01507085	.02001604	.06000000
365	.00596625	.01027186	.01506575	.02001453	.06000000

ANSWERS TO ODD-NUMBERED EXERCISES

CHAPTER 1

EXERCISES 1.1, page 10

1, 3, 5.

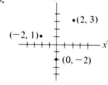

7.

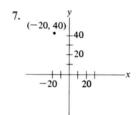

9. $m = 5, b = 8$ **11.** $m = 0, b = 3$

13. $y = -2x + 3$ **15.** $x = \frac{5}{3}$ **17.** $(2, 0), (0, 8)$ **19.** $(7, 0)$, none

21.

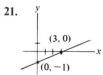

23.

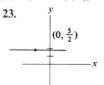

25.

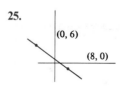

27. a, b, c, e **29.** (a) L_3 (b) L_1 (c) L_2 **31.** (a) 4 min., 40 sec (b) 72° water was placed into the kettle (c) No
33. Up   **35.** $y = 0$ **37.** No   **39.** When the line passes through the origin

EXERCISES 1.2, page 20

1. False   **3.** True   **5.** $x \geq 4$ **7.** $x \geq 3$ **9.** $y \leq -2x + 5$ **11.** $y \geq 15x - 18$ **13.** $x \geq -\frac{3}{4}$ **15.** True
17. False   **19.** True   **21.** True   **23.**

25.

27.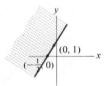

29.

31.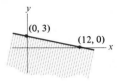

33.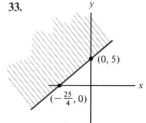

35.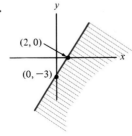

37.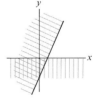

39.

41.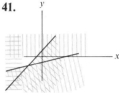

43. In **45.** Not in **47.** Below **49.** Above **51.** $\begin{cases} y \geq 2x - 1 \\ y \leq 2x \end{cases}$

53. $x \geq -2, \ x \leq 4, \ 2x + 3y \leq 6, \ y \geq -3$

EXERCISES 1.3, page 27

1. $(2, 3)$ **3.** $(2, 1)$ **5.** $(12, 3)$ **7.** Yes **9.** $(\frac{10}{3}, \frac{1}{3})$ **11.** $(-\frac{7}{9}, -\frac{22}{9})$ **13.** $A = (3, 4), B = (6, 2)$
15. $A = (0, 0), B = (2, 4), C = (5, 5\frac{1}{2}), D = (5, 0)$ **17.** **19.**

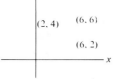

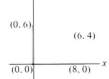

21.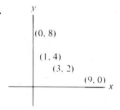

23. (a) 19,500 (b) 5 cents **25.** (a) $3 (b) 29,500

27. Working: 32; supervising; 8

1. $\frac{2}{3}$ 3. 5 5. $\frac{5}{4}$

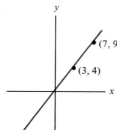

(7, 9)
(3, 4)

7. $\frac{4}{5}$

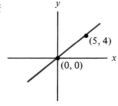

(5, 4)
(0, 0)

9. Not defined

11.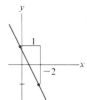

13.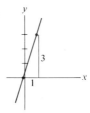

15. $y = -2x + 7$

17. $y = -2x + 4$ 19. $y = \frac{1}{4}x + \frac{2}{3}$ 21. $y = -x$ 23. $y = 3$ 25. (0, 3)

27. Each unit sold yields a commission of \$5. In addition, she receives \$60 per week base pay.

29. (a) \$1200; at \$1200 no one will buy the item (b) 400 items; even if the item is given away, only 400 people will want it (c) -3; to sell an additional item, the price must be reduced by \$3 (d) \$150 (e) 300 items (f)

p
(0, 1200)
(400, 0)
q

31. (a) $y = 90x + 5000$ (b) \$5000 (c) \$90 (d)

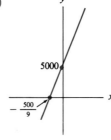

5000
$-\frac{500}{9}$

33. (a) \$30,000 (b) 60 coats (c) (0, 0); if no coats are sold, there is no revenue (d) 100; each additional coat sold yields an additional \$100 in revenue. Therefore, \$100 is the selling price of a coat.

35

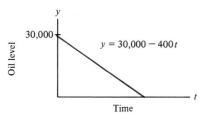

30,000
$y = 30,000 - 400t$
Oil level
Time

37. 12,000 gallons

39. (75, 0). The tank will be empty after 75 days.   **41.** $50,000
43. ($15\frac{1}{3}$, 0). The cash reserves will be depleted after $15\frac{1}{3}$ days; that is, during July 16.
45. Morning of July 11; that is, after 10 days.   **47.** 3   **49.** 1400   **51.** $260.00   **53.** $y = 3x - 1$
55. $y = x + 1$ **57.** $y = -7x + 35$ **59.** $y = 4$ **61.** $y - 1 = \frac{1}{2}(x - 2)$ **63.** $y = -2x$ **65.** 5; 1; -1
67. $-\frac{5}{4}$; $-\frac{3}{2}$; $-\frac{3}{4}$ **69.** l_1 **71.** $F = \frac{9}{5}C + 32$ **73.** Counterclockwise   **75.** $y \geq 4x + 3$
77. $x \geq 0, y \geq 0, x - y \leq 3, x + y \leq 5, x + 2y \leq 8$ **79.** 9   **81.** $-.05$

CHAPTER 1: SUPPLEMENTARY EXERCISES, page 44

1. $x = 0$ **3.** $(2, -\frac{4}{5})$ **5.** $y = -\frac{1}{2}x + 5$ **7.** Yes   **9.** $y = \frac{1}{5}x + 13$ **11.** (5, 0)   **13.** $x = 7, y = 10$ **15.** (0, 7)

17. No   **19.** $b = \dfrac{a}{2}$ **21.** $y \geq 2.4x - 5.8$ **23.** $y - 1 = -\frac{2}{5}(x - 1)$ **25.**

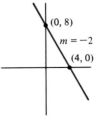

(0, 8)
$m = -2$
(4, 0)

29. (a) L_3 (b) L_1 (c) L_2 **31.** $(\frac{8}{3}, 300)$ **33.** $y = x + \frac{5}{2}$
35. (a) A: $y = .10x + 50$; B: $y = .20x + 40$ (b) B (c) A (d) 100 miles
37.

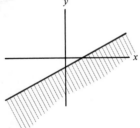

CHAPTER 2

EXERCISES 2.1, page 58

1. $\begin{cases} x - 6y = 4 \\ 5x + 4y = 1 \end{cases}$ **3.** $\begin{cases} x + 2y = 3 \\ \quad\;\; 14y = 16 \end{cases}$ **5.** $\begin{cases} x - 2y + \;\; z = 0 \\ \quad\;\; y - 2z = 4 \\ \quad\; 9y - \;\; z = 5 \end{cases}$ **7.** $\begin{bmatrix} 1 & 0 & | & 5 \\ 0 & 1 & | & 4 \end{bmatrix}$ **9.** $[2] + 2[1]$
$\quad\;\; 2[1]$ $\qquad\qquad\quad [2] + 5[1]$ $\qquad\qquad\; [3] + (-4)[1]$ $\qquad\; [1] + (\frac{1}{2})[2]$

11. $[1] + (-2)[2]$ **13.** Interchange rows 1 and 2 or rows 1 and 3.   **15.** $[1] + (-3)[3]$ **17.** $x = -1, y = 1$
19. $x = -\frac{8}{7}, y = -\frac{9}{7}, z = -\frac{3}{7}$ **21.** $x = -1, y = 1$ **23.** $x = 1, y = 2, z = -1$ **25.** $x = -2.5, y = 15$
27. $(1, -6, 2)$ **29.** $(-1, -2, 5)$ **31.** 137 adults, 463 children   **33.** $x = \$25,000, y = \$50,000, z = \$25,000$

EXERCISES 2.2, page 67

1. $\begin{bmatrix} 1 & -2 & 3 \\ 0 & 13 & -8 \end{bmatrix}$ **3.** $\begin{bmatrix} 9 & -1 & 0 & -7 \\ -\frac{1}{2} & \frac{1}{2} & 1 & 3 \\ 5 & -1 & 0 & -3 \end{bmatrix}$ **5.** $\begin{bmatrix} 1 & \frac{3}{2} \\ 0 & -9 \\ 0 & \frac{7}{2} \end{bmatrix}$ **7.** $\begin{bmatrix} 4 & 3 & 0 \\ 1 & 1 & 0 \\ \frac{1}{6} & \frac{1}{2} & 1 \end{bmatrix}$ **9.** $y =$ any value, $x = 3 + 2y$

11. $x = 1, y = 2$ **13.** No solution **15.** $z =$ any value, $y = 5, x = -6 - z$ **17.** No solution
19. $z =$ any value, $w =$ any value, $x = 2z + w, y = 5 - 3w$ **21.** $x = 5, y = 7$.
23. $x = -13, y = 9, z = 0; x = 2, y = 0, z = 3; x = -3, y = 3, z = 2$
25. $x = 23, y = 0, z = 5; x = 2, y = 3, z = 5; x = 9, y = 2, z = 5$
27. $z =$ any value, $x = 300 - z, y = 100 - z$. Of course, to be realistic, we must have $0 \le z \le 100$.
29. $x = 1, y = 2, z = 3; x = -1, y = 2, z = 3; x = 1, y = -2, z = 3; x = 1, y = 2, z = -3;$
$\quad x = 1, y = -2, z = -3; x = -1, y = 2, z = -3; x = -1, y = -2, z = 3; x = -1, y = -2, z = -3.$
31. No solution if $k \ne -12$; infinitely many solutions if $k = -12$.

EXERCISES 2.3, page 80

1. 2×3 **3.** 1×3, row matrix **5.** 1×1, square matrix **7.** $\begin{bmatrix} 9 & 3 \\ 7 & -1 \end{bmatrix}$ **9.** $\begin{bmatrix} 1 & 3 \\ 1 & 2 \\ 4 & -2 \end{bmatrix}$ **11.** $[11]$ **13.** $[10]$

15. 3×5 **17.** Not defined **19.** 3×1 **21.** $\begin{bmatrix} 6 & 17 \\ 6 & 10 \end{bmatrix}$ **23.** $\begin{bmatrix} 21 \\ -4 \\ 8 \end{bmatrix}$ **25.** $\begin{bmatrix} 5 & 6 \\ 7 & 8 \end{bmatrix}$ **27.** $\begin{bmatrix} .48 & .39 \\ .52 & .61 \end{bmatrix}$

29. $\begin{bmatrix} 25 & 17 & 2 \\ 3 & -1 & 2 \\ 1 & 1 & 4 \end{bmatrix}$ **31.** $\begin{cases} 2x + 3y = 6 \\ 4x + 5y = 7 \end{cases}$ **33.** $\begin{cases} x + 2y + 3z = 10 \\ 4x + 5y + 6z = 11 \\ 7x + 8y + 9z = 12 \end{cases}$ **35.** $\begin{bmatrix} 3 & 2 \\ 7 & -1 \end{bmatrix}\begin{bmatrix} x \\ y \end{bmatrix} = \begin{bmatrix} -1 \\ 2 \end{bmatrix}$

37. $\begin{bmatrix} 1 & -2 & 3 \\ 0 & 1 & 1 \\ 0 & 0 & 1 \end{bmatrix}\begin{bmatrix} x \\ y \\ z \end{bmatrix} = \begin{bmatrix} 5 \\ 6 \\ 2 \end{bmatrix}$ **39.** $\begin{bmatrix} 4 & 24 \\ 20 & 24 \end{bmatrix}$ **43.** (a) $\begin{bmatrix} 340 \\ 265 \end{bmatrix}$ (b) Mike's clothes cost \$340.
Don's clothes cost \$265.
45. (a) I: 2.75, II: 2, III: 1.3 (b) A: 74, B: 112, C: 128, D: 64, F: 22.
47. Number voting Democratic $= 10,100$; number voting Republican $= 7900$.
49. (a) Shift 1: \$1000, shift 2: \$1050, shift 3: \$600: (b) Carpenters: \$1000, bricklayers: \$1050, plumbers: \$600.
51. $x = 3, y = 4, z = 5$

EXERCISES 2.4, page 93

1. $x = 2, y = 0$ **3.** $\begin{bmatrix} 1 & -2 \\ -3 & 7 \end{bmatrix}$ **5.** $\begin{bmatrix} 1 & -1 \\ -2.5 & 3 \end{bmatrix}$ **7.** $\begin{bmatrix} 1.6 & -.4 \\ -.6 & 1.4 \end{bmatrix}$ **9.** $[\tfrac{1}{3}]$ **11.** $x = 4, y = -\tfrac{1}{2}$
13. $x = 32, y = -6$

15. (a) $\begin{bmatrix} .8 & .3 \\ .2 & .7 \end{bmatrix}\begin{bmatrix} x \\ y \end{bmatrix} = \begin{bmatrix} m \\ s \end{bmatrix}$ (b) $\begin{bmatrix} x \\ y \end{bmatrix} = \begin{bmatrix} 1.4 & -.6 \\ -.4 & 1.6 \end{bmatrix}\begin{bmatrix} m \\ s \end{bmatrix}$ (c) 110,000; 40,000 (d) 130,000; 20,000
17. $x = 9, y = -2, z = -2$ **19.** $x = 1, y = 5, z = -4, w = 9$

23. (a) $\begin{bmatrix} 1 & 2 \\ \tfrac{9}{10} & 0 \end{bmatrix}\begin{bmatrix} x \\ y \end{bmatrix} = \begin{bmatrix} a \\ b \end{bmatrix}$ (b) After 1 year: 1,170,000 in group I, 405,000 in group II. After 2 years: 1,980,000
in group I, 1,053,000 in group II. (c) 700,000 in group I, 55,000 in group II

EXERCISES 2.5, page 100

1. $\begin{bmatrix} -2 & 3 \\ 5 & -7 \end{bmatrix}$ **3.** $\begin{bmatrix} \tfrac{1}{19} & \tfrac{3}{19} \\ \tfrac{3}{76} & -\tfrac{5}{38} \end{bmatrix}$ **5.** No inverse **7.** $\begin{bmatrix} -1 & 2 & -4 \\ 1 & -1 & 3 \\ 0 & 0 & 1 \end{bmatrix}$ **9.** No inverse **11.** $\begin{bmatrix} -5 & 6 & 0 & 0 \\ 1 & -1 & 0 & 0 \\ 0 & 0 & -\tfrac{1}{46} & \tfrac{1}{46} \\ 0 & 0 & \tfrac{25}{46} & -\tfrac{1}{23} \end{bmatrix}$
13. $x = 2, y = -3, z = 2$ **15.** $x = 4, y = -4, z = 3, w = -1$

17. Either no solution or infinitely many solutions **19.** $\begin{bmatrix} -3 & 5 \\ 10 & -16 \end{bmatrix}$

EXERCISES 2.6, page 107

1. Coal: $8.84 billion, steel: $3.73 billion, electricity: $9.90 billion

3. Computers: $354 million, semiconductors: $172 million   **7.** Plastics: $955,000; industrial equipment: $590,000

CHAPTER 2: SUPPLEMENTARY EXERCISES, page 109

1. $\begin{bmatrix} 1 & -2 & \frac{1}{3} \\ 0 & 8 & \frac{16}{3} \end{bmatrix}$ **3.** $x = 4, y = 5$ **5.** $x = -1, y = \frac{2}{3}, z = \frac{1}{3}$ **7.** $z =$ any value, $x = 1 - 3z, y = 4z, w = 5$

9. $\begin{bmatrix} 5 \\ 3 \\ 7 \end{bmatrix}$ **11.** $x = -2, y = 3$ **13.** $\begin{bmatrix} -1 & 3 \\ \frac{1}{2} & -1 \end{bmatrix}$ **15.** Industry I: 20 units, Industry II: 20 units

CHAPTER 3

EXERCISES 3.1, page 116

1. Yes   **3.** No

5. (a)

	A	B	Truck capacity
Volume	4	3	300
Weight	100	200	10,000
Earnings	13	9	

(b) $4x + 3y \leq 300$
$100x + 200y \leq 10,000$

(c) $y \leq 2x, x \geq 0, y \geq 0$

(d) $13x + 9y$ (e)

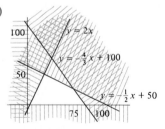

7. (a)

	Alfalfa	Corn	Requirements
Protein	.13	.065	4,550
TDN	.48	.96	26,880
Vitamin A	2.16	0	43,200
Cost	.01	.016	

(b) $x \geq 0, y \geq 0, .13x + .065y \geq 4550, .48x + .96y \geq 26,880, 2.16x \geq 43,200$ (c)
(d) $C = .01x + .016y$

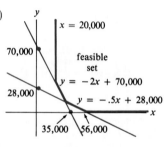

$-y \geq -300$

$y \leq 300$

$-x \geq -60$

$x \leq 60$

A14   *Answers to Odd-Numbered Exercises*

1. $x = 20$, $y = 0$ **3.** $x = 6$, $y = 0$ **5.** 75 crates of cargo A, 0 crates of cargo B   **7.** Produce 16 chairs and 0 sofas

9.

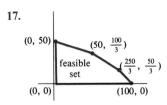

The minimum value is 40 and occurs at the point (4.3).

11.

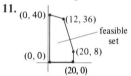

The maximum value is 6600 and occurs at the point (12, 36).

13.

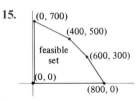

80 homes of first type, 60 homes of second type.

15.

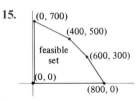

400 cans of Fruit Delight, 500 cans of Heavenly Punch.

17.

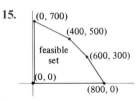

$\frac{250}{3}$ acres of oats
$\frac{50}{3}$ acres of corn
Profit = \$5,283.33

19.

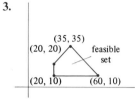

(a) 8 of item I_1, 4 of item I_2
(b) \$88
(c) 40 ounces of M_1, 20 ounces of M_2, 28 ounces of M_3
(d) 10 of item I_1, 0 of item I_2

21. The feasible set contains no points.

1.

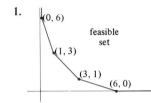

1 can of brand A, 3 cans of brand B.

3.

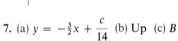

35 crates of oranges, 35 crates of grapefruit, 30 crates of avocados.

5.

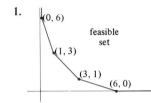

Numbers in hundreds

In Detroit make 100 cars, 300 trucks. In Cleveland make 500 cars and 0 trucks.

7. (a) $y = -\frac{3}{2}x + \dfrac{c}{14}$ (b) Up (c) B

175 + 210 – 240

100 + 120 – 120

300 + 60 –

100 + 60 +

9.

(5000, 50,000) $y = 50,000$ (45,000, 50,000)

$x = 5000$ feasible set $y = 95,000 - x$

(5000, 15,000)

$y = 20,000 - x$ (90,000, 5000)

$y = 5000$

(15,000, 5000)

90,000 gallons of gasoline, 5000 gallons of jet fuel, 5000 gallons of diesel fuel.

11. (a)

	Tom	Dick	Harry
Gardening	8	1	0
Painting	0	7	8

(b)

	Tom	Dick	Harry
Gardening	0	1	8
Painting	8	7	0

(c)

	Tom	Dick	Harry
Gardening	2	0	7
Painting	6	8	1

(d) same as (a)

(e) $A \le D \le C \le B$

13. $\frac{1}{4} \le k \le 3$

CHAPTER 3: SUPPLEMENTARY EXERCISES, page 142

1.

(6, 6) feasible set

(4, 4) (10, 3)

10 planes of type A, 3 planes of type B

3.

(0, 20)

(9, 16)

feasible set (15, 10)

(0, 0) (20, 0)

9 hardtops, 16 sports cars

5.

(51, 17) feasible set

(48, 16)

(60, 8)

60 elementary, 8 intermediate, 4 advanced.

CHAPTER 4

EXERCISES 4.1, page 153

1.
$$\begin{cases} 20x + 30y + u & = 3500 \\ 50x + 10y \quad\; + v & = 5000 \\ -8x - 13y \qquad\quad + M = 0 \end{cases}$$

Find a solution to the linear system for which $x \ge 0, y \ge 0, u \ge 0, v \ge 0$ and M is as large as possible.

3.
$$\begin{cases} x + \quad y + \; z + u & = 100 \\ 3x \qquad\; + \; z \quad\; + v & = 200 \\ 5x + 10y \qquad\quad + w & = 100 \\ -x - \; 2y + 3z \qquad\qquad + M = 0 \end{cases}$$

Find a solution to the linear system for which $x \ge 0, y \ge 0, z \ge 0, u \ge 0, v \ge 0, w \ge 0$ and M is as large as possible.

$$
\begin{array}{c}
\begin{array}{ccccc} x & y & u & v & M \end{array} \\
5.\ \left[\begin{array}{ccccc|c}
20 & 30 & 1 & 0 & 0 & 3500 \\
50 & 10 & 0 & 1 & 0 & 5000 \\
-8 & -13 & 0 & 0 & 1 & 0
\end{array}\right]
\end{array}
\qquad
\begin{array}{c}
\begin{array}{ccccccc} x & y & z & u & v & w & M \end{array} \\
7.\ \left[\begin{array}{ccccccc|c}
1 & 1 & 1 & 1 & 0 & 0 & 0 & 100 \\
3 & 0 & 1 & 0 & 1 & 0 & 0 & 200 \\
5 & 10 & 0 & 0 & 0 & 1 & 0 & 100 \\
-1 & -2 & 3 & 0 & 0 & 0 & 1 & 0
\end{array}\right]
\end{array}
$$

$x = 0, y = 0, u = 3500, v = 5000, M = 0$ $x = 0, y = 0, z = 0, u = 100, v = 200, w = 100, M = 0$

9. $x = 15, y = 0, u = 10, v = 0, M = 20$ **11.** $x = 10, y = 0, z = 15, u = 23, v = 0, w = 0, M = -11$

$$
\begin{array}{c}
\begin{array}{ccccc} x & y & u & v & M \end{array} \\
13.\ (a)\ \left[\begin{array}{ccccc|c}
1 & \frac{3}{2} & \frac{1}{2} & 0 & 0 & 6 \\
0 & -\frac{1}{2} & -\frac{1}{2} & 1 & 0 & 4 \\
0 & -5 & 5 & 0 & 1 & 60
\end{array}\right]
\end{array}
\qquad
\begin{array}{c}
\begin{array}{ccccc} x & y & u & v & M \end{array} \\
(b)\ \left[\begin{array}{ccccc|c}
\frac{2}{3} & 1 & \frac{1}{3} & 0 & 0 & 4 \\
\frac{1}{3} & 0 & -\frac{1}{3} & 1 & 0 & 6 \\
\frac{10}{3} & 0 & \frac{20}{3} & 0 & 1 & 80
\end{array}\right]
\end{array}
$$

$x = 6, y = 0, u = 0, v = 4, M = 60$ $x = 0, y = 4, u = 0, v = 6, M = 80$

$$
\begin{array}{c}
\begin{array}{ccccc} x & y & u & v & M \end{array} \\
(c)\ \left[\begin{array}{ccccc|c}
0 & 1 & 1 & -2 & 0 & -8 \\
1 & 1 & 0 & 1 & 0 & 10 \\
0 & -10 & 0 & 10 & 1 & 100
\end{array}\right]
\end{array}
\qquad
\begin{array}{c}
\begin{array}{ccccc} x & y & u & v & M \end{array} \\
(d)\ \left[\begin{array}{ccccc|c}
-1 & 0 & 1 & -3 & 0 & -18 \\
1 & 1 & 0 & 1 & 0 & 10 \\
10 & 0 & 0 & 20 & 1 & 200
\end{array}\right]
\end{array}
$$

$x = 10, y = 0, u = -8, v = 0, M = 100$ $x = 0, y = 10, u = -18, v = 0, M = 200$

15. (d)

EXERCISES 4.2, page 165

$$
1.\ (a)\ 3\ (b)\
\begin{array}{c}
\begin{array}{ccccc} x & y & u & v & M \end{array} \\
\left[\begin{array}{ccccc|c}
\frac{16}{3} & 0 & 1 & -\frac{2}{3} & 0 & 6 \\
\frac{1}{3} & 1 & 0 & \frac{1}{3} & 0 & 2 \\
0 & 0 & 0 & 4 & 1 & 24
\end{array}\right]
\end{array}
\ (c)\ x = 0, y = 2, u = 6, v = 0, M = 24
$$

$$
3.\ (a)\ 10\ (b)\
\begin{array}{c}
\begin{array}{ccccc} x & y & u & v & M \end{array} \\
\left[\begin{array}{ccccc|c}
-13 & 0 & 1 & -\frac{6}{5} & 0 & 6 \\
\frac{3}{2} & 1 & 0 & \frac{1}{10} & 0 & \frac{1}{2} \\
7 & 0 & 0 & \frac{1}{5} & 1 & 1
\end{array}\right]
\end{array}
\ (c)\ x = 0, y = \frac{1}{2}, u = 6, v = 0, M = 1
$$

5. $x = 0, y = 5, M = 15$ **7.** $x = 12, y = 20, M = 88$ **9.** $x = 0, y = \frac{19}{3}, z = 5, M = 44$
11. $x = 0, y = 30, M = 90$ **13.** $x = 50, y = 100, M = 1300$ **15.** 98 chairs, 4 sofas, 21 tables; $10,640 profit
17. 75 small sofas, 15 large sofas, 30 chairs; max. profit = $6900 **19.** $x = 100, y = 50$; max. value = 45,000

EXERCISES 4.3, page 175

1. $156; x = \frac{3}{5}, y = \frac{22}{5}$ **3.** $8; x = \frac{5}{2}, y = \frac{1}{2}$ **5.** $59; x = 3, y = 5$
7. 1 serving of food A, 3 servings of food B; cost is $7.50 **9.** 100 brand A, 50 brand B, 450 brand C

EXERCISES 4.4, page 183

1. $x = 4, y = 22$; profit = $122 **3.** $x = 25, y = 25$; cost = $250 **5.** $-15 \le h \le 12$

$$
7.\ \left[\begin{array}{ccc}
9 & 1 & 1 \\
4 & 8 & -3
\end{array}\right]
\qquad
9.\ \left[\begin{array}{c}
7 \\ 6 \\ 5 \\ 1
\end{array}\right]
\qquad
11.\ \text{Yes}
$$

$6-$

13. Minimize $\begin{bmatrix} 7 & 5 & 4 \end{bmatrix} \begin{bmatrix} x \\ y \\ z \end{bmatrix}$ subject to

$$\begin{bmatrix} 3 & 8 & 9 \\ 1 & 2 & 5 \\ 4 & 1 & 7 \end{bmatrix} \begin{bmatrix} x \\ y \\ z \end{bmatrix} \geq \begin{bmatrix} 75 \\ 80 \\ 67 \end{bmatrix} \quad \text{and} \quad \begin{bmatrix} x \\ y \\ z \end{bmatrix} \geq \begin{bmatrix} 0 \\ 0 \\ 0 \end{bmatrix}.$$

15. Maximize $\begin{bmatrix} 3 & 5 \end{bmatrix} \begin{bmatrix} x \\ v \end{bmatrix}$ subject to the constraints

$$\begin{bmatrix} 3 & 6 \\ 7 & 5 \\ 4 & 3 \end{bmatrix} \begin{bmatrix} x \\ y \end{bmatrix} \leq \begin{bmatrix} 90 \\ 138 \\ 120 \end{bmatrix} \quad \text{and} \quad \begin{bmatrix} x \\ y \end{bmatrix} \geq \begin{bmatrix} 0 \\ 0 \end{bmatrix}$$

17. Minimize the objective function $2x + 3y$ subject to the constraints

$$\begin{cases} 7x + 4y \geq 33 \\ 5x + 8y \geq 44 \\ x + 3y \geq 55 \\ x \geq 0, \quad y \geq 0 \end{cases}$$

EXERCISES 4.5, page 196

1. Minimize $80u + 76v$ subject to the constraints

$$\begin{cases} 5u + 3v \geq 4 \\ u + 2v \geq 2 \\ u \geq 0, \quad v \geq 0. \end{cases}$$

3. Maximize $u + 2v + w$ subject to the constraints

$$\begin{cases} u - v + 2w \leq 10 \\ 2u + v + 3w \leq 12 \\ u \geq 0, \quad v \geq 0, \quad w \geq 0. \end{cases}$$

5. Maximize $-7u + 10v$ subject to the constraints

$$\begin{cases} -2u + 8v \leq 3 \\ 4u + v \leq 5 \\ 6u + 9v \leq 1 \\ u \geq 0, \quad v \geq 0. \end{cases}$$

7. $x = 12, y = 20, M = 88; u = \frac{2}{7}, v = \frac{6}{7}, M = 88$ **9.** $x = 0, y = 2, M = 24; u = 0, v = 12, M = 24$
11. Maximize $3u + 5v$ subject to the constraints

$$\begin{cases} u + 2v \leq 3 \\ u \leq 1 \\ u \geq 0, \quad v \geq 0 \end{cases}$$

Solution: $x = \frac{5}{2}, y = \frac{1}{2}, M = 8; u = 1, v = 1, M = 8$

13. Minimize $6u + 9v + 12w$ subject to

$$\begin{cases} u + 3v \geq 10 \\ -2u + w \geq 12 \\ v + 3w \geq 10 \\ u \geq 0, \quad v \geq 0, \quad w \geq 0 \end{cases}$$

Solution: $x = 3, y = 12, z = 0, M = 174; u = 0, v = \frac{10}{3}, w = 12, M = 174$

15. Suppose we could hire out our workers at a profit of u dollars per hour, and sell our supply of steel and wood at a profit of v and w dollars per unit, respectively. Then the least acceptable profits are determined by minimizing the objective function $M = 90u + 138v + 120w$ subject to the constraints $3u + 7v + 4w \geq 3, 6u + 5v + 3w \geq 5, u \geq 0, v \geq 0, w \geq 0$.

17. Suppose we could buy coal from another miner at the costs of $u, v,$ and w dollars per ton for ordinary coal, bituminous coal, and anthracite, respectively, and supply it to our customer. The other miner would want to maximize $M = 80u + 60v + 75w$ subject to the constraints $4u + 4v + 7w \leq 150, 10u + 5v + 5w \leq 200, u \geq 0, v \geq 0, w \geq 0$.

19. \$3.63

CHAPTER 4: SUPPLEMENTARY EXERCISES, page 199

1. $x = 2, y = 3, M = 18$ **3.** $x = 4, y = 5, M = 23$ **5.** $6; x = 5, y = 1$ **7.** $110; x = 4, y = 1$
9. $x = 1, y = 6, z = 8, M = 884$
11. Minimize $14u + 9v + 24w$ subject to the constraints.

$$\begin{cases} u + v + 3w \geq 2 \\ 2u + v + 2w \geq 3 \\ u \geq 0, \quad v \geq 0, \quad w \geq 0. \end{cases}$$

13. $x = 4, y = 5, M = 23; u = 1, v = 1, w = 0, M = 23.$

15. $A = \begin{bmatrix} 1 & 2 \\ 1 & 1 \\ 3 & 2 \end{bmatrix}, B = \begin{bmatrix} 14 \\ 9 \\ 24 \end{bmatrix}, C = \begin{bmatrix} 2 & 3 \end{bmatrix}, X = \begin{bmatrix} x \\ y \end{bmatrix}.$ Maximize CX subject to $AX \leq B, X \geq \mathbf{0}.$ Dual: $U = \begin{bmatrix} u & v & w \end{bmatrix}.$
Minimize UB subject to $UA \leq C, U \geq \mathbf{0}.$
17. \$11.00

CHAPTER 5

EXERCISES 5.1, page 207

1. (a) $\{5, 6, 7\}$ (b) $\{1, 2, 3, 4, 5, 7\}$ (c) $\{1, 3\}$ (d) $\{5, 7\}$ **3.** (a) $\{a, b, c, d, e, f\}$ (b) $\{c\}$ (c) $\emptyset$ **5.** $\{1, 2\}, \{1\}, \{2\}, \emptyset$
7. (a) {all male college students who like football} (b) {all female college students} (c) {all female college students who dislike football} (d) {all college students who are either male or like football}
9. (a) $\{1976, 1975, 1967, 1963, 1958, 1951, 1950\}$ (b) $\{1976, 1975, 1967, 1963, 1961, 1958, 1955, 1954, 1951, 1950\}$
(c) $\{1976, 1975, 1967, 1963, 1958, 1951, 1950\}$ (d) $\{1961, 1955, 1954\}$ (e) $\emptyset$
11. From 1950 to 1977, whenever the Standard and Poor's Index increased by 2 or more percent during the first five days of a year, it always increased by at least 16% for that year.
13. (a) $\{d, f\}$ (b) $\{a, b, c, e, f\}$ (c) $\emptyset$ (d) $\{a, c\}$ (e) $\{e\}$ (f) $\{a, c, e, f\}$ (g) $\{a, b, c, e\}$ (h) $\{a, c\}$ (i) $\{d\}$
15. S **17.** U **19.** $\emptyset$ **21.** $L \cup T$ **23.** $P \cap L$ **25.** $P \cap L \cap T$ **27.** S' **29.** $S \cup D \cup A$ **31.** $(A \cap S)' \cap D$
33. Mount College's male students **35.** Mount College's teachers who are also students
37. Mount College's males and students **39.** Mount College's females **41.** S' **43.** $(V \cup C) \cap S'$
45. $V' \cap C'$ **47.** (a)$\{B, C, D, E\}$ (b)$\{C, D, E, F\}$ (c)$\{A, D, E, F\}$ (d)$\{A, C, D, E, F\}$ (e)$\{A, F\}$ (f)$\{D, E\}$ **49.** $\{2\}$
51. S is a subset of T

EXERCISES 5.2, page 216

1. 7 **3.** 0 **5.** 11 **7.** *S* is a subset of *T* **9.** 14 million **11.** 19,000 **13.** 452

15.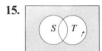

17.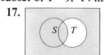

19.

21.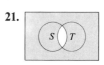

23.

25.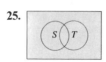

27.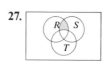

29.

31.

33.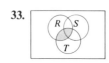

35.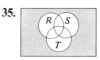

37.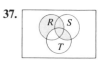

39. $S' \cup T'$ **41.** $S \cap T'$ **43.** U **45.** S' **47.** $R \cap T$ **49.** $T \cap S \cap R'$ **51.** $T \cup (R \cap S')$
53. $(R \cap S \cap T) \cup (R' \cap S' \cap T')$

EXERCISES 5.3, page 222

1.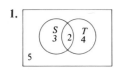

3.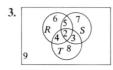

5.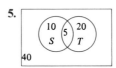

7.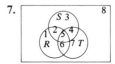

9. 25 **11.** 2, 5 **13.** 30 **15.** 4 **17.** 4, 2, 1 **19.** 28 **21.** 12 **23.** 2000 **25.** 200 **27.** 1600 **29.** 90 **31.** 6
33. 140 **35.** 30 **37.** 190 **39.** 180 **41.** 210 **43.** 35 **45.** Between 47% and 67%

EXERCISES 5.4, page 229

1. 15 **3.** 676 **5.** 380 **7.** 20 **9.** 120 **11.** 64 **13.** 6840 **15.** 870 **17.** 24 **19.** 32 **21.** 360,000
23. (a) 2401 (b) 840 (c) 343 (d) 240 **25.** (a) 362,880 (b) 40,320 (c) 720 **27.** 256 **29.** 3^6 **31.** 168 **33.** 63,973

EXERCISES 5.5, page 236

1. 12 **3.** 120 **5.** 120 **7.** 5 **9.** 720 **11.** 36 **13.** 24 **15.** 35 **17.** 15,600 **19.** $C(100, 15)$ **21.** $C(52, 5)$
23. Yes; Moe 36, Joe 35 **25.** 120

EXERCISES 5.6, page 242

1. (a) 64 (b) 20 (c) 22 (d) 57 **3.** 126 **5.** (a) 120 (b) 56 (c) 64 **7.** $C(100, 25) \cdot C(75, 40)$ **9.** 60
11. $C(50, 15) \cdot 2^{15}$ **13.** 24 **15.** 3744 **17.** 3050 **19.** 45^5

EXERCISES 5.7, page 249

1. 15 **3.** 1 **5.** 816 **7.** 1 **9.** $x^7 + 7x^6y + 21x^5y^2 + 35x^4y^3 + 35x^3y^4 + 21x^2y^5 + 7xy^6 + y^7$
11. $x^{10}, 10x^9y, 45x^8y^2$ **13.** 64 **15.** 16 **17.** 3050 **19.** 5040 **21.** (a) 14,520 (b) $C(12, 5) \cdot 2^5 = 25,344$ **27.** 32

EXERCISES 5.8, page 255

1. 20 **3.** 180 **5.** 210 **7.** 34,650 **9.** 166,320 **11.** 1,401,400 **13.** 2,858,856 **15.** $\dfrac{20!}{7!5!8!}$ **17.** $\dfrac{30!}{10!2!18!}$

19. $\dfrac{1}{4!} \cdot \dfrac{20!}{(5!)^4}$ **21.** 12 **23.** 135,135

CHAPTER 5: SUPPLEMENTARY EXERCISES, page 257

1. $\{a, b\}, \{a\}, \{b\}, \varnothing$ **3.** 120   **5.**   **7.** 840   **9.** 35   **11.** 0   **13.** 6   **15.** 47   **17.** 46   **19.** 22

21. 126   **23.** 550   **25.** 628   **27.** $\binom{4}{3}\binom{48}{2}$ **29.** 6   **31.** 12   **33.** The second teacher   **35.** 5

CHAPTER 6

EXERCISES 6.2, page 269

1. (a) {RS, RT, RU, RV, ST, SU, SV, TU, TV, UV} (b) {RS, RT, RU, RV} (c) {TU, TV, UV}
3. (a) {HH, HT, TH, TT} (b) {HH, HT}
5. (a) {(Urn I, red),(urn I, white), (urn II, red), (urn II, white)} (b) {(Urn I, red), (urn I, white)}
7. (a) {all positive numbers} (b) "More than 5 but less than 8 minutes," $\varnothing$, "5 minutes or less," "8 minutes or more," "5 minutes or less," "less than 4 minutes," S
9. (a) {(1, 1), (1, 2), (1, 3), (1, 4), (2, 1), (2, 2), (2, 3), (2, 4), (3, 1), (3, 2), (3, 3), (3, 4), (4, 1), (4, 2), (4, 3), (4, 4)} (b) (i) {(2, 2), (2, 4), (4, 2), (4, 4)} (ii) {(1, 2), (1, 4), (2, 1), (2, 2), (2, 3), (2, 4), (3, 2), (3, 4), (4, 1), (4, 2), (4, 3), (4, 4)} (iii) {(3, 3), (3, 4), (4, 3), (4, 4)} (iv) {(2, 4), (3, 3), (4, 2)} (v) {(1, 4), (2, 3), (2, 4), (3, 2), (3, 3), (3, 4), (4, 1), (4, 2), (4, 3), (4, 4)} (vi) {(1, 1), (2, 2), (3, 3), (4, 4)} (vii) {(1, 2), (1, 3), (2, 1), (2, 2), (2, 4), (3, 1), (3, 3), (3, 4), (4, 2), (4, 3)} (viii) {(1, 1), (1, 2), (1, 3), (2, 1), (2, 2), (2, 3), (3, 1), (3, 2), (3, 3)}
11. (a) No (b) Yes   **13.** $S, \{a, b\}, \{a, c\}, \{b, c\}, \{a\}, \{b\}, \{c\}, \varnothing$ **15.** Yes   **17.** (a) {0, 1, 2, . . . , 10} (b) {6, 7, 8, 9, 10}
19. (a) No (b) Yes (c) Yes   **21.** The set of nonnegative integers   **23.** The set of nonnegative numbers

EXERCISES 6.3, page 280

1. (a) $\dfrac{46{,}277}{774{,}746}$ (b) $\dfrac{48{,}132}{774{,}746}$ (c) $\dfrac{726{,}614}{774{,}746}$ **3.** (a) $\frac{5}{36}$ (b) $\frac{1}{6}$ **5.** $\frac{1}{19}$ **7.** $\frac{1}{6}$ **9.** (a) .7 (b) .7   **11.** (a) $\frac{10}{11}$ (b) $\frac{1}{3}$ (c) $\frac{4}{9}$

13. 9 to 91   **15.** .61, .39   **17.** (a) .7 (b) .2   **19.** (a) .7 (b) 7000   **21.**

Failures in:	Prob.
Month 1	.05
Month 2	.05
Month 3	.10
Month 4	.05
Month 5	.05
Month 6	.02
No failures in months 1–6	.68

EXERCISES 6.4, page 288

1. $\frac{5}{11}$ **3.** $\frac{25}{42}$ **5.** $1 - \dfrac{30 \cdot 29 \cdot 28 \cdot 27}{30^4} = .188$ **7.** $\frac{1}{2}$ **9.** (a) .25 (b) .75 (c) .8   **11.** 0   **13.** $\frac{1}{11}$ **15.** $\frac{2}{5}$

17. (a) $\frac{15}{28}$ (b) $\frac{15}{56}$ (c) $\frac{9}{56}$ (d) $\frac{9}{14}$ **19.** $\frac{5}{6}$ **21.** $\frac{1}{3}$ **23.** $2 \Big/ \binom{40}{6}$ **25.** (a) $\frac{3}{55}$ (b) $\frac{18}{55}$ (c) $\frac{16}{55}$ (d) $\frac{30}{55}$

27. (a) $\frac{7}{128}$ (b) .43 or $1 - (\frac{121}{128})^{10}$ (c) $\frac{121}{128}$ **29.** $.5165 = 1 - C(12, 3)/C(15, 3)$ **31.** 55.6%

EXERCISES 6.5, page 301

1. $\frac{1}{3}, \frac{1}{5}$ 3. $\frac{4}{7}$ 5. No 7. (a) .36 (b) .81 9. $(.99)^5(.98)^5(.975)^3$ 11. $\frac{3}{4}, \frac{1}{2}$ 13. (a) $\frac{2}{5}$ (b) $\frac{3}{5}$ (c) $\frac{3}{4}$ 15. $\frac{1}{4}$ 17. .94
21. .5, .25, .25 25. (a) .28 (b) .07 (c) .6065 27. (a) $\frac{40}{250}$ (b) $\frac{160}{250}$ (c) $\frac{90}{250}$ (d) $\frac{50}{120}$ (e) $\frac{70}{110}$ (f) $\frac{40}{130}$ 29. (a) .4 (b) .56 (c) .1818
(d) .7 (e) .45 (f) .67

EXERCISES 6.6, page 310

1. 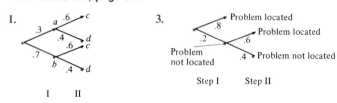   3. 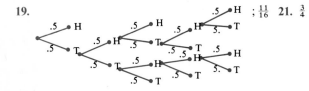   5. .08 7. .295

9. $\frac{7}{12}$ 11. $\frac{4}{5}$ 13. (a) .026 (b) $\frac{9}{13}$ 15. .00473 17. $1 - (.9999)^3$

19. ; $\frac{11}{16}$ 21. $\frac{3}{4}$

EXERCISES 6.7, page 315

1. $\frac{40}{265}$ 3. $\frac{3}{7}$ 5. .075 7. .92 9. (a) .1325 (b) $\approx$.23 11. $\frac{115}{2475} \sim$ .046 13. (a) $\frac{1}{4}$ (b) 76% (c) 13%

CHAPTER 6: SUPPLEMENTARY EXERCISES, page 319

1. $\frac{31}{32}$ 3. $\frac{4}{9}$ 5. $\frac{2}{5}$ 7. (a) $\frac{1}{12}$ (b) $\frac{1}{2}$ 9. $\frac{1}{12}$ 11. No 13. $\frac{1}{3}$ 15. $\frac{1}{6}$ 17. $\frac{2}{3}$ 19. .52 21. .0000083

CHAPTER 7

EXERCISES 7.1, page 332

1. Grade	Rel. freq.
0	.08
1	.12
2	.40
3	.24
4	.16

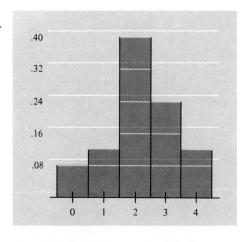

3. No. Calls	Rel. freq.
20	.05
21	.05
22	.00
23	.10
24	.30
25	.20
26	.00
27	.15
28	.10
29	.05

5.

No. Heads	Rel. freq.
0	$\frac{1}{8}$
1	$\frac{3}{8}$
2	$\frac{3}{8}$
3	$\frac{1}{8}$

7.

No. Shots	Prob.
1	$\frac{1}{3}$
2	$\frac{2}{9}$
3	$\frac{4}{27}$
4	$\frac{8}{27}$

9.

Earnings	Prob.
$5	$\frac{1}{15}$
$1	$\frac{8}{15}$
$-$1	$\frac{2}{5}$

11. .6

13.

k	$\Pr(X^2 = k)$
0	.1
1	.2
4	.3
9	.2
16	.2

15.

k	$\Pr(X - 1 = k)$
-1	.1
0	.2
1	.3
2	.2
3	.2

17.

k	$\Pr(\frac{1}{5} Y = k)$
1	.3
2	.4
3	.1
4	.1
5	.1

19.

k	$\Pr((X + 1)^2 = k)$
1	.1
4	.2
9	.3
16	.2
25	.2

EXERCISES 7.2, page 343

1. 2.35   **3.** (a) 2.9 (b)

Grade	Prob.
4	.3
3	.4
2	.2
1	.1

(c) 2.9   **5.** A

7.

Earnings	Prob.
$35	$\frac{1}{38}$
$-$1	$\frac{37}{38}$

, $\mu = -5.26$ cents

9.

Earnings	Prob.
$-50¢$	$\frac{1}{3}$
0¢	$\frac{4}{15}$
50¢	$\frac{1}{5}$
$1.00	$\frac{2}{15}$
$1.50	$\frac{1}{15}$

, $\mu = 16.67$ cents

11. $1000   **13.** $\frac{161}{36}$

EXERCISES 7.3, page 353

1. 1.4 **3.** B **5.** (a) $\mu_A = 15$, $\sigma_A^2 = 160$, $\mu_B = 13$, $\sigma_B^2 = 141$ (b) A (c) B
7. (a) $\mu_A = 103$, $\sigma_A^2 = 4.6$, $\mu_B = 104$, $\sigma_B^2 = 3.4$ (b) B (c) B **9.** (a) $\geq \frac{3}{4}$ (b) $\geq \frac{8}{9}$ (c) $\geq \frac{11}{36}$ **11.** ≥ 4688
13. 8 **15.** (a) 7, $\frac{35}{6}$ (b) $\frac{5}{6}$ (c) $\geq \frac{19}{54}$

17.

k	$\Pr(X^2 = k)$	$k \cdot \Pr(X^2 = k)$
0	.1	0
1	.3	.3
4	.5	2
9	.1	.9
		$3.2 = E(X^2)$

$$E(X^2) - \mu^2 = 3.2 - (1.6)^2$$
$$= 3.2 - 2.56$$
$$= .64$$

19.

k	$\Pr(2X = k)$	$(k - \mu)^2$
-2	$\frac{1}{8}$	4
-1	$\frac{3}{8}$	1
0	$\frac{1}{8}$	0
1	$\frac{1}{8}$	1
2	$\frac{2}{8}$	4

$$\mu = -2(\tfrac{1}{8}) + (-1)\tfrac{3}{8} + 0(\tfrac{1}{8}) + 1(\tfrac{1}{8}) + 2(\tfrac{2}{8})$$
$$= 0$$
$$\text{Variance of } 2X = 4(\tfrac{1}{8}) + 1(\tfrac{3}{8}) + 0(\tfrac{1}{8}) + 1(\tfrac{1}{8}) + 4(\tfrac{2}{8})$$
$$= 2 = 4(\tfrac{1}{2})$$
$$= 4(\text{variance of } X)$$

EXERCISES 7.4, page 362

1. $\frac{25}{216}$ **3.** $\frac{3}{64}$ **5.** $\frac{992}{3125}$ **7.** .087; selection procedure might have been biased **9.** .8507 **11.** .5583 **13.** $\frac{50}{51}$

EXERCISES 7.5, page 375

1. .8944 **3.** .4013 **5.** .2417 **7.** .6170 **9.** $z = 1.75$ **11.** $z = 0.75$ **13.** $\mu = 6$, $\sigma = 2$ **15.** $\mu = 9$, $\sigma = 1$
17. $-\frac{8}{3}$ **19.** $15\frac{1}{2}$ **21.** .9772 **23.** .6247 **25.** .0002 **27.** .9876 **29.** .0122 **31.** (a) 618 (b) between 397 and 643
33. 19,750 miles **35.** (a) 5.35 ounces (b) 4.76 ounces

EXERCISES 7.6, page 384

1. (a) .1974 (b) .7888 (c) .9878 **3.** .0062 **5.** .0062 **7.** .0013 **9.** .6368 **11.** .1056 **13.** .2; .0401

CHAPTER 7: SUPPLEMENTARY EXERCISES, page 386

1. (a), (b)

k	$\Pr(X = k)$	$k \cdot \Pr(X = k)$	$(k - \mu)^2$	$(k - \mu)^2 \Pr(X = k)$
0	$\frac{8}{27}$	0	1	$\frac{8}{27}$
1	$\frac{12}{27}$	$\frac{12}{27}$	0	0
2	$\frac{6}{27}$	$\frac{12}{27}$	1	$\frac{6}{27}$
3	$\frac{1}{27}$	$\frac{3}{27}$	4	$\frac{4}{27}$
		$\mu = 1$		$\sigma^2 = \frac{2}{3}$

3. .2857 **5.** $\geq \frac{8}{9}$ **7.** 10.56% **9.** 0.0122 **11.** $z = 102.5$ **13.** .13, .25, .31, .31

CHAPTER 8

EXERCISES 8.1, page 398

1. Stochastic **3.** Not stochastic **5.** Stochastic **7.** 36, 34 (precisely 34.4) **9.** Same as original distribution

11. (a) $\begin{array}{c} \\ S \\ M \\ L \end{array} \begin{array}{c} S \quad M \quad L \\ \begin{bmatrix} .4 & .5 & .3 \\ .6 & 0 & .2 \\ 0 & .5 & .5 \end{bmatrix} \end{array}$ (b) 44% **13.** (a) $\begin{array}{c} \\ D \\ R \end{array} \begin{array}{c} D \quad R \\ \begin{bmatrix} .7 & .4 \\ .3 & .6 \end{bmatrix} \end{array}$ (b) $\begin{bmatrix} .61 & .52 \\ .39 & .48 \end{bmatrix}, \begin{bmatrix} .583 & .556 \\ .417 & .444 \end{bmatrix}$ (c) 58.3%

15. All are same as given matrix

17. $\begin{bmatrix} .1 & .3 \\ .9 & .7 \end{bmatrix}, \begin{bmatrix} .28 & .24 \\ .72 & .76 \end{bmatrix}, \begin{bmatrix} .24 & .25 \\ .76 & .75 \end{bmatrix}, \begin{bmatrix} .25 & .25 \\ .75 & .75 \end{bmatrix}, \begin{bmatrix} .25 & .25 \\ .75 & .75 \end{bmatrix}$

19. All are same as given matrix **21.** Regular (A^2 has no zero entries)

EXERCISES 8.2, page 407

1. Regular **3.** Regular **5.** Regular **7.** $\begin{bmatrix} \frac{1}{6} \\ \frac{5}{6} \end{bmatrix}$ **9.** $\begin{bmatrix} \frac{3}{5} \\ \frac{2}{5} \end{bmatrix}$ **11.** $\begin{bmatrix} \frac{5}{14} \\ \frac{6}{14} \\ \frac{3}{14} \end{bmatrix}$ **13.** 40% **15.** 25%

EXERCISES 8.3, page 416

1. No **3.** Yes **5.** $R = [.5], S = \begin{bmatrix} .3 \\ .2 \end{bmatrix}; [I - R]^{-1} = [2]; \begin{bmatrix} 1 & 0 & .6 \\ 0 & 1 & .4 \\ 0 & 0 & 0 \end{bmatrix}$

7. $R = \begin{bmatrix} .3 & .6 \\ .1 & .2 \end{bmatrix}, S = \begin{bmatrix} .1 & 0 \\ .5 & .2 \end{bmatrix}; [I - R]^{-1} = \begin{bmatrix} 1.6 & 1.2 \\ .2 & 1.4 \end{bmatrix}; \begin{bmatrix} 1 & 0 & .16 & .12 \\ 0 & 1 & .84 & .88 \\ 0 & 0 & 0 & 0 \\ 0 & 0 & 0 & 0 \end{bmatrix}$

9. $R = \begin{bmatrix} .5 & 0 \\ .1 & .6 \end{bmatrix}, S = \begin{bmatrix} .1 & .2 \\ .3 & 0 \\ 0 & .2 \end{bmatrix}, [I - R]^{-1} = \begin{bmatrix} 2 & 0 \\ .5 & 2.5 \end{bmatrix}; \begin{bmatrix} 1 & 0 & 0 & .3 & .5 \\ 0 & 1 & 0 & .6 & 0 \\ 0 & 0 & 1 & .1 & .5 \\ 0 & 0 & 0 & 0 & 0 \\ 0 & 0 & 0 & 0 & 0 \end{bmatrix}$

11. (a) $\begin{array}{c} \\ D \\ G \\ F \\ S \end{array} \begin{array}{c} D \quad G \quad F \quad S \\ \begin{bmatrix} 1 & 0 & .2 & .1 \\ 0 & 1 & 0 & .9 \\ 0 & 0 & 0 & 0 \\ 0 & 0 & .8 & 0 \end{bmatrix} \end{array}$ (b) $\begin{bmatrix} 1 & 0 & .28 & .1 \\ 0 & 1 & .72 & .9 \\ 0 & 0 & 0 & 0 \\ 0 & 0 & 0 & 0 \end{bmatrix}$ (c) .72

1. Stochastic, neither   **3.** Stochastic, regular   **5.** Not stochastic   **7.** $\begin{bmatrix} \frac{5}{9} \\ \frac{4}{9} \end{bmatrix}$ **9.** (a) $\begin{bmatrix} .5 & .4 & .3 \\ .4 & .3 & .5 \\ .1 & .3 & .2 \end{bmatrix}$ (b) 38% (c) $\frac{19}{97}$

11. $\begin{bmatrix} 1 & 0 & \frac{11}{12} & 1 & \frac{1}{2} \\ 0 & 1 & \frac{1}{12} & 0 & \frac{1}{2} \\ 0 & 0 & 0 & 0 & 0 \\ 0 & 0 & 0 & 0 & 0 \\ 0 & 0 & 0 & 0 & 0 \end{bmatrix}$ **13.** (c)

CHAPTER 9

EXERCISES 9.1, page 428

1. R: row 2, C: column 1   **3.** R: row 2, C: column 1   **5.** R: row 1, C: column 1   **7.** (a) 0: row 1, column 2 (b) 0

9. $\begin{array}{c} \\ H \\ T \end{array} \begin{array}{cc} H & T \\ \begin{bmatrix} 2 & -1 \\ -1 & -4 \end{bmatrix} \end{array}$; strictly determined; R shows heads, C shows tails

11. $\begin{array}{c} \\ F \\ A \\ N \end{array} \begin{array}{ccc} F & A & N \\ \begin{bmatrix} 8000 & -1000 & 1000 \\ -7000 & 4000 & -2000 \\ 3000 & 3000 & 2000 \end{bmatrix} \end{array}$; strictly determined; both should be neutral

13. $\begin{array}{c} \\ 5 \\ 10 \end{array} \begin{array}{ccc} 6 & 7 & 8 \\ \begin{bmatrix} 1 & -5 & -5 \\ -6 & 3 & 2 \end{bmatrix} \end{array}$; not strictly determined

EXERCISES 9.2, page 436

1. (a) 0 (b) 1 (c) -1.12 (d) .5 (The most advantageous is (b).)   **3.** $29,200   **5.** 0

EXERCISES 9.3, page 447

1. $\begin{bmatrix} \frac{1}{2} & \frac{1}{2} \end{bmatrix}$; $\begin{bmatrix} \frac{1}{4} \\ \frac{3}{4} \end{bmatrix}$ **3.** $\begin{bmatrix} \frac{1}{2} & \frac{1}{2} \end{bmatrix}$; $\begin{bmatrix} \frac{5}{9} \\ \frac{4}{9} \end{bmatrix}$ **5.** $\begin{bmatrix} \frac{2}{5} & \frac{3}{5} \end{bmatrix}$; $\begin{bmatrix} \frac{3}{5} \\ \frac{2}{5} \end{bmatrix}$ **7.** $\begin{bmatrix} \frac{7}{13} & \frac{6}{13} \end{bmatrix}$ **9.** $\begin{bmatrix} 0 \\ 1 \end{bmatrix}$ **11.** (a) $\begin{bmatrix} \frac{8}{17} & \frac{9}{17} \end{bmatrix}$ (b) $\begin{bmatrix} \frac{8}{17} \\ \frac{9}{17} \end{bmatrix}$ (c) $\approx$ $2765

13. $\begin{bmatrix} \frac{5}{8} & \frac{3}{8} & 0 \end{bmatrix}$; $\begin{bmatrix} \frac{1}{2} \\ \frac{1}{2} \end{bmatrix}$

1. Strictly determined; R: row 3, C: column 3; 2   **3.** Not strictly determined   **5.** 7   **7.** 1.4   **9.** $\begin{bmatrix} \frac{4}{11} & \frac{7}{11} \end{bmatrix}$; $\begin{bmatrix} \frac{6}{11} \\ \frac{5}{11} \end{bmatrix}$

11. $\begin{bmatrix} 0 & 1 \end{bmatrix}$ **13.** (a) $\begin{bmatrix} \frac{5}{14} & \frac{9}{14} \end{bmatrix}$, $\begin{bmatrix} \frac{9}{14} \\ \frac{5}{14} \end{bmatrix}$ (b) Carol

CHAPTER 10

EXERCISES 10.1, page 457

1. (a) $i = .01, n = 24$ (b) $i = .02, n = 20$ (c) $i = .05, n = 40$
3. (a) $i = .06, n = 4, P = \$500, F = \$631.24$ (b) $i = .02, n = 120, P = \$800, F = \$8612.13$
(c) $i = .06, n = 19, P = \$2974.62, F = \$9000$
5. \$1127.16 **7.** \$5053.45 **9.** \$8584.61; \$2584.61 **11.** (a) \$3548.74 (b) \$451.26 (c)

Month	Interest	Balance
0		\$3548.74
1	17.74	3566.48
2	17.83	3584.31
3	17.92	3602.23

13. \$42,918.71 **15.** \$8874.49 **17.** (a) 12% (b) \$10.20, \$1030.30 (c) \$12.57, \$1269.73
19. \$1700 in 9 years **21.** 30% compounded annually **23.** \$7884.93
25. (a) $r = .07, n = \frac{1}{2}, P = \$500, A = \$517.50$ (b) $r = .08, n = 2, P = \$500, A = \$580$ **27.** \$1270
29. \$2000 **31.** 50% **33.** 10 years **35.** $P = A/(1 + rn)$ **37.** \$126.25, 26.25% **39.** 12.36% **41.** 6.168%
43. $a + (a^2/4)$ **45.** $(1 + a/n)^n - 1$

EXERCISES 10.2, page 467

1. (a) 31 (b) 11 (c) $\frac{31}{16}$ (d) $\frac{93}{16}$
3. (a) $i = .005, n = 120, R = \$50, F = \$8193.97$ (b) $i = .06, n = 20, R = \$1767, F = \$65,000$
5. \$8166.97 **7.** \$4698.97 **9.** (a) \$30,611.30 (b) \$6,611.30 (c)

Month	Interest	Balance
1		500.00
2	5.00	1005.00
3	10.05	1515.05

11. \$253.83; \$9137.88 deposited; \$2862.12 interest **13.** \$200 per month is worth \$323.79 more **15.** \$12,648.91
17. \$877.91 **19.** \$1000 at end of each month for 10 years **21.** \$7239.33 **23.** \$17,584.61 **25.** \$10,000
27. \$4339.35 **29.** \$59,189.85 **31.** (a) \$1 (b) $i \cdot s_{\overline{n}|i}$ **35.** \$3623.52 **37.** (a) \$1269.67 (b) \$65,134.88

EXERCISES 10.3, page 475

1. \$287.68 **3.** \$11,469.92 **5.** (a) \$583.31 (b) \$16.69 (c) \$58,314.31 (d) \$26,973.02 (e) \$4188.65 (f) \$269.73
7. (a) 265.71 (b) \$9565.56 (c) \$1565.56 (d) \$5644.58 (e) \$2990.59 (f) \$534.53

(g)

Payment number	Amount	Interest	Applied to Principal	Unpaid balance
1	\$265.71	\$80.00	\$185.71	\$7814.29
2	265.71	78.14	187.57	7626.72
3	265.71	76.27	189.44	7437.28
4	265.71	74.37	191.34	7245.94

9. \$9000 **11.** \$14,042.36 **15.** \$2001.60 **17.** \$65,900 **19.** \$2039.89
21. (a) \$183,927.96 (b) \$24,989.92 (c) \$105,266.63 **23.** \$2.5 billion
25. Yes. The sinking fund will have a surplus of \$12.26 million.

CHAPTER 10: SUPPLEMENTARY EXERCISES, page 479

1. $347.77 **3.** $26,541.30 **5.** 10% compounded annually **7.** (a) $2400.34 (b) $167,304.95 **9.** $15,149.74
11. $293.42 **13.** $1200.17 **15.** $349,496.41 **17.** Investment A **19.** 10.25% **21.** $146,861.85
23. $151,843.34 **25.** 265,720

CHAPTER 11

EXERCISES 11.1, page 488

1. $4, -6; 2$ **3.** $-\frac{1}{2}, 0; 0$ **5.** $-\frac{2}{3}, 15; 9$

7. (a) $10, 4, 1, -\frac{1}{2}, -\frac{5}{4}$ (b) (c) $y_n = -2 + 12(\frac{1}{2})^n$

9. (a) $3.5, 4, 5, 7, 11$ (b) (c) $y_n = 3 + (.5)2^n$

11. (a) $17.5, 0, 7, 4.2, 5.32$ (b) 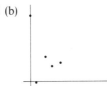 (c) $y_n = 5 + (12.5)(-.4)^n$

13. (a) $15, 14, 12, 8, 0$ (b) 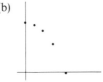 (c) $y_n = 16 - 2^n$ **15.** $1, 5, 5.8, 5.96, 5.992$

17. $y_{n+1} = 1.05y_n, y_0 = 1000$ **19.** $y_{n+1} = 1.05y_n + 100, y_0 = 1000$

21. (a) $1, 3, 5, 7, 9$ (b) (c) Since $a = 1$, $1 - a = 0$ and so $\dfrac{b}{1-a} = \dfrac{2}{0}$, which is not defined

23. $30.

EXERCISES 11.2, page 496

1. $y_n = 1 + 5n$ **3.** $80(1.0075)^{60}$ **5.** $80(1 + \frac{1}{365})^{1825}$ **7.** 108 **9.** $A\left(1 + \dfrac{r}{k}\right)^{kt}$

11.

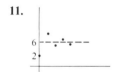

Notice that $\dfrac{b}{1-a} = \dfrac{-10}{1-2} = 10$ and that when $y_0 \neq 10$, the terms are repelled (i.e., move away) from the line $y = 10$.

13. $y_n = 5 + 2(.4)^n$; y_n approaches 5
15. $y_n = 2(-5)^n$; y_n gets arbitrarily large and alternates between being positive and negative
17. $y_{n+1} = 1.0075y_n - 350$, $y_0 = 38,900$ **19.** $y_{n+1} = y_n - 2000$, $y_0 = 50,000$; $y_n = 50,000 - 2000n$

EXERCISES 11.3, page 506

1. A, B, D, F, H **3.** B, D, E, F **5.** B, D, E, F **7.** A, C, H, (possibly G) **9.**

11.

13.

15.

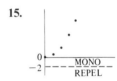

17.

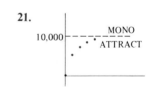

19.

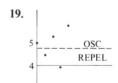

21.

23.

25. Less than $60,000 **27.** (a) $y_{n+1} = 1.06y_n - 120$ (b) At least $2000

EXERCISES 11.4, page 513

1. $y_{n+1} = 1.0075y_n - 261.50$, $y_0 = 32,500$ **3.** $y_{n+1} = 1.015y_n + 200$, $y_0 = 4000$ **5.** $46,000 **7.** $11,000
9. $2000 **11.** $133.02

1.

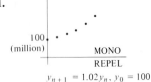

$y_{n+1} = 1.02y_n, y_0 = 100$

3.

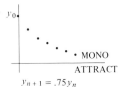

$y_{n+1} = .75y_n$

5.

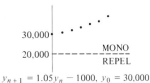

$y_{n+1} = .92y_n + 8, y_0 = 0$

7.

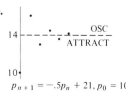

$y_{n+1} = .7y_n + 3.6, y_0 = 0$

9.

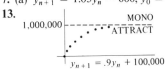

$y_{n+1} = 1.05y_n - 1000, y_0 = 30,000$

11.

$y_{n+1} = .8y_n + 14, y_0 = 40$

13.

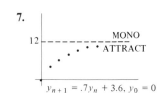

$p_{n+1} = -.5p_n + 21, p_0 = 10$

CHAPTER 11: SUPPLEMENTARY EXERCISES, page 520

1. (a) $5, -7, 29$ (b) $y_n = 2 - (-3)^n$ (c) -79 **3.** \$3000 **5.**

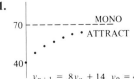

7. (a) $y_{n+1} = 1.03y_n - 600, y_0 = 120,000$ (b) $200,000$ **9.** \$20 **11.** \$22,500

13.

$y_{n+1} = .9y_n + 100,000$

CHAPTER 12

EXERCISES 12.1, page 527

1. statement **3.** statement **5.** not a statement **7.** not a statement **9.** statement **11.** not a statement **13.** not a statement **15.** statement

17. (a) Ozone is opaque to ultraviolet light and life on earth requires ozone. (b) Ozone is not opaque to ultraviolet light or life on earth requires ozone. (c) Ozone is not opaque to ultraviolet light or life on earth does not require ozone. (d) Life on earth requires ozone. **19.** (a) $p \vee q$ (b) $p \wedge \sim q$ (c) $q \wedge \sim p$ (d) $\sim p \wedge \sim q$

EXERCISES 12.2, page 535

1.

p	q	$p \wedge \sim q$
T	T	F
T	F	T
F	T	F
F	F	F

3.

p	q	r	$(p \wedge \sim r) \vee q$
T	T	T	T
T	T	F	T
T	F	T	F
T	F	F	T
F	T	T	T
F	T	F	T
F	F	T	F
F	F	F	F

5.

p	q	r	$\sim[(p \wedge r) \vee q]$
T	T	T	F
T	T	F	F
T	F	T	F
T	F	F	T
F	T	T	F
F	T	F	F
F	F	T	T
F	F	F	T

7. always TRUE

9.

p	q	r	$p \oplus (q \vee r)$
T	T	T	F
T	T	F	F
T	F	T	F
T	F	F	T
F	T	T	T
F	T	F	T
F	F	T	T
F	F	F	F

11.

p	q	r	$(p \vee q) \wedge (p \vee r)$
T	T	T	T
T	T	F	T
T	F	T	T
T	F	F	T
F	T	T	T
F	T	F	F
F	F	T	F
F	F	F	F

13. always FALSE

15. TRUE in the case where p is F, q is F, and r is T, and FALSE otherwise.

17.

p	q	r	$\sim p \vee (q \wedge r)$
T	T	T	T
T	T	F	F
T	F	T	F
T	F	F	F
F	T	T	T
F	T	F	T
F	F	T	T
F	F	F	T

19. They are identical.   **21.** They are identical.

23. $(p \wedge q) \vee r$ is T and $p \wedge (q \vee r)$ is F when p is F and r is T. Otherwise, the tables are identical.

25. (a)

p	$p \vert p$
T	F
F	T

(b–d)

p	q	$(p\vert p)\vert(q\vert q)$	$(p\vert q)\vert(p\vert q)$	$p\vert((p\vert q)\vert q)$
T	T	T	T	F
T	F	T	F	F
F	T	T	F	T
F	F	F	F	T

27. p has truth value T and q has truth value F. (a) T (b) F (c) T (d) F (e) F (f) F
29. p has truth value F and q has truth value T. (a) F (b) F (c) F (d) F

31. $(p \oplus q) \wedge r$ **33.** $(p \vee q) \wedge (p \vee \sim r)$

 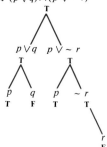

EXERCISES 12.3, page 543

1.
p	q	$p \rightarrow \sim q$
T	T	F
T	F	T
F	T	T
F	F	T

3.
p	q	$(p \oplus q) \rightarrow q$
T	T	T
T	F	F
F	T	T
F	F	T

5.
p	q	r	$(\sim p \wedge q) \rightarrow r$
T	T	T	T
T	T	F	T
T	F	T	T
T	F	F	T
F	T	T	T
F	T	F	F
F	F	T	T
F	F	F	T

7. tautology   **9.**
p	q	r	$(p \rightarrow q) \rightarrow r$
T	T	T	T
T	T	F	F
T	F	T	T
T	F	F	T
F	T	T	T
F	T	F	F
F	F	T	T
F	F	F	F

11. $\sim(p \vee q) \rightarrow (\sim p \wedge r)$ is true except in case p, q, and r are false.

13.
p	q	$(p \vee q) \leftrightarrow (p \wedge q)$
T	T	T
T	F	F
F	T	F
F	F	T

15. This is a tautology. In Exercises 17–25, p is true and q is false.

17. T **19.** T **21.** T **23.** F **25.** T **27.** $p \leftrightarrow q$ **29.** $q \rightarrow p$ **31.** $q \rightarrow p$ **33.** $\sim p \rightarrow \sim q$ **35.** $\sim p \rightarrow \sim q$ (T)
37. $\sim q \rightarrow \sim p$ (T)
39. (a) hyp: A person is healthy. con: A person lives a long life. (b) hyp: The train stops at the station. con: A passenger requests the stop. (c) hyp: The plant grows: con: The plant is exposed to sunlight. (d) hyp: I will go to the store. con: Jane goes to the store.
41. (a) If City Sanitation collects the garbage, then the mayor calls. (b) The price of beans goes down if there is no drought. (c) If Lake Erie is fresh water, then goldfish swim in Lake Erie. (d) If tap water boils slowly, it is not salted.
43. (a) 4 (b) 4 (c) 4 (d) 4 (e) 6 (f) 6 **45.** (a) 7 (b) 7 (c) 0 (d) 7 (e) 0 (f) 0
47. (a) 6 (b) -12 (c) 0 (d) 0 (e) 28 (f) 0 **49.** (a) 3 (b) 3 (c) -37 (d) -7 (e) 43 (f) -7.

EXERCISES 12.4, page 555

1. When p is F and q is T, the statement is FALSE.

5. Use truth tables to show the tautologies in (a) to (c). (d) $(p \to q) \Leftrightarrow p|(q|q)$ (e) $p|q \Leftrightarrow \sim(p \land q)$
(f) $(p \oplus q) \Leftrightarrow [[(p|p)|(q|q)]|(p|q)]|[[(p|p)|(q|q)]|(p|q)]$

7. $p \quad q \quad c \quad (p \to q) \Leftrightarrow [(p \land \sim q) \to c]$ **9.** False **11.** $p \oplus q \Leftrightarrow \sim[\sim(p \lor q) \lor \sim(\sim p \lor \sim q)]$

p	q	c	$(p \to q)$	$\Leftrightarrow$	$[(p \land \sim q)$	$\to$	$c]$
T	T	F	T	T	F	T	
T	F	F	F	T	T	F	
F	T	F	T	T	F	T	
F	F	F	T	T	F	T	
			(1)	(4)	(2)	(3)	

13. (a) Arizona does not border California or Arizona does not border Nevada. (b) There are no tickets available and the agency cannot get tickets. (c) The killer's hat was neither white nor gray.

15. (a) I have a ticket to the theater and I did not spend a lot of money. (b) Basketball is played on an indoor court and the players do not wear sneakers. (c) The stock market is going up and the interest rates are not going down. (d) Man has enough water and man is not staying healthy.

17. (a) contrapositive: (F) If a bird is not a hummingbird, then it is not small. converse: (T) If a bird is a hummingbird, then it is small. (b) contrapositive: (T) If two lines are not parallel, they do not have the same slope. converse: (T) If two lines are parallel they have the same slope. (c) contrapositive: (T) If we are not in France, then we are not in Paris. converse: (F) If we are in France, then we are in Paris. (d) contrapositive: (T) If you can legally make a U-turn, the road is not one-way. converse: (F) If you cannot legally make a U-turn, the road is one-way.

19. Ask either guard, "Would you say your door is the door to freedom?"

EXERCISES 12.5, page 562

1. s = "Sue goes to the movies."   1. $s \lor r$ hyp.
    r = "She reads."   2. $\sim s$ hyp.
                                    3. r disj. syllogism (1, 2)

3. a = "My allowance comes this week."   1. $a \land r \to b$ hyp.
    r = "I pay rent."   2. $\sim r \to e$ hyp.
    b = "My bank acct. is in the black."   3. $\sim e \land a$ hyp.
    e = "I am evicted."   4. $\sim e$ subtr. (3)
                                    5. r mod. tollens (2, 4)
                                    6. a subtr. (3)
                                    7. b mod. ponens (5, 6, 1)

5. p = "Price of oil increases."   1. $p \to a$ hyp.
    a = "OPEC countries agree."   2. $\sim u \to p$ hyp.
    u = "There is a UN debate."   3. $\sim a$ hyp.
                                  4. $\sim p$ mod. tollens (1, 3)
                                  5. u mod. tollens (2, 4)

7. g = "The germ is present."   1. $g \to r \land f$ hyp.
    f = "The fever is present."   2. f hyp.
    r = "The rash is present."   3. $\sim r$ hyp.
                                  4. $\sim r \lor \sim f$ addition (3)
                                  5. $\sim(r \land f)$ De Morgan (4)
                                  6. $\sim g$ mod. tollens (1, 5)

9. c = "The material is cotton."
r = "The material is rayon."
d = "The material can be made into a dress."

1. $c \lor r \to d$ hyp.
2. $\sim d$ hyp.
3. $\sim(c \lor r)$ mod. tollens (1, 2)
4. $\sim c \land \sim r$ De Morgan (3)
5. $\sim r$ subtr. (4)

11. s = "The salaries go up."
a = "More people apply."

$(s \to a) \land (s \lor a)$ does not logically imply s. Invalid argument. (Suppose s is false, a is true.)

13. y = "The balloon is yellow."
p = "The ribbon is pink."
h = "The balloon is filled with helium."

1. $y \lor p$ hyp.
2. $h \to \sim y$ hyp.
3. h hyp.
4. $\sim y$ mod. ponens (2, 3)
5. p disj. syllogism (1, 4)
The argument is valid.

15. p = "Papa bear sits."
m = "Mama bear stands."
b = "Baby bear crawls on the floor."

1. $p \to m$ hyp.
2. $m \to b$ hyp.
3. $\sim b$ hyp.
4. $\sim m$ mod. tollens (2, 3)
5. $\sim p$ mod. tollens (1, 4)
The argument is valid.

17. w = "Wheat prices are steady."
e = "Exports increase."
g = "GNP is steady."

The argument is invalid because $[w \to (e \lor g)] \land (w \land g)$ does not logically imply e. (Suppose w and g are true, e is false.)

19. i = "Tim is industrious."
p = "Tim is in line for promotion."
q = "Tim is thinking of leaving."

The argument is invalid because $(i \to p) \land (p \lor q)$ does not logically imply $(q \to i)$. (Suppose i is false, q and p are true.)

21. s = "Sam goes to the store."
m = "Sam needs milk."

1. s by way of contradiction
2. $s \to m$ hyp.
3. $\sim m$ hyp.
4. m mod. ponens (1, 2)
5. Contradiction (3, 4) $\therefore \sim s$.

23. n = "The newspaper reports a crime."
r = "TV reports a crime."
s = "A crime is serious."
k = "A person is killed."

1. $\sim s$ by way of contradiction
2. $n \land r \to s$ hyp.
3. $k \to n$ hyp.
4. k hyp.
5. r hyp.
6. $\sim(n \land r)$ mod. tollens (1, 2)
7. $\sim n \lor \sim r$ De Morgan (6)
8. n mod. ponens (3, 4)
9. $\sim r$ disj. syllogism (7, 8)
10. Contradiction (5, 9) $\therefore s$.

EXERCISES 12.6, page 573

1. (a) F (b) T (c) T (d) T (e) F   **3.** (a) $\forall x\, p(x)$ (b) $\sim\forall x\, p(x)$ (c) $\forall x\, [\sim p(x)]$ (d) (c) $\rightarrow$ (b)
5. $\forall x\, [\sim p(x)] \Leftrightarrow \sim\exists x\, [p(x)]$ Abby meant to say, "Not all men cheat on their wives."
7. (a) $\forall x\, p(x)$ (b) $\exists x\, [\sim p(x)]$ (c) $\exists x\, p(x)$ (d) $\sim[\forall x\, p(x)]$ (e) $\forall x\, [\sim p(x)]$ (f) $\sim[\exists x\, p(x)]$ or $\forall x\, [\sim p(x)]$ (g) (b) $\Leftrightarrow$ (d)
and (e) $\Leftrightarrow$ (f)
9. (a) T (b) F   **11.** (a) T (b) F (c) T (d) F (e) F (f) T (g) F (h) T
13. (a) Some dogs don't have their day. (b) All men do not fight wars, or No men fight wars. (c) Some mothers are
unmarried. (d) There exists a pot without a cover. (e) All children have pets. (f) Every month has 30 days.
15. (a) For every x and y in U, $x + y$ is greater than 12. FALSE: let $x = 0$ and $y = 3$. (b) For all x in U, there
exists a y in U such that $x + y$ is greater than 12. TRUE. (c) There exists x in U such that for all y in U, $x + y$ is
greater than 12. TRUE. (d) There exist x and y in U such that $x + y$ is greater than 12. TRUE.
17. (a) FALSE; let $x = 5$ and $y = 6$. (b) TRUE; given x in U, we let $y = x$. Then $p(x, y)$ is true. (c) TRUE; let
$x = 1$. Then $p(x, y)$ is true for all y in U. (d) FALSE; there is no y in U such that x divides y regardless of the value
of x in U. (e) TRUE; given y in U, let $x = y$. Then $p(x, y)$ is true. (f) TRUE; x divides itself for every x in U.
19. (a) $x \in S \rightarrow x \in T$ (b) No; $13 \in S$ but $13 \notin T$.
21. $S = \{2, 4, 6, 8\}$ and $T = \{2, 3, 4, 6, 8\}$. Therefore, $x \in S \rightarrow x \in T$.   **23.** $S = T = \{3\}$.

CHAPTER 12: SUPPLEMENTARY EXERCISES, page 578

1. (a), (c), (e) are statements.
3. (a) Contrapositive: If the Yankees are not playing in Yankee Stadium, then they are not in New York City.
Converse: If the Yankees are playing in Yankee Stadium, then they are in New York City. (b) Contrapositive: If
the quake is not considered major, then the Richter scale does not indicate the earthquake is a 7. Converse: If the
quake is considered major, then the Richter scale indicates the earthquake is a 7. (c) Contrapositive: If the coat is
not warm, it is not fur. Converse: If the coat is warm, it is made of fur. (d) Contrapositive: If Jane is not in Moscow,
she is not in the USSR. Converse: If Jane is in Moscow, she is in the USSR.
5. (a) and (b) are tautologies.   **7.** (a) is true (b) is false   **9.** (a) false (b) true.   **11.** (a) 50 (b) -25 (c) -15 (d) 10
13. (a) cannot determine truth value (b) cannot determine truth value (c) TRUE   **15.** (a) indeterminate
(b) indeterminate (c) TRUE
17. m = "I study mathematics."

b = "I study business."	1. $m \wedge b$	hyp.
p = "I can write poetry."	2. $b \rightarrow \sim p \vee \sim m$	hyp.
	3. b	subtr. (1)
	4. $\sim p \vee \sim m$	mod. ponens (2, 3)
	5. m	subtr. (1)
	6. $\sim p$	disj. syllogism (4, 5)

19. a = "Asters grow in the garden."

d = Dahlias grow in the garden."	1. $a \vee d$	hyp.
s = "It is spring."	2. $s \rightarrow \sim a$	hyp.
	3. s	hyp.
	4. $\sim a$	mod. ponens (2, 3)
	5. d	disj. syllogism (1, 4)

CHAPTER 13

EXERCISES 13.1, page 582

1. G_1: (a) parallel edges cd, dc, no loops (b) deg $a = 1$, deg $b = 2$, deg $c = 3$, deg $d = 3$, deg $e = 1$, deg $f = 0$,
   total $= 10$; (c) 4 (d) 5 = number of edges
   G_2: (a) no parallel edges, loop at e (b) deg a = deg b = deg c = deg $d = 4$, deg $e = 6$ (c) 0 (d) 11 = number
   of edges
   G_3: (a) no parallel edges, no loops (b) all have deg $= 2$ (c) 0 (d) 6 = number of edges

3. **5.** The graph would have an odd number of vertices of odd degree.

7. (a) 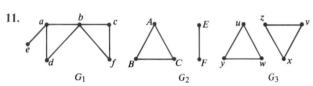 (b) (c) none **9.** 10

11. 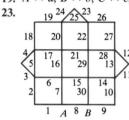 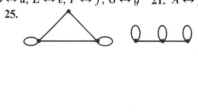 **13.** In G_1, deg $d = 4$. No vertex in G_2 has deg 4.

G_1 G_2 G_3

15. $a \leftrightarrow z, b \leftrightarrow y, c \leftrightarrow u, d \leftrightarrow v, e \leftrightarrow x, f \leftrightarrow w$ **17.** Not equivalent. G_2 has no vertex of deg 1; G_1 does.
19. $A \leftrightarrow a, B \leftrightarrow b, C \leftrightarrow c, D \leftrightarrow d, E \leftrightarrow e, F \leftrightarrow f, G \leftrightarrow g$ **21.** $A \leftrightarrow y, B \leftrightarrow v, C \leftrightarrow u, D \leftrightarrow x, E \leftrightarrow w$
23. **25.**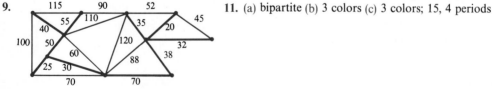

	19	24	25	23	26	
18	20		22		27	
4	17		21		28	12
5	16		29		13	
3						11
2	6		15		14	
	7		30		10	
	1	A	8	B	9	

EXERCISES 13.2, page 607

1. (a) 3 (b) 5 (c) 7 (d) 5 (e) 4 (f) 7
3. (a) 4, none (b) 3, simple path (c) 5, none (d) 5, closed path, circuit, simple circuit (e) 4, none (f) 6, closed path, circuit, simple circuit
5. (a) *ABCD, ACD, ACFD, ABCFED* (b) *CABC, CFDC, CFEDC*, yes **7.** (a) connected (b) (c) not connected
9. *G* cannot be disconnected; we could use at most 6 edges for 4 vertices; graph with 5 vertices and 6 edges might be disconnected
11. (a) **13.** (a) $v\ u\ y\ u\ w\ y\ v$ (b) $v\ r\ s\ w\ z\ t\ u\ r\ t\ y\ x\ z\ s\ v$ **15.** $a\ f\ c\ b\ e\ f\ d\ c\ a$ **17.** *ABEDCBFGCA*
19. *DABEFBCDE*; each Euler path begins or ends at *D* and *E*
21. (a) $b\ c\ d$; length $= 2$ (b) $d\ e\ a\ f\ g$; length $= 4$ (c) $c\ a\ f\ g$; length $= 3$

EXERCISES 13.3, page 622

1. (a) *SRTWVUYXZS* (b) $a\ g\ i\ j\ f\ e\ d\ c\ b\ h\ a$
3. (a) no Euler circuit, no Hamiltonian circuit (b) Euler circuit, $y\ r\ s\ y\ t\ u\ v\ w\ x\ s\ t\ x\ y$; Hamiltonian circuit, $r\ s\ t\ u\ v\ w\ x\ y\ r$ (c) no Euler circuit; Hamiltonian circuit, $a\ b\ c\ d\ e\ f\ g\ h\ a$
5. no conclusion; G might have a Hamiltonian circuit although one vertex has degree less than 3 **7.** 23
9. **11.** (a) bipartite (b) 3 colors (c) 3 colors; 15, 4 periods

115	90	52	
55	110	35	45
40		20	
100	50	120	32
	60	88	38
25	30		
70	70		

EXERCISES 13.4, page 644

1. connected: (b) (c); strongly connected: (c)
3. (a)

vertex	A	B	C	D	E	F
id	1	2	2	1	2	0
od	1	2	1	2	0	2

(b) id of a vertex = the number of procedures called by that procedure; od of a vertex = the number of procedures that call that procedure
5. (a) cycle: ABA (b) acyclic (c) acyclic
7. (a) $u\ v\ w\ u\ p\ s\ w\ p\ r\ s\ z\ y\ x\ z\ u$ (b) neither; id $(C) \ne$ od (C), id $(G) \ne$ od (G), id $(F) \ne$ od (F)
9. Euler circuit: $AHGAFCEDCBFGEBA$ 11. 25 days, critical path: $SA_1A_2A_3A_6A_7A_8A_{10}A_{11}E$
13. 65 days critical path: $ADEF$ 15.

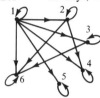   17.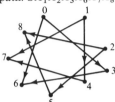

19. (a)

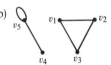

(b) error message (c) C

EXERCISES 13.5, page 659

1. (a) $A = \begin{bmatrix} 0 & 1 & 0 & 0 & 0 \\ 1 & 0 & 1 & 1 & 0 \\ 0 & 1 & 0 & 0 & 0 \\ 0 & 1 & 0 & 0 & 1 \\ 0 & 0 & 0 & 1 & 0 \end{bmatrix}$ (b) $A^2 = \begin{bmatrix} 1 & 0 & 1 & 1 & 0 \\ 0 & 3 & 0 & 0 & 1 \\ 1 & 0 & 1 & 1 & 0 \\ 1 & 0 & 1 & 2 & 0 \\ 0 & 1 & 0 & 0 & 1 \end{bmatrix}$. There is one path of length 2 from v_1 to v_4. (c) R is a

matrix of 1's.
3. (a) The graph is not connected. The fourth and fifth rows and columns will always have 0's. (b)

5. (a) $A = \begin{bmatrix} 0 & 1 & 1 & 0 & 0 \\ 1 & 0 & 1 & 0 & 0 \\ 1 & 1 & 0 & 1 & 0 \\ 0 & 0 & 1 & 0 & 0 \\ 0 & 0 & 0 & 0 & 1 \end{bmatrix}$ (b) The entry in the first row and fourth column of A^3 is 1.

(c) R has 1 in every position except in the last row, where all entries but the fifth column are 0, and the last column, where all entries but the last row are 0.

7. (a) $A = \begin{bmatrix} 0 & 1 & 1 & 0 & 0 \\ 1 & 0 & 1 & 0 & 0 \\ 1 & 1 & 0 & 1 & 0 \\ 0 & 0 & 1 & 0 & 1 \\ 0 & 0 & 0 & 1 & 0 \end{bmatrix}$ (b) 6 (c) We are guaranteed a path of length 6 from any vertex to any other.

9.

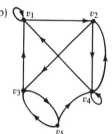

11. (a) $A = \begin{bmatrix} 0 & 1 & 1 & 0 \\ 1 & 0 & 0 & 0 \\ 0 & 1 & 0 & 0 \\ 0 & 1 & 0 & 0 \end{bmatrix}$ (b) 2 (c) $a\,c\,b\,a\,b,\ a\,b\,a\,c\,b$ (d) $R = \begin{bmatrix} 1 & 1 & 1 & 0 \\ 1 & 1 & 1 & 0 \\ 1 & 1 & 1 & 0 \\ 1 & 1 & 1 & 1 \end{bmatrix}$ (e) connected, not strongly connected

13. (a)

Vertex	v_1	v_2	v_3	v_4	v_5
id	3	2	2	3	1
od	2	2	2	3	2

(b) (c) 2 (d) $v_1 v_2 v_4 v_4,\ v_1 v_1 v_2 v_4$

EXERCISES 13.6, page 670

1. **3.**

5. (a) d, e, i, j, k, h (b) g, h, e, f, c, d (c) $x, 6, y$ **7.** (a) a (b) b (c) $*$
9. A tree with 7 vertices must have 6 edges. This tree cannot be drawn.
11. By Property 2, if G is connected, has n vertices, and has $n - 1$ edges, it is a tree.
13. A tree on 3 vertices has 2 edges.

15. **17.** $\dfrac{a}{b + c} - b^c$ **19.**

21. 7 questions, 10 questions

CHAPTER 13: SUPPLEMENTARY EXERCISES, page 673

1. G_1 vertex a b c d G_2 vertex r s t u v w x

         deg   2   3   2   5   deg   3   2   3   3   4   4   3

         sum of degrees = 12   sum of degrees = 22

         number of edges is 6   number of edges = 11

3. G_1: weight = 9 G_2: weight = 15

 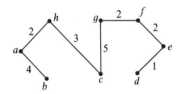

5. G_1 vertex a b c d e f g h G_2 vertex a b c d e f g h

         deg   2   3   3   4   2   3   5   4   deg   2   4   2   4   4   2   4   4

         There are 4 vertices of odd degree.   There are 0 vertices of odd degree.

7. G_1: weight = 19 G_2: weight = 37

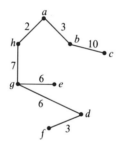

9. There are 9 paths of length 3 from v_1 to v_4 in G_1.

$$A(G_1) = \begin{bmatrix} 0 & 1 & 0 & 1 & 1 \\ 1 & 0 & 1 & 1 & 1 \\ 0 & 1 & 0 & 1 & 0 \\ 1 & 1 & 1 & 0 & 1 \\ 1 & 1 & 0 & 1 & 0 \end{bmatrix}$$

In G_2 there are 6 paths of length 3 from v_1 to v_4.

$$A(G_2) = \begin{bmatrix} 0 & 1 & 0 & 1 & 0 \\ 1 & 0 & 1 & 1 & 0 \\ 0 & 1 & 0 & 1 & 0 \\ 1 & 1 & 1 & 0 & 1 \\ 0 & 0 & 0 & 1 & 0 \end{bmatrix}$$

11. 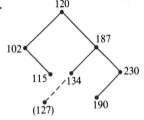 G and G' are equivalent. Associate a to w, b to u, c to x, and d to v.

13. 3 colors **15.**

120

102

187

115 134 230

(127) 190

17. Start, *ABDFG*, END **19.** 3 colors.

INDEX

A

Absorbing state, 410
Absorbing stochastic matrix, 410
Acquaintance table, 593
Activity matrix, 191
Acyclic:
    digraph, 633
    graph, 599, 616
Addition, in logic, 554, 559
Adjacency matrix, 648–649
    of a digraph, 654
Algebraic expression, tree structure of, 668
Algorithm, 603
    Fleury's, 604
Allocation problem, 191
Amortization, 471ff
AND, 533–534
$a_{\overline{n}|i}$, 465
Annuity, 461ff, 511ff
Annuity due, 466
Appel, Kenneth, 618
Argument, definition of, 558
Associative laws of logic, 549
Average, 336

B

Balloon payment, 474
Basic region of Venn diagram, 218
Bayes' theorem, 312ff
Bernoulli trials, 356
Beta probability distribution, 630
Biconditional connective, 540
Binary search, 667

Binary tree, 667
Binomial coefficient, 246
Binomial theorem, 246
Binomial trials, 356ff
    expected value, 356
    probabilities, 356
    probabilities via normal curve, 360, 378ff
Bipartite graph, 619
Birthday problem, 286
Break-even point, 44

C

Cartesian coordinate system:
    line, 4
    plane, 4–5
Chebychev inequality, 351
Child, in a tree, 667
Chromatic number, 620
Circuit, 599
    Euler, 602, 604
    Euler, in a digraph, 635
    Hamiltonian, 613, 614, 619
    Hamiltonian, in a digraph, 636
    in a digraph, 633
    simple, 599
Closed path, 599
    in a digraph, 633
Coloring problems, 618
Combinations, 232–234, 244, 245
Combinatorics, 211
Commutative laws of logic, 549
Complement of a set, 206
Complementary slackness, principle of, 192
Compound amount, 452

Compound statement, 524
Conclusion, in logic, 537
  definition of, 558
Conditional connective, 537
Conditional probability, 292–295
Conditional statement, 537
Conjunction, 529
Connected digraph, 633, 657
Connected graph, 600, 602, 654
Connective, logical:
  $\wedge$, 525
  $\leftrightarrow$, 540
  $\rightarrow$, 526, 537
  $\vee$, 525, 532
  $\oplus$, 532
  conditional, 537
Constructive dilemma, 559, 554
Contradiction, 531, 548
Contrapositive, 549, 550
Converse, 539
Coordinate axes, 4
Cost curve, 34
Cost matrix, 192
Counterexample, 571
Critical path, 627
Cycle, 599

D

Dandolas, Nicholas, 342
Dantzig, George B., 145
De Morgan's laws, 215, 550, 568, 570, 572, 549
Deferred annuity, 469
Degree of a vertex, 585
  in a digraph, 634
Demand curves, 25
Depreciation, 9, 34
Detachment, law of, 554
Difference equation, 482ff
  graph, 484
  solution ($a \neq 1$), 485
  solution ($a = 1$), 489
Difference, symmetric, 571
Digraph, 584, 626, 654
  activity, 627
  acyclic, 633
  connected, 633, 657
  strongly connected, 633–634, 657
Dilemma, constructive, 554, 559
Directed graph, (see Digraph)
Disconnected graph, 600
Disjunction, 530
Disjunctive syllogism, 554, 559
Distance, in a digraph, 639

Distributive laws, 81, 549
Double negation, 549, 550
Dual problem, 186, 190, 194
Duality, 185ff

E

Edge, graph theory definition, 584
Elementary row operations, 52
Empty set, 206
Equally likely outcomes, 283
Equation:
  graph, 6
  linear, 8
Equivalence, logical, 547, 549
Equivalency of simple graphs, 587, 589, 650
Euler circuit, 602, 604
  in a digraph, 635
Euler path, 602
  in a digraph, 635
  property of, 605
Euler's Theorem, 603
Even vertex, 585
Event(s), 263–264
  certain, 267
  complement, 268
  disjoint, 269
  elementary, 275
  impossible, 267
  independent, 297, 299
  intersection, 268
  mutually exclusive, 269
  probability, 275
  union, 267
Exclusive OR, 532
Existential quantifier, 566
Expected value, 339
Experiment, 262

F

Factorial, 233
Feasible set, 19
Final demand of an economy, 103
Finishing time:
  earliest, 631
  latest, 632
Fixed cost, 34
Point-slope formula, 35–36
Fleury's Algorithm, 604
Four-color problem, 618
Frequency distribution, 334
Fundamental Theorem of Duality, 189

Fundamental Theorem of Game Theory, 445
Furniture manufacturing problem, 112–114,
        120–121, 136, 146, 151, 177, 190, 195
Future value of an annuity, 462

G

Games, 422ff
    fundamental problem, 423
    fundamental theorem, 445
    strictly determined, 426
    value of strictly determined, 426
    zero-sum, 424
Gauss-Jordan method, 96–97
Gaussian elimination method, 51, 62, 64
General annuity, 466
Generalized multiplication principle, 226
Geometric progression, 462
Graph coloring, 618
Graph of a difference equation, 484
    asymptotic, 498
    constant, 499
    decreasing, 498
    increasing, 497
    long-run behavior, 498
    monotonic, 498
    oscillating, 498
    technique for sketching, 503
    unbounded, 498
    vertical direction, 497
Graph of an equation, 6
Graph Property 1, 585
Graph Property 2, 586
Graph theory, 582
Graphs, 582
    acyclic, 599, 616
    bipartite, 619
    connected, 600, 602, 654
    disconnected, 600
    graph theory definition, 584
    in matrix form, 649
    planar, 585
    simple, 587
    weighted, 616

H

Haken, Wolfgang, 618
Hamilton's puzzle, 613
Hamilton, Sir William Rowan, 613
Hamiltonian circuit, 613, 614, 619
    in a digraph, 636

Hamiltonian path:
    in a digraph, 636
    in a tournament, 639
Histogram, 326, 329
Hypothesis, 537
    definition of, 558
Hypothetical syllogism, 554, 559

I

Idempotent laws of logic, 549
Identity laws of logic, 549
Identity matrix, 78
If . . . THEN . . . ELSE, 541
Implication, 537ff, 549
    rule, 551
    logical, 553, 554
Implies, 537
Imputed value matrix, 192
Inclusion-exclusion principle:
    for counting, 211
    for probability, 278
Indegree of a vertex, 634
Independent events, 297, 299
Indirect proof, 561
Inequalities:
    addition property, 14
    multiplication property, 15
    signs, 14
    subtraction property, 14
Inequality signs, 14
Inference, rules of, 558–559
Initial value, 484
Input-output analysis, 102ff
Intercepts, 8
Interest:
    compound, 452, 457, 490, 491
    simple, 456, 490, 491
Intersection:
    of lines, 23ff
    of sets, 204
Inverse of a matrix, 85ff
    2×2 case, 88
    existence, 89
    Gauss-Jordan method, 96–97
Isolated vertex, 585

K

Kantorovich, 114
Kirkman, Thomas R., 613
Königsberg bridge problem, 606
Koopmans, 114

L

Leaf, 665
Length, of a path, 599
   in a digraph, 639
Leontieff, Vassily, 102
Leontieff model:
   closed, 106
   open, 106
LET statement, 541
Linear depreciation, 9, 34
Linear equation, 8
   graph, 9
   graphing procedure, 9
   intercepts, 8
   standard form, 8
Linear inequalities:
   feasible set for system, 19
   graphs, 16–17
   standard forms, 15
Linear polynomial, 145
Linear programming, 112ff
   duality, 185ff
   economic interpretation of dual problem, 190
   feasible set, 19
   fundamental theorem, 119
   geometric solution procedure, 121
   marginal analysis, 176ff
   matrix formulation, 181, 183
   problem in standard form, 145
   simplex method, 162, 170
Logic, 523
   relation to sets, 571–572
   rules of, 549, 570
Logical connectives and operators:
   $\wedge$, 525
   $\leftrightarrow$, 540
   $\rightarrow$, 526, 537
   $\sim$, 525
   $\vee$, 525, 532
   $\exists$, 566
   $\forall$, 566
   $\oplus$, 532
   precedence of, 540
Logical equivalence, 547, 549
Logical implication, 553, 554
Logically equivalent, 548
Loop, 585

M

Marginal analysis of linear programming
   problems, 176ff
Marginal cost of production, 34
Marginal revenue of production, 43

Markov process, 390ff
Matrix, 56, 69, 649
   absorbing stochastic, 410
   activity, 191
   addition, 71–72
   adjacency, 648–649
   adjacency, of a digraph, 654
   capacity, 191
   coefficient, 56, 79
   column, 70
   cost, 192
   distribution, 393
   entries, 71
   equality, 71
   equation, 90
   formulation of linear programming problem,
      181–182
   identity, 78
   ijth entry (a_{ij}), 71
   imputed value, 192
   input-output, 102
   inverse, 85
   multiplication, 72–75, 84
   payoff, 424
   profit, 192
   reachability, 654
   reachability, of a digraph, 656
   regular stochastic, 403
   requirements, 192
   row, 70
   size, 70
   square, 70
   stable absorbing, 413–414
   stable regular, 404
   stochastic, 392
   subtraction, 71–72
   symmetric, 649
   transition, 391
   transpose, 182
Mean, 336, 337
Minimal-length path, 601
Modus ponens, 554, 559
Modus tollens, 554, 559
Mortgage, 472, 509
Multinomial coefficient, 251, 252
Multiplication principle, 226

N

NAND, (*see* Sheffer stroke)
Necessary condition, 538
Negation:
   double, 549, 550
   logical, 525
Normal curve, 360, 363, 366

Normal curve table, 367
Normal distribution, 632
Normally distributed outcomes, 363, 371
NOT, 533–534
Not, logical operator, 525

O

Objective function, 121
Odd vertex, 585
Odds, 279
Open sentence, 565
Operator, logical negation ($\sim$), 525
Optimal scheduling with randomness, 629
Optimum point, 119
OR, 533–534
Origin, 4
Outcome, 262–263
Outdegree of a vertex, 634

P

Parallel edges, graph theory definition, 585
Parallel property of slope, 37
Parameter, 335
Parent, in a tree, 667
Partitions:
  ordered, 251
  unordered, 253
Path, 598
  closed, 599
  closed, in a digraph, 633
  critical, 627
  critical, with randomness, 631
  Euler, 602
  Euler, in a digraph, 635
  Euler, property of, 605
  Hamiltonian, in a digraph, 636
  Hamiltonian, in a tournament, 639
  in a digraph, 633
  length, 599
  length, in a digraph, 639
  of minimal length, 601
  simple, 599
Paths, number of, 651–652
Payoff matrix, 424
Percentiles, 368–369, 374
Permutations, 232–233
Perpendicular property of slope, 37
Perpetuity, 469
Pivoting, 61
  pivot element, 162, 170
Planar graph, 585

Population, 335
Population mean, 337
Precedence of logical connectives, 540
Predicate calculus, 565ff
Present value, 455
Present value of an annuity, 465
Principal, 452
Principle of complementary slackness, 192
Probability, 272
  addition principle, 275
  complement rule, 286
  conditional, 292–295
  distribution, 273, 327
  fundamental properties, 275
  product rule, 296
Product rule, 296
Production schedule, 115
Profit matrix, 192
Proof:
  definition of, 558
  indirect, 561
Proportional, 514
Propositional calculus, (see Logic)

Q

Quantifier:
  universal, 566
  existential, 566

R

Radioactive decay, 515
Random variable, 330
Probability distribution, 330
Reachability matrix, 654
  of a digraph, 656
Reductio ad absurdum, 549
Regular stochastic matrix, 404
Relative frequency distribution, 324
Rent of an annuity, 462, 465
Requirements matrix, 192
Resources matrix, 191
Revenue curve, 43
Root, of a tree, 665

S

Saddle point, 426
Salvage value, 10
Sample, 335
  mean, 336, 337, 349
  variance, 349

Sample space, 262–263
Scheduling, optimal, 626
Search, binary, 667
Sentence, open, 565
Set(s), 203ff
  complement, 206
  element of, 203
  empty, 206
  intersection, 204
  number of elements, 211
  number of subsets, 248
  relation to logic, 571–572
  sub-, 206
  union, 204
  universal, 206
Shadow prices, 191
Sheffer stroke, 536, 555
Simple circuit, 599
Simple graphs, 587
  equivalency of, 587, 588, 650
Simple path, 599
Simple statement, 524
Simplex method, 145, 162, 170
Simplex tableau, 150
Simplification, in logic, 554
Sinking fund, 477
Slack time, 632
Slack variable, 146
Slope of a line, 29
  geometric definition, 30
  parallel property, 37
  perpendicular property, 37
  point-slope formula, 35–36
  steepness property, 31–34
$s_{\overline{n}|i}$, 463
Solution of matrix equation:
  via Gauss elimination method, 51–52, 62, 64
  via inverse matrix, 90
Span, 616
Stable distribution, 403
Stable matrix of absorbing stochastic matrix, 413–417
Stable matrix of regular stochastic matrix, 403, 404
Standard deviation, 347
Standard form:
  linear equation, 8
  linear inequality, 15
  linear programming problem, 145
Standard normal curve, 366
Starting time:
  earliest, 631
  latest, 632
Statement, 523
  compound, 524

Statement (cont.):
  conditional, 537
  simple, 524
Statement form, 528
Statistic, 335, 337
Statistics, 323
Steepness property of slope, 31–34
Straight-line depreciation, 34
Strategies:
  expected value of mixed, 431
  expected value of pair, 433
  mixed, 430
  optimal mixed for C, 425, 442
  optimal mixed for R, 425, 440
  optimal pure for C, 438
  optimal pure for R, 437
  pure, 424
Strictly determined game, 426
Strongly connected digraph, 633–634, 657
Subgraph, 616
Substitution principles, 552
Subtraction, logical, 559
Sufficient condition, 538
Supply curves, 25
Syllogism, disjunctive, 554, 559
Syllogism, hypothetical, 554, 559
Symmetric difference, 571
Symmetric matrix, 649
System of linear equations:
  coefficient matrix, 79
  diagonal form, 51
  general solution, 64
  solution, 50ff
  unique solution, 62
  with inconsistent equation, 66

T

Table, transition, 647
Tautology, 531, 548
Tournament, 636–637
  transitive, 638
Transition matrix, 391
Transition table, 647
Transitive tournament, 638
Transpose of a matrix, 182
Tree, 616, 663
Tree Diagrams, 305ff
Tree Property 1, 664
Tree Property 2, 664
Tree:
  algebraic expression, structure of, 668
  binary, 667
  child in, 667

Tree (cont.):
    minimal spanning, 616
    minimal spanning, algorithm for, 616
    parent in, 667
    rooted, 665
    statement form, 534
Trial, 262
Truth table, 528
Truth value, 523

U

Unbiased estimate, 349
Union of sets, 204, 205
Universal quantifier, 566
Universal set, 206
Universe, in predicate calculus, 565

V

Valid, definition of, 558
Valuation problem, 191
Variance, 347, 349, 350
    of finishing time, 632
Venn diagram, 213

Vertex, 119
    degree, 585
    degree, in a digraph, 634
    even, 585
    graph theory definition, 584
    indegree, 634
    isolated, 585
    odd, 585
    outdegree, 634
    pairings, 588

W

Weighted graph, 616

X

x-axis, 4
x-coordinate of a point, 4
XOR, 533–534

Y

y-axis, 4
y-coordinate of point, 4